Modeling the Adaptations of Agricultural Production to Climate Change

Modeling the Adaptations of Agricultural Production to Climate Change

Editors

Dengpan Xiao
Wenjiao Shi

MDPI • Basel • Beijing • Wuhan • Barcelona • Belgrade • Manchester • Tokyo • Cluj • Tianjin

Editors
Dengpan Xiao
Hebei Normal University
China

Wenjiao Shi
Institute of Geographic
Sciences and Natural
Resources Research, CAS
China

Editorial Office
MDPI
St. Alban-Anlage 66
4052 Basel, Switzerland

This is a reprint of articles from the Special Issue published online in the open access journal *Agriculture* (ISSN 2077-0472) (available at: https://www.mdpi.com/journal/agriculture/special_issues/modeling_agriculture_climate).

For citation purposes, cite each article independently as indicated on the article page online and as indicated below:

LastName, A.A.; LastName, B.B.; LastName, C.C. Article Title. *Journal Name* **Year**, *Volume Number*, Page Range.

ISBN 978-3-0365-6802-7 (Hbk)
ISBN 978-3-0365-6803-4 (PDF)

Contents

About the Editors

Dengpan Xiao

Dengpan Xiao, Professor at College of Geography Science, Hebei Normal University. His research interests include climate change and its impacts on agricultural production, the efficient use of agricultural water resources, and the adjustment of agricultural systems and their impacts.

Wenjiao Shi

Wenjiao Shi, Professor at Institute of Geographic Sciences and Natural Resources Research, CAS. Her research interests include land use/cover change and its environmental effects; the impacts of global change on agriculture and ecology; ecosystem services; and spatial information analysis.

Preface to "Modeling the Adaptations of Agricultural Production to Climate Change"

Globally, climate change and its impacts on agricultural production and food security are a significant public concern. Modeling is a key tool for exploring the impacts of climate change on agriculture and proposing adaptation strategies. Generally, establishing closer links between experiments and statistical and/or eco-physiological crop models may facilitate the necessary methodological advances. It is expected that insights derived from this book will be helpful for relevant decision makers in the areas of agricultural adaptation and food security.

We would like to sincerely thank all the authors who submitted papers to the Special Issue of *Agriculture* entitled "Modeling the Adaptations of Agricultural Production to Climate Change", the reviewers of these papers for their constructive comments and thoughtful suggestions, and the editorial staff of *Agriculture*.

Dengpan Xiao and Wenjiao Shi
Editors

agriculture

Editorial

Modeling the Adaptation of Agricultural Production to Climate Change

Dengpan Xiao [1,2,*] and Wenjiao Shi [3,*]

1 College of Geography Science, Hebei Normal University, Shijiazhuang 050024, China
2 Hebei Laboratory of Environmental Evolution and Ecological Construction, Shijiazhuang 050024, China
3 Key Laboratory of Land Surface Pattern and Simulation, Institute of Geographic Sciences and Natural Resources Research, Chinese Academy of Sciences, Beijing 100101, China
* Correspondence: xiaodp@sjziam.ac.cn (D.X.); shiwj@lreis.ac.cn (W.S.)

Climate change and its impacts on agricultural production and food security are a significant source of public concern around the world. In order to reduce the negative impacts of climate change on agriculture, maintain crop production levels, and even discover opportunities in agricultural intensification, researchers have made great efforts to assess changes in agricultural climate resources and develop adaptation measures in different growing areas of the world experiencing climate change. Modeling is a key tool for exploring the impacts of climate change on agriculture and proposing adaptation strategies. Currently, the two main fields where further progress is required include a more mechanistic understanding of climate impacts and management options for adaptation and mitigation, and a focus on cropping systems and integrative multiscale assessments instead of single season and crops. Therefore, establishing closer links between experiments and statistical and/or eco-physiological crop models may not only facilitate the necessary methodological advances but also achieve the above goals.

With these goals in mind, we have organized this Special Issue "Modeling the Adaptation of Agricultural Production to Climate Change (MAAPCC)". The Special Issue of MAAPCC has a total of 21 papers [1–21], and papers were submitted from five countries: China, Japan, Thailand, Iran, and South Africa. Moreover, the Special Issue covers a wide range of plants, including not only grain crops such as maize [15], rice [6,10], wheat [2,4,16], and soybean [14], and cash crops such as cotton [7] and sugarcane [3], but also the apple tree [1] and traditional Chinese medicinal plants such as Rheum nanum (*R. nanum*) [17]. In terms of the research time scale, the Special Issue not only focused on climate change and its impact in the historical period, but also analyzed the impacts of different climate scenarios on plant distribution, crop production, and climatic resources in the future in 10 papers [1,6,8,12,14–17,19,20], and some of the papers adopted the latest climate scenario data, namely Shared Socioeconomic Pathways (SSPs) from the Coupled Model Comparison Program (CMIP6). As for the research methods, these papers not only use the traditional statistical analysis methods, but also involve the widely used crop mechanism models, including the Agricultural Production Systems Simulator (APSIM) [2], the Crop Environment Resource Synthesis (CERES)-Wheat [4], CERES-Rice [6], the denitrification-decomposition (DNDC) [12], the integrated climate–hydrological–economic model [13], CROPWAT [14], and Environmental Policy Integrated Climate (EPIC) [15]. In addition, part of the papers also used the current, more popular, machine learning method for analysis and prediction [2,18]. Overall, the papers in the Special Issue of MAAPCC were grouped into three categories: assessment of climate resources in the context of climate change [5,7,8,14,19,20], assessment of the impact of climate change on crop production [1,3,10,13,15–17], and some methodological studies related to climate change [2,4,6,9,11,12,18,21].

The first category has six papers under the following sub-heading: assessment of climate resources in the context of climate change.

Citation: Xiao, D.; Shi, W. Modeling the Adaptation of Agricultural Production to Climate Change. *Agriculture* **2023**, *13*, 414. https://doi.org/10.3390/agriculture13020414

Received: 6 February 2023
Accepted: 9 February 2023
Published: 10 February 2023

Generally, evaluating the response of evapotranspiration to climatic change can provide theoretical support on the optimal allocation of regional water resources and agricultural water management under climate change. In the two papers in this section, the trends and climatic causes of potential evapotranspiration (ET_0) in Heilongjia Province from 1960 to 2019 [5] and in the Xinjiang Autonomous Region of China from 1957 to 2017 [7] have been quantitatively assessed, and the results suggested that historical climate change has had significant impacts on regional ET_0 and will further affect crop water demand and consumption. Moreover, the paper by Li et al. [14] investigated the spatial and temporal distribution of ET_0, crop water requirement (ET_C), irrigation water requirement (I_r), effective precipitation (P_e), and the coupling degree of ET_C and P_e for soybean during the growth period for the future period from 2021 to 2080 in Heilongjiang Province, China.

Assessing the climatic suitability of crops is critical for mitigating and adapting to the negative impacts of climate change on crop production. The paper by Zhao et al. [20] developed a climate suitability model of maize and investigated the climate suitability of summer maize during the past and future periods in the North China Plain. The paper by Nooni et al. [19] investigated the future changes in drought events for four SSPs in the African continent. The projected wetter trends in humid areas may benefit agricultural production and ecological conservation, and the drier trends in non-humid areas may require appropriate drought adaptation strategies and development plans to minimize impacts [19]. The last paper in this section was from Shi et al. [8], which proposed a land use/land cover (LULC) simulation framework from 2000 to 2030 for four different development scenarios in the Xinjiang region. This study stated that both the supply and demand of carbon stock in Xinjiang would increase in 2025 and 2030, with the demand exceeding the supply [8].

The second category is the assessment of the impacts of climate change on crop production.

There were seven papers exploring the effects of climate change on crop (or plant) distribution and production. In general, climate change plays an important role in the distribution of suitable zones of plant cultivation. In the two papers in this section, the potential distribution of suitable habitats and range shifts of apple trees in the near present and near future (i.e., the 2030s and the 2050s) under two climate scenarios (i.e., SSP126 and SSP585) was simulated based on three pieces of software (the maximum entropy model, IDRISI, and ArcGIS) [1], and the potential distribution of *Rheum nanum* (*R. nanum*), a famous traditional Chinese medicinal plant, was developed for three periods (current, 2050s: 2041–2060, and 2070s: 2061–2080) using MaxEnt and ArcGIS [17]. These two studies may improve our understanding of the effects of climate warming on plant distribution and could be useful for relevant agricultural decision-making. In addition, the paper by Wu et al. [16] determined the planting boundary of winter wheat in north China for the future period based on four critical parameters of percentages of extreme minimum temperature years (POEMTYs), first day of the overwintering period (FD), sowing date (SD), and precipitation before winter (PBW).

Currently, a large number of studies focus on the effects of climate change on crop yields and/or production. The paper by Choruma et al. [15] assessed the effects of future climate change on maize yield in the Eastern Cape Province of South Africa, and indicated a decrease in maize production for two future periods (mid-century (2040–2069) and late century (2070–2099)). The paper by Shayanmehr et al. [13] constructed a new integrated climate–hydrological–economic model to assess the impact of future climate change on water resources and crop production. The findings noted that in the majority of cases, crop production will reduce in response to climate scenarios so that rainfed wheat will experience the greatest decline (approximately 59.95%) [13]. Zhang [10] analyzed the spatiotemporal change in heat stress and its impacts on rice growth in the middle and lower reaches of the Yangtze River, China, and indicated that the change in heat stress is attributed to climate changes and extreme meteorological events. The last paper in this section was from Yao et al. [3], which comprehensively assessed multiple sugarcane agrometeorological disasters with regard to sugarcane yield in Southern China. The results suggested that the yield-reducing effect of sugarcane flood was more obvious than that of drought [3].

The last category is some methodological studies related to climate change.

To meet the challenges of climate change and the increasing food demand, an accurate, timely, and dynamic yield estimation of regional or global crop yield is critical to food trade and policy-making. In the two papers in this section, the winter wheat yield in the North China Plain [2] and in China [18] was accuracy predicted by coupling the crop model with machine learning algorithms based on multi-source data. These findings indicated that the prediction model can be used to develop adaptation strategies to mitigate the negative effects of climate change on crop productivity and provide the data support for food security. Moreover, the paper by Zheng et al. [6] used a process-based crop model (CERES-Rice) which was calibrated and validated based on experimental data from the Songnen Plain of China, and driven by multiple global climate models (GCMs) from the CMIP6 to predict rice growth period, yield, and light and heat resource utilization efficiency under future climate change conditions. The results showed that optimizing the sowing date could make full use of climate resources to improve rice yield and light and heat resource utilization indexes under future climate conditions [6]. In addition, the CERES-Wheat model was applied to investigate the optimal irrigation amount for high yield, water saving, and the trade-off between high yield and water saving of winter wheat in the North China Plain [4]. Therefore, crop mechanism models play an important role in assessing the effects of climate change on crop production and proposing effective coping strategies. In the future, we need to continuously develop crop models to improve the effectiveness and versatility of their simulation.

In this section, three papers developed models or devices to simulate or observe greenhouse gas emissions from agricultural processes, which have made a significant contribution to climate warming [9,11,12]. In the paper by Salehi et al. [11], two types of data-driven models were proposed to predict biogas production from the anaerobic digestion of spent mushroom compost supplement with wheat straw used as a nutrient source. The paper by Yi et al. [12] evaluated crop yields, nitrous oxide (N_2O) emission, and soil organic carbon (SOC) in a typical wheat–corn rotation system field on the North China Plain on a 50-year scale using the denitrification-decomposition (DNDC) model, and proposed adaptive strategies for each climate scenario. Moreover, a gas diffusion analysis method for simulating N_2O surface flux from soil gas measured in a soil-interred silicone diffusion cell using a low-cost device was developed by Bandara et al. [9]. The last paper in this section was from Huang et al. [21], which evaluated the accuracy of three reanalysis temperature data systems (e.g., the China Meteorological Administration Land Data Assimilation System (CLDAS), the U.S. Global Land Data Assimilation System (GLDAS), and the European Centre for Medium-Range Weather Forecasts (ECMWF) Reanalysis Version 5 (ERA5)-Land) across China. The results indicated that the CLDAS product demonstrated a relatively high reliability, which was of great significance for the study of climate change and forcing crop models [21].

In summary, this Special Issue focuses on the quantitative assessment of the impact of climate change on agricultural production based on multi-source model simulation and reveals the role and mechanism of improved management measures in adapting to climate change. It is expected that insights derived from this Special Issue will be helpful for relevant decision-makers in the areas of agricultural adaptation and food security.

Author Contributions: Conceptualization, D.X. and W.S.; investigation, D.X. and W.S.; writing—original draft preparation, D.X.; writing—review and editing, D.X. and W.S.; visualization, D.X. and W.S. All authors have read and agreed to the published version of the manuscript.

Funding: This material is based upon work that is supported by the Hebei Provincial Science Foundation for Distinguished Young Scholars (No. D2022205010), and the Science Fund for Creative Research Groups of the National Natural Science Foundation of China (Grant No. 72221002).

Acknowledgments: We would like to sincerely thank all of the authors who submitted papers to the Special Issue of *Agriculture* entitled "Modeling the Adaptations of Agricultural Production to Climate Change", the reviewers of these papers for their constructive comments and thoughtful suggestions, and the editorial staff of *Agriculture*.

Conflicts of Interest: The authors declare no conflict of interest.

References

1. Xu, W.; Miao, Y.; Zhu, S.; Cheng, J.; Jin, J. Modelling the Geographical Distribution Pattern of Apple Trees on the Loess Plateau, China. *Agriculture* **2023**, *13*, 291. [CrossRef]
2. Zhao, Y.; Xiao, D.; Bai, H.; Tang, J.; Liu, D.L.; Qi, Y.; Shen, Y. The Prediction of Wheat Yield in the North China Plain by Coupling Crop Model with Machine Learning Algorithms. *Agriculture* **2023**, *13*, 99. [CrossRef]
3. Yao, P.; Qian, L.; Wang, Z.; Meng, H.; Ju, X. Assessing Drought, Flood, and High Temperature Disasters during Sugarcane Growth Stages in Southern China. *Agriculture* **2022**, *12*, 2117. [CrossRef]
4. Shi, X.; Shi, W.; Dai, N.; Wang, M. Optimal Irrigation under the Constraint of Water Resources for Winter Wheat in the North China Plain. *Agriculture* **2022**, *12*, 2057. [CrossRef]
5. Nie, T.; Yuan, R.; Liao, S.; Zhang, Z.; Gong, Z.; Zhao, X.; Chen, P.; Li, T.; Lin, Y.; Du, C.; et al. Characteristics of Potential Evapotranspiration Changes and Its Climatic Causes in Heilongjiang Province from 1960 to 2019. *Agriculture* **2022**, *12*, 2017. [CrossRef]
6. Zheng, E.; Qin, M.; Chen, P.; Xu, T.; Zhang, Z. Climate Change Affects the Utilization of Light and Heat Resources in Paddy Field on the Songnen Plain, China. *Agriculture* **2022**, *12*, 1648. [CrossRef]
7. Jia, Y.; Shen, X.; Yi, R.; Song, N. Spatial and Temporal Variability of *ETo* in Xinjiang Autonomous Region of China during 1957–2017. *Agriculture* **2022**, *12*, 1380. [CrossRef]
8. Shi, M.; Wu, H.; Jiang, P.; Shi, W.; Zhang, M.; Zhang, L.; Zhang, H.; Fan, X.; Liu, Z.; Zheng, K.; et al. Cropland Expansion Mitigates the Supply and Demand Deficit for Carbon Sequestration Service under Different Scenarios in the Future—The Case of Xinjiang. *Agriculture* **2022**, *12*, 1182. [CrossRef]
9. Bandara, K.M.T.S.; Sakai, K.; Nakandakari, T.; Yuge, K. A Gas Diffusion Analysis Method for Simulating Surface Nitrous Oxide Emissions in Soil Gas Concentrations Measurement. *Agriculture* **2022**, *12*, 1098. [CrossRef]
10. Zhang, S. Spatiotemporal Change of Heat Stress and Its Impacts on Rice Growth in the Middle and Lower Reaches of the Yangtze River. *Agriculture* **2022**, *12*, 1097. [CrossRef]
11. Salehi, R.; Yuan, Q.; Chaiprapat, S. Development of Data-Driven Models to Predict Biogas Production from Spent Mushroom Compost. *Agriculture* **2022**, *12*, 1090. [CrossRef]
12. Yi, X.; Chang, N.; Ding, W.; Xu, C.; Zhang, J.; Zhang, J.; Li, H. Modeling Adaptive Strategies on Maintaining Wheat-Corn Production and Reducing Net Greenhouse Gas Emissions under Climate Change. *Agriculture* **2022**, *12*, 1089. [CrossRef]
13. Shayanmehr, S.; Porhajašová, J.I.; Babošová, M.; Sabouhi Sabouni, M.; Mohammadi, H.; Rastegari Henneberry, S.; Shahnoushi Foroushani, N. The Impacts of Climate Change on Water Resources and Crop Production in an Arid Region. *Agriculture* **2022**, *12*, 1056. [CrossRef]
14. Li, N.; Nie, T.; Tang, Y.; Lu, D.; Wang, T.; Zhang, Z.; Chen, P.; Li, T.; Meng, L.; Jiao, Y.; et al. Responses of Soybean Water Supply and Requirement to Future Climate Conditions in Heilongjiang Province. *Agriculture* **2022**, *12*, 1035. [CrossRef]
15. Choruma, D.J.; Akamagwuna, F.C.; Odume, N.O. Simulating the Impacts of Climate Change on Maize Yields Using EPIC: A Case Study in the Eastern Cape Province of South Africa. *Agriculture* **2022**, *12*, 794. [CrossRef]
16. Wu, M.; Xu, Y.; Zheng, J.; Hao, Z. North Expansion of Winter Wheat Planting Area in China under Different Emissions Scenarios. *Agriculture* **2022**, *12*, 763. [CrossRef]
17. Xu, W.; Zhu, S.; Yang, T.; Cheng, J.; Jin, J. Maximum Entropy Niche-Based Modeling for Predicting the Potential Suitable Habitats of a Traditional Medicinal Plant (*Rheum nanum*) in Asia under Climate Change Conditions. *Agriculture* **2022**, *12*, 610. [CrossRef]
18. Sun, Y.; Zhang, S.; Tao, F.; Aboelenein, R.; Amer, A. Improving Winter Wheat Yield Forecasting Based on Multi-Source Data and Machine Learning. *Agriculture* **2022**, *12*, 571. [CrossRef]
19. Nooni, I.K.; Hagan, D.F.T.; Ullah, W.; Lu, J.; Li, S.; Prempeh, N.A.; Gnitou, G.T.; Lim Kam Sian, K.T.C. Projections of Drought Characteristics Based on the CNRM-CM6 Model over Africa. *Agriculture* **2022**, *12*, 495. [CrossRef]
20. Zhao, Y.; Xiao, D.; Bai, H.; Tang, J.; Liu, D. Future Projection for Climate Suitability of Summer Maize in the North China Plain. *Agriculture* **2022**, *12*, 348. [CrossRef]
21. Huang, X.; Han, S.; Shi, C. Multiscale Assessments of Three Reanalysis Temperature Data Systems over China. *Agriculture* **2021**, *11*, 1292. [CrossRef]

agriculture

Article

Multiscale Assessments of Three Reanalysis Temperature Data Systems over China

Xiaolong Huang [1,2], Shuai Han [3,*] and Chunxiang Shi [3]

1 Sichuan Meteorological Observation and Data Centre, Chengdu 610072, China; xiaolonghuang116@163.com
2 Heavy Rain and Drought-Flood Disasters in Plateau and Basin Key Laboratory of Sichuan Province, Chengdu 610072, China
3 National Meteorological Information Center, China Meteorological Administration, Beijing 100081, China; shicx@cma.gov.cn
* Correspondence: hans@cma.gov.cn

Abstract: Temperature is one of the most important meteorological variables for global climate change and human sustainable development. It plays an important role in agroclimatic regionalization and crop production. To date, temperature data have come from a wide range of sources. A detailed understanding of the reliability and applicability of these data will help us to better carry out research in crop modelling, agricultural ecology and irrigation. In this study, temperature reanalysis products produced by the China Meteorological Administration Land Data Assimilation System (CLDAS), the U.S. Global Land Data Assimilation System (GLDAS) and the European Centre for Medium-Range Weather Forecasts (ECMWF) Reanalysis version5 (ERA5)-Land are verified against hourly observations collected from 2265 national automatic weather stations (NAWS) in China for the period 2017–2019. The above three reanalysis systems are advanced and widely used multi-source data fusion and re-analysis systems at present. The station observations have gone through data Quality Control (QC) and are taken as "true values" in the present study. The three reanalysis temperature datasets were spatial interpolated using the bi-linear interpolation method to station locations at each time. By calculating the statistical metrics, the accuracy of the gridded datasets can be evaluated. The conclusions are as follows. (1) Based on the evaluation of temporal variability and spatial distribution as well as correlation and bias analysis, all the three reanalysis products are reasonable in China. (2) Statistically, the CLDAS product has the highest accuracy with the root mean square error (RMSE) of 0.83 °C. The RMSEs of the other two reanalysis datasets produced by ERA5-Land and GLDAS are 2.72 °C and 2.91 °C, respectively. This result indicates that the CLDAS performs better than ERA5-Land and GLDAS, while ERA5-Land performs better than GLDAS. (3) The accuracy of the data decreases with increasing elevation, which is common for all of the three products. This implies that more caution is needed when using the three reanalysis temperature data in mountainous regions with complex terrain. The major conclusion of this study is that the CLDAS product demonstrates a relatively high reliability, which is of great significance for the study of climate change and forcing crop models.

Keywords: temperature; evaluation; CLDAS; GLDAS; ERA5-Land

Citation: Huang, X.; Han, S.; Shi, C. Multiscale Assessments of Three Reanalysis Temperature Data Systems over China. *Agriculture* **2021**, *11*, 1292. https://doi.org/10.3390/agriculture11121292

Academic Editors: Dengpan Xiao and Wenjiao Shi

Received: 21 November 2021
Accepted: 17 December 2021
Published: 19 December 2021

1. Introduction

Climate change and its impact on agricultural regionalization and crop production is one of the most important fields of study around the world. Temperature is an important indicator of the energy balance of the earth's surface and directly affects global climate change. Accurate temperature data can reasonably drive crop models to simulate the impact of climate change on agricultural production, which is a key tool to explore planting management systems and put forward adaptation strategies [1–3]. To date, temperature data come from a wide range of sources. Conventional temperature observations at ground-based weather stations are single-point observations. Although a single-point observation may have high accuracy, its spatial coverage is quite limited, especially in mountainous areas with complex terrain, where the observation stations are often sparsely and unevenly distributed due to multiple constraints, such as the topography and environment conditions and maintenance difficulties. As a result, observations at such areas often have certain limitations in representativeness and applicability [4–6]. Temperature retrievals from remote sensing can provide spatially continuous observations, yet the retrieval accuracy is low [7–9]. Numerical model simulations have certain advantages regarding the spatial-temporal resolution. However, the model results are severely affected by various physical parameterization schemes [10,11], which often lead to large uncertainties in the output.

In recent years, various real-time analysis or reanalysis datasets (hereafter referred to as gridded dataset) on regular grids with high spatial resolution and temporal continuity have been produced by different multi-source data fusion and assimilation systems. A data fusion and assimilation system can take advantage of various data, analyze and process data from different sources, like direct observations, retrievals and model simulations, and output gridded dataset. A gridded dataset can cover a large spatial area over a long period, and thus effectively makes up for the lack of observations in areas where observation stations are sparse [12]. These gridded datasets provide basic data support for gridded forecasting, climate analysis and application services in meteorological agencies [13]. They can also be used as input data for numerical models to drive land surface, hydrological and ecological models to obtain more reliable results [10].

At present, various global and regional reanalysis datasets have been released. These datasets, including atmospheric and surface datasets, are produced by different data assimilation and fusion. The atmospheric reanalysis datasets include meteorological variables, such as temperature, relative humidity and UV winds on various pressure levels. The surface reanalysis datasets are composed of surface air temperature, and soil moisture, etc. By far, the National Centers for Environmental Prediction (NCEP) and National Centers for Atmospheric Research (NCAR) reanalysis products [14,15], the ECMWF reanalysis product (ERA5) [16–18], and the Japan Meteorological Agency (JMA) product JRA-55 [19] are the most widely used in the meteorological field. In practical applications, however, the requirement for spatial resolution of the near-surface elements are higher than that for the atmospheric elements in the upper air. Therefore, the land surface fusion system has been developing rapidly in recent years, and the fusion datasets that include near-surface meteorological elements and soil variables have been widely used in weather and climate prediction, water resources management and water cycle studies. In the beginning of the 2000s, the National Oceanic and Atmospheric Administration (NOAA) established the Global Land Data Assimilation System (GLDAS) [20]. In 2019, ECMWF released the high-accuracy ERA5-Land gridded surface dataset [21]. In 2015, the China Meteorological Administration Land Data Assimilation System (CLDAS-V2.0) was successfully developed [12,22,23].

Following the continuous improvement of observational systems, assimilation systems and numerical models, spatial and temporal resolutions of gridded temperature datasets also increased. They provide a rich data source for the mechanism study of regional atmospheric circulation and climate change studies. However, due to differences in the input data sources, as well as fusion models and assimilation systems, the simulation effect of temperature can be good or bad [24]. Therefore, the accuracy and applicability of temperature reanalysis datasets have always been a big concern of meteorologists. Many studies have evaluated the applicability of surface air temperature in several reanalysis products, such as ERA-40, JRA-25, NCEP/NCAR and NCEP/DOE, etc. It is found that these gridded reanalysis datasets can, to a certain degree, reflect spatial and temporal distribution characteristics of the observations [25–31], yet the differences between them demonstrate obvious regional and seasonal changes. Although the near surface air temperatures from GLDAS, ERA5-Landand CLDAS have been respectively evaluated over limited areas, comprehensive and detailed evaluation and comparison of these data over land areas of China have not been conducted. Note that the evaluation of gridded surface air temperature datasets is an important component of climate change study. Results of the evaluation provide a valuable reference for understanding regional temperature changes and promoting sustainable development.

Based on observations collected at automatic weather stations in China, this study analyzes the accuracy of near-surface air temperature in GLDAS, ERA5-Land and CLDAS gridded datasets over mainland China from different temporal and spatial perspectives. Results of the evaluation will be helpful for researchers to understand the applicability of these gridded datasets in China and provide a reference for the selection of appropriate temperature datasets in the studies of climate change, extreme weather, the Earth's energy and various numerical models. Meanwhile, this study will also help research institutions to further improve the algorithms used for producing these gridded datasets.

2. Data and Methods

2.1. Data

Table 1 lists of the spatial and temporal resolutions of the datasets used in this study and their coverage areas. It contains in-site data (NAWS) and three grid datasets (GLDAS, ERA5-Land and CLDAS).

Table 1. Characteristics of datasets.

Datasets	Spatial Coverage	Spatial Resolution	Temporal Resolution [1]	Data Type	Unit
NAWS	Over major land areas of China	2265 Stations	Hourly	Point	°C
GLDAS	180° W–180° E; 60° S–90° N	0.25°	Every 3 Hours	Grid	K
ERA5-Land	180° W–180° E; 60° S–90° N	0.1°	Hourly	Grid	K
CLDAS	70°–140° E; 0°–60° N	0.05°	Hourly	Grid	K

[1] Instantaneous value of temperature at current time.

2.1.1. GLDAS Data

GLDAS is evolved from the land information system [32], which is a land surface data assimilation system that consists of multiple land surface models. It is applied to integrate observation-based data and produce surface state (such as soil moisture, surface temperature) and flux (such as evaporation, latent heat and sensible heat flux) variables. GLDAS includes four land surface models [33], i.e., Noah, Mosaic, CLM and VIC. Driven by Princeton University's global meteorological dataset, GLDAS-2 created a more climatologically consistent dataset that covers the period from 1948 to 2010 [34]. The horizontal resolution of this dataset is 0.25° and the temporal resolution is 3-h. It covers the area of (60° S–90° N, 180° W–180° E). In the present study, GLDAS gridded temperature data are downloaded from NASA Goddard Earth Science Data and Information Service Center (GES DISC) (http://disc.sci.gsfc.nasa.gov/hydrology/data-holdings, accessed on 5 September 2021).

2.1.2. ERA5-Land Data

ERA5-Land is a repeat of the ERA5 climate reanalysis, while a series of improvements have been made to make it better meet application requirements [21,35–37]. In particular, ERA5-Land runs at enhanced resolution (9 km vs. 31 km in ERA5). The temporal frequency of the output is hourly and the fields are masked for all oceans, making them lighter to handle. ERA5-Land is produced by a single model simulation that is not incorporated into the ECMWF Integrated Forecasting System (IFS). The ERA5-Land historical dataset for the period 1950–1980 was released in September 2021, and the dataset for the period since 1981 was initially released to users in 2019 and is being updated in real-time. The spatial and temporal resolutions of the dataset are 1-h and 0.1°, respectively. In the present study, the ERA5-Land surface air temperature gridded dataset is downloaded from ECMWF Copernicus Climate Change Service Climate Data Store (C3S CDS) (https://cds.climate.copernicus.eu/cdsapp#!/dataset/reanalysis-era5-land?tab=form, accessed on 5 September 2021).

2.1.3. CLDAS Data

The version2 China Meteorological Administration Land Assimilation System (CLDAS-V2.0) is developed by the National Meteorological Information Center of the China Meteorological Administration. It runs four physical parameterization schemes [23] (Noah, CLM3.5, Common Land Model CoLM and Noah MP) to simulate various soil variables, such as soil temperature, soil moisture, evapotranspiration, surface heat flux, etc. The CLDAS datasets mainly include the forcing dataset and the land surface dataset. The forcing dataset is produced from fusion of observations collected at more than 60,000 automatic weather stations with numerical model predictions, as well as satellite remote sensing data using multi-grid variational analysis technology, discrete ordinates radiation model, hybrid radiation estimation model and terrain correction algorithm [12,23]. CLDAS provides gridded 2 m air temperature, 2 m humidity, 10 m UV wind, ground pressure, ground incident solar radiation, precipitation and other elements. In this study, the hourly gridded temperature dataset of the second edition of CLDAS is obtained from the National Meteorological Information Center of China Meteorological Administration. The spatial and temporal resolutions of the dataset are 1-h and 0.05°, respectively, and it covers the area of (0° N–60° N, 70° E–140° E) (http://data.cma.cn/, accessed on 5 September 2021).

2.1.4. NAWS Observation Data

NAWS observations are obtained from the CIMISS database in Sichuan Meteorological Observation Data Center. In total, 2281 national weather stations in mainland China with data integrity above 98% are selected (Figure 1). The observation instruments at the above weather stations are regularly calibrated, upgraded, and maintained by professionals in the meteorological field. The data collected at these weather stations have passed the national, provincial, and station quality controls and all data are marked with the QC flags [38]. Due to the lack of representativeness of the data observed by alpine stations in the plain area, the observation stations are excluded. Finally, data collected at the remaining 2265 stations are considered to be the most reliable observational data, which can be used as the benchmark data for the evaluation of CLDAS, ERA5-Land and GLDAS.

2.2. Data Processing

In this study, hourly 2 m temperature data collected at 2265 NAWS for the period 2017–2019 are used to evaluate the accuracy of CLDAS, ERA5-Land and GLDAS gridded datasets. Only those data indicated by QC flag as "correct" are selected to produce the "true value" dataset for the evaluation of reanalysis datasets. To address possible impacts caused by the displacement of weather stations during the evaluation period, GLDAS, ERA5-Land and CLDAS temperatures are spatially interpolated using the bi-linear interpolation method [39] to station locations according to the latitude and longitude information of the stations at each time to obtain comparative sequences. A total of 19,642,844 samples have been obtained. By calculating the statistical metrics defined in Section 2.3, the accuracy of the gridded datasets can be evaluated.

2.3. Metrics Used for Evaluation

Correlation coefficient (COR), mean error (ME) and root mean square error (RMSE) are used to compare CLDAS, GLDAS and ERA5-Land data with NAWS observations. They are defined as follows.

$$\mathrm{COR} = \frac{\sum_{i=1}^{n} (G_i - \bar{G})(O_i - \bar{O})}{\sqrt{\sum_{i=1}^{n} (G_i - \bar{G})^2} \sqrt{\sum_{i=1}^{n} (O_i - \bar{O})^2}} \quad (1)$$

$$BIAS = \frac{1}{N}\sum_{i=1}^{n} G_i - O_i \tag{2}$$

$$RMSE = \sqrt{\frac{1}{N}\sum_{i=1}^{n} (G_i - O_i)^2} \tag{3}$$

where O_i is the weather station observation, G_i is the gridded temperature data interpolated to the station locations, N is the total number of samples used in the evaluation (number of stations). COR varies within [−1.0–1.0]; the closer the value is to 1, the better the data consistency and closer it is to −1, the stronger the opposite relationship. When COR is 0, it means that there is no linear relationship between product and observation. BIAS reflects the degree of deviation of the gridded temperature data from the observation at the station. A negative value indicates that the gridded temperature data is underestimated, while a positive value indicates that the temperature is overestimated in the reanalysis dataset. The closer the RMSE is to 0, the more accurate the gridded temperature data set is. During the evaluation and inspection period, all the samples used for evaluation are calculated based on the cumulative results of hourly observations.

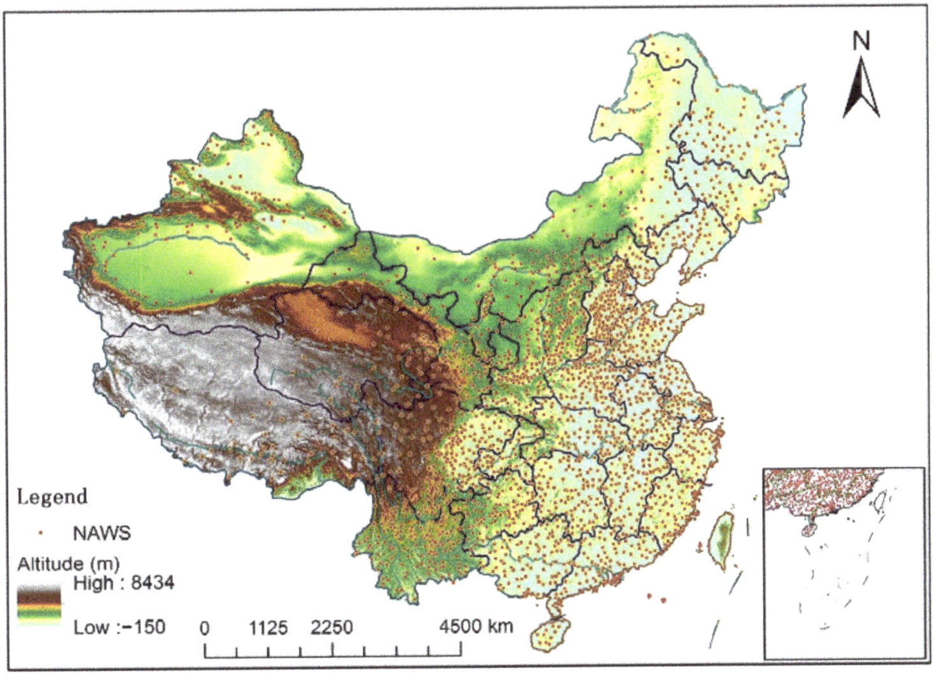

Figure 1. Distribution of National Automatic Weather Stations (NAWS) in China.

3. Results Analysis

3.1. Evaluation of Overall Accuracy

3.1.1. Overall Accuracy during the Study Period

For the evaluation period of 2017–2019, the overall accuracy results are listed in Table 2. The average temperature observations at the weather stations in mainland China is 13.93 °C, and the averages CLDAS, ERA5-Land and GLDAS are 13.88 °C, 13.22 °C and 13.55 °C, respectively, and are 0.06 °C, 0.71 °C and 0.38 °C lower than the station observation. Note that the average of CLDAS is very close to that of the observations. From the perspective of correlation, the highest value of 0.998 is found between CLDAS and observations and the lowest value of 0.970 is found between GLDAS and observations, while that between the ERA5-Land and observations is insignificant. The biases of the three reanalysis datasets all are negative in mainland China, indicating that temperatures in these gridded datasets are underestimated compared to station observations, and the underestimation is most severe in ERA5-Land with a value of 0.71 °C. The RMSEs for CLDAS, ERA5-Land and GLDAS are 0.83 °C, 2.72 °C and 2.91 °C, respectively. Overall, both correlation and bias metrics indicate that the accuracy of CLDAS is obviously higher than that of the other two gridded datasets for the evaluation period. The accuracy of ERA5-Land is better than that of GLDAS, although the difference between them is small.

Table 2. Statistics of CLDAS, ERA5-Land and GLDAS temperatures for the period 2017–2019.

Gridded Dataset	Average (°C)	COR	BIAS (°C)	RMSE (°C)
CLDAS	13.875	0.998	−0.053	0.833
ERA5-Land	13.221	0.976	−0.706	2.717
GLDAS	13.549	0.970	−0.378	2.911

Temperatures at the 4 h of 00, 06, 12, and 18 UTC on the 15th day of each month from January to December 2019 are used to represent the annual mean temperature in 2019. Scatterplots of temperature from the NAWS observations and the three gridded datasets, as well as their linear fittings, are displayed in Figure 2. The goodness of fit (R^2) are 0.995, 0.95 and 0.945, respectively, for CLDAS, ERA5-Land and GLDAS. Intuitively, it can be seen that the CLDAS has a higher accuracy.

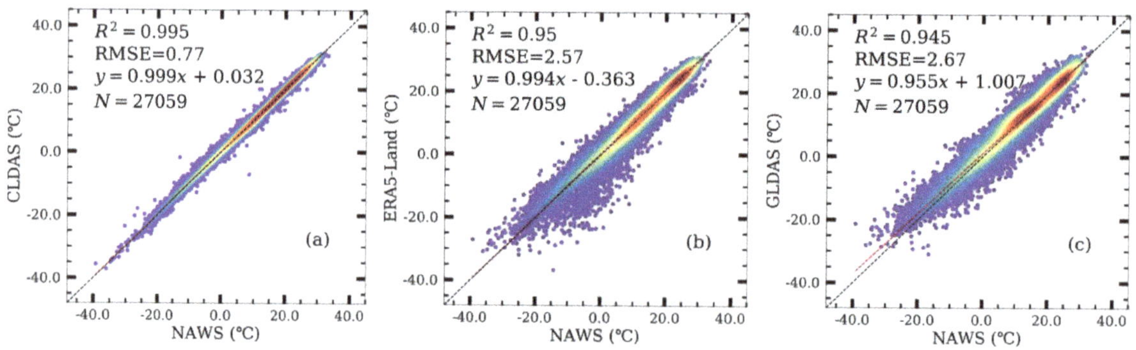

Figure 2. Fitting deviations of CLDAS (**a**), ERA5-Land (**b**) and GLDAS (**c**) data from NAWS observations.

3.1.2. Evaluation at Individual Stations

Figure 3 displays the spatial distributions of correlation coefficients between the three gridded datasets and observations in mainland China. For most stations, the COR values for CLDAS are higher than that for GLDAS and ERA5-Land. For the three datasets, the COR values all decrease from east to west. As shown in Figure 3a for CLDAS, the COR values are greater than 0.99 at most stations, except a few individual stations over the Tibetan Plateau, the Hengduan Mountains and other high elevation areas. For ERA5-Land (Figure 3b), stations with a COR larger than 0.99 are concentrated in Northeast China, North China, and the middle and lower reaches of the Yangzi River. The COR values decrease from 0.98 to 0.95 over inland China and are largely below 0.96 in West China. As shown in Figure 3c for GLDAS, the spatial pattern of COR is similar to that for ERA5-Land, while the COR values are largely smaller than that for ERA5-Land over inland China and Sichuan Basin. Figure 3d presents the Kernel Distribution Estimation (KDE) of the density of stations corresponding to COR values for the three gridded datasets. Note that the greater the number of stations with COR close to 1.0, the better the correlation of the gridded dataset with station observations. Figure 3d clearly indicates that CLDAS is the best among the three datasets, while ERA5-Land is better than GLDAS.

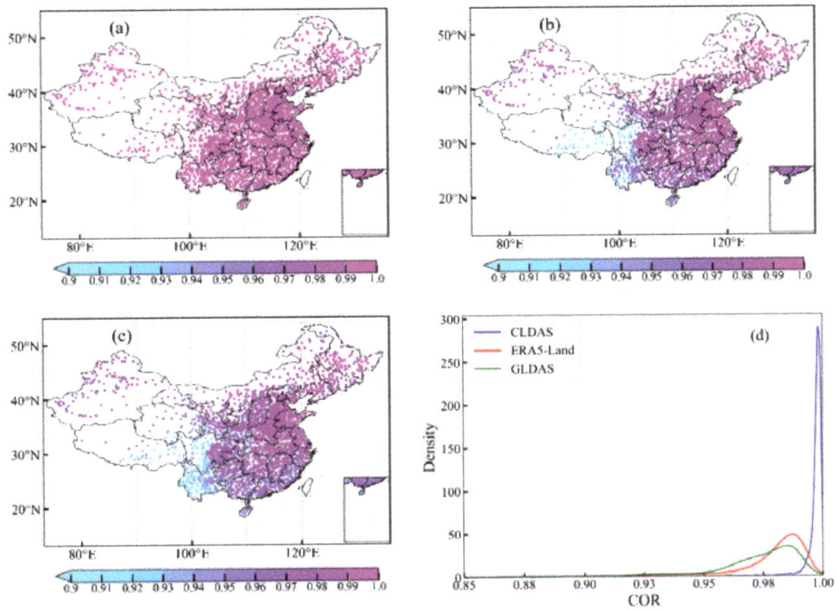

Figure 3. Spatial distributions of COR ((**a**): CLDAS; (**b**): ERA5-Land; (**c**): GLDAS); (**d**): KDE of stations on COR for CLDAS, ERA5-Land and GLDAS datasets.

Spatial distributions of RMSEs for the three datasets are presented in Figure 4, which indicates that the RMSEs for CLDAS are smaller than those for GLDAS and ERA5-Land at most stations. The value of RMSE increases from east to west for all three datasets. As shown in Figure 4a for CLDAS, the RMSE values are smaller than 0.5 °C at all of the stations, except for those in Xinjiang, Yunnan, the Tibetan Plateau and the high elevation areas in western Sichuan. The spatial distributions of RMSEs for ERA5-Land (Figure 4b) and GLDAS (Figure 4c) tend to be similar, with the values concentrated over 1.0–3.0 °C for ERA5-Land and over 1.5–4.0 °C for GLDAS. The KDE of density of stations with RMSE for the three datasets are presented in Figure 4d, which shows clearly that CLDAS is better than the other two datasets.

Figure 5 displays the spatial distributions of BIAS for the three datasets in mainland China. For CLDAS (Figure 5a), the biases at most stations vary between −1.0–1.0 °C with positive values in the east and negative values in the west, where the terrain elevation is relatively high. The spatial distributions of bias for ERA5-Land (Figure 4b) and GLDAS (Figure 4c) are basically consistent. Large positive values occur in the North China Plain and the Taklimakan Desert in Xinjiang, while negative values mostly occur in Fujian and southwest China (except Sichuan Basin). Figure 4d displays the KDE of density of stations with BIAS for the three datasets. It shows that positive and negative biases each account for half of the total stations for CLDAS, while negative biases prevail for ERA5-Land and the opposite is true for GLDAS.

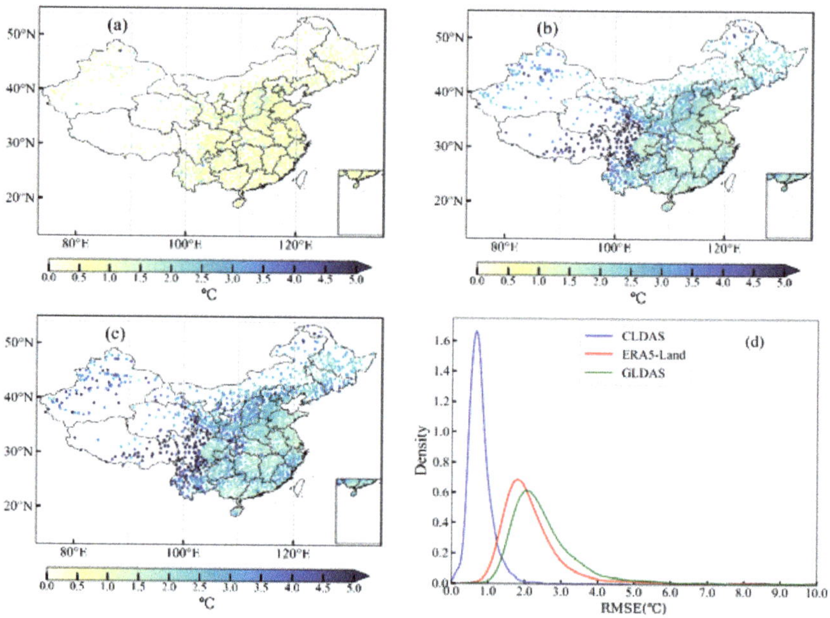

Figure 4. Spatial distributions of RMSE ((**a**): CLDAS; (**b**): ERA5-Land; (**c**): GLDAS); (**d**): KDE of stations on RMSE for CLDAS, ERA5-Land and GLDAS.

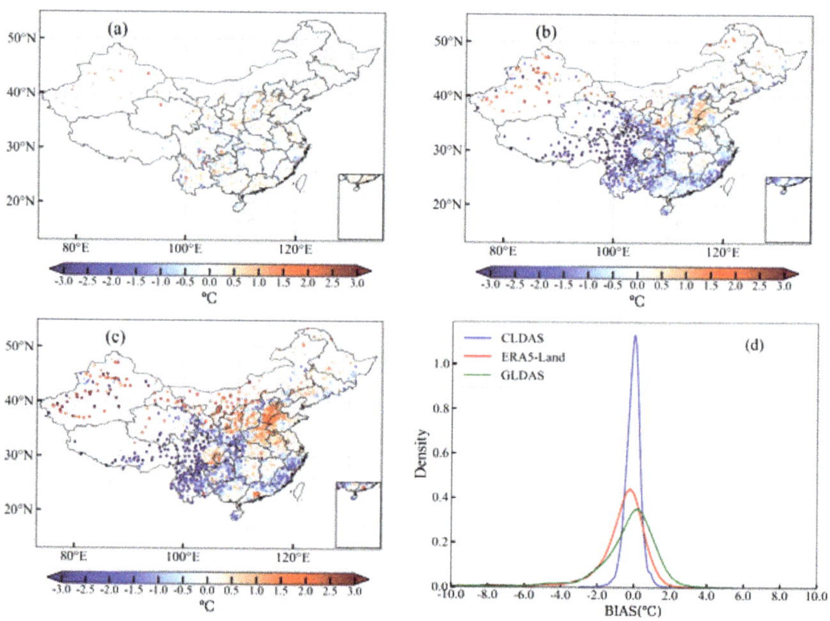

Figure 5. Spatial distributions of BIAS ((**a**): CLDAS; (**b**): ERA5-Land; (**c**): GLDAS); (**d**): KDE of stations on BIAS for CLDAS, ERA5-Land and GLDAS.

3.2. Evaluation at Various Time Scales

3.2.1. At Different Times of the Day

Figure 6 displays diurnal features of the four statistics metrics for the evaluation of the three datasets over the period 2017–2019. Multi-year averages of temperature at different times of the day are shown in Figure 6a, which indicates that the three gridded datasets exhibit consistent diurnal temperature variation with observations. COR at different times of the day are displayed in Figure 6b, which suggests that the COR for CLDAS changes little, with a value around 0.997. The COR value for ERA5-Land first decreases from 0.978 at 00 UTC to 0.97 at 06 UTC, then gradually increases and reaches the peak at 12 UTC, and then slowly decreases again. The COR for GLDAS first increases from 0.974 at 00 UTC to 0.979 at 03 UTC, and then gradually decreases. Figure 6c presents RMSEs at different times of the day. The largest RMSEs occur at 15, 06 and 12 UTC for CLDAS, ERA5-Land and GLDAS, respectively with the values of 0.94 °C, 2.86 °C and 3.27 °C. Biases at different times of the day for the three datasets are displayed in Figure 6d, which shows positive biases for CLDAS at 00, 03 and 06 UTC with the largest positive bias of 0.22 °C at 03 UTC and the largest negative bias of −0.29 °C at 12 UTC. GLDAS has positive biases at 00 and 03 UTC and the bias is up to 0.66 °C at 00 UTC, and negative biases occur at all other times with the largest negative bias of −1.43 °C at 12 UTC. ERA5-Land exhibits negative biases at all times of the day with the largest negative bias of −0.98 °C at 09 UTC. Overall, negative bias prevails at different times of the day for all the three datasets.

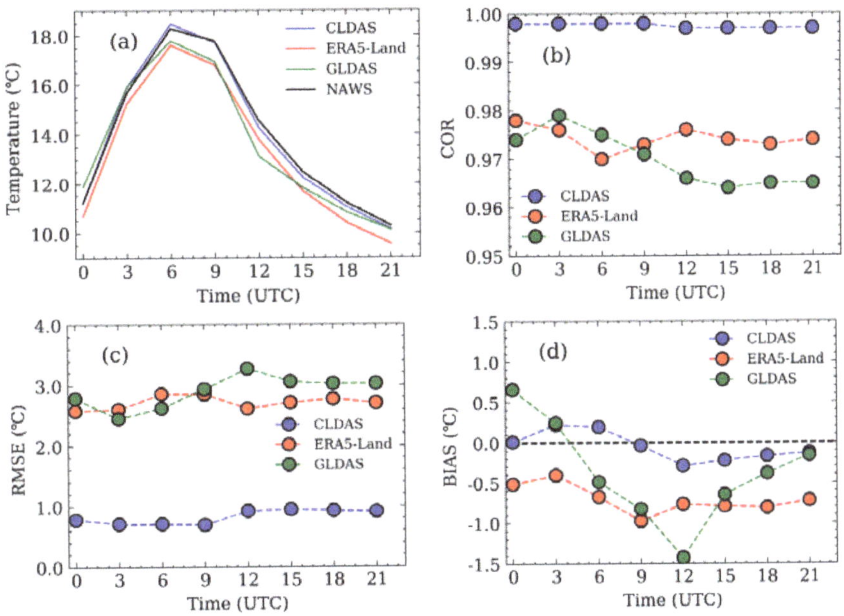

Figure 6. Diurnal variations of temperature during 2017–2019. (**a**): Multi-year average temperature, (**b**): COR; (**c**): RMSE; (**d**): BIAS.

3.2.2. Daily Evaluation

Figure 7 presents daily variations of the evaluation metrics during 2017–2019. Daily mean temperatures for the three datasets and observations are shown in Figure 7a, which indicates that the daily temperature variation during the study period for the three datasets is consistent with station observations. Figure 7b presents daily CORs during the study period. The daily COR between CLDAS and station observations shows little changes, whereas the CORs of ERA5-Land and GLDAS with station observations exhibit large daily variations with the values ranging between 0.891–0.977 and 0.848–0.979, respectively. Daily RMSEs of the three datasets are displayed in Figure 7c, which indicates that the RMSEs of CLDAS are obviously smaller than those of the other two datasets. The RMSEs of ERA5-Land and GLDAS are close, yet the RMSEs of ERA5-Land are smaller than those of GLDAS in most days. Daily RMSEs of CLDAS, ERA5-Land and GLDAS vary between 0.61–2.35 °C, 1.97–3.80 °C and 2.43–3.76 °C, respectively. Figure 7d presents daily biases of the three datasets. It is obvious that the biases of CLDAS basically are negative but very close to 0.0. The biases of ERA5-Land are also negative almost all of the time except for a few days. GLDAS is dominated by negative biases in autumn and winter, while positive biases mainly occur in spring and summer.

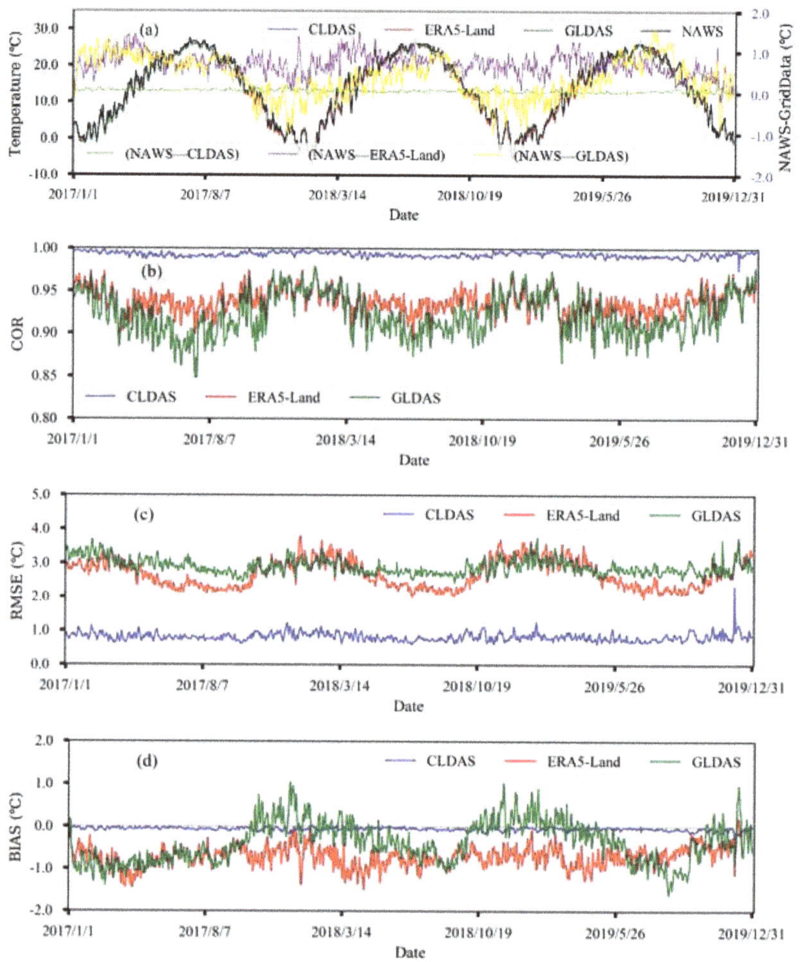

Figure 7. Daily evaluation of temperature during 2017–2019. (**a**): Daily average temperature, (**b**): COR; (**c**): RMSE; (**d**): BIAS.

3.2.3. Monthly Changes

Figure 8 shows monthly changes of temperature and evaluation metrics for the three datasets over 2017–2019. The curves of monthly average temperature of the three grid datasets are basically consistent with the observed values (Figure 8a). Monthly CORs are presented in Figure 8b, which shows that the three gridded datasets exhibit similar monthly variation patterns, i.e., the COR gradually decreases from January to June, and gradually increases from July to December. The possible reason is that the average temperature in China gradually increases because it is located in the northern hemisphere, and the change range of hourly temperature increases relatively. However, the reanalysis temperature products are affected by spatial resolution, and there is a certain regional smoothness in the response to this temperature change, which results in a decreasing COR. Similarly, COR gradually increased from June to December. The CORs of GLDAS are significantly lower than those of ERA5-Land during April-October, but no obvious differences can be found in other months. Figure 8c shows monthly RMSEs. The RMSEs of ERA5-Land overall are smaller than those of GLDAS but are slightly higher in November and February. Monthly biases of CLDAS and ERA5-Land all are negative, while negative biases prevail in GLDAS (Figure 8d) with positive biases only occurring from November 2017 to February 2018 and from October 2018 to February 2019. Among the three gridded datasets, the monthly biases of GLDAS vary the most and those of the CLDAS vary the least.

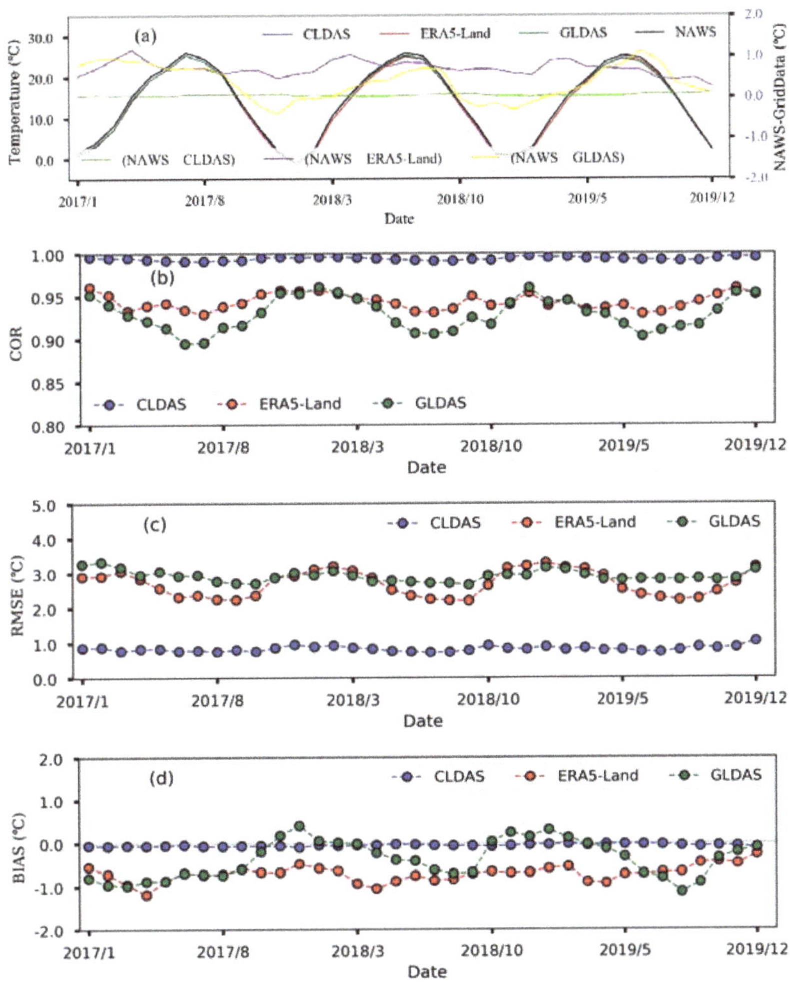

Figure 8. Monthly changes of temperature during 2017–2019. (**a**): Monthly average temperature, (**b**): COR; (**c**): RMSE; (**d**): BIAS.

3.2.4. Seasonal Changes

Figure 9 shows seasonal characteristics of the evaluation metrics over the study period. Seasonal average temperatures of the three gridded datasets and station observations are presented in Figure 9a. Seasonal mean temperatures of station observations and CLDAS, ERA5-Land and GLDAS datasets are 15.0 °C, 14.96 °C, 14.06 °C and 14.57 °C, respectively, in spring. Summer mean temperatures of the above four datasets are 24.61 °C, 24.56 °C, 23.87 °C and 23.88 °C, respectively. Autumn mean temperatures of these datasets are 14.23 °C, 14.16 °C, 13.63 °C and 13.95 °C, respectively. Seasonal mean temperatures in winter are 1.64 °C, 1.58 °C, 1.08 °C and 1.55 °C for the four datasets, respectively. In all the four seasons, the seasonal mean temperature of CLDAS is the closest to the observations, followed by that of GLDAS, and the result of ERA5-Land is the worst. Figure 9b shows the seasonal correlation between the three gridded datasets and station observations. It is found that the correlation is the lowest in summer, the highest in autumn and is higher in winter than in spring. This is a common feature for all three gridded datasets. RMSEs are displayed in Figure 9c, which shows that the RMSEs of the three datasets are the largest in winter with the value of 0.91 °C for CLDAS and 3.1 °C for both ERA5-Land and GLDAS. Figure 9d presents seasonal biases. Compared to station observations, temperature in all four seasons is underestimated in the three gridded datasets. The largest negative bias of CLDAS is −0.06 °C, which appears in autumn. The largest negative bias of ERA5-Land appears in winter with a value of −0.93 °C. GLDAS has the largest negative bias of −0.73 °C that appears in summer.

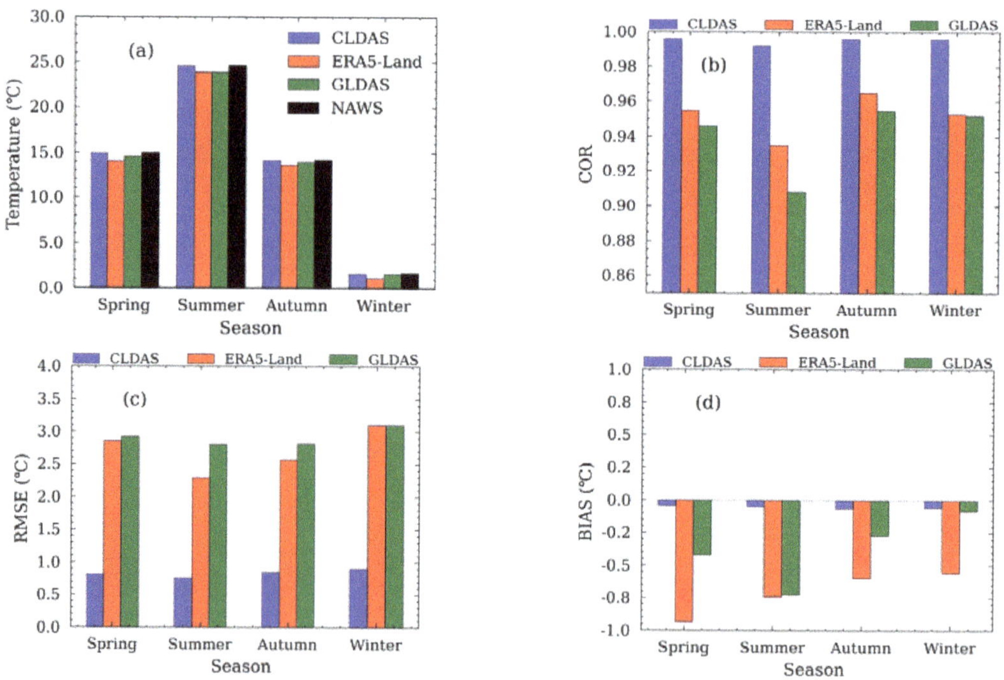

Figure 9. Seasonal changes of temperature during 2017–2019. (**a**): Seasonal average temperature, (**b**): COR; (**c**): RMSE; (**d**): BIAS.

3.3. Evaluation over Subregions

3.3.1. Evaluation over Subregions Divided according to Climate Regimes

With reference to previous studies [40,41], China is divided into eight subregions for evaluation based on topographic and climatic characteristics. Figure 10 shows the eight subregions: subregion I (33°–50° N, 72°–105° E), subregion II (25°–33° N, 76°–95° E), subregion III (17°–25° N, 95°–105° E), subregion IV (25°–33° N, 95°–105° E), subregion V (7°–30° N, 105°–125° E), subregion VI (30°–40° N, 105°–125° E), subregion VII (40°–55° N, 118°–135° E), and subregion VIII (40°–55° N, 105°–118° E).

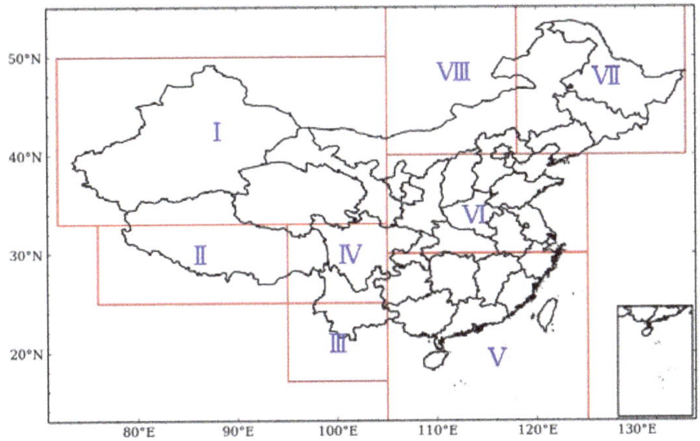

Figure 10. Subregions of China according to climate regimes.

Results of evaluation over climate regimes are listed in Table 3, which shows that the RMSEs of ERA5-Land and GLDAS are the largest in subregion II. This subregion is located in the Tibetan Plateau, where negative biases prevail. This is also the region with the largest negative bias. In contrast, the RMSE of CLDAS is the smallest in subregion II, where positive biases appear. The RMSEs of ERA5-Land and GLDAS are smaller in subregions V, VI and VII than in other subregions. Subregions V, VI and VII are located in eastern China, where the terrain is relatively flat. The RMSEs of the three datasets are larger in subregion IV than in other subregions, which is attributed to the fact that subregion IV is located in the transitional zone from the Tibetan Plateau to Sichuan Basin, where the terrain is extremely complex.

Table 3. Evaluation results over subregions of different climate regimes.

Area	CLDAS			ERA5-Land			GLDAS		
	COR	BIAS (°C)	RMSE (°C)	COR	BIAS (°C)	RMSE (°C)	COR	BIAS (°C)	RMSE (°C)
I	0.997	0.023	1.000	0.963	−0.883	3.689	0.958	−0.011	3.734
II	0.998	0.058	0.586	0.891	−5.156	6.76	0.896	−3.93	5.62
III	0.987	−0.139	1.004	0.935	−1.195	2.526	0.89	−1.858	3.433
IV	0.995	−0.097	0.851	0.919	−3.247	5.18	0.909	−3.201	5.072
V	0.996	−0.073	0.752	0.977	−0.735	1.926	0.967	−0.564	2.223
VI	0.997	−0.044	0.828	0.984	−0.183	1.973	0.978	0.292	2.324
VII	0.998	−0.042	0.842	0.989	0.011	2.271	0.985	0.047	2.634
VIII	0.998	−0.087	0.797	0.987	−0.109	2.265	0.981	0.299	2.733

3.3.2. Evaluation over Administrative Regions

The province is the second-level administrative unit in China, which possesses certain geographical and human attributes. Most operational meteorological services or scientific research projects are conducted according to the territorial principle. Therefore, it is necessary to evaluate the gridded datasets from the perspective of the provinces. For this reason, all of the national automatic weather stations in China are grouped according to their provincial attributes, and the biases in each province are calculated individually Results are listed in Table 4, which shows that, except for Tibet and Guizhou for GLDAS, and except Tibet for ERA5-Land, the CORs of the two gridded datasets with station observations are above 0.90, while the CORs of CLDAS are above 0.99 in all of the provinces The RMSEs of CLDAS are below 1.0 °C in all of the provinces, except for Shanxi and Xinjiang, where the values are 1.012 °C and 1.088 °C, respectively. The RMSEs of ERA5-Land and GLDAS are below 3.0 °C in all of the provinces except for Gansu, Xinjiang, Qinghai, Guizhou, Sichuan and Tibet, and the largest RMSEs of the two datasets both occur in Tibet, with the values of 7.86 °C and 6.292 °C, respectively. The numbers of provinces with negative biases in CLDAS, ERA5-Land and GLDAS respectively account for 61%, 81%, and 55% of the total number of provinces in mainland China. The evaluation results show that the quality of CLDAS, ERA5-Land and GLDAS is significantly better in the eastern provinces than in the western provinces of China. Compared with ERA5-Land and GLDAS, CLDAS is closer to observations in each individual province. ERA5-land is better than GLDAS in all of the provinces except for Sichuan, Qinghai and Tibet, where the biases of ERA5-Land are slightly larger than those of GLDAS.

Table 4. Evaluation results over provinces in mainland China.

Province	CLDAS			ERA5-Land			GLDAS		
	COR	BIAS (°C)	RMSE (°C)	COR	BIAS (°C)	RMSE (°C)	COR	BIAS (°C)	RMSE (°C)
Anhui	0.997	−0.013	0.735	0.987	−0.164	1.596	0.983	0.226	1.841
Beijing	0.997	0.023	0.881	0.983	−0.638	2.355	0.979	0.132	2.541
Chongqing	0.994	−0.115	0.924	0.973	−0.79	2.101	0.954	−0.49	2.556
Fujian	0.996	−0.136	0.731	0.974	−1.113	2.066	0.96	−1.296	2.495
Gansu	0.996	−0.06	0.947	0.963	−1.13	3.216	0.956	−0.426	3.228
Guangdong	0.996	0.187	0.612	0.97	−0.842	1.823	0.955	−0.231	2.039
Guangxi	0.996	0.03	0.66	0.968	−0.856	2.009	0.958	−0.507	2.127
Guizhou	0.99	−0.107	0.992	0.933	−1.678	3.035	0.896	−2.481	4.053
Hainan	0.991	−0.311	0.725	0.948	−0.965	1.807	0.932	−1.202	2.15
Hebei	0.998	−0.039	0.858	0.987	0.1	2.023	0.982	0.67	2.408
Heilongjiang	0.999	0.029	0.863	0.99	0.124	2.282	0.987	0.085	2.592
Henan	0.997	−0.149	0.782	0.985	0.083	1.876	0.982	0.554	2.093
Hubei	0.996	−0.098	0.812	0.981	−0.497	1.933	0.97	−0.36	2.337
Hunan	0.997	−0.151	0.746	0.983	−0.43	1.75	0.973	−0.323	2.109
Inner Mongolia	0.999	−0.107	0.717	0.989	0.093	2.202	0.983	0.506	2.727
Jiangsu	0.998	0.114	0.656	0.99	−0.237	1.42	0.987	0.203	1.63
Jiangxi	0.997	−0.113	0.716	0.983	−0.219	1.661	0.976	−0.17	1.954
Jilin	0.998	−0.122	0.866	0.988	−0.186	2.317	0.983	−0.092	2.731
Liaoning	0.998	−0.041	0.884	0.988	−0.025	2.121	0.983	0.191	2.528
Ningxia	0.997	0.039	0.97	0.983	−0.031	2.131	0.974	0.931	2.805
Qinghai	0.996	0.1	0.901	0.94	−2.986	4.948	0.944	−1.697	3.896
Shaanxi	0.996	0.038	0.906	0.975	−0.455	2.399	0.963	−0.151	2.895
Shandong	0.998	−0.055	0.744	0.989	0.028	1.652	0.986	0.515	1.942
Shanghai	0.999	0.001	0.442	0.991	−0.448	1.339	0.986	0.769	1.704
Shanxi	0.996	0.021	1.012	0.98	−0.25	2.365	0.974	0.165	2.72
Sichuan	0.996	−0.153	0.782	0.932	−2.536	4.49	0.915	−2.127	4.423
Tianjing	0.998	0.088	0.766	0.99	−0.131	1.72	0.987	0.879	2.131
Tibet	0.998	−0.013	0.61	0.865	−6.112	7.856	0.883	−4.638	6.292
Xinjiang	0.997	0.011	1.088	0.973	0.25	3.292	0.966	0.95	3.796
Yunnan	0.995	−0.097	0.813	0.974	−1.133	2.155	0.962	−0.913	2.401
Zhejiang	0.996	−0.212	0.867	0.981	−0.861	1.936	0.972	−1.09	2.39

4. Discussion

The present study reveals some important issues that are different to previous studies [30,41,42]. For example, the biases of CLDAS, ERA5-Land and GLDAS at night are larger than that in daytime, and all three datasets have negative biases in the nighttime. Monthly biases of the three gridded datasets demonstrate certain regularities. From January to June, their correlations with station observations gradually decrease, and the biases increase. From July to December, the correlations gradually increase, and the biases decrease. Seasonal correlations of the three datasets with observations are the lowest in summer and the highest in autumn, while the correlation in winter is higher than that in spring. Similar assessment results also found a monthly variation in the GLDAS evaluation results, but the deviation was the lowest in August [43], which may be due to different time periods of evaluation.

In addition, the change in temperature is significantly related to geographical locations and variations, such as altitude and slope. The Integrated Nowcasting through Comprehensive Analysis (INCA) [44] were used in the fine lattice simulation and application of temperature over complex terrain, and this method compared and analyzed the other three interpolation methods (inverse distance weighting method, inverse distance weighting method and ordinary Kriging method) [45]. The altitude of the station will have a great impact on the results of the four grid methods, and the error increases gradually with the increase of the elevation of the verification station. However, it is mainly aimed at Zhejiang Province in eastern China [46]. There are few reports that provide a detailed evaluation of site classification according to terrain across the whole of China.

The topography in China is high in the west and low in the east, showing a staircase-like distribution with multiple terrain patterns and large mountainous areas. The 2065 observation stations used in this study are located at different elevations. The highest station is the Amdo Station in Tibet, the elevation of which is 4800 m. The lowest station is the Turpan Station in Xinjiang, western China, and its elevation is −48.7 m. Evaluation of the present study at individual stations and over various regions indicate that the accuracy of the three datasets is, to a certain degree, related to topography. This is because surface air temperature in gridded datasets is simulated at each fixed grid, where the elevation is the grid-average value. However, the elevation of a weather station may not be able to well represent the average elevation of its nearby area, which may possibly lead to biases in the gridded dataset. Next, we will further classify the slope and elevation of the observational stations, and discuss the influences of the two main terrain features on the accuracy of the gridded datasets.

4.1. Impact of Terrain Elevation on the Accuracy of Gridded Dataset

According to their elevations, the stations are divided into eight categories, i.e., elevation < 500 m, ≥500–1000 m, ≥1000–1500 m, ≥1500–2000 m, ≥2000–2500 m, ≥2500–3000 m, ≥3000–3500 m and ≥3500 m. Figures 11 and 12 show the bias characteristics of the three gridded datasets at different elevations. The correlations of ERA5-Land and GLDAS with station observations both show a downward trend with increasing elevation, while their average biases gradually increase with more severe underestimation and the bias range at a single station becomes more divergent. Compared to ERA5-Land and GLDAS, the CLDAS dataset is less affected by elevation.

Several previous studies have also found that elevation differences between stations and model grids are a major reason for the biases in reanalysis datasets [47–49]. Specifically, weather stations over the Hengduan Mountain in western Sichuan are concentrated in the river valley, where the elevation is greatly different to the surrounding areas. Large cold biases in this area are found in gridded datasets because the station elevations there are lower than the heights of corresponding model grids. For those stations located at the top of mountains, their elevations probably are higher than the heights of model grids at the same place. As a result, warm biases are found at these stations in the gridded datasets. The above discussion indicates that the elevation correction of temperature in the gridded dataset can effectively reduce the biases and improve the applicability of the dataset [31,50,51]. In addition, possible input data errors, model system errors, and interpolation errors (from Gaussian grid to latitude-longitude grid) of the fusion system are also sources of biases.

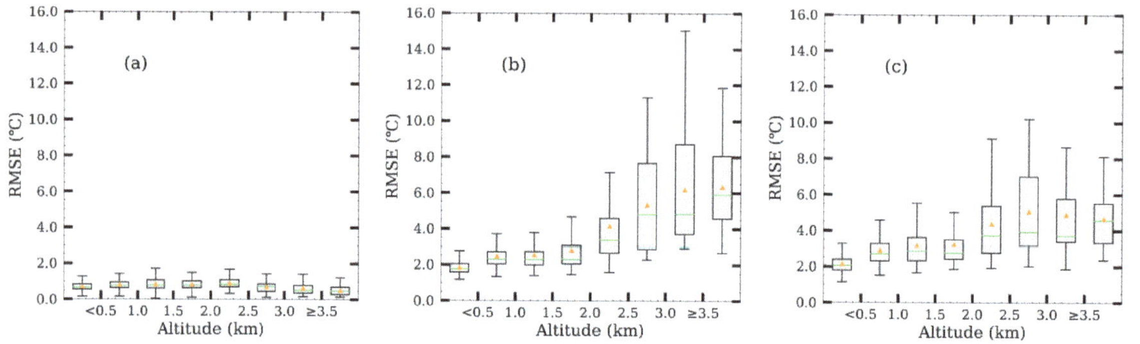

Figure 11. RMSE changes with elevation for ((**a**): CLDAS, (**b**): ERA5-Land, (**c**): GLDAS) datasets.

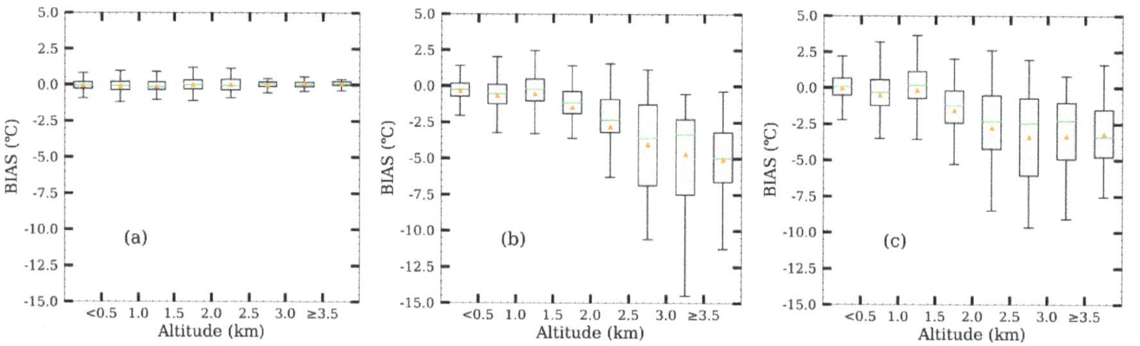

Figure 12. BIAS changes with elevation for ((**a**): CLDAS, (**b**): ERA5-Land, (**c**): GLDAS) datasets.

4.2. Impact of Slope on the Accuracy of Gridded Dataset

According to the classification of slopes proposed by the International Geographical Union and the Geomorphological Mapping Committee for the application of detailed geomorphological maps [52], the slope grades are divided into: plain (0°–0.5°), slight slope (>0.5°–2°), gentle slope (>2°–5°), slope (>5°–15°), steep slope (>15°–35°), steep slope cliffs slope (>35°–55°), vertical slope (>55°–90°). Figures 13 and 14 displays the RMSE and BIAS characteristics of the three datasets over different types of slope. It is found that RMSEs and BIASs of ERA5-Land and GLDAS both increase with increasing slope, while the correlations of the two datasets with observations decrease and the mean errors gradually increase. The underestimation of temperature in the two gridded datasets also gradually intensifies, with a wider spread of biases at individual stations. Compared with the above two datasets, CLDAS is less affected by the terrain slope.

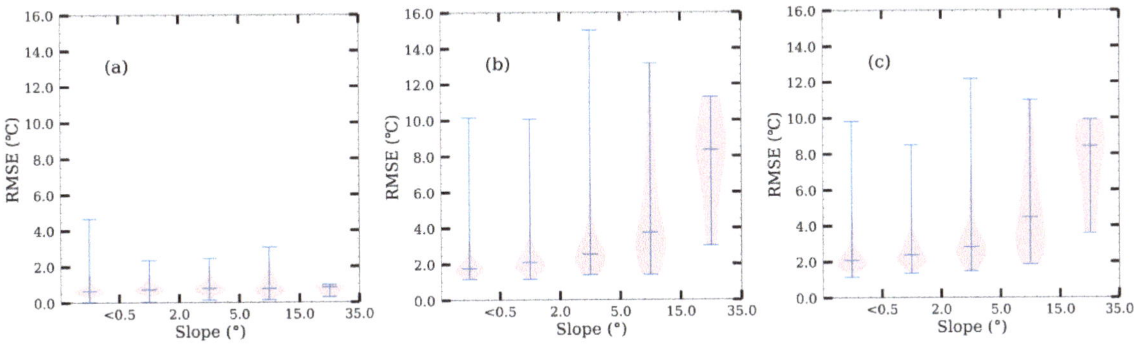

Figure 13. RMSE changes with slope for ((**a**): CLDAS, (**b**): ERA5-Land, (**c**): GLDAS) datasets.

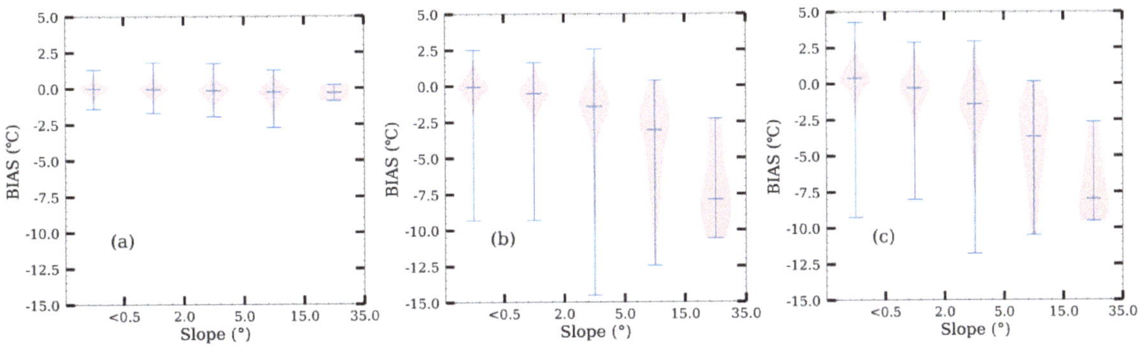

Figure 14. BIAS changes with slope for ((**a**): CLDAS, (**b**): ERA5-Land, (**c**): GLDAS) datasets.

5. Conclusions

In the present study, the gridded temperature datasets (CLDAS, ERA5-Land and GLDAS) that have been widely used in mainland China are evaluated for the past three years (2017–2019) on multiple times scales from hours of the day to daily, monthly and seasonal, etc. Spatially, the evaluation is conducted at single stations and over various climate regimes and administrative regions, etc. The results indicate that the three gridded datasets can represent the near surface air temperature in mainland China and realistically reflect the overall characters of temperature over major land areas of China. Compared to station observations, temperatures in the three datasets all are underestimated to varying degrees. The underestimation is most severe in ERA5-Land, followed by that in GLDAS. Overall, CLDAS exhibits the highest accuracy in mainland China, and ERA5-Land shows the second highest accuracy. GLDAS is the worst. However, note that the accuracy of ERA5-Land and GLDAS are only slightly different, and the two datasets demonstrate their own advantages and disadvantages in different regions.

In summary, differences in evaluation results can be attributed to various factors, including different resolutions of the gridded datasets, different remapping methods used to match gridded data with station observations and different evaluation metrics, etc. The present study compares the evaluation results of three gridded temperature datasets from different perspectives and finds that CLDAS has the highest accuracy in mainland China, followed by ERA5-Land, with GLDAS being the worst. However, CLDAS dataset mainly covers China and the surrounding areas, whereas ERA5-Land and GLDAS datasets are long-term, global datasets. Therefore, appropriate datasets should be selected based on different applications.

Author Contributions: Conceptualization, X.H., S.H. and C.S.; methodology, X.H., S.H. and C.S.; validation, S.H.; data curation, S.H. and C.S.; writing—original draft preparation, X.H.; writing—review and editing, X.H., S.H. and C.S.; visualization, X.H.; supervision, S.H.; funding acquisition, S.H. and C.S. All authors have read and agreed to the published version of the manuscript.

Funding: This research was funded by the National Key Research and Development Program of China (No. 2018YFC1506601), Key Techniques and Data Sets of Land Surface Reanalysis in Qinghai Xizang Plateau (No. NMICJY202106), Study on the Fusion of Precipitation and Soil Moisture with Multi-Source Data (No. 2011DFG23150), the Key Technology Development Project of Weather Forecasting (No. YBGJXM(2020)1A-08) and the Innovative Development Project of the China Meteorological Administration (No. CXFZ2021Z007).

Institutional Review Board Statement: Not applicable.

Informed Consent Statement: Not applicable.

Conflicts of Interest: The authors declare no conflict of interest.

References

1. Chen, S.; Gong, B. Response and adaptation of agriculture to climate change: Evidence from China. *J. Dev. Econ.* **2021**, *148*, 102557. [CrossRef]
2. Rolland, C. Spatial and seasonal variations of air temperature lapse rates in Alpine regions. *J. Clim.* **2003**, *16*, 1032–1046. [CrossRef]
3. Yang, J.; Duan, S.-B.; Zhang, X.; Wu, P.; Huang, C.; Leng, P.; Gao, M. Evaluation of seven atmospheric profiles from reanalysis and satellite-derived products: Implication for single-channel land surface temperature retrieval. *Remote Sens.* **2020**, *12*, 791. [CrossRef]
4. Chronopoulos, K.; Tsiros, I.; Alvertos, N.; Dimopoulos, I. Estimation of microclimatic data in remote mountainous areas using an artificial neural network model-based approach. *Glob. NEST J.* **2010**, *12*, 384–389.
5. Shi, G.; Sun, Z.; Qiu, X.; Zeng, Y.; Chen, P.; Liu, C. Comparison of two air temperature gridding methods over complex terrain in China. *Theor. Appl. Climatol.* **2018**, *133*, 1009–1019. [CrossRef]
6. Willmott, C.J.; Robeson, S.M.; Feddema, J. Influence of spatially variable instrument networks on climatic averages. *Geophys. Res. Lett.* **1991**, *18*, 2249–2251. [CrossRef]
7. Lipton, A.E.; Ward, J.M. Satellite-view biases in retrieved surface temperatures in mountain areas. *Remote Sens. Environ.* **1997**, *60*, 92–100. [CrossRef]
8. Neteler, M. Estimating daily land surface temperatures in mountainous environments by reconstructed MODIS LST data. *Remote Sens.* **2010**, *2*, 333–351. [CrossRef]
9. Vogt, J.V.; Viau, A.A.; Paquet, F. Mapping regional air temperature fields using satellite-derived surface skin temperatures. *Int. J. Climatol.* **1997**, *17*, 1559–1579. [CrossRef]
10. Han, S.; Shi, C.; Xu, B.; Sun, S.; Zhang, T.; Jiang, L.; Liang, X. Development and Evaluation of Hourly and Kilometer Resolution Retrospective and Real-Time Surface Meteorological Blended Forcing Dataset(SMBFD) in China. *J. Meteorol. Res.* **2019**, *33*, 1168–1181. [CrossRef]
11. Yang, F.; Lu, H.; Yang, K.; He, J.; Wang, W.; Wright, J.S.; Li, C.; Han, M.; Li, Y. Evaluation of multiple forcing data sets for precipitation and shortwave radiation over major land areas of China. *Hydrol. Earth Syst. Sci.* **2017**, *21*, 5805–5821. [CrossRef]
12. Shi, C.; Pan, Y.; Gu, J.; Xu, B.; Han, S.; Zhu, Z.; Zhang, L.; Sun, S.; Jiang, Z. A review of multi-source meteorological data fusion products. *Acta Meteorol. Sin.* **2019**, *77*, 774–783. (In Chinese) [CrossRef]
13. Xia, Y.; Hao, Z.; Shi, C.; Li, Y.; Meng, J.; Xu, T.; Wu, X.; Zhang, B. Regional and global land data assimilation systems: Innovations, challenges, and prospects. *J. Meteorol. Res.* **2019**, *33*, 159–189. [CrossRef]
14. Kalnay, E.; Kanamitsu, M.; Kistler, R.; Collins, W.; Deaven, D.; Gandin, L.; Iredell, M.; Saha, S.; White, G.; Woollen, J. The NCEP/NCAR 40-year reanalysis project. *Bull. Am. Meteorol. Soc.* **1996**, *77*, 437–472. [CrossRef]
15. Kanamitsu, M.; Ebisuzaki, W.; Woollen, J.; Yang, S.-K.; Hnilo, J.; Fiorino, M.; Potter, G. Ncep–doe amip-ii reanalysis (r-2). *Bull. Am. Meteorol. Soc.* **2002**, *83*, 1631–1644. [CrossRef]
16. Hersbach, H.; Bell, B.; Berrisford, P.; Hirahara, S.; Horányi, A.; Muñoz-Sabater, J.; Nicolas, J.; Peubey, C.; Radu, R.; Schepers, D. The ERA5 global reanalysis. *Q. J. R. Meteorol. Soc.* **2020**, *146*, 1999–2049. [CrossRef]
17. Hersbach, H.; Bell, B.; Berrisford, P.; Horányi, A.; Sabater, J.M.; Nicolas, J.; Radu, R.; Schepers, D.; Simmons, A.; Soci, C. Global reanalysis: Goodbye ERA-Interim, hello ERA5. *ECMWF Newsl.* **2019**, *159*, 17–24.
18. Hoffmann, L.; Günther, G.; Li, D.; Stein, O.; Wu, X.; Griessbach, S.; Heng, Y.; Konopka, P.; Müller, R.; Vogel, B. From ERA-Interim to ERA5: The considerable impact of ECMWF's next-generation reanalysis on Lagrangian transport simulations. *Atmos. Chem. Phys.* **2019**, *19*, 3097–3124. [CrossRef]
19. Kobayashi, S.; Ota, Y.; Harada, Y.; Ebita, A.; Moriya, M.; Onoda, H.; Onogi, K.; Kamahori, H.; Kobayashi, C.; Endo, H. The JRA-55 reanalysis: General specifications and basic characteristics. *J. Meteorol. Soc. Jpn.* **2015**, *93*, 5–48. [CrossRef]
20. Rodell, M.; Houser, P.; Jambor, U.; Gottschalck, J.; Mitchell, K.; Meng, C.-J.; Arsenault, K.; Cosgrove, B.; Radakovich, J.; Bosilovich, M. The global land data assimilation system. *Bull. Am. Meteorol. Soc.* **2004**, *85*, 381–394. [CrossRef]

21. Muñoz-Sabater, J.; Dutra, E.; Agustí-Panareda, A.; Albergel, C.; Arduini, G.; Balsamo, G.; Boussetta, S.; Choulga, M.; Harrigan, S.; Hersbach, H. ERA5-Land: A state-of-the-art global reanalysis dataset for land applications. *Earth Syst. Sci. Data Discuss.* **2021**, *13*, 4349–4383. [CrossRef]
22. Shi, C.; Xie, Z.; Qian, H.; Liang, M.; Yang, X. China land soil moisture EnKF data assimilation based on satellite remote sensing data. *Sci. China Earth Sci.* **2011**, *54*, 1430–1440. [CrossRef]
23. Han, S.; Shi, C.; Jiang, Z.; Xu, B.; Li, X.; Zhang, T.; Jiang, L.; Liang, X.; Zhu, Z.; Liu, J. Development and progress of high resolution CMA land surface data assimilation system. *Adv. Meteorol. Sci. Technol.* **2018**, *8*, 116. (In Chinese) [CrossRef]
24. Zhao, T.; Congbin, F. Applicability Evaluation of Surface Air Temperature from Several Reanalysis Datasets in China. *Plateau Meteorol.* **2009**, *28*, 594–606. (In Chinese) [CrossRef]
25. Zhao, T.; Fu, C.; Ke, Z.; Guo, W. Global atmosphere reanalysis datasets: Current status and recent advances. *Adv. Earth Sci.* **2010**, *25*, 241–254. (In Chinese) [CrossRef]
26. Zhao, S.; He, W.; Jiang, Y. Evaluation of NCEP-2 and CFSR reanalysis seasonal temperature data in China using detrended fluctuation analysis. *Int. J. Climatol.* **2018**, *38*, 252–263. [CrossRef]
27. Xie, X.; He, J.; Qi, L. A review on applicability evaluation of four reanalysis datasets in China. *J. Meteorol. Environ.* **2011**, *27*, 58–65. (In Chinese)
28. Jiang, Y.; Han, S.; Shi, C.; Gao, T.; Zhen, H.; Liu, X. Evaluation of HRCLDAS and ERA5 Datasets for Near-Surface Wind over Hainan Island and South China Sea. *Atmosphere* **2021**, *12*, 766. [CrossRef]
29. He, W.-p.; Zhao, S.-s. Assessment of the quality of NCEP-2 and CFSR reanalysis daily temperature in China based on long-range correlation. *Clim. Dyn.* **2018**, *50*, 493–505. [CrossRef]
30. Han, S.; Liu, B.; Shi, C.; Liu, Y.; Qiu, M.; Sun, S. Evaluation of CLDAS and GLDAS datasets for Near-surface Air Temperature over major land areas of China. *Sustainability* **2020**, *12*, 4311. [CrossRef]
31. Bao, X.; Zhang, F. Evaluation of NCEP–CFSR, NCEP–NCAR, ERA-Interim, and ERA-40 reanalysis datasets against independent sounding observations over the Tibetan Plateau. *J. Clim.* **2013**, *26*, 206–214. [CrossRef]
32. Kumar, S.V.; Peters-Lidard, C.D.; Tian, Y.; Houser, P.R.; Geiger, J.; Olden, S.; Lighty, L.; Eastman, J.L.; Doty, B.; Dirmeyer, P. Land information system: An interoperable framework for high resolution land surface modeling. *Environ. Model. Softw.* **2006**, *21*, 1402–1415. [CrossRef]
33. Hua-lan, R.; Beaudoing, H.J.G.D.; Greenbelt, M.D. README Document for Global Land Data Assimilation System Version 1 (GLDAS-1) Products. 2011. Available online: ftp://hydro1.sci.gsfc.nasa.gov/data/s4pa/GLDAS_V1/README.GLDAS.pdf (accessed on 5 September 2021).
34. Rui, H.; Beaudoing, H. Readme Document for Global Land Data Assimilation System Version 2 (GLDAS-2) Products. GES DISC. 2011; Volume 2011, pp. 1–22. Available online: ftp://hydro1.sci.gsfc.nasa.gov/data/s4pa/GLDAS/README.GLDAS2.pdf (accessed on 5 September 2021).
35. Wu, Z.; Feng, H.; He, H.; Zhou, J.; Zhang, Y. Evaluation of Soil Moisture Climatology and Anomaly Components Derived From ERA5-Land and GLDAS-2.1 in China. *Water Resour. Manag.* **2021**, *35*, 629–643. [CrossRef]
36. Pelosi, A.; Terribile, F.; D'Urso, G.; Chirico, G.B. Comparison of ERA5-Land and UERRA MESCAN-SURFEX reanalysis data with spatially interpolated weather observations for the regional assessment of reference evapotranspiration. *Water* **2020**, *12*, 1669. [CrossRef]
37. Camargo, L.R.; Schmidt, J. Simulation of multi-annual time series of solar photovoltaic power: Is the ERA5-land reanalysis the next big step? *Sustain. Energy Technol. Assess.* **2020**, *42*, 100829. [CrossRef]
38. Ren, Z.; Zhang, Z.; Sun, C.; Liu, Y.; Li, J.; Ju, X.; Zhao, Y.; Li, Z.; Zhang, W.; Li, H. Development of Three-Step Quality Control System of Real-Time Observation Data from AWS in China. *Meteorol. Mon.* **2015**, *41*, 1268–1277. (In Chinese)
39. Wang, Q.; Li, W.; Xiao, C.; Ai, W. Evaluation of High-Resolution Crop Model Meteorological Forcing Datasets at Regional Scale: Air Temperature and Precipitation over Major Land Areas of China. *Atmosphere* **2020**, *11*, 1011. [CrossRef]
40. Chen, Y.; Yuan, H. Evaluation of nine sub-daily soil moisture model products over China using high-resolution in situ observations. *J. Hydrol.* **2020**, *588*, 125054. [CrossRef]
41. Liu, Y.; Shi, C.; Wang, H.; Han, S. Applicability assessment of CLDAS temperature data in China. *Trans. Atmos. Sci.* **2021**, *44*, 540–548. (In Chinese) [CrossRef]
42. Ji, L.; Senay, G.B.; Verdin, J.P. Evaluation of the Global Land Data Assimilation System (GLDAS) air temperature data products. *J. Hydrometeorol.* **2015**, *16*, 2463–2480. [CrossRef]
43. Cui, W.; Wang, W.; Wang, X.; Chen, X. Evaluation of GLDAS-1 and GLDAS-2 Forcing Data and Noah Model Simulations over China at the Monthly Scale. *J. Hydrometeorol.* **2016**, *17*, 2815–2833. [CrossRef]
44. Kann, A.; Pistotnik, G.; Bica, B. Research. INCA-CE: A Central European initiative in nowcasting severe weather and its applications. *Adv. Sci. Res.* **2012**, *8*, 67–75. [CrossRef]
45. Haiden, T.; Kann, A.; Wittmann, C.; Pistotnik, G.; Bica, B.; Gruber, C.J.W. Forecasting. The Integrated Nowcasting through Comprehensive Analysis (INCA) system and its validation over the Eastern Alpine region. *Weather Forecast.* **2011**, *26*, 166–183. [CrossRef]
46. Chen, F.; Yang, X.; Ji, C.; Li, Y.; Deng, F.; Dong, M. Establishment and assessment of hourly high-resolution gridded air temperature data sets in Zhejiang, China. *Meteorol. Appl.* **2019**, *26*, 396–408. [CrossRef]

47. You, Q.; Kang, S.; Pepin, N.; Flügel, W.-A.; Yan, Y.; Behrawan, H.; Huang, J.J.G. Relationship between temperature trend magnitude, elevation and mean temperature in the Tibetan Plateau from homogenized surface stations and reanalysis data. *Glob. Planet. Chang.* **2010**, *71*, 124–133. [CrossRef]
48. You, Q.; Fraedrich, K.; Ren, G.; Pepin, N.; Kang, S. Variability of temperature in the Tibetan Plateau based on homogenized surface stations and reanalysis data. *Int. J. Climatol.* **2013**, *33*, 1337–1347. [CrossRef]
49. Xie, A.; Ren, J.; Qin, X.; Kang, S. Feasibility comparison of reanalysis data from NCEP-I and NCEP-II in the Himalayas. *J. Mt. Sci.* **2009**, *6*, 56–65. [CrossRef]
50. Gao, L.; Bernhardt, M.; Schulz, K. Elevation correction of ERA-Interim temperature data in complex terrain. *Hydrol. Earth Syst. Sci.* **2012**, *16*, 4661–4673. [CrossRef]
51. Frauenfeld, O.W.; Zhang, T.; Serreze, M.C. Climate change and variability using European Centre for Medium-Range Weather Forecasts reanalysis (ERA-40) temperatures on the Tibetan Plateau. *J. Geophys. Res. Atmos.* **2005**, *110*. [CrossRef]
52. Kun, Z.; Yan, X.; Zhenfang, H.; Min, G. Topography features of Qilian Mountains nature reserve based on SRTM DEM. *Arid Land Geogr.* **2020**, *43*, 1559–1566. (In Chinese)

agriculture

MDPI

Article

Future Projection for Climate Suitability of Summer Maize in the North China Plain

Yanxi Zhao [1,2,3], Dengpan Xiao [1,2,3,*], Huizi Bai [3], Jianzhao Tang [3] and Deli Liu [4]

1 College of Geography Science, Hebei Normal University, Shijiazhuang 050024, China; 18233181223@163.com
2 Hebei Laboratory of Environmental Evolution and Ecological Construction, Shijiazhuang 050024, China
3 Engineering Technology Research Center, Geographic Information Development and Application of Hebei, Institute of Geographical Science, Hebei Academy of Sciences, Shijiazhuang 050011, China; baihz272@126.com (H.B.); tjzcau@163.com (J.T.)
4 NSW Department of Primary Industries, Wagga Wagga Agricultural Institute, Wagga Wagga, NSW 2650, Australia; de.li.liu@dpi.nsw.gov.au
* Correspondence: xiaodp@sjziam.ac.cn

Abstract: Climate change has and will continue to exert significant effects on social economy, natural environment, and human life. Research on the climatic suitability of crops is critical for mitigating and adapting to the negative impacts of climate change on crop production. In the study, we developed the climate suitability model of maize and investigated the climate suitability of summer maize during the base period (1981–2010) and two future periods of 2031–2060 (2040s) and 2071–2100 (2080s) in the North China Plain (NCP) based on BCC-CSM2-MR model (BCC) from the Coupled Model Comparison Program (CMIP6) under two Shared Socioeconomic Pathways (SSP) 245 and SSP585. The phenological shift of maize under future climate scenarios was simulated by the Agricultural Production Systems Simulator (APSIM). The results showed that the root mean square errors (*RMSE*) between observations and projections for sunshine suitability (S_S), temperature suitability (S_T), precipitation suitability (S_P), and integrated climate suitability (S_Z) during the whole growth period were 0.069, 0.072, 0.057, and 0.040, respectively. Overall, the BCC projections for climate suitability were in suitable consistency with the observations in the NCP. During 1981–2010, the S_P, S_T, and S_Z were high in the north of the NCP and low in the south. The S_P, S_T, and S_Z showed a downward trend under all the future climate scenarios in most areas of NCP while the S_S increased. Therein, the change range of S_P and S_S was 0–0.1 under all the future climate scenarios. The S_T declined by 0.1–0.2 in the future except for the decrease of more than 0.3 under the SSP585 scenario in the 2080s. The decrease in S_Z in the 2040s and 2080s under both SSP scenarios varied from 0 to 0.2. Moreover, the optimum area decreases greatly under future scenarios while the suitable area increases significantly. Adjusting sowing data (SD) would have essential impacts on climate suitability. To some extent, delaying SD was beneficial to improve the climate suitability of summer maize in the NCP, especially under the SSP585 scenario in the 2080s. Our findings can not only provide data support for summer maize production to adapt to climate change but also help to propose agricultural management measures to cope with future climate change.

Keywords: adaptation; climate change; summer maize; phenology shift; GCM

Citation: Zhao, Y.; Xiao, D.; Bai, H.; Tang, J.; Liu, D. Future Projection for Climate Suitability of Summer Maize in the North China Plain. *Agriculture* **2022**, *12*, 348. https://doi.org/10.3390/agriculture12030348

Academic Editor: Claudia Di Bene

Received: 20 January 2022
Accepted: 27 February 2022
Published: 28 February 2022

1. Introduction

Over the past 100 years, global warming has become more and more significant, and it has become one of the major issues affecting the sustainable development of human society [1]. Global warming has exerted a significant impact on the natural environment, social economy, and human life, among which the impact on agricultural production on which human survival depends has attracted widespread attention [2–9]. Generally, different crops have different demands for climate resources, and more or less, climate resources are not conducive to the normal growth and development of crops [10,11]. The quantity variation of key climatic factors (i.e., sunshine hours, temperature, and precipitation) can be transformed into the climate suitability of crop growth and development based on the membership function method in fuzzy mathematics [12,13]. Moreover, the crop climate suitability can play a role in predicting final yield [14,15]. The study of climate suitability not only helps to make specific divisions of regions according to climate conditions, e.g., unsuitable area (unsuitable area for crop growing), less suitable area (less suitable area for crop growing), suitable area (suitable area for crop growing) and optimum area (optimum area for crop growing) [16,17], but also can rationally plan the planting areas of crops [17–20]. Moreover, the climate suitability for different varieties of the crop is different. By comparing the climate suitability for three varieties of early maturity, medium maturity, and late maturity of spring soybean in North China, the medium maturity variety was the most suitable variety for planting [21]. The climate suitability model is a useful tool to investigate the sensitivity of crops to climatic factors such as temperature, precipitation, and sunshine, analyze the response mechanism of crops to climate change, and optimize the selection of crop varieties [22,23].

Climate change exerted an important impact on the crop growth process, and the study of future climate suitability is crucial for taking effective adaptive measures to cope with the adverse effects of climate change on crop production. Global climate model (GCM) is an effective tool to explore the mechanisms of climate change and predict future climate change, and the simulation results of GCMs can provide important data support to study the impacts of future climate change on agricultural production at different scales [24]. The World Climate Research Program (WCRP) initiated a new round of the International Coupled Model Comparison Program (CMIP6). Since the implementation of CMIP was more than 20 years, the number of participating models for CMIP6 was the largest, the scientific experiment design was the most complete, and the largest amount of model data was provided [25]. Compared with previous climate models such as CMIP5, the simulations of CMIP6 models for climate systems were closer to the observations and had less uncertainty, so the simulating ability of climate change has significantly improved [26,27]. For example, in terms of simulation for extreme climate at global scale, CMIP6 models had a general improvement in simulating the changing trend of extreme climate compared to CMIP5 models [28]. In the study, the BCC-CSM2-MR model (BCC) from CMIP6 was used to explore the crop climate suitability in the future. In contrast with the previous versions of the BCC model from CMIP5, the physical mechanism of BCC, such as atmospheric radiation and deep convection process, was improved to make it more suitable for the simulation of climate distribution [29–31].

Maize is an important food and feed crop in the world and has a considerable impact on agricultural economy [32]. In China, maize is one of three major food crops while it ranks first among miscellaneous grain crops, widely distributed in Northeast, North, Northwest, and Southwest China [33]. The North China Plain (NCP) is an important grain production base in China, with maize production accounting for more than 30% of the country's total output [34]. Over the past few decades, climate change has had a significant impact on maize production in the NCP [35]. From 1980 to 2009, the contribution of climate change to maize yield reduction in the NCP was 15–30%, among which the contribution from the reduction in solar radiation was 12–24%, and that from temperature increment was 3–9% [36]. Therefore, It is of great significance to study the climate suitability of maize in the NCP and evaluate the impact of climate resources on agricultural production. The evaluation of agricultural climate suitability is beneficial to cope with the impact of climate change on maize production, rationally use agricultural climate resources, improve the agricultural management level and ensure agricultural production safety [14]. Tang and Liu [15] analyzed the spatial-temporal characteristics of maize climate suitability during the current period and future period in the NCP based on 30 CMIP5 GCMs. However, the phenology shift of maize in the above study was obtained by calculating the active accumulated temperature, with a lack of mechanism. During the historical period, the observed records of maize phenology at the agro-meteorological stations should be used for developing the climate suitability model. In addition, the crop model can be used to simulate crop growth under various environmental conditions and agricultural management measures. Therefore, this study took advantage of the crop model to simulate the future phenology of maize. The Agricultural Production Systems Simulator (APSIM) was selected to simulate the phenological shift of maize under future climate scenarios.

In the study, we mainly investigated the climate suitability of maize in the NCP under future climate scenarios on the basis of daily climate data of BCC from CMIP6 and future phenology conditions simulated by the APSIM model. The objectives of the study were (1) to develop the climate suitability model of maize based on the regional climatic conditions and fuzzy mathematics; (2) to analyze the spatial and temporal change characteristics of climate suitability for maize in the NCP under future scenarios; and (3) to evaluate the effect of adjusting sowing date on maize climate suitability.

2. Materials and Methods

2.1. Study Area

The NCP (113.7–122.7° E, 32.9–40.5° N) is delimitated in the east by the sea, the west by Taihang Mountains, the south by the main stream of the Huaihe River, the north by the Yan Mountains, accounting for approximately 1.4×10^5 km^2 of arable land (Figure 1) [37]. The region has a warm temperate monsoon climate with plenty of light and heat resources [37]. The mean annual temperature across the study area ranged from 9.6 to 16.0 °C in nearly fifty years [38]. The annual precipitation is not evenly distributed, with over 70% of precipitation appearing in July through September. The main soil type in the NCP is the loam of Aeolian origin, a soil type deposited by rivers over geological periods. The NCP is an important grain production region in China, where the main cropping system is the double-cropping systems of winter wheat-summer maize [36]. Summer maize is usually planted in middle June and harvested in September [39].

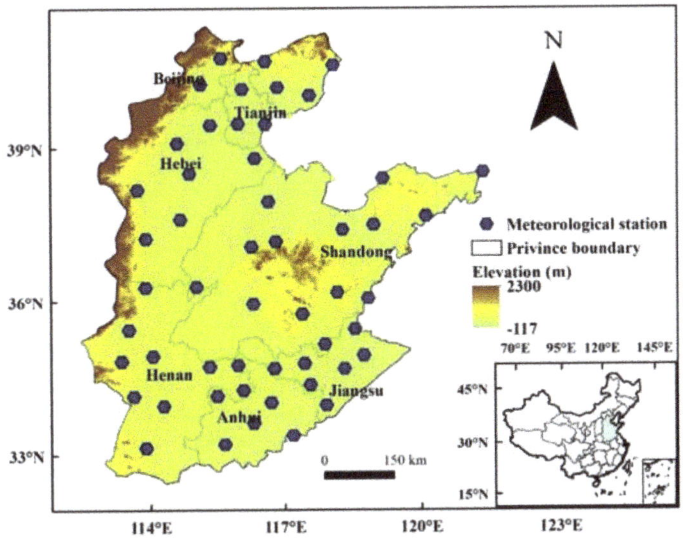

Figure 1. The spatial distribution of 52 meteorological stations in the North China Plain.

2.2. Climate Data

The historical records about daily climate data, including mean temperature (*Tmean*), maximum temperature (*Tmax*), minimum temperature (*Tmin*), precipitation (*Prec*), and sunshine hours (*Sh*) during 1981 to 2010 for 52 meteorological stations across the NCP, were obtained from China's Meteorological Administration (CMA).

Future climate scenario data were obtained from GCMs, which is provided by the World Climate Research Program (WCRP) of Coupled Model Inter-comparison Project phase 6 (CMIP6, https://esgf-node.llnl.gov/search/cmip6/ (accessed on 20 February 2020)). CMIP6 integrated climate change information from the CMIP5 simulations of the representative concentration pathway (RCP) and future societal development pathways (SSPs) [40]. The SSPs describe alternative evolutions of future society under climate change and/or climate policy. SSPs 1 and 5 envision relatively optimistic trends for human development, with substantial investments in education and health, rapid economic growth, and well-functioning institutions [40]. However, SSP5 assumes an energy-intensive, fossil-based economy, while SSP1 assumes an increasing shift toward sustainable practices. Further details of the SSPs can be found in O'Neill et al. [40]. In our study, we focused on fossil-fueled development trends (i.e., SSP5) and the highest forcing pathway (i.e., RCP8.5, the radiative forcing of 8.5 W m^{-2} by 2100), defined by SSP5-85 (SSP585). Additionally, the combination of medium social vulnerability and medium radiative forcing, defined by SSP2-45 (SSP245), was also used for comparison. SSP245 is the updated RCP4.5 scenario, and the radiative forcing will stabilize at 4.5 W/m^2 in 2100. A GCM named BCC-CSM2-MR (BCC) under SSP245 and SSP585 with a time span of 2031–2060 (2040s) and 2071–2100 (2080s) developed by Beijing Climate Center, China, was selected. In the study, the 2040s under SSP245, 2080s under SSP245, 2040s under SSP585, and 2080s under SSP585 were defined as S1, S2, S3, and S4, respectively.

The statistically downscaled method developed by Liu and Zuo [41] was used to generate daily climate data at each station from the monthly data of the BCC-CSM2-MR model (BCC). This method was mainly divided into two steps: spatial downscaling and temporal downscaling. The spatial downscaling was to transform monthly GCMs on the grid scale into monthly stations data using the inverse distance-weighted interpolation (IDW). The formula is as follows:

$$S_i = \sum_{k=1}^{4}\left[\frac{1}{d_{i,k}^{m}}\left(\sum_{j=1}^{4}\frac{1}{d_{i,j}^{m}}\right)^{-1}P_k\right] \quad (1)$$

where S_i is the downscaled site-specific GCMs projection at site i, P_k is the GCMs projection at cell k, $d_{i,k}$ ($d_{i,j}$) is the distance between site i and the center of cell k (j), m is the control parameter.

In the spatial downscaling process, the qq-mapping bias correction method was applied to correct the bias of the spatial downscaling data to match with the observations. The bias-corrected spatial downscaling data were calculated by

$$x_k^f = y_i^o + \frac{y_{i+1}^o - y_i^o}{x_{i+1}^h - x_i^h}\left(x_k^r - x_i^h\right),\ x_i^h \le x_k^r \le x_{i+1}^h \quad (2)$$

where x_k^f is the bias-corrected spatial downscaling data, $k = 1, 2, \ldots, n$, y_i^o is the monthly observed data in the baseline period, x_i^h is the GCMs monthly data in the baseline period, x_k^r is the future GCMs projected data before bias correction.

Then the spatial downscaling monthly climate data for each station was transformed to daily climate data using a modified stochastic weather generator (WGEN) [42].

2.3. *Phenology Data and Future Phenology Simulation*

The observed data of maize phenology, including sowing data (SD), flowering data (FD), and maturity data (MD) for 49 agro-meteorological stations across the NCP during 1981–2010, were obtained from China's Meteorological Administration (CMA). To investigate the climate suitability of maize at different growth periods, the whole growth period (WGP) was divided into two periods: vegetative growth period from SD to FD (VGP) and reproductive growth period from FD to MD (RGP). The phenology data of the adjacent agro-meteorological stations were used to calculate climate suitability for the three meteorological stations without phenology data. With future climate change, the phenology of maize across the NCP will also change. In order to more accurately calculate the climate suitability under future climate scenarios, the APSIM, which is a comprehensive model developed to simulate biophysical processes in agricultural systems [43,44], was selected to simulate the maize phenology under future climate scenarios. Generally, the APSIM model can provide an acceptable prediction of crop productivity under the combined influences of climate change, soil condition, and management measures and was widely used in agricultural research [36,45–48]. The APSIM model had been calibrated and validated based on observed phenological data for the selected stations in the previous study [49]. In addition, SD adjustments were considered in the validated APSIM model to simulate the maize phenology shift under future climate scenarios. A total of 5 sowing dates were set up, including observed SD during the historical period (S_Base), observed SD advanced by 30 days (S_A30) and 15 days (S_A15), and observed SD delayed by 15 days (S_D15) and 30 days (S_D30). The results can be used to evaluate the impacts of adjusting SD on the phenology and climate suitability of maize.

2.4. Climate Suitability Model

To characterize the climatic adaptability of maize to key climatic factors (i.e., sunshine hours, temperature, and precipitation) and integrated climatic conditions in the NCP, the climate suitability model was built with reference to related studies [12,50–52].

2.4.1. Sunshine Suitability (S_S) Model

The sunshine hours had a great influence on the growth and development of crops. The S_S of maize was calculated as follows [23,52,53]:

$$S_S = \begin{cases} e^{-[(S_i - S_0)/r]^2} & S_i < S_0 \\ 1 & S_i \geq S_0 \end{cases} \quad (3)$$

where S_0 is the daily sunshine hours when the percentage of daily sunshine hours reaches 70%; S_i is daily sunshine hours (h); r is a constant that can be determined according to the climatic conditions across the NCP and relevant studies [52,54]. The values for r at different growth periods are shown in Table 1. S_S at the VGP, RGP, and WGP are referred to as S_{S_VGP}, S_{S_RGP}, and S_{S_WGP}, respectively.

Table 1. Values of related parameters for calculating the suitability of sunshine, temperature, and precipitation in vegetative growth period (VGP) and reproductive growth period (RGP) of summer maize.

Parameters	VGP	RGP
r	5.01	5.19
T_1	20	10
T_0	25	22
T_2	35	32
Kc	0.7	0.9

2.4.2. Temperature Suitability (S_T) Model

S_T is related to three base point temperatures at different growth periods of crops. Three basis point temperatures include the optimal temperature, lower limit temperature, and upper limit temperature of the crop life process. Under the optimal temperature, crops grow quickly and well, while crops cease to grow and develop above upper limit temperature or below lower limit temperature [23]. The S_T of maize was calculated as follows [23,52,53]:

$$S_T = \frac{\left[(T_i - T_1)(T_2 - T_i)^B\right]}{\left[(T_0 - T_1)(T_2 - T_0)^B\right]} \quad (4)$$

$$\text{Among}\left\{B = \frac{(T_2 - T_0)}{(T_0 - T_1)}\right\}$$

where T_i is daily mean temperature (°C); T_0 is the optimal temperature (°C) at different growth periods; T_1 and T_2 is the lower limit temperature (°C) and the upper limit temperature (°C) during various growth periods. The specific values of T_0, T_1, and T_2 refer to the climatic conditions across the NCP and relevant studies [50,52,55] and are shown in Table 1. The S_T during the VGP, RGP and WGP were defined by S_{T_VGP}, S_{T_RGP}, and S_{T_WGP}, respectively.

2.4.3. Precipitation Suitability (S_P) Model

It is very important for crop growth if precipitation can match the physiological water requirement of crops. S_P was defined as the ratio of precipitation to physiological water requirement when precipitation is less or greater than physiological water requirement during the crop growth period. The S_P of maize is calculated as follows [51]:

$$S_P = \begin{cases} R/\ R_0\ \ R_i < R_0 \\ R_0/\ R\ R_i \geq R_0 \end{cases} \tag{5}$$

where R is precipitation (mm); R_0 is the physiological water requirement of crops, which can be calculated as follows:

$$R_0 = Kc \cdot ET_0 \tag{6}$$

where Kc is the crop coefficient and ET_0 is the reference crop evapotranspiration (mm). The Kc values of maize during various growth stages were determined according to relevant studies [56,57] and listed in Table 1. The ET_0 values of maize are calculated based on the Penman–Monteith formula [57].

$$ET_0 = \frac{0.408\Delta(R_n - G) + \gamma\frac{900}{T+273}U_2(e_s - e_a)}{\Delta + \gamma(1 + 0.34U_2)} \tag{7}$$

where Δ is the slope on the saturation vapor pressure temperature curve (kPa °C^{-1}); T is the daily mean temperature (°C); R_n represents net radiation (MJ·m^{-2}·d^{-1}); G is the soil heat flux (MJ m^{-2} d^{-1}); γ is the hygrometer constant (kPa °C^{-1}); U_2 is the wind speed 2 m above the ground (m s^{-1}); e_s and e_a are respectively the saturated vapor pressure and the actual vapor pressure (kPa) at temperature T. The S_P at the VGP, RGP and WGP were defined by S_{P_VGP}, S_{P_RGP}, and S_{P_WGP}, respectively.

2.4.4. Crop Climate Suitability during Different Crop Growth Stages

The sunshine and temperature suitability during different growth stages are calculated by the arithmetical average method according to the following formula:

$$S_c = \frac{1}{m}\sum_{i}^{m} S_{ci} \tag{8}$$

where S_c represents S_S (S_T) at different growth stages of maize; I is the corresponding day sequence at each growth stage; S_{ci} is the S_S (S_T) suitability of i day; m is the total number of days at the corresponding growth stage.

2.4.5. Integrated Climate Suitability (S_Z) Model

Crop growth and development are jointly affected by sunshine, temperature, and precipitation together. S_S, S_T, and S_P can only reflect the influence of a single climatic factor on crop growth. Therefore, S_Z is developed to make a comprehensive reflection of the impact of sunshine, temperature, and precipitation on maize growth and development. The formula is as follows:

$$S_Z = aS_S + bS_T + cS_R \tag{9}$$

Based on related studies and characteristics of crop growth and development in the NCP [50,52,58], the coefficients *a*, *b*, and *c* are taken as values 0.20, 0.32, and 0.48, respectively in this study. The S_Z during the VGP, RGP, and WGP were defined by S_{Z_VGP}, S_{Z_RGP}, and S_{Z_WGP}, respectively.

2.5. Spatial Interpolation

In the study, the inverse distance weighting (IDW) in ArcGIS 10.3 was used to map the spatial distribution characteristics and to analyze spatial variation trends of S_S, S_T, S_P, and S_Z during the 2040s and 2080s compared with the baseline period (1981–2010) in the NCP.

3. Results

3.1. Comparison between Observations and Simulations for Climate Suitability during 1981–2010

In the study, it can be found from Figure 2 that the Pearson correlation coefficient (*R*) calculated between observations and projections for S_S, S_T, S_P and S_Z at different stages ranged from 0.10 to 0.46. Meantime, the root mean square error (*RMSE*) calculated between observations and projections for S_S, S_T, S_P and S_Z at different stages was relatively small. In detail, the *RMSE* of S_{S_WGP}, S_{T_WGP}, S_{P_WGP} and S_{Z_WGP} was 0.069, 0.072, 0.057, and 0.040, respectively. There was small uncertainty for the BCC projection in the inter-annual variation during 1981–2010, which can reasonably reproduce the temporal variation trend of the observed climate suitability. As was shown in Figure 2, the BCC projection displayed similar trends of observations for climate suitability. These results showed that the BCC projections for climate suitability were in suitable consistency with the observations and suitable for the assessment and prediction of climate suitability in the NCP.

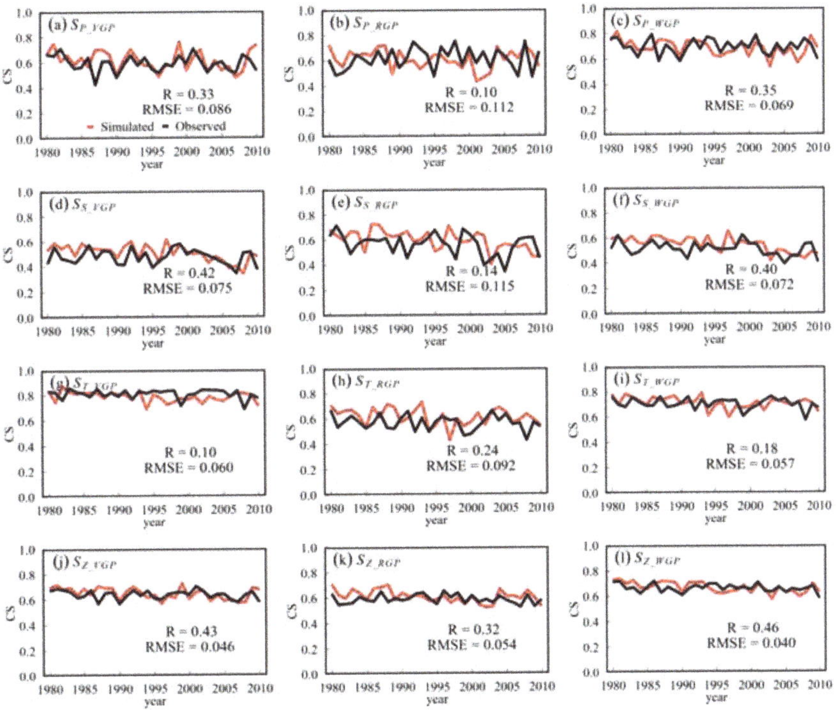

Figure 2. Observed and simulated maize climate suitability for the vegetative growth period (**a**,**d**,**g**,**j**), reproductive growth period (**b**,**e**,**h**,**k**) and whole growth period (**c**,**f**,**i**,**l**) during 1981–2010 across the North China Plain.

3.2. Temporal Changes of Climate Suitability under the Future Climate Scenarios

The changes of BCC projected S_S, S_T, S_P, and S_Z in the 2040s and 2080s under SSP245 and SSP585 compared to the baseline period of 1981–2010 are shown in Figure 3.

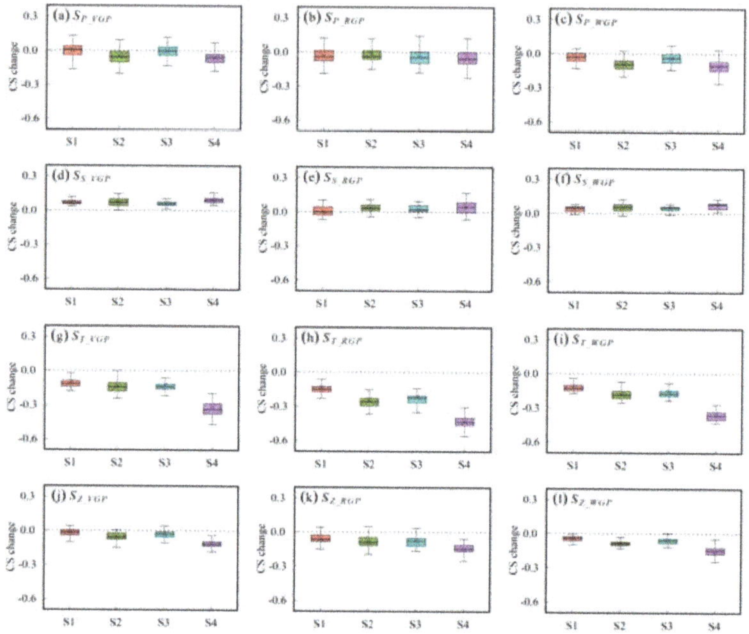

Figure 3. Changes in projected maize climate suitability for the vegetative growth period (**a**,**d**,**g**,**j**), reproductive growth period (**b**,**e**,**h**,**k**) and whole growth period (**c**,**f**,**i**,**l**) during the 2040s and 2080s compared to 1981–2010 (S1, S2, S3, and S4 are the SSP245_2040s, SSP245_2080s, SSP585_2040s, and SSP585_2080s).

The change of BCC projected S_{P_VGP} under both scenarios in the 2040s compared to the baseline period was small (less than 0.01), while the decrease in BCC projected S_{P_VGP} in the 2080s was from 0.05 to 0.06 (Figure 3a). S_{P_RGP} in the 2040s and 2080s showed a uniform downward trend, and the range of decrease was 0.02–0.04 (Figure 3b). On the whole, the decline of S_{P_WGP} in the future compared to the baseline period was even more pronounced. The S_{P_WGP} under both scenarios in the 2040s decreased by 0.03–0.04, and the decline of S_{P_WGP} in the 2080s was up to 0.09–0.11 (Figure 3c).

The increase in BCC projected S_{S_VGP} under both scenarios during the 2040s and 2080s compared to the baseline period was from 0.06 to 0.09 (Figure 3d). The increase in BCC projected S_{S_RGP} was relatively small, less than 0.05 (Figure 3e). The overall rise of S_{S_WGP} for the future periods under SSP245 and SSP585 was 0.04–0.07 (Figure 3f). The change of S_S in the future across the NCP represented a slight increasing trend.

The decrease in S_{T_VGP}, S_{T_RGP} and S_{T_WGP} during the SSP245_2040s, SSP245_2080s and SSP585_2040s ranged from 0.1 to 0.3, while the decline of S_{T_VGP}, S_{T_RGP} and S_{T_WGP} in the SSP585_2080s was between 0.3 and 0.5 (Figure 3g–i). The downtrend of the S_T during the 2040s and 2080s across the NCP was significant, especially under SSP585_2080s.

The S_Z was affected by sunshine, temperature, and precipitation together. The variation trend of S_Z for the future across the NCP was consistent with the S_T, while the magnitude of decrease for the S_Z was smaller than the S_T on account of the S_S and the S_P. The decline of S_{Z_VGP}, S_{Z_RGP}, and S_{Z_WGP} during the SSP245_2040s, SSP245_2080s, and SSP585_2040s was less than 0.1, and the S_{Z_VGP}, S_{Z_RGP}, and S_{Z_WGP} in the SSP585_2080s decreased by 0.1–0.2 (Figure 3j–l).

3.3. Spatial Distribution of Climate Suitability in the Baseline Period (1981–2010) and Future Periods (2040s and 2080s)

The spatial distribution of the S_S, S_T, S_P, and S_Z during the baseline period (1981–2010) in the NCP is shown in Figure 4. S_{P_VGP} in the northwest of the NCP was higher than 0.6, while S_{P_VGP} in the southeast of the NCP was lower than 0.6 (Figure 4a). There were most areas of the NCP with S_{P_RGP} higher than 0.6 (Figure 4b). The areas with S_{P_WGP} lower than 0.7 were distributed in the southeast of the NCP, and S_{P_WGP} in the northwest of the NCP exceeded 0.7 (Figure 4c). The spatial difference of S_S across the NCP was relatively small. The values of S_{S_VGP}, S_{S_RGP}, and S_{S_WGP} during 1981–2010 in most areas of the NCP ranged from 0.4 to 0.6 (Figure 4d–f). The spatial distribution of S_T across the NCP was low in the south and high in the north. The ranges of S_{T_VGP}, S_{T_RGP}, and S_{T_WGP} in Hebei and Shandong during 1981–2010 were 0.8–0.9, 0.6–0.7, and 0.7–0.8, respectively, while the values of S_{T_VGP}, S_{T_RGP}, and S_{T_WGP} in the south of the NCP were less than the values in the north (Figure 4g–i). There was a spatial characteristic with high in the north and low in the south for the S_Z. S_{Z_VGP} and S_{Z_WGP} in most areas were higher than 0.6, and the values of S_{Z_RGP} ranged from 0.5 to 0.7 (Figure 4j–l).

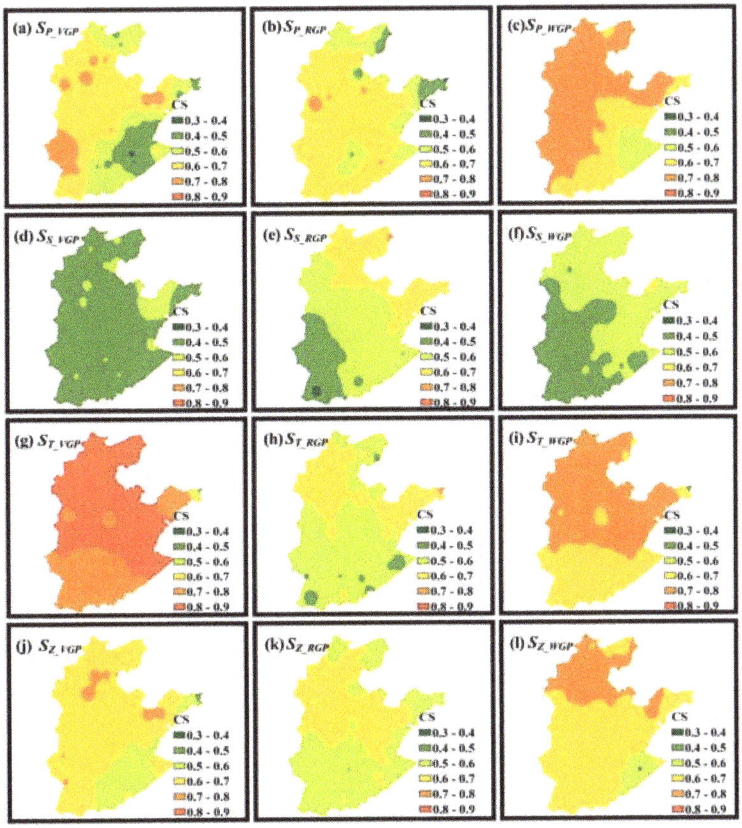

Figure 4. The spatial distribution of maize climate suitability for the vegetative growth period (**a**,**d**,**g**,**j**), reproductive growth period (**b**,**e**,**h**,**k**) and whole growth period (**c**,**f**,**i**,**l**) during the baseline period (1981–2010) in the North China Plain.

The spatial change characteristics of climate suitability in the NCP under the SSP245 scenario during the 2040s and 2080s are shown in Figures S1 and S2. Relative to the baseline period, S_{P_VGP} increased by 0–0.1 in the central region of the NCP during the 2040s and

2080s (Figures S1a and S2a). The areas with S_{P_RGP} and S_{P_WGP} decreasing by 0–0.1 were distributed in most areas of the NCP for two future periods (Figures S1b–c and S2b–c). The S_S increased by 0–0.1 at most parts of the NCP except S_{S_RGP} in the 2040s (Figures S1d–f and S2d–f). In contrast to the S_S, the S_T mainly presented a spatial characteristic of decline with a magnitude over 0.1 across the NCP in the 2040s and 2080s, particularly the decreasing amplitude of S_{T_RGP} during 2080s exceeding 0.2 (Figures S1g–i and S2g–i). Similar to the S_T, there was a spatial characteristic of decreasing for the S_Z in most parts of the NCP. However, the extent of decline for the S_Z was significantly smaller than that of the S_T (Figures S1j–l and S2j–l).

The spatial change characteristics of climate suitability in the NCP under the SSP585 scenario in the 2040s and 2080s are presented in Figures 5 and 6. Compared to the baseline period, the spatial change characteristic of S_P under the SSP585 scenario coincided with that under the SSP245 scenario. In addition, the decline for S_{P_WGP} in the north of the NCP during the 2080s surpassed 0.1 (Figures 5a–c and 6a–c). The S_S increased by 0–0.1 at most parts of the NCP, while the increase in S_S in the south of the NCP exceeded 0.1 (Figures 5d–f and 6d–f). The decreasing trend of S_T under the SSP585 scenario was more significant than that under the SSP245 scenario. The S_T decreased by 0.1–0.3 during the 2040s across the NCP (Figure 5g–i). Furthermore, the magnitude of decline was over 0.3 during the 2080s (Figure 6g–i). The change characteristic of S_Z under SSP585 in the 2040s was consistent with that of S_Z under SSP245 in the 2040s and 2080s, with a decrease of 0–0.1 (Figure 5j–l). The range of decline for the S_Z under SSP585 in the 2080s was up to 0.1–0.2 (Figure 6j–l).

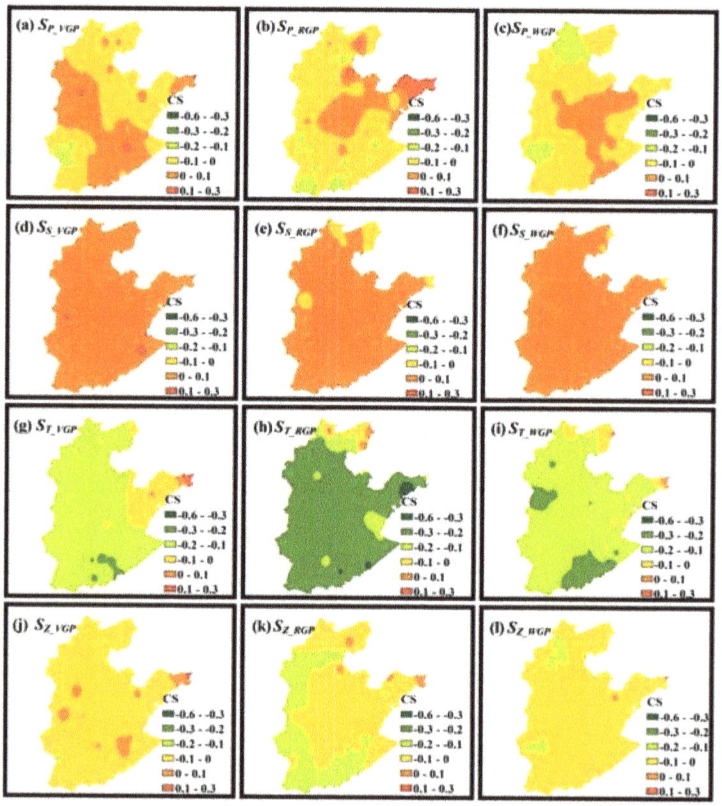

Figure 5. The spatial distribution of the change of maize climate suitability for the vegetative growth period (**a**,**d**,**g**,**j**), reproductive growth period (**b**,**e**,**h**,**k**) and whole growth period (**c**,**f**,**i**,**l**) under SSP585 during the 2040s compared to the baseline period (1981–2010) in the North China Plain.

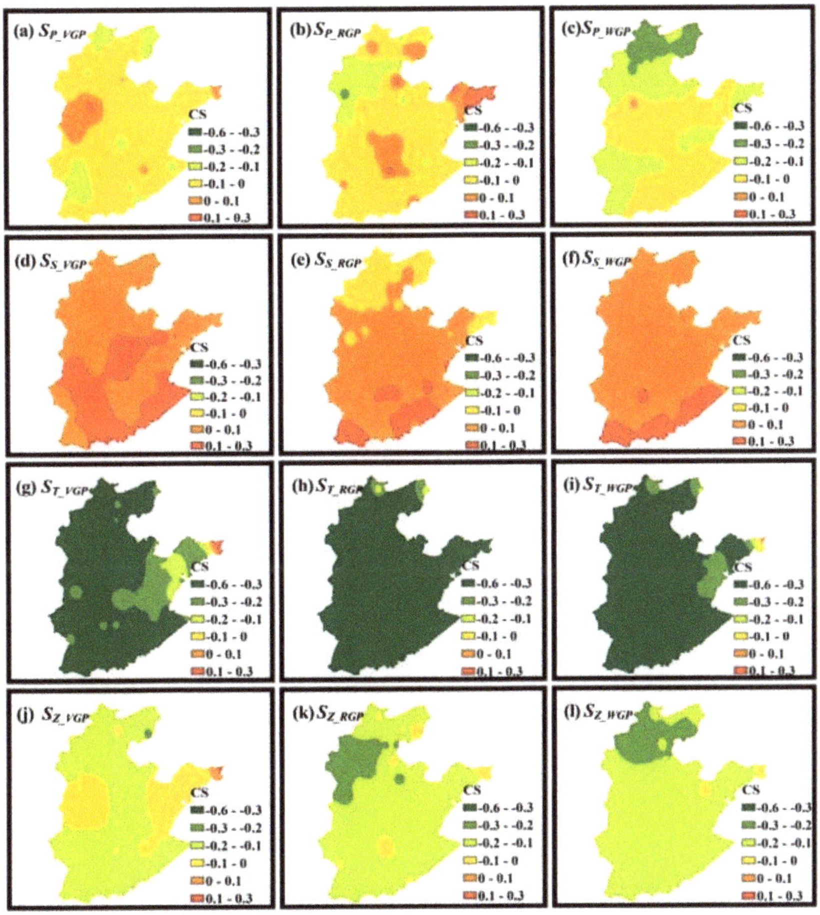

Figure 6. The spatial distribution of change of maize climate suitability for the vegetative growth period (**a**,**d**,**g**,**j**), reproductive growth period (**b**,**e**,**h**,**k**) and whole growth period (**c**,**f**,**i**,**l**) under SSP585 during the 2080s compared to the baseline period (1981–2010) in the North China Plain.

3.4. Regional Division of Climate Suitability for Maize in the Baseline Period (1981–2010) and Future Periods (2040s and 2080s)

According to the statistical principle and referring to the expression of possibility in the fourth assessment report of the Intergovernmental Panel on Climate Change (IPCC), the climate suitability of maize planting area was set as four grades: unsuitable area (CS < 0.05, unsuitable area for maize growing), less suitable area ($0.05 \leq$ CS < 0.33, less suitable area for maize growing), suitable area ($0.33 \leq$ CS < 0.66, suitable area for maize growing) and optimum area (CS ≥ 0.66, optimum area for maize growing) [16,17].

The regional division of climate suitability for maize during the baseline period (1981–2010) in the NCP is shown in Figure 7a. As an important maize production base in China, the climate in the NCP was generally conducive to the growth of maize. The optimum area and the suitable area accounted for about 50%, respectively. The optimum area is distributed in the north of the NCP, and the suitable area is mainly distributed in the south (Figure 7a). This is similar to the research results of He and Zhou [17]. The regional division of climate suitability for maize under four future scenarios in the NCP is shown in Figure 7b–e. With future climate warming, the overall climate suitability in the NCP shows a downward trend. The optimum area will decrease greatly under four future scenarios in the NCP, while the suitable area will increase significantly (Figure 7b–e).

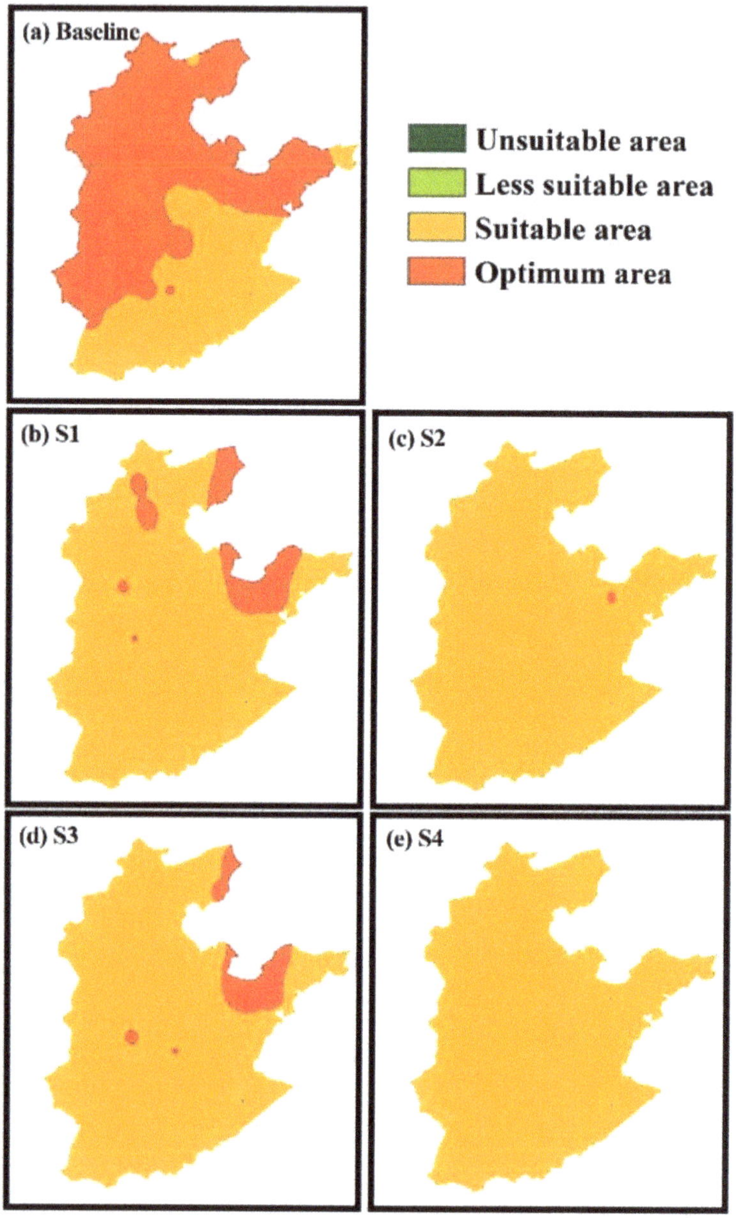

Figure 7. Regional division of climate suitability for maize during the baseline period (1981–2010) (**a**) and four future scenarios (**b–e**) in the NCP.

3.5. Effects of Sowing Date Adjustment on the Climate Suitability of Maize in the NCP

Adjustment of SD has a significant impact on the phenology of maize, while the FD and MD of maize changed with the adjustment of SD (Figure S3). Moreover, the climate suitability was closely related to the growth periods. So there was a significant difference for the climate suitability based on various SDs (Figure 8). The S_S, S_T, S_P and S_Z at the VGP declined with the delay of SD under all future scenarios, while the climate suitability went up with the advance of SD (Figure 8a,d,g,j). Additionally, the climate suitability for the RGP increased due to the delay of SD except the S_P (Figure 8b,e,h,k). Overall, the delay of SD can effectively increase the climate suitability at the WGP, especially in the S4 (Figure 8c,f,i,l). The adjustment of SD led to the change of maize growth period, while there were great differences in climatic resources at different growth periods of maize. The temporal changes of main climate factors (precipitation, sunshine duration, mean temperature, maximum temperature, minimum temperature) from May to October in the baseline period, S1, S2, S3, and S4, are shown in Figure 9. The climate factors varied greatly from month to month, which contributed to the change of climate suitability for different SDs.

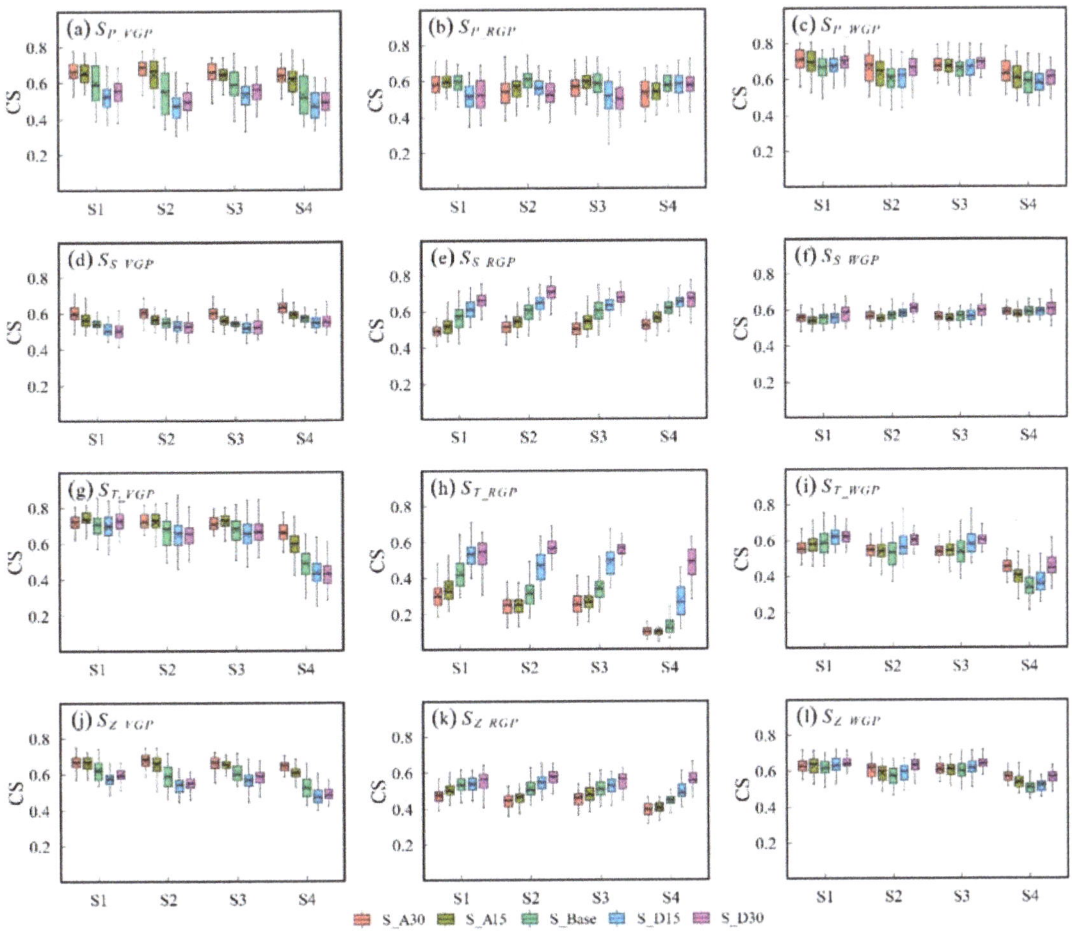

Figure 8. The climate suitability of maize for the vegetative growth period (**a**,**d**,**g**,**j**), reproductive growth period (**b**,**e**,**h**,**k**) and whole growth period (**c**,**f**,**i**,**l**) under different sowing dates in the NCP during the 2040s and 2080s under SSP245 and SSP585 (S1, S2, S3, and S4 are the SSP245_2040s, SSP245_2080s, SSP585_2040s, and SSP585_2080s).

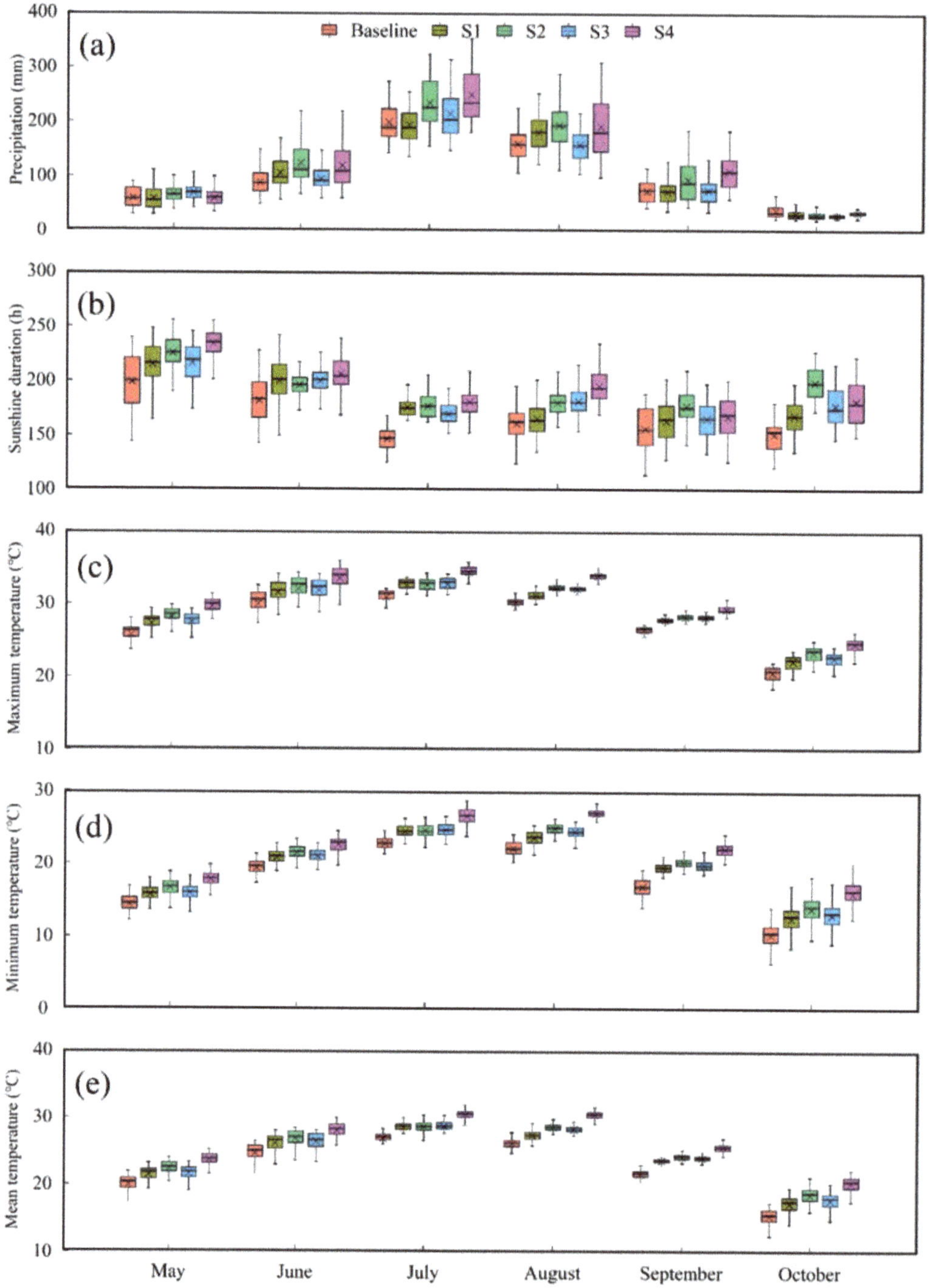

Figure 9. The temporal change of precipitation (**a**), sunshine duration (**b**), maximum temperature (**c**), minimum temperature(**d**) and, mean temperature (**e**) from May to October in the North China Plain during the baseline period, the 2040s and 2080s.

4. Discussion

Climate change has a significant impact on the growth process and yield formation of maize in China. In recent decades, climate warming not only brought forward flowering date and maturity date but also shortened RGP and WGP [46,59,60]. With future climate change, the phenological period of maize in the future will also change further [61]. In order to study the influence of future climate suitability on maize better, the phenology shift of maize should also be taken into account. The previous study analyzed the spatial-temporal characteristics of maize climate suitability in the future period across the NCP, while the phenology of maize under future scenarios was obtained by calculating the active accumulated temperature [15]. Nevertheless, the process of crop growth is complicated. There is a deficiency of mechanism to determine crop growth period by calculating effective accumulated temperature. Consequently, we used the APSIM model to simulate the flowering date and maturity date of maize. The flowering date and maturity date simulated by the APSIM model under future scenarios in our study present the trend of significant advance across the NCP (Figure S4).

Overall, the S_T across the NCP during the baseline period was higher than 0.5, while the values of S_{T_VGP} and S_{T_WGP} in the north of NCP exceeded 0.7 (Figure 4g–i). However, the decrease in S_{T_VGP}, $S_{T\_RGP,}$ and S_{T_WGP} in the future was remarkable, especially the decline of S_{T_VGP}, $S_{T\_RGP,}$ and S_{T_WGP} under SSP585_2080s was between 0.3 and 0.5 (Figure 3g–i). This may be largely associated with rising temperature. As shown in Figure 9c–e, the mean temperature, maximum temperature, and minimum temperature across the NCP will increase markedly under four future scenarios compared to the baseline period. The warm temperatures can improve the growth of crops before the threshold is reached, while yields will abruptly diminish subsequently [62,63]. Furthermore, the sensitivity to extreme temperature changes at different growth stages of crops [64] is particularly significant during the reproductive growth period [65]. The heat stress during the reproductive growth period can affect pollination, reduce male fertility and seed quality, ultimately lead to the loss of kernel weight and yield [65–67]. Our results indicated that S_{T_RGP} was lower than S_{T_VGP} during 1981–2010 across the NCP (Figure 2g–h), while the decrease in S_{T_RGP} was significantly higher than S_{T_VGP} under future scenarios (Figure 3g–h). The risk of high temperature for summer maize during future periods in the NCP will become an important field of climate change-related research [65].

In order to ensure crop yield, some adaptive agricultural management measures should be taken to counteract the adverse effects of climate change, including variety renovation, adjustment of sowing date, improvement of fertilization and irrigation conditions, and so on [68,69]. For example, the renovation of maize varieties delayed the heading date and maturity date and prolonged the whole growth period at more than 90% of stations in China, while appropriate late sowing can also prolong the whole growth period [5]. These adaptive agricultural management measures offset the impacts of climate change on maize to some extent and further ensured maize yield [5]. In this study, we investigated the climate suitability of maize for different sowing dates in the NCP during the 2040s and 2080s under SSP245 and SSP585 (Figure 8). Compared to the observed SD during the historical period, the S_S, S_T, $S_{P,}$ and S_Z for the VGP declined with the delay of SD in the future, while these for the RGP increased due to the delay of SD except the precipitation suitability (Figure 8a,b,d,e,g,h,j,k). On the whole, the delay of SD can effectively increase the climate suitability during the whole growth period of maize (Figure 8c,f,i,l). Due to the adjustment of SD, the overall growth process of maize will change. The FD and MD will postpone delaying the SD of maize (Figure S3). The grain filling stage of maize is sensitive to high temperatures. The delay of SD can postpone the grain filling stage to a relatively cool period, which can reduce the heat damage stress on maize to a certain extent [37,70]. Our study indicated the temporal changes of main climate factors from May to October during the present period, the 2040s and 2080s, in the NCP. In the future period, the increase in temperature and sunshine hours in September and October will better meet the growing demand for sunshine and temperature resources in the late growth period of maize and avoid the invasion of high temperature.

Based on the global climate model (GCM), we can investigate the climate suitability under future scenarios and further deal with the risks of ecological environment protection and social development from climate change. Compared with CMIP5, CMIP6 models have a certain degree of improvement and development in terms of resolution, physical parameterization, experimental design, and simulation computing capability. The simulation of the climate models from CMIP6 was closer to the observed value, while the uncertainty of simulation was smaller and the accuracy of the simulation was higher [26–28]. Taking the study of extreme climate indices as an example, the climate models from CMIP6 had stronger effects on the extreme temperature indices and extreme precipitation indices than the climate models from CMIP5, which can well reproduce the changing trend of the extreme climate indices [71,72]. In our study, the BCC-CSM2-MR model from CMIP6 processed by the statistically downscaled method developed by Liu and Zuo [41] was used to estimate the spatio-temporal variation characteristics of future climate suitability under different climate scenarios. The statistically downscaled method could reproduce the climate statistics at multiple time scales for historical periods and correct the stationary errors effectively [44], while the statistically downscaled data using the method has been applied in previous research [38,73,74].

The study of climate suitability can be used to cope with the impact of climate change on crop production. We can not only make specific divisions of regions based on the climate suitability model [17–20] but also optimize the selection of crop variety [21–23]. We set the regional division of climate suitability for maize in the baseline and future periods according to the statistical principle and referring to the expression of possibility in the IPCC4 (Figure 7a). The regional division of climate suitability during 1981–2010 in the study was close to the results of relevant research [17]. With climate warming, ST and SZ will decrease significantly under four future scenarios, which causes the change in the regional division of climate suitability (Figure 7b–e). Under four future scenarios, the most area of the optimum area in the NCP will change into the suitable area (Figure 7b–e). In the future, the adoption of specific adaptations or mitigation measures against the risk of heatwaves needs to be taken seriously [70]. Moreover, there was a certain correlation between climate suitability and climatic yield [14,15]. The climate suitability combined with climate information and phenology information can be used as an improved climate index in the yield prediction model to improve the effect of prediction.

5. Conclusions

The study of crop climate suitability can enhance the ability to cope with the impact of climate change on crop production. The study developed a climate suitability model of maize and investigated the climate suitability of maize under the historical period and two future periods in the NCP based on the BCC-CSM2-MR model from CMIP6. APSIM model was used to simulate the phenology data of maize under future climate scenarios to improve accuracy and reliability. The results showed that the BCC projections for climate suitability were in suitable consistency with the observations. In 1981–2010, the S_P, S_T, and S_Z were high in the north and low in the south. The S_P, S_T, and S_Z decreased under all the future climate scenarios in most areas of NCP while the Ss presented an increasing trend. Therefore, the optimum area decreases greatly under four future scenarios in the NCP while the suitable area increases significantly. Moreover, the delay of SD can effectively increase the climate suitability during the whole growth period, especially under the SSP585 scenario in 2071–2100. Thus the adjustment of SD had essential impacts on the climate suitability, which was advantageous to adapt to climate change and promoted agricultural production in the NCP.

Supplementary Materials: The following supporting information can be downloaded at: https://www.mdpi.com/article/10.3390/agriculture12030348/s1, Figure S1: The spatial distribution of the change of maize climate suitability for the vegetative growth pe-riod (a,d,g,j), reproductive growth period (b,e,h,k) and whole growth period (c,f,i,l) under SSP245 during 2040s compared to the baseline period (1980–2010) in the NCP, Figure S2: The spatial distribution of the change of maize climate suitability for the vegetative growth pe-riod (a,d,g,j), reproductive growth period (b,e,h,k) and whole growth period (c,f,i,l) under SSP245 during 2080s compared to the baseline period (1980–2010) in the NCP, Figure S3: The simulated flowering date (a) and maturity date (b) of maize for different sowing dates in the NCP during 2040s and 2080s under SSP245 and SSP585, Figure S4: The simulated changes of flowering date and maturity date across the North China Plain during 2040s and 2080s under SSP245 and SSP585 compared to the baseline (1981–2010).

Author Contributions: Conceptualization, D.X.; methodology and data analysis, Y.Z.; writing of original draft preparation, Y.Z.; providing the future climate scenario data, D.L.; writing of review and editing, D.X., H.B., J.T. and D.L. All authors have read and agreed to the published version of the manuscript.

Funding: National Natural Science Foundation of China, grant/award number: 41901128.

Institutional Review Board Statement: Not applicable.

Informed Consent Statement: Not applicable.

Data Availability Statement: The climatological data that support the findings of this study during manuscript preparation were available from the corresponding author on reasonable request. The detailed access to the data: The historical records about daily climate data from China's Meteorological Administration (CMA) (http://data.cma.cn/, accessed on 1 February 2020); Future climate scenario data were provided by the World Climate Research Program (WCRP) of Coupled Model Inter-comparison Project phase 6 (CMIP6, https://esgf-node.llnl.gov/search/cmip6/ (accessed on 20 February 2020)).

Acknowledgments: We are grateful to the NSW Department of Primary Industries, Australia, for providing the statistically downscaled climate data for the BCC-CSM2-MR model (BCC) from Coupled Model Comparison Program (CMIP6).

Conflicts of Interest: The authors declare no conflict of interest.

References

1. IPCC. Summary for Policymakers. In *Global Warming of 1.5 °C. An IPCC Special Report on the Impacts of Global Warming of 1.5 °C above Pre-Industrial Levels and Related Global Greenhouse Gas Emission Pathways, in the Context of Strengthening the Global Response to the Threat of Climate Change, Sustainable Development, and Efforts to Eradicate Poverty*; IPCC: Geneva, Switzerland, 2018.
2. Lobell, D.B.; Asner, G.P. Climate and management contributions to recent trends in U.S. agricultural yields. *Science* **2003**, *299*, 1032. [CrossRef] [PubMed]
3. Lobell, D.B.; Burke, M.B.; Tebaldi, C.; Mastrandrea, M.D.; Falcon, W.P.; Naylor, R.L. Prioritizing climate change adaptation needs for food security in 2030. *Science* **2008**, *319*, 607–610. [CrossRef] [PubMed]
4. Tao, F.; Zhang, Z.; Shi, W.; Liu, Y.; Xiao, D.; Zhang, S.; Zhu, Z.; Wang, M.; Liu, F. Single rice growth period was prolonged by cultivars shifts, but yield was damaged by climate change during 1981–2009 in China, and late rice was just opposite. *Glob. Chang. Biol.* **2013**, *19*, 3200–3209. [CrossRef] [PubMed]
5. Tao, F.; Zhang, S.; Zhang, Z.; Rotter, R.P. Maize growing duration was prolonged across China in the past three decades under the combined effects of temperature, agronomic management, and cultivar shift. *Glob. Chang. Biol.* **2014**, *20*, 3686–3699. [CrossRef]
6. Liu, B.; Asseng, S.; Müller, C.; Ewert, F.; Elliott, J.; Lobell, D.B.; Martre, P.; Ruane, A.C.; Wallach, D.; Jones, J.W.; et al. Similar estimates of temperature impacts on global wheat yield by three independent methods. *Nat. Clim. Chang.* **2016**, *6*, 1130–1136. [CrossRef]
7. Fletcher, A.L.; Chen, C.; Ota, N.; Lawes, R.A.; Oliver, Y.M. Has historic climate change affected the spatial distribution of water-limited wheat yield across Western Australia? *Clim. Chang.* **2020**, *159*, 347–364. [CrossRef]
8. Ye, Z.; Qiu, X.; Chen, J.; Cammarano, D.; Ge, Z.; Ruane, A.C.; Liu, L.; Tang, L.; Cao, W.; Liu, B.; et al. Impacts of 1.5 °C and 2.0 °C global warming above pre-industrial on potential winter wheat production of China. *Eur. J. Agron.* **2020**, *120*, 126149. [CrossRef]
9. Salman, S.A.; Shahid, S.; Sharafati, A.; Salem, G.S.A.; Bakar, A.A.; Farooque, A.A.; Chung, E.-S.; Ahmed, Y.A.; Mikhail, B.; Yaseen, Z.M. Projection of Agricultural Water Stress for Climate Change Scenarios: A Regional Case Study of Iraq. *Agriculture* **2021**, *11*, 1288. [CrossRef]
10. Mall, R.K.; Lal, M.; Bhatia, V.S.; Rathore, L.S.; Singh, R. Mitigating climate change impact on soybean productivity in India: A simulation study. *Agric. For. Meteorol.* **2004**, *121*, 113–125. [CrossRef]
11. Wang, P.; Wu, D.; Yang, J.; Ma, Y.; Feng, R.; Huo, Z. Summer maize growth under different precipitation years in the Huang-Huai-Hai Plain of China. *Agric. For. Meteorol.* **2020**, *285–286*, 107927. [CrossRef]

12. Ma, S.Q. The climatic and ecological suitability of central Jilin Province for developing maize zones. *J. Ecol.* **1990**, *54*, 40–45.
13. Gong, L.J.; Yang-Hui, J.I.; Wang, P.; Zhu, H.X.; Wang, L.L.; Wang, Q.J. Variation of Climate Suitability of Maize in Northeast of China. *J. Maize Sci.* **2013**, *21*, 140–146.
14. Cao, Y.; Qi, J.; Wang, F.; Li, L.; Lu, J. Analysis of climate suitability of spring maize in Liaoning Province based on modulus and mathematics. *Sci. Geogr. Sin.* **2020**, *40*, 1210–1220.
15. Tang, X.; Liu, H. Climate suitability for summer maize on the North China Plain under current and future climate scenarios. *Int. J. Climatol.* **2020**, *41*, 1–18. [CrossRef]
16. He, Q.; Zhou, G. Climatic suitability of potential summer maize planting zones in China. *Acta Geogr. Sin.* **2011**, *66*, 1443–1450.
17. He, Q.; Zhou, G. The climatic suitability for maize cultivation in China. *Chin. Sci. Bull.* **2012**, *57*, 395–403. [CrossRef]
18. Sun, J.; Zhou, G.; Sui, X. Climatic suitability of the distribution of the winter wheat cultivation zone in China. *Eur. J. Agron.* **2012**, *43*, 77–86.
19. Irimia, L.M.; Patriche, C.V.; Quenol, H.; Sfîcă, L.; Foss, C. Shifts in climate suitability for wine production as a result of climate change in a temperate climate wine region of Romania. *Theor. Appl. Climatol.* **2017**, *131*, 1069–1081. [CrossRef]
20. He, Q.; Zhou, G.; Lü, X.; Zhou, M. Climatic suitability and spatial distribution for summer maize cultivation in China at 1.5 and 2.0 °C global warming. *Sci. Bull.* **2019**, *64*, 690–697. [CrossRef]
21. He, Y.; Yao, Y.; Tang, H.; Chen, Y.; Li, J.; Yang, P.; Chen, Z.; Xin, X.; Wang, L.; Li, D.; et al. Using an Integrated Response-Function Method to Explore Agro-Climatic Suitability for Spring Soybean Growth in North China. *J. Appl. Meteorol. Climatol.* **2011**, *50*, 1354–1361. [CrossRef]
22. Holzkämper, A.; Calanca, P.; Fuhrer, J. Identifying climatic limitations to grain maize yield potentials using a suitability evaluation approach. *Agric. For. Meteorol.* **2013**, *168*, 149–159. [CrossRef]
23. Zhao, J.; Guo, J.; Xu, Y.; Mu, J. Effects of climate change on cultivation patterns of spring maize and its climatic suitability in Northeast China. *Agric. Ecosyst. Environ.* **2015**, *202*, 178–187. [CrossRef]
24. Ding, Y.; Wang, H. Newly acquired knowledge on the scientific issues related to climate change over the recent 100 years in China. *Chin. Sci. Bull.* **2015**, *61*, 1029–1041. [CrossRef]
25. Zhou, T.J.; Zou, L.W.; Chen, X.L. Commentary on the Coupled Model Intercomparison Project Phase 6 (CMIP6). *Clim. Chang. Res.* **2019**, *15*, 445–456.
26. Fu, Y.; Lin, Z.; Guo, D. Improvement of the simulation of the summer East Asian westerly jet from CMIP5 to CMIP6. *Atmos. Ocean. Sci. Lett.* **2020**, *13*, 550–558. [CrossRef]
27. Kim, Y.-H.; Min, S.-K.; Zhang, X.; Sillmann, J.; Sandstad, M. Evaluation of the CMIP6 multi-model ensemble for climate extreme indices. *Weather Clim. Extrem.* **2020**, *29*, 100269. [CrossRef]
28. Chen, H.; Sun, J.; Lin, W.; Xu, H. Comparison of CMIP6 and CMIP5 models in simulating climate extremes. *Sci. Bull.* **2020**, *65*, 1415–1418. [CrossRef]
29. Wu, T.; Lu, Y.; Fang, Y.; Xin, X.; Li, L.; Li, W.; Jie, W.; Zhang, J.; Liu, Y.; Zhang, L.; et al. The Beijing Climate Center Climate System Model (BCC-CSM): Main Progress from CMIP5 to CMIP6. *Geosci. Model Dev. Discuss.* **2018**, *12*, 1573–1600. [CrossRef]
30. Xin, X.G.; Wu, T.; Zhang, J.; Zhang, F.; Wei-Ping, L.I.; Zhang, Y.W.; Yi-Xiong, L.U.; Fang, Y.J.; Jie, W.H.; Zhang, L. Introduction of BCC models and its participation in CMIP6. *Clim. Chang. Res.* **2019**, *15*, 533–539.
31. Sang, Y.; Ren, H.-L.; Shi, X.; Xu, X.; Chen, H. Improvement of Soil Moisture Simulation in Eurasia by the Beijing Climate Center Climate System Model from CMIP5 to CMIP6. *Adv. Atmos. Sci.* **2020**, *38*, 237–252. [CrossRef]
32. Haarhoff, S.J.; Swanepoel, P.A. Plant Population and Maize Grain Yield: A Global Systematic Review of Rainfed Trials. *Crop Sci.* **2018**, *58*, 1819–1829. [CrossRef]
33. Zhao, Y.; Xiao, D.; Bai, H.; Tao, F. Research progress on the response and adaptation of crop phenology to climate change in China. *Prog. Geogr.* **2019**, *38*, 224–235.
34. National Bureau of Statistics of China. *China Rural Statistical Yearbook*; China Statistics Press: Beijing, China, 2018.
35. Huang, S.; Lv, L.; Zhu, J.; Li, Y.; Tao, H.; Wang, P. Extending growing period is limited to offsetting negative effects of climate changes on maize yield in the North China Plain. *Field Crops Res.* **2018**, *215*, 66–73. [CrossRef]
36. Xiao, D.; Tao, F. Contributions of cultivar shift, management practice and climate change to maize yield in North China Plain in 1981–2009. *Int. J. Biometeorol.* **2016**, *60*, 1111–1122. [CrossRef] [PubMed]
37. Xiao, D.; Liu, D.L.; Wang, B.; Feng, P.; Waters, C. Designing high-yielding maize ideotypes to adapt changing climate in the North China Plain. *Agric. Syst.* **2020**, *181*, 102805. [CrossRef]
38. Bai, H.; Xiao, D.; Wang, B.; Liu, D.L.; Feng, P.; Tang, J. Multi-model ensemble of CMIP6 projections for future extreme climate stress on wheat in the North China plain. *Int. J. Climatol.* **2020**, *41*, 1–16. [CrossRef]
39. Xiao, D.; Liu, D.L.; Wang, B.; Feng, P.; Bai, H.; Tang, J. Climate change impact on yields and water use of wheat and maize in the North China Plain under future climate change scenarios. *Agric. Water Manag.* **2020**, *238*, 106238. [CrossRef]
40. O'Neill, B.C.; Tebaldi, C.; van Vuuren, D.P.; Eyring, V.; Friedlingstein, P.; Hurtt, G.; Knutti, R.; Kriegler, E.; Lamarque, J.-F.; Lowe, J.; et al. The Scenario Model Intercomparison Project (ScenarioMIP) for CMIP6. *Geosci. Model Dev.* **2016**, *9*, 3461–3482. [CrossRef]
41. Liu, D.L.; Zuo, H. Statistical downscaling of daily climate variables for climate change impact assessment over New South Wales, Australia. *Clim. Chang.* **2012**, *115*, 629–666. [CrossRef]

42. Richardson, C.; Wright, D. *WGEN: A Model for Generating Daily Weather Variables*; USA Department of Agriculture, Agricultural Research Service: Springfield, VA, USA, 1984; Volume ARS-8.
43. Asseng, S.; Keating, B.A.; Fillery, I.R.P.; Gregory, P.J.; Abrecht, D.G. Performance of the APSIM-wheat model in Western Australia. *Field Crops Res.* **1998**, *57*, 163–179. [CrossRef]
44. Asseng, S.; Keulen; Stol, W. Performance and application of the APSIM Nwheat model in the Netherlands. *Eur. J. Agron.* **2000**, *12*, 37–54. [CrossRef]
45. Keating, B.A.; Carberry, P.S.; Hammer, G.L.; Probert, M.E.; Robertson, M.J.; Holzworth, D.; Huth, N.I.; Hargreaves, J.N.G.; Meinke, H.; Hochman, Z. An overview of APSIM, a model designed for farming systems simulation. *Eur. J. Agron.* **2003**, *18*, 267–288. [CrossRef]
46. Xiao, D.; Qi, Y.; Shen, Y.; Tao, F.; Moiwo, J.P.; Liu, J.; Wang, R.; Zhang, H.; Liu, F. Impact of warming climate and cultivar change on maize phenology in the last three decades in North China Plain. *Theor. Appl. Climatol.* **2016**, *124*, 653–661. [CrossRef]
47. Teixeira, E.I.; Brown, H.E.; Michel, A.; Meenken, E.; Hu, W.; Thomas, S.; Huth, N.I.; Holzworth, D.P. Field estimation of water extraction coefficients with APSIM-Slurp for water uptake assessments in perennial forages. *Field Crops Res.* **2018**, *222*, 26–38. [CrossRef]
48. Arshad, A.; Raza, M.A.; Zhang, Y.; Zhang, L.; Wang, X.; Ahmed, M.; Habib-ur-Rehman, M. Impact of Climate Warming on Cotton Growth and Yields in China and Pakistan: A Regional Perspective. *Agriculture* **2021**, *11*, 97. [CrossRef]
49. Xiao, D.; Bai, H.; Liu, D.L.; Tang, J.; Wang, B.; Shen, Y.; Cao, J.; Feng, P. Projecting future changes in extreme climate for maize production in the North China Plain and the role of adjusting sowing date. *Mitig. Adapt. Strateg. Glob. Chang.* **2022**, *27*, 21. [CrossRef]
50. Pu, J.Y.; Yao, X.Y.; Yao, R.X. Variations of summer and autumn grain crops' climatic suitability in the areas east of Yellow River in Gansu in recent 40 years. *Agric. Res. Arid. Areas* **2011**, *29*, 253–258.
51. Hou, Y.Y.; Zhang, Y.H.; Wang, L.Y.; Hou-Quan, L.; Song, Y.B. Climatic suitability model for spring maize in Northeast China. *Chin. J. Appl. Ecol.* **2013**, *24*, 3207–3212.
52. Zhao, Y.; Xiao, D.; Bai, H.; Tang, J.Z. Climatic suitability degrees of winter wheat and summer maize in the North China Plain. *Chin. J. Ecol.* **2020**, *39*, 1–12.
53. Zhao, F.; Qian, H.S.; Jiao, S.X. The climatic suitability model of crop: A case study of winter wheat in Henan province. *Resour. Sci.* **2003**, *25*, 77–82.
54. Huang, H. A study on the climatic ecology adaptability of the crop production in the red and yellow soils region of China. *J. Nat. Resour.* **1996**, *11*, 340–346.
55. Xu, L.; Lv, H.; Fang, L. Effect of Climate Change on the Climate Suitability of Summer Maize on the Huang-Huai-Hai Plain. *Resour. Sci.* **2014**, *36*, 0782–0787.
56. Wang, E.; Han, X. The productivity evaluation and its application of winter wheat and summer maize in Huang-Huai-Hai region. *Chin. J. Agrometeorol.* **1990**, *11*, 41–46.
57. Allen, R.; Pereira, L.; Dirk, R.; Smith, M. *Crop Evapotranspiration Guidelines for Computing Crop Water Requirements*; FAO Irrigation and Drainage Paper, No. 56; FAO—Food and Agriculture Organization of the United Nations: Rome, Italy, 1998; Volume 300, pp. 53–62.
58. Wang, L.X.; Ren, Z.Y. Appraisal and analysis of agriculture eco-climatic amenity in Datong city of Shanxi province. *Geogr. Res.* **2007**, *11*, 319–324.
59. Wang, Z.; Chen, J.; Li, Y.; Li, C.; Zhang, L.; Chen, F. Effects of climate change and cultivar on summer maize phenology. *Int. J. Plant Prod.* **2016**, *10*, 509–526.
60. Xiao, D.; Zhao, Y.; Bai, H.; Hu, Y.; Cao, J. Impacts of climate warming and crop management on maize phenology in northern China. *J. Arid. Land* **2019**, *11*, 892–903. [CrossRef]
61. Xiao, D.; Bai, H.; Liu, D. Impact of Future Climate Change on Wheat Production: A Simulated Case for China's Wheat System. *Sustainability* **2018**, *10*, 1277. [CrossRef]
62. Lobell, D.B.; Roberts, M.J.; Schlenker, W.; Braun, N.; Little, B.B.; Rejesus, R.M.; Hammer, G.L. Greater sensitivity to drought accompanies maize yield increase in the U.S. Midwest. *Science* **2014**, *344*, 516–519. [CrossRef]
63. Zhao, C.; Liu, B.; Piao, S.; Wang, X.; Lobell, D.B.; Huang, Y.; Huang, M.; Yao, Y.; Bassu, S.; Ciais, P.; et al. Temperature increase reduces global yields of major crops in four independent estimates. *Proc. Natl. Acad. Sci. USA* **2017**, *114*, 9326–9331. [CrossRef]
64. Bailey-Serres, J.; Parker, J.E.; Ainsworth, E.A.; Oldroyd, G.E.D.; Schroeder, J.I. Genetic strategies for improving crop yields. *Nature* **2019**, *575*, 109–118. [CrossRef]
65. Bita, C.E.; Gerats, T. Plant tolerance to high temperature in a changing environment: Scientific fundamentals and production of heat stress-tolerant crops. *Front. Plant Sci.* **2013**, *4*, 273. [CrossRef] [PubMed]
66. Wang, N.; Wang, E.; Wang, J.; Zhang, J.; Zheng, B.; Huang, Y.; Tan, M. Modelling maize phenology, biomass growth and yield under contrasting temperature conditions. *Agric. For. Meteorol.* **2018**, *250–251*, 319–329. [CrossRef]
67. Siddik, M.A.; Zhang, J.; Chen, J.; Qian, H.; Jiang, Y.; Raheem, A.K.; Deng, A.; Song, Z.; Zheng, C.; Zhang, W. Responses of indica rice yield and quality to extreme high and low temperatures during the reproductive period. *Eur. J. Agron.* **2019**, *106*, 30–38. [CrossRef]
68. Owusu Danquah, E.; Beletse, Y.; Stirzaker, R.; Smith, C.; Yeboah, S.; Oteng-Darko, P.; Frimpong, F.; Ennin, S.A. Monitoring and Modelling Analysis of Maize (*Zea mays* L.) Yield Gap in Smallholder Farming in Ghana. *Agriculture* **2020**, *10*, 420. [CrossRef]

69. Zhai, L.; Zhang, L.; Yao, H.; Zheng, M.; Ming, B.; Xie, R.; Zhang, J.; Jia, X.; Ji, J. The Optimal Cultivar × Sowing Date × Plant Density for Grain Yield and Resource Use Efficiency of Summer Maize in the Northern Huang-Huai-Hai Plain of China. *Agriculture* **2021**, *12*, 7. [CrossRef]
70. Lv, Z.; Li, F.; Lu, G. Adjusting sowing date and cultivar shift improve maize adaption to climate change in China. *Mitig. Adapt. Strateg. Glob. Chang.* **2020**, *25*, 87–106. [CrossRef]
71. Lin, W.; Chen, H. Assessment of model performance of precipitation extremes over the mid-high latitude areas of Northern Hemisphere: From CMIP5 to CMIP6. *Atmos. Ocean. Sci. Lett.* **2020**, *13*, 598–603. [CrossRef]
72. Luo, N.; Guo, Y.; Gao, Z.B.; Chen, K.; Chou, J. Assessment of CMIP6 and CMIP5 model performance for extreme temperature in China. *Atmos. Ocean. Sci. Lett.* **2020**, *13*, 589–597. [CrossRef]
73. Wang, B.; Liu, D.L.; Macadam, I.; Alexander, L.V.; Abramowitz, G.; Yu, Q. Multi-model ensemble projections of future extreme temperature change using a statistical downscaling method in south eastern Australia. *Clim. Chang.* **2016**, *138*, 85–98. [CrossRef]
74. Wang, B.; Feng, P.; Li Liu, D.; O'Leary, G.J.; Macadam, I.; Waters, C.; Asseng, S.; Cowie, A.; Jiang, T.; Xiao, D. Sources of uncertainty for wheat yield projections under future climate are site-specific. *Nat. Food* **2020**, *1*, 720–728. [CrossRef]

agriculture

MDPI

Article

Projections of Drought Characteristics Based on the CNRM-CM6 Model over Africa

Isaac Kwesi Nooni [1,2,3], Daniel Fiifi Tawia Hagan [2,*], Waheed Ullah [2], Jiao Lu [2], Shijie Li [2], Nana Agyemang Prempeh [4], Gnim Tchalim Gnitou [1] and Kenny Thiam Choy Lim Kam Sian [1,3]

1 Binjiang College, Nanjing University of Information Science and Technology, Wuxi 214105, China; nooni25593@alumni.itc.nl (I.K.N.); patrickgnitou@yahoo.fr (G.T.G.); kennlks@gmail.com (K.T.C.L.K.S.)
2 School of Geographical Sciences, Nanjing University of Information Science and Technology, Nanjing 210044, China; waheed.khan@nuist.edu.cn (W.U.); jiao_lu@nuist.edu.cn (J.L.); lishijie@nuist.edu.cn (S.L.)
3 Wuxi Institute of Technology, Nanjing University of Information Science & Technology, Wuxi 214105, China
4 Department of Geographical Sciences, School of Geosciences, University of Energy and Natural Resources, Sunyani P.O. Box 214, Ghana; agyemang.prempeh@uenr.edu.gh
* Correspondence: dhagan@ucar.edu; Tel.: +86-13813369428

Abstract: In a warming climate, drought events are projected to increase in many regions across the world, which would have detrimental impacts on water resources for agriculture activity and human life. Thus, projecting drought changes, especially the frequency of future drought events, is very important for the African continent. This study investigates the future changes in drought events based on the France Centre National de Recherches Météorologiques (CNRM-CM6) model in the Coupled Model Intercomparison Project phase six (CMIP6) datasets for four shared socio-economic pathways (SSP): SSP1-2.6, SSP2-4.5, SSP3-7.0, and SSP5-8.5; and three time slices: near future (2020–2039), mid-century (2050–2069), and end-of-century (2080–2099), relative to a historical baseline period (1995–2014). The interannual variability and trends of the self-calibrating Palmer Drought Severity Index (scPDSI) based on the Penman–Monteith methods for measuring potential evapotranspiration (PET) are used to estimate future droughts. The temporal analysis shows that the drought frequency, intensity, and affected area will increase throughout the 21st century. Among the scenarios, SSP3-7.0 and SSP5-8.5 project a larger upward trend in drought characteristics than SSP1-2.6 and SSP2-4.5. The spatial pattern shows drought frequency decreases in humid regions and increases in non-humid regions across Africa. For all SSP scenarios, the projected wetting trend per decade ranges from 0.05 to 0.25, while the drying trend per decade ranges from −0.05 to 0.25. A regional trend analysis revealed key differences in spatial pattern, with varied trend projections of wetter and drier conditions in humid and non-humid regions under all SSP scenarios. Drier conditions are expected to intensify in Southern Africa under all SSP scenarios but are projected to be more intense under either SSP3-7.0 and SSP5-8.5. In general, the projected wetter trends in humid areas may favor agricultural production and ecological conservation, and drier trends in non-humid regions may call for the possible adoption of tailor-made drought adaptation strategies and development programmes to minimize impacts.

Keywords: CNRM-CM6; PET; climate change; IPCC-AR6; SSP scenarios

Citation: Nooni, I.K.; Hagan, D.F.T.; Ullah, W.; Lu, J.; Li, S.; Prempeh, N.A.; Gnitou, G.T.; Lim Kam Sian, K.T.C. Projections of Drought Characteristics Based on the CNRM-CM6 Model over Africa. *Agriculture* **2022**, *12*, 495. https://doi.org/10.3390/agriculture12040495

Academic Editors: Dengpan Xiao and Wenjiao Shi

Received: 25 February 2022
Accepted: 30 March 2022
Published: 31 March 2022

1. Introduction

Under a warming climate, the frequency of droughts is expected to increase in many regions due to the increase in projected temperature (TEMP) [1]. Drought is a complex natural process with adverse effects that ripple through multiple sectors of society, especially water resources for agricultural activities and human livelihood [2]. Droughts may be classified as meteorological, hydrological, agricultural, or socio-economic, based on their physical characteristics (see [2,3] for more details).

An increase in TEMP is expected to significantly affect hydrological processes [1,4] and may considerably change regional climates, leading to more frequent extreme events (e.g., droughts, heat stress) [1]. Skoulikaris et al. [4], among others, investigated the heat stress on agriculture due to climate change. Drought indices are commonly used to quantify drought events at any spatial–temporal scale. According to the World Meteorological Organisation (WMO), there are over 50 indices designed to compute droughts [2,3]. However, the most widely used and recommended drought indices are the Standardized Precipitation Index (SPI) [5], the Standardized Precipitation Evapotranspiration Index (SPEI) [6], the Palmer Drought Severity Index (PDSI) [7], and the self-calibrating PDSI (scPDSI) [8]. In a historical context, a wide range of studies based on independent observation data sets has examined drought indices and characterized drought parameters, such as the frequency, intensity, and spatial extent over the last three decades [2]. Among the indices mentioned above, the SPEI and the PDSI/scPDSI are the most widely used because their design incorporates two key components of the water cycle—precipitation (i.e., water supply) and evapotranspiration (i.e., water demand)—to represent a drought condition [2,3].

When comparing SPEI and scPDSI, the latter is preferred due to its better physical representation of drought condition in tropical regions, particularly in non-humid regions [9–12]. The advantage of using scPDSI, for example in a tropical region such as Africa, is its ability to represent drought conditions in non-humid regions (e.g., the Saharan or Kalahari deserts) and humid regions (e.g., the equatorial regions of Africa). In scPDSI, the change in the water balance is based on the difference between precipitation (PRE) and potential evapotranspiration (PET) together with parameters related to the soil/surface characteristics at each geographic region [8]. For example, the response to actual ET in water-limited regions (such as arid or semi-arid climates) is related to PRE changes rather than PET. On the contrary, in an energy-limited region (such as the equatorial region of Africa), PET, rather than PRE, is a driver to actual ET changes [13,14]. Thus, the scPDSI has been suggested by many studies [8,15,16], despite the index not being multi-scalar [15]. However, characterizing drought events at an interannual scale makes scPDSI comparable to SPEI at similar timescales [13], thus improving our understanding of drought events over the past century [16,17] and those documented in [8,16,18,19]. Recently, the scPDSI based on the Penman–Monteith methods for calculating PET, provided reasonable estimates of drought characteristics over the tropical climates of Africa compared to the Thornwaite method [20], a pattern which is well documented in historical drought study literature [15,17,21].

Good knowledge of the evolution of drought characteristics in the near and distant future can assist in early and efficient preparation for a drought event. Outputs from global climate models (GCMs) used in the Coupled Model Intercomparison Project (CMIP) framework allow us to understand the evolution of the climate under different emission scenarios [1]. Many studies have delved into drought characteristics using older CMIP versions. The effect of climate warming on drought intensifications (aridity) is also well studied and reported in [1,18,19,22]. The release of the new CMIP datasets [21,23] with improved quality and resolution makes further drought studies [24] of great interest, as updates in CMIP6 large-scale physics and dynamics are expected to introduce differences in how they perform in different climate regions. A typical example is presented by Voldoire et al. [25], where updates of several schemes, such as those in the France Centre National de Recherches Météorologiques (CNRM-CM) model, improved the simulation outputs of tropical climates, which is of great interest to Africa's climate studies.

Also, an understanding of interannual variability and long-term changes in future droughts is further motivated by recent studies [26,27], which demonstrate potential shifts in climate zones under a future global warming scenario. According to these studies, different climate regions are likely to be influenced by a warming climate at the end of the century. This could suggest that in a future climate, an altered energy/water-limited regional response to actual ET will be related to PRE (PET) changes rather than PET (PRE) [14,28]. The new demarcation of the African sub-regions based on climate zones largely puts this study into proper context [29], thus providing meaningful information needed to achieve effective regional drought mitigation strategies under climate warming. However, knowledge of future droughts interannual variability and long-term changes based on the CMIP6 data is limited. A review of previous studies showed that relatively significant studies have used the CMIP3 or the CMIP5 to study drought in different African regions [30–34].

In a multiplicity of global datasets coupled with improved data representation, many studies have adopted the multi-mean ensemble (MME) technique to study drought events. On the other hand, single model studies have gained significant attention and have advanced in recent times [25,35–40]. Most single model studies compared the two different versions of the same models [25,36–40]. Unlike those studies, the focus of the present study is to examine the climatology of drought events and their parameters based on the CNRM-CM6 GCM. This study follows a related study [35] that examined the future ET climatology for different SSP scenarios using the CNRM-CM6 model across climate regions and indicated that the ET variability may likely influence the distribution of extreme events, such as droughts, in both space and time, especially across Africa. This study investigates the temporal variability of future drought characteristics under four emissions scenarios. Moreover, the spatial pattern of drought event frequency and the wetting and drying trends from the CNRM-CM6 model using the scPDSI is examined based on the Penman-Monteith (PM) methods for measuring PET to identify droughts. The choice of scPDSI to represent future drought is documented in CMIP (phase 5 and 6 based drought analysis) [1,24,26].

The remainder of the paper is structured as follows: Section 2 describes the study area, and introduces the data and methods used in the study. Section 3 presents the results of the projected change in drought characteristics and projected trends in wetting and drying conditions. The discussions are presented in Section 4 and the conclusion in Section 5.

2. Data and Methods

2.1. Study Area

Africa is located between 32° N and 35° S and 14° W and 52° E (Figure 1). The entire African land area is nearly 30.37 million km^2, and the equator divides the continent into two, with more states in the Northern than in the Southern Hemispheres. Africa is the second-largest continent after Asia, in land size and population growth. Its vulnerability to climate variabilities is highly noticeable when extreme events occur, as three-quarters of the continent's Gross Domestic Product is heavily dependent on rain-fed agriculture [41], which is tied to climate variability [41]. As climate change is expected to reshape the spatiotemporal pattern of climate zones in the future climate [26,27], we present a Köppen–Geiger map overlaid on the latest Intergovernmental Panel on Climate Change (IPCC) regional demarcations for Africa region [29] (Figure 1). The IPCC regional demarcations for Africa [29] is divided into seven regions: the Sahara (SAH), West Africa (WAF), Central Africa (CAF), Northern East Africa (NEAF), Southern East Africa (SEAF), Western South Africa (WSAF), and Eastern South Africa (ESAF) (Figure 1) and are used in [42,43].

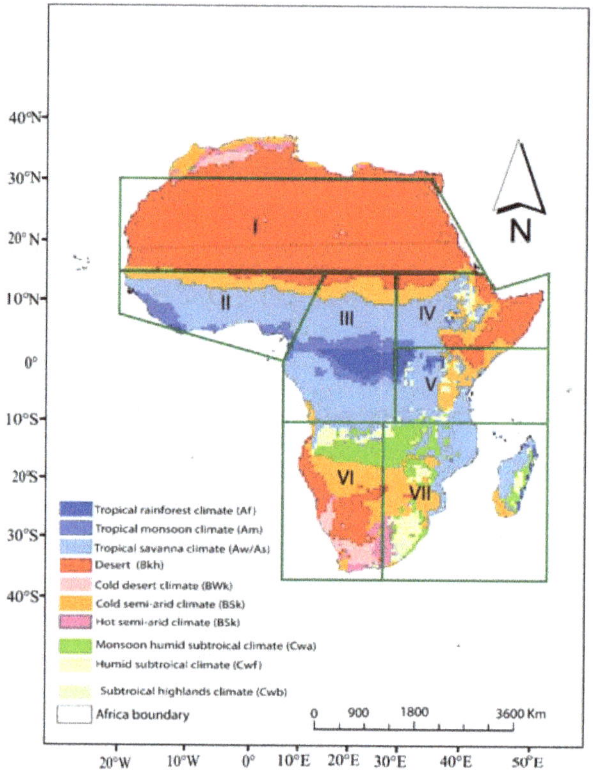

Figure 1. Map of Köppen–Geiger climate classification for 2071–2100 [27] overlaid with the updated IPCC sub-regions for Africa climate studies [29]. Abbreviations: (I) SAH: Sahara, (II) WAF: West Africa, (III) CAF: Central Africa; (IV) NEAF: Northern East Africa, (V) SEAF: Southern East Africa; (VI) WSAF: Western South Africa, (VII) ESAF: Eastern South Africa.

2.2. Data

Following the recommendation from a previous study of the region [35], the present study uses the France Centre National de Recherches Météorologiques (CNRM-CM6) dataset produced by [25,44]. The dataset is jointly developed by the Centre National de Recherches Météorologiques—Groupe d'Étude de l'Atmosphère Météorologique (CNRM-GAME) and the Centre Européen de Recherche et de Formation Avancée en Calcul Scientifique (CERFACS). In CNRM, the atmosphere model is represented by ARPEGE-Climat (v6.3) [45], which incorporates the land surface scheme ISBA-CTRIP [46,47]. The lake areas uses the revised FLAKE model, which is incorporated in the SURFES v8.0 [48] externalized surface system as well as being fully-coupled with the NEMO version 3.6 ocean model [49]. The sea ice model GELATO scheme [50] through the OASIS-MCT coupling system [51,52] and the Total Runoff Integrating Pathways (TRIP) river routing scheme [53] were used. An output system called the XIOS server is added to the system to allow online output processing [54]. Readers are directed to Voldoire et al. [25] for more details of the CNRM-CM6 GCM. The CNRM-CM selection was influenced partly as a result of past studies on PRE [55,56], ET [35] and TEMP [55]. Further, an evaluation of CNRM-CM5 and CNRM-CM6 by Voldoire et al. [25] highlighted significant improved simulation of tropical climates.

The spatiotemporal resolution of CNRM-CM6 is $1.4^{\circ} \times 1.4^{\circ}$ and extends from 1995–2014 for historical data and 2015–2100 for projections. The study considered projections for three time slices: near future (2020–2039), mid-century (2050–2069), and end-of-century (2080–2099). We use the first ensemble member (r1i1p1f1 (r1: realization index; i1: initialization index; p1: physics index; and f1: forcing index). The CNRM-CM6 datasets are publicly available at [57]. The projections are studied for four Shared Socio-economic Pathway (SSP): SSP1-2.6, SSP 2-4.5, SSP3-7.0, and SSP5-8.5, representing the low forcing (i.e., sustainability pathway), medium forcing (i.e., middle-of-the-road pathway) medium-to-high forcing pathway (i.e., a medium challenge to mitigation and adaptation pathway) and high-end forcing pathway (i.e., the worst possible scenario), respectively [58].

2.3. Methods

2.3.1. Potential Evapotranspiration (PET) Computation Using the Penman–Monteith Model

PET is a key component of the scPDSI. We used the Penman–Monteith (PM) model to compute PET based on the Food and Agriculture (FAO) recommendation [41]. The choice of PM is based on a previous study in the region [20] and it has been documented in many studies across the globe [15,16,18,19]. The PET computation with the PM model uses relative humidity, wind, temperature, and short and long-wave radiation. See for more details about the PM approach [59].

2.3.2. Self-Calibrated Palmer Drought Severity Index (scPDSI) Model

The scPDSI is widely used to quantify future droughts [60]. We computed the scPDSI with PRE, PET, and available water capacity (AWC) following Wells et al. [8] and Dai [59]. The drought is computed for the baseline (1995–2014) and future (2015–2100) periods. To compute the projected drought for each time window, the difference between the future time window and the baseline period is estimated and projected from 2015 to 2100 under the four SSP scenarios following [60]. Details of scPDSI formulations and calculation is found in [59] and others related studies [15,16,18,19,61].

2.3.3. Drought Characteristics

The run theory is used [62] to extract drought events and describe their basic characteristics (i.e., drought frequency (D_F), drought intensity (D_I) and mean drought-affected area (D_A)). For each month, grids with scPDSI values lower than −2 are considered as a drought event.

The mathematical expression below (Equation (1)) is used to calculate drought frequency (D_F). D_F is the ratio of the number of drought months to the total number of months.

$$D_F = \frac{n}{N} \tag{1}$$

where D_F denotes the frequency of droughts, n denotes the number of drought months, and the N represents the total number of months.

Drought area is the total area of grids affected by droughts. The drought area coverage (D_A) is expressed as (Equation (2)):

$$D_A = \frac{\sum_{i=1}^{n} d_a}{n_a} \tag{2}$$

where D_A represents the drought area coverage, i is a month, n is the nth month, n_a is the total number of pixels under drought condition, and d_a denotes the number of pixels with scPDSI < -2 for a specific intensity in month i.

Drought intensity is the average drought index of grids experiencing droughts. Drought intensity is computed by the averaging the intensity of all drought events on each grid during the reference period and the three future periods (Equation (3)):

$$D_I = \left| \frac{1}{n} \sum_{i=1}^{n} scPSDI_i \right| \tag{3}$$

where D_I represents the drought intensity and n denotes the total number of grids with drought conditions in months with $scPSDI < -2$.

2.3.4. Mann–Kendall Test and Theil–Sen's Slope Test

The trends in drought are examined using the Mann–Kendall tau-b nonparametric technique [63,64]. The study also used the Theil–Sen formula, to estimate and characterize linear trends [65]. The mathematical formulation for both the Mann–Kendall tau-b nonparametric technique and the Theil–Sen formula is well known in the literature. The computation procedures are presented in many studies [61,66,67].

2.3.5. Unit of Analysis

The unit of analysis used in this study is based on annual and decadal scales for future climate change analysis. A flowchart of the paper is illustrated in Figure S1. In addition, to better characterize the drought events over the African region, the projected drought changes are performed for six spatial domains defined by Iturbide et al. [29] and are adopted in the IPCC AR6 [1] and other studies [42,43]. The new demarcation provides a possible scientific basis for describing drought events under different climate zones and proposing tailormade adaptations and mitigation policies for different regions.

Based on lessons from past studies [16,18,68], this work defines drought episodes as periods with a monthly drought index (less than −2) under the thresholds shown in Table 1. Thus, we calculate the drought index for each grid and the four SSP scenarios of the CNRM-CM6 data. The drought properties are then spatially averaged for each 20-year period. The projected droughts are computed by subtracting the historical mean (1995–2014) from the entire projected time series: near future (2020–2039), mid-century (2050–2069), and end-of-century (2080–2099). All data processing is performed using the Climate Data Operation (CDO).

Table 1. Classifications of droughts based on scPDSI.

Categories	scPDSI
Extremely dry	≤−4.0
Severely dry	−3.99 to −3.0
Moderately dry	−2.99 to −2.0
Near normal	−1.99 to 1.99
Moderately wet	2.0–2.99
Severely wet	3.0–3.99
Extremely wet	≥4.0

3. Results

3.1. Projected Climatological Changes in Drought Characteristics

Figure 2 illustrates the projected drought frequency for the four SSPs scenarios during 2015–2100 relative to the baseline period (1995–2014). In general, the frequency of future drought events shows an increasing trend for all SSP scenarios. Moreover, the magnitude of the trend increases with an increase in radiative forcing.

SSP5-8.5 (red) and SSP3-7.5 (orange) illustrate an increasing trend throughout the century, with the frequency of drought events estimated to range between 3–8 yr^{-1}. The magnitude of the frequency of drought events is slightly lower (3–6 yr^{-1}) in SSP1-2.6 (deep blue) and SSP2-4.5 (light blue). Overall, all the SSP scenarios show an increasing trend in drought frequency throughout the century.

Figure 3 illustrates the time series of projected drought intensity for the four SSP scenarios. The projected time series presents different drought intensities for the different SSPs. Overall, the projected intensity varies significantly across all SSP scenarios with values ranging from −2.5 to −4 for different periods. These changes in drought intensities are more distinguishable in the mid- (2050–2069) to end-of-century (2080–2099), as projected intensity changes from severe (i.e., −2.5–−3.5) to extreme (i.e., ≥3.5) droughts under SSP3-7.0 (orange) and SSP5-8.5 (red) at the end of the century. In the SSP1-2.6 (deep blue) and SSP2-4.5 (light blue) scenarios, the drought intensities are moderate (i.e., ≤2.5) in the near future (2020–2039), changing to severe in the mid-century (2050–2069), and back to moderate drought intensity at the end of the century (2080–2099).

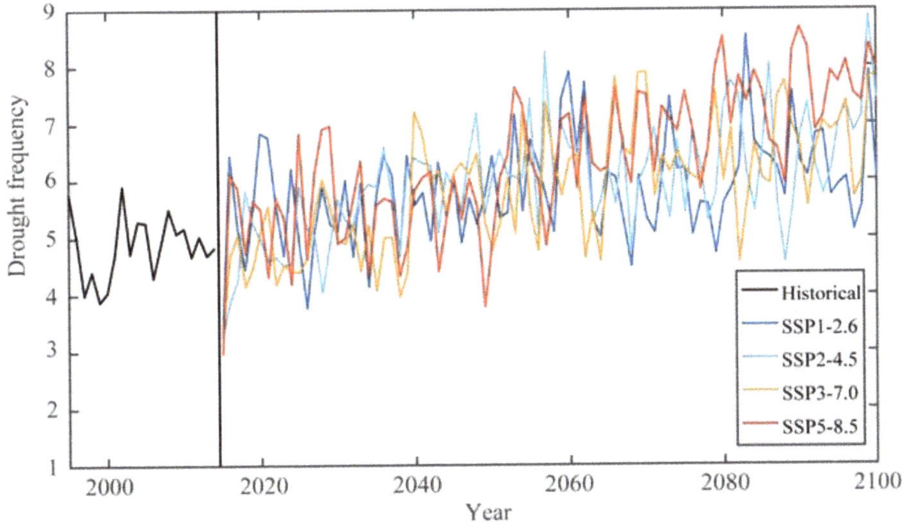

Figure 2. Annual drought frequency averaged over Africa during the baseline (1995–2014; black line) and future (2015–2100) periods. The colored lines represent results under the four SSP: SSP1-2.6, SSP2-4.5, SSP3-7.0, and SSP5-8.5.

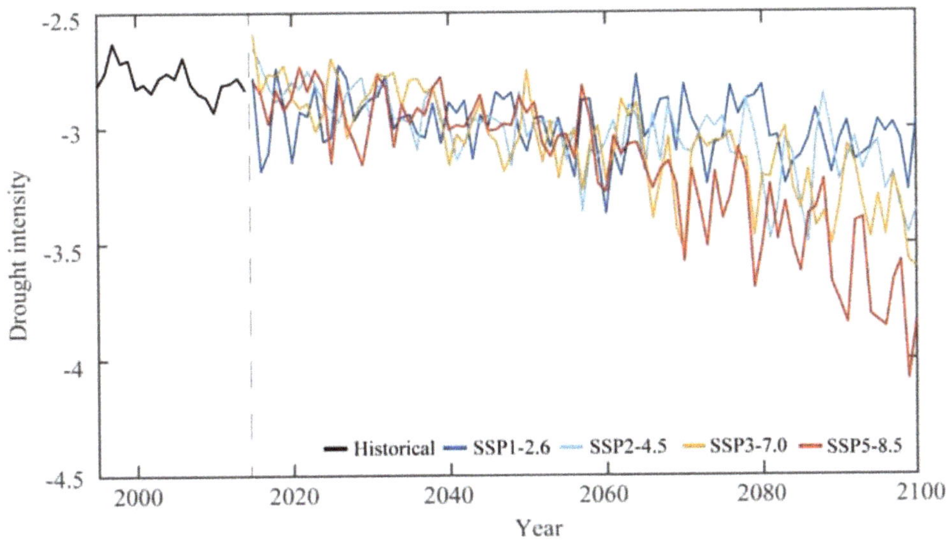

Figure 3. Annual drought intensity averaged over Africa during the baseline (1995–2014, black line) and future (2015–2100) periods. The color lines represent the results under the four SSP: SSP1-2.6, SSP2-4.5, SSP3-7.0, and SSP5-8.5.

Figure 4 shows the time series of the projected drought-affected area for the different SSP scenarios during the 21st century. The figure shows mixed upward and downward trends during the different time periods. Quantitatively, the affected areas range from 900–1800 (10^4 km^{-2}) in the near future (2020–2039) under all SSP scenarios, followed by an increase in coverage from 2000 (10^4 km^{-2}) in the 2060s and a decrease to 1700 (10^4 km^{-2}) at the end of mid-century (i.e., 2070). The end-of-century projects upwards trends for all SSP scenarios from 1600 to 2099 (10^4 km^{-2}). The affected areas for SSP3-7.0 (orange color) and SSP5-8.5 (red) are slightly closer to each other in magnitudes, while SSP1-2.5 (deep blue) and SSP 2-4.5 (light blue) scenarios show a similar increase in range of magnitudes.

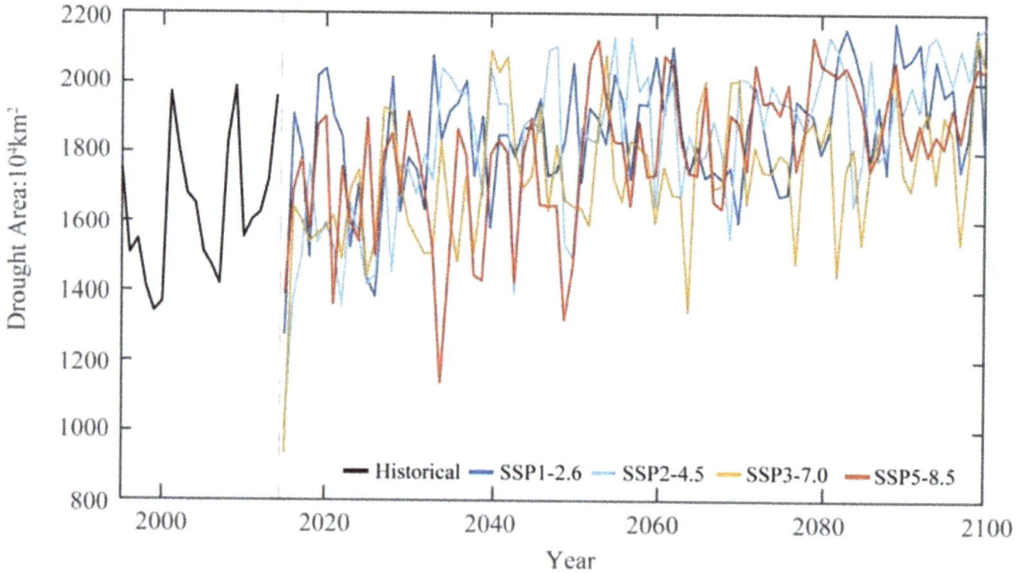

Figure 4. Annual drought spatial coverage averaged over Africa during the baseline (1995–2014; black line) and future (2015–2100) periods. The color lines represent the results under the four SSP: SSP1-2.6, SSP2-4.5, SSP3-7.0, and SSP5-8.5.

3.2. Projected Changes in Drought Frequency

Figure 5 illustrates the spatial distributions of climatological drought frequency across Africa for the different SSPs scenarios during 2020–2099. The drought frequency is computed by subtracting each grid of the baseline period from that of the future periods. Overall, the spatial pattern of drought frequency is projected to increase for different periods and SSP scenarios. The spatial trend of drought frequency across Africa complements the temporal trend by indicating the regions of possible increase or decrease in drought frequency.

The analysis of the projected drought frequency across the continent shows striking differences in spatial patterns. Regional differences are observed when considering the different sub-regions (Figure 5). In SAH, high drought frequency values (6 to >10) are observed for each scenario (Figure 5). The drought frequency is projected to increase with an increase in radiative forcing scenarios in this region over the period. Similar patterns are noted for WSAF and ESAF, with the drought frequency varying from 6 to >10. However, we observed higher frequency values over the arid regions of WSAF than over the semi-arid region of ESAF, except under SSP5-8.5. The drought frequency pattern of SAH and WSAF is related to the similarity in PRE pattern and the amount of the regions.

In contrast, the equatorial region shows different drought frequencies under each scenario. In WAF, the drought frequency is relatively lower and ranges from 2 to 4, with the frequency slightly reducing with increasing scenarios. In CAF, drought frequency varies from <2 to 6 and the value reduces from SSP1-2.6 to SSP5-8.5. We observe values of <2 in drought frequency in CAF, especially over the Congo Basin. In the EAF, we observe slightly mixed results. For example, NEAF presents slightly higher drought frequency values ranging from 2 to >8, while SEAF shows values ranging from 2 to 6. Drought frequency of 4–6 yr^{-1} is observed over parts of Sudan and Ethiopia but reduces from 6 to 2 over the Somalia and Eritrea region for each scenario. In summary, drought frequency is projected to decrease in CAF and WAF and slightly in SEAF for each scenario while an increase is expected in SAH, WSAF, and ESAF.

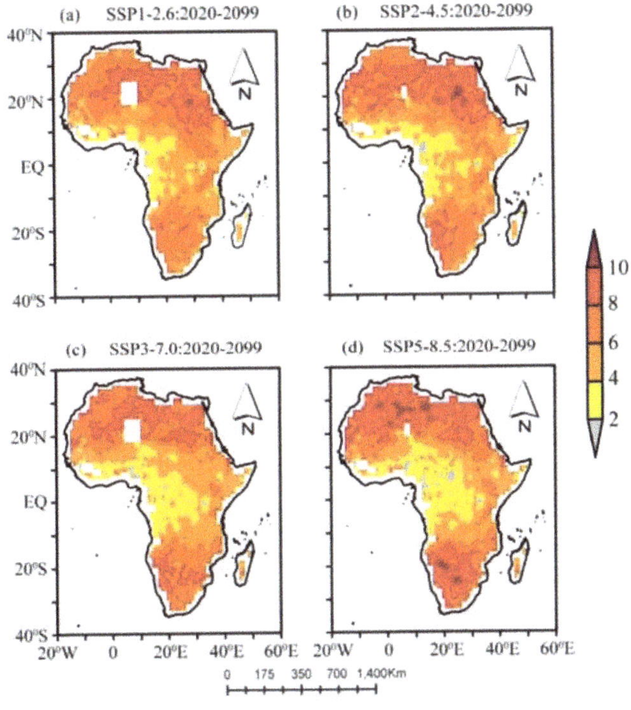

Figure 5. Geographic distribution of projected annual mean drought frequency (yr^{-1}) during 2020–2099 under the four SSP scenarios. (**a**) SSP1-2.6; (**b**) SSP2-4.5; (**c**) SSP3-7.0; and (**d**) SSP5-8.5. The anomaly is calculated as projection minus baseline period. The white background shows areas with no values.

We further investigate the spatial distributions of projected climatological changes in drought frequency in the three time slices (Figure 6). Overall, the spatial pattern shows nearly similar value ranges in drought frequencies across the continent. However, sub-regional analysis shows an interesting pattern with a distinguishable reduction in the number of drought occurrences in WAF and CAF. WAF shows progressively decreasing values in drought frequency throughout the century. The number of drought occurrences decreases at a much lower rate, with a sharper decrease noted in CAF. In general, in WAF and CAF, drought frequencies is lower under SSP3-7.0 and SSP5-8.5 than under SSP1-2.5 and SSP2-4.5.

On the contrary, an increase in the number of drought occurrences is noted in SAH, WSAF, and ESAF. On the other hand, the drought frequency in NEAF and SEAF is relatively similar for each scenario. The drought frequency is likely to increase in SAH, WSAF, and ESAF under SSP3-7.0 and SSP5-8.5 than under SSP1-2.5 and SSP2-4.5.

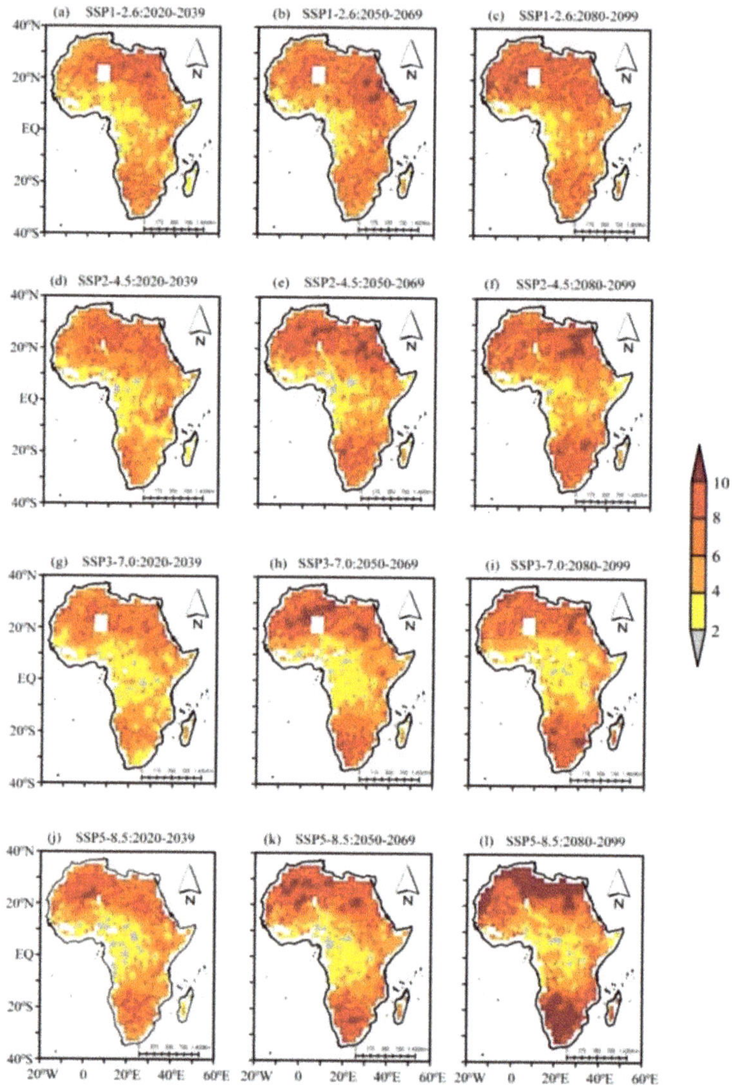

Figure 6. Geographic distribution of projected mean drought frequency (yr^{-1}) for the near future (2020–2039), mid-century (2050–2069), and end-of-century (2080–2099) under **(a–c)** SSP1-2.6, **(d–f)** SSP2-4.5, **(g–i)** SSP3-7.0, and **(j–l)** SSP5-8.5 scenarios. The white background shows areas with no values.

3.3. Projected Wetting and Drying Trends

This section focuses on the projected scPDSI trends for the different scenarios during the 2020–2099. For this purpose, we consider an event as dry and wet when $scPDSI_{PM}$ is <-2 and $scPDSI_{PM} >+2$, respectively. Figure 7 illustrates the spatial distribution of $scPDSI_{PM}$ linear trends based on the Mann–Kendall test. The results are tested at 5% significance level.

The spatial patterns clearly illustrate that those mixed drying and wetting signals are likely to dominate many parts of Africa throughout the 21st century. The spatially complex trends show that drying conditions are likely to increase from 0.05 to 0.25 $decade^{-1}$ under

SSP1-2.5 and SSP2-4.5 across Africa, with packets of wetting conditions scattered in the continent (Figure 7a,b).

In contrast, the spatial patterns in Figure 7c,d, clearly show that wetting conditions are likely to increase throughout the 21st century, with distinguished variations observed in the equatorial region (i.e., WAF and CAF), with trends of SSP3-7.0 and SSP5-8.5 scenarios increasing from 0.05 to 0.25 decade^{-1}. Moreover, a clear distinction is observed in SAH, WSAF, and ESAF, showing that future drying conditions are likely to increase from −0.05 to −0.25 decade^{-1} for SSP3-7.0 and SSP5-8 scenarios. It is worth observing that NEAF and SEAF present mixed results of wetting and drying in different scenarios.

The spatial distribution in SEAF shows a slightly larger increase in wetting than in NEAF under all scenarios. In SEAF, a striking pattern is the wetting trends before 30–40° E and the drying trend along 40–45° E in the SSP3-7.0 and 5–8.5 scenarios. However, in Figure 7a,b, a distinct spatial pattern between the two scenarios in NEAF and SEAF shows no trends in SSP1-2.5 and wetting trends in SSP 2-4.5 along 40–45° E. In WSAF and ESAF, a drying trend gradually increases from SSP1-2.5 to the SSP5-8.5 scenario. Our analysis shows more pronounced drying conditions in SSP3-7.0 and SSP5-8 scenarios over WSAF and ESAF.

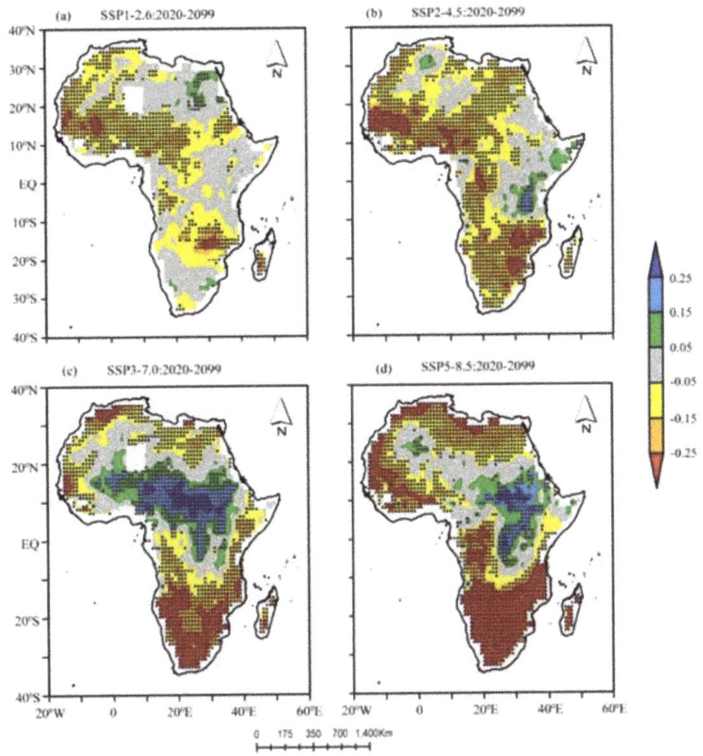

Figure 7. Pixel-wise linear trends for $scPDSI_{PM}$ <(−2.0, during 2020–2099. (a) SSP1-2.6, (b) SSP2-4.5, (c) SSP3-7.0, and (d) SSP5-8.5 scenarios. The values are expressed in changes per decade (the dots denote passing a 5% significance test). The white background shows areas with no values.

Figure 8 illustrates the spatial distribution of linear trends for $scPDSI_{PM}$ for all scenarios during the near future ((2020–2039), mid-century (2050–2069), and end-of-century (2080–2099), based on the Mann–Kendall test at 5% significance level. In general, a distinguishable spatial pattern of wetting and drying trends is shown for each SSP scenario and time slice. The sub-regional analysis shows different spatial patterns of wet and dry conditions with differences in trend values for the three time slices under all SSP scenarios. SAH show mixed results of wetting and drying trends in all SSPs. In WAF and CAF, we observed a more distinct drying trend under SSP2-4.5 (Figure 8d) than SSP1-2.6 (Figure 8a) in the near future (2020–2039), which reverts to a wetting trend in the mid-century (2050–2069) with magnitudes >0.25 decade^{-1} in Figure 8b,d, and mixed trends at the end-of-century (2080–2099) (Figure 8c,f). On the contrary, NEAF and SEAF are relatively similar, but they have mixed wetting and drying trends in three SSPs except for SSP5-8.5. A strong wetting trend under SSP5-8.5 is observed in the near future and end-of-century in NEAF and SEAF but with mixed trends in mid-century (2020–2039). On the other hand, WSAF and ESAF show a mixed trend in all SSPs but more pronounced drying conditions in the SSP3-7.0 and SSP5-8 scenarios at ESAF than WSAF at the end of century.

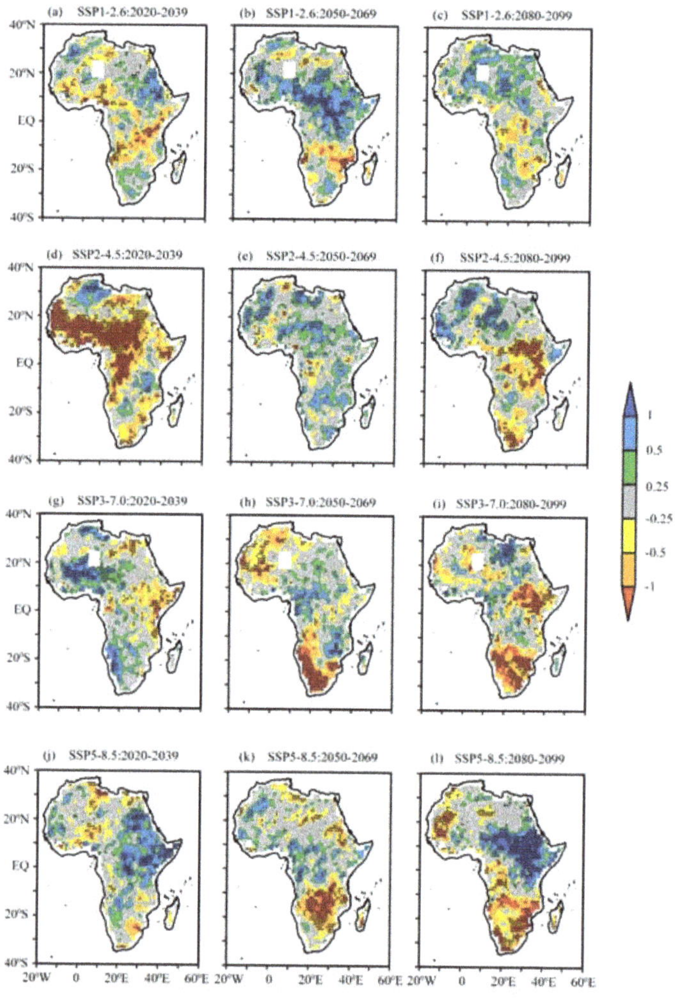

Figure 8. Pixel-wise linear trends for $scPDSI_{PM} < -2.0$, for the near future (2020–2039), mid-century (2050–2069), and end-of-century (2080–2099) under (**a–c**) SSP1-2.6, (**d–f**) SSP2-4.5, (**g–i**) SSP3-7.0, and (**j–l**) SSP5-8.5 scenarios, respectively. The values are expressed in changes per decade (dots denote passing a 5% significance test). The white background shows areas with no values.

4. Discussion

This study uses the CNRM-CM6 data to investigate future drought characteristics and trends during the 21st century (near future, mid-century, and end-of-century) for four SSP scenarios (SSP1-2.6, SSP2-4.5, SSP3-7.0, and SSP5-8.5). Using these scenarios provides the scientific community with an opportunity to investigate changes in droughts across the African continent against the backdrop of the continent's vulnerability to weather and climate variability, which is coupled with population growth.

The time series of change in drought intensity between the future and the baseline periods are computed and presented in Figure 3. An increase in drought event intensity is experienced under all SSP scenarios and different time periods. The extent of the drought-affected areas significantly increases over time and the upward trend increases with increasing SSPs (Figure 4). This indicates that the frequency of future drought events will be higher and drought-event severity will intensify over time, thus increasing the spatial extent of the affected area. Overall, higher emissions (i.e., SSP370 and SSP585) scenarios will exhibit higher frequency of drought events, more severe drought over significantly larger areas than lower emissions scenarios (SSP1.2.6 and SSP2-4.5). This result is consistent with the IPCC AR6 [1].

Geographically, the drought-event frequency shows some mixed results under different climate conditions. In absolute terms, drought events will be more frequent in areas under arid and semi-arid climate conditions in Africa (Figure 5) for different future periods. These results are consistent with a previous study over the region, but it is based on SPI and SPEI [69]. Moreover, this pattern is comparable to RCP2-4.5 and RCP8.5 scenarios from CMIP5 [32]. The spatial pattern shows similar value ranges in drought frequencies across the continent (Figure 6). The WAF region shows progressively decreasing values in drought frequency across all three time slices (Figure 6a–i). The number of drought occurrences decreases at a much lower rate, with a sharper decrease noted in the CAF (Figure 6a–i). This result is consistent with previous studies in the WAF and CAF regions [70,71]. On the contrary, an increase in the number of drought in SAH, WSAF, and ESAF (Figure 6a–i) is in agreement with Shongwe et al. [72]. On the other hand, the number of drought frequencies in NEAF and SEAF is relatively similar for different time slices and under each SSP-RCP scenario (Figure 6a–i), in agreement with Makula, and Zhou [73] and Ayugi et al. [68]. A general spatial pattern shows a likely increase in drought frequencies over SAH, WSAF, and ESAF under SSP3-7.0 and SSP5-8.5 than under SSP1-2.5 and SSP2-4.5.

To investigate future trends in wetting and drying conditions, the Mann–Kendall test [64] and the Theil–Sen slope estimator were used [65]. The scPDSI present significant trends for all different scenarios. The trends are mostly negative (indicating likely increases in drying conditions) across Africa. Overall, lower emissions (SSP1-2.5 and 2-4, Figure 7a,b) present a larger area of drying trends than higher emission (SSP3-7.0 and SSP5-8.5, Figure 7c,d). The regional analysis of projected wetting and drying trends shows spatial pattern differences across the continent. A larger area of negative trends obtained in SAH, WSAF, and ESAF indicate that future drought events may be further intensified in arid regions. Similar drying conditions were observed in Bellprat et al. [74] in WSAF and ESAF, and in SAH [75]. This may be partly attributed to the spatial pattern of future PRE in this region due to water-limited conditions. Droughts respond more to PRE than PET, which is consistent with Munday and Washington [76], or model uncertainties [77]. Considering the SSP scenarios, lower emissions (i.e., SSP1-2.6 and SSP2-4.5) are expected to witness more drying conditions than higher emissions (SSP3-7.0 and SSP5-8.5). The WSAF and ESAF point to drying conditions, with trends significant under SSP2-4.5, SSP3-7.0, and SSP5-8.5. This result is consistent with a study by Iyakaremye et al. [78], who projected that SAH and WSAF would warm faster, relative to other parts of the continent with changes in PRE [13,79]. The magnitudes of negative trends under SSP3-7.0 and SSP5-8.5 signal severe aridification trends in the arid region of WSAF and semi-arid conditions of ESAF than in SAH. The striking difference in drying trends in SAH relative to WSAF and ESAF is that the projected increase in PRE in the Northern Hemisphere (NH) is slightly higher than that in the Southern Hemisphere (SH), consistent with Almazroui et al. [55], Lim Kam Sian et al. [56], and Babaousmail et al. [80]. These results indicate that the arid conditions in the SH (i.e., WSAF and ESAF) exhibit more pronounced drying conditions than the NH (i.e., SAH). Similar results were noted in Lee & Wang [81]. The plausible reason for these results may be related to the interhemispheric difference in the warming rate documented in Kitoh et al. [82]. This signal has a potential implication for this region, as these impacts may affect regional socio-economic stability and ecological security for countries located in the SAF.

In WAF and CAF, SSP1-2.5 and SSP2-4.5 (Figure 7a,b) show opposite indicators to SSP3-7.0 and SSP5-8.5 (Figure 7c,d), as the magnitude of change cannot be ignored. Large changes are expected in the WAF and CAF regions, as large areas of significant positive trends under a high emissions rate and negative trends under lower emissions rate are observed. The possibility of an increase in negative trends under SSP1-2.5 and SSP2-4.5 (Figure 7a,b) and positive trends under SSP3-7.0 and SSP5-8.5 (Figure 7c,d) provides an interesting result, since, in humid conditions, drought responds to PET rather than PRE. A lower emissions rate likely will present a drying trend, and wetting trends in higher emissions scenarios in a humid environment such as WAF and CAF may be related to a weakening of the land–atmosphere coupling [11,83–85]. Similar results were reported in Dosio et al. [28] using both regional climate models (RCM) and GCMs over WAF but for PRE projections. A further study is recommended to examine this phenomenon of low emission scenarios (i.e., SSP1-2.6 and SSP2-4.5) that shows a stronger wetting trend with high magnitudes than high emission scenarios (i.e., SSP3-7.0 and SSP5-8.5).

The NEAF and SEAF regions present distinct spatial patterns between the four scenarios (Figure 7a–d). The SSP2-4.5 illustrates a positive trend in the NEAF and SEAF region, while the SSP1-2.5 exhibits no trend in large areas, with scattered packets of negative trends. However, the drought conditions under the SSP3-7.0 and SSP5-8.5 in the NEAF and SEAF generally present negative and positive trends. The negative trend is related to a decrease in PRE [79,80] and the increase in ET [35] or TEMP [55] in the NEAF and SEAF regions in the future. The drying conditions identified here agree with previous studies in this region, reflecting the complex patterns of PRE and ET trends [86,87]. The results noted that the negative trends are located in already vulnerable states, such as Kenya and Somalia, with arid conditions and poor adaptation mitigation. This pattern in NEAF and SEAF shows that the results are comparable in SSP2-4.5 and SSP5-8.5 scenarios from CMIP6 in Haile et al. [30]. Overall, the CNRM-CM6 model indicates that the possible future wetting and drying patterns are changing in different regions across Africa. We urge readers to interpret the results with caution, as the results are based on non-bias adjusted CNRM-CM6 data. Different studies have noted that non-bias adjusted data may over- or underestimate regions with significant variation in local features, such as topography and water bodies [88–90]. Future studies should consider the impact of bias adjustment on the historical and projected droughts events over Africa. This information may provide insight into the ongoing climate discussion and improve our understanding of drought events over the African region.

In summary, stakeholders have reiterated the need for an evidence-based approach to studing extreme events to inform policymaking at local scales. Many countries in Africa are highly dependent on rain-fed agriculture. The projected wetting and drying trend throughout the century for all SSP scenarios may likely impact the future agricultural production and ecological stability of humid, arid and semi-arid climates, as documented in past studies [30,91]. The regional analysis of projected drought climatology shows significant spatial differences. The spatial differences suggest that drought impact may vary with locations, and so will a region's capacity to respond to drought events. Future climate adaptation policies should be tailored to specific regional needs. The results form a wider network of previous related studies published in FAO [41] to inform national policymakers of the identified future drought-prone regions to develop adaptation policies across Africa.

5. Conclusions

This study describes the long-term changes in drought characteristics using the scPDSI based on the CNRM-CM6 model of the CMIP6 datasets. The following conclusions are drawn from the model projections across Africa for all four SSP scenarios.

- The spatio-temporal pattern and trends reveal that Africa is likely to experience changes in drought characteristics under all SSP scenarios.
- The spatial pattern of drought frequency across the continent reveals regional differences, as arid and semi-arid regions are to likely to have more droughts.
- The CNRM-CM6 model projections indicate a regional difference in wetting and drying trends over Africa for different SSP scenarios.
- Overall, regions in Africa located below the equator are likely to experience a general drying trend with droughts intensifying over time, while arid regions above the equator are likely to show moderate drying conditions in all SSP scenarios.
- The CNRM-CM6′s ability to satisfactorily identify the extent of drought parameters and trends over Africa is consistent with previous studies and further increases the confidence of the CMIP6 datasets for future studies of extreme events.

Within the context of the Paris Agreement, the Agenda 20,230 of the United Nations Sustainable Development Goals (SDGs), and the Malabo 2025 declaration, the findings of this study are significant and provides a basis for stakeholders in the region to further explore the changing trends of projected drought episodes and its potential impact on various sectors of the society. Readers are urged to interpret the results with caution as the objective of this study is not to confirm the superiority of the CNRM-CM6 datasets over other CMIP6 datasets or the ensemble approach, but rather to demonstrate its potential use in a local context. A future study plans to explore the implication of future drought climatology on direct (e.g., water use efficiency and crop yields) and indirect costs in African countries whose economy is tied to climate variability.

Supplementary Materials: The following supporting information can be downloaded at: https://www.mdpi.com/article/10.3390/agriculture12040495/s1, Figure S1: Methodological Flowchart.

Author Contributions: Conceptualization, I.K.N. and D.F.T.H.; methodology, I.K.N. and D.F.T.H.; software, J.L.; validation, D.F.T.H. and W.U.; formal analysis, I.K.N., J.L. and D.F.T.H.; investigation, I.K.N. and D.F.T.H.; resources, S.L. and J.L.; data curation, S.L. and J.L.; writing—original draft preparation, I.K.N.; writing—review and editing, W.U., N.A.P., G.T.G. and K.T.C.L.K.S.; visualization, S.L. and J.L.; supervision, D.F.T.H. All authors have read and agreed to the published version of the manuscript.

Funding: This research received no external funding.

Institutional Review Board Statement: Not applicable.

Informed Consent Statement: Not applicable.

Data Availability Statement: The CNRM-CM6 datasets used here are publicly available at the France Centre National de Recherches Météorologiques website (https://esg1.umr-cnrm.fr/ (accessed on 20 May 2021)) or CMIP6: https://esgf-node.llnl.gov/search/cmip6 (accessed on 20 May 2021).

Acknowledgments: The authors are thankful to the World Climate Research Programme (WCRP)-Working Group on Coupled Modelling (WGCM) for making the CMIP6 data publicly available.

Conflicts of Interest: The authors declare no conflict of interest.

References

1. IPCC. Summary for Policymakers. In *Climate Change 2021, The Physical Science Basis Contribution of Working Group I to the Sixth Assessment Report of the Intergovernmental Panel on Climate Change*; Cambridge University Press: Cambridge, UK, 2021.
2. Mishra, A.K.; Singh, V.P. A review of drought concepts. *J. Hydrol.* **2010**, *391*, 202–216. [CrossRef]
3. Heim, R.R.J. A review of twentieth-century drought indices used in the united states. *Bull. Am. Meteorol. Soc.* **2002**, *83*, 1149–1166. [CrossRef]

4. Skoulikaris, C.; Makris, C.; Katirtzidou, M.; Baltikas, V.; Krestenitis, Y. Assessing the Vulnerability of a Deltaic Environment due to Climate Change Impact on Surface and Coastal Waters: The Case of Nestos River (Greece). *Environ. Model. Assess.* **2021**, *26*, 459–486. [CrossRef]
5. Mckee, T.B.; Doesken, N.J.; Kleist, J. The relationship of drought frequency and duration to time scales. In Proceedings of the 8th Conference on Applied Climatology, Boston, MA, USA, 17–22 January 1993; Volume 17, pp. 179–183.
6. Vicente-Serrano, S.M.; Beguería, S.; López-Moreno, J.I. A multiscalar drought index sensitive to global warming: The Standardized Precipitation Evapotranspiration Index. *J. Clim.* **2010**, *23*, 1696–1718. [CrossRef]
7. Palmer, W.C. *Meteorological Drought*; U.S. Department of Commerce: Washington, DC, USA, 1965. Available online: https://www.droughtmanagement.info/literature/USWB_Meteorological_Drought_1965.pdf (accessed on 20 May 2021).
8. Wells, N.; Goddard, S.; Michaelsen, J. A self-calibrating Palmer Drought Severity Index. *J. Clim.* **2004**, *17*, 2335–2351. [CrossRef]
9. Ayantobo, O.O.; Wei, J. Appraising regional multi-category and multi-scalar drought monitoring using Standardized Moisture Anomaly Index (SZI): A water-energy balance approach. *J. Hydrol.* **2019**, *579*, 124139. [CrossRef]
10. Chen, S.; Gan, T.Y.; Tan, X.; Shao, D.; Zhu, J. Assessment of CFSR, ERA-Interim, JRA-55, MERRA-2, NCEP-2 reanalysis data for drought analysis over China. *Clim. Dyn.* **2019**, *53*, 737–757. [CrossRef]
11. Berg, A.; Findell, K.; Lintner, B.; Alessandra, G.; Sonia, I.S.; van den Hurk, B.; Ruth, L.; Pitman, A.; Hagemann, S.; Meier, A.; et al. Land—Atmosphere feedbacks amplify aridity increase over land under global warming. *Nat. Clim. Chang.* **2016**, *6*, 869–874. [CrossRef]
12. Yang, Y.; Roderick, M.L.; Zhang, S.; McVicar, T.R.; Donohue, R.J. Hydrologic implications of vegetation response to elevated CO_2 in climate projections. *Nat. Clim. Chang.* **2018**, *9*, 44–48. [CrossRef]
13. Dai, A.; Zhao, T.; Chen, J. Climate change and drought: A precipitation and evaporation perspective. *Curr. Clim. Chang. Rep.* **2018**, *4*, 301–312. [CrossRef]
14. Seneviratne, S.I.; Corti, T.; Davin, E.L.; Hirschi, M.; Jaeger, E.B.; Lehner, I.; Orlowsky, B.; Teuling, A.J. Investigating soil moisture–climate interactions in a changing climate: A review. *Earth Sci. Rev.* **2010**, *99*, 125–161. [CrossRef]
15. Van der Schrier, G.; Barichivich, J.; Briffa, K.R.; Jones, P.D. A scPDSI-based global data set of dry and wet spells for 1901–2009. *J. Geophys. Res. Atmos.* **2013**, *118*, 4025–4048. [CrossRef]
16. Seneviratne, S.I. Historical drought trends revisited. *Nature* **2012**, *491*, 338–339. [CrossRef]
17. Trenberth, K.E.; Dai, A.; van der Schrier, G.; Jones, P.D.; Barichivich, J.; Briffa, K.R.; Sheffield, J. Global warming and changes in drought. *Nat. Clim. Chang.* **2014**, *4*, 17–22. [CrossRef]
18. Douville, H.; Krishnan, R.; Renwick, J.; Allan, R.; Arias, P.; Barlow, M.; Cerezomota, R.; Cherchi, A.; Gan, T.Y. Water cycle change. In *Climatic Change, 2021: The Physical Science Basis Contribution of Working Group I to 6th Assessment Report of IPCC*; Cambridge University Press: Cambridge, UK, 2021.
19. IPCC. *Climate Change 2007, The Physical Science Basis, Contribution of Working Group I to the Fourth Assessment Report of the Intergovernmental Panel on Climate Change*; Cambridge University Press: Cambridge, UK, 2007.
20. Nooni, I.K.; Hagan, D.F.T.; Wang, G.; Ullah, W.; Li, S.; Lu, J.; Bhatti, A.S.; Shi, X.; Lou, D.; Prempeh, N.A.; et al. Spatiotemporal Characteristics and Trend Analysis of Two Evapotranspiration-Based Drought Products and Their Mechanisms in Sub-Saharan Africa. *Remote Sens.* **2021**, *13*, 533.
21. O'Neill, B.C.; Tebaldi, C.; van Vuuren, D.P.; Eyring, V.; Friedlingstein, P.; Hurtt, G.; Knutti, R.; Kriegler, E.; Lamarque, J.F.; Lowe, J.; et al. The Scenario Model Intercomparison Project (ScenarioMIP) for CMIP6. *Geosci. Model Dev.* **2016**, *9*, 3461–3482. [CrossRef]
22. IPCC. *Climate Change 2014: Impacts, Adaptation, and Vulnerability*; Cambridge University Press: Cambridge, UK, 2014.
23. Eyring, V.; Bony, S.; Meehl, G.A.; Senior, C.A.; Stevens, B.; Stouffer, R.J.; Taylor, K.E. Overview of the Coupled Model Intercomparison Project Phase 6 (CMIP6) experimental design and organization. *Geosci. Model Dev.* **2016**, *9*, 1937–1958. [CrossRef]
24. Stouffer, R.J.; Eyring, V.; Meehl, G.A.; Bony, S.; Senior, C.; Stevens, B.; Taylor, K.E. CMIP5 Scientific Gaps and Recommendations for CMIP6. *Bull. Am. Meteorol. Soc.* **2017**, *98*, 95–105. [CrossRef]
25. Voldoire, A.; Saint-Martin, D.; Sénési, S.; Decharme, B.; Alias, A.; Chevallier, M.; Colin, J.; Guérémy, J.-F.; Michou, M.; Moine, M.-P.; et al. Evaluation of CMIP6 DECK Experiments With CNRM-CM6-1. *J. Adv. Model. Earth Syst.* **2019**, *11*, 2177–2213. [CrossRef]
26. Li, M.; Wu, P.; Sexton, D.M.H.; Ma, Z. Potential shifts in climate zones under a future global warming scenario using soil moisture classification. *Clim. Dyn.* **2021**, *56*, 2071–2092. [CrossRef]
27. Beck, H.E.; Zimmermann, N.E.; McVicar, T.R.; Vergopolan, N.; Berg, A.; Wood, E.F. Present and future Köppen-Geiger climate classification maps at 1-km resolution. *Sci. Data* **2018**, *5*, 180214. [CrossRef]
28. Dosio, A.; Turner, A.G.; Tamoffo, A.T.; Sylla, M.B.; Lennard, C.; Jones, R.G.; Terray, L.; Nikulin, G.; Hewitson, B. A tale of two futures: Contrasting scenarios of future precipitation for West Africa from an ensemble of regional climate models. *Environ. Res. Lett.* **2020**, *15*, 064007. [CrossRef]
29. Iturbide, M.; Gutiérrez, J.; Alves, L.; Bedia, J.; Cerezo-Mota, R.; Cimadevilla, E.; Cofiño, A.; Di Luca, A.; Faria, S.; Gorodetskaya, I.; et al. An update of IPCC climate reference regions for subcontinental analysis of climate model data: Definition and aggregated datasets. *Earth Syst. Sci. Data* **2020**, *12*, 2959–2970. [CrossRef]
30. Haile, G.G.; Tang, Q.; Hosseini-Moghari, S.M.; Liu, X.; Gebremicael, T.G.; Leng, G.; Leng, G.; Kebede, A.; Xu, X.; Yun, X. Projected impacts of climate change on drought patterns over East Africa. *Earths Future* **2020**, *8*, e2020EF001502. [CrossRef]
31. Chami, D.E.; Moujabber, M.E. Drought, climate change and sustainability of water in agriculture: A roadmap towards the NWRS2. *S. Afr. J. Sci.* **2016**, *112*, 1–4. [CrossRef]

32. Ahmadalipour, A.; Moradkhani, H.; Castelletti, A.; Magliocca, N. Future drought risk in Africa: Integrating vulnerability, climate change, and population growth. *Sci. Total Environ.* **2019**, *662*, 672–686. [CrossRef]
33. Cook, B.I.; Mankin, J.S.; Anchukaitis, K.J. Climate change and drought: From past to future. *Curr. Clim. Chang. Rep.* **2018**, *4*, 164–179. [CrossRef]
34. Gan, T.Y.; Ito, M.; Hülsmann, S.; Qin, X.; Lu, X.X.; Liong, S.Y.; Rutschman, P.; Disse, M.; Koivusalo, H. Possible climate change/variability and human impacts, vulnerability of drought-prone regions, water resources and capacity building for Africa. *Hydrol. Sci. J.* **2016**, *61*, 1209–1226. [CrossRef]
35. Nooni, I.K.; Hagan, D.F.T.; Wang, G.; Ullah, W.; Lu, J.; Li, S.; Dzakpasu, M.; Prempeh, N.A.; Lim Kam Sian, K.T.C. Future Changes in Simulated Evapotranspiration across Continental Africa Based on CMIP6 CNRM-CM6. *Int. J. Environ. Res. Public Health* **2021**, *18*, 6760. [CrossRef]
36. Brient, F.; Roehrig, R.; Voldoire, A. Evaluating Marine Stratocumulus Clouds in the CNRM-CM6-1 Model Using Short-Term Hindcasts. *J. Adv. Model. Earth Syst.* **2019**, *11*, 127–148. [CrossRef]
37. Volodin, E.M.; Mortikov, E.V.; Kostrykin, S.V.; Galin, V.Y.; Lykossov, V.N.; Gritsun, A.S.; Diansky, N.A.; Gusev, A.V.; Iakovlev, N.G.; Shestakova, A.A.; et al. Simulation of the modern climate using the INM-CM48 climate model. *Russ. J. Numer. Anal. Math. Model.* **2018**, *33*, 367–374. [CrossRef]
38. Wu, T.; Lu, Y.; Fang, Y.; Xin, X.; Li, L.; Li, W.; Jie, W.; Zhang, J.; Liu, Y.; Zhang, L.; et al. The Beijing Climate Center Climate System Model (BCC-CSM): The main progress from CMIP5 to CMIP6. *Geosci. Model Dev.* **2019**, *12*, 1573–1600. [CrossRef]
39. Wyser, K.; van Noije, T.; Yang, S.; von Hardenberg, J.; O'Donnell, D.; Döscher, R. On the increased climate sensitivity in the EC-Earth model from CMIP5 to CMIP6. *Geosci. Model Dev.* **2020**, *13*, 3465–3474. [CrossRef]
40. Wyser, K.; Kjellström, E.; Koenigk, T.; Martins, H.; Döscher, R. Warmer climate projections in EC-Earth3-Veg: The role of changes in the greenhouse gas concentrations from CMIP5 to CMIP6. *Environ. Res. Lett.* **2020**, *15*, 054020. [CrossRef]
41. World Health Organization. *The State of Food Security and Nutrition in the World—Transforming Food Systems for Food Security, Improved Nutrition and Affordable Healthy Diets for All*; FAO: Rome, Italy, 2021.
42. Teichmann, C.; Jacob, D.; Remedio, A.R.; Remke, T.; Buntemeyer, L.; Hoffmann, P.; Kriegsmann, A.; Lierhammer, L.; Bülow, K.; Weber, T.; et al. Assessing mean climate change signals in the global CORDEX-CORE ensemble. *Clim. Dyn.* **2021**, *57*, 1269–1292. [CrossRef]
43. Gnitou, G.T.; Tan, G.; Niu, R.; Nooni, I.K. Assessing Past Climate Biases and the Added Value of CORDEX-CORE Precipitation Simulations over Africa. *Remote Sens.* **2021**, *13*, 2058.
44. Voldoire, A.; Sanchez-Gomez, E.; Salas y Mélia, D.; Decharme, B.; Cassou, C.; Sénési, S.; Valcke, S.; Beau, I.; Alias, A.; Chevallier, M.; et al. The CNRM-CM5.1 global climate model: Description and basic evaluation. *Clim. Dyn.* **2013**, *40*, 2091–2121. [CrossRef]
45. Déqué, M.; Dreveton, C.; Braun, A.; Cariolle, D. The ARPEGE/IFS atmosphere model: A contribution to the French community climate modelling. *Clim. Dyn.* **1994**, *10*, 249–266. [CrossRef]
46. Noilhan, J.; Planton, S. A Simple Parameterization of Land Surface Processes for Meteorological Models. *Mon. Weather Rev.* **1989**, *117*, 536–549. [CrossRef]
47. Decharme, B.; Delire, C.; Minvielle, M.; Colin, J.; Vergnes, J.-P.; Alias, A.; Saint-Martin, D.; Séférian, R.; Sénési, S.; Voldoire, A. Recent Changes in the ISBA-CTRIP Land Surface System for Use in the CNRM-CM6 Climate Model and in Global Off-Line Hydrological Applications. *J. Adv. Model. Earth Syst.* **2019**, *11*, 1207–1252. [CrossRef]
48. Masson, V.; Le Moigne, P.; Martin, E.; Faroux, S.; Alias, A.; Alkama, R.; Belamari, S.; Barbu, A.; Boone, A.; Bouyssel, F.; et al. The SURFEXv7.2 land and ocean surface platform for coupled or offline simulation of earth surface variables and fluxes. *Geosci. Model Dev.* **2013**, *6*, 929–960. [CrossRef]
49. Madec, G.; Bourdallé-Badie, R.; Bouttier, P.A.; Bricaud, C.; Bruciaferri, D.; Calvert, D.; Jérôme, C.; Clementi, E.; Coward, A.; Delrosso, D.; et al. *NEMO Ocean Engine*; Notes du Pôle de Modélisation de l'Institut Pierre-Simon Laplace (IPSL): Singapore, 2017. [CrossRef]
50. Salas Mélia, D. A global coupled sea ice–ocean model. *Ocean. Model.* **2002**, *4*, 137–172. [CrossRef]
51. Craig, A.; Valcke, S.; Coquart, L. Development and performance of a new version of the OASIS coupler, OASIS3-MCT_3.0. *Geosci. Model Dev.* **2017**, *10*, 3297–3308. [CrossRef]
52. Valcke, S.; Guilyardi, E.; Larsson, C. PRISM and ENES: A European approach to Earth system modelling. *Concurr. Comput. Pract. Exp.* **2006**, *18*, 247–262. [CrossRef]
53. Oki, T.; Sud, Y.C. Design of Total Runoff Integrating Pathways (TRIP)—A Global River Channel Network. *Earth Interact.* **1998**, *2*, 1–37. [CrossRef]
54. Meurdesoif, Y. *Xios Fortran Reference Guide*; IPSL: Singapore, 2018.
55. Almazroui, M.; Saeed, F.; Saeed, S.; Nazrul Islam, M.; Ismail, M.; Klutse, N.A.B.; Siddiqui, M.H. Projected Change in Temperature and Precipitation Over Africa from CMIP6. *Earth Syst. Environ.* **2020**, *4*, 455–475. [CrossRef]
56. Lim Kam Sian, K.T.C.; Wang, J.; Ayugi, B.O.; Nooni, I.K.; Ongoma, V. Multi-Decadal Variability and Future Changes in Precipitation over Southern Africa. *Atmosphere* **2021**, *12*, 742. [CrossRef]
57. WCRP CMIP6. WCRP Coupled Model Intercomparison Project (Phase 6) Homepage. Available online: https://esgf-node-.llnl.gov/projects/cmip6/ (accessed on 10 January 2020).
58. Van Vuuren, D.P.; Edmonds, J.; Kainuma, M.; Riahi, K.; Thomson, A.; Hibbard, K.; Hurtt, G.C.; Kram, T.; Krey, V.; Lamarque, J.-F.; et al. The representative concentration pathways: An overview. *Clim. Chang.* **2011**, *109*, 5. [CrossRef]

59. Dai, A. Characteristics and trends in various forms of the Palmer Drought Severity Index during 1900–2008. *J. Geophys. Res. Atmos.* **2011**, *116*, D12115. [CrossRef]
60. Taylor, I.H.; Burke, E.; McColl, L.; Falloon, P.D.; Harris, G.R.; McNeall, D. The impact of climate mitigation on projections of future drought. *Hydrol. Earth Syst. Sci.* **2013**, *17*, 2339–2358. [CrossRef]
61. Nooni, I.K.; Wang, G.; Hagan, D.F.T.; Lu, J.; Ullah, W.; Li, S. Evapotranspiration and its Components in the Nile River Basin Based on Long-Term Satellite Assimilation Product. *Water* **2019**, *11*, 1400.
62. Yevjevich, V. An objective approach to definitions and investigations of continental hydrologic droughts. *Hydrol. Pap.* **1969**, *23*, 1–18.
63. Mann, H.B. Non-parametric tests against trend. *Econometrica* **1945**, *13*, 245–259. [CrossRef]
64. Kendall, M. *Rank Correlation Measures*; Charles Griffin: London, UK, 1975.
65. Sen, P.K. Estimates of the regression coefficients based on Kendall's tau. *J. Am. Stat. Assoc.* **1968**, *63*, 1379–1389. [CrossRef]
66. Ullah, W.; Wang, G.; Ali, G.; Tawia Hagan, D.F.; Bhatti, A.S.; Lou, D. Comparing Multiple Precipitation Products against In-Situ Observations over Different Climate Regions of Pakistan. *Remote Sens.* **2019**, *11*, 628.
67. Ullah, W.; Wang, G.; Lou, D.; Ullah, S.; Bhatti, A.S.; Ullah, S.; Karim, A.; Hagan, D.F.T.; Ali, G. Large-scale atmospheric circulation patterns associated with extreme monsoon precipitation in Pakistan during 1981–2018. *Atmos. Res.* **2021**, *253*, 105489. [CrossRef]
68. Ayugi, B.; Dike, V.; Ngoma, H.; Babaousmail, H.; Mumo, R.; Ongoma, V. Future Changes in Precipitation Extremes over East Africa Based on CMIP6 Models. *Water* **2021**, *13*, 2358. [CrossRef]
69. Ajayi, V.O.; Ilori, O.W. Projected Drought Events over West Africa Using RCA4 Regional Climate Model. *Earth Syst. Environ.* **2020**, *4*, 329–348. [CrossRef]
70. Akinsanola, A.A.; Zhou, W. Projection of West African summer monsoon rainfall in dynamically downscaled CMIP5 models. *Clim. Dyn.* **2019**, *53*, 81–95. [CrossRef]
71. Faye, A.; Akinsanola, A.A. Evaluation of extreme precipitation indices over West Africa in CMIP6 models. *Clim. Dyn.* **2022**, *58*, 925–939. [CrossRef]
72. Shongwe, M.E.; van Oldenborgh, G.J.; van den Hurk, B.J.J.M.; de Boer, B.; Coelho, C.A.S.; van Aalst, M.K. Projected Changes in Mean and Extreme Precipitation in Africa under Global Warming. Part I: Southern Africa. *J. Clim.* **2009**, *22*, 3819–3837. [CrossRef]
73. Makula, E.K.; Zhou, B. Coupled Model Intercomparison Project phase 6 evaluation and projection of East African precipitation. *Int. J. Climatol.* **2022**, *42*, 2398–2412. [CrossRef]
74. Bellprat, O.; Lott, F.; Gulizia, C.; Young, H.; Pampuch, L.; Pinto, I.; Ciavarella, A.; Stott, P. Unusual past dry and wet rainy seasons over Southern Africa and South America from a climate perspective. *Weather Clim. Extrem.* **2015**, *9*, 36–46. [CrossRef]
75. Driouech, F.; ElRhaz, K.; Moufouma-Okia, W.; Arjdal, K.; Balhane, S. Assessing Future Changes of Climate Extreme Events in the CORDEX-MENA Region Using Regional Climate Model ALADIN-Climate. *Earth Syst. Environ.* **2020**, *4*, 477–492. [CrossRef]
76. Munday, C.; Washington, R. Controls on the Diversity in Climate Model Projections of Early Summer Drying over Southern Africa. *J. Clim.* **2019**, *32*, 3707–3725. [CrossRef]
77. Murphy, J.M.; Sexton, D.M.H.; Barnett, D.N.; Jones, G.S.; Webb, M.J.; Collins, M.; Stainforth, D.A. Quantification of modelling uncertainties in a large ensemble of climate change simulations. *Nature* **2004**, *430*, 768–772. [CrossRef]
78. Iyakaremye, V.; Zeng, G.; Yang, X.; Zhang, G.; Ullah, I.; Gahigi, A.; Vuguziga, F.; Asfaw, T.G.; Ayugi, B. Increased high-temperature extremes and associated population exposure in Africa by the mid-21st century. *Sci. Total Environ.* **2021**, *790*, 148162. [CrossRef]
79. Donat, M.G.; Lowry, A.L.; Alexander, L.V.; O'Gorman, P.A.; Maher, N. More extreme precipitation in the world's dry and wet regions. *Nat. Clim. Chang.* **2016**, *6*, 508–513. [CrossRef]
80. Babaousmail, H.; Hou, R.; Ayugi, B.; Ojara, M.; Ngoma, H.; Karim, R.; Rajasekar, A.; Ongoma, V. Evaluation of the Performance of CMIP6 Models in Reproducing Rainfall Patterns over North Africa. *Atmosphere* **2021**, *12*, 475. [CrossRef]
81. Lee, J.Y.; Wang, B. Future change of global monsoon in CMIP5. *Clim. Dyn.* **2014**, *42*, 101–119. [CrossRef]
82. Kitoh, A.; Endo, H.; Kumar, K.K.; Cavalcanti, I.F.A.; Goswami, P.; Zhou, T. Monsoons in a changing world: A regional perspective in a global context. *J. Geophys. Res. Atmos.* **2013**, *118*, 3053–3065. [CrossRef]
83. Dirmeyer, P.A. The terrestrial segment of soil moisture-climate coupling. *Geophys. Res. Lett.* **2011**, *38*, L16702. [CrossRef]
84. Koster, R.D.; Dirmeyer, P.A.; Guo, Z.; Bonan, G.; Chan, E.; Cox, P.; Gordon, C.T.; Kanae, S.; Kowalczyk, E.; Lawrence, D.; et al. Regions of strong coupling between soil moisture and precipitation. *Science* **2004**, *305*, 1138–1140. [CrossRef]
85. Berg, A.; Sheffield, J. Soil Moisture-Evapotranspiration Coupling in CMIP5 Models: Relationship with Simulated Climate and Projections. *J. Clim.* **2018**, *31*, 4865–4878. [CrossRef]
86. Dosio, A.; Jury, M.W.; Almazroui, M.; Ashfaq, M.; Diallo, I.; Engelbrecht, F.A.; Klutse, N.A.B.; Lennard, C.; Pinto, I.; Sylla, M.B.; et al. Projected future daily characteristics of African precipitation based on global (CMIP5, CMIP6) and regional (CORDEX, CORDEX-CORE) climate models. *Clim. Dyn.* **2021**, *57*, 3135–3158. [CrossRef]
87. Pinto, I.; Lennard, C.; Tadross, M.; Hewitson, B.; Dosio, A.; Nikulin, G.; Panitz, H.-J.; Shongwe, M.E. Evaluation and projections of extreme precipitation over southern Africa from two CORDEX models. *Clim. Chang.* **2016**, *135*, 655–668. [CrossRef]
88. Skoulikaris, C.; Venetsanou, P.; Lazoglou, G.; Anagnostopoulou, C.; Voudouris, K. Spatio-Temporal Interpolation and Bias Correction Ordering Analysis for Hydrological Simulations: An Assessment on a Mountainous River Basin. *Water* **2022**, *14*, 660. [CrossRef]

89. Chen, J.; Brissette, F.P.; Chaumont, D.; Braun, M. Finding Appropriate Bias Correction Methods in Downscaling Precipitation for Hydrologic Impact Studies over North America. *Water Resour. Res.* **2013**, *49*, 4187–4205. [CrossRef]
90. Teutschbein, C.; Seibert, J. Bias correction of regional climate model simulations for hydrological climate-change impact studies: Review and evaluation of different methods. *J. Hydrol.* **2012**, *456*, 12–29. [CrossRef]
91. Shi, Y.; Shen, Y.; Kang, E.; Li, D.; Ding, Y.; Zhang, G.; Hu, R. Recent and future climate change in northwest China. *Clim. Chang.* **2007**, *80*, 379–393. [CrossRef]

agriculture

Article

Improving Winter Wheat Yield Forecasting Based on Multi-Source Data and Machine Learning

Yuexia Sun [1,2], Shuai Zhang [1,2,*], Fulu Tao [1,2,3], Rashad Aboelenein [4] and Alia Amer [5]

1 Key Laboratory of Land Surface Pattern and Simulation, Institute of Geographic Sciences and Natural Resources Research, CAS, Beijing 100101, China; sunyx.20s@igsnrr.ac.cn (Y.S.); taofl@igsnrr.ac.cn (F.T.)
2 College of Resources and Environment, University of Chinese Academy of Sciences, Beijing 100049, China
3 Natural Resources Institute Finland (Luke), 00790 Helsinki, Finland
4 Barley Research Department, Field Crops Research Institute, Agricultural Research Center, Giza 583121, Egypt; rashadaboelenein@yahoo.com
5 Medicinal and Aromatic Plants Department, Horticulture Research Institute, Agricultural Research Center, Giza 583121, Egypt; aliaamer@arc.sci.eg
* Correspondence: zhangshuai@igsnrr.ac.cn

Abstract: To meet the challenges of climate change, population growth, and an increasing food demand, an accurate, timely and dynamic yield estimation of regional and global crop yield is critical to food trade and policy-making. In this study, a machine learning method (Random Forest, RF) was used to estimate winter wheat yield in China from 2014 to 2018 by integrating satellite data, climate data, and geographic information. The results show that the yield estimation accuracy of RF is higher than that of the multiple linear regression method. The yield estimation accuracy can be significantly improved by using climate data and geographic information. According to the model results, the estimation accuracy of winter wheat yield increases dramatically and then flattens out over months; it approached the maximum in March, with R^2 and RMSE reaching 0.87 and 488.59 kg/ha, respectively; this model can achieve a better yield forecasting at a large scale two months in advance.

Keywords: solar induced chlorophyll fluorescence (SIF); winter wheat; yield forecast; random forest; enhanced vegetation index (EVI)

Citation: Sun, Y.; Zhang, S.; Tao, F.; Aboelenein, R.; Amer, A. Improving Winter Wheat Yield Forecasting Based on Multi-Source Data and Machine Learning. *Agriculture* **2022**, *12*, 571. https://doi.org/10.3390/agriculture12050571

Academic Editor: William A. Payne

Received: 21 March 2022
Accepted: 16 April 2022
Published: 19 April 2022

1. Introduction

As the world's largest producer and consumer of wheat [1], China faces great challenges of food security. Winter wheat production, one of the most important summer grain production of China [2], stagnated in 56% of China from 1961 to 2008 [3]. Therefore, a timely and an accurate winter wheat yield forecasting in China is of great importance for food trade and policymakers. Recently, there has been increasing research on winter wheat yield estimation where the yield prediction models are based on the physiological and the ecological processes of crops. These have been developed constantly, such as WOFOST [4], DSSAT [5], APSIM [6], STICS [7], and MONICA [8]. Such models mostly simulate daily crop development, growth, and yield formation as well as climate variables that are used as the main inputs to describe environmental conditions during the period of crop growth. However, the growth state of crops is not only affected by abiotic factors (growth environment) but also by biological factors (such as plant diseases) [9–11]. Therefore, using climate data alone may not be sufficient to estimate yield. Meanwhile, due to the high spatial heterogeneity of crop varieties, farmer management policies, and environments, there is significant uncertainty in the practical application of the model on a large scale [12,13].

Satellite remote sensing can continuously monitor crop growth across various spectral bands and provide useful additional information for crop yield estimations [14–16]. In the past decades, remote sensing monitoring technology has been successfully applied to crop yield estimations [17,18]. Such research was mostly about the empirical relationship between vegetation indexes (VIs), which is based on visible light and near infrared (e.g., NDVI data, are important in the models of crop yield estimation [19–22], radiation data EVI, GCVI) [23–26], and observed yield [27]. EVI, which is sensitive to a higher canopy leaf area index and less affected by the atmospheric aerosol, is most commonly used in crop yield estimation. However, VIs are based on greenness and not sensitive to the physiological changes of vegetation caused by meteorological factors such as temperature, vapor pressure, and absorbed radiation. In recent years, extensive studies have shown [28–31] that sun-induced chlorophyll fluorescence (SIF) can directly reflect the respiration of crops, respond timely and accurately to environmental stress, and it is directly related to biomass [32–35]. Currently some studies use SIF directly to estimate crop yield and to obtain better results at the field scale than the vegetation index [29,36–38]. It has been shown that the yield estimation accuracy of crop models based on climate data and on satellite data is generally better than that of the models based on data alone [39–41]. Water-supply-related variables (such as precipitation), temperature-related variables (such as maximum temperature), and water-demand-related variables (such as potential evapotranspiration), among all climate are also important variables for crop growth simulation [42]. Recognizing that various satellite products have common overlapping and complementary information is conducive to yield estimation [43]. However, better combining satellite data with other environmental factors in crop yield estimation needs to be further studied [44]. It is also unclear how multi-source climate data (i.e., climate and satellite data) promote the formation of the final yield estimation models and how their contributions to the models vary with the growing season. Furthermore, more and more approaches based on machine learning (ML) or deep learning are applied to agricultural applications, such as crop type classification [45,46], disease prediction [47], crop growth monitoring [48], and yield estimation [49–52]. Frausto-Solis [53] estimated the yield of many kinds of crops based on data of daily minimum and maximum temperature, and precipitation by using the decision tree (DT), while Jeong [54] used climate data to estimate the yields of wheat and corn and other crops in the world and some regions based on RF. At present, the research on the estimation of winter wheat yield in China based on RF are mostly concentrated in the North China Plain, such as Anhui Province, Henan Province, and so on [54–56].

To avoid the randomness of an individual model, this paper used multiple linear regression (MLR), a most common method for forecasting, and random forest (RF), a typical machine learning method, to build a crop yield estimation model in which the crop growth environment, agricultural policies, and spatial heterogeneity of yield are considered. This model combines the spatial information data, the climate data, and the satellite data. This paper aims to understand the relationship between different input elements and crop yield and to compare the effects of different input data combinations and time series data of different growth periods on the model's performance. Besides, this paper intends to unravel and to quantify the contributions of climate and satellite data on growing seasons to the crop yield estimation. This study mainly used EVI and SIF to find out whether satellite data can improve climate-based yield estimation methods on a large scale and also to explore whether SIF still maintains the advantage of the sensitive capture of photosynthetic activity of crops on a large scale [57,58].

2. Materials and Methods

2.1. Study Area

As the target area in this study, the winter wheat planting areas in China (Figure 1) are mainly in the North China Plain and also include the winter wheat planting area in ten provinces—Inner Mongolia, Shandong, Hubei, Chongqing, Sichuan, Guizhou, Yunnan, Shaanxi, Gansu, and Qinghai—and two autonomous regions: Ningxia and Xinjiang. It is thus clear that the study area is widely distributed and the proportion of planting areas in other provinces are relatively small. Winter wheat in China is generally sown at the end of September or at the beginning of October, and it is harvested by the middle of June of the following year [59]. Generally, irrigation and fertilization are available in these areas. In this study, January to June is defined as the major growth period of winter wheat on which the analysis and the modeling are focused in view of the effect of cold and frost injury on crops in January [60].

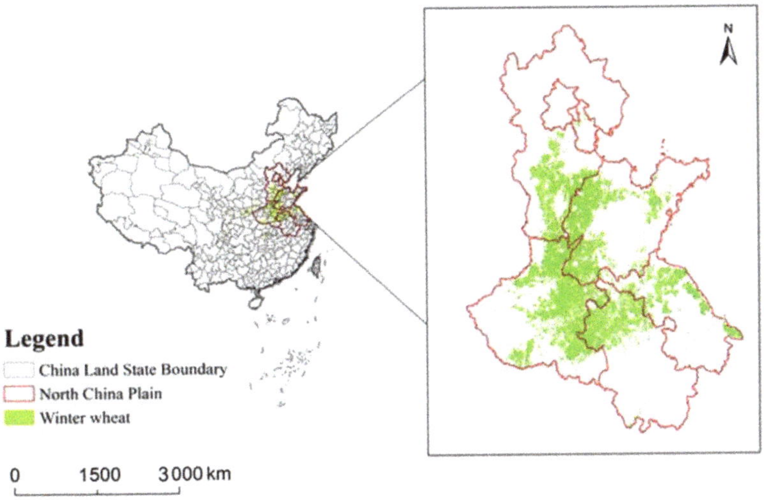

Figure 1. Winter wheat growing area in China.

2.2. *Data and Preprocessing*

2.2.1. Dataset

The study uses data on crop yield, planting area, satellite data, climate data, and spatial information (Table 1). The multi-source data collected in the study have various temporal and spatial resolutions. Therefore, firstly the raster data were resampled to a spatial resolution of 1 km and the climate and satellite data are unified into monthly interval. Then, monthly climate and satellite are aggregated at prefecture-level by using the crop map generated.

Table 1. Details of dataset.

Category	Variable(s)	Temporal Resolution	Spatial Resolution
Crop yield and area	Winter wheat yield Winter wheat area	yearly	prefecture-level
Satellite	EVI, SIF	monthly	0.05 degree
Climate Spatial information	Precipitation (pre); wet day frequency (wet); near-surface average temperature (tmp); near-surface temperature minimum (tmn); near-surface temperature maximum (tmx); potential evapotranspiration (pet); vapour pressure (vap); air specific humidity (shum); surface downward shortwave radiation (srad); surface downward longwave radiation (lrad) latitude (lat); longitude (lon)	monthly	0.5 degree, 0.1 degree

Crop yield and planting area: the winter wheat yield of prefecture-level cities from 2014 to 2018 (unit: kg/ha) was collected from local agricultural statistical yearbooks. Based on the previous work, winter wheat planting distribution of China at 1 km resolution from 2000 to 2015 are identified [61] (https://doi.org/10.6084/m9.figshare.8313530 (accessed on 21 August 2019)). The main planting areas of crops remain almost unchanged in a short period, therefore in this study the data of 2014 is used to represent the winter wheat planting distribution in China from 2014 to 2018.

Satellite data: satellite data includes EVI (MOD13C2 V6) (https://search.earthdata.nasa.gov (accessed on 1 February 2000)) and SIF reanalysis datasets (GOSIF) of OCO-2 satellite (http://data.globalecology.unh.edu (accessed on 27 November 2019)). Compared with the normalized vegetation index (NDVI), EVI is closely related to biomass and crop yield [62], and it can better represent the leaf and the chlorophyll content of crop canopy; GOSIF is a reanalysis dataset based on SIF data derived from OCO-2, MODIS data, and climate data. Compared with SIF data with coarse resolution and calculated directly from OCO-2, GOSIF has a better spatio-temporal resolution (0.05°, 8 days), continuous global coverage, and longer records.

Climate data: a total of 10 climate variables are collected from CRU_TS 4.04 (Climatic Research Unit Timeseries 4.04) series and CMFD (The China Meteorological Forcing Dataset) series (Table 1). The CRU_TS series is based on the record analysis of more than 4000 independent weather stations and it is gridded at a resolution of $0.5 \times 0.5°$, including monthly precipitation, daily maximum and minimum temperatures, cloud cover, and other variables covering the terrestrial region of the earth from 1901 to 2020. The CMFD dataset, with a spatial resolution of $0.1 \times 0.1°$, mainly includes precipitation per 3 h, surface radiation, wind speed, air specific humidity, and other variables covering 1979 to 2018.

Geographical basic data: the crops growth status and growth environment have spatial heterogeneity. Studies have indicated that crop yields in neighboring counties are usually similar in a certain year. The spatial autocorrelation can be explained by coding geographical coordinates (lat, lon) in feature space [62,63]. In this study, all data, including EVI, SIF, climate variables, and geographical coding, are covered by the raster data of winter wheat distribution and they were collected to prefecture-level cities with an average value.

2.2.2. Data Preprocessing

Selecting input variables was indispensable before machine learning and linear regression, which can not only reduce the input dimension, i.e., integrating expert knowledge to select the most appropriate input but also quantify the correlation between different potential independent variables and dependent variables to help to explain the results of machine learning algorithms. Based on previous studies on the relation between climate and crop yield [19–22], ten climate variables were selected for the study. To facilitate variable selection and interpretation, the best variable combinations of yield estimation were chosen without wasting information. Firstly, the 10 climate variables were divided into four groups according to prior knowledge: (1) water-supply-related, including precipitation (pre), wet day frequency (wet), and air specific humidity (shum); (2) temperature-related, including near-surface average temperature (tmp), near-surface temperature minimum (tmn), and near-surface temperature maximum (tmx); (3) water-demand-related, including potential evapotranspiration (pet) and vapor pressure (vap); and (4) radiation-related, including surface downward shortwave radiation (srad) and surface downward longwave radiation (lrad). The correlation analysis was carried out based on the mean value of the variables of the growing season (January–June) to eliminate the influence of the seasonal cycle. This study selected appropriate dependent variable inputs from the climate variables based on the following criteria: 1) selecting the variables which have the maximum absolute correlation with the yield in each group; and 2) selecting the variables whose value of correlation with the previously selected climate variables in the same group is not greater than 0.5.

2.3. Research Methods

2.3.1. Multiple Linear Regression

Multiple linear regression (MLR) is one of the most widely used methods of crop yield estimation, and it is easy to use. Based on the principle of Ordinary Least Square (OLS) and the stepwise regression method, the independent variables were selected with significant effects and they constructed the optimal regression model for winter wheat yield estimation by using the climate, satellite, and space information data. The yield estimation model is calculated by Equation (1)

$$Y = a_1x_1 + a_2x_2 + a_3x_3 \cdots a_nx_n + \beta + \varepsilon \quad (1)$$

where Y represents the winter wheat yield of prefecture-level cities; $x_1 \ldots x_n$ represent different independent variable factors used to predict Y; $a_1 \ldots a_n$ represent partial regression coefficient; β is a random variable and a constant term; and ε represents random error. The criterion for the stepwise regression model to pass the significance test is that the equation of linear relation model passes the F test and all the coefficients of the equation passes the t test.

2.3.2. Random Forest

Random Forest (RF) is an integrated learning technology, which classifies or regresses by combining a group of CART decision trees. Due to the introduction of randomness, RF is not prone to over-fitting, and it has good learning stability [56]. In this paper, the scikit-learn, an ML library of Python, is used to develop the RF model, which includes three steps: (1) normalizing all the selected variables and yield and randomly dividing the whole data set into training data with 70% and test data with 30% [64,65]; (2) for the training data set only, optimizing the key parameters of each model based on the highest R^2 and the lowest RMSE by ten-fold cross-verification; and (3) conducting the "leave one year out" experiment from 2014 to 2018, and R^2 and RMSE are used to evaluate the performance and generalization of the model. Considering the climate data, satellite data, and spatial information, this study counts the yield data of 187 out of 385 prefecture-level cities in China from 2014 to 2018.

2.4. Experiment Design

Two groups of experiments were designed (Figure 2) to answer the research questions raised in this paper. The purpose of the first group of experiments was to explore the effect of different input combinations on crop yield estimation models and to compare the potential of SIF in crop yield estimation. There are 11 data input combinations for the experiment, namely: (1) only SIF; (2) only EVI; (3) only climate; (4) SIF combined with spatial information; (5) EVI combined with spatial information; (6) climate combined with spatial information; (7) SIF combined with climate; (8) EVI combined with climate; (9) SIF combined with climate and spatial information; (10) EVI combined with climate and spatial information; and (11) SIF combined with EVI, climate, and spatial information. To assess the practicality of these models, based on the most suitable selected input, we recursively performed hindcasting for each year from 2014 to 2018 to evaluate whether the models can be promoted in different years; for example, the data for 2014–2017 was collected as training data to predict winter wheat yield in 2018. Certainly, future data cannot be used to predict current data. However, more verification samples can be provided for these hypotheses to increase the understanding of the model's performance. The RMSE (root mean square error) and R^2 (determination coefficient) between the predicted yield and the actual yield of winter wheat were calculated to verify the accuracy of the model.

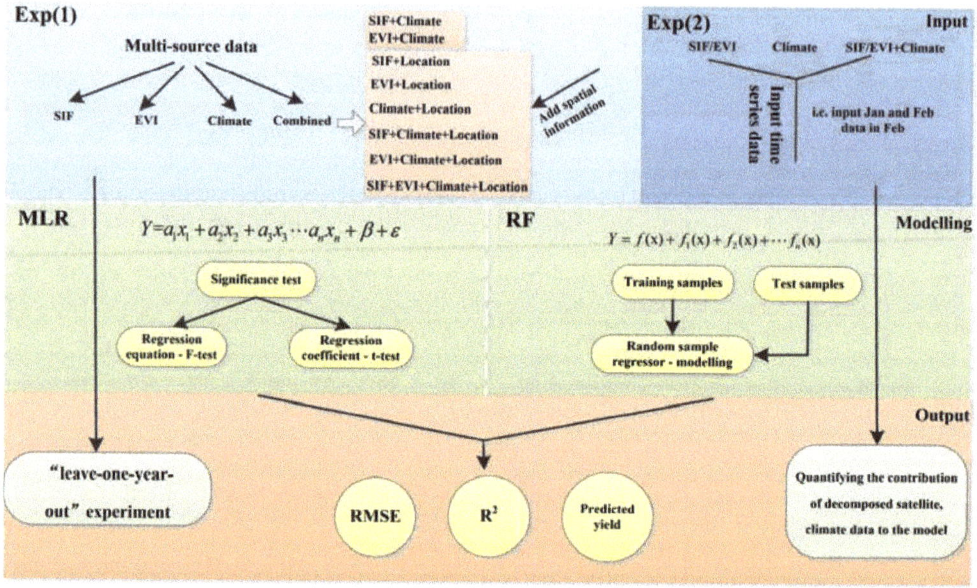

Figure 2. Experimental flow chart (Related parameter description: Location represents spatial information, including latitude and longitude; $a_1 \ldots a_n$ represent partial regression coefficient; β is a random variable and a constant term; ε represents random error; Jan is short for January; Feb is short for February).

The second group of experiments explored the effect of time series data on yield estimation models and the contribution of climate data and satellite data to crop yield prediction at different growth stages. In this experiment, the location information was not added since by default the spatial information of crops remained unchanged in this study in the short term. In the experiments, the climate and the satellite data were added and compared the change of R^2 and RMSE based on two methods of modeling to evaluate the change of performance of winter wheat estimation models. During the growing season (January–June), the input data of all months were used to predict winter wheat yield. The experiments were based on the three input combinations (namely climate, satellite, as well as climate and satellite). According to the experiment results, the added value of climate or satellite data to the estimation model in any period can be determined, and through different methods and input combinations the time for the model to achieve the best performance of estimation can be tested.

3. Results

3.1. Selection of Climate Variables Combination

Figure 3 shows the correlation analysis results of 10 climate variables, which demonstrated that water-supply-related variables (shum, pre, and wet) are all positively correlated with yield, while tmx and tmp among the temperature-related variables are negatively correlated with yield. Among the water-demand-related variables, pet is negatively correlated with yield, while vap is positively correlated with yield. Among radiation-related variables, srad is negatively correlated with yield, and lrad is positively correlated with yield. To select appropriate variables from each group as the input of yield estimation, 5 out of 10 climate variables are selected according to the method in Section 2.4, namely wet, tmx, pet, srad, and vap.

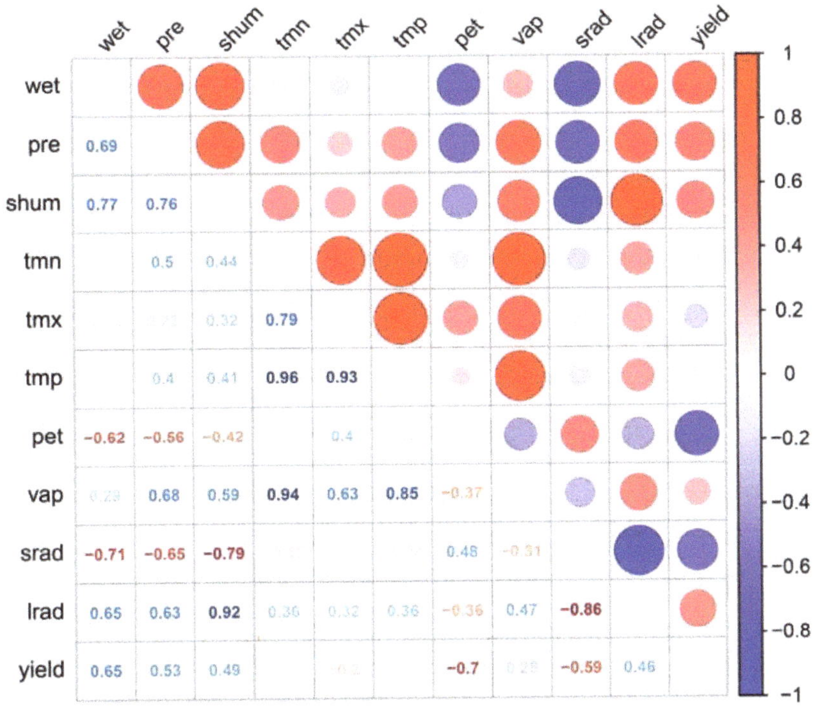

Figure 3. Correlations among the 10 climate variables and correlations between each climate variable and yield.

3.2. *The Influence of Different Input Data Combinations on the Simulation of the Model*

The results of the first group of experiments given in (Figure 4) that two models have the following similar characteristics with different combinations of data inputs: in the single data, the yield estimation performance of climate data is better. It may be because the climate variables can better simulate the growing environment of crops. There is an obvious spatial pattern of winter wheat yield at the prefecture scale, which indicates that the addition of spatial information is helpful to improve the prediction accuracy of the model. A better simulated result of yield is obtained by combining satellite data, climate data, and spatial information (MLR: R^2~0.68; RF: R^2~0.95). What is noteworthy is that on a monthly scale, compared with the addition of EVI, that of SIF does not significantly improve the yield estimation accuracy. The estimation effect of RF by combining SIF with other environmental factors is even lower than that of EVI. The result indicates that at the seasonal and the prefecture scale, SIF cannot provide much additional information that is different from EVI in crop yield estimation, and it shows no advantages of yield estimation on the small scale in the field. This may be related to the low signal-to-noise ratio, coarse resolution, and complex extraction algorithm of SIF [38]. However, the resolution of the SIF dataset used in this study (0.05°) has been improved compared with previous datasets (0.5~1°) [38]. Yet, the performance of the model remains the same, indicating that the downscale SIF dataset based on statistical methods alone cannot significantly enhance the effect of seasonal-scale crop yield estimation, which is consistent with Lindsey [66].

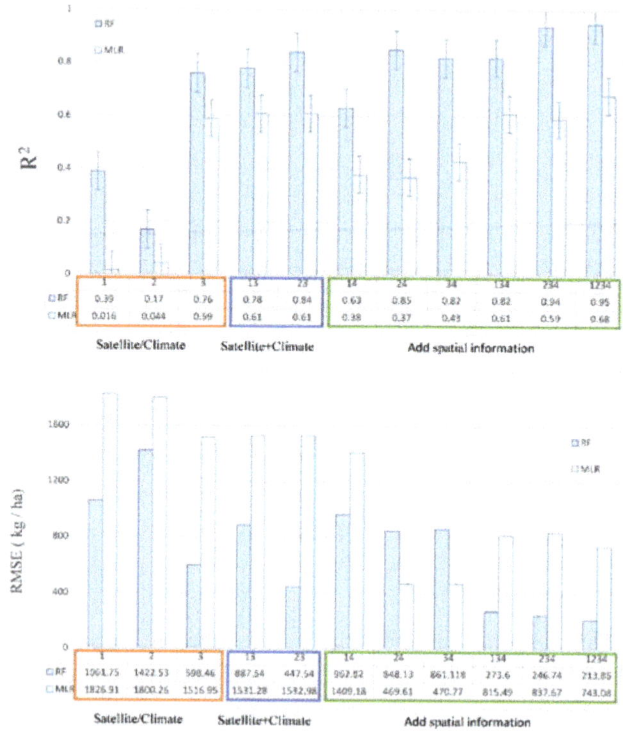

Figure 4. The model performance (predicted R^2, RMSE) of two methods using different combinations of inputs for the whole growing season, each number representing a different input combination, which can be expressed as 1: SIF, 2: EVI, 3: Climate, 4: Location, 13: SIF + Climate, 23: EVI + Climate, 14: SIF + Location, 24: EVI + Location, 34: Climate + Location, 134: SIF + Climate + Location, 234: EVI + Climate + Location, 1234: SIF + EVI + Climate + Location.

3.3. Comparison of Yield Estimation Performance of the Model

The results show that the yield estimation performance of RF is generally higher than that of MLR, which may be because the relationships between crop yield and variables are mostly nonlinear while nonlinear methods capture these relationships better than linear methods. Besides comparing the performance and generalization of the models, we conduct a "leave-one-year-out" experiment to verify the extrapolation potential of the models, which is establishing models based on all data and the crop yields in four out of five years and then separately verifying the estimated yield result of the year left. The result is shown in Table 2 in which each row represents the model performance of one year. The RMSE and R^2 between the estimated and the actual yield of winter wheat are compared. The results show that R^2 and RMSE are fairly stable in each year of winter wheat yield estimation at the prefecture scale, except for 2015. For example, the spatial distribution of yield prediction of 2014 based on two models (Figure 5) shows that RF can well reflect the spatial difference of winter wheat yield, especially in the North China Plain, in addition to having a high potential of yield estimation. In 2014, the errors of RF are mainly in Henan Province, while MLR generally underestimates the crop yield in high-yield areas, and the errors are concentrated in Henan, Hebei, and Shandong provinces.

Table 2. The validation results of "leave one-year out" experiment.

Year	MLR		RF	
	R^2	RMSE (kg/ha)	R^2	RMSE (kg/ha)
2014	0.74	1100.92	0.91	363.15
2015	0.61	1250.06	0.82	529.11
2016	0.82	964.75	0.87	441.01
2017	0.77	1306.90	0.89	491.89
2018	0.71	1527.23	0.83	501.43
Median	0.73	1229.97	0.85	465.32

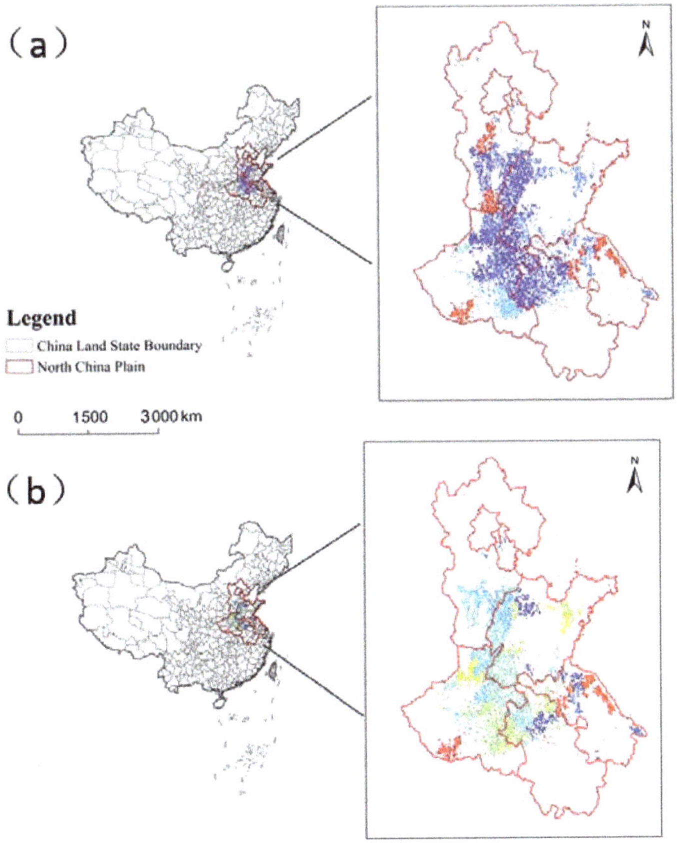

Figure 5. *Cont.*

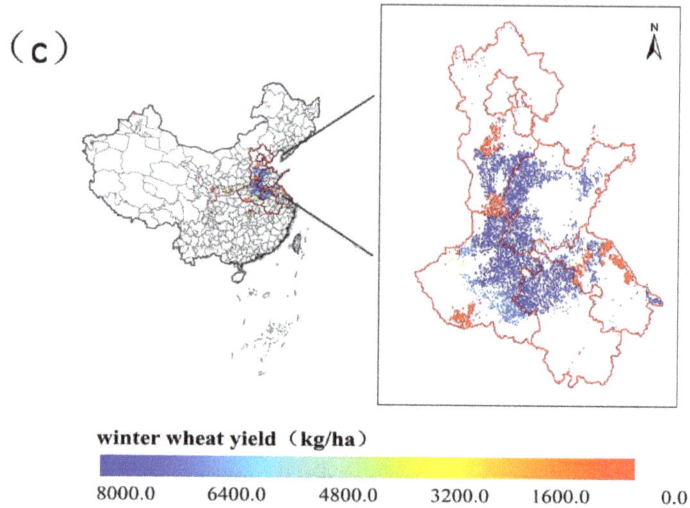

Figure 5. Spatial pattern of winter wheat yield forecast in 2014: (**a**) RF, (**b**) MLR, (**c**) actual yield.

3.4. *The Influence of Time Series Data on the Simulation Ability of the Model*

Since SIF has no advantage in large-scale yield estimation, EVI is used as an example in this experiment. The results (Figure 6) illustrate that the two models have the following similar characteristics: (1) for any particular inputs, the yield estimation accuracy of the model increases rapidly with the increase of acquired data and the growth rate slows down and gradually reaches saturation at a later stage of the growing season; (2) the combination can significantly improve the performance of the yield estimation model which can be significantly improved through combining climate data with satellite data, and climate data plays an essential role in the model. However, there are significant differences in the trajectory of model prediction performance produced by different inputs. With only satellite data as inputs, the model generally starts from very poor performance (R^2~0.1–0.2), and then it improves relatively by much (R^2~0.4–0.5), while with only climate data and combined data as inputs, the model starts with relatively good performance (R^2~0.4–0.6) and has a small increase during the growing season. To understand more clearly the effect of multi-resource data on the model, we assume that climate and satellite data have independent and overlapping contributions to the yield estimation model. We quantify the contributions of data from different sources. For example, the independent influence of climate data on the model is the difference between the combination R^2 and satellite R^2. The results indicate that (Figure 7) climate data always play a vital role in the performance of the model. With the advance of the growing season, the proportion of overlapping information becomes higher, and the contribution of climate data to the model decreases gradually. The results depict that satellite data gradually absorbs climate information as time goes by. In addition, models with only climate data and only satellite data can generally achieve a high simulation performance in May, while the performance of those with multi-source data can generally get close to the maximum in March (RF) and April (MLR). Therefore, the combination of multi-source data can achieve a high estimation accuracy of the crop yield one or two months in advance.

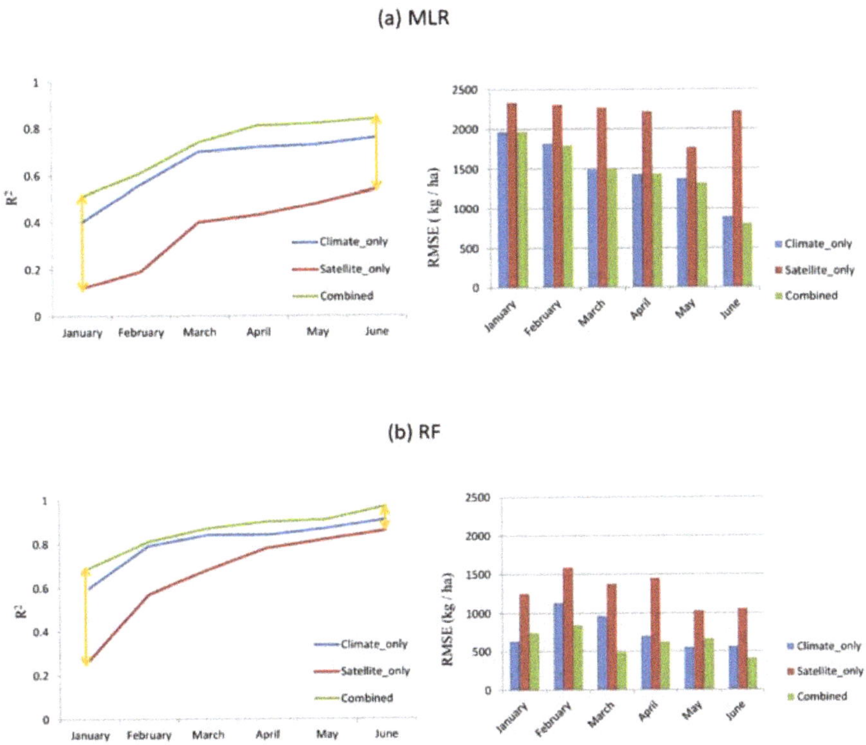

Figure 6. Cumulative input effect on the model performance. In each set of figures, the left figure represents the changing trend of R^2 with the increase of inputs; the figure on the right shows the changing trend of RMSE.

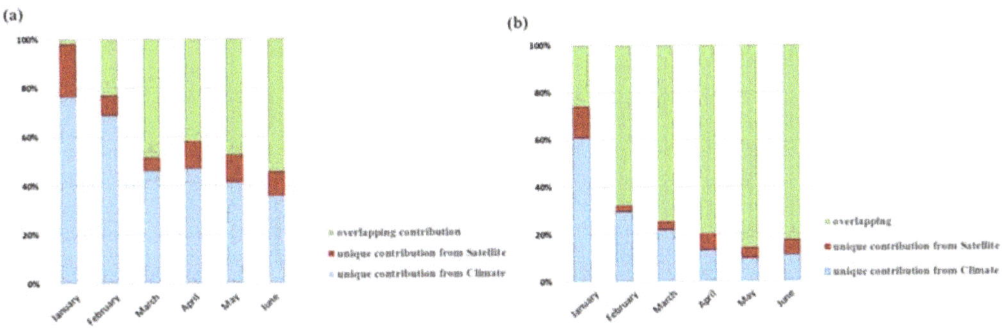

Figure 7. Independent and overlapping contributions of satellite and climate data to the model: (**a**) MLR, (**b**) RF.

4. Discussion

The results show the effects of different input combinations on yield estimation accuracy, and different yield estimation models have similar results. During the whole crop growing season, climate data always provide important information for crop yield estimation, which is consistent with the previous conclusions [20–22]. The performance of two yield estimation models with satellite data has been significantly improved, which is consistent with the view of Guan et al., who indicate that satellite data can provide for crop growth additional information different from climate data [43]. It is worth noting that the addition of SIF in the regional range does not significantly improve the yield estimation accuracy of the model in this case. Compared with EVI, SIF has no significant advantage in performance of yield estimation. It has not provided much additional information to the yield estimation model. It agrees with previous research results on whether SIF has an advantage in capturing crop growth state on the regional scale [38,66], which may be related to the coarse resolution and the complex extraction algorithm and thus more uncertainties of SIF. Random Forest can better predict crop yield. This is consistent with the literature [39], as both temperature and precipitation have a nonlinear response to yield [67,68]; and the nonlinear yield estimation model was more in line with the actual situation.

As time goes on in the growing season, the amount of the input satellite and the climate data increases, and the changes of yield estimation accuracy of different models show a similarity. In accordance with the previous conclusions [38], in the early stage of crop growth climate data play an important role in the model, the satellite gradually absorbs crop growth information, and the yield estimation accuracy of the model increases significantly, while the yield estimation accuracy reaches the maximum in the late stage of growth. It is worth noting that the time when the estimation accuracy reaches the maximum varies slightly between estimation models. While in the regression model the accuracy of yield estimation peaks only one month before harvest, RF achieves a high performance of the yield prediction two months in advance.

The winter wheat planting areas are widely distributed in China, with obvious spatial differences. The main planting areas of winter wheat are in the North China Plain, about 15,309.1 kha. Most of the previous studies have not considered the situation of Inner Mongolia, Ningxia, Xinjiang, and other regions, which accounts for approximately 27.4 percent of the total winter wheat planting area in China. This study establishes models of yield estimation at the national scale with consideration of the spatial heterogeneity of winter wheat yield by adding extra basic geographic data. The results show that adding spatial information data can improve the yield estimation accuracy of models [3] and it is helpful to establish a unified model on a large scale.

In this paper, RF and MLR are used to build yield estimation models of winter wheat in China, which avoids the randomness of single model analysis. However, due to the availability of data, the research is mainly on the prediction of crop yield at the prefecture scale. Furthermore, the spatial resolution of the satellite data used is relatively coarse, which leads to small training samples, and it limits the ability of machine learning methods [38,64,69]. The newly launched satellite provides a variety of data (such as EVI, SIF, and climate variables) and it has a higher spatiotemporal resolution (such as Landsat, Sentinel, and Fluorescence EXplorer) [70–72], which can provide the potential for future improvement. Besides, although the machine learning model performs well in the yield estimation in the prefecture, the process-based explanation is limited, which weakens the traceability and the interpretability of the model. How to better combine the process-based model with machine learning algorithm to realize more efficiently the extrapolation beyond the training conditions [54,69] and special migration can be investigated in the future. This is to improve the crop yield estimation of models in areas where there are not enough historical yield records, such as Africa [44]. At the same time, some key factors have not been considered in this study, such as biological factors other than those captured by satellite, namely soil characteristics [26,40], which will also help to explain more yield variability. In addition, due to the data limitation of the spatial distribution of winter wheat, the spatial distribution of winter wheat from 2014 to 2018 was represented by 2014 data in this paper, which may also lead to errors in the statistics of remote sensing data. For future research, it is suggested that the crop interannual spatial distribution information data should be generated from satellite data to reduce potential errors.

5. Conclusions

To avoid the randomness of a single model, this study conducts yield estimation of winter wheat in China at the prefecture scale based on a MLR model and a RF model combining climate data, satellite data, and spatial information. The effects of different input combinations and time-series data on the performance of the model have been discussed. The main conclusions are as follows:

(1) By decomposing and quantifying the contribution of satellite data and climate data to the model's performance in different growth periods, we find that satellite data can gradually capture the changes of crop growth and with accumulation of information can absorb part of the climate data. Spatial information and climate data have made a unique contribution to the yield forecasting of winter wheat in the whole growing season.

(2) By comparing the satellite data from two sources (i.e., SIF and EVI), it was found that the downsized SIF products do not perform better than EVI on the yield forecasting at the prefecture scale in China, which may be largely owing to the low signal-to-noise ratio of SIF products and the difficulty of extraction algorithm.

(3) By comparing the extrapolation and the spatial generalization ability of two models, RF can generally better capture the spatiotemporal heterogeneity of crop growth and thus is expected to better understand the impact of meteorology on agricultural production.

This study demonstrated a new scalable, simple, and inexpensive framework in estimating winter wheat yields over a wide range of areas based on publicly available data, which is applicable to other crops and geographical environments.

Author Contributions: S.Z. led the project and developed the framework; F.T. and S.Z. conceptualized and designed this research strategy; Y.S. carried out the field work and was responsible for data processing and manuscript writing; S.Z., R.A. and A.A. revised the manuscript. All authors have read and agreed to the published version of the manuscript.

Funding: This study was supported by the National Key Research and Development Program of China (Project No. 2016YFD0300201) and the National Science Foundation of China (Project No. 41801078).

Institutional Review Board Statement: Not applicable.

Informed Consent Statement: Not applicable.

Data Availability Statement: The data presented in this study are available on request from the authors.

Conflicts of Interest: The authors declare no conflict of interest.

References

1. Wei, L. Spatial-temporal evolution of Wheat production in China and its influencing factors. *Chin. J. Agric. Resour. Reg. Plan.* **2019**, *40*, 49–57.
2. Huang, J.; Tian, L.; Liang, S.; Ma, H.; Becker-Reshef, I.; Huang, Y.; Su, W.; Zhang, X.; Zhu, D.; Wu, W. Improving winter wheat yield estimation by assimilation of the leaf area index from Landsat TM and MODIS data into the WOFOST model. *Agric. For. Meteorol.* **2015**, *204*, 106–121. [CrossRef]
3. Chen, Y.; Zhang, Z.; Tao, F.; Wang, P.; Wei, X. Spatio-temporal patterns of winter wheat yield potential and yield gap during the past three decades in North China. *Field Crops Res.* **2017**, *206*, 11–20. [CrossRef]
4. Van Diepen, C.V.; Wolf, J.V.; Van Keulen, H.; Rappoldt, C. WOFOST: A simulation model of crop production. *Soil Use Manag.* **2010**, *5*, 16–24. [CrossRef]
5. Jones, J.W.; Hoogenboom, G.; Porter, C.H.; Boote, K.J.; Batchelor, W.D.; Hunt, L.A.; Wilkens, P.W.; Singh, U.; Gijsman, A.J.; Ritchie, J.T. The DSSAT cropping system model. *Eur. J. Agron.* **2003**, *18*, 235–265. [CrossRef]
6. Keating, B.A.; Carberry, P.S.; Hammer, G.L.; Probert, M.E.; Smith, C.J. An overview of APSIM, a model designed for farming systems simulation. *Eur. J. Agron.* **2003**, *18*, 267–288. [CrossRef]
7. Brisson, N.; Gate, P.; Lorgeou, J.; Nicoullaud, B.; Justes, E. STICS: A generic model for simulating cropsand their water and nitrogen balances.II. Model validation for wheat and maize. *Agronomie* **2002**, *22*, 69–92. [CrossRef]

8. Nendel, C.; Berg, M.; Kersebaum, K.C.; Mirschel, W.; Specka, X.; Wegehenkel, M.; Wenkel, K.O.; Wieland, R. The MONICA model: Testing predictability for crop growth, soil moisture and nitrogen dynamics. *Ecol. Model.* **2011**, *222*, 1614–1625. [CrossRef]
9. Hatfield, J.L.; Gitelson, A.A.; Schepers, J.S.; Walthall, C.L. Application of Spectral Remote Sensing for Agronomic Decisions. *Agron. J.* **2008**, *100*, S-117–S-131. [CrossRef]
10. Mahlein, A.K.; Oerke, E.C.; Steiner, U.; Dehne, H.W. Recent advances in sensing plant diseases for precision crop protection. *Eur. J. Plant Pathol.* **2012**, *133*, 197–209. [CrossRef]
11. Lichtenthaler, H.K. Vegetation Stress: An Introduction to the Stress Concept in Plants. *J. Plant Physiol.* **1996**, *148*, 4–14. [CrossRef]
12. Ginaldi, F.; Bajocco, S.; Bregaglio, S.; Cappelli, G. *Spatializing Crop Models for Sustainable Agriculture*; Springer International Publishing: Cham, Switzerland, 2019; pp. 599–619.
13. Kamir, E.; Waldner, F.; Hochman, Z. Estimating wheat yields in Australia using climate records, satellite image time series and machine learning methods. *ISPRS J. Photogramm. Remote Sens.* **2020**, *160*, 124–135. [CrossRef]
14. Anderson, M.C.; Hain, C.; Otkin, J.; Zhan, X.; Mo, K.; Svoboda, M.; Wardlow, B.; Pimstein, A. An Intercomparison of Drought Indicators Based on Thermal Remote Sensing and NLDAS-2 Simulations with U.S. Drought Monitor Classifications. *J. Hydrometeorol.* **2013**, *14*, 1035–1056. [CrossRef]
15. Rodriguez, D.; Sadras, V.O.; Christensen, L.K.; Belford, R. Spatial assessment of the physiological status of wheat crops as affected by water and nitrogen supply using infrared thermal imagery. *Aust. J. Agric. Res.* **2005**, *56*, 983–993. [CrossRef]
16. Zhang, L.; Qiao, N.; Huang, C.; Wang, S. Monitoring Drought Effects on Vegetation Productivity Using Satellite Solar-Induced Chlorophyll Fluorescence. *Remote Sens.* **2019**, *11*, 378. [CrossRef]
17. Doraiswamy, P.C.; Hatfield, J.L.; Jackson, T.J.; Akhmedov, B.; Prueger, J.; Stern, A. Crop condition and yield simulations using Landsat and MODIS. *Remote Sens. Environ.* **2004**, *92*, 548–559. [CrossRef]
18. Huang, J.; Ma, H.; Wei, S.; Zhang, X.; Wu, W. Jointly Assimilating MODIS LAI and et Products into the SWAP Model for Winter Wheat Yield Estimation. *IEEE J. Sel. Top. Appl. Earth Obs. Remote Sens.* **2015**, *8*, 4060–4071. [CrossRef]
19. Fitzpatrick, E. Estimates of pan evaporation from mean maximum temperature and vapour pressure. *J. Appl. Meteorol.* **1963**, *2*, 780–792. [CrossRef]
20. Fitzpatrick, E.A.; Nix, H.A. A model for simulating soil water regime in alternating fallow-crop systems. *Agric. Meteorol.* **1969**, *6*, 303–319. [CrossRef]
21. Gouache, D.; Bouchon, A.-S.; Jouanneau, E.; Le Bris, X. Agrometeorological analysis and prediction of wheat yield at the departmental level in France. *Agric. For. Meteorol.* **2015**, *209*, 1–10. [CrossRef]
22. Landau, S.; Mitchell, R.A.C.; Barnett, V.; Colls, J.J.; Craigon, J.; Payne, R.W. A parsimonious, multiple-regression model of wheat yield response to environment. *Agric. For. Meteorol.* **2000**, *101*, 151–166. [CrossRef]
23. Hui, Q. *Assessment of Net Primary Productivity (NPP) of Vegetation in North China Plain Using MODIS Remote Sensing Information*; Jilin University: Changchun, China, 2004.
24. Guo, Y.S.; Liu, Q.S.; Liu, G.H.; Huang, C. Extraction of main crop planting information based on MODIS Time series NDVI. *J. Nat. Resour.* **2017**, *32*, 1808–1818.
25. Holzman, M.E.; Rivas, R.; Piccolo, M.C. Estimating soil moisture and the relationship with crop yield using surface temperature and vegetation index. *Int. J. Appl. Earth Obs. Geoinf.* **2014**, *28*, 181–192. [CrossRef]
26. Potgieter, A.B.; Lawson, K.; Huete, A.R. Determining crop acreage estimates for specific winter crops using shape attributes from sequential MODIS imagery. *Int. J. Appl. Earth Obs. Geoinf.* **2013**, *23*, 254–263. [CrossRef]
27. Cao, J.; Zhang, Z.; Tao, F.; Zhang, L.; Luo, Y.; Zhang, J.; Han, J.; Xie, J. Integrating Multi-Source Data for Rice Yield Prediction across China using Machine Learning and Deep Learning Approaches. *Agric. For. Meteorol.* **2021**, *297*, 108275. [CrossRef]
28. Yongguang, Z. *Passive Remote Sensing of Chlorophyll Fluorescence in Plants and Its Application*; Zhejiang University: Zhejiang, China, 2006.
29. Guan, K.; Berry, J.A.; Zhang, Y.; Joiner, J.; Guanter, L.; Badgley, G.; Lobell, D.B. Improving the monitoring of crop productivity using spaceborne solar-induced fluorescence. *Glob. Chang. Biol.* **2016**, *22*, 716–726. [CrossRef] [PubMed]
30. Guanter, L.; Zhang, Y.; Jung, M.; Joiner, J.; Voigt, M.; Berry, J.A.; Frankenberg, C.; Huete, A.R.; Zarco-Tejada, P.; Lee, J.E.; et al. Global and time-resolved monitoring of crop photosynthesis with chlorophyll fluorescence. *Proc. Natl. Acad. Sci. USA* **2014**, *111*, E1327–E1333. [CrossRef]
31. Zhang, Y.; Guanter, L.; Berry, J.A.; Joiner, J.; van der Tol, C.; Huete, A.; Gitelson, A.; Voigt, M.; Köhler, P. Estimation of Vegetation Photosynthetic Capacity from Space-Based Measurements of Chlorophyll Fluorescence for Terrestrial Biosphere Models. *Glob. Change Biol.* **2014**, *20*, 3727–3742. [CrossRef]
32. Berry, J.; Frankenberg, C.; Wennberg, P. New Methods for Measurements of Photosynthesis from Space. 2013. Available online: https://authors.library.caltech.edu/92893/ (accessed on 1 April 2013).
33. Frankenberg, C.; Fisher, J.B.; Worden, J.; Badgley, G.; Saatchi, S.S.; Lee, J.E.; Toon, G.C.; Butz, A.; Jung, M.; Kuze, A.; et al. New global observations of the terrestrial carbon cycle from GOSAT: Patterns of plant fluorescence with gross primary productivity. *Geophys. Res. Lett.* **2011**, *38*, 17. [CrossRef]
34. Zhang, Y.; Guanter, L.; Joiner, J.; Song, L.; Guan, K. Spatially-explicit monitoring of crop photosynthetic capacity through the use of space-based chlorophyll fluorescence data. *Remote Sens. Environ.* **2018**, *210*, 362–374. [CrossRef]

35. Zhou, H.; Wu, D.; Lin, Y. The relationship between solar-induced fluorescence and gross primary productivity under different growth conditions: Global analysis using satellite and biogeochemical model data. *Int. J. Remote Sens.* **2020**, *41*, 7660–7679. [CrossRef]
36. Somkuti, P.; Bosch, H.; Feng, L.; Palmer, P.; Parker, R.J.; Quaife, T. A new space-borne perspective of crop productivity variations over the US Corn Belt. *Agric. For. Meteorol.* **2020**, *281*, 107826. [CrossRef]
37. Wei, J.; Tang, X.; Gu, Q.; Wang, M.; Ma, M.; Han, X. Using Solar-Induced Chlorophyll Fluorescence Observed by OCO-2 to Predict Autumn Crop Production in China. *Remote Sens.* **2019**, *11*, 1715. [CrossRef]
38. Cai, Y.; Guan, K.; Lobell, D.; Potgieter, A.B.; Wang, S.; Peng, J.; Xu, T.; Asseng, S.; Zhang, Y.; You, L.; et al. Integrating satellite and climate data to predict wheat yield in Australia using machine learning approaches. *Agric. For. Meteorol.* **2019**, *274*, 144–159. [CrossRef]
39. Franch, B.; Vermote, E.F.; Becker-Reshef, I.; Claverie, M.; Huang, J.; Zhang, J.; Justice, C.; Sobrino, J.A. Improving the timeliness of winter wheat production forecast in the United States of America, Ukraine and China using MODIS data and NCAR Growing Degree Day information. *Remote Sens. Environ.* **2015**, *161*, 131–148. [CrossRef]
40. Newlands, N.K.; Zamar, D.S.; Kouadio, L.A.; Zhang, Y.; Chipanshi, A.; Potgieter, A.; Toure, S.; Hill, H.S. An integrated, probabilistic model for improved seasonal forecasting of agricutral crop yield under environmental uncertainty. *Front. Environ. Sci.* **2014**, *2*, 7–8. [CrossRef]
41. Balaghi, R.; Tychon, B.; Eerens, H.; Jlibene, M. Empirical regression models using NDVI, rainfall and temperature data for the early prediction of wheat grain yields in Morocco. *Int. J. Appl. Earth Obs. Geoinf.* **2008**, *10*, 438–452. [CrossRef]
42. Tao, F.; Zhang, Z.; Liu, J.; Yokozawa, M. Modelling the impacts of weather and climate variability on crop productivity over a large area: A new super-ensemble-based probabilistic projection. *Agric. For. Meteorol.* **2009**, *149*, 1266–1278. [CrossRef]
43. Guan, K.; Wu, J.; Kimball, J.S.; Anderson, M.C.; Frolking, S.; Li, B.; Hain, C.R.; Lobe, D.B. The shared and unique values of optical, fluorescence, thermal and microwave satellite data for estimating large-scale crop yields. *Remote Sens. Environ.* **2017**, *199*, 333–349. [CrossRef]
44. Jiang, H.; Hu, H.; Zhong, R.; Xu, J.; Xu, J.; Huang, J.; Wang, S.; Ying, Y.; Lin, T. A deep learning approach to conflating heterogeneous geospatial data for corn yield estimation: A case study of the US Corn Belt at the county level. *Glob. Change Biol.* **2020**, *26*, 1754–1766. [CrossRef]
45. Cai, Y.; Guan, K.; Peng, J.; Wang, S.; Seifert, C.; Wardlow, B.; Li, Z. A high-performance and in-season classification system of field-level crop types using time-series Landsat data and a machine learning approach. *Remote Sens. Environ.* **2018**, *210*, 35–47. [CrossRef]
46. Kussul, N.; Lavreniuk, M.; Skakun, S.; Shelestov, A. Deep Learning Classification of Land Cover and Crop Types Using Remote Sensing Data. *IEEE Geosci. Remote Sens. Lett.* **2017**, *14*, 778–782. [CrossRef]
47. Miotto, R.; Li, L.; Dudley, J.T. Deep Learning to Predict Patient Future Diseases from the Electronic Health Records. In *European Conference on Information Retrieval, Padua, Italy, 20–23 March 2016*; Springer International Publishing: Cham, Switzerland, 2016.
48. Wolanin, A.; Camps-Valls, G.; Gómez-Chova, L.; Mateo-García, G.; van der Tol, C.; Zhang, Y.; Guanter, L. Estimating crop primary productivity with Sentinel-2 and Landsat 8 using machine learning methods trained with radiative transfer simulations. *Remote Sens. Environ.* **2019**, *225*, 441–457. [CrossRef]
49. Khaki, S.; Wang, L. Crop yield prediction using deep neural networks. *Front. Plant Sci.* **2019**, *10*, 621. [CrossRef] [PubMed]
50. Kuwata, K.; Shibasaki, R. Estimating crop yields with deep learning and remotely sensed data. In Proceedings of the 2015 IEEE International Geoscience and Remote Sensing Symposium (IGARSS), Milan, Italy, 26–31 July 2015.
51. Yang, Q.; Shi, L.; Han, J.; Zha, Y.; Zhu, P. Deep convolutional neural networks for rice grain yield estimation at the ripening stage using UAV-based remotely sensed images. *Field Crops Res.* **2019**, *235*, 142–153. [CrossRef]
52. Johnson, D.M. An assessment of pre- and within-season remotely sensed variables for forecasting corn and soybean yields in the United States. *Remote Sens. Environ.* **2014**, *141*, 116–128. [CrossRef]
53. Frausto-Solis, J.; Gonzalez-Sanchez, A.; Larre, M. A New Method for Optimal Cropping Pattern. In *Mexican International Conference on Artificial Intelligence*; Springer: Berlin/Heidelberg, Germany, 2009.
54. Cao, J.; Zhang, Z.; Tao, F.; Zhang, L.; Luo, Y.; Han, J.; Li, Z. Identifying the contributions of multi-source data for winter wheat yield prediction in China. *Remote Sens.* **2020**, *12*, 750. [CrossRef]
55. Xinhua, Y.; Weiqing, Z.; Zaichun, Z.; Xubao, D.; Zhaozhi, Z. Multi-scale winter wheat yield estimation based on remote sensing and crop growth model. *Spectrosc. Spectr. Anal.* **2021**, *41*, 2205–2211.
56. Ying, L.; Huaiyong, S. Study on optimal time window and influencing factors of winter wheat yield prediction in Henan Province based on stochastic forest algorithm. *J. Triticeae Crops* **2020**, *40*, 874–880.
57. Sun, Y.; Fu, R.; Dickinson, R.; Joiner, J.; Frankenberg, C.; Gu, L.; Xia, Y.; Fernando, N. Drought onset mechanisms revealed by satellite solar-induced chlorophyll fluorescence: Insights from two contrasting extreme events. *J. Geophys. Res. Biogeosciences* **2015**, *120*, 2427–2440. [CrossRef]
58. Joiner, J.; Guanter, L.; Lindstrot, R.; Voigt, M.; Vasilkov, A.P.; Middleton, E.M.; Huemmrich, K.F.; Yoshida, Y.; Frankenberg, C. Global monitoring of terrestrial chlorophyll fluorescence from moderate-spectral-resolution near-infrared satellite measurements: Methodology, simulations, and application to GOME-2. *Atmos. Meas. Tech.* **2013**, *6*, 2803–2823. [CrossRef]
59. Chen, Y.; Zhang, Z.; Tao, F. Improving regional winter wheat yield estimation through assimilation of phenology and leaf area index from remote sensing data. *Eur. J. Agron.* **2018**, *101*, 163–173. [CrossRef]

60. Tao, F.; Xiao, D.; Zhang, S.; Zhang, Z.; Rötter, R.P. Wheat yield benefited from increases in minimum temperature in the Huang-Huai-Hai Plain of China in the past three decades. *Agric. For. Meteorol.* **2017**, *239*, 1–14. [CrossRef]
61. Luo, Y.; Zhang, Z.; Chen, Y.; Li, Z.; Tao, F. ChinaCropPhen1km: A high-resolution crop phenological dataset for three staple crops in China during 2000–2015 based on leaf area index (LAI) products. *Earth Syst. Sci. Data* **2020**, *12*, 197–214. [CrossRef]
62. Ma, Y.; Kang, Y.; Ozdogan, M.; Zhang, Z. County-Level Corn Yield Prediction Using Deep Transfer Learning [Z]. 2019: B54D-02. Available online: https://ui.adsabs.harvard.edu/abs/2019AGUFM.B54D..02M/abstract (accessed on 4 December 2019).
63. You, J.; Li, X.; Low, M.; Lobell, D.; Ermon, S. Deep gaussian process for crop yield prediction based on remote sensing data. In Proceedings of the Thirty First AAAI conference on artificial intelligence, San Francisco, CA, USA, 4–9 February 2017; Springer International Publishing: Cham, Switzerland, 2017.
64. Cao, J.; Zhang, Z.; Wang, C.; Liu, J.; Zhang, L. Susceptibility assessment of landslides triggered by earthquakes in the Western Sichuan Plateau. *CATENA* **2019**, *175*, 63–76. [CrossRef]
65. Youssef, A.M.; Pourghasemi, H.R.; Pourtaghi, Z.S.; Al-Katheeri, M.M. Landslide susceptibility mapping using random forest, boosted regression tree, classification and regression tree, and general linear models and comparison of their performance at Wadi Tayyah Basin, Asir Region, Saudi Arabia. *Landslides* **2016**, *13*, 839–856. [CrossRef]
66. Sloat, L.L.; Lin, M.; Butler, E.E.; Johnson, D.; Holbrook, N.M.; Huybers, P.J.; Lee, J.-E.; Mueller, N.D. Evaluating the benefits of chlorophyll fluorescence for in-season crop productivity forecasting. *Remote Sens. Environ.* **2021**, *260*, 112478. [CrossRef]
67. Alvarez, R. Predicting average regional yield and production of wheat in the Argentine Pampas by an artificial neural network approach. *Eur. J. Agron.* **2009**, *30*, 70–77. [CrossRef]
68. Stockle, C.O.; Donatelli, M.; Nelson, R. CropSyst, a cropping systems simulation model. *Eur. J. Agron.* **2003**, *18*, 289–307. [CrossRef]
69. Zhang, L.; Zhang, Z.; Luo, Y.; Cao, J.; Tao, F. Combining Optical, Fluorescence, Thermal Satellite, and Environmental Data to Predict County-Level Maize Yield in China Using Machine Learning Approaches. *Remote Sens.* **2020**, *12*, 21. [CrossRef]
70. Guanter, L.; Aben, I.; Tol, P.; Krijger, J.M.; Hollstein, A.; Köhler, P.; Damm, A.; Joiner, J.; Frankenberg, C.; Landgraf, J. Potential of the TROPOspheric Monitoring Instrument (TROPOMI) onboard the Sentinel-5 Precursor for the monitoring of terrestrial chlorophyll fluorescence. *Atmos. Meas. Tech.* **2015**, *8*, 1337–1352. [CrossRef]
71. Stark, H.R.; Moller, H.L.; Courrèges-Lacoste, G.B.; Koopman, R.; Mezzasoma, S.; Veihelmann, B. The Sentinel-4 Mission and its implementation. In *ESA Living Planet Symposium*; Springer International Publishing: Cham, Switzerland, 2013.
72. Drusch, M.; Moreno, J.; Del Bello, U.; Franco, R.; Goulas, Y.; Huth, A.; Kraft, S.; Middleton, E.; Miglietta, F.; Mohammed, G.; et al. The FLuorescence EXplorer Mission Concept-ESA's Earth Explorer 8. *IEEE Trans. Geosci. Remote Sens.* **2016**, *55*, 1273–1284. [CrossRef]

agriculture

MDPI

Article

Maximum Entropy Niche-Based Modeling for Predicting the Potential Suitable Habitats of a Traditional Medicinal Plant (*Rheum nanum*) in Asia under Climate Change Conditions

Wei Xu [1,†], Shuaimeng Zhu [2,†], Tianli Yang [3], Jimin Cheng [4] and Jingwei Jin [5,*]

1 Humanities and Social Sciences, Jiangsu University of Science and Technology, Zhenjiang 212000, China; xuwei12@mails.ucas.ac.cn
2 School of Surveying and Land Information Engineering, Henan Polytechnic University, Jiaozuo 454000, China; zhusm@hpu.edu.cn
3 College of Forestry, Northwest Agriculture and Forestry University, Xianyang 712100, China; yangtianli@nwafu.edu.cn
4 Institute of Soil and Water Conservation, Chinese Academy of Sciences and Ministry of Water Resources, Xianyang 712100, China; gyzcjm@ms.iswc.ac.cn
5 Institute of Soil and Water Conservation, Northwest Agriculture and Forestry University, Xianyang 712100, China
* Correspondence: jjw@ms.iswc.ac.cn
† These authors contributed equally to this work.

Abstract: *Rheum nanum*, a perennial herb, is a famous traditional Chinese medicinal plant that has great value in modern medicine. In order to determine the potential distribution of *R. nanum* in Asia, we specifically developed the potential distribution maps for three periods (current, 2050s: 2041–2060, and 2070s: 2061–2080) using MaxEnt and ArcGIS, and these were based on the current and future climate data under two climate scenarios (RCP2.6 and RCP6.0). To predict the potential impacts of global warming, we measured the area of suitable habitats, habitat suitability changes, and habitat core changes. We found that bio16 (i.e., the precipitation of the wettest quarter) and bio1 (i.e., the annual mean temperature) were the most important climate factors that influenced the distribution of *R. nanum*. The areas of high suitable habitats (HH) and middle suitable habitats (MH) in the current period were 156,284.7 ± 0.99 km^2 and 361,875.0 ± 3.61 km^2, respectively. The areas of HH and MH in 2070RCP6.0 were 27,309.0 ± 0.35 km^2 and 123,750 ± 2.36 km^2, respectively. The ranges of 82.0–90.3° E, 43.8–46.5° N were the mostly degraded areas of the 2050s and 2070s, and RCP6.0 had a larger decrease in habitable area than that found in RCP2.6. All the HH cores shifted south, and the shift distance of HH in 2070RCP6.0 was 115.65 km. This study provides a feasible approach for efficiently utilizing low-number occurrences, and presents an important attempt at predicting the potential distribution of species based on a small sample size. This may improve our understanding of the impacts of global warming on plant distribution and could be useful for relevant agricultural decision-making.

Keywords: geographic distribution; suitable habitat; *Rheum nanum*; MaxEnt; ArcGIS; range shifts; climate scenario

Citation: Xu, W.; Zhu, S.; Yang, T.; Cheng, J.; Jin, J. Maximum Entropy Niche-Based Modeling for Predicting the Potential Suitable Habitats of a Traditional Medicinal Plant (*Rheum nanum*) in Asia under Climate Change Conditions. *Agriculture* **2022**, *12*, 610. https://doi.org/10.3390/agriculture12050610

Academic Editor: Nándor Fodor

Received: 15 April 2022
Accepted: 22 April 2022
Published: 26 April 2022

1. Introduction

In China, *Rheum* L. plants are traditional medicines distributed in the Greater Khingan Range, the Taihang Mountains region, the Qinling Mountains region, the Daba Mountains region, west of the Yunnan–Guizhou Plateau, and the Qinghai-Tibet Plateau [1–3]. At present, studies on *Rheum* L. plants mainly focus on its functional extracts [4–8]. As one of the most heat-tolerant and drought-tolerant species of *Rheum* L. plants [1], *Rheum nanum* Siev. ex Pall has been found in Mongolia, Kazakhstan, Russia, and China (i.e., Xinjiang, Gansu, and Inner Mongolia), which mainly inhabit hillsides, valleys, and gravel lands below 1000 m [9]. The latest study showed that chrysophanol, an active constituent of *R. nanum*, has been effective for obesity and may cure certain skin disorders such as acne vulgaris [10]. Emodin, another active constituent of *R. nanum*, has suppressed the growth and the invasion of colorectal cancer [11] and has been used to cure acute kidney injury [12]. Previous studies have shown that precipitation and temperature are the dominate factors affecting the geographic distribution of *Rheum* L. plants [1,3]. *R. nanum* extracts have played an important role in modern medicine; however, to the best of our knowledge, the exact geographic distribution and the suitable habitats of this ancient medicine remain unexplored.

With the development of computer technology, it has become possible to predict the distribution pattern of the species' niche by using occurrence data and associated environmental variables [13] with the help of species distribution models (SDMs). According to their dependence of occurrence data, SDMs can be roughly divided in two groups: occurrence-absence data groups (e.g., generalized additive models, generalized linear models, generalized boosting models) and occurrence-data-only groups (e.g., domain model, ecological niche-factor analysis model) [14,15]. The occurrence-data-only models typically have greater advantages as the absence data are usually difficult to collect in the real world, especially for poorly known species [16]. Among various occurrence-data-only SDMs, the maximum entropy model (MaxEnt) [17] has been used extensively because of its strong statistical foundation (i.e., high accuracy and robustness [14,18–21]) and ability to simulate ecological relationships [22]. Until now, MaxEnt has had a wide range of applications with relevant predictions [23–29]. Moreover, MaxEnt has better performance for species with a small sample size [16], especially for those species with typically little occurrence records [30].

Over the past 100 years, the global average temperature has risen by 0.6 °C [31,32]. By the end of this century, with a maximum increase of 2.6–4.8 °C in global average temperature [33], global precipitation will increase by approximately 31% [34]. Throughout the evolutionary history of diverse species [31,35], climate change has had significant impacts on the spatial distribution patterns of organisms [32,36]. Global warming has also been confirmed to have the ability to change species' survival [35] and conservation situations both spatially and temporally [37,38]. With the growing concern of global warming and its potential effects, there have been many useful studies on ecological modeling and conservations using multiple factors and methods [39]. In sub-Saharan Africa, under future climate scenarios, rice production will experience relevant losses in different regions [40]. For alpine plant communities in mountain ecosystems, elevation-dependent warming presents additional challenges and may result in the extinction of cold-adapted flora species [41,42]. In addition, the herbaceous and woody species in mountain basins and the Taklamakan Desert are more sensitive to climate change than plant species in meadows and steppe regions [36]. For *Stipa purpurea*, a perennial herbaceous plant widespread throughout the Tibetan Plateau, its number of suitable habitats is expected to increase from 1990 to 2050 [27]. Both the herbaceous species and dwarf shrubs will shift upwards in the alpine tundra region of the Changbai Mountains [43]. Hence, it is expected that the possible biogeographic patterns of *R. nanum* could change with global warming. Based on the effects of increasing temperatures on the global distribution of flora species, we developed two hypotheses: (1) high suitable habitats of *R. nanum* will increase and move northward; (2) the higher emission scenario may cause a larger increase in the habitat area of *R. nanum* than the lower emission scenario. The aim of this study was to predict the potential distribution of *R. nanum* in Asia under climate change and to verify our two hypotheses.

2. Materials and Methods

2.1. Occurrence Data and Study Area

The study area range was 70.9–136.5° N, 17.8–55.5° E in Asia. We initially selected 58 region names with relevant habitat descriptions (e.g., text, photo, or GPS points) from online resources (http://www.cvh.ac.cn/ accessed on 21 April 2022, http://www.cfh.ac.cn/ accessed on 21 April 2022 and http://www.plantphoto.cn/ accessed on 29 June 2016) and the published literature [1,3]. At an approximate position, we conducted a two-and-a-half year (2016–2018) field survey in north and northwest China, and collected 246 occurrence data with GPS. To improve the accuracy of prediction and reduce overfitting [44], SDMs toolbox version 2.4 for ArcGIS 10.2 was used to rarefy the occurrence data. Since the resolution of environmental variables used in the MaxEnt model was 2.5 arc-minutes (approximately 5 km), to ensure that each variable grid covered, at most, one occurrence point, we filtered the occurrence data with a buffer distance of 5 km and finally selected 16 occurrence points (Figure 1) for the MaxEnt model. The occurrence points are usually divided into two types according to their usage: the training sample layer and the test sample layer. The rarefied sample size was too small and the widely used method of setting "random test percentages" (i.e., 20–30% of occurrence data [45,46]) would further reduce the sample size. Therefore, to overcome the limitations of a small sample size, we decided to set the 16 occurrence points (i.e., a 5 km buffer) as the training sample layer, and also sought other occurrence points that differed from the training samples as the test sample layer. We rarified the 246 occurrence data again with a 1 km buffer, and then obtained 27 points, which covered the 16 occurrence points (i.e., a 5 km buffer) previously selected. After removing the 16 samples from the 27 newly acquired samples (i.e., a 1 km buffer), we obtained 11 sample points that could be set as the test sample layer for the MaxEnt model.

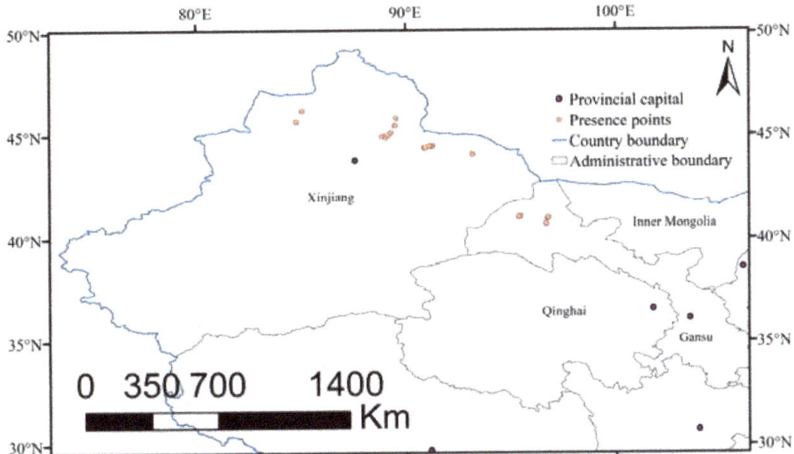

Figure 1. Occurrence data of *Rheum nanum* in China.

2.2. Environmental Variables

Nineteen climate variables (2.5 arc-minutes resolution, Table S1) for the current (1970–2000) and future (2050s: 2041–2060 and 2070s: 2061–2080) periods were downloaded from the WorldClim version 1.4 dataset (https://www.worldclim.org accessed on 21 April 2022). The current climate data are interpolations from the observed data, and the future climate data are projections from the Intergovernmental Panel on Climate Change (IPCC [47]) Fifth Assessment Reports (AR5). The future projection data has four "representative concentration pathways" (RCPs. 2.6, 4.5, 6.0, and 8.5 [48]) and examines four possible future radiative forcing levels (2.6, 4.5, 6.0, and 8.5 W/m^2 [46]). The emission scenarios consider the potential impacts of policies on future greenhouse gas emissions and provide more scientific descriptions of possible future climate changes [17,49]. Among them, RCP2.6 is the lowest emission scenario (i.e., the most optimistic case) with ~490 parts per million (ppm) CO_2 concentrations until 2100, RCP4.5 and RCP6.0 are middle of the road (i.e., the more optimistic case, ~650 ppm and 850 ppm, respectively), and RCP8.5 is the highest emission scenario (i.e., the worst case, ~1370 ppm, https://www.carbonbrief.org accessed on 21 April 2022). RCP2.6 is the only case that could generally limit the global mean temperature increase below 2 °C by 2100 [49], because previous studies have indicated that without proper control, the CO_2 concentrations will likely reach 560 ppm (double the pre-industrial level) and 800 ppm by 2060 and 2080, respectively [50]. Hence, in this study, we chose RCP2.6 and RCP6.0 of the Beijing Climate Center–Climate System Model version 1.1 (BCC–CSM 1.1 [51,52]) as the future climate change scenarios.

2.3. Model Processing and Evaluation

For SDMs, rare samples may lead to an imperfect match between the environmental variables and the occurrence data, which brings great difficulties to model evaluation and easily leads to model overfitting [53]. To our knowledge, MaxEnt is one of the few models that perform well in small sample simulations [16,53]. Over the past two decades, MaxEnt, an occurrence-data-only machine-learning program [54], has had extremely extensive procedures on predicting the potential distribution of species [44,55]. Previous studies indicates that 15 might be the minimum number of occurrence points for ecological niche modeling [16]. For Asian species which have >15 available occurrences, the MaxEnt model may provide a broad spectrum of predictions [16]. Moreover, the jackknife test reflects the contributions and the importance of environmental factors in the MaxEnt model [56–58]. The area under the curve (AUC) value, based on the receiver-operating characteristic (ROC), ranges from 0 to 1 and reflects the accuracy of the MaxEnt model's prediction [14,56,59]. Values of 0.8–0.9 indicate a good prediction while values of 0.9–1.0 represent an excellent prediction [17,32,54].

We converted 19 bioclimate environmental variables (bio1–bio19) to ASCII format using Toolbox of ArcGIS 10.2, and then imported the file along with the species occurrence data (in ASCII format) into MaxEnt 3.4.1 for pre-selection. Based on the default settings, we set 16 occurrence data (i.e., at 5 km) as the model training dataset and 11 occurrence data (i.e., at 1 km) as the model testing dataset. A correlation analysis ($r > 0.8$) was conducted to avoid the potential over-fitting [60,61] of environmental variables, and we obtained two factors (more details in [62]): the annual mean temperature (bio1) and the annual precipitation (bio12). In order to measure the contribution and importance of the 19 variables, we conducted a pre-selection with a set of 1000 iterations, a jackknife test, and an analysis of the response curves, and used default values for other settings. According to the pre-selection result (Table S1), the precipitation of the wettest quarter (bio16) and bio1 contributed to the model the most. Previous studies have indicated that, for *Rheum* L. plants, seasonal precipitation and relevant temperature during the growth period (March to September [9], and March to May for some species in extreme conditions) are the key factors affecting its spatial distribution [1–3]. Hence, after comprehensively considering the results of the correlation analysis, the pre-selection, and the possible environmental needs of *R. nanum*, we finally chose 6 bioclimatic factors (Table 1) as the environmental layers of the MaxEnt in this study. To improve the projection accuracy and reduce uncertainty, we launched the MaxEnt with the previous settings and ran it 10 times (i.e., 10 "replicates" with basic settings). We calculated the contributions and importance of the 6 factors (Table 1) with Excel and the obtained raster layers in ASCII format were then imported into ArcGIS for further analysis.

Table 1. Environmental variables used for MaxEnt modeling.

Code	Bioclimatic Data	Unit	Contribution (%)	Importance (%)
bio1	Annual mean temperature	°C	13.9 ± 0.8	54.1 ± 2.7
bio9	Mean temperature of driest quarter	°C	7.4 ± 0.6	3.1 ± 0.6
bio14	Precipitation of driest month	mm	9.7 ± 1.8	0.1 ± 0.1
bio15	Precipitation seasonality		6.2 ± 1.1	9.3 ± 2.1
bio16	Precipitation of wettest quarter	mm	55.9 ± 2.1	20.0 ± 5.3
bio19	Precipitation of coldest quarter	mm	6.9 ± 2.2	13.4 ± 2.6

We defined the logistic threshold of "balance training omission, predicted area, and threshold value" as TH, and divided the obtained float raster maps into four classes using the reclassify tool [54,55]: UH (unsuitable habitat, 0–TH), LH (low suitable habitat, TH–0.3), MH (middle suitable habitat, 0.3–0.6), and HH (high suitable habitat, 0.6–1.0). The area of suitable habitats as well as habitat maps showing the increases and decreases in habitat range were measured using a raster calculator tool. The cores of habitats and range shifts of habitats were calculated by the zonal geometry (i.e., by calculating the area and centroid of the input raster or feature zone data using a specific zone field, which must be an integer field) and the mean center tool (i.e., by identifying the geographic center for a set of features with case field and weight field, which determine the rank and weight value, respectively; the rank represents the suitability classes, and the weight represents the floating value of MaxEnt.). To quantify the level of increase and decrease in habitat suitability, we defined UH, LH, MH, and HH as 1, 2, 3, and 4, respectively. The raster calculator tool of ArcGIS was used to test and compare the changes in habitat suitability over two periods. The changes in the habitat suitability degree of −1, −2, and −3, respectively, represent a slight, moderate, and dramatic decrease, while +1, +2, and +3, respectively, represent a slight, moderate, and dramatic increase. In addition, we chose the ratio of the area corresponding to the habitat suitability change degree and the total area of suitability change (i.e., increase or decrease) to quantify the changes in habitat suitability.

3. Results

3.1. Contributions and Importance of Environmental Variable in the MaxEnt Model

The mean AUC value of the training data was 0.992 ± 0.003 (Figure S1a), indicating that MaxEnt had excellent prediction performance. The Jackknife importance test showed that the mean value of the regularized training gain of the six factors was 3.127 (Figure S1b,c). When the environmental variables were used in isolation, bio16 (i.e., the precipitation of the wettest quarter) had the highest gain of 1.825, bio1 (i.e., the annual mean temperature) had the second highest gain of 1.455, and bio14 (i.e., the precipitation of the driest month) had the lowest gain of 0.291. Moreover, bio1 (2.751) had the largest decrease in gains when it was omitted, followed by bio16 (3.021) and bio15 (3.047) (Figure S1c).

The MaxEnt model results showed that the variable that contributed the most was bio16 (55.9%), followed by bio1 (13.9%) and bio14 (9.7%); the most important factor in the MaxEnt model was bio1 (54.1%), and the second and third most important factors were bio16 (20.0%) and bio19 (i.e., the precipitation of the coldest quarter, 13.4%), respectively.

3.2. Potential Geographic Distribution and Suitable Habitat Area of R. nanum

Under the RCP2.6 and RCP6.0 scenarios, *R. nanum*'s potential geographic distribution of suitable habitats over the three periods (current, 2050s and 2070s) modeled by MaxEnt was visualized by ArcGIS (Figure 2). The distribution maps showed that, under current climate conditions, the range of MH was bounded at 81.7–97.3° E, 39.4–48.6° N, and the range of HH was bounded at 82.4–93.4° E, 42.5–47.9° N.

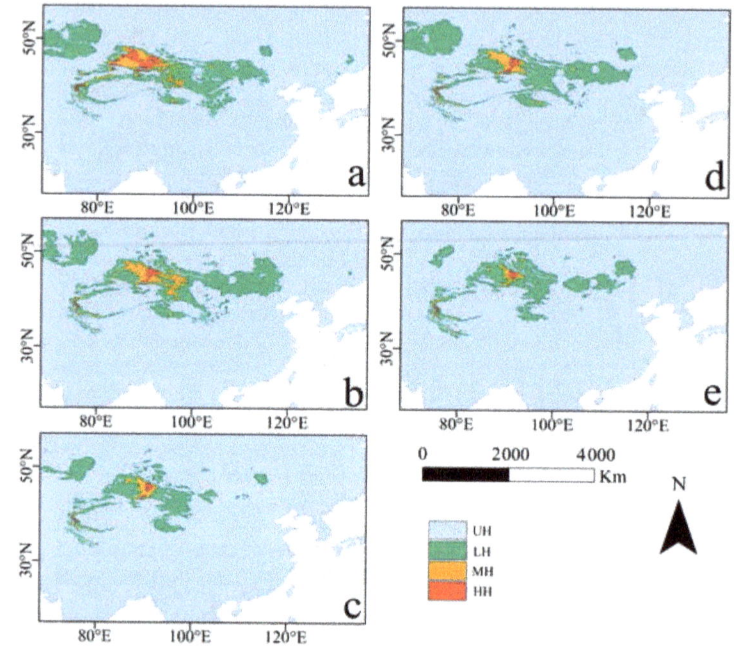

Figure 2. Potential suitable habitat maps of the current period, 2050s, and 2070s: (**a**) current; (**b**) 2050RCP2.6; (**c**) 2050RCP6.0; (**d**) 2070RCP2.6; (**e**) 2070RCP6.0. UH, LH, MH, and HH represent unsuitable habitat, low suitable habitat, middle suitable habitat, and high suitable habitat, respectively.

The areas of MH and HH were 361,875.0 ± 3.61 km^2 and 156,284.7 ± 0.99 km^2 in the current climate conditions, respectively (Figure 3). The area of suitable habitats (LH, MH, and HH) under the RCP6.0 scenario was smaller than that under the RCP2.6 scenario, in both the 2050s and the 2070s. Under the RCP2.6 scenario and the RCP6.0 scenario, the area of MH in the 2070s would be 225,972.2 ± 3.12 km^2 and 123,750 ± 2.36 km^2, respectively, while that of the HH in the 2070s would be 48,194.4 ± 0.54 km^2 and 27,309.0 ± 0.35 km^2. For MH, the area that was reduced the most was 2070RCP60 (65.80%), followed by 2050RCP60 (62.15%) and 2070RCP26 (37.56%) (Figure 2 and Table S2). For HH, the area that was reduced the most was 2070RCP60 (128,975.7 km^2, 82.53%), followed by 2050RCP60 (128,368.1 km^2, 82.14%), 2050RCP26 (110,920.1 km^2, 70.97%), and 2070RCP26 (108,090.3 km^2, 69.16%) (Figure 3 and Table S2).

3.3. Changes of Potential Suitable Habitats in the Future Distribution Pattern

Environment variables have a significant influence on the distribution of *R. nanum*. The increase and decrease maps (Figure 4) described the changes in their potential suitable habitats for three periods (current, 2050s, and 2070s). The predicted maps showed that, under the RCP2.6 scenario, the habitat suitability drastically decreased in the ranges of 81.6–89.6° E, 43.6–46.8° N (Figure 4a) and 82.0–90.3° E, 43.8–46.5° N (Figure 4b) in 2070–current and 2050–current period, respectively; under the RCP6.0 scenario, the drastically decreased area of habitat suitability in 2070–current period was at 81.9–91.9° E, 42.1–48.3° N (Figure 4c), while that of 2050–current period was at 81.9–90.6° E, 43.6–48.1° N (Figure 4d). Under the RCP2.6 and RCP6.0 scenarios, the changes in habitat suitability were varied by different degrees: a larger proportion of suitability change was found between adjacent levels (e.g., MH and HH), and a smaller proportion of change was found between further apart habitat levels (e.g., LH and HH). The change range of habitat suitability at −1 and +1 was 51.84–59.31% and 65.64–76.91%, while that of −3 and +3 was 14.82–18.63% and

4.55–7.37%, respectively (Figure 4 and Table S3). Among all conditions of the 2050s and the 2070s, compared with the current distribution pattern, RCP6.0 led to a larger change in the ratio of −1 than RCP2.6, and led to a smaller change in +1 (Table S3).

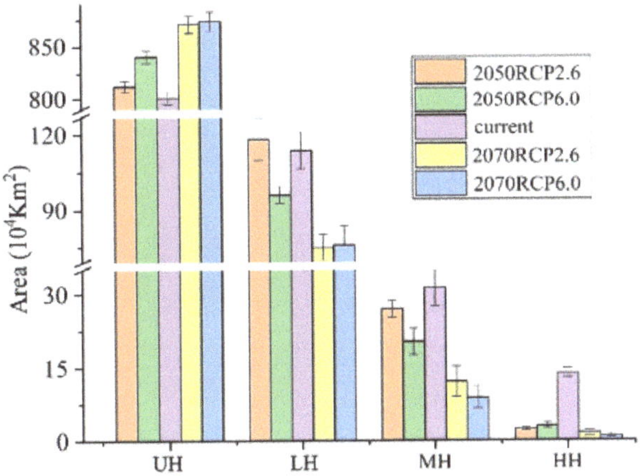

Figure 3. The area of potential suitable habitats over three periods (current, 2050s, and 2070s) under RCP2.6 and RCP6.0 scenarios. UH, LH, MH, and HH represent unsuitable habitat, low suitable habitat, middle suitable habitat, and high suitable habitat, respectively.

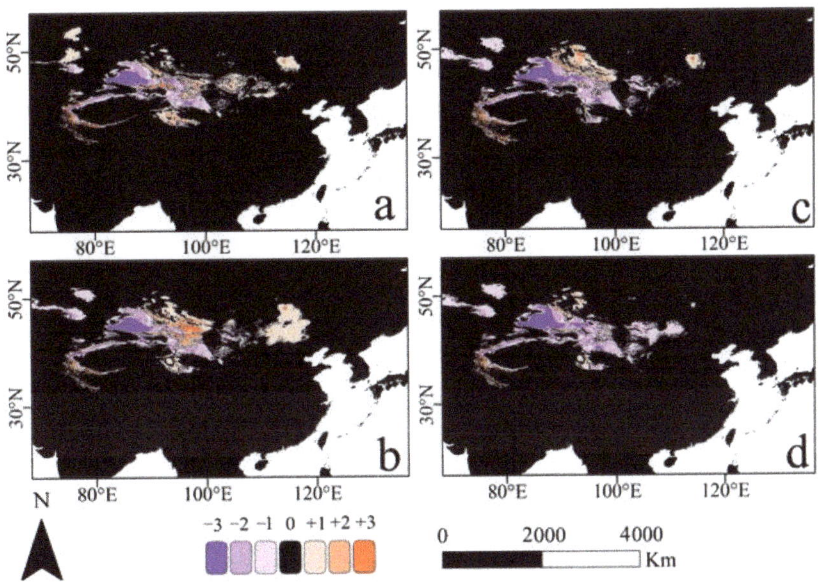

Figure 4. The increase and decrease maps of suitable habitats area: (**a**,**b**) RCP2.6 scenario; (**c**,**d**) RCP6.0 scenario; (**a**,**c**) 2070–current; (**b**,**d**) 2050–current. Zero represents no change of suitability, negative numbers represent a decrease in habitat suitability, and positive numbers represent an increase in habitat suitability. The degree of habitat suitability changes increases from −1 to −3 and +1 to +3.

3.4. Range Shifts of Suitable Habitat Cores under Two Climate Scenarios

Under the RCP2.6 and RCP6.0 scenarios, all cores of *R. nanum*'s UH were located in the area of 98.7–99.6° E, 38.3–39.0° N, the border of Gansu (GS) and Qinghai (QH), and did not have significant changes for three periods (current, 2050s, and 2070s) (Figure 5). However, the cores of LH, MH, and HH in future periods (i.e., 2050s and 2070s) had significant changes under RCP2.6 and RCP6.0, compared with their current distribution. Notably, under the RCP2.6 and the RCP6.0 scenarios, all cores of LH, MH, and HH in the 2050s and the 2070s were located in Xinjiang (XJ), China (Figure 5). Compared to the habitat range of MH in the current period, in the west–east direction, the MH cores of 2050RCP60, 2070RCP26, and 2070RCP60 shifted toward the west, while the MH core of 2050RCP26 moved east; in the north–south direction, except for 2070RCP60, the MH range of 2050RCP60, 2070RCP26, and 2050RCP26 shifted toward the north (Figure 5). The shift distances of MH cores between the current distribution and that of 2070RCP60, 2070RCP26, 2050RCP60, and 2050RCP26 were 118.68 km, 63.52 km, 142.73 km, and 148.49 km, respectively. Compared to the core of HH in the current distribution, all cores of HH shifted toward the south under the RCP2.6 and the RCP6.0 future scenarios (Figure 5). Compared with the range of HH in the current distribution, the cores of HH shifted by 100.18 km and 115.65 km in 2050RCP60 and 2070RCP60, respectively.

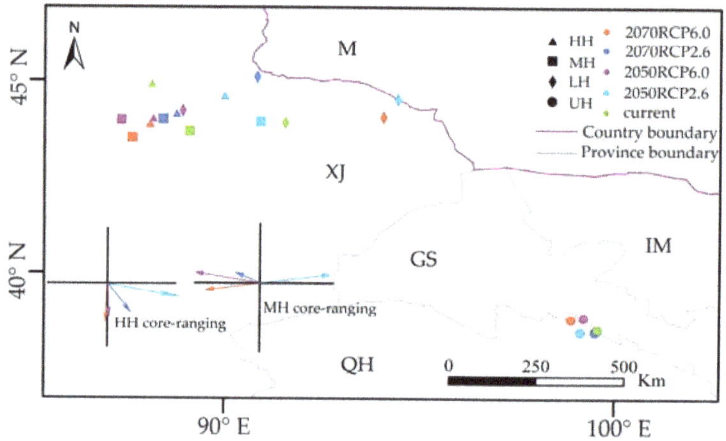

Figure 5. Range shifts of four habitats under RCP2.6 and RCP6.0 scenarios. Lines with color show the direction and distance of core change from current (start of line) to specific time (end of arrow) in MH and HH, and the color of the line depicts the specific time. UH: unsuitable habitat; LH: low suitable habitat; MH: middle suitable habitat; HH: high suitable habitat; M: Mongolia; IM: Inner Mongolia; XJ: Xinjiang; GS: Gansu; QH: Qinghai.

4. Discussion

4.1. Effects of Climate Change on Suitable Habitat Range

Our simulation of the potential distribution of *R. nanum* in the current period, the 2050s, and the 2070s had an accurate performance (AUC > 0.9), and the results indicated that *R. nanum*'s distribution was largely influenced by three precipitation factors (i.e., bio14, bio16, and bio19) and one temperature factor (bio1). For flora species, precipitation affects soil moisture and infiltration [15] and plays a key role in determining the distribution of plants. Moreover, under the growing pressure of changing environments, an increase in temperature [34] has resulted in a reduction in precipitation in Himachal Pradesh, and has impacted the productivity of agricultural crops [63]. In this study, we compared the range changes of the potential suitable habitat of three periods (i.e., current, 2050s, and 2070s) under two climate scenarios (i.e., RCP2.6 and RCP6.0) to quantify the impacts of temperature increase on the potential distribution of *R. nanum*. Compared to the current distribution, almost every suitable habitat (LH, MH, and HH) under RCP2.6 and RCP6.0 scenarios would decrease by different degrees in the future except for LH (increase of 2.58%) and MH (increase of 2.32%) under the 2050RCP2.6 scenario (Figure 3 and Table S2). We believe that, under the 2050RCP2.6 scenario, the unbalanced changes of suitable habitats (e.g., for MH, the degraded area from HH could not offset the newly added suitable habitats from UH) may be responsible for the increase of LH and MH. The comparison of the potential distribution maps of the 2050s and the 2070s (Figure 4 and Table S2) revealed that, compared to the RCP2.6 scenario, RCP6.0 led to a greater reduction in the area of suitable habitats (LH, MH, and HH), which was contrary to previous studies [27] and our former hypotheses. The ecosystems of arid and semi-arid regions are too sensitive to global warming [64], and we believe that this may be the cause of this opposite result.

As the optimal habitat of *R. nanum*, the range of HH was predicted to shift south in the future, as compared to its current range. This result is contrary to the moving direction of the other species in previous studies [27,65]. We inferred that species in arid and semi-arid lands may have different survival strategies [36] when facing global warming, as compared to species in other ecosystems [26], and this may have caused the opposite range shifts. Murray et al. [66] suggested that species with a limited distribution always have a narrow ecological tolerance, and that even slight climate changes may affect their distribution patterns. In addition to the poleward shift, the cores of HH also shifted to higher altitudes with complex terrain (Figure S2), which was consistent with our original hypothesis. Previous studies have also indicated that species would shift poleward and upward to adapt to climate changes [67–69]. However, the poleward migration is essentially the same strategy as the high-elevation migration. For species with little elevation change in their typical habitats, range shifts north would provide an environment similar to their previous habitat in terms of temperature, and species could also shift to a colder environment by increasing living elevation by a small range [65]. Notably, unlike the simple horizon or elevation shift shown on maps, species' range shifts in nature are typically a combination of two migration strategies [67].

With a continued increase in greenhouse gas emissions [48], the higher-emission scenario (RCP6.0) had higher rates of temperature increases than those found in the lower-emission scenario (RCP2.6). Until the end of this century, MH and HH had further range shifts in RCP6.0 than in RCP2.6. One possible reason was that, as an arid and semi-arid species [36], *R. nanum* is highly sensitive to climate changes (e.g., temperature and precipitation). Notably, the geometric comparison of the suitable habitat range maps of the three periods under the two climate scenarios indicated that the HH range of 2050RCP26, 2050RCP60, 2070RCP26, and 2070RCP60 were all included in the HH area in the present day (Figure 2). In addition, the global temperature and mountain ranges largely led to the range shifts of plant species [68]. Therefore, combined with the range of suitable habitats and their range shifts, we assumed that the irregular shrinkage of their suitable habitats may be the main reason behind the range shifts in the future. For insight into the essentials of species range shifts, quantifying and evaluating the importance and contribution of the irregular shrinkage of their habitats should be a focus in future studies.

4.2. Conservation of Species in Ecologically Fragile Areas

The complex terrains of the high mountains in Asia, such as the Kunlun Mountains, the Altai Mountains, the Qilian Mountains, and the Tianshan–Pamir–Hindu Kush–Karakoram mountain ranges, have created a large region with relatively stable arid and semi-arid environments [70]. Previous studies have also proven that the valleys and the rivers of the Hengduan Mountains have provided refuge for plants to survive and evolve in the Last Interglacial and the Last Glacial Maximum [71]. For the habitat suitability of *R. nanum*, the drastically increased range was commonly bounded by 74.1–81.4° E and 36.0–37.6° N while the drastically decreased range was commonly bounded by 81.6–89.6° E and 43.6–46.8° N (Figure 4). We speculate that the mountain regions of Central Asia provide a relatively stable habitat for *R. nanum* to survive, and previous studies have provided some consistent clues [27,70,71]. Under two climate scenarios of the 2050s and the 2070s, the decreased range of habitat suitability coincides with the major distribution range (MH and HH) of *R. nanum* under current climate conditions (Figures 2a and 4). We support the belief that the ecosystems in arid and semi-arid regions are relatively fragile and sensitive to environmental changes [36], and believe that temperature and corresponding precipitation changes [34,35] may be the dominant factors causing the decrease in *R. nanum*'s habitat suitability in Central Asia. The growing pressure of potential evaporation caused by the increasing global warming will accelerate the decrease in soil moisture [72], which may trigger severe droughts [36] with increasing frequency in arid and semi-arid regions. The severe drought stress could lead to low richness patterns of plant species [73], which would likely offer less opportunities [74] for the adaption and distribution of species in this region.

The variations (e.g., seasonality changes) in environmental factors (i.e., temperature and precipitation [41]) have an obvious effect on flora plants in arid and semi-arid areas. Compared with the herbaceous species distributed in this region, the living strategies (e.g., deeper root systems [36]) of woody species play an important role in ensuring survival. However, they also limit the speciation during their life history and may reduce their potential distribution range [75]. In terms of the decrease in the habitat suitability and irregular shrinkage of habitat range of *R. nanum*, compared to species with similar habitats, we cannot state with certainty whether this is an individualistic response [41] by the species to climate change. When it comes to conservation, for *R. nanum* and other plants with medicinal properties [48], we hold the opinion that appropriate human intervention is necessary. Since the extinction of species due to severe climate change is most likely to occur in sensitive and fragile ecosystems (e.g., the Taklamakan Desert [36]), it is recommended that protection areas [76] are set up with suitable habitats for target species in order to prevent excessive digging [77] and illegal trade [48]. Additionally, to ensure the management and protection of agricultural crops of economic value, regular collection of field germplasm resources and corresponding artificial cultures are also necessary and deserve special attention.

4.3. Dominant Environmental Factors and Limitations in Predicting Species-Distribution Ranges

Our study succeeded in predicting the geographic distribution of species living in arid and semi-arid regions. Objectively speaking, however, our results had several limitations. Firstly, MaxEnt's niche simulation of the target species is based on the premise that the species would be extensively present in the sites where their environmental conditions (i.e., temperature and precipitation) have maximum similarity to the sites of known occurrence data [59,62,78]. Some geographic barriers (e.g., monsoons, mountains, and rivers) that affect the distribution ranges of species are typically ignored [20]. Secondly, we only obtained a few occurrence data in this study, and the prediction did not consider more relative environment factors (i.e., soil category, light, terrain). For more accurate predictions, future studies should consider as many relevant factors (e.g., abiotic and biotic factors) as possible and integrate them with more adaptable algorithms [53]. Thirdly, we predicted the potential distribution of *R. nanum* using MaxEnt alone. However, previous studies have shown that crosslinked models have had a higher prediction accuracy compared to a single model [68]. In addition, as an important complement and confirmation of our work, studies of the Last Glacial Maximum [45] and the Mid-Holocene will further refine species' response theory to climate change.

5. Conclusions

In this study, with the help of ArcGIS and the MaxEnt model, we successfully predicted *R. nanum*'s potential distribution and evaluated suitable habitats in the current period, the 2050s, and the 2070s under the RCP2.6 and RCP6.0 scenarios based upon the relevant environmental factors (i.e., temperature and precipitation). In contrast to the simulation of other species with abundant occurrence data, we categorized the selected occurrence data into the test layer and training layer manually, and realized MaxEnt's effective utilization of our small sample size. Our results suggest that the potential distribution habitat of *R. nanum*'s range was 81.7–97.3° E and 39.4–48.6° N in the current period. The key environmental factors that affected the distribution of *R. nanum* were bio1 and bio16. Under the two climate scenarios, the areas of suitable habitats (i.e., LH, MH, and HH) had different degrees of decreasing in both the 2050s and the 2070s, and RCP6.0 led to larger habitat range reductions than those found with RCP2.6. Moreover, the suitable habitats for *R. nanum* will shift toward the south with different distances in the future. In particular, we found that the irregular shrinkage of suitable habitats may be an ignored reason that led to the movement of habitat cores. To prevent the illegal digging and trade of agricultural crops with economic value, it is feasible to establish protection areas and management standards. We believe our study can provide a vital reference for the habitat simulation of species with a small sample size, and may provide supports for species conservation in arid and semi-arid regions.

Supplementary Materials: The following supporting information can be downloaded at: https://www.mdpi.com/article/10.3390/agriculture12050610/s1, Figure S1: The curves of the MaxEnt model (a) and bio-climate factors (b) and Jackknife test results (c); Figure S2: The terrain remote sensing image (300 km) of high suitable habitat cores; Table S1: Contribution of environmental variables used in MaxEnt; Table S2: Area changes of suitable habitats in different climate scenarios; Table S3: The area (10^4 km^2) and ratio (%) of habitat suitability change at different period interval situations.

Author Contributions: W.X. designed the study, collected occurrence data, and wrote the main body of the manuscript; W.X., S.Z. and T.Y. drew the map, tables, and figures; J.J. and J.C. acquired the funding and revised the manuscript and figures. All authors have read and agreed to the published version of the manuscript.

Funding: This study was financially supported by the Key Research and Development Program of Shaanxi Province (2021NY-006), Natural Science Foundation of China (31601987), China Agriculture Research System (CARS-34), Doctoral Fund of Henan Polytechnic University (B2019-4), and Deployment Program of the Chinese Academy of Sciences (KJZD-EW-TZ-G10).

Acknowledgments: We thank the national and international organizations that provided data and associated software for this work. The authors are indebted to Yixian Chen, Hanqi Liu, and Kai Jin for their kind help with ArcGIS operation and data processing. Junru Wang assisted in confirming species' Latin names. During the two and a half years of fieldwork, Weiwei Ren and numerous anonymous and kind individuals helped collect the species' field occurrence data. Moreover, we would like to thank Shuo Wang for her suggestions on the color selection of the full-text Figures. Importantly, the spiritual support of Runzhi Mao was the indispensable base for the completion of this work. We also thank the anonymous reviewers who made many important suggestions for improving the paper and others who contributed to the improvement of this manuscript.

Conflicts of Interest: The authors declare no conflict of interest.

References

1. Xie, Z. Ecogeographical distribution of the species from Rheum.L (polygonaceae) in China. In Proceedings of the Third National Symposium on the Conservation and Sustainable Use of Biological Diversity, Kunming, China, 11−13 December 1998.
2. Zhu, R. The Survey of Rhubarb Resources in Guyuan Country and Its Geographical Distribution. Master's Thesis, Lanzhou University, Lanzhou, China, 2011.
3. Zhe, W. A Study on the Pharmacophylogeny of the Rheum L. in China. Ph.D. Thesis, Chinese Academy of Medical Scienses & Peking Union Medical College, Beijing, China, 2011.
4. Shang, X.-F.; Zhao, Z.-M.; Li, J.-C.; Yang, G.-Z.; Liu, Y.-Q.; Dai, L.-X.; Zhang, Z.-J.; Yang, Z.-G.; Miao, X.-L.; Yang, C.-J.; et al. Insecticidal and antifungal activities of Rheum palmatum L. anthraquinones and structurally related compounds. *Ind. Crops Prod.* **2019**, *137*, 508–520. [CrossRef]
5. Yuan, S.; Jian, T.; Li, W.; Huang, Y. Extraction process optimization and activity assays of antioxidative substances from Rheum officinale. *J. Food Meas. Charact.* **2019**, *14*, 176–184. [CrossRef]
6. Zhang, Q.; Hu, F.; Guo, F.; Zhou, Q.; Xiang, H.; Shang, D. Emodin attenuates adenosine triphosphate-induced pancreatic ductal cell injury in vitro via the inhibition of the P2X7/NLRP3 signaling pathway. *Oncol. Rep.* **2019**, *42*, 1589–1597. [CrossRef] [PubMed]
7. Kalkan, S.; Otag, M.R.; Engin, M.S. Physicochemical and bioactive properties of edible methylcellulose films containing Rheum ribes L. extract. *Food Chem.* **2020**, *307*, 125524. [CrossRef]
8. Qin, T.; Wu, L.; Hua, Q.; Song, Z.; Pan, Y.; Liu, T. Prediction of the mechanisms of action of Shenkang in chronic kidney disease: A network pharmacology study and experimental validation. *J. Ethnopharmacol.* **2020**, *246*, 112128. [CrossRef]
9. Li, A.; Gao, Z.; Mao, Z. (Eds.) *Flora of China*; Science Press: Beijing, China, 1998; Volume 25.

10. Kwon, H.C.; Kim, T.Y.; Lee, C.M.; Lee, K.S.; Lee, K.K. Active compound chrysophanol of Cassia tora seeds suppresses heat-induced lipogenesis via inactivation of JNK/p38 MAPK signaling in human sebocytes. *Lipids Health* **2019**, *18*, 135. [CrossRef]
11. Dai, G.; Ding, K.; Cao, Q.; Xu, T.; He, F.; Liu, S.; Ju, W. Emodin suppresses growth and invasion of colorectal cancer cells by inhibiting VEGFR2. *Eur. J. Pharmacol.* **2019**, *859*, 172525. [CrossRef]
12. Sun, J.; Luo, J.-W.; Yao, W.-J.; Luo, X.-T.; Wei, Y.-H. Effect of emodin on gut microbiota of rats with acute kidney failure. *China J. Chin. Mater. Med.* **2019**, *44*, 758–764. [CrossRef]
13. Guo, Y.; Li, X.; Zhao, Z.; Wei, H.; Gao, B.; Gu, W. Prediction of the potential geographic distribution of the ectomycorrhizal mushroom Tricholoma matsutake under multiple climate change scenarios. *Sci. Rep.* **2017**, *7*, 46221. [CrossRef]
14. Phillips, S.J.; Anderson, R.P.; Schapire, R.E. Maximum entropy modeling of species geographic distributions. *Ecol. Model.* **2006**, *190*, 231–259. [CrossRef]
15. Abolmaali, S.M.-R.; Tarkesh, M.; Bashari, H. MaxEnt modeling for predicting suitable habitats and identifying the effects of climate change on a threatened species, *Daphne mucronata*, in central Iran. *Ecol. Inform.* **2018**, *43*, 116–123. [CrossRef]
16. Papes, M.; Gaubert, P. Modelling ecological niches from low numbers of occurrences: Assessment of the conservation status of poorly known viverrids (Mammalia, Carnivora) across two continents. *Divers. Distrib.* **2007**, *13*, 890–902. [CrossRef]
17. Tang, J.; Cheng, Y.; Luo, L.; Zhang, L.; Jiang, X. Maxent-based prediction of overwintering areas of *Loxostege sticticalis* in China under different climate changes scenarios. *Acta Ecol. Sin.* **2017**, *37*, 4852–4863. [CrossRef]
18. Araujo, M.B.; Pearson, R.G.; Thuiller, W.; Erhard, M. Validation of species-climate impact models under climate change. *Glob. Chang. Biol.* **2005**, *11*, 1504–1513. [CrossRef]
19. Peterson, A.T. Uses and Requirements of Ecological Niche Models and Related Distributional Models. *Biodivers. Inform.* **2006**, *3*, 59–72. [CrossRef]
20. Phillips, S.J.; Dudik, M. Modeling of species distributions with Maxent: New extensions and a comprehensive evaluation. *Ecography* **2008**, *31*, 161–175. [CrossRef]
21. Guisan, A.; Zimmermann, N.E. Predictive habitat distribution models in ecology. *Ecol. Model.* **2000**, *135*, 147–186. [CrossRef]
22. Elith, J.; Graham, C.H.; Anderson, R.P.; Dudik, M.; Ferrier, S.; Guisan, A.; Hijmans, R.J.; Huettmann, F.; Leathwick, J.R.; Lehmann, A.; et al. Novel methods improve prediction of species' distributions from occurrence data. *Ecography* **2006**, *29*, 129–151. [CrossRef]
23. Garcia, K.; Lasco, R.; Ines, A.; Lyon, B.; Pulhin, F. Predicting geographic distribution and habitat suitability due to climate change of selected threatened forest tree species in the. Philippines. *Appl. Geogr.* **2013**, *44*, 12–22. [CrossRef]
24. Illoldi-Rangel, P.; Sanchez-Cordero, V.; Peterson, A.T. Predicting distributions of Mexican mammals using ecological niche modeling. *J. Mammal.* **2004**, *85*, 658–662. [CrossRef]
25. Remya, K.; Ramachandran, A.; Jayakumar, S. Predicting the current and future suitable habitat distribution of *Myristica dactyloides Gaertn.* using MaxEnt model in the Eastern Ghats, India. *Ecol. Eng.* **2015**, *82*, 184–188. [CrossRef]
26. Zhang, K.; Yao, L.; Meng, J.; Tao, J. Maxent modeling for predicting the potential geographical distribution of two peony species under climate change. *Sci. Total Environ.* **2018**, *634*, 1326–1334. [CrossRef] [PubMed]
27. Ma, B.; Sun, J. Predicting the distribution of *Stipa purpurea* across the Tibetan Plateau via the MaxEnt model. *BMC Ecol.* **2018**, *18*, 10. [CrossRef] [PubMed]
28. Ciss, M.; Biteye, B.; Fall, A.G.; Fall, M.; Gahn, M.C.B.; Leroux, L.; Apolloni, A. Ecological niche modelling to estimate the distribution of Culicoides, potential vectors of bluetongue virus in Senegal. *BMC Ecol.* **2019**, *19*, 45. [CrossRef]
29. Guo, Y.L.; Li, X.; Zhao, Z.F.; Nawaz, Z. Predicting the impacts of climate change, soils and vegetation types on the geographic distribution of *Polyporus umbellatus* in China. *Sci. Total Environ.* **2019**, *648*, 1–11. [CrossRef] [PubMed]
30. Qin, A.; Liu, B.; Guo, Q.; Bussmann, R.W.; Ma, F.; Jian, Z.; Xu, G.; Pei, S. Maxent modeling for predicting impacts of climate change on the potential distribution of *Thuja sutchuenensis* Franch., an extremely endangered conifer from southwestern China. *Glob. Ecol. Conserv.* **2017**, *10*, 139–146. [CrossRef]
31. Root, T.L.; Price, J.T.; Hall, K.R.; Schneider, S.H.; Rosenzweig, C.; Pounds, J.A. Fingerprints of global warming on wild animals and plants. *Nature* **2003**, *421*, 57–60. [CrossRef] [PubMed]
32. Xu, W.; Jin, J.; Cheng, J. Predicting the Potential Geographic Distribution and Habitat Suitability of Two Economic Forest Trees on the Loess Plateau, China. *Forests* **2021**, *12*, 747. [CrossRef]
33. Xu, X.; Zhang, H.; Yue, J.; Xie, T.; Xu, Y.; Tian, Y. Predicting Shifts in the Suitable Climatic Distribution of Walnut (Juglans regia L.) in China: Maximum Entropy Model Paves the Way to Forest Management. *Forests* **2018**, *9*, 103. [CrossRef]
34. Bracegirdle, T.J.; Krinner, G.; Tonelli, M.; Haumann, F.A.; Naughten, K.A.; Rackow, T.; Roach, L.A.; Wainer, I. Twenty first century changes in Antarctic and Southern Ocean surface climate in CMIP6. *Atmos. Sci. Lett.* **2020**, *14*, e984. [CrossRef]
35. Kellermann, V.; van Heerwaarden, B. Terrestrial insects and climate change: Adaptive responses in key traits. *Physiol. Entomol.* **2019**, *44*, 99–115. [CrossRef]
36. Sun, Y.; Sun, Y.; Yao, S.R.; Akram, M.A.; Hu, W.G.; Dong, L.W.; Li, H.L.; Wei, M.H.; Gong, H.Y.; Xie, S.B.; et al. Impact of climate change on plant species richness across drylands in China: From past to present and into the future. *Ecol. Indic.* **2021**, *132*, 108288. [CrossRef]
37. Sun, J.; Qin, X.; Yang, J. The response of vegetation dynamics of the different alpine grassland types to temperature and precipitation on the Tibetan Plateau. *Environ. Monit. Assess.* **2016**, *188*, 20. [CrossRef] [PubMed]

38. Hughes, L. Biological consequences of global warming: Is the signal already apparent? *Trends Ecol. Evol.* **2000**, *15*, 56–61. [CrossRef]
39. Piao, S.L.; Liu, Q.; Chen, A.P.; Janssens, I.A.; Fu, Y.S.; Dai, J.H.; Liu, L.L.; Lian, X.; Shen, M.G.; Zhu, X.L. Plant phenology and global climate change: Current progresses and challenges. *Glob. Chang. Biol.* **2019**, *25*, 1922–1940. [CrossRef]
40. Iannella, M.; De Simone, W.; D'Alessandro, P.; Biondi, M. Climate change favours connectivity between virus-bearing pest and rice cultivations in sub-Saharan Africa, depressing local economies. *Peerj* **2021**, *9*, e12387. [CrossRef]
41. Alexander, J.M.; Chalmandrier, L.; Lenoir, J.; Burgess, T.I.; Essl, F.; Haider, S.; Kueffer, C.; McDougall, K.; Milbau, A.; Nunez, M.A.; et al. Lags in the response of mountain plant communities to climate change. *Glob. Chang. Biol.* **2018**, *24*, 563–579. [CrossRef]
42. Pepin, N.; Bradley, R.S.; Diaz, H.F.; Baraer, M.; Caceres, E.B.; Forsythe, N.; Fowler, H.; Greenwood, G.; Hashmi, M.Z.; Liu, X.D.; et al. Elevation-dependent warming in mountain regions of the world. *Nat. Clim. Chang.* **2015**, *5*, 424–430. [CrossRef]
43. Wang, L.; Wang, W.J.; Wu, Z.F.; Du, H.B.; Zong, S.W.; Ma, S. Potential Distribution Shifts of Plant Species under Climate Change in Changbai Mountains, China. *Forests* **2019**, *10*, 498. [CrossRef]
44. Shankhwar, R.; Bhandari, M.S.; Meena, R.K.; Shekhar, C.; Pandey, V.V.; Saxena, J.; Kant, R.; Barthwal, S.; Naithani, H.B.; Pandey, S.; et al. Potential eco-distribution mapping of *Myrica esculenta* in northwestern Himalayas. *Ecol. Eng.* **2019**, *128*, 98–111. [CrossRef]
45. Wang, S.; Xu, X.; Shrestha, N.; Zimmermann, N.E.; Wang, Z. Response of spatial vegetation distribution in China to climate changes since the Last Glacial Maximum (LGM). *PLoS ONE* **2017**, *12*, e0175742. [CrossRef] [PubMed]
46. Wu, Y.M.; Shen, X.L.; Tong, L.; Lei, F.W.; Zhang, Z.X. Impact of Past and Future Climate Change on the Potential Distribution of an Endangered Montane Shrub Lonicera oblata and Its Conservation Implications. *Forests* **2021**, *12*, 125. [CrossRef]
47. Tian, B.J.; Dong, X.Y. The Double-ITCZ Bias in CMIP3, CMIP5, and CMIP6 Models Based on Annual Mean Precipitation. *Geophys. Res. Lett.* **2020**, *47*, 11. [CrossRef]
48. Shrestha, B.; Tsiftsis, S.; Chapagain, D.J.; Khadka, C.; Bhattarai, P.; Kayastha Shrestha, N.; Alicja Kolanowska, M.; Kindlmann, P. Suitability of Habitats in Nepal for Dactylorhiza hatagirea Now and under Predicted Future Changes in Climate. *Plants* **2021**, *10*, 467. [CrossRef]
49. van Vuuren, D.P.; Stehfest, E.; den Elzen, M.G.J.; Kram, T.; van Vliet, J.; Deetman, S.; Isaac, M.; Goldewijk, K.K.; Hof, A.; Beltran, A.M.; et al. RCP2.6: Exploring the possibility to keep global mean temperature increase below 2 degrees C. *Clim. Chang.* **2011**, *109*, 95–116. [CrossRef]
50. O'Neill, B.C.; Tebaldi, C.; van Vuuren, D.P.; Eyring, V.; Friedlingstein, P.; Hurtt, G.; Knutti, R.; Kriegler, E.; Lamarque, J.F.; Lowe, J.; et al. The Scenario Model Intercomparison Project (ScenarioMIP) for CMIP6. *Geosci. Model Dev.* **2016**, *9*, 3461–3482. [CrossRef]
51. Rana, S.K.; Rana, H.K.; Ghimire, S.K.; Shrestha, K.K.; Ranjitkar, S. Predicting the impact of climate change on the distribution of two threatened Himalayan medicinal plants of Liliaceae in Nepal. *J. Mt. Sci.* **2017**, *14*, 558–570. [CrossRef]
52. Xin, X.G.; Zhang, L.; Zhang, J.; Wu, T.W.; Fang, Y.J. Climate Change Projections over East Asia with BCC_CSM1.1 Climate Model under RCP Scenarios. *J. Meteorol. Soc. Jpn.* **2013**, *91*, 413–429. [CrossRef]
53. Breiner, F.T.; Guisan, A.; Bergamini, A.; Nobis, M.P. Overcoming limitations of modelling rare species by using ensembles of small models. *Methods Ecol. Evol.* **2015**, *6*, 1210–1218. [CrossRef]
54. Hu, X.; Wu, F.; Guo, W.; Liu, N. Identification of Potential Cultivation Region fie *Santalum album* in China by the MaxEnt Ecologic Niche Model. *Sci. Silvae Sin.* **2014**, *50*, 27–33. [CrossRef]
55. Fu, G.Q.; Xu, X.Y.; Ma, J.P.; Xu, M.S.; Jiang, L.; Ding, A.Q. Responses of Haloxylon ammodendron potential geographical distribution to the hydrothermal conditions under MaxEnt model. *Pratacult. Sci.* **2016**, *33*, 2173–2179. [CrossRef]
56. Elith, J.; Phillips, S.J.; Hastie, T.; Dudik, M.; Chee, Y.E.; Yates, C.J. A statistical explanation of MaxEnt for ecologists. *Divers. Distrib.* **2011**, *17*, 43–57. [CrossRef]
57. Lv, W.; Li, Z.; Wu, X.; Ni, W.; Qv, W. *Maximum Entropy Niche-Based Modeling (Maxent) of Potential Geographical Distributions of Lobesia Botrana (Lepidoptera: Tortricidae) in China*; Springer: Berlin/Heidelberg, Germany, 2012.
58. Hernandez, P.A.; Graham, C.H.; Master, L.L.; Albert, D.L. The effect of sample size and species characteristics on performance of different species distribution modeling methods. *Ecography* **2006**, *29*, 773–785. [CrossRef]
59. Guisan, A.; Graham, C.H.; Elith, J.; Huettmann, F.; Distri, N.S. Sensitivity of predictive species distribution models to change in grain size. *Divers. Distrib.* **2007**, *13*, 332–340. [CrossRef]
60. Graham, M.H. Confronting multicollinearity in ecological multiple regression. *Ecology* **2003**, *84*, 2809–2815. [CrossRef]
61. Vaughan, I.P.; Ormerod, S.J. The Continuing Challenges of Testing Species Distribution Models. *J. Appl. Ecol.* **2005**, *42*, 720–730. [CrossRef]
62. Xu, W.; Sun, H.; Jin, J.; Cheng, J. Predicting the Potential Distribution of Apple Canker Pathogen (Valsa mali) in China under Climate Change. *Forests* **2020**, *11*, 1126. [CrossRef]
63. Rana, R.S.; Singh, M.; Pathania, R.; Upadhyay, S.K.; Kalia, V. Impact of changes in climatic conditions on temperate fruit production of Himachal Pradesh. *Mausam* **2017**, *68*, 655–662. [CrossRef]
64. Yan, X.; Zhang, Q.; Yan, X.; Wang, S.; Ren, X.; Zhao, F. An Overview of Distribution Characteristics and Formation Mechanisms in Global Arid Areas. *Adv. Earth Sci.* **2019**, *34*, 826–841. [CrossRef]

65. Ribeiro, M.M.; Roque, N.; Ribeiro, S.; Gavinhos, C.; Castanheira, I.; Quinta-Nova, L.; Albuquerque, T.; Gerassis, S. Bioclimatic modeling in the Last Glacial Maximum, Mid-Holocene and facing future climatic changes in the strawberry tree (*Arbutus unedo* L.). *PLoS ONE* **2019**, *14*, 15. [CrossRef]
66. Murray, K.A.; Rosauer, D.; McCallum, H.; Skerratt, L.F. Integrating species traits with extrinsic threats: Closing the gap between predicting and preventing species declines. *Proc. Biol. Sci.* **2011**, *278*, 1515–1523. [CrossRef]
67. Lenoir, J.; Gegout, J.C.; Marquet, P.A.; de Ruffray, P.; Brisse, H. A Significant Upward Shift in Plant Species Optimum Elevation During the 20th Century. *Science* **2008**, *320*, 1768–1771. [CrossRef] [PubMed]
68. Yu, F.Y.; Wang, T.J.; Groen, T.A.; Skidmore, A.K.; Yang, X.F.; Ma, K.P.; Wu, Z.F. Climate and land use changes will degrade the distribution of Rhododendrons in China. *Sci. Total Environ.* **2019**, *659*, 515–528. [CrossRef] [PubMed]
69. Chen, I.C.; Hill, J.K.; Ohlemuller, R.; Roy, D.B.; Thomas, C.D. Rapid Range Shifts of Species Associated with High Levels of Climate Warming. *Science* **2011**, *333*, 1024–1026. [CrossRef] [PubMed]
70. Li, J.; McCarthy, T.M.; Wang, H.; Weckworth, B.V.; Schaller, G.B.; Mishra, C.; Lu, Z.; Beissinger, S.R. Climate refugia of snow leopards in High Asia. *Biol. Conserv.* **2016**, *203*, 188–196. [CrossRef]
71. Hu, Z.-j.; Zhang, Y.-l.; Yu, H.-b. Simulation of *Stipa purpurea* disttibution pattern on Tibetan Plateau based on MaxEnt model. *Chin. J. Appl. Ecol.* **2015**, *26*, 505–511. [CrossRef]
72. Sato, T.; Kimura, F.; Kitoh, A. Projection of global warming onto regional precipitation over Mongolia using a regional climate model. *J. Hydrol.* **2007**, *333*, 144–154. [CrossRef]
73. Brown, S.C.; Wigley, T.M.L.; Otto-Bliesner, B.L.; Rahbek, C.; Fordham, D.A. Persistent Quaternary climate refugia are hospices for biodiversity in the Anthropocene. *Nat. Clim. Change* **2020**, *10*, 244–248. [CrossRef]
74. Ron, R.; Fragman-Sapir, O.; Kadmon, R. The role of species pools in determining species diversity in spatially heterogeneous communities. *J. Ecol.* **2018**, *106*, 1023–1032. [CrossRef]
75. Albuquerque, F.S.; Olalla-Tarraga, M.A.; Montoya, D.; Rodriguez, M.A. Environmental determinants of woody and herb plant species richness patterns in Great Britain. *Ecoscience* **2011**, *18*, 394–401. [CrossRef]
76. Kaky, E.; Gilbert, F. Using species distribution models to assess the importance of Egypt's protected areas for the conservation of medicinal plants. *J. Arid. Environ.* **2016**, *135*, 140–146. [CrossRef]
77. Bhattarai, P.; Pandey, B.; Gautam, R.K.; Chhetri, R. Ecology and Conservation Status of Threatened Orchid Dactylorhiza hatagirea (D. Don) Soo in Manaslu Conservation Area, Central Nepal. *Am. J. Plant Sci.* **2014**, *5*, 3483–3491. [CrossRef]
78. Guisan, A.; Thuiller, W. Predicting species distribution: Offering more than simple habitat models. *Ecol. Lett.* **2005**, *8*, 993–1009. [CrossRef] [PubMed]

agriculture

MDPI

Article

North Expansion of Winter Wheat Planting Area in China under Different Emissions Scenarios

Maowei Wu [1], Yang Xu [1], Jingyun Zheng [1,2] and Zhixin Hao [1,2,*]

1 Key Laboratory of Land Surface Pattern and Simulation, Institute of Geographic Sciences and Natural Resources Research, Chinese Academy of Sciences, Beijing 100101, China; wumw.13b@igsnrr.ac.cn (M.W.); xy7912@stu.syau.edu.cn (Y.X.); zhengjy@igsnrr.ac.cn (J.Z.)
2 University of Chinese Academy of Sciences, Beijing 100049, China
* Correspondence: haozx@igsnrr.ac.cn

Abstract: Suitable planting areas for winter wheat in north China are expected to shift northwardly due to climate change, however, increasing extreme events and the deficient water supply are threatening the security of planting systems. Thus, based on predicted climate data for 2021–2050 under the Shared Socioeconomic Pathways (SSP1-2.6, SSP3-7.0, and SSP5-8.5) emission scenarios, as well as historical data from 1961–1990, we use four critical parameters of percentages of extreme minimum temperature years (POEMTY), first day of the overwintering period (FD), sowing date (SD), and precipitation before winter (PBW), in order to determine the planting boundary of winter wheat. The results show that the frequency of extreme minimum temperature occurrences is expected to decrease in the North winter wheat area, which will result in a northward movement of the western part of northern boundary by 73, 94, and 114 km on average, in addition to FD delays ranging from 6.0 to 10.5 days. Moreover, agrometeorological conditions in the Huang-Huai winter wheat area are expected to exhibit more pronounced changes than the rest of the studied areas, especially near the southern boundary, which is expected to retreat by approximately 213, 215, and 233 km, northwardly. The north boundary is expected to move 90–140 km northward. Therefore, the change in southern and northern boundaries will lead the potential planting areas of the entire North winter wheat area to increase by 10,700 and 28,000 km^2 on average in the SSP3-7.0 and SSP5-8.5 scenarios, respectively, but to decrease by 38,100 km^2 in the SSP1-2.6 scenario; however, the lack of precipitation remains a limitation for extending planting areas in the future.

Keywords: climate change; agriculture; food security; planting boundary; winter wheat

Citation: Wu, M.; Xu, Y.; Zheng, J.; Hao, Z. North Expansion of Winter Wheat Planting Area in China under Different Emissions Scenarios. *Agriculture* **2022**, *12*, 763. https://doi.org/10.3390/agriculture12060763

Academic Editor: Danilo Scordia

Received: 29 April 2022
Accepted: 25 May 2022
Published: 27 May 2022

1. Introduction

Wheat is the third-largest crop in the world, and provides 20% of human dietary protein and caloric intake globally [1]. Its broad adaptability to climatic conditions and variety diversity accounts for its unparalleled cultivation range, from 67° N in Scandinavia and Russia, to 45° S in Argentina [2]. As stated by the Food and Agriculture Organization of the United Nations (FAO), the production of wheat has increased from 222 million tons in 1961 to 732 million tons in 2013 [3]. Nonetheless, with the continually increasing global mean surface temperature since the Industrial Revolution [4], climate change and increasing extreme climate events disturb the agricultural ecosystem, and result in changes in local suitable agrometeorological conditions, which affects wheat growth. Thus, climate change is expected to substantially expand the suitable regions for winter wheat cultivation in North America northwardly into Canada, and extend the fall-sown spring wheat region northwardly and eastwardly [5,6]. In northern Europe, suitable areas for winter wheat cultivation have expanded almost into the Arctic Circle (66.5° N) [7]. Crops in southern Europe, such as maize, sunflower, and soya beans, could also expand further north and occur at higher altitudes [8,9]. Although warmer temperatures benefit wheat cultivation at high latitudes by reducing cold-temperature constraints on agricultural development, typical wheat planting areas in the tropics will be gradually reduced [10]. Therefore, the negative impacts of climate change on global wheat production will likely become a critical issue to address in the future. As predicted by Balkovic et al. [11], global wheat production under current conventional management methods would decrease under all Representative Concentration Pathways (RCPs) by 37–52 Mt, and by 54–103 Mt in the 2050s and 2090s, respectively.

China is the largest wheat producer in the world, possessing 11% of the global wheat cultivation areas and contributing to 17% of global wheat production [12]. Winter wheat accounts for more than 90% of the total wheat yields in China [13]. The winter wheat cultivation zone climatic indices in China are based on tolerable low temperatures for winter wheat growth [14,15], as varying degrees of freeze damage during the overwintering stage may have different negative impacts on winter survival rate, crop vigor, and therefore final yields [16,17]. However, temperature change spatial distribution patterns in China exhibit strong similarities with global changes, with temperature increases occurring throughout the entire region, albeit more noticeably in the northern region [18]. This northward expansion over the past few decades has been largely attributed to the longer growing seasons and decreased temperature-related constraints on crop growth that have resulted from warmer temperatures. Increasing attention has begun to focus on the changes in winter wheat cultivation distribution and possible planting boundaries in China, and substantial progress in the characterization of this phenomenon has been achieved. For instance, an observed significant relationship is that the sowing date is delayed for 4 days when the temperature increases by 1 °C [19]. The planting boundaries for different winter wheat varieties in China moved significantly northward in 1981–2010, compared to the 1951–1980 period [20]. Moreover, the overall potential planting areas for winter wheat increased as well. The strong winterness-variety winter wheat had the largest change in both the movement of the planting boundary and in planting area increase. Hao et al. analyzed changes in suitable winter wheat planting boundaries along its production areas in China under the RCP4.5 scenario, and predicted a northward shift of the northern winter wheat boundary by 1–2° N [21]. Planting area increased 1420 km^2 in the year 2019 compared to that in the year 2000, as measured by Landsat image mapping [22].

So far, most previous studies have focused on the northern boundary shifts, rather than on those occurring in both the southern and northern boundaries in the coming decades. Moreover, some studies have been limited to the provincial or regional scale, and thus cannot represent the general impact of climate change on agriculture. Therefore, this study aimed to assess the impacts of climate change on agrometeorological index trends associated with wheat safe overwintering in the winter wheat region of China under different Shared Socioeconomic Pathways of low (SSP1-2.6), medium-high (SSP3-7.0), and high (SSP5-8.5) emission scenarios, as well as further, in order to explore potential wheat planting boundaries in the future. Thus, the findings of this study could provide reference for other agriculture planting regions and scientific data for climate change adaptation and responses in food security.

2. Materials and Methods

2.1. Study Area

The winter wheat planting region in China is divided into the North winter wheat area and the South winter wheat area border by the Huaihe River, based on the geographical environment, natural conditions, climatic factors, farming system, and wheat varieties [14]. In this study, we focus on the North winter wheat area (I), which is located south of the Great Wall and north of the Huaihe River, and includes the Shandong, Henan, Hebei, Shanxi, Shaanxi, southeastern Gansu, and the northern parts of Jiangsu and Anhui Provinces (Figure 1). Crops ripen twice per year, or three times every two years [23,24]. The climatic conditions are suitable for winter or strong winter varieties of winter wheat growing, with an annual mean temperature of 9–15 °C, extreme minimum temperature of −30.0 to −13.2 °C from north to south, and annual active accumulated temperature ranges of 2750−4900 °C. Area I can be divided into the Northern winter wheat subregion (Ia) and

Huang-Huai winter wheat subregion (Ib), according to latitudes, terrains, and climatic conditions. For subregion Ia, the winter wheat is sown from the end of September to early-October, and harvested until mid-June to late-June, but for subregion Ib, the sowing date is the same as for Ia, and the harvest time is advanced to early-June. In addition, since the possible winter wheat planting boundaries might move northward as a result of climate warming, we also plot the Spring wheat area (II) in the north of area I in Figure 1.

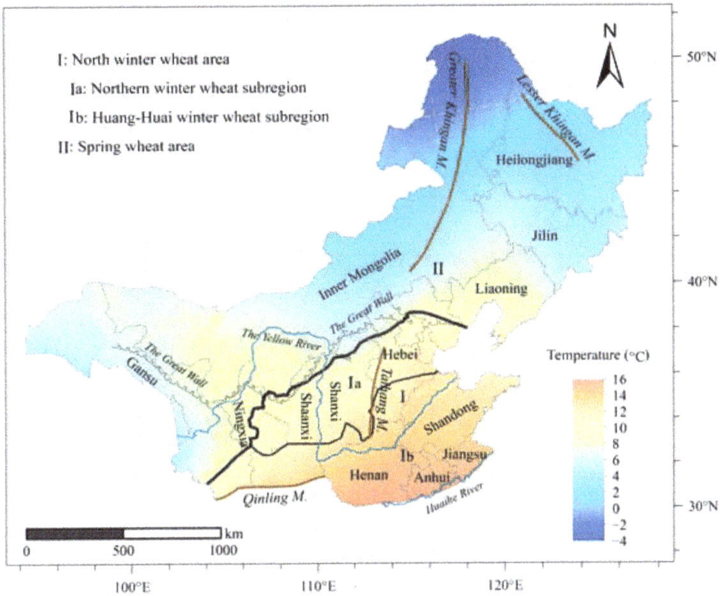

Figure 1. Map of North winter wheat area; shaded colors indicate the temperature spatial pattern (mean value from 1991–2019), using the daily meteorological data set of basic meteorological elements of China National Surface Weather Station (V3.0).

2.2. Data

The Shared Socioeconomic Pathways (SSPs) are five distinctly different scenarios determined by an international team of climate scientists, economists and energy systems modelers, and adopted by the Sixth Assessment Report of the Intergovernmental Panel on Climate Change, which describe how global societies, populations, and economies will change under the context of climate change adaptation and mitigation [25]. The SSPs provide a framework to describe alternative socioeconomic developments between and within countries, and represent five scenarios, including sustainable pathway (SSP1), middle pathway (SSP2), regional rivalry pathway (SSP3), divided pathway (SSP4), and fossil-fueled development pathway (SSP5). For the analysis of socioeconomic and climate systems, this study combines SSPs and RCPs to form a set of future global change scenarios determined by socioeconomics, emissions, climate response, and anthropogenic forcing of climate systems, which makes future scenarios more reasonable for social development [26]. Three combined SSP-RCP scenarios are selected in this study for future wheat overwintering indices in model prediction: (1) a low forcing and sustainability pathway (SSP1-2.6), which represents the combined scenario of a lower challenge of mitigation with low radiation forcing which peaks at 2.6 W/m^2 by 2100; (2) a new forcing scenario (SSP3-7.0), which represents a combination of high social vulnerability and relatively high radiative forcing that stabilizes at 7.0 W/m^2 by 2100; and (3) a high forcing scenario (SSP5-8.5), which represents a highly energy-intensive socioeconomic development pathway whereby radiative forcing reaches 8.5 W/m^2 by 2100.

Climate scenario data for 2021–2051 and historical data for 1961–1991 were provided by the Inter-Sectoral Impact Model Intercomparison Project (ISI–MIP; https://data.isimip.org/search/tree/ISIMIP3b/; accessed on 1 May 2021). The data included five climate model simulation outputs: GFDL-ESM4 (NOAA-GFDL), UKESM1-0-LL (MOHC), MPI-ESM1-2-HR (MPI-M), IPSL-CM6A-LR (IPSL), and MRI-ESM2-0 (MRI), which have been bias-corrected based on the raw data from the five aforementioned Coupled Model Intercomparison Project (CMIP6) models. The monthly mean values of the simulated data were adjusted to match the observed data, in order to preserve the long-term absolute or relative trends of the simulated data. Afterward, these bias-adjusted data were bilinearly interpolated at 0.5° × 0.5° spatial resolution [27]. In this study, the data used were daily precipitation, daily mean temperature, and daily minimum temperature. We used the daily gridded climate data set from 1961 to 1990 as observation data, in order to evaluate the simulation capability of the climate model. The daily gridded climate data set with spatial resolution of 0.5° × 0.5° is converted from observation data, and can be downloaded from the National Meteorological Information Center (http://101.200.76.197/data/detail/dataCode/SURF_CLI_CHN_PRE_DAY_GRID_0.5.html; accessed on 1 January 2021).

The study area includes approximately 1425 grid points. A Taylor diagram is used to evaluate the simulation capability of extreme minimum temperature (≤−22 °C) days (EMTD), accumulated temperature (AT), and accumulated precipitation (AP) from 1 September to 31 December over the entire study area (Figure 2). The spatial correlation coefficients of EMTD, AT, and AP between the multi-model ensemble (MME) and the observation data were 0.968, 0.974 and 0.970, respectively, and all coefficients passed the significance test at the 99% confidence level. The respective ratios of standard deviations between MME and observation data were 1.070, 1.036 and 1.018, and the normalized root mean square differences (RMSD) between simulation and observation data for all three climatic indices were less than 0.5. These results indicate that MME can effectively capture the temperature and precipitation characteristics of the study area. Therefore, we used MME to analyze the meteorological conditions related to wheat overwintering under different emissions scenarios relative to 1961–1990.

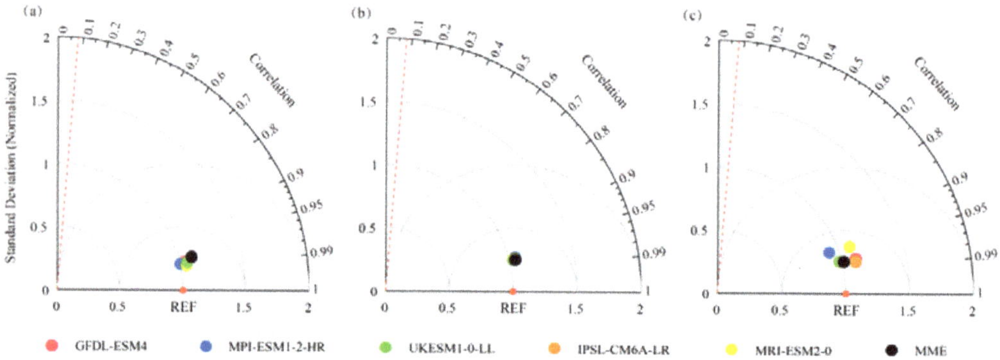

Figure 2. Taylor diagrams of (**a**) EMTD (≤−22 °C), (**b**) AT, and (**c**) AP from September 1st to December 31st over the entire study area for the period 1961–1990. The red dotted line corresponds to the 99% confidence level; REF: observation; azimuthal position: spatial correlation coefficient; radial distance: ratio of standard deviation; distance from REF point: normalized root mean square difference.

2.3. Methodology

In the North winter wheat Area, whether the winter wheat can be grown safely or not is determined by climate conditions during the winter, and thus, four critical indices with significant effects on winter wheat safe planting boundaries are used here: (1) the percentage of extreme minimum temperature years occurrence, (2) the first day of the overwintering period, (3) the sowing date, and (4) the precipitation before overwintering. The percentage of extreme minimum temperature years occurrence determines whether winter wheat can resist freezing injury in severe winter. The first day of the overwintering period and sowing date both account for the accumulated temperature of winter wheat prior to the overwintering stage, as well as its overwintering ability; if the accumulated temperature between sowing and overwintering is too low or high, weak wheat seedlings may encounter difficulties in surviving through winter. The precipitation before overwintering influences the strength of wheat seedlings before entering the overwintering stage. Here, since the sowing dates are influenced by complex factors such as wheat variety and terrain (e.g., mountain area microclimates), we obtain the sowing date through the required accumulative temperature from sowing date to the overwinter period (e.g., 450, 550, 700 °C), which can ensure winter wheat seedling survival during the winter. It is also worth noting that the accumulated temperature was calculated until December 31st over the south region of the whole winter wheat area, where no obvious overwintering period is observed. The calculations of the four indices were performed as follows:

(i) Percentage of extreme minimum temperature (−22 °C) years occurring(POEMTY) in a given period:

$$\text{POEMTY} = \sum_{k=1}^{ty} I\{tmin_k \leq -22\}/ty \tag{1}$$

where ty is the total years of a study period, $tmin_k$ is the daily minimum temperature for a certain year k, and I is a sign function, which is 1 if $tmin_k$ is lower than or equal to −22 °C; the specified −22 °C is the lowest temperature that winter wheat can endure safely through the winter [28,29].

(ii) First day of the overwintering period (FD): defined as the first day at which the daily mean temperature (i.e., based on a 5-day moving average) was below 0 °C [30].

(iii) Sowing date (SD): the date when the cumulative temperature reaches the required accumulated temperature before winter (Ta), calculating back from the FD. Ta (i.e., 450, 550, 700 °C) is positive accumulated temperature calculation for daily average temperature greater than 0 before overwintering, according to the method described by Cui et al. [31].

(iv) Precipitation before overwintering (PBW): total precipitation from the SD to the FD.

For each grid, the FD and SD are determined with an 80% guarantee rate (i.e., an agricultural climate criterion), and the PBW was calculated via the mean value over the referenced period and forecasting period.

3. Results

3.1. Percentage of Extreme Minimum Temperature (−22 °C) Years Occurrence

The lowest critical temperature for winter wheat cultivation in the regions near the north boundary (along the Great Wall) was demonstrated to be −22 °C [29]. Therefore, POEMTY is among the most important indices to reflect climatic conditions during the winter wheat overwintering period, which directly impacts seedling survival rate. Figure 3 illustrates the spatial distribution of POEMTY for 1961–1990 and 2021–2050 under SSP1-2.6, SSP3-7.0, and SSP5-8.5 scenarios. The POEMTYs are mainly concentrated in 0–20% and 80–100% intervals, and the 0% areas are marked with white lattices in Figure 3, where no lower than −22 °C extreme minimum temperatures occurred. However, values of POEMTY above 50% have cold injury risk for agriculture, and farmers would no longer choose these areas to plant winter wheat. Therefore, colored areas with above 50% of POEMTY were defined as high-risk region for winter wheat growing, and low-risk (0 < POEMTY ≤ 20%) and medium-risk regions (20% < POEMTY ≤ 50%) were also defined, as shown in Figure 3. In particular, the medium-risk region was considered as the potential extended winter wheat area along the northern boundary.

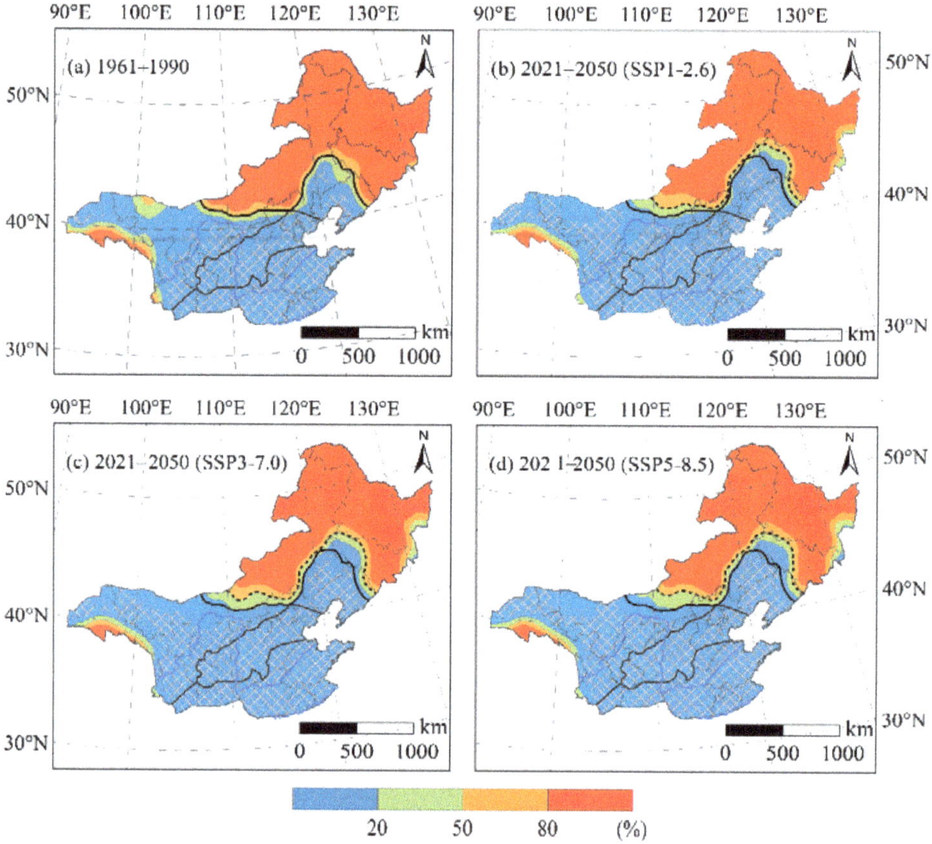

Figure 3. Spatial distribution of POEMTYs for the 1961–1990 and the 2021–2050 periods. (**a**) Historical period, (**b**) SSP1-2.6, (**c**) SSP3-7.0, and (**d**) SSP5-8.5 scenarios. White lattices indicate that no extreme minimum temperature ≤ -22 °C occurred. The black solid and dotted lines represent the northern border of winter wheat cultivation in the 1961–1990 and 2021–2050 periods, respectively.

During the historical period 1961–1990 (Figure 3a), the POEMTY in the North winter wheat area (I) was below 20%, which meets the 80% assurance rate of guaranteed minimum temperature. In fact, the potential winter wheat planting area northern boundary was further north than the current boundary in the eastern and western portions of area I. At the end of the 20th century, the experiments of winter wheat northward migration were successfully carried out in Liaoning Province and Inner Mongolia [32].

For 2021−2050, the potential safe overwintering areas for winter wheat are projected to expand under the SSP1-2.6, SSP3-7.0, and SSP5-8.5 scenarios compared to the 1961−1990 period (Figure 3b–d), and the high-risk area will move northward as a result of intended climate warming. Under the SSP1-2.6 scenario, the safe overwintering areas for winter wheat will increase by 11.8% relative to 1961–1990. The risk-free area will increase by 24.7%, and the wheat in 94.5% of the current area I will no longer experience extremely low temperatures. The northern boundary of the potential planting area in northeastern China (i.e., the eastern region) will extend to the western part of Jilin Province, and the boundary in central Inner Mongolia (i.e., the western region) will move slightly north as well. The western part of the northern boundary could move northwardly on average by approximately 73 km, and the northernmost tip of the eastern region could move 111 km northward. Under the SSP3-7.0 scenario, the low-risk area will increase by 14.7%, and the potential northern planting boundary will move northwardly by approximately 94 km on average in the western region, and by 152 km in the northernmost tip of the eastern region. Moreover, 99.0% of the current area I will no longer experience extreme low temperatures, and the risk-free area will increase by 32.5%. Under the SSP5-8.5 scenario, the area for safe wheat overwintering will have an increase of 16.6%, the risk-free area will increase by 34.8%, and the northern boundaries in the western and eastern regions will both substantially move to the north (i.e., 114 and 158 km, respectively).

3.2. First Day of the Overwintering Period

During the 1961–1990 reference period (Figure 4a), the FD varied from mid-November to late December in most of the North winter wheat area (I), while moving from north to south along the latitudinal gradient. In the Spring wheat area (II), the FD occurred in early November in the southern part of northeastern China and western Inner Mongolia, and in October in the northern and western parts of northeastern China. However, most of the northern parts of area II were unsuitable for winter wheat, given the low-temperature limitations mentioned in Section 3.1. The FD was from early November to early December in the northern winter wheat subregion (Ia), and concentrated in mid-to-late December in the Huang-Huai winter wheat subregion (Ib). However, some areas along the Huaihe River in the southernmost part of subregion (Ib) did not exhibit noticeable overwintering periods, since they are in the temperate-subtropical transition region, which is represented with dark grey dots in Figure 4a.

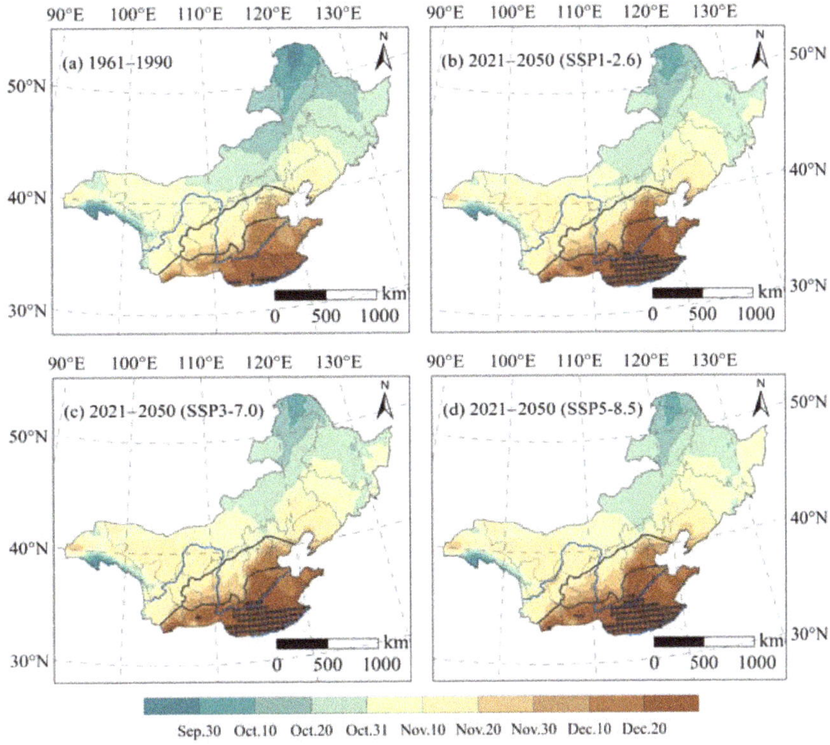

Figure 4. Spatial distribution of FDs for the 1961–1990 and 2021–2050 periods. (**a**) Historical period, (**b**) SSP1-2.6, (**c**) SSP3-7.0, and (**d**) SSP5-8.5 scenarios. The grid points marked by dark grey dots indicate that there are no obvious overwintering periods.

For the period of 2021–2050, the FD exhibited a relatively consistent spatial distribution under the three SSP-RCP scenarios, albeit with delays towards late November and later (Figure 4b–d). Taking the SSP1-2.6 scenario as an example, the winter wheat in area II will enter the overwintering period as early as early October, and the boundaries of the overwintering periods will move northwardly. Moreover, the largest changes will occur between late October to mid-November in northeastern China, and from early- to mid-November in western Inner Mongolia. In subregion Ia, the FD will range from mid-November to late December from north to south, respectively. Specifically, it is mainly from mid-November to early December in the western high elevations, and from mid-November to late December in the east. Winter wheat will begin to enter the overwintering period in early to mid-December in the western regions, and in late-December to later in most of the eastern plains. There will be no obvious overwintering periods in the south part of subregion Ib, and the areas without obvious overwintering periods will increase and account for approximately 36.7%, 37.0%, and 39.7% in subregion Ib under the SSP1-2.6, SSP3-7.0, and SSP5-8.5 scenarios, respectively, and will be mainly distributed in some of the southern provinces, such as Shandong, Henan, Jiangsu, and Anhui, which will cause the southern boundaries of area I to move approximately 213, 215, and 233 km to the north (i.e., as determined by the SSP1-2.6, SSP3-7.0, and SSP5-8.5 scenarios, respectively). Compared with the reference period, the FD is delayed by 6.0, 6.6, and 7.1 days in area II; by 8.9, 7.7, and 10.4 days in subregion Ia; and by 10.5, 9.0, and 10.3 days in subregion Ib under the SSP1-2.6, SSP3-7.0, and SSP5-8.5 scenarios, respectively. Importantly, the MME projects higher temperatures for more than half of the days from overwintering periods in 63.0% of the grids over area I under the SSP1-2.6 scenario compared to the SSP3-7.0 scenario. Therefore, the changing trends of some overwintering meteorological indices in this region are larger under the SSP1-2.6 scenario, such as the area without obvious overwintering periods and the FD delays.

3.3. *Sowing Date*

The accumulated temperature (AT) before entering the overwintering period is important factor affecting the ability of winter wheat to resist freezing conditions, and the SD in this study was calculated when the pre-winter positive accumulated temperature reached a certain value. According to previous studies [21], the AT for viable seedlings was approximately 570–720 °C, and the lower limit for safe overwintering of wheat was approximately 420 °C. However, wheat seedlings tended to grow excessively before winter if the AT was excessive, which also led to poor cold resistance ability. In this study, values of 450, 550, and 700 °C were used as references to analyze SD changes in the North winter wheat area (I).

Figure 5 illustrates the spatial distribution features of SD during the 1961–1990 historical period, as well as under the SSP1-2.6, SSP3-7.0, and SSP5-8.5 scenarios in 2021–2050 with AT values of 450, 550, and 700 °C, respectively. The SD exhibited a gradually delayed spatial distribution pattern from south to north, but occurred earlier if the AT increased for each scenario. In 1961–1990, when the AT was 450 °C, the SD began in mid-September and earlier in the spring wheat area (II), in August in the Greater Khingan Mountains and Lesser Khingan Mountains, and in mid-September in the southern part of the northeast plain and western part of Inner Mongolia. The SD was from mid-September to early-October in the Qinling Mountain area, the western Taihang Mountain, the northern part of the North China Plain, and the Liaoning area; mid-October in the lower and middle reaches of the Yellow River; and late October in the southern part of the North China Plain. The SD spatial distributions for values of 550 and 700 °C were similar to those for 450 °C; however, the SD advanced as AT requirements increased, as illustrated in Figure 5(a2,a3). Additionally, the SDs of different regions changed to varying degrees over the 1961–1990 period, exhibiting average delays of 0.3–0.4 days/decade in area II, 0.8–0.9 days/decade in the Northern winter wheat subregion (Ia), 0.6–0.8 days/decade in the Huang-Huai winter wheat subregion (Ib), and 0.7–0.8 days/decade in the entirety of area I.

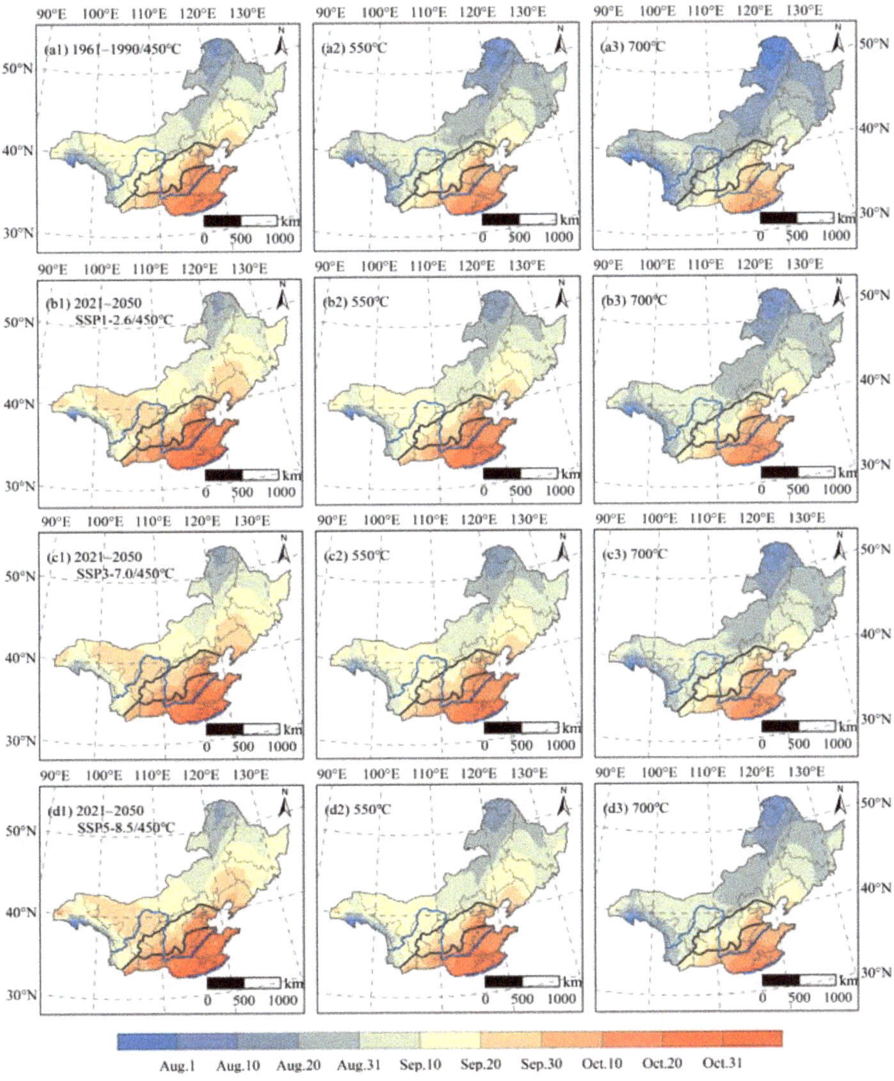

Figure 5. Spatial distribution of SD in the 1961–1990 and 2021–2050 periods for accumulated temperature before overwintering periods at 450, 550, and 700 °C. (**a1–a3**) historical period, (**b1–b3**) SSP1-2.6, (**c1–c3**) SSP3-7.0, and (**d1–d3**) SSP5-8.5 scenarios.

For the period of 2021–2050, the SDs exhibited relatively consistent spatial distributions under the three emissions scenarios for 450, 550, and 700 °C, and all showed delayed dates compared with 1961–1990. For example, when the AT reaches 450 °C under the SSP1-2.6 scenario, the SDs are delayed, from mid and late August in the reference period, to late August and early September in most areas of the Greater Khingan Mountains and the Lesser Khingan Mountains, and from early-mid September to mid-to-late September in the northern boundary and its surrounding areas. In subregion Ia, the SDs range from mid-September and early October, to late September and early October in the western regions, and concentrate in early and mid-October in the eastern regions. In subregion Ib, the SDs are delayed from late September and early October, to early and mid-October in the areas surrounding Qinling Mountain, and from late October to November in the areas along the Huaihe River. On average, the SD is delayed by 8.1 days in area II, 7.5 days in subregion Ia, and 8.7 days in subregion Ib. For the AT of 550 and 700 °C, the SD spatial distributions are shown in Figure 5(b2,b3), with a delay of 7.7–8.8 days and 7.8–8.9 days for the three sub-districts. The SD would be further delayed under the SSP3-7.0 and SSP5-8.5 scenarios, with an average delay of 7.5 (450 °C) to 8.0 (700 °C) days for SSP3-7.0, and 8.8 (450 °C) to 9.1 (700 °C) days for SSP5-8.5 across the entirety of area I. Furthermore, the SD trends over a given region were similar for different AT requirements under the same emissions scenario over the predicted 2021–2050 period, and the SD delay rate accelerated for all conditions compared to 1961—1990. The SDs under SSP1-2.6, SSP3-7.0, and SSP5-8.5 were respectively delayed by 1.3–1.4, 1.6–1.7, and 2.2–2.4 days/decade on average for the entirety of area I, suggesting that the impact is greater under higher emissions scenarios.

3.4. Precipitation before Winter

The historical spatial distribution features of PBW in 1961–1990, as well as in 2021–2050 under the SSP1-2.6, SSP3-7.0, and SSP5-8.5 radiative forcing scenarios are illustrated, with AT values of 450, 550, and 700 °C (Figure 6). During the 1961–1990 reference period, when the AT was 450 °C, PBW reached more than 60 mm in the northern Greater Khingan Mountains, the Lesser Khingan Mountains, Qinling Mountain, and Changbai Mountain, which was followed by 40–60 mm in Taihang Mountain and the Huaihe River area, and 20–40 mm in the Northeast Plain and North China Plain. The PBW in the plateau areas in the west of Taihang Mountain was generally less than 60 mm, and decreased from east to west. Moreover, there was an increase in PBW with higher AT, especially at high altitudes. For example, at 700 °C of AT, the PBW was generally above 100 mm in the Greater Khingan Mountains, Lesser Khingan Mountain, Changbai Mountain, and Qinling Mountain area; 60–80 mm in the Taihang Mountain area; 20–80 mm in the Northeast Plain and Northern China Plain; and 80–100 mm in the areas along the Huaihe River.

During the period of 2021–2050, the PBWs in most areas of the North winter wheat area (I) show a decreasing trend relative to the reference period under the SSP1-2.6, SSP3-7.0, and SSP5-8.5 scenarios. No substantial differences were observed between the spatial distribution pattern of PBW for different emissions scenarios and AT requirements. When the AT reached 450 °C under the SSP1-2.6 scenario, the regions with more than 60 mm of PBW almost reached the Greater Khingan Mountains, Lesser Khingan Mountains, and Changbai Mountain, while the western regions of Inner Mongolia received little precipitation. The western regions of the Northern winter wheat subregion (Ia) exhibited more precipitation than the eastern regions, with 40–60 mm and less than 40 mm of PBW, respectively. The PBWs in most regions of the Huang-Huai winter wheat subregion (Ib) only received 20–40 mm, but the area near the Qinling Mountain exceeded 40 mm. On average, the PBW decreased by 13.8% in the spring wheat area, 13.7% in subregion (Ia), and 26.7% in subregion (Ib), relative to the reference period. For AT values of 550 and 700 °C, the PBW spatial distributions are illustrated in Figure 5(b2,b3), with 13.1–24.8% and 10.1–22.6% decreases for the three sub-districts, respectively. The PBWs would still decrease with larger reductions under the SSP5-8.5 scenario than under the SSP1-2.6 and SSP3-7.0 scenarios, which indicated that the issue of water deficiency during the historical period was still intractable [33]; development and management of irrigation facilities may require more attention over this region [34].

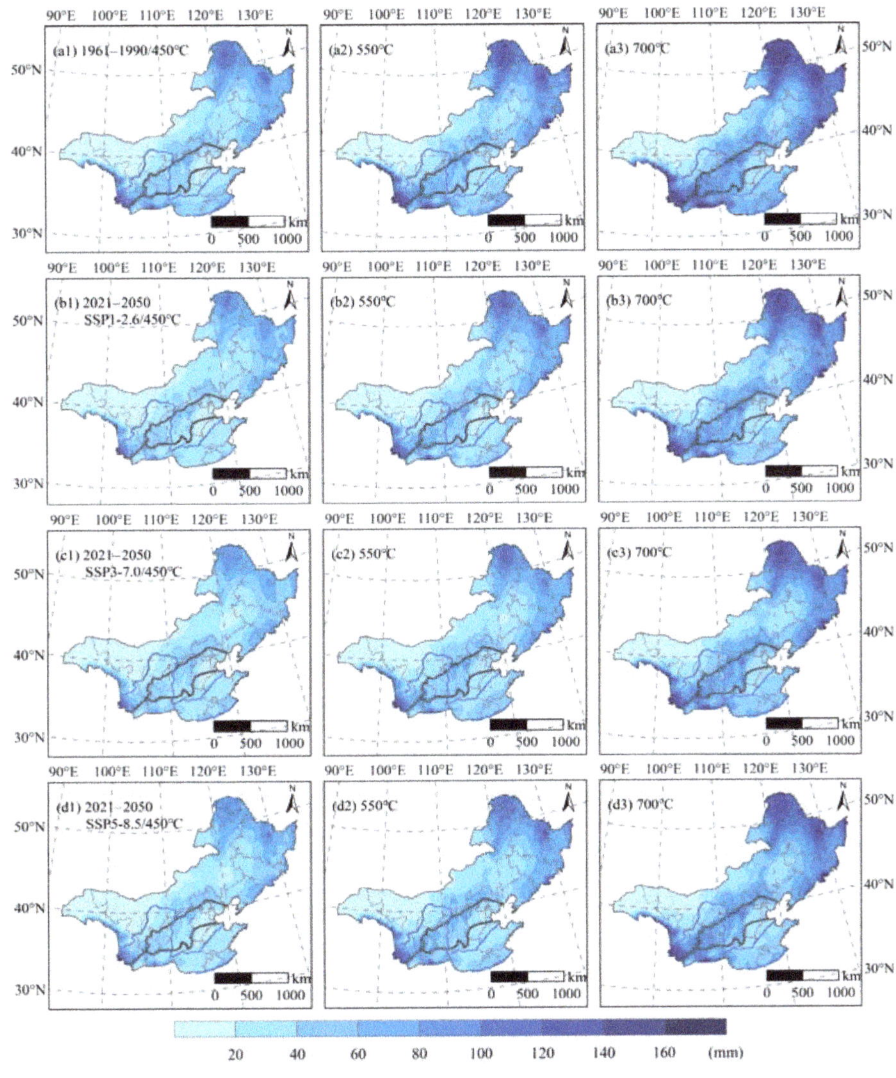

Figure 6. Spatial distribution of PBW in the 1961–1990 and 2021–2050 periods for accumulated temperature before overwintering periods at 450, 550, and 700 °C. (**a1–a3**) historical scenario, (**b1–b3**) SSP1-2.6, (**c1–c3**) SSP3-7.0, and (**d1–d3**) SSP5-8.5 scenarios.

3.5. Planting Boundaries under the Different Scenarios

Based on above analysis from 3.1 to 3.4, the safe planting areas are illustrated in Figure 7 for each scenario. The western region (west of 115° E) of the potential northern planting boundaries will move northward by approximately 73, 94, and 114 km on average under the SSP1-2.6, SSP3-7.0, and SSP5-8.5 scenarios, respectively, and the northernmost tip of the eastern part will respectively move northwardly by 111, 152, and 158 km. Due to climate change, almost 40% the Huang-Huai winter wheat subregion (i.e., 36.7%, 37.0%, and 39.7% under the three radiative forcing scenarios) will exhibit no obvious overwintering periods, causing the southern boundaries of the North winter wheat area (I) to retract approximately 213, 215, and 233 km to the north. This indicates that some provinces in southern area I, such as Shandong, Henan, Jiangsu, and Anhui, would become unsuitable for winter wheat cultivation. However, the potential planting areas of the entirety of area I will increase by 10.7 and 28.0 thousand km^2 on average in the SSP3-7.0 and SSP5-8.5 scenarios, respectively, and decrease 38.1 thousand km^2 in the SSP1-2.6 scenario. It is worth noting that although the radiative forcing of SSP3-7.0 is higher than that of SSP1-2.6, obvious warming of SSP3-7.0 exists as a regional difference.

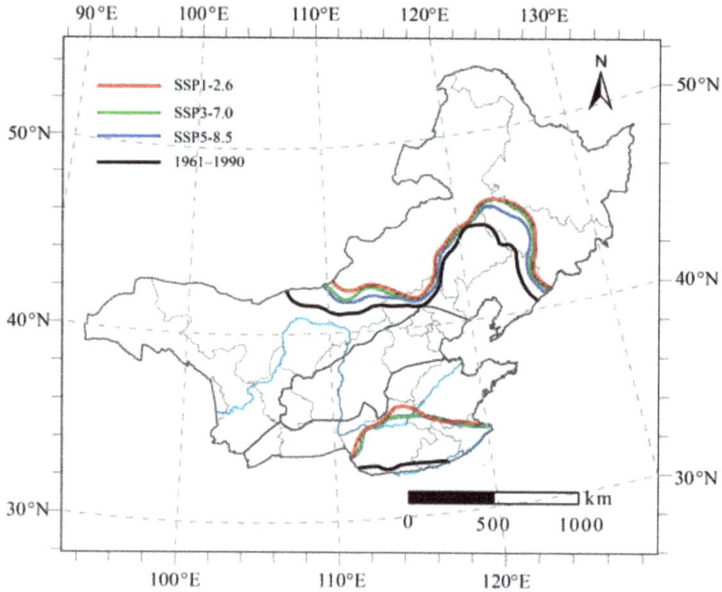

Figure 7. The possible planting boundaries for the North winter wheat area for 1961–1990 and 2021–2050 (SSP1-2.6, SSP3-7.0, and SSP5-8.5 scenarios).

4. Discussion

Winter wheat is generally planted in China as a result of its broad climatic adaptability, and the distribution of its cultivation zones has become the focus of a growing number of scientists in the search to maintain high and stable yields. Previous studies mainly focused on the northward displacement of the northern planting boundary [19,32,35]. However, we predicted that areas without obvious overwintering periods are expected to increase significantly by reaching the Yellow River within the Huang-Huai winter wheat subregion. This indicates that most of the Huang-Huai winter wheat subregion will no longer be suitable for the currently cultivated winter wheat variety in the future. It is also worth noting that when 'extreme spring cold spells' (ESCSs), occur in northern China, continuous negative temperature anomalies can have a catastrophic impact on wheat yields, resulting in yield losses of20% or more. Without an overwintering period, winter wheat will likely grow excessively fast during the winter and subsequently encounter difficulties in resisting cold injury in the early spring. We assumed that ESCS will occur when the daily temperature remains at least 3 °C lower than the climatological daily mean during 5 consecutive days, and analyzed the probability of ESCS in the area with no obvious overwintering periods. We found that the probability of ESCS will increase from 12.5% in the 1961–1990 to 20.5%–25.5% in the forecast period (i.e., 25.5%, 20.5%, and 21.1% under the three scenarios). Additionally, plant diseases and insect pests may increase as well as a result of the climate change, which would further reduce winter wheat yields. Climate change also impacts the winter wheat sowing date. Here, we found that the sowing date was delayed by 0.7–0.8 days/decade on average for the entire North winter wheat area in

1961–1990. Similar results were found in some previous studies; for instance, Xiao et al. observed 1.5 days/decade average delays in the North China Plain during the 1981–2009 period [36]; Liu et al. reported a larger value (3.1 days/decade) for the same period [37]. These large variations in sowing date delays were largely due to differences in study areas and data sources. Nonetheless, our study determined that sowing dates will be delayed by 1.1–1.3, 1.9–2.0, and 2.2–2.4 days/decade in 2021–2050 under the SSP1-2.6, SSP3-7.0, and SSP5-8.5 scenarios, respectively, suggesting that the sowing date delay rate will further increase in the coming three decades.

In order to predict the northern winter wheat planting boundary, we mainly focused on whether winter wheat could safely survive winter conditions, and calculated an index based on the minimum temperature that could be tolerated by winter wheat. However, this threshold (−22 °C) was established for the winter wheat variety of 1980; current varieties are known to tolerate lower temperatures. For example, the strong winter wheat varieties "Dongnong winter wheat No.1," which was introduced and bred in Heilongjiang Province, can withstand a minimum temperature range of −30 to −35 °C [38]. Considering that the winter wheat varieties planted in most areas of the North winter wheat area (I) do not have such strong resistance to freezing temperatures, this study still conservatively assumed a −22 °C minimum temperature in the winter wheat cultivation northern boundary.

Climate change plays an important role in the distribution of suitable winter wheat cultivation zones in China. This study analyzed the changes of potential winter wheat planting areas in the North winter wheat area (I) from a strict meteorological perspective. However, our results do not necessarily imply that winter wheat can be grown in any area where the meteorological conditions are favorable, as local soil conditions, productivity levels, and agricultural policies are also critical factors that determine the suitability of a region for winter wheat cultivation. Additionally, farmers may adjust to the local conditions by choosing the correct sowing depth, adjusting the sowing date, improving crop varieties, and expanding irrigation infrastructure, all of which may lead to further variations in the actual planting boundaries for winter wheat relative to the meteorological boundaries. Furthermore, farmers will no longer engage in production activities in areas where winter wheat planting has failed repeatedly. As a result, wheat cultivation in the northern boundary has changed very little since 2000 to reduce or avoid climate risks. Nonetheless, the winter wheat meteorological boundary is still an important reference boundary for the exploration of new winter wheat cultivation areas, as winter wheat would be difficult to produce profitably beyond these boundaries. Therefore, actual changes in winter wheat cultivation areas should be considered comprehensively in combination with regional meteorological conditions and human activities in future studies.

5. Conclusions

This study predicted the potential changes in the northern and southern winter wheat cultivation boundaries during 2021–2050 under the SSP1-2.6, SSP3-7.0, and SSP5-8.5 radiative forcing scenarios in the North winter wheat area and its Spring wheat area. The findings of this study indicate that the occurrence of extremely low temperature years will decline in the North winter wheat area due to climate change, which will result in an increase in the potential safe overwintering areas for winter wheat cultivation in 2021–2050 by 11.8%, 14.7%, and 16.6% under the SSP1-2.6, SSP3-7.0, and SSP5-8.5 scenarios, respectively, compared to the 1961–1990 reference period. The north boundary will move 0.8–1.3° northward on average, and the south boundary will retract 1.9–2.1° in latitude.

SD and FD, the two phenological stages before winter, are projected to be delayed in the forecasted period. Compared with the 1961–1990 reference period, the SD is delayed by 8.2–8.5, 7.5–8.0, and 8.8–9.1 days, and FD is delayed by 9.9, 8.5, and 10.3 days in the entire North winter wheat area under the SSP1-2.6, SSP3-7.0, and SSP5-8.5 scenarios, respectively. Moreover, PBW is projected to experience large decreases across the entire region under the three scenarios, especially in the Huang-Huai winter wheat subregion, with reduction rates above 20%. This increases the likelihood that the southern winter wheat cultivation boundary will recede northwardly.

Author Contributions: M.W. and Y.X. analyzed and processed the data. J.Z. and Z.H. were major contributors to the drafting of the manuscript. Z.H. provided financing throughout the experiment and generated the outlines of the research. All authors have read and agreed to the published version of the manuscript.

Funding: This study was funded by The Strategic Priority Research Program of Chinese Academy of Sciences (XDA23100403) and National Natural Science Foundation China (42171030).

Institutional Review Board Statement: Not applicable.

Informed Consent Statement: Not applicable.

Data Availability Statement: Data are contained within the article.

Conflicts of Interest: The authors declare no conflict of interest.

References

1. Shiferaw, B.; Smale, M.; Braun, H.; Duveiller, E.; Reynolds, M.; Muricho, G. Crops that feed the world 10. Past successes and future challenges to the role played by wheat in global food security. *Food Secur.* **2013**, *5*, 291–317. [CrossRef]
2. Shewry, P.; Hey, S. The contribution of wheat to human diet and health. *Food Energy Secur.* **2015**, *4*, 178–202. [CrossRef] [PubMed]
3. Tadesse, W.; Amri, A.; Ogbonnaya, F.C.; Sanchez-Garcia, M.; Sohail, Q.; Baum, M. *Genetic and Genomic Resources for Grain Cereals Improvement*; National Academic Press: Queensland, Australia, 2016. [CrossRef]
4. Masson-Delmotte, V.; Zhai, P.; Pirani, A.; Connors, S.L.; Péan, C.; Berger, S.; Caud, N.; Chen, Y.; Goldfarb, L.; Gomis, M.I.; et al. IPCC, 2021. Summary for Policymakers. In *Climate Change 2021: The Physical Science Basis. Contribution of Working Group I to the Sixth Assessment Report of the Intergovernmental Panel on Climate Change*; Cambridge University Press: Cambridge, UK, 2021.
5. Rosenberg, N. The increasing CO_2 concentration in the atmosphere and its implication on agricultural productivity. II. Effects through CO_2-induced climatic change. *Clim. Chang.* **1982**, *4*, 239–254. [CrossRef]
6. Rosenzweig, C. Potential CO_2-induced climate effects on North American wheat producing regions. *Clim. Chang.* **1985**, *7*, 367–389. [CrossRef]
7. Kenny, G.; Harrison, P.; Olesen, J.; Parry, M. The effects of climate change on land suitability of grain maize, winter wheat and cauliflower in Europe. *Eur. J. Agron.* **1993**, *2*, 325–338. [CrossRef]
8. Audsley, E.; Pearn, K.; Simota, C.; Cojocaru, G.; Koutsidou, E.; Rounsevell, M.D.A.; Trnka, M.; Alexandrov, V. What can scenario modelling tell us about future European scale agricultural land use, and what not? *Environ. Sci. Policy* **2006**, *9*, 148–162. [CrossRef]
9. Olesen, J.; Carter, T.; Díaz-Ambrona, C.; Fronzek, S.; Heidmann, T.; Hickler, T.; Holt, T.; Minguez, M.I.; Morales, P.; Palutikof, J.; et al. Uncertainties in projected impacts of climate change on European agriculture and terrestrial ecosystems based on scenarios from regional climate models. *Clim. Chang.* **2007**, *81*, 123–143. [CrossRef]
10. Rosenzweig, C.; Parry, M. Potential impacts of climate change on world food supply. *Nature* **1994**, *367*, 133–138. [CrossRef]
11. Balkovic, J.; van der Velde, M.; Skalsky, R.; Xiong, W.; Folberth, C.; Khabarov, N.; Smirnov, A.; Mueller, N.; Obersteiner, M. Global wheat production potentials and management flexibility under the representative concentration pathways. *Glob. Planet. Chang.* **2014**, *122*, 107–121. [CrossRef]
12. United States Department of Agriculture (USDA), World Agricultural Production. Circular Series WAP. 2016, pp. 7–16. Available online: https://downloads.usda.library.cornell.edu/usda-esmis/files/5q47rn72z/6w924c19w/44558d80d/worldag-production-07-12-2016.pdf (accessed on 1 May 2021).
13. National Bureau of Statistics of China, National Bureau of Statistics of China. National Data. 2017. Available online: http://data.stats.gov.cn/easyquery.htm?cn=C01 (accessed on 1 May 2021).
14. Zhao, G. Study on Chinese wheat planting regionalization. *J. Triticeae Crop.* **2010**, *30*, 886–895. (In Chinese)
15. Sun, J.; Zhou, G.; Sui, X. Climatic suitability of the distribution of the winter wheat cultivation zone in China. *Eur. J. Agron.* **2012**, *43*, 77–86.
16. Vico, G.; Hurry, V.; Weih, M. Snowed in for survival: Quantifying the risk of winter damage to overwintering field crops in northern temperate latitudes. *Agric. For. Meteorol.* **2014**, *197*, 65–75. [CrossRef]
17. Zheng, D.; Yang, X.; Minguez, M.; Mu, C.; He, Q.; Xia, W. Effect of freezing temperature and duration on winter survival and grain yield of winter wheat. *Agric. For. Meteorol.* **2018**, *260*, 1–8. [CrossRef]
18. Guo, W.; Shi, H.; Ma, J.; Zhang, Y.; Wang, J.; Shu, W.; Zhang, Z. Basic Features of Climate Change in North China during 1961–2010. *Adv. Clim. Chang. Res.* **2013**, *4*, 73–83. [CrossRef]
19. Li, Y.; Liang, H.; Wang, P. Effects of Climate Warming on the Planting Boundary and Developmental Stages of Winter Wheat. *J. Triticeae Crop.* **2013**, *33*, 382–388.

20. Li, K.; Yang, X.; Mu, C.; Xu, H.; Chen, F. The possible effects of global warming on cropping system in China VIII-The effects of climate change on planting boundaries of different winter-spring varieties of winter wheat in China. *Sci. Agric Sin.* **2013**, *46*, 1583–1594.
21. Hao, Z.; Geng, X.; Wang, F.; Zheng, J. Impacts of climate change on agrometeorological indices at winter wheat overwintering stage in northern China during 2021–2050. *Int. J. Climatol.* **2018**, *38*, 5576–5588. [CrossRef]
22. Zhang, W.; Brandt, M.; Prishchepov, A.V.; Li, Z.; Lyu, C.; Fensholt, R. Mapping the Dynamics of Winter Wheat in the North China Plain from Dense Landsat Time Series (1999 to 2019). *Remote Sens.* **2021**, *13*, 1170. [CrossRef]
23. Hu, Q.; Ma, X.; Pan, X.; Huang, B. Climate Warming Changed the Planting Boundaries of Varieties of Summer Corn with Different Maturity Levels in the North China Plain. *J. Appl. Meteorol. Climatol.* **2019**, *12*, 2605–2615. [CrossRef]
24. Li, J.; Lei, H. Tracking the spatio-temporal change of planting area of winter wheat-summer maize cropping system in the North China Plain during 2001–2018. *Comput. Electron. Agric.* **2021**, *187*, 106222. [CrossRef]
25. Riahi, K.; Vuuren, D.; Kriegler, E.; Edmonds, J.; O'Neill, B.; Fujimori, S.; Bauer, N.; Calvin, K.; Dellink, R.; Fricko, O.; et al. The Shared Socioeconomic Pathways and their energy, land use, and greenhouse gas emissions implications: An overview. *Glob. Environ. Chang.* **2017**, *42*, 153–168. [CrossRef]
26. O'Neill, B.C.; Tebaldi, C.; Vuuren, D.P.v.; Eyring, V.; Friedlingstein, P.; Hurtt, G.; Knutti, R.; Kriegler, E.; Lamarque, J.-F.; Lowe, J. The scenario model intercomparison project (ScenarioMIP) for CMIP6. *Geosci. Model Dev.* **2016**, *9*, 3461–3482. [CrossRef]
27. Hempel, S.; Frieler, K.; Warszawski, L.; Schewe, J.; Piontek, F. A trend-preserving bias correction—The ISI-MIP approach. *Earth Syst. Dyn.* **2013**, *4*, 219–236. [CrossRef]
28. Cooperative Agricultural and Forest Crop Regionalization Group in China. *Agricultural and Forest Crop Climate Regionalization in China*; China Meteorological Press: Beijing, China, 1987. (In Chinese)
29. Jin, S. *Wheat Science in China*; China Agriculture Press: Beijing, China, 1996. (In Chinese)
30. Wang, S. A statistical method for the first and last date with the daily temperature steadily passing through the threshold. *Meteorol. Mon.* **1982**, *8*, 29–30. (In Chinese)
31. Cui, Y.; Han, J.; Cao, G.; Jiang, M.; Zhang, J. Effect of per-winter positive accumulated temperature on suitable planting dates of winter wheat in south centre area of Hebei. *Chin. Sci. Bull.* **2008**, *24*, 195–198. (In Chinese)
32. Hao, Z.; Zheng, J.; Tao, X. A study on northern boundary of winter wheat during climate warming: A case study in Liaoning Province. *Prog. Geogr.* **2001**, *20*, 254–261. (In Chinese)
33. Gao, J.; Yang, X.; Zheng, B.; Liu, Z.; Zhao, J.; Sun, S.; Li, K.; Dong, C. Effects of climate change on the extension of the potential double cropping region and crop water requirements in Northern China. *Agric. For. Meteorol.* **2019**, *268*, 146–155. [CrossRef]
34. Yang, G.; Li, S.; Wang, H.; Wang, L. Study on agricultural cultivation development layout based on the matching characteristic of water and land resources in North China Plain. *Agric. Water Manag.* **2022**, *259*, 107272. [CrossRef]
35. Zou, L.; Zhang, J.; Jiang, Q.; Qing, Z.; Wang, G.; Zhao, H. Research and development of winter wheat growing in northern region. *Chin. J. Agrometeorol.* **2001**, *22*, 54–58. (In Chinese)
36. Xiao, D.; Tao, F.; Liu, Y.; Shi, W.; Wang, M.; Liu, G.; Zhang, S.; Zhu, Z. Observed changes in winter wheat phenology in the North China Plain for 1981–2009. *Int. J. Biometeorol.* **2013**, *57*, 275–285. [CrossRef] [PubMed]
37. Liu, Y.; Chen, Q.; Ge, Q.; Dai, J. Spatiotemporal differentiation of changes in wheat phenology in China under climate change from 1981 to 2010. *Sci. China Earth Sci.* **2018**, *61*, 1088–1097. [CrossRef]
38. Wang, D.; Zeng, Y.; Mou, Y.; Yu, J.; Cang, J. Research on antifreeze proteins of Dongnongdongmai 1 in high–cold area. *J. Triticeae Crop.* **2009**, *29*, 823–826. (In Chinese)

agriculture

MDPI

Article

Simulating the Impacts of Climate Change on Maize Yields Using EPIC: A Case Study in the Eastern Cape Province of South Africa †

Dennis Junior Choruma [1,2,*], Frank Chukwuzuoke Akamagwuna [2] and Nelson Oghenekaro Odume [2]

1 African Studies Centre, Rhodes University, P.O. Box 94, Grahamstown 6139, South Africa
2 Institute for Water Research, Rhodes University, P.O. Box 94, Grahamstown 6139, South Africa; f.akamagwuna@ru.ac.za (F.C.A.); n.odume@ru.ac.za (N.O.O.)
* Correspondence: d.choruma@ru.ac.za
† This article is part of the Ph.D. thesis of Dennis Junior Choruma.

Abstract: Climate change has been projected to impact negatively on African agricultural systems. However, there is still an insufficient understanding of the possible effects of climate change on crop yields in Africa. In this study, a previously calibrated Environmental Policy Integrated Climate (EPIC) model was used to assess the effects of future climate change on maize (*Zea mays* L.) yield in the Eastern Cape Province of South Africa. The study aimed to compare maize yields obtained from EPIC simulations using baseline (1980–2010) weather data with maize yields obtained from EPIC using statistically downscaled future climate data sets for two future periods (mid-century (2040–2069) and late century (2070–2099)). We used three general circulation models (GCMs): BCC-CSM1.1, GFDL-ESM2M and MIROC-ES under two Representative Concentration Pathways (RCPs), RCP 4.5 and RCP 8.5, to drive the future maize yield simulations. Simulation results showed that for all three GCMs and for both future periods, a decrease in maize production was projected. Maize yield was projected to decrease by as much as 23.8% for MIROC, RCP 8.5, (2070–2099). The temperature was projected to rise by over 50% in winter under RCP 8.5 for both future periods. For both future scenarios, rainfall was projected to decrease in the summer months while increasing in the winter months. Overall, this study provides preliminary evidence that local farmers and the Eastern Cape government can utilise to develop local climate change adaptation strategies.

Keywords: climate change; agriculture; crop modelling; yield; future climate scenarios

Citation: Choruma, D.J.; Akamagwuna, F.C.; Odume, N.O. Simulating the Impacts of Climate Change on Maize Yields Using EPIC: A Case Study in the Eastern Cape Province of South Africa. *Agriculture* **2022**, *12*, 794. https://doi.org/10.3390/agriculture12060794

Academic Editors: Dengpan Xiao and Wenjiao Shi

Received: 7 May 2022
Accepted: 12 May 2022
Published: 31 May 2022

1. Introduction

Climate change is anticipated to significantly impact the resilience of agricultural systems in semi-arid developing countries such as South Africa. The Intergovernmental Panel on Climate Change (IPCC) has projected that increases in greenhouse gases, particularly carbon dioxide (CO_2), are expected to modify global climate by increasing surface air temperature, altering rainfall patterns and increasing the occurrence of extreme weather events [1]. While the increased temperature may boost the yields of some crops in some regions by increasing the rate of biomass accumulation [2], the negative effects of climate change such as increased rainfall variability and droughts are expected to far outweigh the positive benefits of climate change [3]. Several studies have predicted a decline in agricultural productivity in most parts of Southern Africa due to increased rainfall variability and elevated temperatures [4–6].

Maize (*Zea mays* L.) is a staple food in South Africa and vital for food security in the country [7]. However, climate change threatens agricultural productivity in South Africa and hence food security and the livelihoods of many subsistence farmers who rely on maize production for their livelihoods [8,9]. A review by [10] showed that maize was projected to decrease by as much as 8–38% under RCP 4.5 and RCP 8.5 scenarios by the end of the

21st century. Several studies have investigated the impacts of climate change on maize production in South Africa. A study by [11] in Southern Africa using a process-based crop model (APSIM) combined with 17 general circulation models (GCMs) predicted a decrease in future maize yields. However, many of these studies focused on the traditional maize growing areas such as KwaZulu-Natal with limited studies focusing on the Eastern Cape. However, many people rely on maize production for their livelihoods in the Eastern Cape [9]. While the Eastern Cape has been predominantly a livestock producing area due to the semi-arid climate, the government is driving efforts to increase cereal production, especially maize, in an effort to increase the region's food security [9,12,13].

Lately, predicting and evaluating the possible impacts of climate change on crop yields has become important in order to develop effective climate change adaptation strategies in agricultural systems. Early knowledge and understanding of potential climate change effects on crops may help farmers and decision makers to make informed decisions that minimise agricultural production risks and take advantage of opportunities arising from climate change [11]. This knowledge of how the future climate may affect agricultural production is important in semi-arid regions such as South Africa, where water scarcity and increasing frequencies of droughts are already limiting crop production [14] and threatening food security.

One way of predicting and evaluating the effects of future climate conditions on agricultural production is by using crop models. Crop models have gained increasing application in agriculture-related research to enhance crop growth, soil water balance and nutrient management under various climate conditions [15,16]. Crop models have also been used to assess the impacts of climate change on crop production and environmental risks [17,18], and explore potential adaptation strategies [19]. In South Africa, studies have applied crop models in the fields of hydrology and agriculture. For example, Warburton et al. [20] investigated the impacts of climate change on the hydrology of catchments. Abraha and Savage [21] assessed the potential effects of climate change on maize yields in the KwaZulu-Natal area of the country. However, most of these studies in South Africa used global climate data to run the crop models. Global climate data may not always be representative of local climate conditions [22].

When simulating future crop yields, variables such as precipitation and temperature are required as model inputs. General circulation models (GCMs) have been created to use different greenhouse gas scenarios and complex earth–atmosphere interactions to project future climate parameters such as precipitation. GCMs are numerical models that use complex mathematical equations to simulate the earth's atmospheric processes and predict climate [23]. GCMs project climate parameters at a resolution of approximately 250 km^2 [24,25]. While accurate predictions can be made at this resolution at the global scale, the resolution is coarse at the local scale to support local decision making and planning [26]. To reduce the uncertainty involved with the use of GCMs, data from GCMs is usually downscaled either statistically or dynamically to produce local climate data or regional climate models (RCMs) that reflect local conditions more accurately [27].

In the dynamic downscaling method, a regional climate model (RCM) is nested into the GCM to represent a given boundary forcing. Statistical downscaling methods use empirical relationships established between large-scale and fine-scale variables using historical data, for example, statistical downscaling uses historically sourced data such as the quantitative links between the state of the larger-scale climatic environment and local variations. In contrast, dynamical downscaling employs boundary conditions (e.g., surface pressure and wind) and an atmospheric circulation system (principle of physics) to generate high-resolution data sets [28]. However, the dynamical downscaling method is computationally and technically complex and expensive [29], limiting the number of institutions employing the approach. In this regard, coupling local and regional baseline climate data with statistically downscaled GCM outputs provides an invaluable way of reducing uncertainty associated with climate projections. In this study, freely available climate data, statistically downscaled to reflect local weather more accurately, were used for the climate simulations.

In South Africa, research groups such as the Council for Scientific and Industrial Research (CSIR) and the Climate Systems Analysis Group (CSAG) have developed local downscaled future climate data. However, despite the availability of these locally developed, downscaled climate data, few studies have used these downscaled climate data to assess the impacts of future climate change on crop yields in South Africa [30,31]. Therefore, this study aims to compare current and future maize yields under different future climate scenarios. While the focus of this study was not on climate uncertainty, three climate models were compared to reduce the uncertainty of climate change projections associated with different models that could affect crop response.

2. Materials and Methods

2.1. Background

This study follows up on our previous study using the EPIC model in the study area. The previous study [32] provides a detailed description of the model calibration and validation using limited data from field trials on maize at the Cradock Research Farm. This present study applies the calibrated and validated EPIC model to simulate future maize yields using future climate data sets. In this study, only a summary of the model performance will be given. A detailed description of the calibration and validation steps can be found in [32] and additional data on model performance can be found in Appendix A.

2.2. Study Area

Biophysical data for model calibration were collected from the Cradock Research Farm (Figure 1) in the Eastern Cape province of South Africa (32°13′11.09″ S, 25°41′11.86″ E, elevation 849 m). The area is predominantly fine-loamy mollic ustifluvent [33], with elevated quantities of Beaufort sediments (alluvial sand and silt and colluvial materials). A description of the major soil characteristics at the Cradock Research Farm is given in Appendix A, Table A1. Rainfall in the area is bimodal, with winter rainfall on the western side of the province and summer rainfall on the eastern side. The region receives an average rainfall amount of 341 mm. The area is drought-prone, and since 2015, most of the Eastern Cape has experienced droughts resulting in water supply shortages [34].

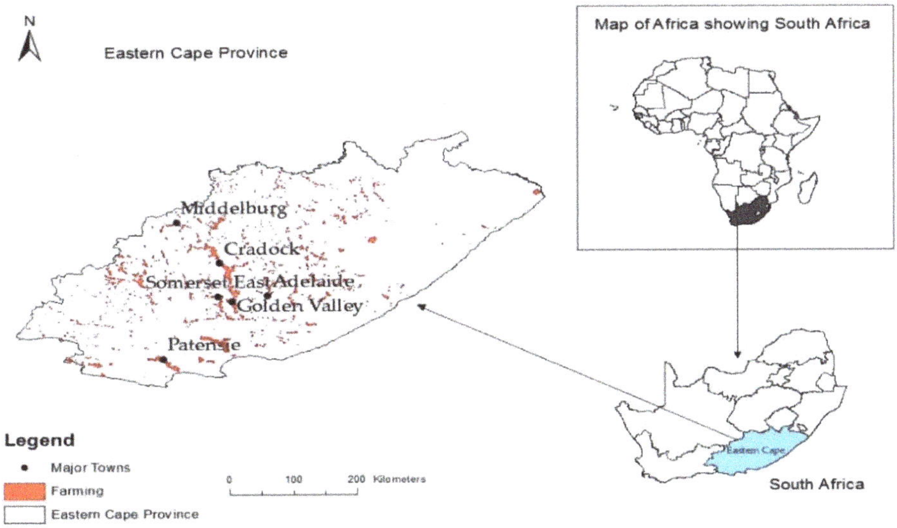

Figure 1. Map of study area indicating the dominant farming towns in the Eastern Cape Province, South Africa. The figure is taken from [32].

The Eastern Cape has been predominantly a livestock production area due to frequent droughts and semi-arid nature of the region. In addition, the soils are inherently infertile and prone to erosion [9]. However, to improve food security in the region, government, through programmes such as the Massive Food Production Programme (MFPP) has been on a drive to increase maize production in the area [12]. Maize is a staple food in the area and key to enhancing the region's food security.

Projections by the South African Department of Environmental Affairs [35] predict significant increases in climate variability for the region. Substantial reductions in both annual and daily precipitation have been forecasted for the area [34,35]. The yearly temperature is also anticipated to rise, accompanied by elevated evapotranspiration rates and the likelihood of droughts. An assessment of mid-century (2040–2060) CMIP5 rainfall predictions by Mahlalela et al. [34] estimate a levelling of the annual rainfall cycle over the Eastern Cape, with summer becoming drier and winter becoming wetter. Generally, the Eastern Cape is projected to have elevated temperatures, a higher frequency of extreme rainfall events and drier conditions, especially in summer [35].

2.3. EPIC Model Description

The EPIC model (version 0810) is an agroecosystem model designed to simulate over 70 crops at the field scale using values characteristic of each crop [36]. Crop yield is estimated based on the biomass accumulated by the plant. Biomass accumulation is affected by model parameters such as planting density (PD), photosynthetic active radiation (PAR), vapor pressure deficit (VPD) and the biomass to energy ratio (WA) [37]. The daily stresses caused by extreme temperature, water and nutrient stress or inappropriate aeration are used to correct the potential daily biomass accumulation to daily actual biomass accumulation. The model also requires weather inputs such as precipitation, minimum and maximum temperature, wind speed and relative humidity. Stresses reduce the biomass accumulation and the harvest index using the value of the most severe stress experienced by the crop [38]. To better reflect the specific site conditions, values of location-specific variables such as potential heat units (PHU) accumulated, HI and optimum temperature (OT) have to be adjusted according to the area or region in which the model is to be used [39].

2.4. Field Work

Field trials on maize were conducted by the Agricultural Research Council (ARC) from 1999 to 2003 at the Cradock Farm to assess the yield potential of hybrid maize cultivars within the Eastern Cape Province of South Africa. Data from these trials were used to calibrate and validate the EPIC model. We selected two fields with similar soil characteristics, one for calibration and one for validation. A randomised block design (RBD) [40], with three replications, was used throughout the field trials. The two fields with similarly performing maize hybrids were managed according to the same agricultural management plan developed by the ARC based on local farmers' management practices. The management plan, including planting and harvesting dates, and irrigation and fertiliser application dates, is shown in Appendix A, Table A2. The management practices were performed around the same time each year. Each year, minor changes to the management plan were carried out based on prevailing weather conditions. In the future climate simulations, management practices including planting dates, fertiliser and irrigation levels used during the maize cultivar evaluation trials were used as the baseline management practices being used in the area.

2.5. Model Inputs

EPIC requires weather inputs such as rainfall, relative humidity, temperature and solar radiation. We obtained weather files for the study area from the AgMERRA [41] climate dataset at 0.5×0.5 arc-degree spatial resolution. Soil parameter values including cation exchange capacity, soil texture, bulk density and electrical conductivity were taken from a previous soil analysis in the Cradock Farm. We selected missing soil parameter values (i.e., soil albedo, organic carbon concentration) from the Harmonized World Soil Database (HWSD) [42] based on the expert opinion given by the Cradock Farm Manager (Mr G. Jordaan 2017, pers. comm).

2.6. EPIC Model Set-Up

2.6.1. Framework

This study used a modelling framework for the EPIC model developed at the International Institute of Applied Systems Analysis (IIASA) [43]. Raster layers on weather, soil and topography were combined and a modelling scheme applied at 5 × 5 arc-min resolution. A grid was set up for the whole Eastern Cape and then divided into homogenous grids that had similar site properties such as soil texture, weather and elevation. We then chose the grid containing Cradock farm and used one soil profile based on the soil characteristics at the farm [43]. The simulation grid containing the Cradock Research Farm was then chosen for the simulations.

The Priestly–Taylor method was used to calculate the potential evapotranspiration (PET). The Priestly–Taylor method was selected due to the method yielding PET values close to the region's reported values by [44]. The model was run for 31 years, corresponding to the length of the weather records available, with the first 19 years serving as a warm-up period for equilibrating EPIC's soil erosion functions. Agricultural land management in the model was set up according to the dates in the management plan (Appendix A, Table A2). Irrigation and fertiliser applications were carried out in the model using the manual setting. One soil profile (see Appendix A, Table A1) was used for all the simulations.

2.6.2. Model Calibration

The calibration and validation of the model were performed using grain yield data from two fields at the Cradock Farm that had similar soil types. Other data such as biomass accumulation rates and nutrient leaching were not available for model calibration and validation as the trials were only designed to evaluate cultivar stability and potential yield. Detailed steps of the calibration process are given in [32]. Model calibration used data from one field and model validation used grain yield data collected from the other field. Grain yield data were for the five-year period from 1999 to 2003.

2.6.3. Model Evaluation

We used four indicators, namely root mean square error (*RMSE*), the coefficient of determination (R^2), Nash–Sutcliffe efficiency (*NSE*) and per cent bias (*PBIAS*) to evaluate model efficiency.

$$RMSE = \left[\frac{1}{n}\sum_{i=1}^{n}(s_i - o_i)^2\right]^{\frac{1}{2}} \quad (1)$$

$$R^2 = \frac{\left[\sum(O_i - O_{mean})(S_i - S_{mean})\right]^2}{\sum(O_i - O_{mean})^2 \sum(S_i - S_{mean})^2} \quad (2)$$

$$NSE = 1 - \frac{\sum_{i=1}^{n}(O_i - S_i)^2}{\sum_{i=1}^{n}(O_i - O_{mean})} \quad (3)$$

$$PBIAS = \frac{\sum_{i=1}^{n} 100(O_i - S_i)}{\sum_{i=1}^{n} O_i} \quad (4)$$

where n represents the sample number, O_{mean} the observed mean value and S_{mean} the simulated mean value. O_i and S_i are the observed and predicted values of the ith observation (i = 1 to n), respectively. Regarding the *RMSE*, values close to zero signify a good fit between observed and simulated yields [45]. An *RMSE* of zero indicates that the model predicts the observations with complete accuracy. The coefficient of determination, R^2, has values ranging from 0 to 1, with higher values denoting less error variance [46]. *NSE* varies from negative infinity to 1, with an *NSE* value of 1 representing perfect model fit between observed and simulated values. In contrast, negative *NSE* values indicate that the mean observed value is a better predictor than the simulated value [46]. The *PBIAS* measures the tendency of simulated data to be larger or smaller than the observed data. *PBIAS* has an ideal value of 0, while positive values indicate model underestimation, and negative values indicate model overestimation [47]. Lastly, the *t*-test evaluated variations between simulated and observed mean values. We considered $R^2 \geq 0.6$, $PBIAS \leq \pm 25\%$ and $NSE \geq 0.4$ as satisfactory model performance criteria following [48].

2.7. Climate Data

We used statistically downscaled climate input data from three general circulation models available from the Coupled Model Intercomparison Project Phase 5 (CMIP5) [49]. The climate data were downloaded from the Climate Systems Analysis Group's (CSAG) Climate Information Portal (CIP) (http://cip.csag.uct.ac.za, accessed 27 July 2019). The climate data come from two primary sources—the Computing Centre for Water Resources located at the University of KwaZulu-Natal and the South African Weather Services. Prior to uploading to the CIP, the data are collated and checked for quality by the CSAG [50]. Due to inherent uncertainties in individual models, three GCMs were used to encompass a range of global mean temperature and precipitation changes and consider a wide range of plausible future scenarios. The selected GCMs have been applied previously in South Africa and found to represent the region accurately in terms of projection signal (see [51] for example). The driving GMCs chosen for this study were the BCC-CSM1.1, GFDL-ESM2M and MIROC-ES models (Table 1).

Table 1. List of driving GCMs and the model abbreviations used in this study.

Driving Regional General Circulation Model	Source	Abbreviation of the Model Used in this Study
BCC-CSM1.1	Beijing Climate Centre, China Meteorological Administration, China	BCC
GFDL-ESM2M	Geophysical Fluid Dynamic Laboratory, USA	GFDL
MIROC-ESM	Atmosphere and Ocean Research Institute (University of Tokyo), National Institute for Environmental Studies and Japan Agency for Marine-Earth Science and Technology	MIROC

For future greenhouse gas emission scenarios, two Regional Concentration Pathways, RCP 4.5 and RCP 8.5, for two future 30-year periods, from 2040–2069 and 2070–2099, were chosen to compare two different possible climate scenarios depending on the level of greenhouse gas emissions. The GCAM modelling team at the Pacific Northwest National Laboratory's Joint Global Change Research Institute (JGCRI) in the United States developed RCP 4.5. It is a stabilisation scenario that stabilises the radiative forcing, overshooting the long-run radiative forcing target level, shortly after 2100 [52,53], whereas RCP 8.5 was created using the MESSAGE model and the IIASA Integrated Assessment Framework by IIASA in Austria. The RCP 8.5 pathway is characterised by increasing greenhouse gas emissions over time and represents a scenario that results in high greenhouse gas levels [54].

We used the weather data for 31 years from 1980 to 2010 for the Cradock Research Farm obtained from the AgMERRA database [41] as input data for the baseline simulation with EPIC. Weather data included daily maximum and minimum temperature and rainfall. In the field trials, the time from physiological maturity to actual harvest date was not recorded. Due to this lack of information on the actual time from physiological maturity to harvest, changes in the length of the growing season under future climate scenarios were not included in the simulations.

2.8. Data Analysis

The model output variables for the simulations analysed included economic yield in tonnes per hectare (t ha^{-1}), seasonal irrigation water applied in millimetres (mm), seasonal evapotranspiration in mm, nitrogen (N) leaching as N lost in percolate in kilogrammes Nitrogen per hectare (kg N ha^{-1}) and water use efficiency (WUE) computed as yield per unit of water use (yield/(rainfall plus irrigation) in kg ha^{-1} mm^{-1}). The means of the output variables for the current scenario were compared to the means of the output variables for the future periods. Model variables were analysed using analysis of variance (ANOVA) computed with the Statistical Package for Social Scientists (SPSS) v21. Prior to ANOVA, Shapiro–Wilks and Levene's tests examined the normality and equality of variance. Tukey's post hoc tests were used to determine the means that significantly varied when ANOVA indicated significant differences. An independent samples t-test was performed to test for mean differences in the output variables between the two future periods, 2040–2069 and 2070–2099. The ANOVA and t-tests were conducted in SPSS v21.

3. Results

3.1. Model Calibration

Before calibration, the following model performance values were observed: $NSE = -3.34$, $RMSE = 3.65$ and $PBIAS = 28.55$. After calibration the following values were observed: $NSE = 0.53$, $RMSE = 1.17$ and $PBIAS = 0.31$. Table 2 summarises model performance after calibration. For the calibration simulation, the model underestimated yields for all years using default parameters. Adjusting the parameters, Parm 20 (microbial decay rate coefficient), Parm 47 (slow humus transformation rate), Parm 52 (tillage effect on residue decay rate) and WSYF (minimum harvest index) decreased the *RMSE*% from 32.4% to 11.4%, while the *NSE* value increased from negative values to 0.47. Adjusting PHU improved model performance with a PHU value of 2480 producing the smallest *RMSE*% (10.7%) value between observed yields and simulated yields. Further adjustments of PHU from 2480 did not produce any improvement in model performance. After PHU adjustment, model performance came within the range set for satisfactory model calibration (i.e., $R^2 > 0.6$ and $PBIAS < \pm 25\%$). Further calibration of the crop parameters HI and WA was therefore not conducted. The relationship between observed and simulated grain yield is given in Appendix A, Figures A1 and A2.

Table 2. Showing Nash–Sutcliff efficiency (*NSE*), root mean square error (*RMSE*) and per cent bias (*PBIAS*) for calibration and validation [32].

	Observed Mean (t ha^{-1})	Simulated Mean (t ha^{-1})	*NSE*	*RMSE* (t ha^{-1})	*PBIAS* %
Calibration	11.26	11.23	0.53	1.17	0.31
Validation	11.12	11.23	0.61	1.018	−0.2

3.2. Validation

Observed yields ranged from 9 t ha^{-1} to 14 t ha^{-1}, while simulated yields ranged from 10 t ha^{-1} to 12 t ha^{-1}. The following model evaluation statistics were observed: $NSE = 0.61$, $RMSE = 10.18$ and $PBIAS = -0.2$. Model performance was within the set criteria and considered satisfactory. Table 2 summarises model performance for the validation simulation. The model overestimated maize yields for three out of the five years used for validation. In the year 2000, there were unusually high observed maize yields (14.01 t ha^{-1}), which were underestimated by the model. In 2003, the trials had low observed yields, which were slightly overestimated by the model. No indications were given in the management records on why there were unusually high observed yields in the year 2000; however, in the year 2003, management records indicated that the trial suffered a heavy weed infestation. No statistical differences were revealed by the Student's *t*-test (alpha = 0.05) between the observed and simulated mean grain yields. The relationship between observed and simulated grain yield is shown graphically in Appendix A, Figures A3 and A4.

3.3. Climate Data Analysis

3.3.1. Temperature and Rainfall

All three GCMs revealed average temperature increases from March to October for both scenarios (Figure 2a,b). For RCP 4.5 scenario (Figure 2a), the increase in average temperature was lower than the RCP 8.5 scenario (Figure 2b). The highest monthly average temperature in the RCP 4.5 scenario was 23.7 °C in January and February for the model MIROC and approximately 21 °C for the GFDL and BCC models. The temperature increase was more prominent in the RCP 8.5 scenario and the MIROC model, where average temperatures in June and July were above 10 °C and approximately 6.8 °C higher than the baseline average for the two months. In the months from September to December, the temperatures were similar across all three models.

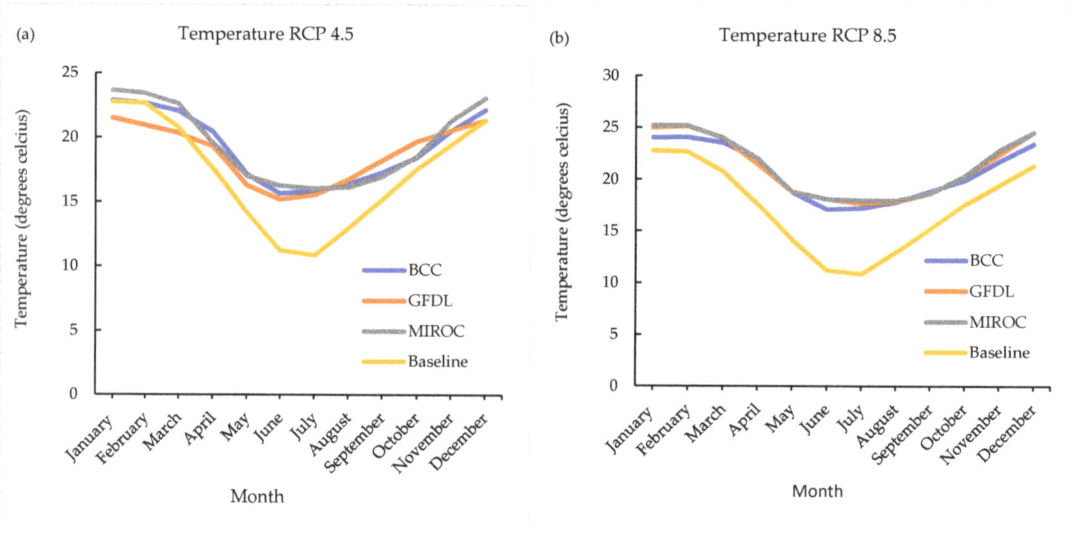

Figure 2. Monthly average temperatures for the two 30-year future periods compared to 31 years of baseline data, (**a**) RCP 4.5 and (**b**) RCP 8.5.

With respect to temperature differences from the baseline (Figure 3), the GCMs that had the highest temperature increase for the RCP 4.5 scenario were the MIROC model, with a monthly percentage difference from the baseline of about 51% in July and the GFDL model with peaks of more than 40% in June and July for the period 2070–2099. RCP 8.5 showed higher temperature differences from the baseline compared to RCP 4.5 for both climate models and future time periods. The highest percentage difference from the baseline in RCP 8.5 was given by the MIROC model, reaching a peak of 71% in July.

Regarding rainfall (Figure 4), an increase in winter rainfall was observed from May to July for both RCPs with higher average rainfall values in RCP 8.5 (Figure 3b). The MIROC model showed a different trend for rainfall from the other models for both the RCP 4.5 (Figure 3a) and RCP 8.5 scenarios (Figure 3b) with higher average monthly rainfall for the months September to December, showing peaks of about 70 mm in November (Figure 3b). The baseline, BCC and GFDL scenarios also showed peaks in November in the RCP 8.5 scenario but with rainfall peaks lower than the MIROC model (Figure 3b).

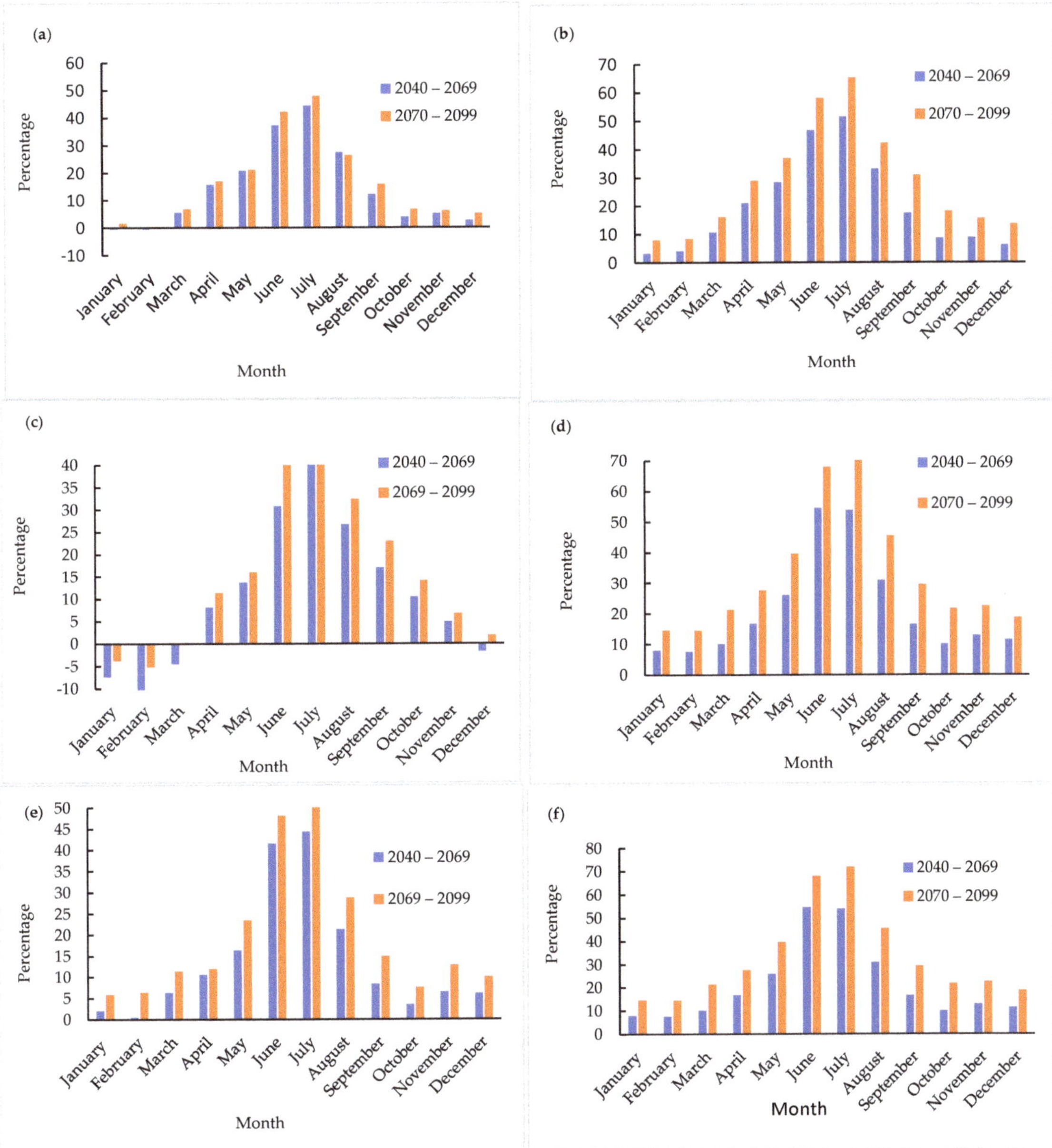

Figure 3. Percentage variations from the baseline of average monthly temperatures for the two thirty-year future periods for all three GCMs under the two RCPs, (**a**) BCC RCP 4.5, (**b**) BCC RCP 8.5, (**c**) GFDL RCP 4.5, (**d**) GFDL RCP 8.5, (**e**) MIROC RCP 4.5 and (**f**) MIROC RCP 8.5).

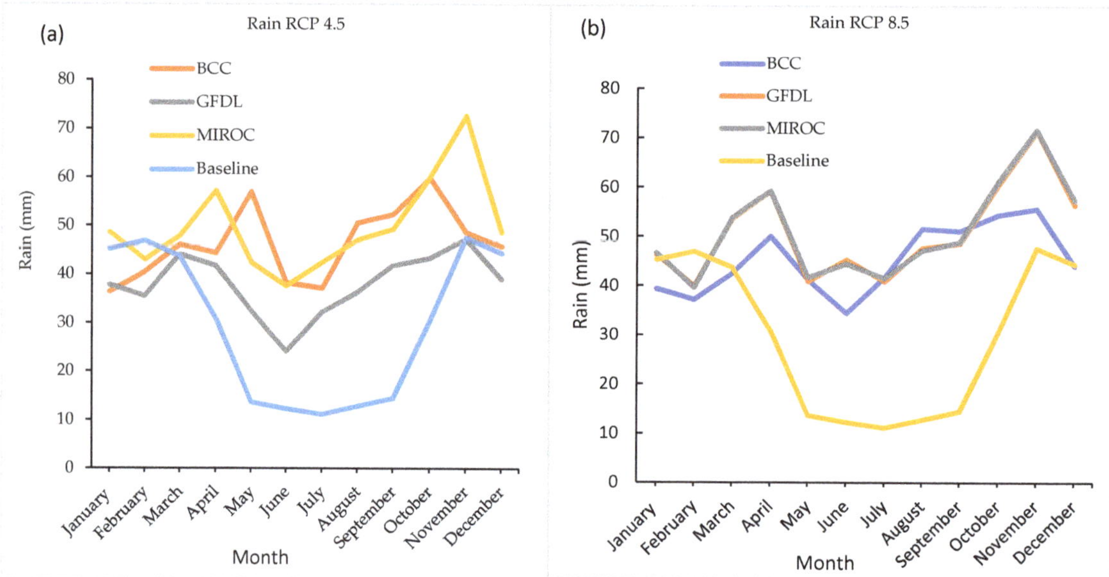

Figure 4. Monthly average rainfall for the two 30-year future periods compared to 31 years of baseline data, (**a**) RCP 4.5 and (**b**) RCP 8.5.

3.3.2. Yield Simulations

Simulation results displayed a similar trend among all the three GCMs used in the RCPs. There was a reduction in maize yield, WUE and seasonal irrigation requirements, and an increase in N leaching and seasonal evapotranspiration for all GCMs under the two future periods (Table 3).

Table 3. Average model output values and mean comparison test for the different scenarios, climate models and future time periods. Different superscript letters on means in the same column indicate significant differences ($p < 0.05$) revealed by a Tukey's post hoc multiple comparison test. Identical superscript letters on means in the same column indicate no significant differences ($p > 0.05$).

	Yield ($t\ ha^{-1}$)	**Irrigation Water Used (mm)**	**WUE ($kg\ ha^{-1}\ mm^{-1}$)**	**N Leaching ($kg\ N\ ha^{-1}$)**	**Seasonal Et (mm)**
Scenario					
Baseline	12.24 ^{A} ± 0.58	562.89 ^{A} ± 82.53	24.13 ^{A} ± 1.33	19.91 ^{B} ± 24.17	907.78 ^{C} ± 46.79
RCP 4.5	11.51 ^{B} ± 1.10	541.09 ^{A} ± 74.29	23.46 ^{A} ± 1.96	36.79 ^{B} ± 34.09	943.10 ^{A} ± 39.08
RCP 8.5	10.20 ^{C} ± 0.81	460.81 ^{B} ± 61.86	22.40 ^{B} ± 1.19	66.13 ^{A} ± 53.58	918.84 ^{B} ± 40.94
General Circulation Model					
BCC-ESM	10.89 ^{A} ± 1.17	509.23 ^{A} ± 66.43	23.24 ^{A} ± 2.34	49.22 ^{A} ± 41.35	922.45 ^{A} ± 32.91
GFDL	11.05 ^{A} ± 1.32	510.82 ^{A} ± 92.8	22.95 ^{B} ± 1.33	47.34 ^{A} ± 52.37	933.88 ^{A} ± 45.78
MIROC	10.62 ^{A} ± 0.95	481.52 ^{A} ± 73.55	22.58 ^{BC} ± 1.09	56.49 ^{A} ± 48.18	936.34 ^{A} ± 44.62
Period					
2040–2069	11.31 * ± 0.73	525.26 *±6 8.77	23.37 * ± 0.76	39.35 * ± 34.14	938.76 * ± 36.59
2070–2099	10.39 * ± 1.33	475.82 * ± 81.78	22.48 * ± 1.33	62.78 * ± 55.77	922.62 * ± 45.13

* Indicates a significant difference at $\alpha = 0.05$ for independent samples t-test. WUE = water use efficiency, Et = evapotranspiration.

Regarding percentage differences between the baseline and future periods, maize yield decreased by up to 23.8% for MIROC, RCP 8.5, (2070–2099). The largest decrease in seasonal irrigation (13.6%) was for GFDL, RCP 8.5 (2040–2069). For WUE, the most significant percentage decrease (22.7%) occurred under MIROC, RCP 8.5, (2070–2099). Concerning N leaching, a significant percentage increase of 375.4% occurred under GFDL, RCP 8.5 (2070–2099). Table 4 shows the percentage differences (future–baseline) between the simulated mean baseline values and simulated mean future values for yield, WUE, seasonal irrigation requirements and N leaching.

Table 4. Percentage differences (future–baseline) between the simulated mean baseline values and simulated mean future values for yield, WUE, seasonal irrigation requirements and N leaching.

Scenario and Period	Yield	Seasonal Irrigation	Water Use Efficiency	N Leaching
		BCC		
RCP 4.5 2040–2069	−8.2	−5.4	−4.3	−26.4
RCP 4.5 2070–2099	−7.4	−5.9	−5.1	108.8
RCP 8.5 2040–2069	−10.7	−5.9	−7.0	148.4
RCP 8.5 2070–2099	−15.6	−8.0	−14.1	215.5
		GFDL		
RCP 4.5 2040–2069	0.0	−1.9	0	17.4
RCP 4.5 2070–2099	−2.5	0.3	−2.8	39.4
RCP 8.5 2040–2069	−14.8	−13.6	−13.4	207.5
RCP 8.5 2070–2099	−20.8	−13.6	−21.7	375.4
		MIROC		
RCP 4.5 2040–2069	−8.2	−13.2	−6.6	113.5
RCP 4.5 2070–2099	−13.1	−8.7	−12.1	178.8
RCP 8.5 2040–2069	−10.7	−12.9	−9.6	153.2
RCP 8.5 2070–2099	−23.8	−13.6	−22.7	373.8

3.3.3. BCC Model

In the second future period, 2070–2099, where the gap from the baseline was more highlighted, maize yield was on average equal to 10.3 t ha^{-1} for RCP 8.5 and 11.3 t ha^{-1} for RCP 4.5. RCP 8.5 2070–2099 gave the most considerable yield difference from the baseline yield (Figure 5a). The seasonal irrigation amount showed a decreasing trend in the future periods compared to the baseline (Figure 5b). The decrease in seasonal irrigation amount was comparable between RCP 4.5 2040–2069, RCP 4.5 2070–2099 and RCP 8.5 2040–2099, with the three periods having similar seasonal irrigation requirements. RCP 8.5 2070–2099 had the largest seasonal irrigation requirement decrease compared to the baseline scenario, with a seasonal irrigation amount 8% lower than the baseline. Future WUE also showed a decreasing trend from the baseline scenario for all future periods (Figure 5c). The largest decrease in WUE was in RCP 8.5 2070–2099, which was 22.7% lower than the baseline WUE. N leaching increased in all future scenarios except in RCP 4.5 2040–2099, where N leaching slightly decreased compared to the baseline scenario (Figure 5d). RCP 8.5 2070–2099 had the largest increase in N leaching compared to the baseline scenario.

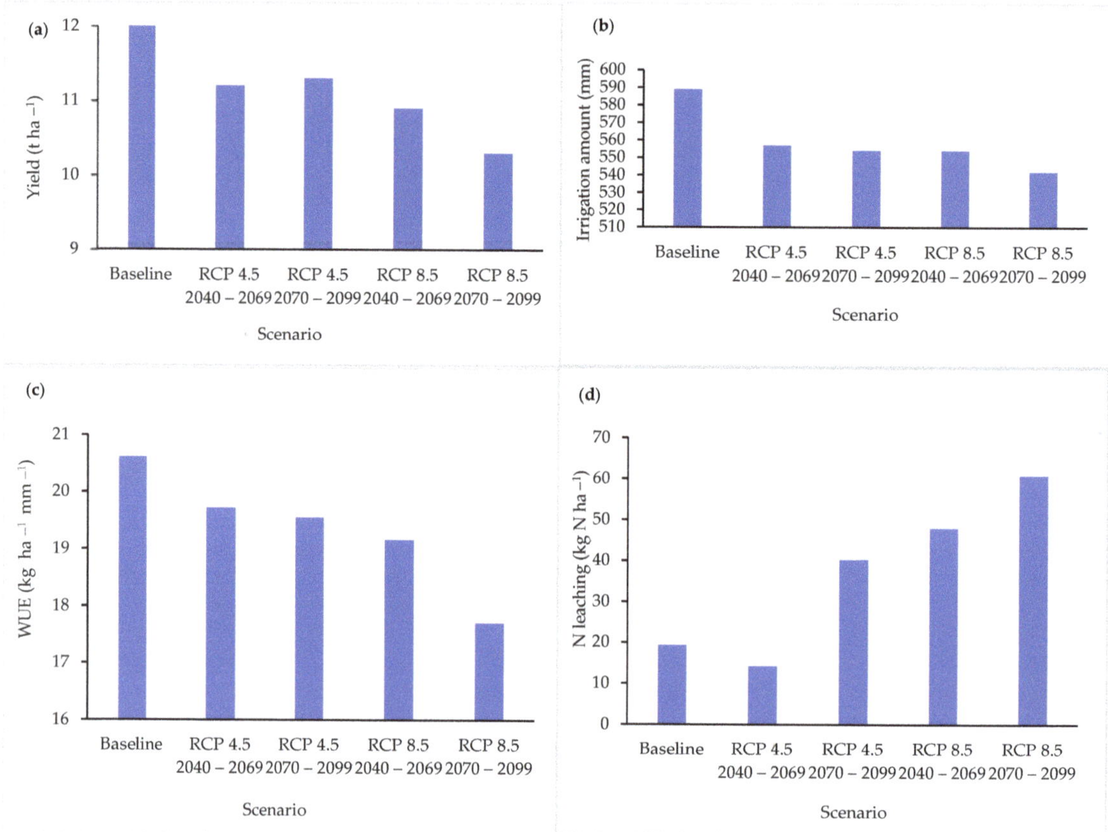

Figure 5. EPIC model outputs from the simulations using BCC-ESM climate data. Values are plotted and shown for the two 30-year periods compared to the baseline simulation; (**a**) yield, (**b**) seasonal irrigation, (**c**) water use efficiency (WUE), (**d**) N leaching.

3.3.4. GFDL Model

For the GFDL model, crop yield was similar to the baseline yields but slightly lower (Figure 6a). For RCP 4.5 and RCP 8.5, there were only slight differences in yield in the two future periods for the GFDL scenario. The two future scenarios for RCP 8.5 showed lower yields compared to both RCP 4.5 and the baseline scenario. Seasonal irrigation was similar to the baseline period for the two future periods in RCP 4.5. However, both future periods for RCP 8.5 showed a marked decrease in seasonal irrigation amount compared to the baseline scenario. The largest decrease in seasonal irrigation compared to the baseline scenario was observed for 2040–2069 in RCP 8.5 (Figure 6b). WUE slightly decreased in the future climate scenarios ranging from 15.91 kg ha^{-1} mm^{-1} in RCP 8.5 2070–2099 to 20.61 kg ha^{-1} mm^{-1} in RCP 4.5 2040–2069 compared to 20.61 kg ha^{-1} mm^{-1} in the baseline scenario (Figure 6c). N leaching increased in all future climate periods for all the scenarios compared to the baseline scenario (Figure 6d). RCP 8.5 2070–2099 had the largest increase in N leaching with an average of 91.64 kg N ha^{-1}.

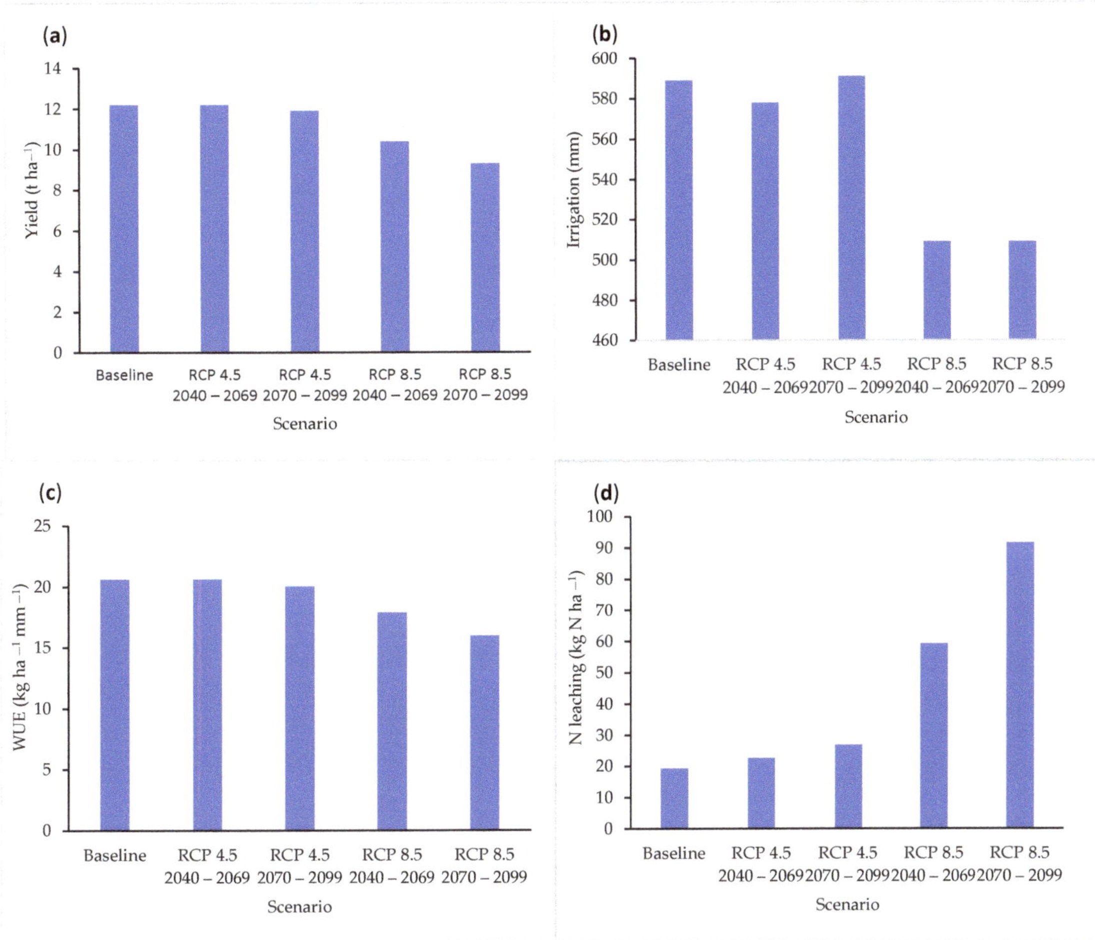

Figure 6. EPIC model outputs from the simulations using GFDL climate data. Values are plotted and shown for the two 30-year periods compared to the baseline simulation; (**a**) yield, (**b**) seasonal irrigation, (**c**) water use efficiency (WUE), (**d**) N leaching.

3.3.5. MIROC Model

The MIROC model showed a similar trend of decreasing yield for all the future periods with respect to the baseline period. Maize yield decreased by up to 23% in RCP 8.5 2070–2099 (Figure 7a). Seasonal irrigation also reduced significantly in the future periods for all RCPs. Seasonal irrigation amount decreased by up to 13% in RCP 4.5 2040–2069 and the two time periods for RCP 8.5 compared to the baseline period (Figure 7b). For WUE, the model simulated a slight decrease over time, particularly in RCP 8.5 2070–2099 (Figure 7c). N leaching increased for all future periods compared to the baseline scenario. RCP 8.5 2070–2099 had the most significant increase in N leaching compared to all the other periods for all three models (Figure 7d).

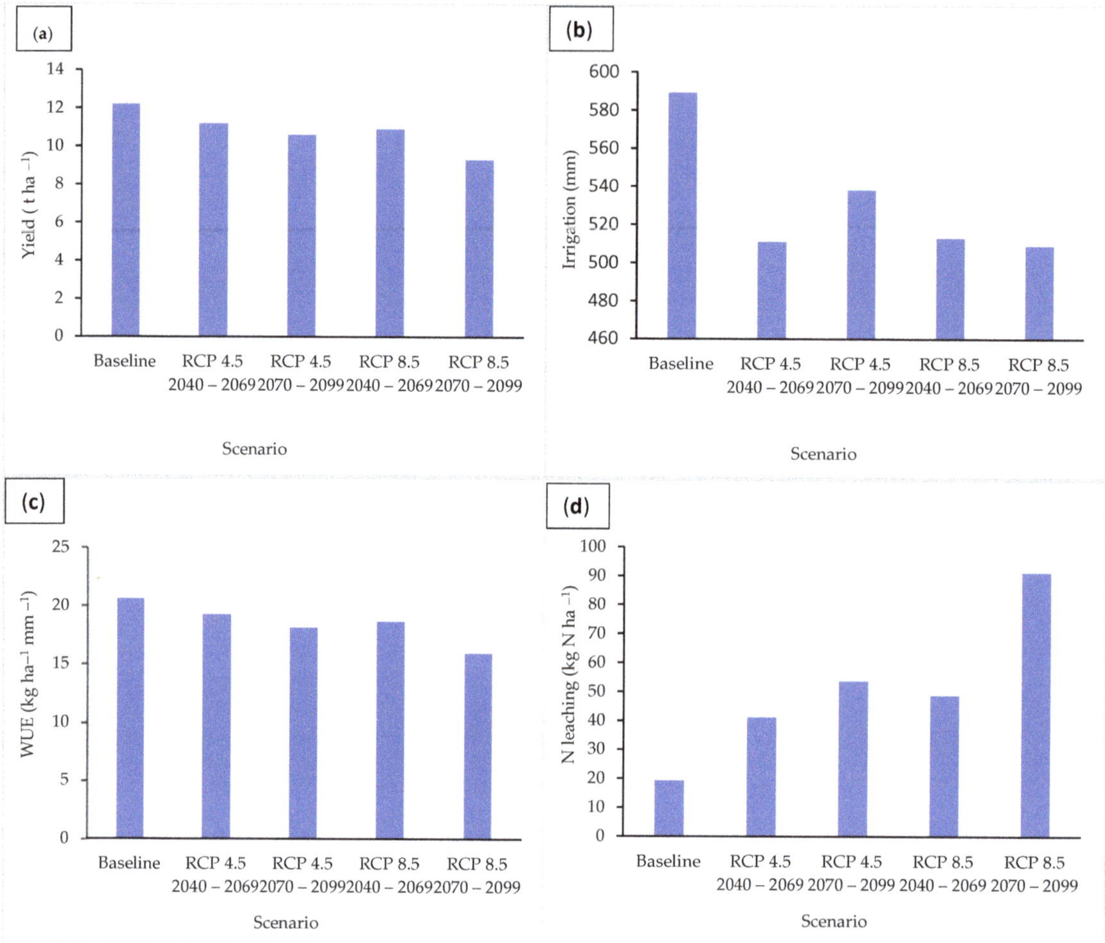

Figure 7. EPIC model outputs from the simulations using MIROC climate data. Values are plotted and shown for the two 30-year periods compared to the baseline simulation; (**a**) yield, (**b**) seasonal irrigation, (**c**) water use efficiency (WUE), (**d**) N leaching.

4. Discussion

4.1. EPIC Model Calibration and Validation

Notwithstanding the limited data available to calibrate and validate the model in this study, the calibration results revealed satisfactory agreement between observed and simulated yields. In the initial simulation with default parameters, the agreement between observed and simulated crop yields was unsatisfactory, suggesting the need for calibration. After adjustment of site-specific model parameters, the model performance improved, showing the value of calibrating models with parameters that are site-specific. Our results provide further evidence to support previous studies that have demonstrated that adjusting parameters with local-scale data increase can increase the accuracy of simulations and reduce model uncertainties considerably [55]. For example, Xiong et al. [56] and Angulo et al. [57] demonstrated that fine-tuning PHUs to local conditions could significantly improve model simulation accuracy. In this study, model simulations improved on adjusting the PHU value. The PHUs are closely related to biomass growth and its final yield allotment, indicating the substantial influence of PHU adjustment on simulated crop yields.

Trials conducted in the USA by Williams et al. [58] showed that the PHUs required for maize to reach maturity ranged between 1000–2900. In this study, 2480 PHUs brought model performance into the range set for satisfactory model calibration. The ARC in South Africa states that maize typically requires 120 days to mature from the day of planting. However, this period is hugely dependent on weather conditions and 120 days is generally for the warmer traditional maize growing regions in South Africa such as KwaZulu Natal [59]. The Cradock area is relatively cooler than the traditional maize growing regions in South Africa, which may account for the higher PHU value found in this study.

Concerning the HI, the default value in the EPIC model is 0.5, which is representative of HI values for improved high yielding maize varieties [60], similar to the varieties used in the field trials for this study. The default HI value of 0.5 used in this study, has been used in studies such as those by [39,61].

In this study, we did not adjust the biomass to energy ratio (WA) since adjusting PHUs improved the model performance considerably to within the range set for satisfactory model calibration. For example, PHU calibration gave an *RMSE* of 1.17 kg ha^{-1} and *PBIAS* of 0.31 between observed and simulated yields. The small *RMSE* and *PBIAS* values suggested that no additional WA and HI adjustments were required since the conditions for satisfactory model performance had been met. Regarding WA, we left WA at the default value of 40 kg ha^{-1} MJ^{-1} m^2. Similar studies have also used the value of 40 kg ha^{-1} MJ^{-1} m^2 for WA (see, e.g., [39,62]). The biomass to energy ratio can significantly influence crop yields [63], and [36] explains that WA can substantially alter crop growth and yield rate. Reference [36] further emphasises that WA should be adjusted only as a final resort and based on experimental data.

The EPIC model potentially overestimates yields, even at low observed yields during calibration and validation (see [32]. Studies conducted by [64,65] also found that the EPIC model tended to overestimate low observed yields. It has been suggested that the overestimation of plant available water at field capacity could potentially lead to the overestimation of yields in the dry years by the EPIC model (see [66]). Thus, Kiniry et al. [66] proposed measuring the maximum depth of water extraction using local cultivars as a solution. However, the solution was not applied as it is beyond the scope of the calibration and validation study. The overestimation observed in this study may be attributed to the influence of weed outbreaks. Agricultural management records used during the field trials note that in 2003 the maize fields were affected by heavy weed outbreaks. At the time of model calibration and validation, the EPIC model had not yet been developed to accurately account for competition from weeds [67]. As such, competition from weeds was not accounted for in the simulations, which may explain why the model overestimated the low yields observed in 2003.

4.2. Climate Change Impacts on Maize Yield

Model ensemble results predicted a decrease in maize yield for all future scenarios with a more pronounced reduction in RCP 8.5 2070–2099. This decrease can be attributed to an increased temperature that would shorten the growth stage of the maize crop. Increased temperature increases the rate of accumulation of growing degree days, thereby influencing growth duration. Several studies have shown that temperature increases lead to early crop maturing, allowing less time to accumulate biomass and form grain yield [68–70]. The projected decrease in maize yield in this study agrees with other studies in Southern Africa. For example, studies by [71] projected decreases in maize yield in Zimbabwe under irrigated and rain-fed agriculture. In their study, [71] used the CERES model driven by GCMs (specifically the GFDL and the Canadian Climate Centre Model). Walker and Schulze [72] also studied the response of smallholder maize production in Potshini village, KwaZulu-Natal, South Africa, up to the late 21st century climates. The study by [72] projected a decrease in average maize yields of approximately 30% and showed that more efficient management of fertiliser and manure applications would be a viable management strategy to adapt to climate change.

A study by [73] in Ethiopia for mid-century maize production projected a shortening of maize maturity period by approximately 9–13% due to elevated temperatures. The reduced maturity period would reduce the amount of time the maize crop was able to capture solar radiation and assimilate carbon dioxide, resulting in a reduction in biomass and yield accumulation [74]. Other studies such as those by [75,76] have reported that photosynthesis is affected by elevated temperatures and low water availability, which in turn can reduce the yield. In this study, projections showed an increase in temperature and decrease in rainfall during the early growing season, leading to a reduction in yield. Although rainfall is predicted to be lower in a portion of months in the growing season, studies have shown that maize requires the right amount and distribution of rainfall [77,78]. In this study, GCM projections predicted low rainfall in the critical growing months for maize. While there was an increase in rainfall in winter, the maize plant would already have been affected by water stress, and hence the reduction in yield.

Rainfall can also influence crop yield as water is key to crop growth and development. In this study, rainfall was predicted to decrease in the early months of the maize growing season. Similar to this study [79], found a shift in precipitation during the growing season. The shift in precipitation may affect yield as studies have shown maize to be sensitive to moisture amount and distribution [80]. Furthermore, the decrease in rainfall projected has implications for food production as rainfall supplements irrigation in the study area. Rainfall is the ultimate source of irrigation water in the study area. A reduction in rainfall would lead to decreased flows in the Great Fish River, leading to further water shortages in an already water-scarce area. Further water shortages would significantly impact food production in the area as the Great Fish River supplies most of the irrigation water used by conventional farmers in the area. A previous study by [81] showed that rain-fed maize yields in the Eastern Cape are very low without irrigation even when sufficient fertiliser is provided.

Regarding nitrate leaching, all future simulations predicted significant increases in N leaching. Generally, increases in temperature accelerate phenological development, leading to a shorter growing period and less nutrient uptake. The shorter growing period, coupled with the increased rainfall towards the end of the growing found in this study, can explain the increased leaching for the future period. The increased N leaching found in this study is similar to the findings of [16]. In the study by [16], under future climate scenarios, nitrate leaching was found to increase significantly compared to the baseline scenario. He et al. [16] attributed the increased leaching to the future high temperature stress and increased precipitations, explaining that the high temperature stress and increased precipitations resulted in low crop N removal and increased drainage. Without matching the amount of fertiliser applied to crop N needs, excess N can be lost to the environment through leaching. This indicates the need to take into consideration the impacts of climate change on N leaching when developing future agricultural land management strategies aimed at maximising the use of N by plants and minimising N losses to the environment.

Considering the predicted impacts of climate change in the study area, farmers may need to obtain financial and technical support to implement on-farm water adaptation strategies such as rainwater harvesting and the use of field water conservation strategies such as mulching. Several studies analysing climate and weather trends in South Africa have shown that average temperatures in the country have increased in the last decades [35,82,83]. A study by [50] on observed and modelled trends for rainfall and temperature for South Africa found significant increases in temperature and rainfall variability in the Eastern Cape. Temperature increases and the decreased rainfall season length predicted in this study suggest that short-term growing maize varieties and drought-tolerant maize varieties may be needed in the Eastern Cape if crop production is to be sustained.

It is worth noting that we did not consider farmers implementing agricultural land management strategies aimed at minimising the effects of climate change in the simulations. This is unlikely to be the case in practice. Agroecosystems are human-managed, and farmers have a variety of possible adaptation options [84,85]. While the study did not show possible yield changes due to the implementation of climate change adaptation measures, the study does provide a clear picture on maize yield and N leaching rates if no climate change adaptation measures are taken. While there is uncertainty associated with climate projections, several studies in sub-Saharan Africa (e.g., [86,87]) have shown that projections of climate impacts appear robust across model ensembles [11].

However, the results of climate impact studies should not be taken in absolute terms but rather as possible pathways for the future of maize production in the Eastern Cape. Decision makers should consider other factors that may influence crop yield. In this study, the combined influence of other factors such as the development of pests and disease on crop yield was assumed to be fully controlled through appropriate management practices. This study's results can be used by farmers and policymakers to plan how to adapt to the projected increases in temperature and decreased rainfall. It is vital to develop adaptation strategies that consider the projected increases in temperature and minimise N leaching. N leaching represents an economic loss to farmers (N fertiliser not utilised by plants) and a potential water pollutant. It is recommended that studies that test the effectiveness of adaptation strategies and current and future climate scenarios using the EPIC model be carried out in the region.

5. Limitations of the Study

Downscaled climate projections inescapably inherit uncertainties from GCMs. Sources of uncertainty arise from internal variability of the model, the greenhouse gas emission scenario used (RCPs), the statistical downscaling process and imperfections in the GCMs from which the downscaled data were derived. Other sources include using only one crop model (EPIC) to project the impacts of climate change on crop yield. Asseng et al. [88] suggested that ensembles of many crop models could give a better estimate of yield than using one model. However, the use of multiple models was beyond the scope of this study.

The results of climate change effects are prone to many uncertainties resulting from the limited knowledge of underlying geophysical processes of global change (GCM uncertainties) and uncertain future scenarios (emission scenario uncertainties) [19]. Uncertainties in climate projections with respect to climate models can have significant impacts on crop model outputs [89,90]. To reduce uncertainties associated with individual climate models, three different models under two contrasting climate scenarios were selected to capture the full range of changes in temperature and precipitation projected by the models. Reference [91] states that emission scenario uncertainties are less relevant until the middle of the 21st century; hence, the 2040–2069 scenario was chosen as the starting period for future climate simulations.

In this study, carbon dioxide (CO_2) fertilisation effects were not considered due to the lack of site-specific annual data on future CO_2 levels for the periods used in the scenarios. Klein [89] explains that model equations are all subject to variability and uncertainty. As a result, processes included in simulation models, such as CO_2 fertilisation effects, may not always be fully understood or well implemented. For example, Free Air Carbon Enrichment (FACE) experiments indicate productivity increases due to increased CO_2 levels but do not address important co-limitations arising from water and nutrient availability [89]. The magnitude of crops' responses to increased CO_2 levels is thus uncertain and the subject of current debates among researchers [2,92–94]. Biernath et al. [95] argue that many crop models are currently unable to capture the complex underlying processes associated with CO_2 fertilization and are therefore unable to reproduce experimental results.

Additionally, we assumed crop management such as fertilisation to be similar across the future periods, which may not be the case in reality as farmers adapt to changing farming conditions. Additionally, by considering one maize cultivar, we assumed the single cultivar would give similar responses to the impacts of climate change as those of different cultivars.

6. Conclusions

EPIC simulations predict that climate change will negatively affect maize production and environmental water quality in the Eastern Cape. Maize yields are projected to decrease, accompanied by an increase in N leaching. Mitigating the future impacts of climate change will be vital to enhancing food security in the region. Models such as EPIC can help predict and anticipate the possible effects of climate change on crop production and help plan appropriate agricultural land management responses that contribute to sustainable food production in South Africa. In this regard, this study's results have demonstrated that the EPIC model can be considered a valuable tool for exploring the future impacts of climate change on crop yields and the environment. Future studies using EPIC should test the effectiveness of various crop rotation and intercropping strategies based on farmers' current crop rotation and intercropping strategies.

Author Contributions: Conceptualization, D.J.C.; methodology, D.J.C.; software, D.J.C.; validation, D.J.C.; formal analysis, D.J.C.; investigation, D.J.C.; resources, N.O.O.; data curation, D.J.C. and N.O.O.; writing—original draft preparation, D.J.C.; writing—review and editing, D.J.C., F.C.A. and N.O.O.; visualization, D.J.C. and F.C.A.; supervision, D.J.C.; project administration, D.J.C. and N.O.O.; funding acquisition, D.J.C. and N.O.O. All authors have read and agreed to the published version of the manuscript.

Funding: This article is the outcome of research conducted within the Rhodes University African Studies Centre (RASC), funded by the Deutsche Forschungsgemeinschaft (DFG, German Research Foundation) under Germany is Excellence Strategy–EXC 2052/1–390713894 and the National Research Foundation (NRF) of South Africa under the Southern African Systems Analysis Centre (SASAC) initiative.

Institutional Review Board Statement: Not applicable.

Informed Consent Statement: Not applicable.

Data Availability Statement: Not applicable.

Conflicts of Interest: The authors declare no conflict of interest. The funders had no role in the design of the study; in the collection, analyses, or interpretation of data; in the writing of the manuscript, or in the decision to publish the results.

Appendix A

Supplementary information on the calibration and validation of the EPIC model.

Table A1. Representative soil characteristics of the Cradock Research Farm used as inputs into the Environmental Policy Integrated Climate (EPIC) model (obtained from [32]). Clay, sand, silt, soil organic carbon units are in percentages, whereas bulk density, soil organic carbon and ion exchange capacity are in g cm^{-3} m and (cmol (+) kg^{-1}), respectively.

Soil Parameters	Soil Layer Number	
	1	2
Bulk density	1.48	1.52
Soil depth	0.3	1.2
Clay	20.4	15.1
Sand	52.8	42.5
Silt	26.8	42.4
pH	6.5	6.5
Soil organic carbon	0.91	0.2
Cation exchange capacity	14.3	13.4

Table A2. Showing the agricultural management plan used during the study period (table obtained from [32]).

Date [1]	Operation	Type	Amount
22 October	Planting	Maize	50,000 plants ha^{-1}
22 October	Fertilizer application	Superphosphate	476 kg ha^{-1}
22 October	Fertilizer application	Ammonium sulfate	330 kg ha^{-1}
22 October	Fertilizer application	Calcium sulfate	120 kg ha^{-1}
22 October	Irrigation	Furrow	75 mm
15 November	Fertiliser application	Ammonium sulfate	300 kg ha^{-1}
26 November	Irrigation	Furrow	75 mm
10 December	Fertiliser application	Ammonium sulfate	300 kg ha^{-1}
17 December	Irrigation	Furrow	75 mm
28 December	Irrigation	Furrow	75 mm
18 January	Irrigation	Furrow	75 mm
8 February	Irrigation	Furrow	75 mm
19 February	Irrigation	Furrow	75 mm
11 March	Irrigation	Furrow	75 mm
5 June	Harvesting	Manual	11 tonnes hectare^{-1} (average)

[1] The dates given in the table are not fixed for each year. They indicate the approximate times of year each management activity was carried out during the trial period.

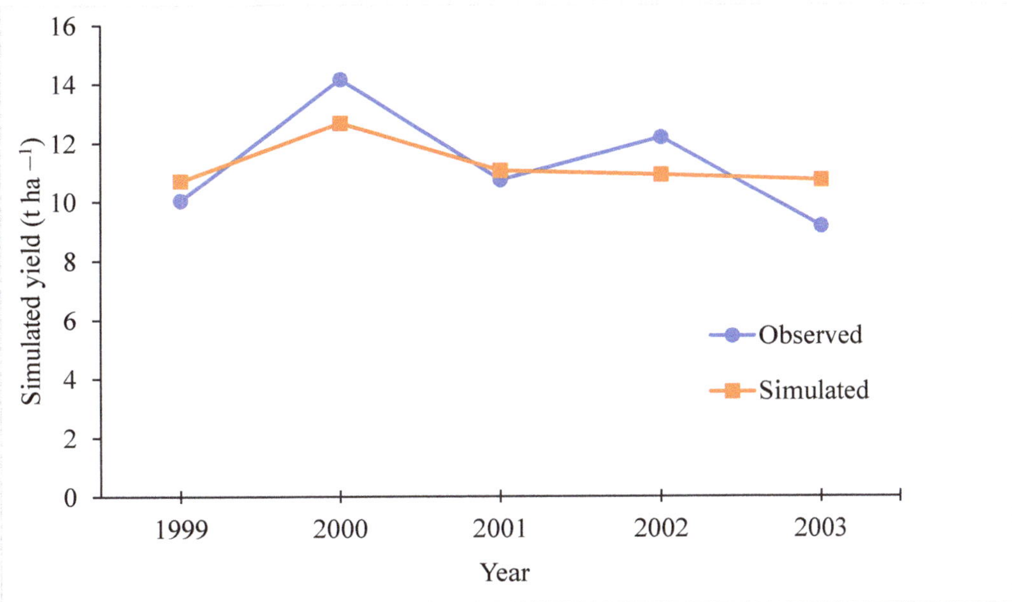

Figure A1. Showing the crop yields (observed and simulated) since the model was in the range set for acceptable model calibration for the study period after PHU calibration (figure obtained from [32]).

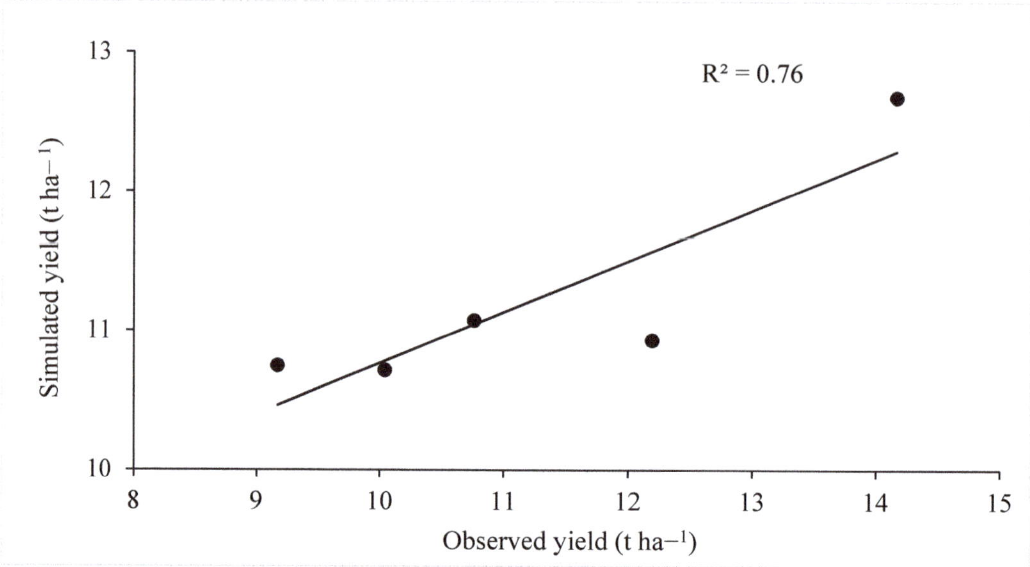

Figure A2. Showing the simulated crop yields on observed maize yields with the calibrated maize crop file (figure obtained from [32]).

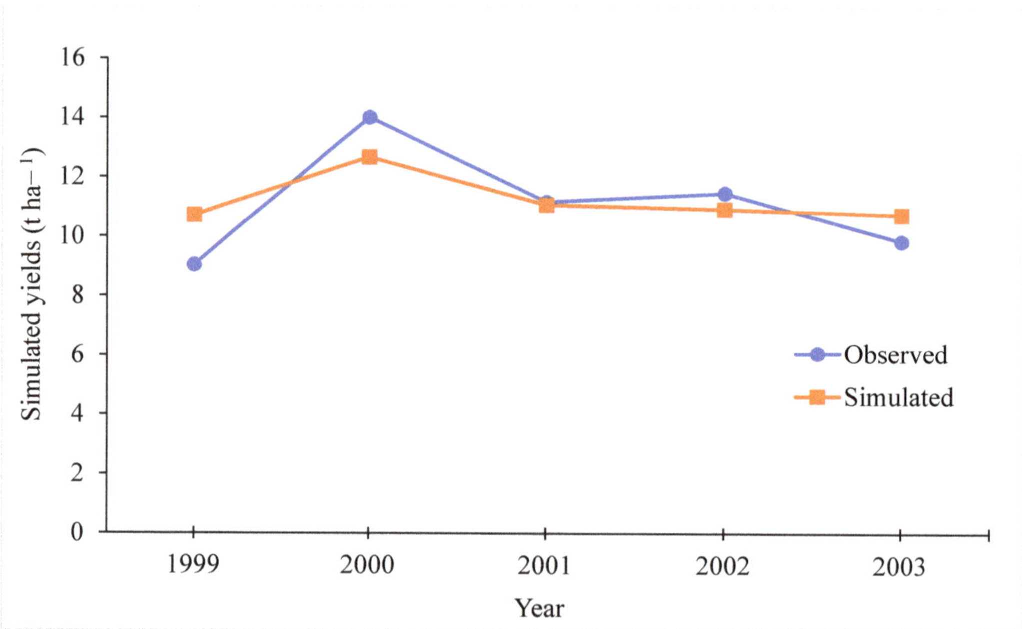

Figure A3. Simulated yields in the validation simulation using the calibrated model (figure obtained from [32]).

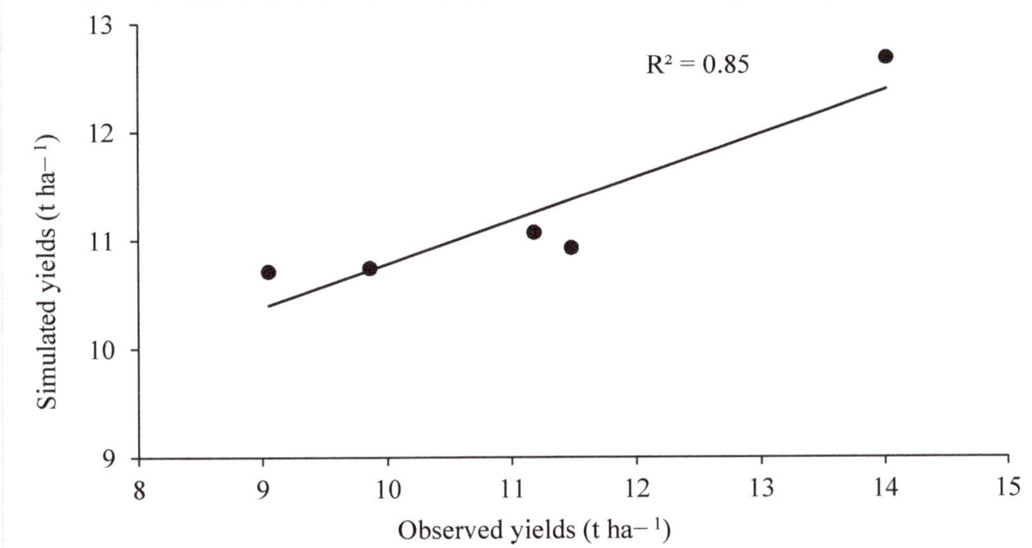

Figure A4. Simulated crop yields (t ha^{-1}) regression result on observed maize yields (figure obtained from [32]).

References

1. Pachauri, R.K.; Allen, M.R.; Barros, V.R.; Broome, J.; Cramer, W.; Christ, R.; Church, J.A.; Clarke, L.; Dahe, Q.; Dasgupta, P.; et al. *Climate Change 2014: Synthesis Report. Contribution of Working Groups I, II and III to the Fifth Assessment Report of the Intergovernmental Panel on Climate Change*; Pachauri, R.K., Meyer, L.A., Eds.; IPCC: Geneva, Switzerland, 2014.
2. Long, S.P.; Ainsworth, E.A.; Leakey, A.D.B.; Noösberger, J.; Ort, D.R. Food for Thought: Lower-Than-Expected Crop Yield Stimulation with Rising CO_2 Concentrations. *Science* **2006**, *312*, 1918–1921. [CrossRef] [PubMed]
3. Webber, H.; Gaiser, T.; Ewert, F. What role can crop models play in supporting climate change adaptation decisions to enhance food security in Sub-Saharan Africa? *Agric. Syst.* **2014**, *127*, 161–177. [CrossRef]
4. Stuart-Hill, S.; Schulze, R.E.; Colvin, J. *Handbook on Adaptive Management Strategies and Options for the Water Sector in S. Afr. under Climate Change*; Water Research Commission: Pretoria, South Africa, 2012; ISBN 9781431202706.
5. Jury, M.R. Climate trends in southern Africa. *S. Afr. J. Sci.* **2013**, *109*, 1–11. [CrossRef]
6. Nhamo, L.; Mabhaudhi, T.; Modi, A. Preparedness or repeated short-term relief aid? Building drought resilience through early warning in southern Africa. *Water SA* **2019**, *45*, 75–85. [CrossRef]
7. Mulungu, K.; Ng'Ombe, J.N. Climate Change Impacts on Sustainable Maize Production in Sub-Saharan Africa: A Review. In *Maize-Production and Use*; IntechOpen Limited: London, UK, 2020. [CrossRef]
8. Agbugba, I.K.; Christian, M.; Obi, A. Economic analysis of smallholder maize farmers: Implications for public extension services in Eastern Cape. *S. Afr. J. Agric. Ext.* **2020**, *48*, 50–63. [CrossRef]
9. Chimonyo, V.G.P.; Mutengwa, C.S.; Chiduza, C.; Tandzi, L. Characteristics of maize growing farmers, varietal use and constraints to increase productivity in selected villages in the Eastern Cape province of South Africa. *S. Afr. J. Agric. Ext.* **2020**, *48*, 64–82. [CrossRef]
10. Kogo, B.K.; Kumar, L.; Koech, R.; Langat, P. Modelling Impacts of Climate Change on Maize (*Zea mays* L.) Growth and Productivity: A Review of Models, Outputs and Limitations. *J. Geosci. Environ. Prot.* **2019**, *7*, 76–95. [CrossRef]
11. Corbeels, M.; Berre, D.; Rusinamhodzi, L.; Lopez-Ridaura, S. Can we use crop modelling for identifying climate change adaptation options? *Agric. For. Meteorol.* **2018**, *256–257*, 46–52. [CrossRef]
12. Jacobson, K. *From Betterment to Bt Maize*; Swedish University of Agricultural Sciences: Uppsala, Sweden, 2013.
13. Khapayi, M.; Celliers, P.R. Factors limiting and preventing emerging farmers to progress to commercial agricultural farming in the King William's Town area of the Eastern Cape Province, South Africa. *S. Afr. J. Agric. Ext. (SAJAE)* **2016**, *44*, 25–41. [CrossRef]
14. Goldblatt, A. *Agriculture: Facts and Trends South Africa*; World Wide Fund for Nature: Gland, Switzerland, 2011; pp. 2–26.
15. Uzoma, K.C.; Smith, W.; Grant, B.; Desjardins, R.L.; Gao, X.; Hanis, K.; Tenuta, M.; Goglio, P.; Li, C. Assessing the effects of agricultural management on nitrous oxide emissions using flux measurements and the DNDC model. *Agric. Ecosyst. Environ.* **2015**, *206*, 71–83. [CrossRef]

16. He, W.; Yang, J.Y.; Qian, B.; Drury, C.F.; Hoogenboom, G.; He, P.; Lapen, D.; Zhou, W. Climate change impacts on crop yield, soil water balance and nitrate leaching in the semiarid and humid regions of Canada. *PLoS ONE* **2018**, *13*, e0207370. [CrossRef] [PubMed]
17. Wang, Z.; Qi, Z.; Xue, L.; Bukovsky, M.; Helmers, M.J. Modeling the impacts of climate change on nitrogen losses and crop yield in a subsurface drained field. *Clim. Chang.* **2015**, *129*, 323–335. [CrossRef]
18. Basche, A.; Archontoulis, S.V.; Kaspar, T.C.; Jaynes, D.B.; Parkin, T.B.; Miguez, F.E. Simulating long-term impacts of cover crops and climate change on crop production and environmental outcomes in the Midwestern United States. *Agric. Ecosyst. Environ.* **2016**, *218*, 95–106. [CrossRef]
19. Folberth, C. Modeling Crop Yield and Water Use in the Context of Global Change with a Focus on Maize in Sub-Saharan Africa. Doctoral Thesis, ETH-Zürich, Zürich, Switzerland, 2013.
20. Warburton, M.; Schulze, R.; Jewitt, G. *Hydrological Responses to Combined Land-Use and Climate Change in Three Diverse South African Catchments*; IAHS Publication: Wallingford, UK, 2013; Volume 359.
21. Abraha, M.; Savage, M. Potential impacts of climate change on the grain yield of maize for the midlands of KwaZulu-Natal, South Africa. *Agric. Ecosyst. Environ.* **2006**, *115*, 150–160. [CrossRef]
22. Navarro-Racines, C.E.; Tarapues, J.; Thornton, P.; Jarvis, A.; Ramirez-Villegas, J. High-resolution and bias-corrected CMIP5 projections for climate change impact assessments. *Sci. Data* **2020**, *7*, 1–14. [CrossRef] [PubMed]
23. Tang, J.; Niu, X.; Wang, S.; Gao, H.; Wang, X.; Wu, J. Statistical downscaling and dynamical downscaling of regional climate in China: Present climate evaluations and future climate projections. *J. Geophys. Res. Atmos.* **2016**, *121*, 2110–2129. [CrossRef]
24. Kaini, S.; Nepal, S.; Pradhananga, S.; Gardner, T.; Sharma, A.K. Representative general circulation models selection and downscaling of climate data for the transboundary Koshi river basin in China and Nepal. *Int. J. Clim.* **2020**, *40*, 4131–4149. [CrossRef]
25. Ziervogel, G.; Shale, M.; DU, M. Climate change adaptation in a developing country context: The case of urban water supply in Cape Town. *Clim. Dev.* **2010**, *2*, 94–110. [CrossRef]
26. Tisseuil, C.; Vrac, M.; Lek, S.; Wade, A. Statistical downscaling of river flows. *J. Hydrol.* **2010**, *385*, 279–291. [CrossRef]
27. Tadross, M.; Jack, C.; Hewitson, B. On RCM-based projections of change in southern African summer climate. *Geophys. Res. Lett.* **2005**, *32*. [CrossRef]
28. Oosthuizen, H.; Schulze, R.; Crespo, O.; Louw, D.; Tadross, M.; Waagsaether, K.; Arowolo, S. *Modelling Impacts of Climate Change on Selected South African Crop Farming Sysytems*; Water Research Commission: Gezina, South Africa, 2016.
29. Siabi, E.K.; Kabobah, A.T.; Akpoti, K.; Anornu, G.K.; Amo-Boateng, M.; Nyantakyi, E.K. Statistical downscaling of global circulation models to assess future climate changes in the Black Volta basin of Ghana. *Environ. Chall.* **2021**, *5*, 100299. [CrossRef]
30. Johnson, K.A.; Smithers, J.C.; E Schulze, R. A review of methods to account for impacts of non-stationary climate data on extreme rainfalls for design rainfall estimation in South Africa. *J. S. Afr. Inst. Civ. Eng.* **2021**, *63*, 55–61. [CrossRef]
31. Ziervogel, G.; New, M.; van Garderen, E.A.; Midgley, G.; Taylor, A.; Hamann, R.; Stuart-Hill, S.; Myers, J.; Warburton, M. Climate change impacts and adaptation in South Africa. *WIREs Clim. Chang.* **2014**, *5*, 605–620. [CrossRef]
32. Choruma, D.J.; Balkovic, J.; Odume, O.N. Calibration and Validation of the EPIC Model for Maize Production in the Eastern Cape, South Africa. *Agronomy* **2019**, *9*, 494. [CrossRef]
33. USDA-Natural Resources Conservation Service. *Soil Survey Staff Keys to Soil Taxonomy*, 12th ed.; USDA-Natural Resources Conservation Service: Washington, DC, USA, 2014.
34. Mahlalela, P.T.; Blamey, R.C.; Hart, N.; Reason, C.J.C. Drought in the Eastern Cape region of South Africa and trends in rainfall characteristics. *Clim. Dyn.* **2020**, *55*, 2743–2759. [CrossRef]
35. Department of Environmental Affairs Long-Term Adaptation Scenarios Flagship Research Programme (LTAS) for South Africa. *Climate Change Implications for the Water Sector in South Africa*; SANBI: Pretoria, South Africa, 2013; Volume 1.
36. Williams, J.R.; Gerik, T.; Dagitz, S.; Magre, M.; Meinardus, A.; Steglich, E.; Taylor, R. *Environmental Policy Integrated Climate Model-Users Manual Version 0810*; Blackland Research and Extension Centre: Temple, TX, USA, 2015.
37. Xiong, W.; Skalský, R.; Porter, C.H.; Balkovič, J.; Jones, J.W.; Yang, D. Calibration-induced uncertainty of the EPIC model to estimate climate change impact on global maize yield. *J. Adv. Model. Earth Syst.* **2016**, *8*, 1358–1375. [CrossRef]
38. Rinaldi, M.; De Luca, D. Application of EPIC model to assess climate change impact on sorghum in southern Italy. *Ital. J. Agron.* **2012**, *7*, 12. [CrossRef]
39. Balkovic, J.; van der Velde, M.; Schmid, E.; Skalský, R.; Khabarov, N.; Obersteiner, M.; Stürmer, B.; Xiong, W. Pan-European crop modelling with EPIC: Implementation, up-scaling and regional crop yield validation. *Agric. Syst.* **2013**, *120*, 61–75. [CrossRef]
40. Clewer, A.G.; Scarisbrick, D.H. *Practical Statistics and Experimental Design for Plant and Crop Science*; Wiley: Hoboken, NJ, USA, 2001; ISBN 9780471899082.
41. Ruane, A.C.; Goldberg, R.; Chryssanthacopoulos, J. Climate forcing datasets for agricultural modeling: Merged products for gap-filling and historical climate series estimation. *Agric. For. Meteorol.* **2015**, *200*, 233–248. [CrossRef]
42. FAO; IIASA; ISRIC; ISS-CAS; JRC. *Harmonized World Soil Database*; Version 1.1; FAO: Rome, Italy; Laxenburg, Austria, 2009.
43. Balkovič, J.; van der Velde, M.; Skalský, R.; Xiong, W.; Folberth, C.; Khabarov, N.; Smirnov, A.; Mueller, N.D.; Obersteiner, M. Global wheat production potentials and management flexibility under the representative concentration pathways. *Glob. Planet. Chang.* **2014**, *122*, 107–121. [CrossRef]

44. Schulze, R.; Maharaj, M. *A-Pan Equivalent Reference Potential Evaporation*; Schulze, R.E., Ed.; University of KwaZulu-Natal: Durban, South Africa, 2007.
45. Bao, Y.; Hoogenboom, G.; McClendon, R.; Vellidis, G. A comparison of the performance of the CSM-CERES-Maize and EPIC models using maize variety trial data. *Agric. Syst.* **2017**, *150*, 109–119. [CrossRef]
46. Moriasi, D.N.; Arnold, J.G.; van Liew, M.W.; Bingner, R.L.; Harmel, R.D.; Veith, T.L. Model Evaluation Guidelines for Systematic Quantification of Accuracy in Watershed Simulations. *Trans. ASABE* **2007**, *50*, 885–900. [CrossRef]
47. Gupta, H.V.; Sorooshian, S.; Yapo, P.O. Status of Automatic Calibration for Hydrologic Models: Comparison with Multilevel Expert Calibration. *J. Hydrol. Eng.* **1999**, *4*, 135–143. [CrossRef]
48. Wang, X.; Williams, J.R.; Gassman, P.W.; Baffaut, C.; Izaurralde, R.; Jeong, J.; Kiniry, J.R. EPIC and APEX: Model Use, Calibration, and Validation. *Trans. ASABE* **2012**, *55*, 1447–1462. [CrossRef]
49. Taylor, K.E.; Stouffer, R.J.; Meehl, G.A. An Overview of CMIP5 and the Experiment Design. *Bull. Am. Meteorol. Soc.* **2012**, *93*, 485–498. [CrossRef]
50. Mackellar, N.; New, M.; Jack, C. Observed and modelled trends in rainfall and temperature for South Africa: 1960–2010. *S. Afr. J. Sci.* **2014**, *110*, 1–13. [CrossRef]
51. Lekalakala, G. Options for Managing Climate Risk and Climate Change Adaptation in Smallholder Farming Systems of the Limpopo Province, South Africa. Ph.D. Thesis, Georg-August-University Göttingen, Göttingen, Germany, 2017.
52. Wise, M.; Calvin, K.; Thomson, A.; Clarke, L.; Bond-Lamberty, B.; Sands, R.; Smith, S.J.; Janetos, A.; Edmonds, J. Implications of Limiting CO_2 Concentrations for Land Use and Energy. *Science* **2009**, *324*, 1183–1186. [CrossRef]
53. Smith, S.J.; Wigley, T. Multi-Gas Forcing Stabilization with Minicam. *Energy J.* **2006**, *27*, 373–391. [CrossRef]
54. Riahi, K.; Rao, S.; Krey, V.; Cho, C.; Chirkov, V.; Fischer, G.; Kindermann, G.E.; Nakicenovic, N.; Rafaj, P. RCP 8.5—A scenario of comparatively high greenhouse gas emissions. *Clim. Chang.* **2011**, *109*, 33–57. [CrossRef]
55. Niu, X.; Easterling, W.; Hays, C.J.; Jacobs, A.; Mearns, L. Reliability and input-data induced uncertainty of the EPIC model to estimate climate change impact on sorghum yields in the U.S. Great Plains. *Agric. Ecosyst. Environ.* **2009**, *129*, 268–276. [CrossRef]
56. Xiong, W.; Balkovič, J.; van der Velde, M.; Zhang, X.; Izaurralde, R.; Skalský, R.; Lin, E.; Mueller, N.; Obersteiner, M. A calibration procedure to improve global rice yield simulations with EPIC. *Ecol. Model.* **2014**, *273*, 128–139. [CrossRef]
57. Angulo, C.; Rötter, R.; Lock, R.; Enders, A.; Fronzek, S.; Ewert, F. Implication of crop model calibration strategies for assessing regional impacts of climate change in Europe. *Agric. For. Meteorol.* **2013**, *170*, 32–46. [CrossRef]
58. Williams, J.R.; Jones, C.A.; Kiniry, J.R.; Spanel, D.A. The EPIC Crop Growth Model. *Trans. ASAE* **1989**, *32*, 0497–0511. [CrossRef]
59. du Plessis, J. *Maize Production*; Department of Agriculture: Pretoria, South Africa, 2003.
60. Folberth, C.; Yang, H.; Gaiser, T.; Abbaspour, K.C.; Schulin, R. Modeling maize yield responses to improvement in nutrient, water and cultivar inputs in sub-Saharan Africa. *Agric. Syst.* **2013**, *119*, 22–34. [CrossRef]
61. Kiniry, J.R.; Williams, J.R.; Vanderlip, R.L.; Atwood, J.D.; Reicosky, D.C.; Mulliken, J.; Cox, W.J.; Mascagni, H.J.; Hollinger, S.E.; Wiebold, W.J. Evaluation of Two Maize Models for Nine U.S. Locations. *Agron. J.* **1997**, *89*, 421–426. [CrossRef]
62. Wang, X.C.; Li, J.; Tahir, M.N.; De Hao, M. Validation of the EPIC model using a long-term experimental data on the semi-arid Loess Plateau of China. *Math. Comput. Model.* **2011**, *54*, 976–986. [CrossRef]
63. Wang, X.; He, X.; Williams, J.R.; Izaurralde, R.; Atwood, J.D. Sensitivity an Uncertainity Analysis of Crop Yields and Soil Organic Carbon Simulated with EPIC. *Trans. ASAE* **2005**, *48*, 1041–1054. [CrossRef]
64. Martin, S.M.; Nearing, M.A.; Bruce, R.R. An Evaluation of the EPIC Model for Soybeans Grown in Southern Piedmont Soils. *Trans. ASAE* **1993**, *36*, 1327–1331. [CrossRef]
65. Warner, G.; Stake, J.; Guillard, K. Evaluation of EPIC for a Shallow New England Soil: I. Maize Yield and Nitrogen Uptake. *Trans. ASABE* **1997**, *40*, 575–583. [CrossRef]
66. Kiniry, J.R.; Williams, J.R.; Major, D.J.; Izaurralde, R.C.; Gassman, P.W.; Morrison, M.; Bergentine, R.; Zentner, R.P. EPIC model parameters for cereal, oilseed, and forage crops in the northern Great Plains region. *Can. J. Plant Sci.* **1995**, *75*, 679–688. [CrossRef]
67. Ko, J.; Piccinni, G.; Steglich, E. Using EPIC model to manage irrigated cotton and maize. *Agric. Water Manag.* **2009**, *96*, 1323–1331. [CrossRef]
68. Craufurd, P.Q.; Wheeler, T.R. Climate change and the flowering time of annual crops. *J. Exp. Bot.* **2009**, *60*, 2529–2539. [CrossRef] [PubMed]
69. Dominguez-Faus, R.; Folberth, C.; Liu, J.; Jaffe, A.M.; Alvarez, P.J.J. Climate Change Would Increase the Water Intensity of Irrigated Corn Ethanol. *Environ. Sci. Technol.* **2013**, *47*, 6030–6037. [CrossRef] [PubMed]
70. Sacks, W.J.; Deryng, D.; Foley, J.A.; Ramankutty, N. Crop planting dates: An analysis of global patterns. *Glob. Ecol. Biogeogr.* **2010**, *19*, 607–620. [CrossRef]
71. Matarira, C.H.; Makadho, J.M.; Mwamuka, F.C. Climate Change Impacts on Maize Production and Adaptive Measures for the Agricultural Sector. In *Interim Report on Climate Change Country Studies*; Ramos-Mane, C., Benioff, R., Eds.; US Country Studies Program: Washington, DC, USA, 1995.
72. Walker, N.; Schulze, R. An assessment of sustainable maize production under different management and climate scenarios for smallholder agro-ecosystems in KwaZulu-Natal, South Africa. *Phys. Chem. Earth Parts A/B/C* **2006**, *31*, 995–1002. [CrossRef]
73. Araya, A.; Kisekka, I.; Lin, X.; Prasad, P.V.; Gowda, P.; Rice, C.; Andales, A. Evaluating the impact of future climate change on irrigated maize production in Kansas. *Clim. Risk Manag.* **2017**, *17*, 139–154. [CrossRef]

74. Bassu, S.; Brisson, N.; Durand, J.-L.; Boote, K.; Lizaso, J.; Jones, J.W.; Rosenzweig, C.; Ruane, A.C.; Adam, M.; Baron, C.; et al. How do various maize crop models vary in their responses to climate change factors? *Glob. Chang. Biol.* **2014**, *20*, 2301–2320. [CrossRef]
75. Islam, A.; Ahuja, L.R.; Garcia, L.A.; Ma, L.; Saseendran, A.S.; Trout, T. Modeling the impacts of climate change on irrigated corn production in the Central Great Plains. *Agric. Water Manag.* **2012**, *110*, 94–108. [CrossRef]
76. Olesen, J.E.; Bindi, M. Consequences of climate change for European agricultural productivity, land use and policy. *Eur. J. Agron.* **2002**, *16*, 239–262. [CrossRef]
77. Yin, G.; Gu, J.; Zhang, F.; Hao, L.; Cong, P.; Liu, Z. Maize Yield Response to Water Supply and Fertilizer Input in a Semi-Arid Environment of Northeast China. *PLoS ONE* **2014**, *9*, e86099. [CrossRef]
78. Durodola, O.S.; Mourad, K.A. Modelling Maize Yield and Water Requirements under Different Climate Change Scenarios. *Climate* **2020**, *8*, 127. [CrossRef]
79. Chisanga, C.B.; Phiri, E.; Chinene, V.R.N.; Chabala, L.M. Projecting maize yield under local-scale climate change scenarios using crop models: Sensitivity to sowing dates, cultivar, and nitrogen fertilizer rates. *Food Energy Secur.* **2020**, *9*. [CrossRef]
80. Mapfumo, P.; Chagwiza, C.; Antwi, M. Impact of Rainfall Variability on Maize Yield in the KwaZulu-Natal, North-West and Free State Provinces of South Africa (1987–2017). *J. Agribus. Rural. Dev.* **2020**, *58*, 359–367. [CrossRef]
81. Mnkeni, P.; Chiduza, C.; Modi, A.T.; Stevens, J.B. *Best Management Practices for Smallholder Farming on Two Irrigation Schemes in the Eastern Cape and Kwazulu-Natal Through Participatory Adaptive Research*; Water Research Commission: Pretoria, South Africa, 2010.
82. Jury, M.R. Climate trends across South Africa since 1980. *Water SA* **2018**, *44*, 297–307. [CrossRef]
83. Davis-Reddy, C.L.; Vincent, K. *Climate Change Handbook for Southern Africa Climate*, 2nd ed.; CSIR: Pretoria, South Africa, 2017; ISBN 9780620765220.
84. Wang, J.; Wang, E.L.; Yang, X.G.; Zhang, F.S.; Yin, H. Increased yield potential of wheat-maize cropping system in the North China Plain by climate change adaptation. *Clim. Chang.* **2012**, *113*, 825–840. [CrossRef]
85. Yang, Y.; Liu, D.L.; Anwar, M.R.; Zuo, H.; Yang, Y. Impact of future climate change on wheat production in relation to plant-available water capacity in a semiaridenvironment. *Arch. Meteorol. Geophys. Bioclimatol. Ser. B* **2013**, *115*, 391–410. [CrossRef]
86. Müller, C.; Waha, K.; Bondeau, A.; Heinke, J. Hotspots of climate change impacts in sub-Saharan Africa and implications for adaptation and development. *Glob. Chang. Biol.* **2014**, *20*, 2505–2517. [CrossRef]
87. Dale, A.L.; Fant, C.W.; Strzepek, K.; Lickley, M.J.; Solomon, S. Climate model uncertainty in impact assessments for agriculture: A multi-ensemble case study on maize in sub-Saharan Africa. *Earth's Futur.* **2017**, *5*, 337–353. [CrossRef]
88. Asseng, S.; Ewert, F.; Rosenzweig, C.; Jones, J.W.; Hatfield, J.L.; Ruane, A.C.; Boote, K.; Thorburn, P.J.; Rötter, R.; Cammarano, D.; et al. Uncertainty in simulating wheat yields under climate change. *Nat. Clim. Chang.* **2013**, *3*, 827–832. [CrossRef]
89. Klein, T.; Holzkämper, A.; Calanca, P.; Seppelt, R.; Fuhrer, J. Adapting agricultural land management to climate change: A regional multi-objective optimization approach. *Landsc. Ecol.* **2013**, *28*, 2029–2047. [CrossRef]
90. Ceglar, A.; Črepinšek, Z.; Kajfež-Bogataj, L.; Pogačar, T. The simulation of phenological development in dynamic crop model: The Bayesian comparison of different methods. *Agric. For. Meteorol.* **2011**, *151*, 101–115. [CrossRef]
91. van der Linden, P.; Mitchell, J. Ensembles: Climate Change and Its Impact: Summary of Research and the Results from the ENSEMBLES Project. *Met Off. Hadley Cent. Exeter* **2009**. [CrossRef]
92. Parry, M.L.; Rosenzweig, C.; Iglesias, A.; Livermore, M.; Fischer, G. Effects of climate change on global food production under SRES emissions and socio-economic scenarios. *Glob. Environ. Chang.* **2004**, *14*, 53–67. [CrossRef]
93. Swann, A.L.S.; Hoffman, F.M.; Koven, C.D.; Randerson, J.T. Plant responses to increasing CO_2 reduce estimates of climate impacts on drought severity. *Proc. Natl. Acad. Sci. USA* **2016**, *113*, 10019–10024. [CrossRef]
94. Mengis, N.; Keller, D.P.; Eby, M.; Oschlies, A. Uncertainty in the response of transpiration to CO_2 and implications for climate change. *Environ. Res. Lett.* **2015**, *10*, 094001. [CrossRef]
95. Biernath, C.; Gayler, S.; Bittner, S.; Klein, C.; Högy, P.; Fangmeier, A.; Priesack, E. Evaluating the ability of four crop models to predict different environmental impacts on spring wheat grown in open-top chambers. *Eur. J. Agron.* **2011**, *35*, 71–82. [CrossRef]

agriculture

MDPI

Article

Responses of Soybean Water Supply and Requirement to Future Climate Conditions in Heilongjiang Province

Na Li [1], Tangzhe Nie [1,2,*], Yi Tang [1], Dehao Lu [1], Tianyi Wang [3], Zhongxue Zhang [2,3], Peng Chen [4], Tiecheng Li [2,3], Linghui Meng [5], Yang Jiao [1] and Kaiwen Cheng [1]

1 School of Water Conservancy and Electric Power, Heilongjiang University, Harbin 150080, China; 2201954@s.hlju.edu.cn (N.L.); 2201943@s.hlju.edu.cn (Y.T.); 2212006@s.hlju.edu.cn (D.L.); 2201957@s.hlju.edu.cn (Y.J.); 20195777@s.hlju.edu.cn (K.C.)

2 Key Laboratory of Efficient Use of Agricultural Water Resources, Ministry of Agriculture and Rural Affairs, Northeast Agricultural University, Harbin 150030, China; zhangzhongxue@neau.edu.cn (Z.Z.); b210101005@neau.edu.cn (T.L.)

3 School of Water Conservancy and Civil Engineering, Northeast Agricultural University, Harbin 150030, China; 2191532@s.hlju.edu.cn

4 College of Agricultural Science and Engineering, Hohai University, Nanjing 210098, China; 190210070001@hhu.edu.cn

5 School of Architecture and Engineering, Liaocheng University, Liaocheng 252000, China; m15762134644@163.com

* Correspondence: 2019036@hlju.edu.cn; Tel.: +86-189-4508-3525

Abstract: Understanding future changes in water supply and requirement under climate change is of great significance for long-term water resource management and agricultural planning. In this study, daily minimum temperature (T_{min}), maximum temperature (T_{max}), solar radiation (*Rad*), and precipitation for 26 meteorological stations under RCP4.5 and RCP8.5 of MIRCO5 for the future period 2021–2080 were downscaled by the LARS-WG model, daily average relative humidity (*RH*) was estimated using the method recommended by FAO-56, and reference crop evapotranspiration (ET_0), crop water requirement (ET_c), irrigation water requirement (I_r), effective precipitation (P_e), and coupling degree of ET_c and P_e (*CD*) for soybean during the growth period were calculated by the CROPWAT model in Heilongjiang Province, China. The spatial and temporal distribution of these variables and meteorological factors were analyzed, and the response of soybean water supply and requirement to climate change was explored. The result showed that the average T_{min}, T_{max}, and *Rad* under RCP4.5 and RCP8.5 increased by 0.2656 and 0.5368 °C, 0.3509 and 0.5897 °C, and 0.0830 and 0.0465 MJ/m^2, respectively, while the average *RH* decreased by 0.0920% and 0.0870% per decade from 2021 to 2080. The annual average ET_0, ET_c, P_e, and I_r under RCP4.5 for 2021–2080 were 542.89, 414.35, 354.10, and 102.44 mm, respectively, and they increased by 1.92%, 1.64%, 2.33%, and −2.12% under the RCP8.5, respectively. The ranges of *CD* under RCP4.5 and RCP8.5 were 0.66–0.95 and 0.66–0.96, respectively, with an average value of 0.84 for 2021–2080. Spatially, the *CD* showed a general trend of increasing first and then decreasing from west to east. In addition, ET_0, ET_c, and P_e increased by 9.55, 7.16, and 8.77 mm per decade, respectively, under RCP8.5, while I_r decreased by 0.65 mm per decade. Under RCP4.5 and RCP8.5, ET_c, P_e, and I_r showed an overall increasing trend from 2021 to 2080. This study provides a basis for water resources management policy in Heilongjiang Province, China.

Keywords: climate change; soybean; CROPWAT; reference crop evapotranspiration (ET_0); crop water requirement (ET_c); irrigation water requirement (I_r)

Citation: Li, N.; Nie, T.; Tang, Y.; Lu, D.; Wang, T.; Zhang, Z.; Chen, P.; Li, T.; Meng, L.; Jiao, Y.; et al. Responses of Soybean Water Supply and Requirement to Future Climate Conditions in Heilongjiang Province. *Agriculture* **2022**, *12*, 1035. https://doi.org/10.3390/agriculture12071035

Academic Editors: Dengpan Xiao and Wenjiao Shi

Received: 23 May 2022
Accepted: 14 July 2022
Published: 15 July 2022

1. Introduction

Global climate change, marked mainly by climate warming, has taken place [1]. Undoubtedly, this change has had and will continue to have an important impact on agricultural water resources on which crop growth depends [2]. In addition, climate factors, such as relative humidity (*RH*), solar radiation (*Rad*), and CO_2 concentration, have a significant effect on crop water requirements (ET_c) [3]. Moreover, the uncertainty of temporal and spatial distribution of precipitation (*P*) and ET_c affects crop irrigation water requirement (I_r) [4]. Therefore, analyzing the impact of future climatic changes on crop water supply and requirement becomes necessary [5].

Future climate change is expected to affect water supply and requirement in a number of ways [6]. Climate change mainly affects the transpiration of plants, evaporation of water from the soil and field surface between plants, and *P* in the agricultural water cycle system [7]. *Rad* is the largest source of energy required for soil water vaporization during evapotranspiration, which converts a large amount of liquid water into water vapor. *Rad* absorbed by the atmosphere and heat emitted from the surface increase the atmospheric temperature [8]. The sensible heat around the atmosphere transmits energy to the crop to control the evapotranspiration rate, and the increase in soil surface temperature promotes evaporation [5]. The water vapor pressure difference between the evapotranspiration surface and the atmosphere is the decisive factor for water vapor movement [9]. The increase in *RH* leads to the saturation of air humidity, forming a protective layer on the field surface, thus reducing the evapotranspiration requirement [10]. However, the increase in CO_2 concentration will also promote the accumulation of crop dry matter, promote plant growth, and increase transpiration [11]. *P* increases soil water content, replenishes the total effective soil water, improves the plant root water absorption rate, and helps to reduce I_r while meeting the needs of crop evapotranspiration [12]. Some researchers found that the *RH* in Zimbabwe areas would decrease in the future, while the average temperature, *Rad* and wind speed would increase, resulting in an increase in ET_0 and ET_c; however, the decrease in *P* would eventually lead to the increase of I_r [13]. In contrast, studies in North China Plain (NCP) found that ET_c and I_r decreased with increasing temperature, *Rad* and *P*, shorten of growth period [14]. In many studies from different regions, the relationships between ET_c, P_e, and I_r varies under climate change. Therefore, more in-depth studies are needed to assess the impact of future climate change on crop water supply and requirement.

ET_c constitutes a major component of regional and global hydrological cycles and, therefore, has important implications in the use of agricultural irrigation water, as well as in analyzing the crop water supply and requirement relationship in agricultural ecosystems [15]. There are many methods for calculating ET_c, such as the empirical estimation method, the Penman–Monteith (P–M) double-crop coefficient method, and the P–M single-crop coefficient method [16]. The empirical formula for estimating ET_c is simple and convenient; however, it is only suitable for local instead of large-scale areas [17]. When using P–M double-crop coefficient method to estimate ET_c, the crop coefficient is divided into basic crop coefficient (K_{cb}) and soil evaporation coefficient (K_e), although the estimation accuracy of ET_c is improved [18]; however, the estimation of K_e is complex and uncertain, which needs the support of a large amount of experimental data [19,20]. The parameters required by the P–M single-crop coefficient method are easy to obtain, which can be directly substituted into the formula for calculation. The calculated ET_c has less difference with the measured ET_c. Generally, the P–M single-crop coefficient method has strong universality in different regions and is considered to be a more efficient, convenient, and accurate method [21,22]. Therefore, most scholars use the P–M single-crop coefficient method recommended by the Food and Agriculture Organization of the United Nations (FAO) to calculate ET_c [23]. Nie et al. estimated rice ET_c in Heilongjiang province using the CROPWAT model based on the P–M single-crop coefficient method [24]; the calculated ET_c was only 21–30 mm different from the measured ET_c in the field experiment. In order to test the practicability and rationality of the P–M single-crop coefficient method for calculating ET_c, Jin et al. calculated wheat ET_c in the Huaihongxinhe Irrigation District using the P–M single-crop coefficient method and found that the average difference in ET_c for the years was only 6 mm [25].

Quantitative estimation of temporal and spatial variability of ET_c, P_e, and I_r under climate change is helpful to maximize the use of rainwater resources and optimize regional water resource allocation [11,14]. P is the main influencing factor of soil moisture content, which provides water for crop evapotranspiration [26]. I_r depends on soil moisture content [27]. Therefore, there is a complex relationship between P_e and I_r, which cannot be fully explained by simple linear equations [28]. In addition, the relationship among ET_c, P_e, and I_r is also affected by P distribution pattern, crop species, and planting area [11]. In the Jayakwadi command area, India, the ET_c of major crops and P_e increased during the growth period, resulting in less I_r under climate change [29]. In the Najafabad plain in Iran, ET_c increased and P_e decreased during the growth period of major crops; therefore, more water needed to be irrigated [30].

As one of the largest developing countries in the world, China constitutes 22% of the world's population and encompasses 9% of the world's arable land [31]. Heilongjiang Province has the largest arable land area in China and is also an important commercial grain base in China [32]. The soybean sowing area and yield in Heilongjiang Province rank first in China, with a sowing area of 4.279×10^6 ha and yield of 7.808×10^6 tons as of 2019 [33]. Soybean sowing area increased by an average of 5×10^5 ha per year in the last 5 years. The climate distribution in Heilongjiang province leads to great differences in temporal and spatial distribution of crop water supply and requirement, and agricultural drought occurs frequently in spring and summer [34]. With the increase in soybean planting area and soybean export share, the study on soybean water supply and requirement under future climatic conditions is of great guidance to ensure soybean production and food security in Heilongjiang Province [35].

The purpose of this study was (1) to clarify the spatial and temporal distribution characteristics of ET_0, ET_c, P_e, I_r, and CD during the soybean growth period for 2021–2080 under RCP4.5 and RCP8.5 in Heilongjiang Province, and (2) to reveal the response of soybean water supply and requirement to climate change for 2021–2080 under RCP4.5 and RCP8.5. This study will provide reasonable planning for water allocation and guide the sustainable development of agricultural irrigation water use in Heilongjiang Province.

2. Materials and Methods

2.1. Study Region and Datasets

The study area is located in Heilongjiang Province, Northeast China, where 26 meteorological stations are located relatively evenly throughout the study area for observations (Figure 1). The area belongs to the cold temperate and temperate continental monsoon climate, with an average annual temperature of 4.52 °C, an average annual solar radiation of 13.72 MJ/m^2, and an average annual P of 511 mm. According to the "Heilongjiang Province Crop Variety Cumulative Temperature Zone Plan" [36] and "Heilongjiang Province 2015 Regional Layout Plan for High-Quality and High-Yielding Major Food Crops" [37] issued by the Heilongjiang Provincial Agriculture Committee, the sixth cumulative temperature zone is not suitable for soybean cultivation; therefore, the sixth cumulative temperature zone is not included in this study.

We used the general circulation model (GCM) of MIRCO5 with a resolution of $1.39° \times 1.41°$ and selected two representative concentration pathways (RCP4.5 for the low-radiation scenario and RCP8.5 for the high-radiation scenario) according to the socioeconomic conditions of the radiative forcing currently faced by humans. The minimum temperature (T_{min}), maximum temperature (T_{max}), Rad, and P data from 26 stations of the China Meteorological Administration (CMA) from 1960–2015 were imported into the LARS-WG stochastic weather generator model to generate future climate datasets. The dataset includes daily T_{min}, T_{max}, Rad, and P for 26 meteorological stations under RCP4.5 and RCP8.5 for the future period (2021–2080). The period 2021–2080 was divided into three time periods: the 2030s (2021–2040), 2050s (2041–2060), and 2070s (2061–2080). Under RCP4.5 and RCP8.5, average RH was estimated using the method recommended by FAO-56.

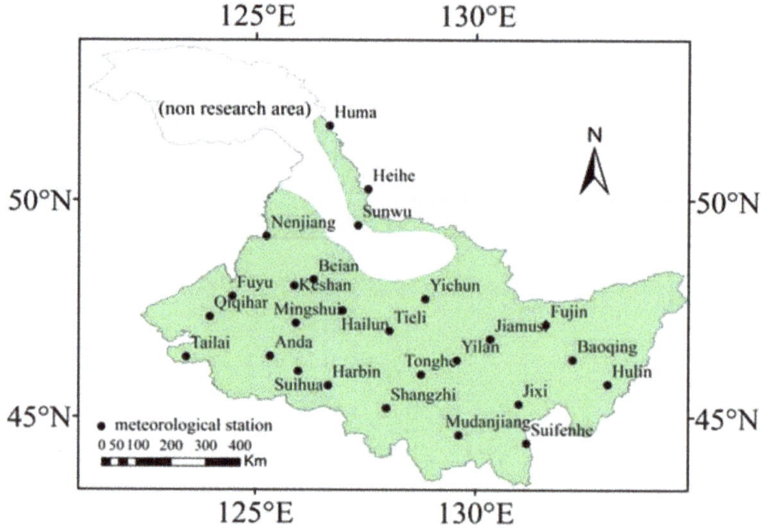

Figure 1. Study area and distribution of 26 meteorological stations in Heilongjiang Province.

2.2. Division of Soybean Growth Period

The FAO divides the crop growth period into four stages: initial stage (L_{ini}), crop development stage (L_{dev}), mid-season stage (L_{mid}), and late stage (L_{late}); the crop coefficients in each growth stage are K_{cini}, K_{cmid} and K_{cend}. In this study, the whole growth period of soybean was divided into sowing to three-leaf stage (L_{ini}), three-leaf stage to flowering stage (L_{dev}), flowering stage to podding stage (L_{mid}), and podding stage to maturity stage (L_{late}). The crop coefficients (K_c) were based on the irrigation series "Crop Guide to Crop Water Requirements" published by FAO-56 and corrected using the method recommended by FAO-56 [38,39]. It was assumed that the soybean variety would not change in the future period. According to the observation data of soybean growth period from 1994 to 2005 at 19 agrometeorological observation stations in Heilongjiang Province, the soybean sowing date and the length of each growth stage were determined. The data of the adjacent agrometeorological observation stations in the same temperature accumulation area were selected as the calculation basis, as shown in Table 1.

Table 1. Average soybean growth period data in 1994–2005.

Agrometeorological Station	L_{ini} (Days)	L_{dev} (Days)	L_{mid} (Days)	L_{Late} (Days)	Total Growth Day (Days)	Meteorological Station
Qinggang	24	31	59	16	130	Anda, Suihua
Hulin	27	30	60	17	134	Hulin
Boli	35	24	16	16	125	Jixi, Mudanjiang, Suifenhe
Bayan	29	24	62	23	138	Tonghe, Shangzhi
Heihe	32	25	59	17	133	Heihe, Sunwu
Harbin	34	32	67	17	150	Haerbin
Nenjiang	29	27	60	17	133	Nenjiang
Longjiang	20	27	63	22	132	Qiqihar, Tailai
Huma	33	18	57	17	125	Huma

Table 1. *Cont.*

Agrometeorological Station	L_{ini} (Days)	L_{dev} (Days)	L_{mid} (Days)	L_{Late} (Days)	Total Growth Day (Days)	Meteorological Station
Qingan	28	21	63	28	140	Tieli
Tangyuan	27	28	65	16	136	Yinlan
Beian	28	26	65	15	134	Keshan, Beian
Baiquan	27	27	55	23	132	Mingshui
Jiayin	30	24	57	16	127	Yichun
Hailun	28	29	63	19	139	Hailun
Jiamusi	28	32	62	17	139	Jiamusi
Fuyu	32	32	57	14	135	Fuyu
Baoqing	22	28	60	19	129	Baoqing
Fujin	29	25	60	21	135	Fujin

2.3. Soil Parameters

Parameters such as soil type, total available soil moisture, maximum rain infiltration rate, and maximum rooting depth were obtained from the Harmonized World Soil Database (HSWD). To improve the accuracy of the model simulation results, the initial soil moisture depletion and initial available soil moisture were adjusted according to the "10 day dataset of crop growth and development and farmland soil moisture in China" from the China Meteorological Data Network (http://data.cma.cn, accessed on 22 May 2022). The obtained soil data were input into the CROPWAT model, and the initial soil water content for each year thereafter was taken as the last day of the previous year.

2.4. Effective Precipitation (P_e)

For upland crops, P_e refers to the total precipitation that can be stored in the crop root layer to meet the crop's water needs, excluding surface runoff and leakage below the crop root layer. In this study, we used the method recommended by the United States Department of Agriculture (USDA) to calculate P_e. The formula was as follows:

$$P_e = \begin{cases} P(125 - 0.6P)/125 & (P \leq 83.3\text{ mm}) \\ 125/3 + 0.1P & (P > 83.3\text{ mm})' \end{cases} \quad (1)$$

where P_e is the effective precipitation (mm), and P is precipitation (mm).

2.5. Crop Water Requirement (ET_c)

Soybean ET_c was calculated using the CROPWAT model as a function of the loading altitude, latitude, longitude, T_{min}, T_{max}, Rad, and RH data from each station into the "climate/ET_0" module to calculate ET_0. The sowing date, harvest date, K_c, and length of each growing period were loaded into the "crop" module to calculate ET_c. Soybean ET_c was calculated using the single-crop coefficient method recommended by FAO-56. ET_c was calculated from ET_0 and K_c using the equation under standard conditions, where ET_0 was considered as the key variable for the estimation of ET_c. Standard conditions mean that there were no limitations to crop growth, including a sufficient supply of water and crops free from diseases and pest infections.

$$ET_c = K_c \times ET_0, \quad (2)$$

where ET_0 is the reference crop evapotranspiration (mm), K_c is the crop coefficient (dimensionless), and ET_c is the crop water requirement (mm).

ET_0 was calculated using the P–M formula recommended by FAO; thus,

$$ET_0 = \frac{0.408\Delta \times (R_n - G) + \frac{900\gamma \times u_2 \times (e_s - e_a)}{(T+273)}}{\Delta + \gamma \times (1 + 0.34u_2)}, \quad (3)$$

where ET_0 is the reference crop evapotranspiration (mm·day^{-1}), Δ is the slope of the vapor pressure curve (kPa·°C^{-1}), R_n is the net radiation at the crop surface (MJ·(m^2·day^{-1})), G is the soil heat flux density (MJ·(m^2·day^{-1})), γ is the psychrometric constant (kPa·°C^{-1}), T is the mean daily air temperature at 2 m height (°C), u_2 is the wind speed at 2 m height (m·s^{-1}), e_s is the saturation vapor pressure, e_a is the actual vapor pressure (kPa), $e_s - e_a$ is the saturation vapor pressure deficit (kPa), and 900 is a conversion factor.

2.6. *Irrigation Water Requirement* (I_r)

The daily soil water balance equation was used to calculate I_r. Irrigation quota should be less than or equal to the root-zone water consumption to avoid deep leakage loss. The calculation formula is as follows:

$$I_{r,i} = D_{r,i-1} + ET_c - D_{r,i} - P_{e_i}, \tag{4}$$

where $I_{r,i}$ is the irrigation water requirement on day i, $D_{r,i-1}$ is the water consumption of the root zone on day $i-1$, ET_c is crop water requirement, $D_{r,i}$ is the water consumption of the root zone on day i, and P_{ei} is the P_e on day i.

2.7. *Climate Tendency Rate*

The climate tendency rate is the changing rate of each variable every 10 years. A positive climate tendency rate indicates an increasing trend of the corresponding variable, while a negative value indicates a decreasing trend. By using the least-square method, the changing trend of variable can be expressed by a linear equation formulas follows:

$$Y_t = at + b, \tag{5}$$

where Y_t is represents the fitted values of each variable, t is the corresponding year, and a and b are regression coefficients.

2.8. *Coupling Degree of* ET_c *and* P_e *(CD)*

During the soybean growth period, the degree to which P_e meets ET_c is called the coupling degree between ET_c and P_e. The calculation equation is as follows:

$$\lambda_i = \begin{cases} 1 & (P_e \geq ET_c) \\ P_e/ET_c & (P_e < ET_c) \end{cases}. \tag{6}$$

2.9. *Mann–Kendall Trend Test*

The Mann–Kendall trend test is a nonparametric statistical method used to reveal how a variable changes with time, introduced by the World Meteorological Organization. Positive and negative values of the statistical variable Z indicate the data changing trend; if the absolute value of Z is greater than 1.64, 2.32, and 2.56, it means that the data have passed the significance test of 95%, 99%, and 99.9% for reliability [40]. This study used this method to test the changing trend of ET_0, ET_c, I_r, P_e, and CD during the soybean growth period.

2.10. *Data Processing*

The reduced-dimension downscaled dataset was processed by Codeblocks20.03 [41] open-source software, which made the data schema acceptable to the CROPWAT model. The CROPWAT8.0 [42] model was used to calculate ET_c, P_e, and I_r under future climate conditions at 26 meteorological stations in Heilongjiang Province. Matlab R2019a [43] was used to perform Mann–Kendall trend tests of ET_0, ET_c, P_e, I_r, and their climate tendency rates under future climatic conditions, and the inverse distance weighting (IDW) method in the spatial analysis toolbox of Arcmap 10.2 was used to spatially interpolate and mapping at a resolution of 0.04° × 0.04°. We used SPSS25.0 [44] to process the correlation analysis of T_{min}, T_{max}, RH, Rad, ET_0, ET_c, P_e, and I_r, as well as the analysis of variance (ANOVA) of T_{min}, T_{max}, P_e, RH, Rad, ET_0, ET_c, I_r, and CD.

3. Results

3.1. Spatial and Temporal Variation of Future Meteorological Factor

ET_0 during the soybean growth period was driven by interacting effects of different climate factors. Therefore, a detailed analysis of changes for each meteorological factor was conducted (Figures 2 and 3). Average T_{max} under RCP4.5 and RCP8.5 showed a significant increasing trend, *Rad* showed an increasing and then decreasing trend, and average *RH* showed a decreasing and then increasing trend (Figure 2). Under RCP4.5 and RCP8.5, the average T_{max} started from 25.19 and 25.36 °C in the 2030s, respectively, and increased significantly to 26.76 and 28.02 °C in the 2070s. Similarly, the average *Rad* increased significantly from 21.04 and 21.15 MJ/m^2 in the 2030s to 21.56 and 21.53 MJ/m^2 in the 2050s, respectively, and then both decreased to 21.42 MJ/m^2 in the 2070s. The average *RH* decreased significantly from 75.07% and 74.92% in the 2030s to 74.36% and 74.42% in the 2050s, before continuing to increase to 74.50% and 74.65% in the 2070s, respectively. Under RCP4.5 and RCP8.5, the highest values of the T_{min} were distributed in the east, and the highest values of the T_{max} were distributed in the south. In addition, the highest *RH* was found in the central part, and the highest *Rad* was found in the western and eastern parts.

Under RCP4.5, the average climate tendency rates of T_{min}, T_{max}, *RH*, and *Rad* for 2021–2080 were 0.2656 °C/(10 years), 0.3509 °C/(10 years), −0.0920%/(10 years), and 0.0830 MJ/m^2/(10 years), respectively (Figure 3). Under RCP8.5, the average climate tendency rates of T_{min}, T_{max}, *RH*, and *Rad* in 2021–2080 were 0.5368 °C/(10 years), 0.5897 °C/(10 years), −0.0870%/(10 years), and 0.0465 MJ/m^2/(10 years), respectively. Under RCP4.5 from 2021–2050, the *RH* declined most quickly, at a rate of 0.3002%/(10 years), while *Rad* increased most quickly, at a rate of 0.2193 MJ/m^2/(10 years).

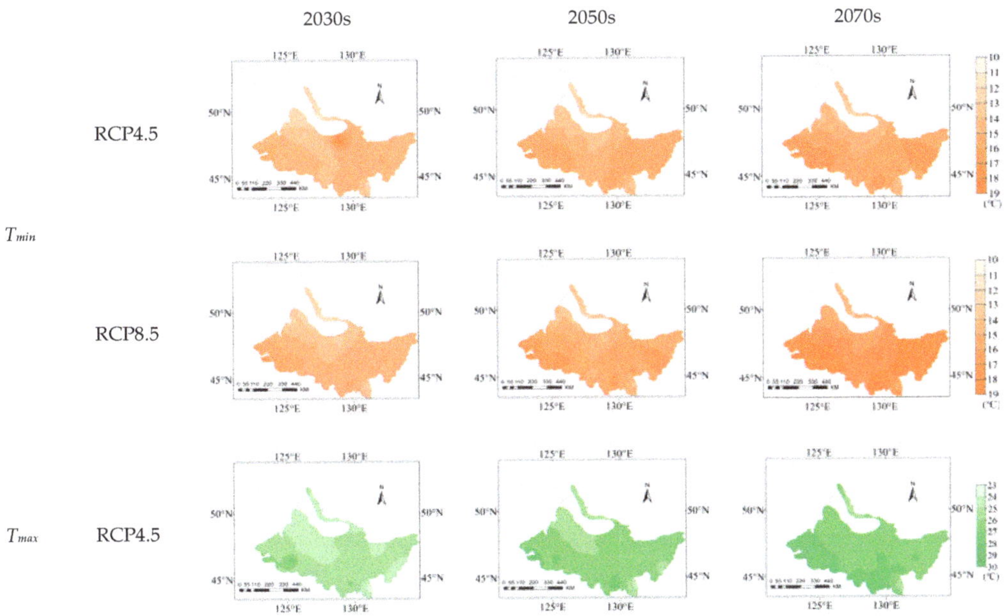

Figure 2. *Cont.*

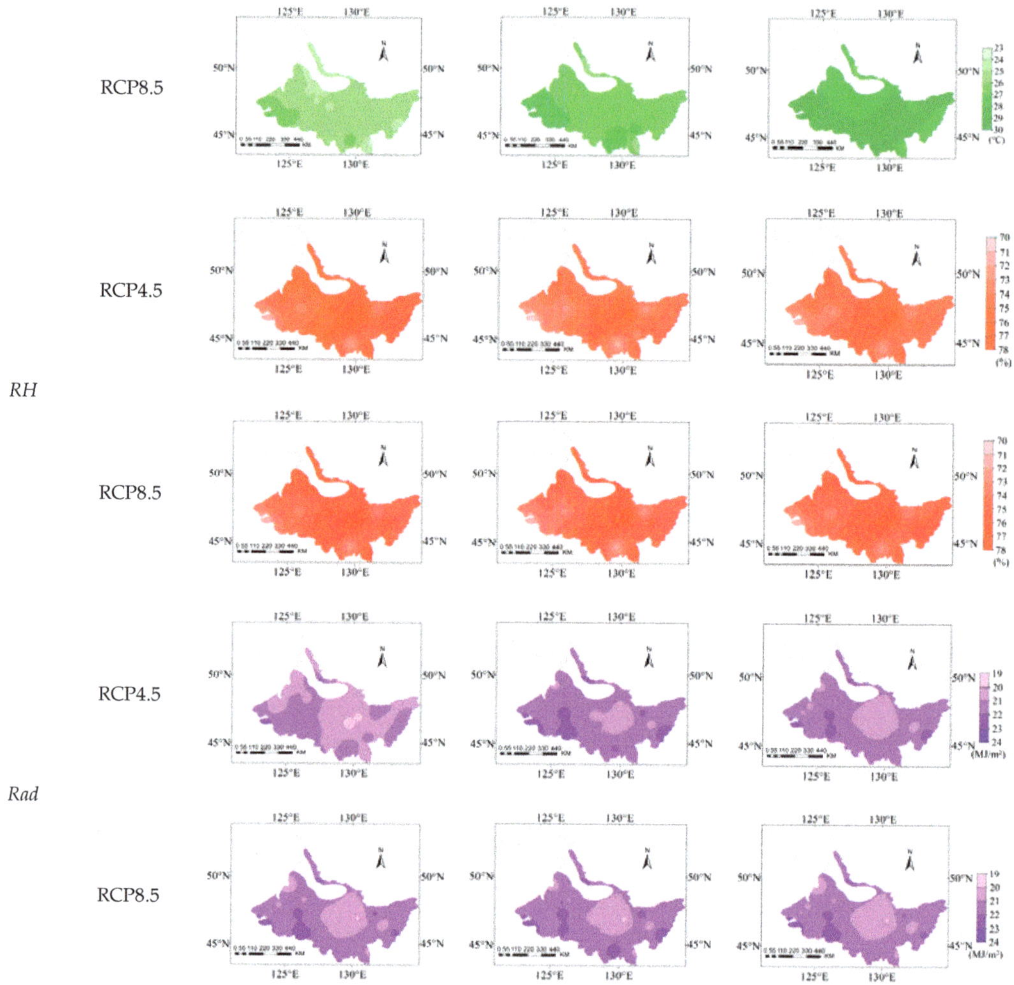

Figure 2. Spatial distribution of average minimum temperature (T_{min}), maximum temperature (T_{max}), relative humidity (*RH*), and solar radiation (*Rad*) during soybean growth period under RCP4.5 and RCP8.5 in the study area in the 2030s, 2050s, and 2070s.

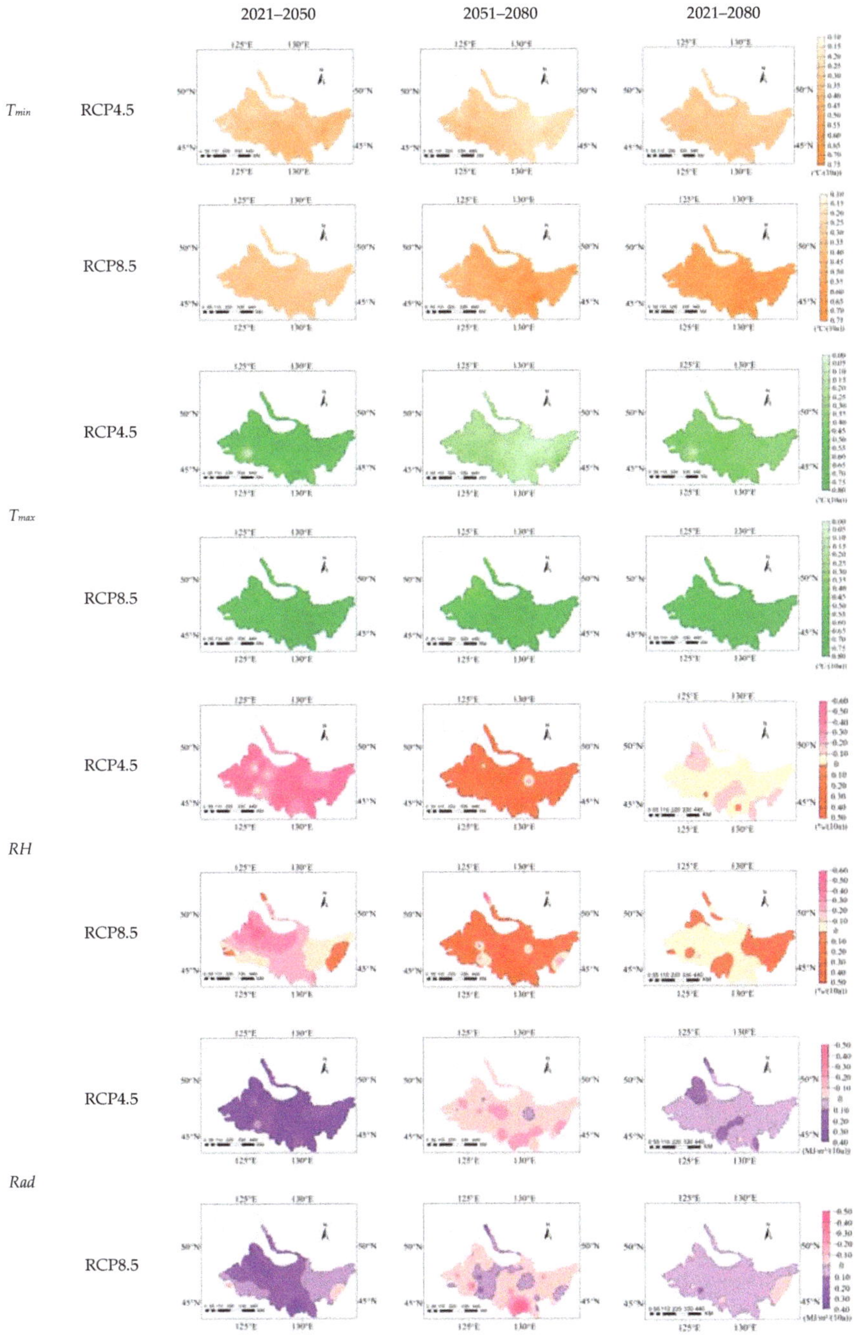

Figure 3. Climate tendency rates of average minimum temperature (T_{min}), maximum temperature (T_{max}), relative humidity (RH), and solar radiation (Rad) during soybean growth period under RCP4.5 and RCP8.5 in the study area in 2021–2050, 2051–2080, and 2021–2080.

3.2. Spatial and Temporal Variation of ET_0

The ET_0 values during the soybean growth period from 2021–2080 under RCP4.5 and RCP8.5 were shown in Figure 4. Under RCP4.5, ET_0 from 2021–2080 was between 409.34 and 621.47 mm, with an average of 542.89 mm. Under RCP8.5, ET_0 was between 492.48 and 642.24 mm, with an average of 553.35 mm. Under RCP4.5 and RCP8.5, ET_0 increased first and then decreased from west to east in the study area.

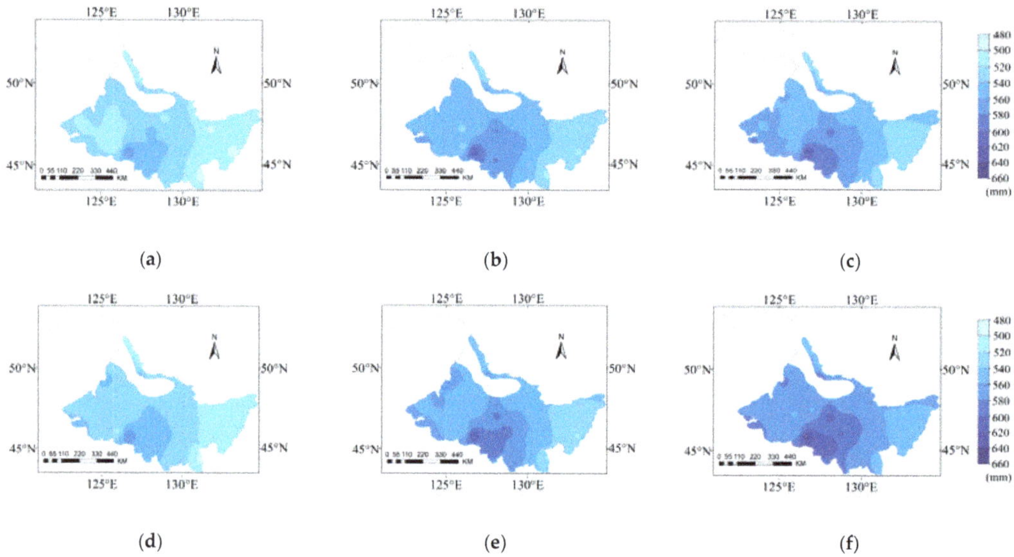

Figure 4. Spatial distribution of reference crop evapotranspiration (ET_0) during the (**a**) 2030s, (**b**) 2050s, and (**c**) 2070s under RCP4.5, and during the (**d**) 2030s, (**e**) 2050s, and (**f**) 2070s under RCP8.5 during the soybean growth period.

The climate tendency rate of ET_0 in the soybean growth period from 2021–2080 under RCP4.5 was 3.71–10.18 mm/(10 years). The climate tendency rates of ET_0 in 2021–2050, 2051–2080, and 2021–2080 were 12.65 mm/(10 years), 1.93 mm/(10 years), and 7.71 mm/(10 years), respectively (Figure 5a–c). Under RCP8.5, the climate tendency rate of ET_0 from 2021–2080 was 7.30–12.07 mm/(10 years), with an average of 9.55 mm/(10 years) (Figure 5d–f). All 26 sites passed the significance test at $\alpha = 0.001$ under both RCP4.5 and RCP8.5.

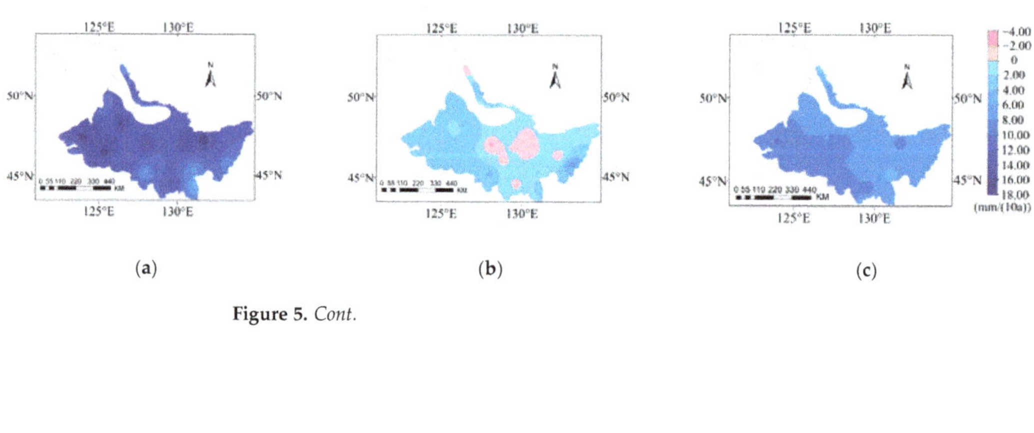

Figure 5. *Cont.*

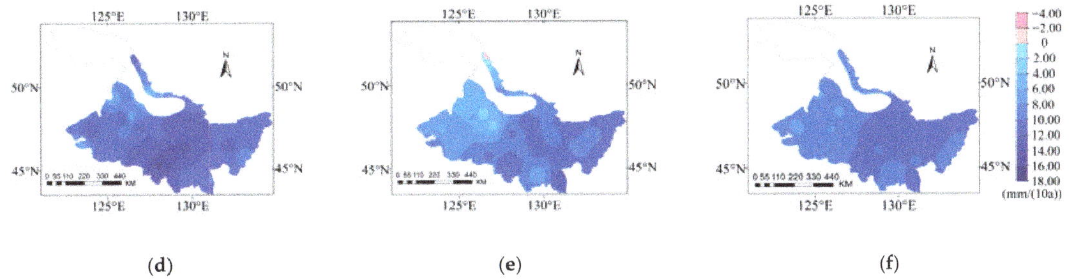

Figure 5. Climate tendency rates of ET_0 in the periods (**a**) 2021–2050, (**b**) 2051–2080, and (**c**) 2021–2080 under RCP4.5, and in the periods (**d**) 2021–2050, (**e**) 2051–2080, and (**f**) 2021–2080 under RCP8.5 during the soybean growth period.

3.3. Spatial and Temporal Variation of ET_c

The spatial distribution of ET_c and its climate tendency rate of soybean growth period for 2021–2080 under RCP4.5 and RCP8.5 are shown in Figures 6 and 7. Under RCP4.5, the ET_c values for 2021–2080 were 356.88–470.45 mm, with an average of 414.35 mm. Under RCP8.5, the average ET_c values for the 2030s, 2050s, and 2070s were 403.94, 423.39, and 436.07 mm, respectively. Under both RCP4.5 and RCP8.5, ET_c increased and then decreased from west to east in the study area.

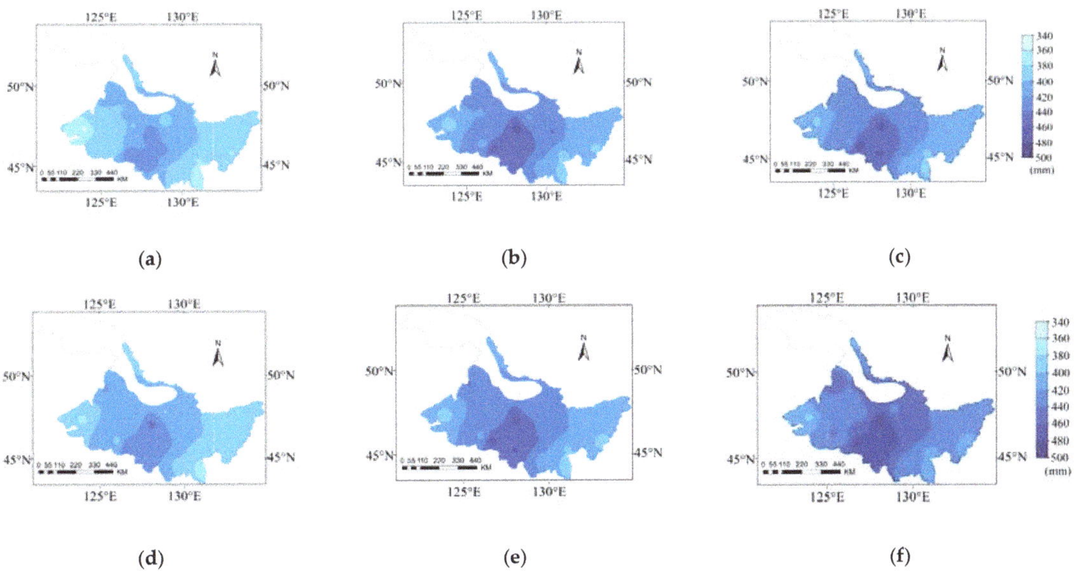

Figure 6. Spatial distribution of crop water requirement (ET_c) during the (**a**) 2030s, (**b**) 2050s, and (**c**) 2070s under RCP4.5, and during the (**d**) 2030s, (**e**) 2050s, and (**f**) 2070s under RCP8.5 during the soybean growth period.

As shown in Figure 7, the climate tendency rates of soybean ET_c for 2021–2080 were 2.92–8.11 mm/(10 years) and 4.08–9.39 mm/(10 years) under RCP4.5 and RCP8.5 with average values of 6.09 mm/(10 years) and 7.16 mm/(10 years), respectively. The ET_c climate tendency rate was higher in the western region than that in the eastern region under RCP4.5, whereas it was higher in the eastern region than that in the western region under RCP8.5. All 26 sites passed the significance test at $\alpha = 0.001$ under both RCP4.5 and

RCP8.5. Specifically, soybean ET_c in Yichun and Suifenhe increased at a rate of more than 11 mm/(10 years) under RCP8.5.

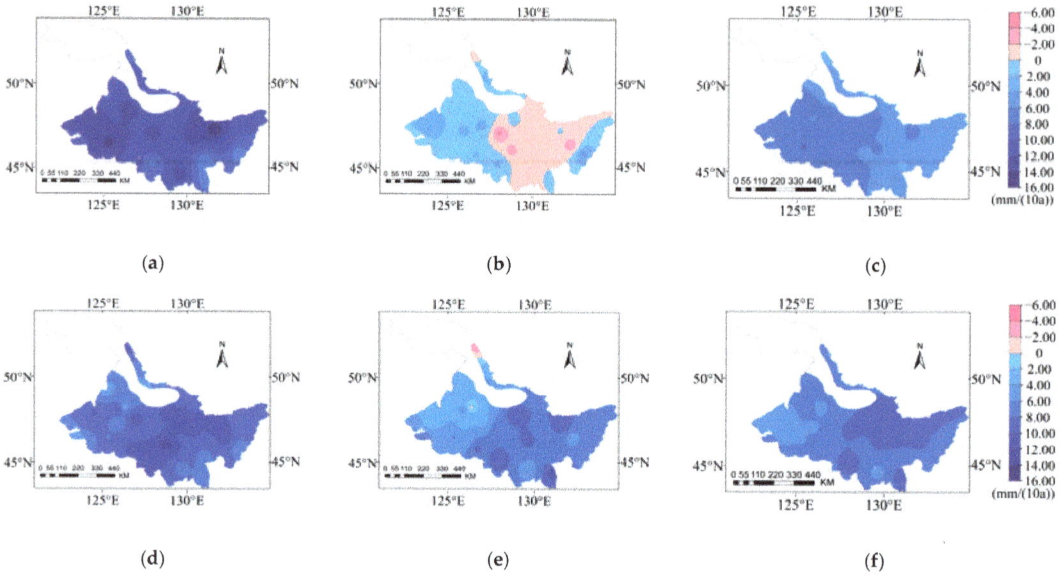

Figure 7. Climate tendency rates of ET_c in the periods (**a**) 2021–2050, (**b**) 2051–2080, and (**c**) 2021–2080 under RCP4.5, and in the periods (**d**) 2021–2050, (**e**) 2051–2080, and (**f**) 2021–2080 under RCP8.5 during the soybean growth period.

3.4. Spatial and Temporal Variation of P_e

The spatial distribution of P_e and its climate tendency rate under RCP4.5 and RCP8.5 during the soybean growth period for 2021–2080 are shown in Figure 8. Under RCP4.5 and RCP8.5, P_e values were 268.41–459.18 mm and 269.53–466.94 mm, with an average of 354.10 and 362.36 mm, respectively. Under RCP8.5, the greatest difference in P_e was 94.99 mm. Under RCP4.5 and RCP8.5, P_e first increased and then decreased from west to east; higher values were mainly distributed in Hailun and Tieli, with an average value greater than 370 mm, while lower values were mainly distributed in Tailai and Huma, with an average value lower than 340 mm.

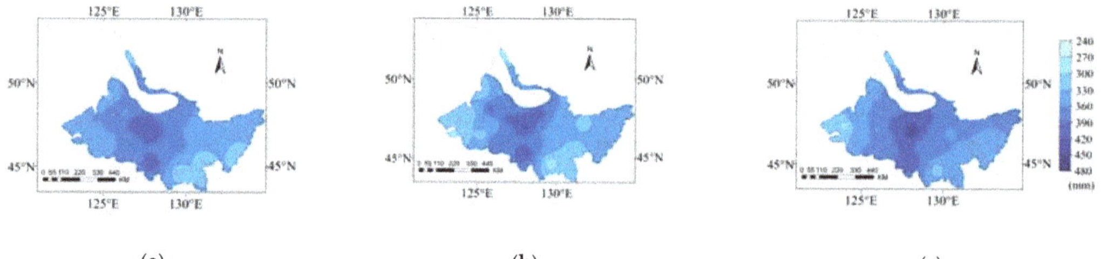

Figure 8. *Cont.*

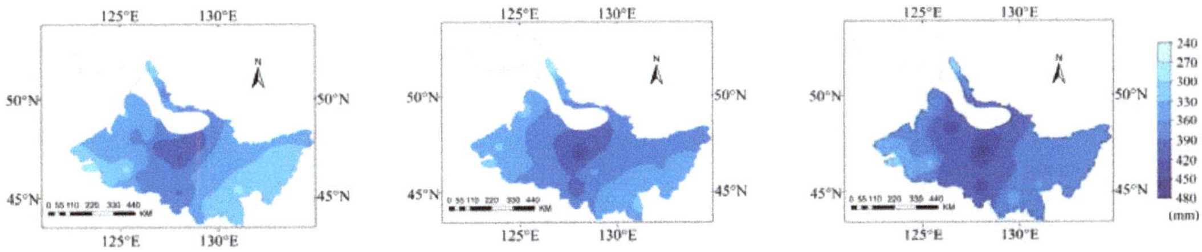

(d) (e) (f)

Figure 8. Spatial distribution of effective precipitation (P_e) during the (**a**) 2030s, (**b**) 2050s, and (**c**) 2070s under RCP4.5, and during the (**d**) 2030s, (**e**) 2050s, and (**f**) 2070s under RCP8.5 during the soybean growth period.

Under RCP4.5, the climate tendency rate of P_e during the soybean growth period from 2021–2080 was −10.81–10.11 mm/(10 years), and the average was 1.37 mm/(10 years) (Figure 9). A total of 16 sites showed an upward trend, while 10 sites showed a downward trend. Bei'an, Harbin, Jixi, Suifenhe, and Tieli passed the significance test at $\alpha = 0.05$, while Hulin, Keshan, and Suihua passed the significance test at $\alpha = 0.1$. Under RCP8.5, the climate tendency rate of Pe for 2021–2080 was −1.16–22.28 mm/(10 years), with an average value of 8.77 mm/(10 years). Bei'an, Mudanjiang, Suifenhe, Suihua, and Tonghe passed the significance test at $\alpha = 0.001$, while Baoqing, Fuyu, Fujin, and Mingshui passed the significance test at $\alpha = 0.05$.

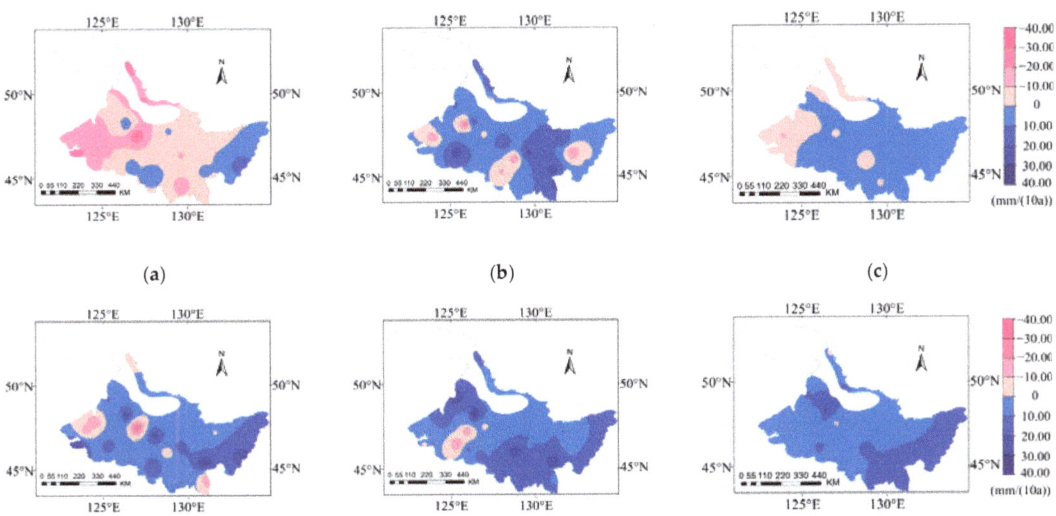

(a) (b) (c)

(d) (e) (f)

Figure 9. Climate tendency rates of P_e in the periods (**a**) 2021–2050, (**b**) 2051–2080, and (**c**) 2021–2080 under RCP4.5, and in the periods (**d**) 2021–2050, (**e**) 2051–2080, and (**f**) 2021–2080 under RCP8.5 during the soybean growth period.

3.5. Spatial and Temporal Variation of CD

The *CD* values during the soybean growth period under RCP4.5 and RCP8.5 from 2021–2080 are shown in Figure 10. Under RCP4.5 and RCP8.5, the *CD* for 2021–2080 ranged from 0.66–0.95 and 0.66–0.96, with average values of 0.84 in both cases, showing a trend of first increasing and then decreasing in the study area.

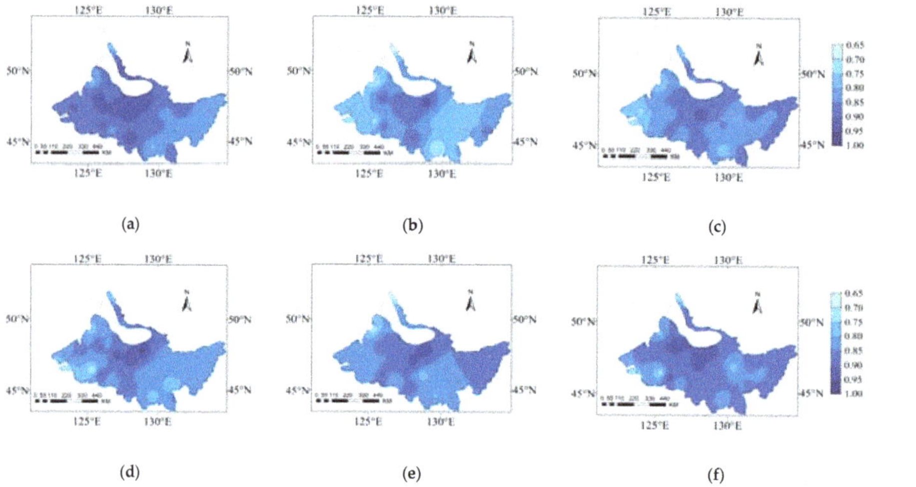

Figure 10. Spatial distribution of *CD* during the (**a**) 2030s, (**b**) 2050s, and (**c**) 2070s under RCP4.5, and during the (**d**) 2030s, (**e**) 2050s, and (**f**) 2070s under RCP8.5 during the soybean growth period.

The climate tendency rate of *CD* during the soybean growth period from 2021–2080 under the RCP4.5 was −0.036–0.014/(10 years), with an average value of −0.007/(10 years), showing an overall downward trend (Figure 11). Among them, Keshan and Qiqihar passed the significance test at α = 0.001, while Fuyu and Shangzhi passed the significance test at α = 0.05. However, under RCP8.5, the *CD* during the growth period of soybean ranged from −0.013 to 0.029/(10 years), with an average of 0.006/(10 years), showing an overall increasing trend. The climate tendency rates of *CD* at the 19 sites were greater than 0, among which Bei'an passed the significance test at α = 0.001, while Mudanjiang and Tonghe passed the significance test at α = 0.05.

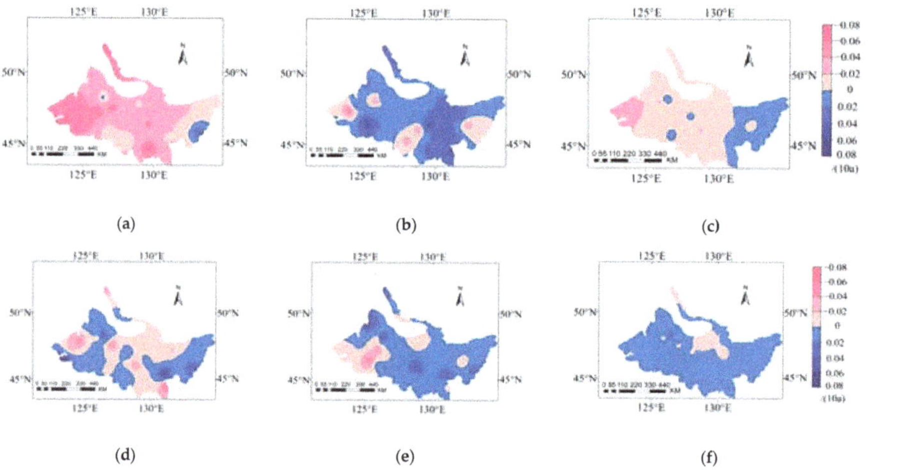

Figure 11. Climate tendency rates of *CD* in the periods (**a**) 2021–2050, (**b**) 2051–2080, and (**c**) 2021–2080 under RCP4.5, and in the periods (**d**) 2021–2050, (**e**) 2051–2080, and (**f**) 2021–2080 under RCP8.5 during the soybean growth period.

3.6. Spatial and Temporal Variation of I_r

The temporal and spatial distributions of I_r under RCP4.5 and RCP8.5 during the soybean growth period for 2021–2080 are shown in Figure 12. Under RCP4.5 and RCP8.5, the I_r values during 2021–2080 were 58.01–159.84 mm and 60.03–166.19 mm, with average values of 102.44 mm and 100.27 mm, respectively, which showed a trend of first decreasing and then increasing from west to east in the study area. Under RCP8.5, the greatest difference in I_r during the 2050s was as high as 43.32 mm, which was higher than that during the 2030s and 2050s (Figure 12d–f).

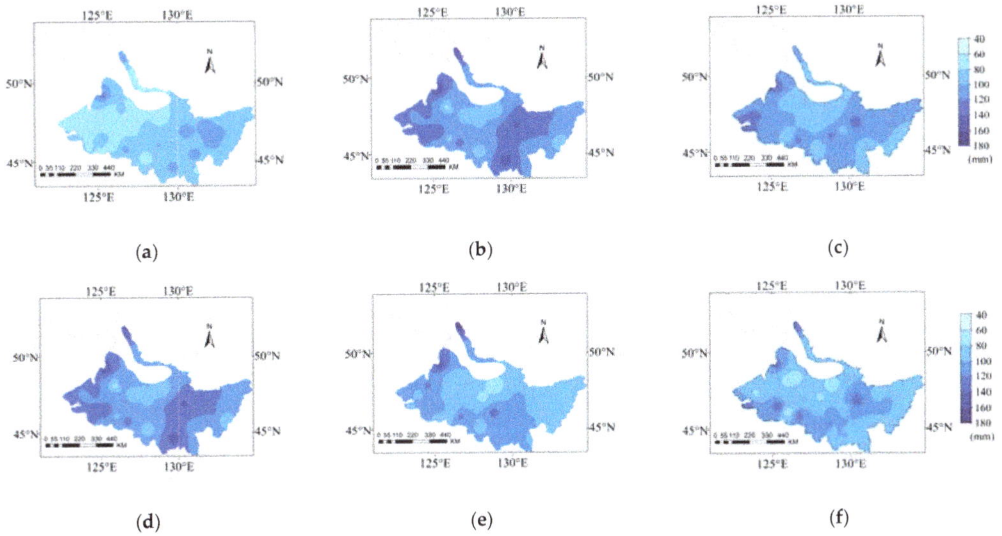

Figure 12. Spatial distribution of irrigation water requirement (I_r) during the (**a**) 2030s, (**b**) 2050s, and (**c**) 2070s under RCP4.5, and during the (**d**) 2030s, (**e**) 2050s, and (**f**) 2070s under RCP8.5 during the soybean growth period.

The average climate tendency rates of I_r during the growth period of soybean under RCP4.5 in 2021–2051, 2051–2080, and 2021–2080 were 14.88, −5.92, and 3.73 mm/(10 years), respectively (Figure 13). Among the 26 sites, Qiqihar increased at a significance of $\alpha = 0.001$. Under RCP8.5, the average climate tendency rate of I_r for 2021–2080 was −0.067 mm/(10 years). I_r showed an overall downward trend (Figure 13d–f). During the whole period of the study, the climate tendency rates of I_r in 16 sites were negative, accounting for 61.54% of the total site number, among which, Bei'an decreased at a significance of $\alpha = 0.001$.

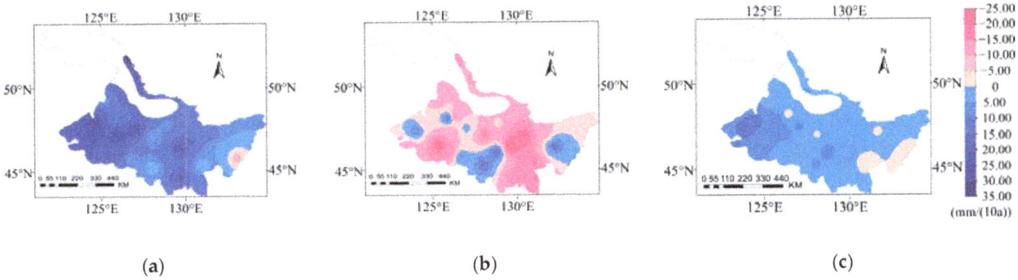

Figure 13. *Cont.*

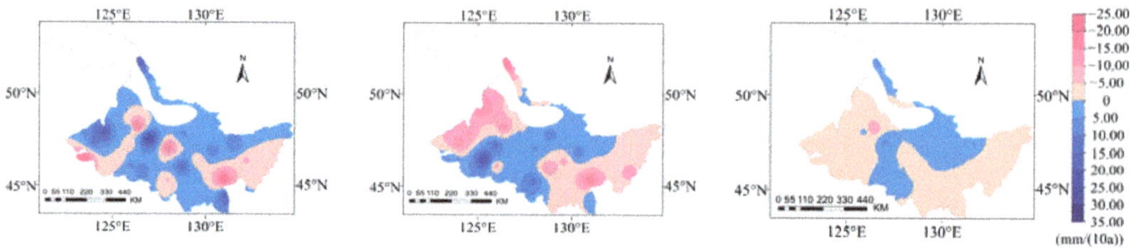

(d) (e) (f)

Figure 13. Climate tendency rates of I_r in the periods (**a**) 2021–2050, (**b**) 2051–2080, and (**c**) 2021–2080 under RCP4.5, and in the periods (**d**) 2021–2050, (**e**) 2051–2080, and (**f**) 2021–2080 under RCP8.5 during the soybean growth period.

3.7. Effect of Climate Change on Water Supply and Requirement

Under RCP4.5 and RCP8.5, soybean ET_c was significantly positively correlated with T_{max} and *Rad* and negatively correlated with *RH* (Table 2). Under RCP8.5, P_e was significantly negatively correlated with T_{max} and *Rad*, significantly positively correlated with T_{min}, and weakly correlated with *RH*. Under RCP4.5, I_r was significantly positively correlated with average T_{min}, T_{max}, and *Rad*, and significantly negatively correlated with average *RH*. The effects of meteorological factors on soybean ET_c in the study area under RCP4.5 and RCP8.5 for 2021–2080 and the relationships among P_e, ET_c, and I_r are shown in Figure 14. Under RCP4.5 and RCP8.5, soybean ET_c was significantly correlated with average T_{min}, T_{max}, *RH*, and *Rad*. The increase in temperature and *Rad* led to an increase in ET_0, further increasing ET_c (Table 2). The correlation coefficient between ET_c and *Rad* was greater than that of T_{max}, indicating that the increasing soybean ET_c was most influenced by *Rad*, followed by T_{max}. The *CD* tended to decrease and then increase under RCP4.5; however, it tended to increase under RCP8.5 from the 2030s to 2070s. Under RCP4.5, ET_c increased by 29.45 and 5.55 mm in the 2030s–2050s and 2050s–2070s, respectively, while P_e decreased by 12.21 mm in 2030s–2050s and then increased by 16.78 mm in 2050s–2080s. The combined effects of ET_c and P_e led to a change in I_r, which first increased by 31.01 mm and then decreased by 13.65 mm from the 2030s to 2070s (Figure 14).

Table 2. Correlation analysis among ET_0, ET_c, P_e, I_r, and meteorological factors during the soybean growth period.

Items	Scenarios	Periods	T_{min}	T_{max}	*RH*	*Rad*
ET_0	RCP4.5	2030s	−0.176	−0.522 *	−0.769 **	0.790 **
		2050s	0.087	0.779 **	−0.727 **	0.892 **
		2070s	0.057	0.926 **	−0.863 **	0.912 **
		2030s–2070s	0.898 **	0.982 **	−0.831 **	0.908 **
	RCP8.5	2030s	−0.059	0.405	−0.450 *	0.508 *
		2050s	−0.386	0.852 **	−0.924 **	0.963 **
		2070s	−0.473 *	0.854 **	−0.885 **	0.929 **
		2030s–2070s	0.971 **	0.991 **	−0.226	0.697 **
ET_c	RCP4.5	2030s	−0.291	0.548 *	−0.826 **	0.849 **
		2050s	−0.089	0.598 **	−0.780 **	0.892 **
		2070s	−0.091	0.790 **	−0.798 *	0.823 **
		2030s–2070s	0.855 **	0.960 **	−0.873 **	0.939 **
	RCP8.5	2030s	−0.214	0.298	−0.595 **	0.649 **
		2050s	−0.395	0.739 **	−0.820 **	0.884 **
		2070s	−0.489 *	0.726 **	−0.841 **	0.890 **
		2030s–2070s	0.962 **	0.984 **	−0.851 **	0.716 **

Table 2. *Cont.*

Items	Scenarios	Periods	T_{min}	T_{max}	RH	Rad
P_e	RCP4.5	2030s	−0.240	−0.338	0.140	−0.114
		2050s	0.390	0.025	−0.487	−0.436
		2070s	−0.149	−0.223	0.255	−0.208
		2030s–2070s	0.167	−0.021	0.387 **	−0.306 *
	RCP8.5	2030s	0.206	0.081	0.136	−0.142
		2050s	0.099	−0.344	0.375	−0.339
		2070s	0.482 *	−0.001	−0.248	−0.306
		2030s–2070s	0.810 **	−0.793 **	−0.110	−0.324 *
I_r	RCP4.5	2030s	0.152	0.287	−0.184	0.167
		2050s	−0.283	0.291	−0.698 **	0.693 **
		2070s	−0.016	0.381	−0.434	0.419
		2030s–2070s	0.409 **	0.590 **	−0.841 **	0.824 **
	RCP8.5	2030s	−0.232	−0.078	−0.209	0.209
		2050s	−0.231	0.434	−0.547 *	0.550 *
		2070s	−0.531 *	0.268	−0.489 *	0.582 **
		2030s–2070s	−0.140	0.094	−0.400 **	0.283 *

Note: * significant correlation at the 0.05 level; ** significant correlation at the 0.01 level.

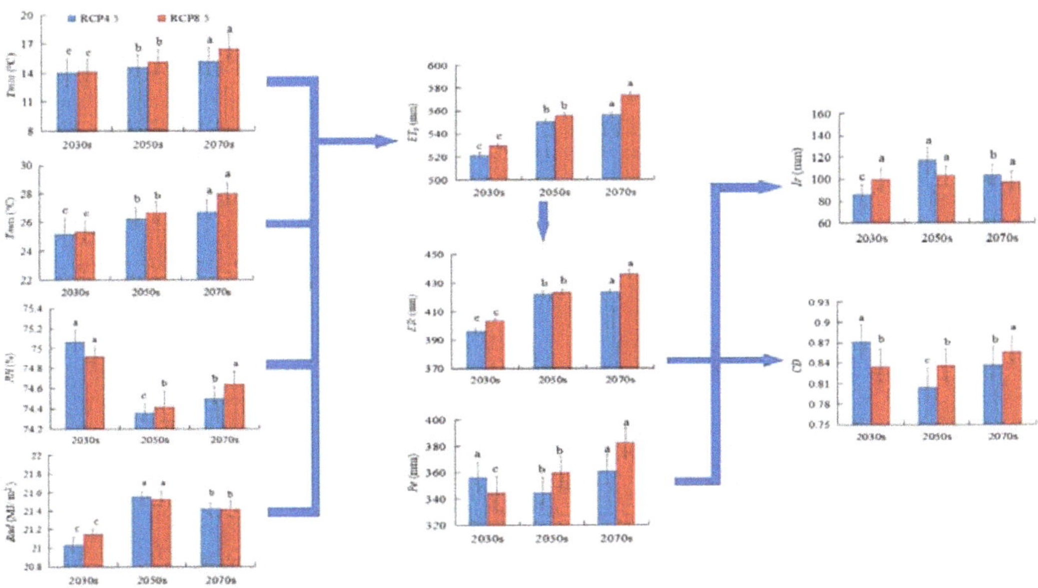

Figure 14. Effects of changes in meteorological factors on soybean ET_c and the relationships among P_e, ET_c, and I_r under RCP4.5 and RCP8.5 for 2021–2080. Bars marked with different lowercase letters indicate significant differences between groups ($p < 0.05$), while those marked with the same lowercase letters indicate insignificant differences between groups ($p > 0.05$).

4. Discussion

4.1. Soybean ET_c and Meteorological Factors

Soybean ET_c showed an increasing trend from the 2030s to 2070s under RCP4.5 and RCP8.5 in Heilongjiang Province of China in this study. An upward trend in ET_c was also observed in previous studies involving different crops under future climate change, including maize in Zimbabwe in the 2020s, 2050s, and 2090s [13], sugarcane in Pakistan in the 2020s, 2050s, and 2080s [45], rice in Kunshan in the 2020s–2080s [7], wheat, maize, and gram in India in the 2020s–2080s [28], wheat in three provinces of northeast China in the 2040s, 2070s, and 2100s [23], and summer maize in Huang-Huai-Hai Plain in 2016–2050 [46].

In the 60 year time series under RCP4.5 and RCP8.5 covered in this study, T_{min}, T_{max}, *Rad*, and *RH* were all strongly related to the increase in ET_c. Yang et al. (2021) found that *Rad*, wind speed, and *P* had the strongest linear correlations with cotton ET_c, with correlation coefficients of 0.410–0.789, 0.361–0.676, and −0.215–−0.410, for 1965–2016 in NCP, respectively. The correlation of ET_c with *RH* and average temperature were weak, in the range of −0.189–−0.047 and −0.102–0.015, respectively [16]. In contrast, under RCP4.5 and RCP8.5, the decline rate of *RH* climate tendency rate in Heilongjiang Province was almost twice that of the NCP. The decreased *RH* in the air increased the evaporation rate, thus increasing ET_c [47]. Nageen reported an increasing trend in ET_c for sugarcane as well, which was due to the forecasted temperature rise in the future Pakistan region, while the increased P_e could not compensate for the increased ET_c [45]. In addition, this study found that the increase in sunshine hours provided more radiation and light energy to the soybean [48], thus promoting the opening of stomata for plant transpiration and leading to an increase in transpiration [49]; accordingly, ET_c showed an increasing trend. Li et al. [46] reported that the temperature would continue to rise in the future in the Huang-Huai-Hai Plain, while the summer maize evaporation would increase, resulting in increased ET_c. However, this study focused more on the impact of the combined effects of T_{min}, T_{max}, *RH*, and *Rad* meteorological factors on the increase in ET_c in soybean.

4.2. Soybean ET_c, P_e, and I_r

In this study, the annual average ET_c for the soybean growing season under RCP4.5 and RCP8.5 for 2021–2080 was higher than that of soybean in Heilongjiang Province for 1966–2015 reported by Li et al. [30]. The higher ET_c indicated soybean in this study area may suggest the need for more water due to the increase in evapotranspiration derived from future climate conditions [50]. Oludare et al. (2020) reported that the average temperature and *Rad* increased, while soybean ET_c increased slightly under RCP4.5 and RCP8.5 for 2021–2099 in the Ogun-Ona River Basin, Nigeria [50]. Similar to the results of this study, the ET_c of soybean in different regions of the world also increased with the same trend of meteorological factors.

Under RCP4.5 and RCP8.5, the P_e and I_r of soybean were higher than reported by Li et al. [30]. Although a small increase in P_e was predicted in the future, more I_r was still needed, which probably increased the pressure on agricultural water, as well as drought frequency [51]. The highest ET_c and P_e under RCP4.5 and RCP8.5 in this study were distributed in the south; however, Li et al. (2020) reported that the highest ET_c and P_e were in the west [34]. Due to the increase in T_{min}, T_{max}, and *Rad* in the southern region and the decrease in *RH*, higher ET_c is expected in the future. Moreover, in the historical period, Li et al. [34] did not consider the influence of *Rad* on ET_c. In addition, ET_c is also affected by the plant itself, such as plant canopy structure and plant physiology [52]. Under RCP4.5, the climate tendency rate of ET_c was much greater in the west than that in the east; however, under RCP8.5, the climate tendency rate of ET_c showed an opposite spatial distribution trend, which differed from the results of Hu et al. [37]. This might be due to the higher values of *Rad* under RCP8.5, resulting in an increase in the climate tendency rate of ET_c in the east. On the other hand, the meteorological factors came from different meteorological stations and time series [53]. Under future climate change, their increasing or decreasing trends and magnitudes are also very different from the past [54]. This study provides long-term information for soybean water and irrigation requirements in Heilongjiang Province of China under future climate change [55].

4.3. Uncertainties and Limitations of the Study

This study indicated that there was a strong relationship among temperature, P_e variability, ET_c, and I_r under climate change. Two limitations should be taken into account in this study. Firstly, due to political and socioeconomic factors, regional climate programs are unable to accurately predict the path of future greenhouse gas emissions [56]. We only considered the RCP4.5 and RCP8.5, in which the concentration of CO_2 is fixed; however, in fact, the concentration of CO_2 varies with time [2]. In addition, the "Special Report on Emission Scenarios" (SRES) also proposed that two other emission scenarios, A1 (emphasizing economic development) and B2 (emphasizing sustainable development), can also predict the future climate [23]. However, anthropogenic-based climate change scenarios are one-sided scenarios. Under the actual climate in the future, the biological and agricultural technological progress of soybean planting will certainly change to reduce the impact of climate change. Secondly, we adjusted the parameters of the CROPWAT crop model for Heilongjiang Province, but there were still some uncertainties in the simulation parameters. For example, K_c and crop phenology are expected to change under future climatic change [57]. Therefore, changes in all meteorological factors caused by global warming and the uncertainties and limitations of the model should be deeply considered in further study.

5. Conclusions

In 2021–2080, T_{min}, T_{max}, and Rad increased while RH decreased under RCP4.5 and RCP8.5. In particular, the climate tendency rates for T_{min}, T_{max}, and Rad were higher under RCP8.5. There was little difference in the climate tendency rate of RH between RCP4.5 and RCP8.5. Affected by the changes in climate factors in the future, the ET_0, ET_c, and P_e during soybean growth period in Heilongjiang Province showed an upward trend under RCP4.5 and RCP8.5. The climate tendency rates of annual ET_c were 6.09 mm/(10 years) and 7.16 mm/(10 years), respectively. The climate tendency rate of annual I_r was 3.73 mm/(10 years) under RCP 4.5, while it was −0.067 mm/(10 years) under RCP 8.5. The results showed that the soybean in whole Heilongjiang province would face water shortage stress in the future, especially in the central and western regions. There would be more P and less ET_c in the eastern region. Therefore, we should appropriately adjust the crop structure, change the planting system, and recommend increasing the soybean planting area in the eastern Heilongjiang Province. This study can guide future irrigation system planning and management policy in Heilongjiang Province.

Author Contributions: Conceptualization, T.N. and N.L.; methodology, T.N. and N.L.; validation, Y.T., Y.J. and Z.Z.; formal analysis, N.L. and T.N.; writing—original draft preparation, N.L. and T.N.; writing—review and editing, T.W., Z.Z., P.C., T.L., L.M. and K.C.; visualization, N.L., Y.T., D.L. and T.W.; supervision, T.N.; project administration, T.N.; funding acquisition, T.N. All authors have read and agreed to the published version of the manuscript.

Funding: This work was fund by Basic Scientific Research Fund of Heilongjiang Provincial Universities (grant number: 2021-KYYWF-0019), the Opening Project of Key Laboratory of Efficient Use of Agricultural Water Resources, Ministry of Agriculture and Rural Affairs of the People's Republic of China in Northeast Agricultural University (grant number: AWR2021002), and the National Natural Science Foundation Project of China (grant number: 51779046).

Institutional Review Board Statement: Not applicable.

Informed Consent Statement: Not applicable.

Data Availability Statement: The data presented in this study are available on request from the corresponding author.

Acknowledgments: We thank the Chinese meteorological data sharing service (http://data.cma.cn, accessed on 22 May 2022) for providing the meteorological data. We thank all the members in the Lab of the Pumping, Hydraulic Teaching, and Experimental Center of Heilongjiang University. Lastly, we thank the anonymous reviewers and the editor for their suggestions, which substantially improved the manuscript.

Conflicts of Interest: The authors declare no conflict of interest.

References

1. Chen, Y.Y. Impact of climate change on grain production in China. *Chin. Agric. Sci. Bull.* **2021**, *12*, 51–57. (In Chinese)
2. Intergovernmental Panel on Climate Change (IPCC). *Global Warming of 1.5 °C; Impacts of 1.5 °C of Global Warming on Natural and Human Systems. Contribution of Working Group I to the Sixth Assessment Report of the Intergovernmental Panel on Climate Change*; Intergovernmental Panel on Climate Change (IPCC): Geneva, Switzerland, 2018; *in press*.
3. Wang, T.; Du, C.; Nie, T.; Sun, Z.; Zhu, S.; Feng, C.; Dai, C.; Chu, L.; Liu, Y.; Liang, Q. Spatiotemporal analysis of maize water requirement in the Heilongjiang Province of China during 1960–2015. *Water* **2020**, *12*, 2472. [CrossRef]
4. Tang, C.X. Influence of rainfall weather on agricultural production and countermeasures. *Agric. Dev. Equip.* **2019**, *10*, 68–69. (In Chinese)
5. Bouras, E.; Jarlan, L.; Khabba, S.; Er-Raki, S.; Dezetter, A.; Sghir, F.; Tramblay, Y. Assessing the impact of global climate changes on irrigated wheat yields and water requirements in a semi-arid environment of Morocco. *Sci. Rep.* **2019**, *9*, 19142. [CrossRef]
6. Ding, Y.; Wang, W.; Zhuang, Q.; Luo, Y. Adaptation of paddy rice in China to climate change: The effects of shifting sowing date on yield and irrigation water requirement. *Agric. Water Manag.* **2020**, *228*, 105890. [CrossRef]
7. Wang, W.G.; Sun, F.Z.; Peng, S.Z.; Xu, J.Z.; Luo, Y.F.; Jiao, X.Y. Simulation of response of rice irrigation water demand to climate change. *Trans. Chin. Soc. Agric. Eng.* **2013**, *14*, 90–98. (In Chinese)
8. Acharjee, T.K.; Ludwig, F.; van Halsema, G.; Hellegers, P.; Supit, I. Future changes in water requirements of Boro rice in the face of climate change in North-West Bangladesh. *Agric. Water Manag.* **2017**, *194*, 172–183. [CrossRef]
9. Cheshmberah, F.; Zolfaghari, A.A. The Effect of Climate Change on Future Reference Evapotranspiration in Different Climatic Zones of Iran. *Pure Appl. Geophys.* **2019**, *176*, 3649–3664. [CrossRef]
10. Ti, J.; Yang, Y.; Yin, X.; Liang, J.; Pu, L.; Jiang, Y.; Wen, X.; Chen, F. Spatio-Temporal Analysis of Meteorological Elements in the North China District of China during 1960–2015. *Water* **2018**, *10*, 789. [CrossRef]
11. Nie, T.; Tang, Y.; Jiao, Y.; Li, N.; Wang, T.; Du, C.; Zhang, Z.; Chen, P.; Li, T.; Sun, Z.; et al. Effects of Irrigation Schedules on Maize Yield and Water Use Efficiency under Future Climate Scenarios in Heilongjiang Province Based on the AquaCrop Model. *Agronomy* **2022**, *12*, 810. [CrossRef]
12. Lyu, J.; He, Q.-Y.; Chen, Q.-W.; Cheng, R.-R.; Li, G.; Otsuki, K.; Yamanaka, N.; Du, S. Distinct transpiration characteristics of black locust plantations acclimated to semiarid and subhumid sites in the Loess Plateau, China. *Agric. Water Manag.* **2022**, *262*, 107402. [CrossRef]
13. Nkomozepi, T.; Chung, S.-O. Assessing the trends and uncertainty of maize net irrigation water requirement estimated from climate change projections for Zimbabwe. *Agric. Water Manag.* **2012**, *111*, 60–67. [CrossRef]
14. Xiao, D.; Liu, D.L.; Wang, B.; Feng, P.; Bai, H.; Tang, J. Climate change impact on yields and water use of wheat and maize in the North China Plain under future climate change scenarios. *Agric. Water Manag.* **2020**, *238*, 106238. [CrossRef]
15. Yang, X.; Jin, X.; Chu, Q.; Pacenka, S.; Steenhuis, T.S. Impact of climate variation from 1965 to 2016 on cotton water requirements in North China Plain. *Agric. Water Manag.* **2021**, *243*, 106502. [CrossRef]
16. Chen, D.W. Advances in Prediction and Calculation of Crop Water Demand. *Hydro Sci. Cold Zone Eng.* **2021**, *4*, 171–174. (In Chinese)
17. Wang, Y.; Shao, X.H.; Li, Y.H. Research progress on real-time prediction method of reference crop water requirement. *J. Anhui Agric. Sci.* **2008**, *36*, 13919–13920. (In Chinese) [CrossRef]
18. Tavakoli, A.R.; Moghadam, M.M.; Sepaskhah, A.R. Evaluation of the AquaCrop model for barley production under deficit irrigation and rainfed condition in Iran. *Agric. Water Manag.* **2015**, *161*, 136–146. [CrossRef]
19. FAO. *OECD-FAO Agricultural Outlook*; OECD Publishing, Food and Agriculture Organization of the United Nations: Rome, Italy, 2009; pp. 2011–2020.
20. Lu, X.P.; Duan, S.Q.; Ma, X.Y.; Bai, S.M. Comparative Study on Calculation of Maize Water Demand by Single and Double Crop Coefficient Method. *Water Sav. Irrig.* **2012**, *11*, 18–21. (In Chinese)
21. Zhang, L.; Tan, F.; Li, S.; Huo, Z. Potential dynamic of irrigation water requirement for rice across Northeast China. *Theor. Appl. Climatol.* **2020**, *142*, 1283–1293. [CrossRef]
22. Wu, J.B.; Ma, Y.F.; Zheng, H.X.; Cao, X.S. Estimation of maize evapotranspiration in semi-arid region based on modified crop coefficient model. *J. Water Resour. Archit. Eng.* **2020**, *18*, 25–29. (In Chinese)
23. Zhang, J.P.; Wang, C.Y.; Yang, X.G.; Zhao, Y.X.; Liu, Z.J.; Wang, J.; Chen, Y.Y. Prediction of the impact of future climate change on maize water demand in Northeast China. *Trans. Chin. Soc. Agric. Eng.* **2009**, *25*, 50–55. (In Chinese)
24. Nie, T.Z.; Zhang, Z.X.; Qi, Z.J.; Chen, P.; Lin, Y.Y.; Sun, Z.Y. Spatial-temporal distribution characteristics of rice water demand in Heilongjiang Province from 1960 to 2015. *Trans. Chin. Soc. Agric. Mach.* **2019**, *50*, 279–290. (In Chinese)
25. Jin, X. Application of Penman Formula in Crop Water Demand Calculation. *Jianghuai Water Resour. Sci. Technol.* **2018**, *1*, 28–30. (In Chinese)
26. Awal, R.; Fares, A.; Bayabil, H. Assessing Potential Climate Change Impacts on Irrigation Requirements of Major Crops in the Brazos Headwaters Basin, Texas. *Water* **2018**, *10*, 1610. [CrossRef]
27. Hong, E.-M.; Nam, W.-H.; Choi, J.-Y.; Pachepsky, Y.A. Projected irrigation requirements for upland crops using soil moisture model under climate change in South Korea. *Agric. Water Manag.* **2016**, *165*, 163–180. [CrossRef]
28. Hossain, M.B.; Roy, D.; Maniruzzaman, M.; Biswas, J.C.; Naher, U.A.; Haque, M.M.; Kalra, N. Response of crop water requirement and yield of irrigated rice to elevated temperature in Bangladesh. *Int. J. Agron.* **2021**, *2021*, 9963201. [CrossRef]

29. Sunil, A.; Deepthi, B.; Mirajkar, A.B.; Adarsh, S. Modeling future irrigation water demands in the context of climate change: A case study of Jayakwadi command area, India. *Model. Earth Syst. Environ.* **2021**, *7*, 1963–1977. [CrossRef]
30. Goodarzi, M.; Abedi-Koupai, J.; Heidarpour, M. Investigating Impacts of Climate Change on Irrigation Water Demands and Its Resulting Consequences on Groundwater Using CMIP5 Models. *GroundWater* **2019**, *57*, 259–268. [CrossRef]
31. Chen, P.; Nie, T.; Chen, S.; Zhang, Z.; Qi, Z.; Liu, W. Recovery efficiency and loss of 15N-labelled urea in a rice-soil system under water saving irrigation in the Songnen Plain of Northeast China. *Agric. Water Manag.* **2019**, *222*, 139–153. [CrossRef]
32. Chen, P.; Xu, J.; Zhang, Z.; Wang, K.; Li, T.; Wei, Q.; Li, Y. Carbon pathways in aggregates and density fractions in Mollisols under water and straw management: Evidence from 13C natural abundance. *Soil Biol. Biochem.* **2022**, *169*, 108684. [CrossRef]
33. Heilongjiang Provincial People's Government. *Heilongjiang Yearbook*; Editorial Department: Harbin, China, 2020; pp. 244–245. ISSN 1008-0791.
34. Li, J.Y.; Wang, X.L.; Wang, Z.B.; Nie, T.Z. Spatial and Temporal Distribution of Soybean Water Demand and Drought in Heilongjiang Province in Recent 50 Years. *Trans. Chin. Soc. Agric. Mach.* **2020**, *51*, 223–237. (In Chinese)
35. Hu, H.J.; Wang, M.; Yin, X.G.; Jiang, Y.L.; Wen, X.Y.; Chen, F. Analysis on Temporal and Spatial Variation Characteristics of Soybean Water Demand in Northeast Agricultural Region under Climate Change. *Zhongguo Nongye Daxue Xuebao* **2017**, *22*, 21–31. (In Chinese)
36. Accumulated Temperature Zone of Crop Varieties in Heilongjiang Province. Available online: http://www.hljagri.gov.cn/ (accessed on 2 September 2020).
37. Area Layout Planning of High-Quality and High-yield Main Food Crops in Heilongjiang Province in 2015. Available online: http://dszz.hljagri.gov.cn/tjxxw/jrtj/tzgg/201501/ (accessed on 2 September 2020).
38. Smith, M. *CROPWAT: A Computer Program for Irrigation Planning and Management*; FAO: Rome, Italy, 1992.
39. FAO. *Crop Evapotranspiration for Computing Crop Water Requirement*; FAO Irrigation & Drainage Paper: Rome, Italy, 1998; p. 56.
40. Gocic, M.; Trajkovic, S. Analysis of changes in meteorological variables using Mann-Kendall and Sen's slope estimator statistical tests in Serbia. *Glob. Planet. Change* **2013**, *100*, 172–182. [CrossRef]
41. Bahmani, S.; Bajic, I.V.; HajShirmohammadi, A. Joint Decoding of Unequally Protected JPEG2000 Bitstreams and Reed-Solomon Codes. *IEEE Trans. Image Process.* **2010**, *19*, 2693–2704. [CrossRef] [PubMed]
42. Tsakmakis, I.D.; Zoidou, M.; Gikas, G.D.; Sylaios, G.K. Impact of Irrigation Technologies and Strategies on Cotton Water Footprint Using AquaCrop and CROPWAT Models. *Environ. Process.* **2018**, *5*, S181–S199. [CrossRef]
43. Dubrau, A.W.; Hendren, L.J. Taming MATLAB. *Acm Sigplan Notice* **2012**, *47*, 503–522. [CrossRef]
44. Xiao, Z.L. A comparative study of using MATLAB and SPSS software to solve regression analysis. *Shanxi Youth* **2016**, *18*, 58–59. (In Chinese)
45. Farooq, N.; Gheewala, S.H. Assessing the impact of climate change on sugarcane and adaptation actions in Pakistan. *Acta Geophys.* **2020**, *68*, 1489–1503. [CrossRef]
46. Li, Y.; Qu, H.J.; Su, W.Y.; Gao, Z.P.; Wu, J.M. Response of Crop Irrigation Water Demand in Typical Area of Huanghuai Plain under HadGEM2-ES Climate Model. *Water Resour. Power* **2017**, *35*, 119–122. (In Chinese)
47. Jalota, S.K.; Sood, A.; Chahal, G.B.S.; Choudhury, B.U. Crop water productivity of cotton (*Gossypium hirsutum* L.)–wheat (*Triticum aestivum* L.) system as influenced by deficit irrigation, soil texture and precipitation. *Agric. Water Manag.* **2006**, *84*, 137–146. [CrossRef]
48. Liu, B.; Zhang, D.; Zhang, H.; Asseng, S.; Yin, T.; Qiu, X.; Ye, Z.; Liu, L.; Tang, L.; Cao, W.; et al. Separating the impacts of heat stress events from rising mean temperatures on winter wheat yield of China. *Environ. Res. Lett.* **2021**, *16*, 124035. [CrossRef]
49. Paul, M.; Dangol, S.; Kholodovsky, V.; Sapkota, A.R.; Negahban-Azar, M.; Lansing, S.L. Modeling the Impacts of Climate Change on Crop Yield and Irrigation in the Monocacy River Watershed, USA. *Climate* **2020**, *8*, 139. [CrossRef]
50. Durodola, O.S.; Mourad, K.A. Modelling the Impacts of Climate Change on Soybeans Water Use and Yields in Ogun-Ona River Basin, Nigeria. *Agriculture* **2020**, *10*, 593. [CrossRef]
51. Soares, D.; Rolim, J.; Fradinho, M.J.; Paço, T.A.D. Climate Change Impacts on Irrigation Requirements of Preserved Forage for Horses under Mediterranean Conditions. *Agronomy* **2020**, *10*, 1758. [CrossRef]
52. Luo, X.; Xia, J.; Yang, H. Modeling water requirements of major crops and their responses to climate change in the North China Plain. *Environ. Earth Sci.* **2015**, *74*, 3531–3541. [CrossRef]
53. Gao, J.; Yang, X.; Zheng, B.; Liu, Z.; Zhao, J.; Sun, S.; Li, K.; Dong, C. Effects of climate change on the extension of the potential double cropping region and crop water requirements in Northern China. *Agric. For. Meteorol.* **2019**, *268*, 146–155. [CrossRef]
54. Mainuddin, M.; Kirby, M.; Chowdhury, R.A.R.; Shah-Newaz, S.M. Spatial and temporal variations of, and the impact of climate change on, the dry season crop irrigation requirements in Bangladesh. *Irrig. Sci.* **2015**, *33*, 107–120. [CrossRef]
55. Bouregaa, T. Impact of climate change on yield and water requirement of rainfed crops in the Setif region. *Manag. Environ. Qual. Int. J.* **2019**, *30*, 851–863. [CrossRef]
56. Chen, X.; Wang, L.; Niu, Z.; Zhang, M.; Li, C.a.; Li, J. The effects of projected climate change and extreme climate on maize and rice in the Yangtze River Basin, China. *Agric. For. Meteorol.* **2020**, *282*, 107867. [CrossRef]
57. Ma, J.Q.; Li, P.F.; Liu, L. Experimental Study on Crop Coefficient of Summer Maize in Henan Province. *Water Sav. Irrig.* **2016**, *4*, 24–27. (In Chinese)

agriculture

MDPI

Article

The Impacts of Climate Change on Water Resources and Crop Production in an Arid Region

Samira Shayanmehr [1], Jana Ivanič Porhajašová [2], Mária Babošová [2], Mahmood Sabouhi Sabouni [1], Hosein Mohammadi [1], Shida Rastegari Henneberry [3] and Naser Shahnoushi Foroushani [1,*]

1 Department of Agricultural Economics, Ferdowsi University of Mashhad, P.O. Box 91775-1111, Mashhad 9177948974, Iran; samira.shayanmehr@mail.um.ac.ir (S.S.); sabouhi@um.ac.ir (M.S.S.); hoseinmohammadi@um.ac.ir (H.M.)

2 Institute of Plant and Environmental Sciences, Slovak University of Agriculture in Nitra, 949 76 Nitra-Chrenová, Slovakia; jana.porhajasova@uniag.sk (J.I.P.); maria.babosova@uniag.sk (M.B.)

3 Department of Agricultural Economics, Oklahoma State University, Stillwater, OK 74078, USA; srh@okstate.edu

* Correspondence: shahnoushi@um.ac.ir

Abstract: Climate change is one of the most pressing global issues of the twenty-first century. This phenomenon has an increasingly severe impact on water resources and crop production. The main purpose of this study is to evaluate the impact of climate change on water resources, crop production, and agricultural sustainability in an arid environment in Iran. To this end, the study constructs a new integrated climate-hydrological-economic model to assess the impact of future climate change on water resources and crop production. Furthermore, the agricultural sustainability is evaluated using the multicriteria decision making (MCDM) technique in the context of climate change. The findings regarding the prediction of climate variables show that the minimum and maximum temperatures are expected to increase by about 5.88% and 6.05%, respectively, while precipitation would decrease by approximately 30.68%. The results of the research reveal that water availability will decrease by about 13.79–15.45% under different climate scenarios. Additionally, the findings show that in the majority of cases crop production will reduce in response to climate scenarios so that rainfed wheat will experience the greatest decline (approximately 59.95%). The results of the MCDM model show that climate change can have adverse effects on economic and environmental aspects and, consequently, on the sustainability of the agricultural system of the study area. Our findings can inform policymakers on effective strategies for mitigating the consequences of climate change on water resources and agricultural production in dry regions.

Keywords: climate change; crop yield; cultivated area; future climate scenarios; water use

Citation: Shayanmehr, S.; Porhajašová, J.I.; Babošová, M.; Sabouhi Sabouni, M.; Mohammadi, H.; Rastegari Henneberry, S.; Shahnoushi Foroushani, N. The Impacts of Climate Change on Water Resources and Crop Production in an Arid Region. *Agriculture* **2022**, *12*, 1056. https://doi.org/10.3390/agriculture12071056

Academic Editors: Dengpan Xiao and Wenjiao Shi

Received: 20 June 2022
Accepted: 17 July 2022
Published: 19 July 2022

1. Introduction

Climate change's impact on agricultural production has raised serious global concerns in the twenty-first century [1,2]. Carbon dioxide (CO_2) emissions have been identified as the primary cause of climate change [3–5]. Human activities, such as the use of fossil fuels, environmental degradation, and land-use changes, have all contributed significantly to rising CO_2 levels in the atmosphere since the industrial revolution [6–9]. The global CO_2 concentration in the atmosphere increased from 288 ppm in 1750 to 415 ppm in 2021 [10]. This has resulted in higher global average temperatures and unpredictability of rainfall patterns [11]. Climate change and variability have far-reaching consequences for natural resources, human communities, and biodiversity [12]. Water resources and agriculture are most affected by climate change because it directly determines the availability of resources in terms of time and space [13]. Some researchers have concluded that climate change has a negative impact on groundwater table recharge, which affects irrigation [12,14]. Others believe that this phenomenon will increase agricultural water demand due to increased evapotranspiration, thus putting more pressure on water resources [12,15]. Aside from influencing the availability of water resources, climate change is expected to reduce crop yield and agricultural efficiency by increasing crop water stress [16]. Climate change has recently had a negative impact on crop production in major agricultural areas; it is also expected to reduce global agricultural production by about 16% by 2030, resulting in widespread food insecurity [17–19]. Therefore, meeting the food needs of the world's rising population has become a major concern around the world [2,20].

Climate change is expected to have the greatest impact on agricultural production in the world's dry and semi-arid regions, such as Iran. Iran's average annual rainfall is around 250 mm, which is less than one-third of the global average; therefore, most parts of the country suffer from a lack of water resources for food production [21]. Nevertheless, the adaptation of supportive polices to achieve self-sufficiency in order to meet the domestic food demand in the presence of climate change has led to increased pressure on water resources in the country's arid and semi-arid regions [22,23]. In this regard, the Mashhad plain in northeastern Iran serves as a good example. Climate data analysis reveals that the phenomenon of climate change has happened in this area as a result of decreasing rainfall and increasing temperature (see Figure 1). As shown in the figure, during the years 1990 to 2016, the total precipitation decreased from 300 mm to 220 mm and the average temperature increased from 11.9 °C to 16.5 °C. Given the importance of this plain in ensuring the country's food security, a thorough understanding of the effects of climate change on water resources and agricultural production in this region is required to adopt accurate and efficient mitigation and adaptation policies.

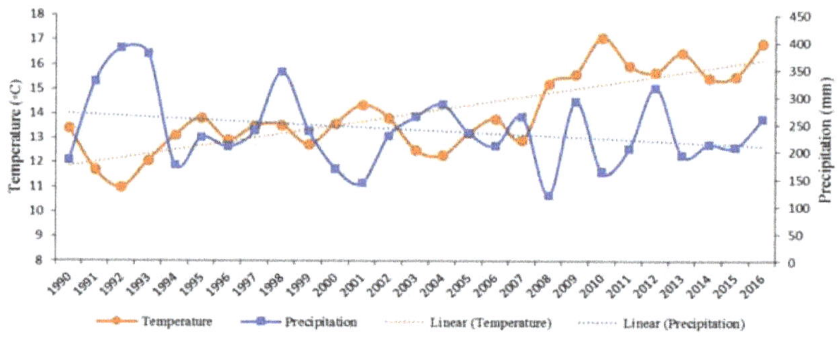

Figure 1. Annual mean temperature and precipitation changes in Mashhad plain.

The literature review indicated that the effects of climate change on water resources and crop production have been studied all over the world. Xiong et al. [24] used climate scenarios of the regional climate model to investigate the consequences of climate change on water availability and cereal production in China in the 2020s and 2040s. The findings of this study revealed that there are insufficient water resources for cereal production, particularly in southern China, due to an increase in nonagricultural water demand and the occurrence of climate change. Palazzoli et al. [16] developed a soil and water assessment tool (SWAT) model to investigate the effects of future climate change on rainfed crop productivity and water resources in Nepal. Based on their results, they predicted significant potential changes in water resources availability (from −26 to +37%) and crop production (rice from −17 to +12%, wheat from −36 to +18% and maize from −17 to +4%). Sinnarong et al. [25] applied an econometric model to estimate the effect of climate change on rice production in Thailand. The results showed that temperature has a negative impact on rice production while precipitation has different regional effects on rice production. Additionally, the findings indicated that rice production under different climatic scenarios would decrease between 4.56% and 33.77%. Mostafa et al. [26] evaluated the impact of climate change on water resources and the agricultural sector of Egypt using climate and irrigation (CROPWAT) models and found the irrigation water requirement for wheat crop would rise by about 6.2% in 2050 and 11.8% in 2100. Furthermore, wheat production would decrease by approximately 8.6% and 11.1% in 2050 and 2100, respectively. Medellín-Azuara et al. [27] estimated the impact of climatic change on crop farming in California using the statewide agricultural production model (SWAP). They found that, by 2050, water supply, agricultural land use, and production of most crops will decrease in California due to climate change, such as rising temperature and declining precipitation. Shahvari et al. [28], using the SWAT model, assessed the impact of climate change on water resources and crop yield in Iran for the future. Their results revealed that future climate scenarios will lead to an increase in runoff in spring and autumn and a decrease in summer and winter. In addition, future climate change will reduce the yield of rainfed crops in the region. Lu et al. [29] constructed a new climate-economic model to analyze the effects of climate change on grain production and water resources. The findings of this study showed that irrigation water consumption has increased by about 100 billion m^3 because of climate change in China. This phenomenon also reduced the grain yield in this area by 1000 kg/hm^2. The current gap in the existing literature is a comprehensive view of all meteorological, hydrological, economic, and sustainability aspects of climate change in the agricultural sector.

The study, therefore, assesses the effect of future climate change on water resources, crop production, and agricultural sustainability in the Mashhad plain under three climate scenarios (RCP 2.6, 4.5, and 8.5). Specifically, the study aims at (1) projecting climate variables using the Long Ashton Research Station Weather Generator (LARS-WG) model alongside HadGEM2-ES outputs and the three RCP scenarios; (2) assessing the impact of future climate change on water resources in the Mashhad plain using panel data model; (3) estimating the relationship between crop yield and climate variables, including minimum temperature, maximum temperature, and precipitation, using the GME (generalized maximum entropy) technique; (4) investigating the impact of climate change on cropping pattern, crop production, and water consumption in the selected area using PMP (positive mathematical programming) model; (5) evaluating the agricultural sustainability under climate scenarios using a MCDM (multicriteria decision making) method. The results of this study are expected to provide policymakers with insights into designing climate change mitigation policies to ensure food security and sustainable production in the region.

The contribution of this study to the literature is twofold. First, to the best of our knowledge, this is the first attempt to apply an integrated climate-hydrological-economic model to evaluate the effect of climate change on water resources, crop production, and cropping pattern in Iran. The second contribution of the study is found in the use of the MCDM approach to investigate the sustainability of agricultural activity at the regional level under different climate scenarios.

The study is structured as follows: Section 2 describes the study area, datasets, and methodology. Section 3 presents the results and discussions, and the last section concludes with the research and policy implications of these findings.

2. Materials and Methods

This study used a variety of methods to achieve its research goals. The LARS-WG model was used to downscale the climate variables (maximum and minimum temperatures and precipitation), and a regression model was used to forecast the groundwater availability in the Mashhad plain. The sensitivity of yield to climate change was estimated using the GME technique. The cropping pattern was then evaluated under climate change using a PMP model. Finally, economic, social, and environmental indicators were ranked using an integrated MCDM method. The complete structure of the framework is presented in Figure 2.

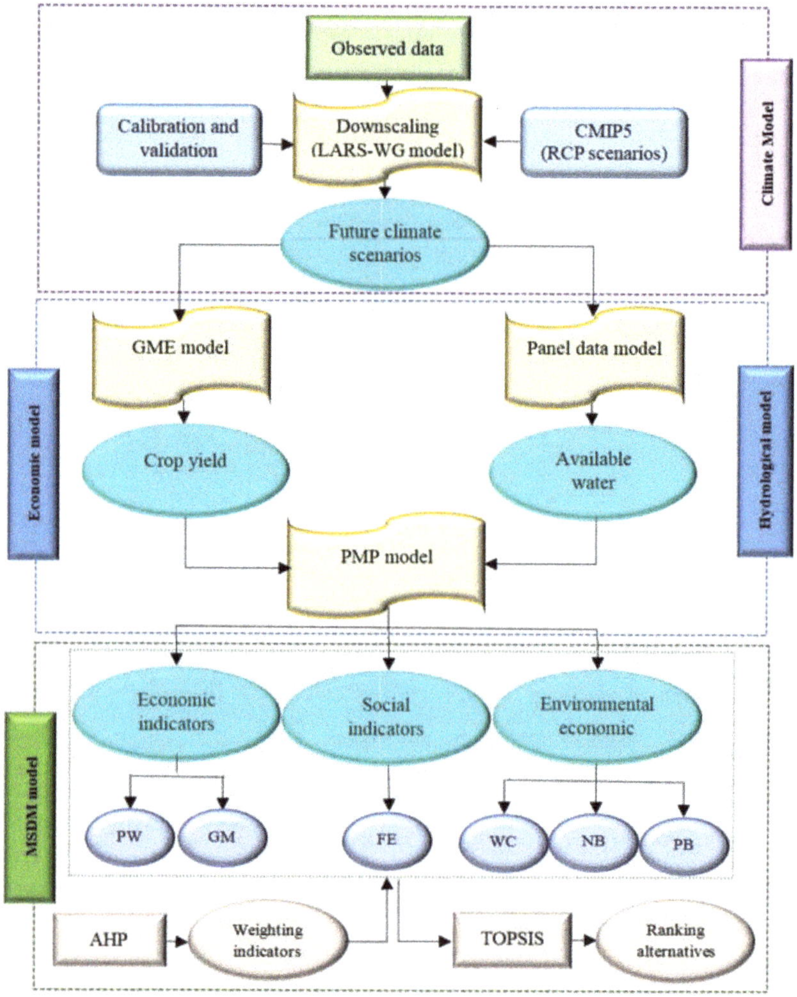

Figure 2. Main steps in the methodological framework.

2.1. Study Area

The present case study is in Northeast Khorasan Razavi Province, Iran, between the latitude 35°59′ to 37°03′ N and longitude 58°22′ to 60°06′ E, covering an area of approximately 9957 km^2. The plain is bounded on the north by Hezar Masjid heights, on the northwest by Atrak river basin, on the south by Binaloud Mountain, and on the southeast by Jamroud river basin [30]. It has a semi-arid to arid climate, with an average annual rainfall from 1991 to 2015 is about 262 mm [31]. The average monthly temperature in this plain is reported to be between 11.6 °C and 26.7 °C. Furthermore, the average annual evapotranspiration ranges from 236 to 310 (mm). The location of the study area is shown in Figure 3. Around 3 million people live in the study basin and rely mostly on groundwater resources for drinking and agricultural cultivation. Climate change and a lack of water resource management in this area have resulted in a 12-m water drop in the water table over a 20-year period [32].

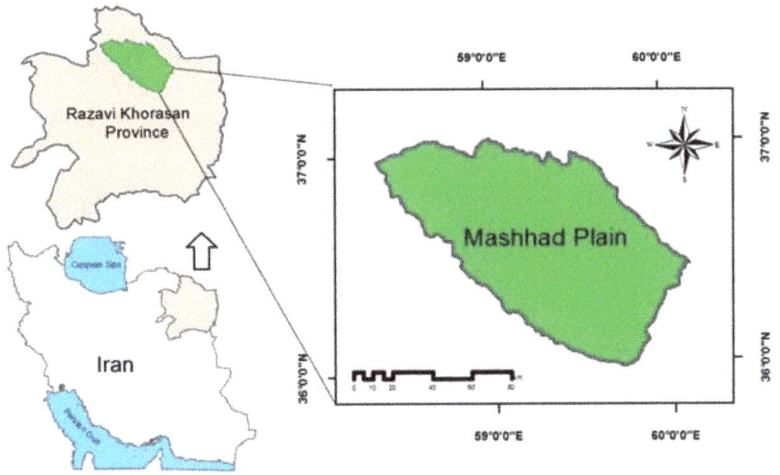

Figure 3. Map of study area.

2.2. Data Collection

The observed daily time series data of minimum temperature, maximum temperature, and precipitation for Mashhad synoptic station during the period 1979–2016 were obtained from Iran's Meteorological Organization. To estimate the model (Equation (6)), the data (2000–2016) on piezometric wells and groundwater depth in the Mashhad plain were provided by Iran Water Resources Management Company. Additionally, the observed monthly temperature and precipitation data (2000–2016) for the synoptic station were obtained from Iran's Meteorological Organization. To estimate the yield response function (Equation (8)), crop yield data (1983–2016) were gathered from Ministry of Agriculture Jihad of Iran. Figure 4 shows the growing seasons of crops (wheat, barley, alfalfa, potato, corn, tomato, melon, onion, sugar beet, and cucumber) in the Mashhad plain. Data and information on outputs prices, inputs costs, technical coefficients, crop yield, and resources availability were gathered through face-to-face interviews with farmers in 2016–2017 cropping season.

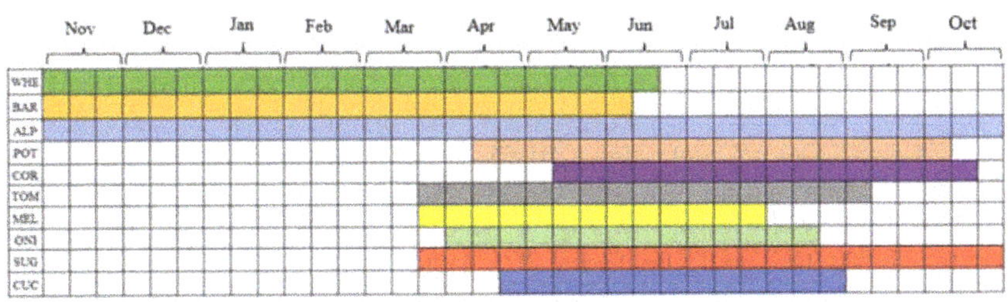

Figure 4. The growing season of crops in the study area. Note: WHE: wheat, BAR: barley, ALP: alfalfa, POT: potato, COR: corn, TOM: tomato, MEL: melon, ONI: onion, SUG: sugar beet, and CUC: cucumber.

2.3. Meteorological Model

The LARS-WG model is a random generator that uses statistical downscaling techniques to generate meteorological data [33]. Because of the repeated calculations, it requires less input data and is simpler and more efficient than other models [34,35]. Racsko et al. [36] proposed this model, which Semenov et al. [37] later revised and developed. This model's sixth version (LARS-WG6) was updated in 2018 for downscaling the coupled model intercomparison project phase (CIMP5) [34]. The HadGEM2-ES universal model data were used in this study to project climate variables during the horizon in 2045 under three climate scenarios, namely RCP 2.6, RCP 4.5, and RCP 8.5. The LARS-WG model was implemented using daily data of maximum and minimum temperature and precipitation from 1979 to 2016.

The performance of the LARS-WG statistical model was evaluated by comparing the simulated and observed maximum and minimum temperatures, as well as precipitation, using the following statistics [2,38,39]: coefficient of determination (R^2), normalized root mean square error (NRMSE), root mean square error (RMSE), mean absolute deviation (MAD), and mean square error (MSE) (see Table 1).

Table 1. The statistical indicators for model validation.

Index	Formulate	Model Performance	
R^2	$R^2 = \frac{(\sum_{n=1}^{N}(S_n-\bar{S})(O_n-\bar{O}))^2}{\sum_{n=1}^{N}(S_n-\bar{S})^2\sum_{n=1}^{N}(O_n-\bar{O})^2}$	≤0.75	
NRMSE	$NRMSE = \frac{RMSE}{\bar{O}} \times 100$	Criteria	performance
		≤10%	Excellent
		10–20%	Good
		20–30%	Fair
		≥30%	Poor
RMSE	$RMSE = \sqrt{\frac{\sum_{n=1}^{N}(S_n-O_n)^2}{N}}$	the lower values show a better model [40]	
MAD	$MAD = \frac{\sum_{n=1}^{N}\lvert S_n-O_n\rvert}{N}$		
MSE	$MSE = \frac{\sum_{n=1}^{N}(S_n-O_n)^2}{N}$		

Note: N is the number of data points, S_n is the simulated values, O_n is the observed values, $\bar{O}$ is the mean of the observed values, and $\bar{S}$ is the mean of the simulated values.

2.4. Hydrological Model

This section investigates the effects of climate change on groundwater resources in the Mashhad plain. The panel data model was used to forecast the groundwater depth in the plain. To this end, the following groundwater conceptual model was estimated using panel data regression [41,42] related to piezometers as cross units from 2000 to 2016.

$$\ln H_t = \alpha_0 + \alpha_1 \ln H_{t-1} + \alpha_2 \ln P_{t-1} + \alpha_3 \ln Tmin_{t-1} + \alpha_4 \ln Tmax_{t-1} \quad (1)$$

where t indicates time, H_t is predicted groundwater depth, H_{t-1} is the initial groundwater depth, P is total monthly precipitation (mm), Tmin and Tmax are monthly minimum and maximum temperature (°C), respectively, and α_0 to α_4 are the model coefficients that should be estimated.

After estimating the amount of changes in groundwater depth because of climate change, Equation (2) was used to calculate the amount of changes in groundwater resources [43–46].

$$\Delta V = A \times S_y \times \Delta H \quad (2)$$

where ΔV is the groundwater storage change (m^3), A is geographical area (m^2), S_y is specific yield (dimensionless), and ΔH is average depth change (m).

2.5. Economic Models

2.5.1. Generalized Maximum Entropy (GME) Model

In this study, the Cobb–Douglas Yield Response (CDYR) model was used to assess the sensitivity of crops yield to climatic variables, such as minimum and maximum temperatures, as well as precipitation [47,48]. The CDYR model, after taking the logarithm of both sides of the equation, can be presented as follows:

$$\log(Y_t) = \beta_0 + \alpha_t \log(Tmin_t) + \lambda_t \log(Tmax_t) + \eta_t \log(P_t) + \nu Trend \quad (3)$$

where t is the set of years; Y is the yield of crops (wheat, barley, alfalfa, potato, corn, tomato, melon, onion, sugar beet, and cucumber); Tmin and Tmax are average minimum and maximum temperature; and P is total growing season precipitation. It is worth noting that the impact of additional factors affecting crop production, which were not incorporated in Equation (3), was covered in the residual terms [49].

Due to the lack of access to crop yields data in the research area, we only analyzed data of 34 years (1983–2016). In this case, because of the limited sample size, traditional estimation approaches, such as ordinary least squares (OLS), may result in parameter estimation with excessive variance [49,50]. To address this issue, following Moreno et al. [51], we used the GME estimator. The GME approaches are founded on Shannon's entropy information measure and the generalized maximum entropy theory [50,52]. Instead of calculating the mean and variance of coefficients directly, GME estimator considers a probability distribution for the coefficients and error terms [52]. Assume that y is dependent on K independent variables $x_k(k = 1, \ldots, k)$:

$$y = X\beta + \varepsilon \tag{4}$$

where y is a $(T \times 1)$ vector of observations for y, X is a $(T \times K)$ matrix of observations for the x_k variables, β is the $(K \times 1)$ vector of explanatory variable coefficients, and ε is a $(T \times 1)$ vector of residual terms. To estimate explanatory variable coefficients using GME, firstly we reparametrize the regression model, and then we recast the coefficients and residual in terms of discrete probability distributions.

In this method, each β_k is assumed to be as a discrete point with M dimension $(M \geq 2)$. Let $Z_k = [Z_{k1}, \ldots, Z_{kM}]$ be support points for parameter β_k, which are symmetrical around zero. Additionally, the probability mass function of z_k is defined as $P_k = [P_{k1}, \ldots, P_{kM}]'$ such that:

$$\beta_k = E_{pk}[z_k] = \acute{z}_k p_k = \sum_{m=1}^{M} z_{km} p_{km}; \forall K = 1, \ldots, K \tag{5}$$

Then, β can be represented as follows:

$$B = \begin{bmatrix} \beta_1 \\ \vdots \\ \beta_K \end{bmatrix} = zp = \begin{bmatrix} \acute{z}_1 & 0 & \ldots & 0 \\ 0 & \acute{z}_2 & \ldots & 0 \\ \vdots & \vdots & \ddots & \vdots \\ 0 & 0 & \ldots & \acute{z}_K \end{bmatrix} \begin{bmatrix} P_1 \\ P_2 \\ \vdots \\ P_K \end{bmatrix} \tag{6}$$

where Z is a $(K \times KM)$ matrix of support values, and P is a $(KM \times 1)$ of vector of unknown weights.

The unknown error is defined as follows:

$$\varepsilon_g = E_{w_g}[v] = \acute{v} w_g = \sum_{j=1}^{J} v_j w_{gj}; \forall g = 1, \ldots, G \tag{7}$$

where $w_g = [w_{g1}, \ldots, w_{gJ}]'$ is a vector of weights, and $\acute{v}_g = [v_{g1}, \ldots, v_{gJ}] (J \geq 2)$ is a set of support points. The error vector is presented as follows:

$$\varepsilon = \begin{bmatrix} \varepsilon_1 \\ \vdots \\ \varepsilon_G \end{bmatrix} = V_w = \begin{bmatrix} \acute{v}_1 & 0 & \ldots & 0 \\ 0 & \acute{v}_2 & \ldots & 0 \\ \vdots & \vdots & \ddots & \vdots \\ 0 & 0 & \ldots & \acute{v}_G \end{bmatrix} \begin{bmatrix} w_1 \\ w_2 \\ \vdots \\ w_G \end{bmatrix} \tag{8}$$

Now, Equation (9) can be extended as follows:

$$y = XZp + Vw \tag{9}$$

In Equation (9), y, Z, and V are known vectors, and P and w are unknown vectors that are estimated using GME, which is defined as follows:

$$\text{Max } H(p, w) = -\sum_{k=1}^{K}\sum_{m=1}^{M} p_{km}\ln(p_{km}) - \sum_{g=1}^{G}\sum_{j=1}^{J} w_{gj}\ln(w_{gj}) \tag{10}$$

and is subject to:

$$\sum_{k=1}^{K}\sum_{m=1}^{M} Z_{km}P_{km} + \sum_{j=1}^{J} v_j w_{gj} = y_g; \forall g = 1, \ldots, G \tag{11}$$

$$\sum_{k=1}^{K} p_{km} = 1; \forall m = 1, \ldots, M \tag{12}$$

$$\sum_{j=1}^{J} w_{gj} = 1; \forall g = 1, \ldots, G \tag{13}$$

The GME techniques are created by solving the optimization problem (Equation (10)) while taking constraint into account (Equations (11–13)). Equation (11) is a condition for the compatibility of the probability of the posterior distribution of the coefficients and the residual terms with the observations. Equations (12) and (13) are normalization constraints for the probabilities.

2.5.2. Positive Mathematical Programming (PMP) Model

The present study used an economic modeling system composed of the PMP model to analyze and assess the effects of climate change (the decline in available water and the changes in crop yields) on the cropping pattern of selected crops and farmers' gross revenues in Mashhad plain.

In recent decades, PMP has been widely used to evaluate the effects of climate change on the agricultural sectors [53–58]. The main objective of the PMP is to improve the accuracy of modeling farmers' behavior in the context of an optimization model utilizing observed values from the baseline year [59]. There are two primary reasons for interest in this approach: firstly, in the presence of incomplete and insufficient data, alternative approaches, such as traditional econometrics, are unable to model farmer behavior; secondly, optimization models cannot properly calibrate farm-level models [60]. In the current study, the impacts of changes in climate variables, such as temperature and rainfall on crop pattern, were simulated in the framework of a developed PMP. The study's empirical model comprises a nonlinear objective function as well as constraints, such as water, labor force, fertilizer, and land. Following Röhm and Dabbert [61] and Radmehr and Shayanmehr [59], PMP is constructed in three stages: (1) solve a linear optimization programming model and obtain shadow prices, (2) use a generalized maximum entropy (GME) approach to calibrate crop yield parameters, (3) solve a nonlinear optimization programming model that includes the objective function and constraints (from step one), as well as calibrated yield functions (obtained in the second step). In order to simplify, the nonlinear optimization model developed in the third step is presented in this section as follows:

$$\text{Max TMG} = \sum_{e}\sum_{r} P_e\left(\alpha_{e,r}X_{e,r} - \beta_{e,r}X_{e,r}^2\right) - \sum_{e}\sum_{r} C_{e,r}X_{e,r} \tag{14}$$

and is subject to:

$$\sum_{r=\text{land}}\sum_{e}\frac{w_e X_{e,r}}{ef} \le b_{r\epsilon \text{ water}} \tag{15}$$

$$\sum_{e} a_{e,r}X_{e,r} \le b_{r\epsilon \text{ land, labor, fertilizer, and machinery}} \tag{16}$$

$$X_{e,r} \geq 0 \quad (17)$$

In this expression, e is the set of different crops; r is the set of production inputs (land, water, labor, machinery, and fertilizers); TMG shows the total gross margin in the region; $X_{e,r}$ is a decision variable that represents the amount of input r used for crop e; $\alpha_{e,r}$ and $\beta_{e,r}$ are coefficients of yield function that calibrated using the GME approach (for more information about the details of the technique, see Paris and Howitt [62]); $C_{e,r}$ is the unit cost of input r for crop e; w_e is the water requirement of crop e; ef is technical efficiency of irrigation water use; $a_{e,r}$ is the technical coefficient of input r for crop e, which shows the amount of input r to produce a unit of crop e; b_r is the total available input r. Equation (14) indicates the objective functions that maximize the total gross margin of production in the irrigated area. Equation (15) is the constraint of water that represented the amount of water allocated for agricultural production and should be less than total water availability for crop production in the region. Equation (16) is the constraint of inputs that shows the amount of input allocated for crop production to be less than total input availability for crop production in the region. Finally, the non-negative constraint (Equation (17)) states that the decision variable ($X_{e,r}$) must be greater than or equal to zero.

2.6. Multi-Criteria Decision-Making Approach

The current study examines the effects of climate change on agricultural sustainability using three types of indicators: social, environmental, and economic indicators. The social index is based on the sub-index of farm employment (FE), while the environmental index includes the sub-indices of nitrogen balance (NB), phosphorus balance (PB), and water consumption (WC). The economic index is introduced using the sub-indices of total gross margin (GM) and profit-to-water consumption ratio (PW) [59,63–65]. Many complex decision-making issues employ MCDM models [66]. Analytic hierarchy process (AHP) and technique for order preferred preference by similarity to ideal solution (TOPSIS) are two approaches of this model that have been used in numerous studies [63,67–69]. The main advantages of AHP method are the ability to (i) depict the rationale of human choice; (ii) evaluate the relative performance of alternatives based on the simple algorithm; and (iii) define flexibly the selection set [70–72].

To select and rank indicators, an integrated AHP and TOPSIS method is used. This method consists of the eight steps listed below:

Step 1. *Build a decision matrix.*

First, a decision matrix is created, which can be expressed as follows:

$$D = \begin{array}{c} \\ A_1 \\ A_2 \\ \vdots \\ A_i \\ \vdots \\ A_m \end{array} \begin{array}{c} \begin{array}{cccccc} F_1 & F_2 & \cdots & F_j & \cdots & F_n \end{array} \\ \begin{bmatrix} f_{11} & f_{12} & \cdots & f_{1j} & \cdots & f_{1n} \\ f_{21} & f_{22} & \cdots & f_{2j} & \cdots & f_{2n} \\ \vdots & \vdots & \cdots & \vdots & \cdots & \vdots \\ f_{i1} & f_{i2} & \cdots & f_{ij} & \cdots & f_{in} \\ \vdots & \vdots & \vdots & \vdots & \cdots & \vdots \\ f_{m1} & f_{m2} & \cdots & f_{mj} & \cdots & f_{mn} \end{bmatrix} \end{array} \quad (18)$$

where A_i is the alternative; F_j is the evaluation indicators; and f_{ij} is the performance value of A_i with respect of F_j.

Step 2. *Construct the normalized decision matrix (r_{ij}) using following formula:*

$$r_{ij} = \frac{f_{ij}}{\sqrt{\sum_{i=1}^{m} f_{ij}^2}} \qquad j = 1, \ldots, n; \; i = 1, \ldots, m \quad (19)$$

Step 3. *Compute the weight (w_{j0}) of the indicators.*

The relative importance of various indicators is determined with respect to the objective, and weights of indicators are given based on their importance.

K indicates an n × n pair-wise comparison matrix:

$$K = \begin{bmatrix} 1 & k_{12} & \dots & k_{1n} \\ k_{21} & 1 & \dots & k_{2n} \\ \dots & \dots & \dots & \dots \\ k_{n1} & k_{n2} & \dots & 1 \end{bmatrix} \quad (20)$$

In an arbitrary random reciprocal matrix K, each criterion k_{ij} is the relative importance of ith alternatives compared to the jth indicators [73]. Therefore, it expresses that the higher values of k_{ij} show stronger preference of k_i over k_j. In the matrix k, $k_{ij} = 1$, when i = j and $k_{ji} = 1/k_{ij}$.

Geometric mean method is employed for normalization and determines the importance degree of the indicators [74]. If W_i indicates the importance degree for the ith attribute, then:

$$W_i = \frac{\prod_{j=1}^{n}(k_{ij})^{1/n}}{\sum_{i=1}^{n}\prod_{j=1}^{n}(k_{ij})^{1/n}} \quad (21)$$

E indicates an n-dimensional column vector, which defines the sum of the weighted values of the importance degree of indicators. Then:

$$E = [e_i]_{n\times 1} = KW^T \; i = 1,\ 2,\ 3, \dots,\ N \quad (22)$$

where

$$KW^T = \begin{bmatrix} 1 & k_{12} & \dots & k_{1n} \\ k_{21} & 1 & \dots & k_{2n} \\ \dots & \dots & \dots & \dots \\ k_{n1} & k_{n2} & \dots & 1 \end{bmatrix} \begin{bmatrix} W_1 & W_2 & \dots & W_n \end{bmatrix} C_n = \begin{bmatrix} C_1 \\ C_2 \\ \dots \end{bmatrix} \quad (23)$$

Consistency values are defined by the following vector:

$EV = [ev_i]_{1\times n}$ with a typical component ev_i calculated as $ev_i = \left(\frac{e_i}{W_i}\right)$, $i = 1,\ 2,\ \dots,\ n$. The CI is the consistency index that is calculated from Equation (24):

$$CI = \left(\frac{\lambda_{max} - n}{n - 1}\right) \quad (24)$$

λ_{max} is maximum Eigen value that can be obtained as follow [74]:

$$\lambda_{max} = \left(\frac{\sum_{i=1}^{n} ev_i}{n}\right) \; i = 1,2,\dots,n \quad (25)$$

The consistency of evaluation in AHP is measured by consistency ratio (CR). Consistency ratio is defined as Equation (26):

$$CR = \frac{CI}{RI} \quad (26)$$

where RI indicates the inconsistency index of a random matrix. If the value of consistency ratio is less than 0.10, the evaluation of the importance of degrees of attributes is acceptable.

Step 4. *Calculate the weighted normalized decision matrix (z_{ij}) using the following formula:*

$$z_{ij} = r_{ij}.w_{j0} \qquad j = 1,\dots,n;\ i = 1,\dots,m \quad (27)$$

Step 5. *Determine the positive* (A^+) *and negative* (A^-) *ideal options.*

$$A^+ = \{z_1^+, z_2^+, \ldots, z_n^+\} = \left[\left(maxz_{ij} \middle| \, j \in J'\right), \left(\text{min}z_{ij} \middle| \, j \in J''\right)\right] \tag{28}$$

$$A^- = \{z_1^-, z_2^-, \ldots, z_n^-\} = \left[\left(\text{min}z_{ij} \middle| j \in J'\right), \left(\text{max}z_{ij} \middle| j \in J''\right)\right] \tag{29}$$

where J' and J'' *are the indicators with positive and negative polarity, respectively.*
Step 6. *Compute the relative distance of each* A_i *from* A^+ *and* A^- [63].

$$D_i^+ = \sqrt{\sum_{j=1}^{n}\left(z_{ij} - z_j^+\right)^2}, \quad i = 1, \ldots, m \tag{30}$$

$$D_i^- = \sqrt{\sum_{j=1}^{n}\left(z_{ij} - z_j^-\right)^2}, \quad i = 1, \ldots, m \tag{31}$$

Step 7. *Determine the relative closeness* (C_i) *to the best alternative* [74].

$$C_i = \frac{D_i^-}{D_i^+ + D_i^-}, \; i = 1, \ldots, m; 0 \le C_i \le 1 \tag{32}$$

Step 8. *Rank the alternatives.*

The alternative that has the highest value of C_i is selected as the best option.

3. Results and Discussion

3.1. Projecting Climate Variables

Using the LARS-WG model, climatic parameters of maximum and minimum air temperature and precipitation were predicted in the Mashhad plain based on data of 1979–2016. To calibrate and ensure the accuracy of the LARS-WG model, the simulated data are compared with the observed data on a monthly scale, as shown in Figure 5. Assessment of the monthly average of maximum air temperature, minimum air temperature, and precipitation shows a good agreement among all three parameters.

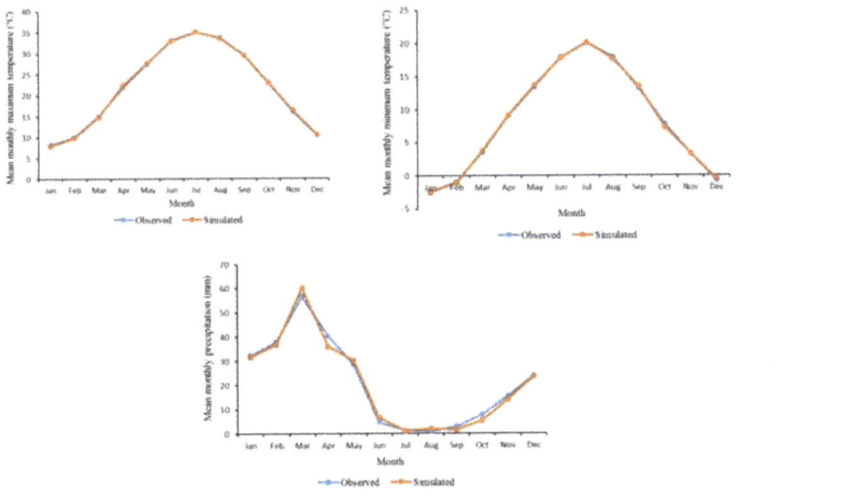

Figure 5. Comparison of the observed and LARS-WG-generated monthly minimum and maximum temperature and precipitation for 1979–2016 in the Mashhad station.

LARS-WG performance also was investigated using R^2, NRMSE, RMSE, MAD, and MSE indicators, as presented in Table 2. The model successfully downscaled the minimum and maximum temperature as well as precipitation, according to the evaluation of these indicators. The higher values of R^2 (>0.98) and the lower values of RMSE (0.21–2.09), MAD (0.17–1.69), MSE (0.04–4.39), and NRMSE (0.95–9.96) for this period reveal that the simulated precipitation and temperature data are acceptable.

Table 2. Results of LARS performance for the observed and simulated data.

Indicators	Minimum Temperature	Maximum Temperature	Precipitation
R^2	0.99	0.99	0.98
NRMSE	2.55	0.95	9.96
RMSE	0.21	0.21	2.09
MAD	0.17	0.17	1.69
MSE	0.04	0.04	4.39

Sources: Research findings.

After evaluating the accuracy of the model, the climate scenarios were generated by downscaling HadGEM2 outputs under climate scenarios of RCP 2.6, 4.5, and 8.5 on the horizons of 2045. (We considered a 30-year period (2016–2045) to investigate the effects of climate change on crop production. Because the term climate is a long-term shift in the weather pattern, it is an average of at least 30 years of weather condition of a particular place [75–77].) The percentage change in climate variables was then calculated and compared to the benchmark year (2016). The obtained findings are shown in Table 3. According to the results, the minimum and maximum temperatures are expected to increase by about 5.88% and 6.05%, respectively, while precipitation would decrease by approximately 30.68%.

Table 3. Forecasting of temperature and precipitation changes under climate scenarios in 2045 horizon compared to 2016 (benchmark year).

Scenario	Minimum Temperature	Maximum Temperature	Precipitation
RCP 2.6	3.40	4.06	−14.46
RCP 4.5	1.26	4.16	−18.14
RCP 8.5	5.88	6.05	−30.68

Note: The unit of numbers is percent.

3.2. Evaluating the Impacts of Climate Change on Water Resources

The groundwater depth was forecasted because of climatic change using a panel data model under three scenarios in the Mashhad plain. The first step in the analysis was to determine whether the variables were stationary. This was performed using the Im-Pesaran-Shin (IPS) and ADF-Fisher-type panel unit root tests. Table 4 shows, for all variables in the model, the null hypothesis of unit roots is rejected with a 99% confidence level. As a result, the model's variables were all stationary. For panel data, the random-effects and fixed-effects models were used. To identify which should be used, the Hausman test was used. As shown in Table 4, the null hypothesis of no correlation between regional effects and independent variables is rejected. As a result, a fixed-effect model with regional-specific effects was used.

The results of the panel data model for sensitivity of the groundwater depth to climate variables are presented in Table 5. The findings indicated that maximum and minimum temperatures had a positive impact on the groundwater depth in the Mashhad plain from 2000 to 2016. Furthermore, precipitation was negatively and significantly related to groundwater depth in the plain. Therefore, as precipitation increased, the groundwater depth decreased, resulting in more water in the well, according to many previous studies, including Shahvari et al. [28] and Izady et al. [41].

Table 4. The results of unit root test and Hausman test.

Variable	IPS		Fisher-ADF	
	Without Trend	With Trend	Without Trend	With Trend
Minimum temperature	−4.57 ***	−5.03 ***	41.63 ***	31.27 ***
Maximum temperature	−4.40 ***	−5.04 ***	39.64 ***	31.31 ***
Precipitation	−18.70 ***	−19.91 ***	360.43 ***	360.43 ***
Initial groundwater depth	−13.48 ***	−13.52 ***	229.55 ***	198.43 ***
Groundwater depth	−13.50 ***	−13.53 ***	230.11 ***	198.84 ***
Fixed effects versus random effects test	**Chi2**		***p*-value**	
Hausman test	7.69 *		0.10	

Note: * and *** show rejection of the unit root hypothesis at the 10% and 1% significance levels, respectively.

Table 5. Sensitivity of the groundwater depth to minimum and maximum temperatures and precipitation in the study area.

Variable	Coefficient	T-Statistic	*P*-value
Minimum temperature	0.003	0.31	0.75
Maximum temperature	0.003	0.26	0.79
Precipitation	−0.001 *	−1.69	0.09
Initial groundwater depth	0.99 ***	249.04	0.00
Constant	0.40 *	1.79	0.07

* and ***, respectively, indicate rejection of the unit root hypothesis at the 10% and 1% significance levels.

The results regarding the percentage changes in groundwater depth and water availability under climatic scenarios compared to the baseline are presented in Table 6. According to the findings, the depth of groundwater in the Mashhad plain is expected to rise about 13.79% in RCP 2.6, 13.50% in RCP 4.5, and 15.45% in RCP 8.5. Furthermore, water availability will increase by approximately 13.25%, 13.26%, and 14.84% in response to RCP 2.6, RCP 4.5, and RCP 8.5, respectively.

Table 6. Percentage change in groundwater depth and water availability in the study area under climate scenarios compared to the baseline.

Scenario	Groundwater Depth	Water Availability
RCP 2.6	13.79	−13.25
RCP 4.5	13.50	−13.26
RCP 8.5	15.45	−14.84

Sources: Research findings.

3.3. *Assessing the Impacts of Climate Change on Crop Yield*

To assess crop yield sensitivity to temperature and precipitation, the yield response function was estimated using the GME model. The estimated results are displayed in Table 7. The Cobb–Douglas functional form was used in the estimation of yield response functions. Therefore, the estimated parameters in Table 5 show the elasticity values. Based on the results obtained from this table, increasing maximum and minimum temperatures reduce crop yield in many crops. In the case of irrigated wheat, for example, the results show that a 1% increase in maximum temperature results in a 1.17% decrease in yield. In addition, a 1% increase in the minimum temperature increases irrigated wheat yield by approximately 0.76%. Precipitation has a negative impact on yield of some crops due to increase humidity or the potential spread of diseases and pests [78]. Irrigated wheat, rainfed wheat, irrigated barley, rainfed barley, alfalfa, corn, sugar beet, melon, cucumber, and tomato yield are positively influenced by precipitation.

Table 7. Estimates of the impact of climatic variables on crop yield using the GME model.

Crop	Minimum Temperature	Minimum Temperature	Precipitation	Trend	Constant
Irrigated wheat	−1.17 *	0.76 ***	0.17 **	0.009 **	9.16 ***
	(0.64)	(0.19)	(0.08)	(0.004)	(1.92)
Rainfed wheat	−0.53	0.93 **	0.49 ***	0.004	4.20
	(1.35)	(0.40)	(0.18)	(0.008)	(4.07)
Irrigated barley	−0.68	0.71 ***	0.13 *	0.002	8.09 ***
	(0.71)	(0.16)	(0.07)	(0.003)	(1.64)
Rainfed barley	−1.22	1.06 ***	0.14	0.003	7.42 **
	(1.12)	(0.34)	(0.15)	(0.007)	(3.35)
Alfalfa	−2.82 ***	−0.62	0.07	0.01 ***	18.61 ***
	(0.77)	(0.31)	(0.06)	(0.004)	(2.41)
Corn	2.37 ***	−0.08	0.06 ***	0.007 *	2.57
	(0.96)	(0.41)	(0.02)	(0.003)	(3.42)
Sugar beet	0.62	−1.19 ***	0.09 **	0.02 ***	10.61 ***
	(0.75)	(0.36)	(0.04)	(0.003)	(2.62)
Potato	0.53	−0.93	−0.10	0.03	10.30 *
	(1.75)	(0.93)	(0.07)	***(0.007)	(5.94)
Onion	−3.16 **	1.55 *	−0.11	0.03 ***	16.26 ***
	(1.47)	(0.89)	(0.09)	(0.006)	(4.77)
Melon	1.61 ***	−0.85 **	0.11 ***	0.01 ***	5.88 ***
	(0.64)	(0.38)	(0.04)	(0.003)	(2.05)
Cucumber	1.99	−2.94 ***	0.11 ***	0.02 ***	10.38 **
	(1.39)	(0.96)	(0.05)	(0.006)	(4.62)
Tomato	1.20	−0.90 **	0.08 *	0.02 ***	7.95 ***
	(0.86)	(0.46)	(0.05)	(0.003)	(2.90)

Note: Numbers in parenthesis indicate standard error. *, **, and ***, respectively, indicate rejection of the unit root hypothesis at the 10%, 5%, and 1% significance levels.

The percentage changes in crop yield in the Mashhad plain under RCP 2.6, RCP 4.5, and RCP 8.5 on the horizon in 2045, as compared to the baseline year, are presented in Table 8. The findings imply that the yield of irrigated wheat, rainfed wheat, irrigated barley, rainfed barley, alfalfa, sugar beet, onion, and cucumber will decrease in response to all three climate scenarios. Alfalfa crop is prone to experience a decrease in yield between 19.07% and 25.20% and is expected to emerge as a highly vulnerable crop in 2045. In addition, climate change will increase the yield of corn, potato, melon, and tomato crops under all scenarios. With changing climate, corn yield will rise more than other crops. This result is in line with the results of Almaraz et al. [79] and Zhang et al. [80].

Table 8. Percentage changes in crop yield under RCP 2.6, RCP 4.5, and RCP 8.5 for 2045, as compared to the base period in the Mashhad plain.

Crop	RCP 2.6	RCP 4.5	RCP 8.5
Irrigated wheat	−8.71	−11.32	−13.57
Rainfed wheat	−14.48	−18.34	−17.29
Irrigated barley	−5.21	−7.90	−8.56
Rainfed barley	−7.17	−11.04	−12.92
Alfalfa	−19.83	−19.07	−25.20
Corn	13.33	15.64	10.11
Sugar beet	−1.75	−0.98	−5.86
Potato	3.53	5.33	4.74
Onion	−8.86	−14.41	−10.46
Melon	3.12	5.92	5.12
Cucumber	−7.71	−3.91	−8.29
Tomato	2.34	4.02	0.15

Sources: Research findings.

3.4. Evaluating the Impacts of Climate Change on Crop Production and Cropping Pattern

In this section, changes in crop yields and water availability under climate change were incorporated into the PMP model to assess the impact of climate change on cropping pattern, crop production, and water consumption. It is worth noting that the cropping pattern is defined as a combination of agricultural crops that are grown in a particular geographical area [81]. The percentage changes in cropland under three climate scenarios in 2045, as compared to the base year, are shown in Figure 6. The results imply that RCP 2.6, RCP 4.5, and RCP 8.5 scenarios will decrease the cultivated area of irrigated wheat, rainfed wheat, irrigated barley, rainfed barley, alfalfa, sugar beet, onion, cucumber, and tomato in 2045 in the Mashhad plain. On the contrary, the cultivated area of corn, potato, and melon will increase under two scenarios of RCP 4.5 and RCP 8.5 in 2045 relative to the base year. As shown in the figure, the most significant reduction in cultivated land is related to rainfed wheat and RCP 4.5, where the area under cultivation will reduce by approximately 51.16%.

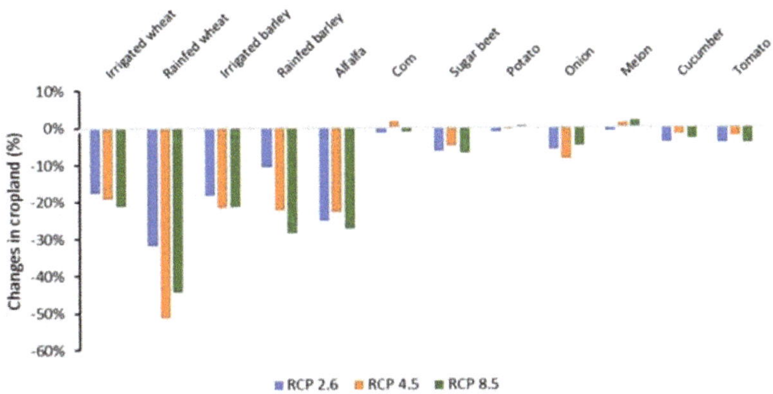

Figure 6. Percentage changes in cropland in the climate scenarios as compared to the baseline in the Mashhad plain.

The percentage changes in crop production under climate scenarios for 2045, as compared to the baseline, are presented in Table 9. The results of this table show that production of crops, such as irrigated wheat, rainfed wheat, irrigated barley, rainfed barley, alfalfa, sugar beet, onion, and cucumber, will decrease in response to all three climate scenarios. Therefore, the biggest decline (59.95%) will occur for rainfed wheat in RCP 4.5 scenario.

Table 9. Percentage changes in crop production under RCP 2.6, RCP 4.5, and RCP 8.5 for 2045 as compared to the base year in the Mashhad plain.

Crop	RCP 2.6	RCP 4.5	RCP 8.5
Irrigated wheat	−24.95	−28.08	−32.36
Rainfed wheat	−41.43	−59.95	−53.75
Irrigated barley	−22.21	−28.05	−28.50
Rainfed barley	−16.98	−30.96	−37.94
Alfalfa	−40.24	−37.66	−45.53
Corn	11.37	17.92	8.61
Sugar beet	−8.28	−5.85	−12.54
Potato	2.64	4.41	5.33
Onion	−14.51	−21.42	−14.40
Melon	2.05	7.14	6.90
Cucumber	−11.77	−5.90	−10.80
Tomato	−2.41	1.40	−4.12

Sources: Research findings.

In addition, corn, potato, and melon production will increase by about 17.92%, 5.33%, and 7.14%, respectively, under climate scenarios. This increase in production is due to an increase in yield or area under cultivation, or both, as discussed in the previous section. Additionally, in the presence of climate change in 2045, the Mashhad plain will experience the biggest increase in crop production in corn and RCP 4.5.

Given the decline in production of most crops that account for a large share of the region's production, it is reasonable to conclude that the occurrence of climate change poses a serious threat to crop production and food security in the region.

Figure 7 shows the water consumption for each crop under current conditions and different climatic scenarios. As expected, under climate scenarios, due to the reduction in water availability and crop yield and, consequently, reduced cropland, water consumption will decrease for crops, such as irrigated wheat (21.40%), irrigated barley (21.88%), alfalfa (27.70%), sugar beet (7.90%), onion (8.71%), cucumber (4.52%), and tomato (4.33%), while there is an increase in water consumption for corn, potato, and melon because of improved yield. Overall, it can be stated that water consumption in climatic scenarios will be reduced by about 13.30%, 13.31%, and 14.90% under RCP 2.6, RCP 4.5, and RCP 8.5, respectively, compared to baseline conditions.

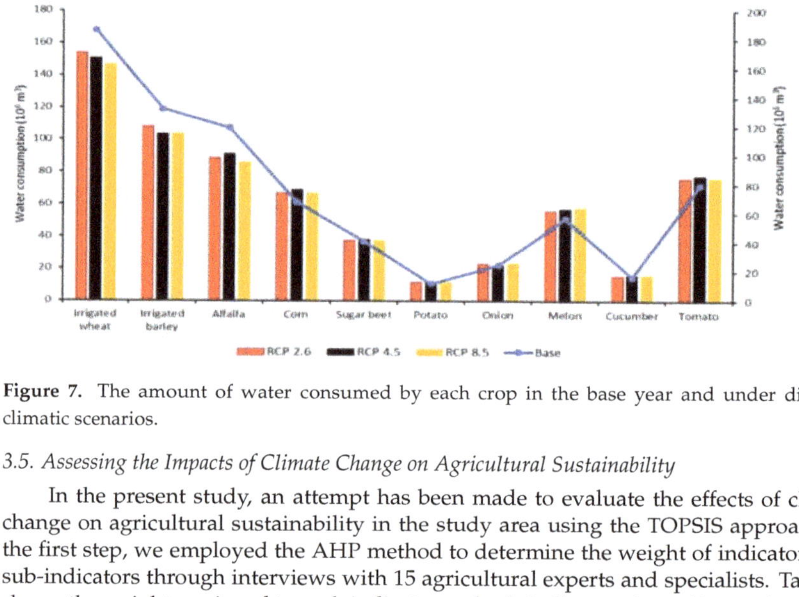

Figure 7. The amount of water consumed by each crop in the base year and under different climatic scenarios.

3.5. Assessing the Impacts of Climate Change on Agricultural Sustainability

In the present study, an attempt has been made to evaluate the effects of climate change on agricultural sustainability in the study area using the TOPSIS approach. In the first step, we employed the AHP method to determine the weight of indicators and sub-indicators through interviews with 15 agricultural experts and specialists. Table 10 shows the weights assigned to each indicator and sub-indicator. According to the results of this table, economic, environmental, and social indicators have the greatest importance with 52%, 33%, and 14%, respectively. Due to the low income of farmers in Iran, improving the profitability of agricultural activities is critical to achieving sustainable development in the agricultural sector [60]. In addition, over the past few decades, the excessive use of chemical fertilizers and the uncontrolled extraction of groundwater resources in the process of food production in agriculture has caused irreversible environmental damage [16]. This highlights the importance of environmental issues in assessing the sustainability of Iran's agricultural sector. In the economic indicator, the importance of the "GM" and "PW" sub-indicators are 58% and 42%, respectively. Because of Iran's high unemployment rate, the FE index was deemed as the mere social indicator, with a weight of 14%. Among the environmental sub-indicators, WC is the most important environmental sub-indicator, followed by the NI and PB.

Table 10. Selected indicators and weights.

Indicator	Weights of Criteria (%)	Sub-Indicator	Measurement Unit	Weights of Indicator (%)	Normalized Weights of Indicator (%)	Polarity of Indicator
Economic	52.43	GM	MT	58.33	30.58	+
		PW	$MT/10^3\ m^3$	41.67	21.85	+
Social	14.10	FE	h/ha	100.00	14.10	+
Environmental	33.47	NB	kg/ha	30.51	10.21	−
		PB	kg/ha	17.16	5.74	−
		WC	$10^3\ m^3/ha$	52.33	17.52	−

Note: MT is million tomans.

In the second step, using the results obtained from the PMP model, the value of each of the sustainability indicators in the base conditions and climatic scenarios was calculated, which forms the decision matrix (see Table 11). Table 11 shows under the climate scenarios that the values of economic indicators are lower and environmental indicators are higher than the baseline conditions. Table 12 indicates the normalized decision matrix. Figure 8 depicts the distribution of positive and negative ideal options across the various sustainability indicators. In the final step, current and climatic conditions were ranked based on sustainability indicators (see Table 13). Based on the results of Table 13, the base condition has the highest Ci values (0.77), indicating that the phenomenon of climate change can be considered a serious threat to the agricultural sustainability in this region. The findings in this section of the research are in line with the study of Karandish et al. [82] and Ghanian et al. [83].

Table 11. Decision matrix.

Alternative	GM	PW	FE	NB	PB	WC
Base	316,270.25	0.43	162.05	153.34	81.55	8.36
RCP 2.6	279,714.25	0.44	173.04	163.50	86.50	8.72
RCP 4.5	278,643.55	0.44	183.86	172.13	91.12	9.26
RCP 8.5	262,996.33	0.42	182.28	171.21	90.53	9.15

Sources: Research findings.

Table 12. Normalized decision matrix.

Alternative	GM	PW	FE	NB	PB	WC
Base	0.55	0.50	0.46	0.46	0.47	0.47
RCP 2.6	0.49	0.51	0.49	0.49	0.49	0.49
RCP 4.5	0.49	0.51	0.52	0.52	0.52	0.52
RCP 8.5	0.46	0.49	0.52	0.52	0.52	0.52

Sources: Research findings.

Table 13. Ranking of alternatives.

Alternative	C_i	Rank
Base	**0.77**	**1**
RCP 2.6	0.38	2
RCP 4.5	0.36	3
RCP 8.5	0.21	4

Sources: Research findings.

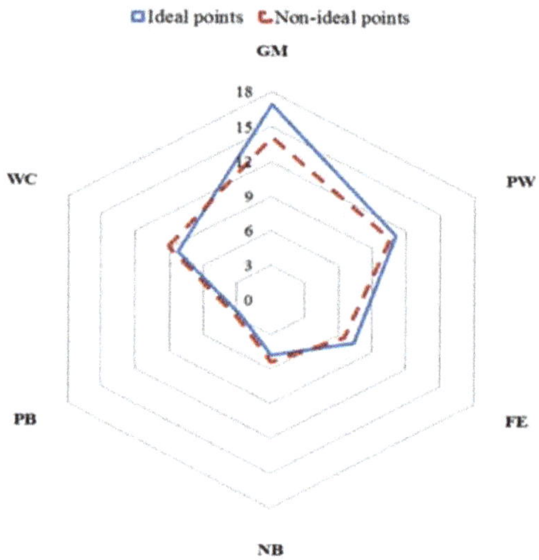

Figure 8. TOPSIS positive ideal and negative ideal options.

To ensure consistency in the results of Table 13, a sensitivity analysis of the economic, social, and environmental criteria weights used in the TOPSIS model is performed. To achieve this goal, a total of seven experiments has been conducted to compare the impact of potential changes in the weights of economic, social, and environmental criteria (see Table 14). In Experiment 1, all sub-indicators have the same weight (16.66%); in Experiment 2–7, the weight of one criterion is higher than the weight of the remaining criteria. The results of the sensitivity analysis described indicate that in all experiments (except Experiment 4), the base condition is selected as the best alternative. In other words, given the importance of economic and environmental indicators in Iran's agricultural sector, it is reasonable to expect that if the harmful effects of climate change are not properly managed, the phenomenon will have a negative impact on agricultural sustainability.

Table 14. Results of the sensitivity analysis.

	Experiment	Best Alternative
1	All Sub-indicators have same weight	Base
2	Weight of GM criterion = 50% Weight of the other criteria = 10%	Base
3	Weight of PW criterion = 50% Weight of the other criteria = 10%	Base
4	Weight of FE criterion = 50% Weight of the other criteria = 10%	RCP 4.5
5	Weight of NB criterion = 50% Weight of the other criteria = 10%	Base
6	Weight of PB criterion = 50% Weight of the other criteria = 10%	Base
7	Weight of WC criterion = 50% Weight of the other criteria = 10%	Base

4. Conclusions

The climate simulation model predicted a 6% increase in minimum and maximum temperatures, as well as a 30% decrease in precipitation. Additionally, the results showed that water availability will decrease between 13% and 15% under different climate scenarios. Crop yields were found to be negatively affected by increasing maximum and minimum temperatures. Precipitation affects crop yield in different ways, with positive effects on irrigated wheat, rainfed wheat, irrigated barley, rainfed barley, alfalfa, corn, sugar beet, melon, cucumber, and tomato yields and negative effects on potato and onion yields. Overall, future climate change is expected to reduce the yield of irrigated wheat, rainfed wheat, irrigated barley, rainfed barley, alfalfa, sugar beet, onion, and cucumber, while the effects will be reversed for corn, potato, melon, and tomato. The results of the PMP model showed that changes in crops yield and water availability will lead to a reduction in the cultivated area of most crops in 2045, among which dryland wheat will experience the greatest decrease (51%). The results of the evaluation of the effects of climate change on agricultural sustainability show that this phenomenon can have adverse economic and environmental effects on the agricultural system of the study area. As a result, it can have a negative impact on agricultural sustainability if not properly managed.

In entirety, the findings of this study reflect the fact that water and food security in the region will be severely adversely affected by climate change in the future. Nevertheless, by continuing to support population growth policies, uncontrolled extraction of groundwater, and expansion of urbanization in the presence of climate change, more severe irreversible effects on water and food resources in the control area are expected. These results underscore the necessity of implementing adaptation policies, such as reforming the cropping pattern and production technologies, as well as the introduction of drought-tolerant varieties to reduce the detrimental effects of climate in the region.

Author Contributions: Conceptualization, S.S. and N.S.F.; methodology, S.S.; software, S.S.; validation, S.S., M.S.S., H.M., S.R.H., and N.S.F.; formal analysis, S.S.; investigation, S.S., N.S.F., J.I.P., M.B., and H.M.; data curation, S.S. and N.S.F.; writing—original draft preparation, S.S.; writing—review and editing, J.I.P., M.B., M.S.S., H.M., S.R.H., and N.S.F.; supervision, N.S.F. All authors have read and agreed to the published version of the manuscript.

Funding: This research received no external funding.

Institutional Review Board Statement: Not applicable.

Informed Consent Statement: Not applicable.

Data Availability Statement: Not applicable.

Acknowledgments: This work was supported by the Ferdowsi University of Mashhad, Iran [No. 57324] and project "Environmental Assessment of Specific Biotopes of Danubian Lowland" [VEGA 1/0604/20].

Conflicts of Interest: The authors declare no conflict of interest.

References

1. Alkaya, E.; Bogurcu, M.; Ulutas, F.; Demirer, G.N. Adaptation to climate change in industry: Improving resource efficiency through sustainable production applications. *Water Environ. Res.* **2015**, *87*, 14–25. [CrossRef] [PubMed]
2. Shayanmehr, S.; Rastegari Henneberry, S.; Sabouhi Sabouni, M.; Shahnoushi Foroushani, N. Climate Change and Sustainability of Crop Yield in Dry Regions Food Insecurity. *Sustainability* **2020**, *12*, 9890. [CrossRef]
3. Chiu, S.-Y.; Kao, C.-Y.; Tsai, M.-T.; Ong, S.-C.; Chen, C.-H.; Lin, C.-S. Lipid accumulation and CO_2 utilization of Nannochloropsis oculata in response to CO_2 aeration. *Bioresour. Technol.* **2009**, *100*, 833–838. [CrossRef] [PubMed]
4. Gao, M.; Yang, H.; Xiao, Q.; Goh, M. A novel fractional grey Riccati model for carbon emission prediction. *J. Clean. Prod.* **2021**, *282*, 124471. [CrossRef]
5. Gao, M.; Yang, H.; Xiao, Q.; Goh, M. A novel method for carbon emission forecasting based on Gompertz's law and fractional grey model: Evidence from American industrial sector. *Renew. Energy* **2022**, *181*, 803–819. [CrossRef]
6. Noh, S.; Son, Y.; Park, J. Life cycle carbon dioxide emissions for fill dams. *J. Clean. Prod.* **2018**, *201*, 820–829. [CrossRef]
7. Radmehr, R.; Henneberry, S.R.; Shayanmehr, S. Renewable energy consumption, CO_2 emissions, and economic growth nexus: A simultaneity spatial modeling analysis of EU countries. *Struct. Change Econ. Dyn.* **2021**, *57*, 13–27. [CrossRef]
8. Kohansal, M.R.; Shayan, M.S. The interplay between energy consumption, economic growth and environmental pollution: Application of spatial panel simultaneous-equations model. *Iran. Energy Econ.* **2016**, *5*, 179–216.
9. Hezareh, R.; Shayanmehr, S.; Darbandi, E.; Schieffer, J. *Energy Consumption and Environmental Pollution: Evidence from the Spatial Panel Simultaneous-Equations Model of Developing Countries*; Southern Agricultural Economics Association (SAEA): Birmingham, AL, USA, 2017.

10. Jin, S.; Hao, Z.; Zhang, K.; Yan, Z.; Chen, J. Advances and challenges for the electrochemical reduction of CO_2 to CO: From fundamentals to industrialization. *Angew. Chem. Int. Ed.* **2021**, *60*, 20627–20648. [CrossRef]
11. Anjum, S.A.; Farooq, M.; Xie, X.-Y.; Liu, X.-J.; Ijaz, M.F. Antioxidant defense system and proline accumulation enables hot pepper to perform better under drought. *Sci. Hortic.* **2012**, *140*, 66–73. [CrossRef]
12. Kiani Ghalehsard, S.; Shahraki, J.; Akbari, A.; Sardar Shahraki, A. Assessment of the impacts of climate change and variability on water resources and use, food security, and economic welfare in Iran. *Environ. Dev. Sustain.* **2021**, *23*, 14666–14682. [CrossRef]
13. Sabbaghi, M.A.; Nazari, M.; Araghinejad, S.; Soufizadeh, S. Economic impacts of climate change on water resources and agriculture in Zayandehroud river basin in Iran. *Agric. Water Manag.* **2020**, *241*, 106323. [CrossRef]
14. Varela Ortega, C.; Esteve, P.; Blanco, I.; Carmona, G.; Ruiz Fernández, J.; Rabah, T. Assessment of Socio-Economic and Climate Change Effects on Water Resources and Agriculture in Southern and Eastern Mediterranean Countries. MEDPRO Technical Report. 2013. Available online: https://ssrn.com/abstract=2276859 (accessed on 10 June 2022).
15. Lehmann, N.; Finger, R.; Klein, T.; Calanca, P.; Walter, A. Adapting crop management practices to climate change: Modeling optimal solutions at the field scale. *Agric. Syst.* **2013**, *117*, 55–65. [CrossRef]
16. Palazzoli, I.; Maskey, S.; Uhlenbrook, S.; Nana, E.; Bocchiola, D. Impact of prospective climate change on water resources and crop yields in the Indrawati basin, Nepal. *Agric. Syst.* **2015**, *133*, 143–157. [CrossRef]
17. Connolly-Boutin, L.; Smit, B. Climate change, food security, and livelihoods in sub-Saharan Africa. *Reg. Environ. Change* **2016**, *16*, 385–399. [CrossRef]
18. Ahmadi, H.; Ghalhari, G.F.; Baaghideh, M. Impacts of climate change on apple tree cultivation areas in Iran. *Clim. Change* **2019**, *153*, 91–103. [CrossRef]
19. Perdinan, P. A Rationale for International Cooperation in Implementing Adaptation Strategies to Climate Change in the Face of Global Inequality. *J. Indones. Focus* **2010**, *1*, 1–8.
20. Jat, M.L.; Stirling, C.M.; Jat, H.S.; Tetarwal, J.P.; Jat, R.K.; Singh, R.; Lopez-Ridaura, S.; Shirsath, P.B. Soil processes and wheat cropping under emerging climate change scenarios in South Asia. *Adv. Agron.* **2018**, *148*, 111–171.
21. Radmehr, R.; Ghorbani, M.; Ziaei, A.N. Quantifying and managing the water-energy-food nexus in dry regions food insecurity: New methods and evidence. *Agric. Water Manag.* **2021**, *245*, 106588. [CrossRef]
22. Mosavi, S.H.; Soltani, S.; Khalilian, S. Coping with climate change in agriculture: Evidence from Hamadan-Bahar plain in Iran. *Agric. Water Manag.* **2020**, *241*, 106332. [CrossRef]
23. Shayanmehr, S.; Shahnoushi, N.; Sabouhi Sabouni, M.; Rastegari, S. Climate Change and Its Consequences on Food Security in Khorasan Region. *Agric. Econ.* **2021**, *15*, 95–128.
24. Xiong, W.; Holman, I.; Lin, E.; Conway, D.; Jiang, J.; Xu, Y.; Li, Y. Climate change, water availability and future cereal production in China. *Agric. Ecosyst. Environ.* **2010**, *135*, 58–69. [CrossRef]
25. Sinnarong, N.; Chen, C.-C.; McCarl, B.; Tran, B.-L. Estimating the potential effects of climate change on rice production in Thailand. *Paddy Water Environ.* **2019**, *17*, 761–769. [CrossRef]
26. Mostafa, S.M.; Wahed, O.; El-Nashar, W.Y.; El-Marsafawy, S.M.; Abd-Elhamid, H.F. Impact of climate change on water resources and crop yield in the Middle Egypt region. *AQUA—Water Infrastruct. Ecosyst. Soc.* **2021**, *70*, 1066–1084. [CrossRef]
27. Medellín-Azuara, J.; Howitt, R.E.; MacEwan, D.J.; Lund, J.R. Economic impacts of climate-related changes to California agriculture. *Clim. Change* **2011**, *109*, 387–405. [CrossRef]
28. Shahvari, N.; Khalilian, S.; Mosavi, S.H.; Mortazavi, S.A. Assessing climate change impacts on water resources and crop yield: A case study of Varamin plain basin, Iran. *Environ. Monit. Assess.* **2019**, *191*, 134. [CrossRef]
29. Lu, S.; Bai, X.; Li, W.; Wang, N. Impacts of climate change on water resources and grain production. *Technol. Forecast. Soc. Chang.* **2019**, *143*, 76–84. [CrossRef]
30. Khashei-Siuki, A.; Sarbazi, M. Evaluation of ANFIS, ANN, and geostatistical models to spatial distribution of groundwater quality (case study: Mashhad plain in Iran). *Arab. J. Geosci.* **2015**, *8*, 903–912. [CrossRef]
31. Toosi, A.S.; Doulabian, S.; Tousi, E.G.; Calbimonte, G.H.; Alaghmand, S. Large-scale flood hazard assessment under climate change: A case study. *Ecol. Eng.* **2020**, *147*, 105765. [CrossRef]
32. Naghibi, S.A.; Vafakhah, M.; Hashemi, H.; Pradhan, B.; Alavi, S.J. Groundwater augmentation through the site selection of floodwater spreading using a data mining approach (case study: Mashhad Plain, Iran). *Water* **2018**, *10*, 1405. [CrossRef]
33. Wilks, D.S.; Wilby, R.L. The weather generation game: A review of stochastic weather models. *Prog. Phys. Geogr.* **1999**, *23*, 329–357. [CrossRef]
34. Tashekaboud, S.H.; Heydari Tashekaboud, A. Investigating the effects of climate change on stream flows of Urmia Lake basin in Iran. *Modeling Earth Syst. Environ.* **2019**, 329–339.
35. Kilsby, C.G.; Jones, P.; Burton, A.; Ford, A.; Fowler, H.J.; Harpham, C.; James, P.; Smith, A.; Wilby, R. A daily weather generator for use in climate change studies. *Environ. Model. Softw.* **2007**, *22*, 1705–1719. [CrossRef]
36. Racsko, P.; Szeidl, L.; Semenov, M. A serial approach to local stochastic weather models. *Ecol. Model.* **1991**, *57*, 27–41. [CrossRef]
37. Semenov, M.A.; Brooks, R.J.; Barrow, E.M.; Richardson, C.W. Comparison of the WGEN and LARS-WG stochastic weather generators for diverse climates. *Clim. Res.* **1998**, *10*, 95–107. [CrossRef]
38. Ahmadi, M.; Etedali, H.R.; Elbeltagi, A. Evaluation of the effect of climate change on maize water footprint under RCPs scenarios in Qazvin plain, Iran. *Agric. Water Manag.* **2021**, *254*, 106969. [CrossRef]

39. Baez-Villanueva, O.M.; Zambrano-Bigiarini, M.; Ribbe, L.; Nauditt, A.; Giraldo-Osorio, J.D.; Thinh, N.X. Temporal and spatial evaluation of satellite rainfall estimates over different regions in Latin-America. *Atmos. Res.* **2018**, *213*, 34–50. [CrossRef]
40. Zarei, A.; Chemura, A.; Gleixner, S.; Hoff, H. Evaluating the grassland NPP dynamics in response to climate change in Tanzania. *Ecol. Indic.* **2021**, *125*, 107600. [CrossRef]
41. Izady, A.; Davary, K.; Alizadeh, A.; Ghahraman, B.; Sadeghi, M.; Moghaddamnia, A. Application of "panel-data" modeling to predict groundwater levels in the Neishaboor Plain, Iran. *Hydrogeol. J.* **2012**, *20*, 435–447. [CrossRef]
42. Arellano, M. *Panel Data Econometrics*; OUP Oxford: Oxford, UK, 2003.
43. Chinnasamy, P.; Maheshwari, B.; Dillon, P.; Purohit, R.; Dashora, Y.; Soni, P.; Dashora, R. Estimation of specific yield using water table fluctuations and cropped area in a hardrock aquifer system of Rajasthan, India. *Agric. Water Manag.* **2018**, *202*, 146–155. [CrossRef]
44. Asoka, A.; Gleeson, T.; Wada, Y.; Mishra, V. Relative contribution of monsoon precipitation and pumping to changes in groundwater storage in India. *Nat. Geosci.* **2017**, *10*, 109–117. [CrossRef]
45. Rohde, M.M.; Edmunds, W.M.; Freyberg, D.; Sharma, O.P.; Sharma, A. Estimating aquifer recharge in fractured hard rock: Analysis of the methodological challenges and application to obtain a water balance (Jaisamand Lake Basin, India). *Hydrogeol. J.* **2015**, *23*, 1573–1586. [CrossRef]
46. Shayannmehr, S. *Climate Change and Its Impacts on Major Crops Production and Market in Iran*; Ferdowsi University of Mashhad: Mashhad, Iran, 2021.
47. Shayanmehr, S.; Rastegari Henneberry, S.; Sabouhi Sabouni, M.; Shahnoushi Foroushani, N. Drought, climate change, and dryland wheat yield response: An econometric approach. *Int. J. Environ. Res. Public Health* **2020**, *17*, 5264. [CrossRef]
48. Attavanich, W.; McCarl, B.A. *The Effect of Climate Change, CO_2 Fertilization, and Crop Production Technology on Crop Yields and Its Economic Implications on Market Outcomes and Welfare Distribution*; Agricultural & Applied Economics Association: Philadelphia, PA, USA, 2011.
49. Baltagi, B.H. *Econometric Analysis of Panel Data*; John Wiley & Sons Ltd.: West Sussex, UK, 2005.
50. Golan, A.; Judge, G.; Karp, L. A maximum entropy approach to estimation and inference in dynamic models or counting fish in the sea using maximum entropy. *J. Econ. Dyn. Control* **1996**, *20*, 559–582. [CrossRef]
51. Moreno, B.; García-Álvarez, M.T.; Ramos, C.; Fernández-Vázquez, E. A General Maximum Entropy Econometric approach to model industrial electricity prices in Spain: A challenge for the competitiveness. *Appl. Energy* **2014**, *135*, 815–824. [CrossRef]
52. Liu, W.; Radmehr, R.; Zhang, S.; Henneberry, S.R.; Wei, C. Driving mechanism of concentrated rural resettlement in upland areas of Sichuan Basin: A perspective of marketing hierarchy transformation. *Land Use Policy* **2020**, *99*, 104879. [CrossRef]
53. Qureshi, M.E.; Whitten, S.M.; Kirby, M. A multi-period positive mathematical programming approach for assessing economic impact of drought in the Murray–Darling Basin, Australia. *Econ. Model.* **2014**, *39*, 293–304. [CrossRef]
54. Fathelrahman, E.; Davies, A.; Davies, S.; Pritchett, J. Assessing climate change impacts on water resources and Colorado agriculture using an equilibrium displacement mathematical programming model. *Water* **2014**, *6*, 1745–1770. [CrossRef]
55. Elouadi, D.; Ouazar, M.; Doukkali, L.; Elyoussfi, A. A mathematical model for assessment of socio-economic impact of climate change on agriculture activities: Cases of the east of Morocco (africa). *Indian J. Sci. Technol.* **2017**, *10*, 1–6. [CrossRef]
56. Zhao, Z.; Wang, G.; Chen, J.; Wang, J.; Zhang, Y. Assessment of climate change adaptation measures on the income of herders in a pastoral region. *J. Clean. Prod.* **2019**, *208*, 728–735. [CrossRef]
57. Henderson, B.; Cacho, O.; Thornton, P.; van Wijk, M.; Herrero, M. The economic potential of residue management and fertilizer use to address climate change impacts on mixed smallholder farmers in Burkina Faso. *Agric. Syst.* **2018**, *167*, 195–205. [CrossRef]
58. Laskookalayeh, S.S.; Najafabadi, M.M.; Shahnazari, A. Investigating the effects of management of irrigation water distribution on farmers' gross profit under uncertainty: A new positive mathematical programming model. *J. Clean. Prod.* **2022**, *351*, 131277. [CrossRef]
59. Radmehr, R.; Shayanmehr, S. The determinants of sustainable irrigation water prces in Iran. *Bulg. J. Agric. Sci* **2018**, *24*, 893–919.
60. Radmehr, R.; Ghorbani, M.; Kulshreshtha, S. Selecting strategic policy for irrigation water management (Case Study: Qazvin Plain, Iran). *J. Agric. Sci. Technol.* **2020**, *22*, 579–593.
61. Röhm, O.; Dabbert, S. Integrating agri-environmental programs into regional production models: An extension of positive mathematical programming. *Am. J. Agric. Econ.* **2003**, *85*, 254–265. [CrossRef]
62. Paris, Q.; Howitt, R.E. An analysis of ill-posed production problems using maximum entropy. *Am. J. Agric. Econ.* **1998**, *80*, 124–138. [CrossRef]
63. Gallego-Ayala, J. Selecting irrigation water pricing alternatives using a multi-methodological approach. *Math. Comput. Model.* **2012**, *55*, 861–883. [CrossRef]
64. Orojloo, M.; Shahdany, S.M.H.; Roozbahani, A. Developing an integrated risk management framework for agricultural water conveyance and distribution systems within fuzzy decision making approaches. *Sci. Total Environ.* **2018**, *627*, 1363–1376. [CrossRef]
65. Mortazavi, S.A.; Hezareh, R.; Ahmadi Kaliji, S.; Shayan Mehr, S. Application of linear and non-linear programming model to assess the sustainability of water resources in agricultural patterns. *Int. J. Agric. Manag. Dev.* **2014**, *4*, 27–32.
66. Kalbar, P.P.; Karmakar, S.; Asolekar, S.R. Technology assessment for wastewater treatment using multiple-attribute decision-making. *Technol. Soc.* **2012**, *34*, 295–302. [CrossRef]
67. Balioti, V.; Tzimopoulos, C.; Evangelides, C. Multi-criteria decision making using TOPSIS method under fuzzy environment. Application in spillway selection. *Multidiscip. Digit. Publ. Inst. Proc.* **2018**, *2*, 637.

68. Durmuşoğlu, Z.D.U. Assessment of techno-entrepreneurship projects by using Analytical Hierarchy Process (AHP). *Technol. Soc.* **2018**, *54*, 41–46. [CrossRef]
69. Rao, C.; Gao, M.; Wen, J.; Goh, M. Multi-attribute group decision making method with dual comprehensive clouds under information environment of dual uncertain Z-numbers. *Inf. Sci.* **2022**, *602*, 106–127. [CrossRef]
70. Shih, H.-S.; Shyur, H.-J.; Lee, E.S. An extension of TOPSIS for group decision making. *Math. Comput. Model.* **2007**, *45*, 801–813. [CrossRef]
71. Othman, M.K.; Fadzil, M.N.; Rahman, N.S.F.A. The Malaysian seafarers psychological distraction assessment using a TOPSIS method. *Int. J. E-Navig. Marit. Econ.* **2015**, *3*, 40–50. [CrossRef]
72. Hsu, T.-K.; Tsai, Y.-F.; Wu, H.-H. The preference analysis for tourist choice of destination: A case study of Taiwan. *Tour. Manag.* **2009**, *30*, 288–297. [CrossRef]
73. Karahalios, H. The application of the AHP-TOPSIS for evaluating ballast water treatment systems by ship operators. *Transp. Res. Part D Transp. Environ.* **2017**, *52*, 172–184. [CrossRef]
74. Jayant, A.; Gupta, P.; Garg, S.; Khan, M. TOPSIS-AHP based approach for selection of reverse logistics service provider: A case study of mobile phone industry. *Procedia Eng.* **2014**, *97*, 2147–2156. [CrossRef]
75. Kumar, V.; Ranjan, D.; Verma, K. Global climate change: The loop between cause and impact. In *Global Climate Change*; Elsevier: Amsterdam, The Netherlands, 2021; pp. 187–211.
76. Nwankwoala, H. Causes of Climate and Environmental Changes: The Need for Environmental-Friendly Education Policy in Nigeria. *J. Educ. Pract.* **2015**, *6*, 224–234.
77. Kumar, S. Climate change and its causes and effects: A review. *IJRSI* **2015**, 106–108.
78. Kokic, P.; Heaney, A.; Pechey, L.; Crimp, S.; Fisher, B.S. Climate change: Predicting the impacts on agriculture: A case study. *Aust. Commod. Forecast. Issues* **2005**, *12*, 161–170.
79. Almaraz, J.J.; Mabood, F.; Zhou, X.; Gregorich, E.G.; Smith, D.L. Climate change, weather variability and corn yield at a higher latitude locale: Southwestern Quebec. *Clim. Change* **2008**, *88*, 187–197. [CrossRef]
80. Zhang, Q.; Zhang, J.; Guo, E.; Yan, D.; Sun, Z. The impacts of long-term and year-to-year temperature change on corn yield in China. *Theor. Appl. Climatol.* **2015**, *119*, 77–82. [CrossRef]
81. Majumder, K. Nature and pattern of crop diversification in West Bengal. *Int. J. Res. Manag. Pharm.* **2014**, *3*, 33–41.
82. Karandish, F.; Mousavi, S.S.; Tabari, H. Climate change impact on precipitation and cardinal temperatures in different climatic zones in Iran: Analyzing the probable effects on cereal water-use efficiency. *Stoch. Environ. Res. Risk Assess.* **2017**, *31*, 2121–2146. [CrossRef]
83. Ghanian, M.; Ghoochani, O.M.; Dehghanpour, M.; Taqipour, M.; Taheri, F.; Cotton, M. Understanding farmers' climate adaptation intention in Iran: A protection-motivation extended model. *Land Use Policy* **2020**, *94*, 104553. [CrossRef]

agriculture

Article

Modeling Adaptive Strategies on Maintaining Wheat-Corn Production and Reducing Net Greenhouse Gas Emissions under Climate Change

Xiaopei Yi [1,†], Naijie Chang [1,†], Wuhan Ding [1,2], Chi Xu [1], Jing Zhang [3], Jianfeng Zhang [1] and Hu Li [1,*]

[1] Institute of Agricultural Resources and Regional Planning, Chinese Academy of Agricultural Sciences, Beijing 100081, China; 82101212143@caas.cn (X.Y.); 82101201035@caas.cn (W.D.); 82101205160@caas.cn (C.X.); zhangjianfeng@caas.cn (J.Z.); changnaijie@caas.cn (N.C.)
[2] Agro-Environmental Protection Institute, Ministry of Agriculture and Rural Affairs, Tianjin 300191, China
[3] Engineering and Technology Research Center for Agricultural Land Pollution Prevention and Control of Guangdong Higher Education Institutes, College of Resources and Environmental Sciences, Zhongkai University of Agriculture and Engineering, Guangzhou 510225, China; zhangjing@zhku.edu.cn
* Correspondence: lihu@caas.cn
† These authors contributed equally to this work.

Abstract: Climate change has posed serious challenges to food production and sustainable development. We evaluated crop yields, N_2O emissions, and soil organic carbon (SOC) in a typical wheat–corn rotation system field on the North China Plain on a 50-year scale using the Denitrification–Decomposition (DNDC) model and proposed adaptive strategies for each climate scenarios. The study showed a good consistency between observations and simulations ($R^2 > 0.95$ and nRMSE < 30%). Among the twelve climate scenarios, we explored ten management practices under four climate scenarios (3 °C temperature change: P/T−3 and P/T+3; 30% precipitation change: 0.7P/T and 1.3P/T), which have a significant impact on crop yields and the net greenhouse effect. The results revealed that changing the crop planting time (CP) and using cold-resistant (CR) varieties could reduce the net greenhouse effect by more than 1/4 without sacrificing crop yields under P/T−3. Straw return (SR) minimized the negative impact on yields and the environment under P/T+3. Fertigation (FG) and Drought-Resistant (DR) varieties reduced the net greenhouse effect by more than 8.34% and maintained yields under 0.7P/T. SR was most beneficial to carbon sequestration, and yields were increased by 3.87% under 1.3P/T. Multiple adaptive strategies should be implemented to balance yields and reduce the environmental burden under future climate change.

Keywords: wheat–corn; DNDC; climate change; crop yield; net greenhouse effect

Citation: Yi, X.; Chang, N.; Ding, W.; Xu, C.; Zhang, J.; Zhang, J.; Li, H. Modeling Adaptive Strategies on Maintaining Wheat-Corn Production and Reducing Net Greenhouse Gas Emissions under Climate Change. *Agriculture* **2022**, *12*, 1089. https://doi.org/10.3390/agriculture12081089

Academic Editors: Dengpan Xiao and Wenjiao Shi

Received: 20 June 2022
Accepted: 18 July 2022
Published: 24 July 2022

Publisher's Note: MDPI stays neutral with regard to jurisdictional claims in published maps and institutional affiliations.

1. Introduction

Climate change has been identified as one of the long-term and severe challenges in the twenty-first century [1]. With a global average surface temperature increase of 0.85 °C during the period from 1880 to 2012 [2], global warming occurred much faster than expected according to a recent study that 1.5 °C temperature increase by 2030 and an increase of 3 °C by the end of this century [3], and changes in the spatial pattern of precipitation also showed intensification across most global land areas between 1950 and 2016 [4]. As a result, the frequency and severity of extreme weather, such as strong heat waves, floods, and droughts, have increased significantly [5,6]. A lot of uncertainty exists in agricultural production activities that are sensitive to climatic conditions, thereby threatening global food security. Researchers agreed that the warmer and wetter conditions may accelerate the growth and development of crops in a few areas [7]. However, several studies highlighted that the climate crisis has a negative impact on agricultural production in most regions, particularly on the North China Plain (NCP) [8], which has a severe drought risk [9–11]. Along with climate change, greenhouse gas (GHG) emissions that contribute to global warming have been increasing. Nitrous oxide (N_2O) is a main source of GHG emissions, and the annual growth rate was between 0.24% and 0.31% in response to climate change [2]. In addition, cropland emit about 43% of global total N_2O emissions [12]. Therefore, we attempt to explore adaptive strategies in agriculture to mitigate food shortages and climate warming.

Cropland management is one possible strategy for addressing climate change [13]. In recent decades, studies have assessed many management mitigation strategies to maintain crop yield and reduce GHG emissions [14,15]. Tillage is a fundamental agriculture management option, and the adoption of conservation tillage practices such as reduced tillage and no tillage were reported to reduce carbon dioxide emissions through soil organic carbon (SOC) sequestration [16]. Li et al. [17] claimed that reducing nitrogen inputs by 13% and increasing straw retention by 20% also promoted the sequestration of SOC, as well as reduced GHGs by 13% and 11%, respectively. In addition, adjustment of the sowing date and the use of different varieties increased attainable maize yields, which will enhance the heat and water utilization efficiency of crops [18,19], and efficient irrigation practices have also alleviated the relationship between yield and climate conditions [20]. Although these strategies have been regarded as effective management practices that regulate GHG emissions, multiple practices have not been compared because of limited treatments in field experiments, and there is also a lack of quantitative analysis of the impact on crop yield and the environment under different possible climate scenarios [21,22]. Due to the uncertainty of future changes in temperature and precipitation, there is a need to identify a model to quantify and compare multiple agricultural strategies when developing optimum options to cope with the foreseeable adverse effect of climate change.

The Denitrification–Decomposition model (DNDC), a process-oriented plant–soil model, links C sequestration to GHG emissions by simulating the soil microbial activities [23,24]. In the past three decades, the model has been widely validated in various areas such as the simulation of greenhouse benefits, crop growth, and soil processes in agricultural fields [23,25,26], and there are also reports on wheat–corn rotation systems on the NCP [27,28]. Wheat and corn are important cereal crops, and wheat–corn rotation is one of the dominate cropping systems on the NCP, providing approximately 68.55% of the wheat and 22.89% of the corn for China [29,30]. In this study, based on the internationally known DNDC model [23] and field monitoring data, we expect to achieve two objectives: (1) Evaluate the impact of temperature and precipitation changes on wheat–corn yields, N_2O emissions, and SOC loss at a 50-year scale. (2) Quantify the relative contributions of different agricultural practices to wheat–corn yields and the net greenhouse effect and select the adaptive strategies according to the performance of 10 strategies under different climate scenarios.

2. Materials and Methods

2.1. Study Area and Experimental Description

The study was conducted in Ye County, Henan Province, China (113°22′ E, 33°31′ N). The field site has a continental monsoon climate with a mean annual air temperature of 14.8 °C, a frost-free period of 212 days, and an average annual precipitation of 819 mm. Soil properties before the experiment are presented in Table 1.

Table 1. Soil properties of the study fields.

Bulk Density	Nitrate N	Ammonium N	SOC	Soil pH
(g cm^{-3})	(mg kg^{-1})	(mg kg^{-1})	(g kg^{-1})	
1.35	0.5	0.05	10	7.0

The study was performed in a field with a typical winter wheat–summer corn double-cropping system in this region. The experimental data were obtained from October 2015 to October 2017. There were two treatments: farmers' conventional practice (FP) and optimal practice (PT). Rotary tillage was applied three times to a depth of 15 cm in FP after corn harvest, and then seeds were applied together with fertilizer. No tillage was carried out in PT, and the wheat seeds were broadcast directly and were mixed in the top 5 cm soil layer. The application of N fertilizer in the wheat season was divided into basal fertilizer (FP: 150 kg N·ha^{-1}; PT: 120 kg N·ha^{-1}) and one additional fertilizer application during the jointing stage (FP: 150 kg N·ha^{-1}; PT: 120 kg N·ha^{-1}). The N fertilizer in the corn season was applied at rates of 300 kg N·ha^{-1} and 240 kg N·ha^{-1} in FP and PT, respectively. Wheat was irrigated in November and April of the next year, and corn was irrigated at the trumpet stage.

The yields of wheat and corn, N_2O fluxes, SOC (0–10 cm), soil temperature, and soil moisture (0–5 cm) were monitored in this study. For details, please refer to a previous study [31].

2.2. DNDC Model

The DNDC model was constructed by two components to describe the process of C and N transformations in an agricultural system [23]. This first component is mainly simplified into soil climate, crop growth, and decomposition sub-models, which predict soil temperature, soil moisture, pH, redox potential (Eh), and substrate concentration driven by ecological factors. The second part simulates C and N transformations mediated by soil microbes and includes nitrification, denitrification, and fermentation sub-models.

In this study, we used DNDC (version 9.5) calibrated by measured data from the FP treatment and verified the model using measured data from the PT treatment. The main inputs of DNDC are climate, soil, and management data, and output results are crop growth and soil physical and chemical data. Table 2 shows a number of user-set parameters in our study, which came from monitoring and calibrated values. Meteorological data were collected from the Meteorological Bureau of Ye country, Henan Province. We used a zero-intercept linear regression between simulations and observations to evaluate the performance of the modified DNDC. The slope and coefficient of determination (R^2) of the regression indicate the consistency and correlation between simulations and observations, respectively. Additionally, the normalized root mean square error (nRMSE) was used for quantitative comparisons between simulations and observations [27].

Table 2. The DNDC Model Input Parameters for Ye County.

Site	Ye Country 2015, 2016 Wheat	Ye Country 2016, 2017 Corn
Latitude (°)	33.5	33.5
Tillage depth (cm)	10	10
Soil depth sampled (cm)	0–10	0–10
Soil pH	7	7
Soil texture	Silty loam	Silty loam
Clay fraction (%)	0.2	0.2
Bulk density (g·cm^{-3})	1.35	1.35
Total organic carbon (kg C·kg^{-1})	0.01	0.01
Field capacity (WFPS)	0.52	0.52
Wilting point (WFPS)	0.34	0.34
Nitrate N (mg N·kg^{-1})	0.5	0.5
Ammonium N (mg N·kg^{-1})	0.05	0.05

2.3. Scenarios of Future Climate Change

The latest research claims that the current global warming rate is far faster than expected and will increase by 1.5 °C by 2030 and by 3 °C by the end of the century [3]. Further, regional precipitation will fluctuate with the continuous warming. Climate change scenarios are used to predict future GHG emissions from agricultural ecosystems and to assess the vulnerability of agricultural production under a future climate. To accurately evaluate the potential of carbon emission reductions from wheat–corn rotation systems under future climate change, the baseline temperature and precipitation datasets were built based on the mean daily climate values from 1998 to 2017, and seven change scenarios for temperature and precipitation were set up in this study (Table 3). The other input parameters of the DNDC model followed FP. Climate change scenario data were set as DNDC meteorological parameters to assess the 50-year effects of different climate change scenarios on the yield, SOC, and greenhouse gas emissions in a typical wheat–corn rotation system.

Table 3. The climate change scenarios.

Temperature Change Scenarios	Explanation	Precipitation Change Scenarios	Explanation
P/T	Baseline	P/T	Baseline
P/T−3	Temperature decrease of 3 °C	0.7P/T	Precipitation decreased by 30%
P/T−2	Temperature decrease of 2 °C	0.8P/T	Precipitation decreased by 20%
P/T−1	Temperature decrease of 1 °C	0.9P/T	Precipitation decreased by 10%
P/T+1	Temperature increase of 1 °C	1.1P/T	Precipitation increased by 10%
P/T+2	Temperature increase of 2 °C	1.2P/T	Precipitation increased by 20%
P/T+3	Temperature increase of 3 °C	1.3P/T	Precipitation increased by 30%

Note: T represents the baseline temperature, P represents the baseline precipitation, and the baseline temperature and precipitation are the daily average of the past 20 years (1998–2017).

2.4. Scenarios of Adaptive Strategies

From the above twelve climate scenarios, the four climate scenarios most sensitive in the wheat–corn rotation system were selected, and we assessed ten strategies for each climate scenario by comparison with the baseline (FP). Ten management practices were selected: (1) Reduce N fertilizer by 20% (0.8N); (2) Increase N fertilizer by 20% (1.2N); (3) Straw return (SR); (4) No tillage (NT); (5) Fertigation (FG); (6) Change the crop planting time (CP); (7) Change the time of irrigation and fertilization (CI, move topdressing irrigation one week ahead); (8) Increase irrigation by 30% (IR); (9) Cold-resistant (CR) varieties; and (10) Drought-resistant (DR) varieties. Except for the management options mentioned above, other factors (such as climate and soil) remained consistent with those used in FP.

2.5. The Net Greenhouse Effect

The net greenhouse effect was evaluated by the global warming potential (*GWP*) by considering the impacts of climate change on GHG and SOC. The *GWP* with a span of 100 years was used to represent the combined effect of N_2O, CO_2, and CH_4, which represented the combined effect of those gases in the atmosphere and the relative effect in causing radiative forcing over different time periods. The net greenhouse effect of the same amount of N_2O is 273 times [32] that of CO_2 over a 100-year period and has been estimated to contribute 7.4% to global warming [33]. The annual CO_2 emissions can be expressed by the annual net SOC (*dSOC*) change. There were no monitoring data of CH_4 emissions to verify the model due to lower emissions under dryland farmland. Thus, the calculation was as follows:

$$GWP(kg\ CO_2 \cdot ha^{-1} \cdot a^{-1}) = 273 \times N_2O - 44/12 \times dSOC \tag{1}$$

where N_2O represents the annual emissions in the output results of the DNDC model, and *dSOC* is the net SOC change in the output results of the model, with a positive or negative value indicating a reduction or increase in CO_2 emissions.

3. Results

3.1. Validation of the DNDC Model

We carried out site verification under two agricultural practices to verify the applicability of the DNDC model in the study area. Modeled crop yields were in good agreement with observations for both the calibration (FP during 2015–2016) and validation (FP during 2016–2017 and PT during 2015–2017) datasets (Figure 1). The calculated statistical indices

(R^2 =0.96, $p < 0.001$, nRMSE = 22.69%) indicated that the simulated crop yields were significantly correlated with observed values. However, the slope of the zero-intercept linear equation was 0.85, indicating that our simulated values were lower than the observed crop yields.

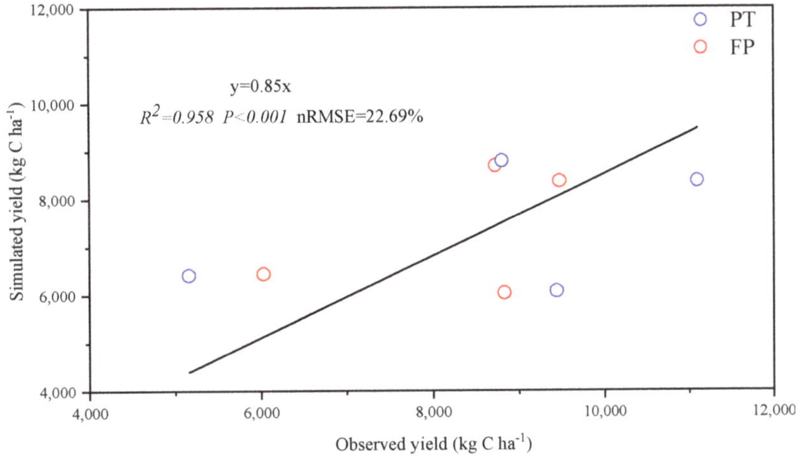

Figure 1. Comparison of simulated and observed yields during 2015–2017. FP represents farmers' conventional practice, PT represents optimal practice.

The modeled soil temperature data were also generally comparable with the observed values (Figure 2). The R^2 of the linear regression of simulated against observed data was 0.98, and nRMSE was 15.97%, indicating that the simulated values had high simulation reliability in the trend and range of soil temperature variation. In addition, the DNDC model overestimated the soil temperature in several periods.

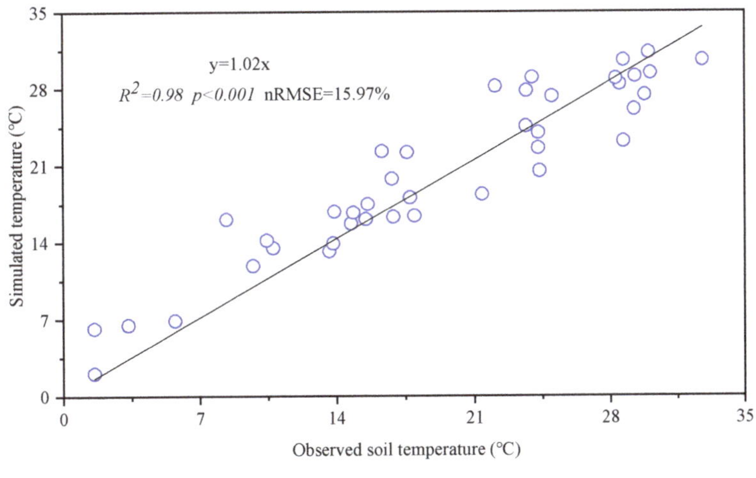

Figure 2. Comparison of simulated and observed soil temperature (0–5 cm) during 2015–2017.

Based on a comparison of the simulated annual cumulative N_2O emissions under FP and PT treatments (2.70 and 1.92 kg N ha^{-1}, respectively), the observed values showed a small discrepancy due to overestimation (Figure 3). However, the nRMSE with a value of 26.51% demonstrated that the simulation deviations were within an acceptable range (nRMSE < 30%) [27].

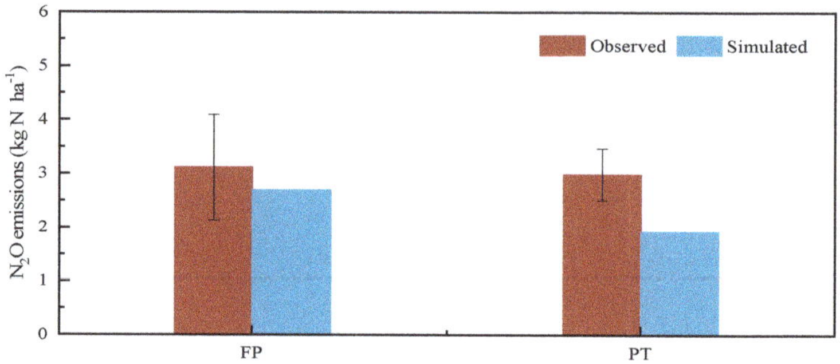

Figure 3. The simulated and observed values of total N_2O emissions in 2016. FP represents farmers' conventional practice; PT represents optimal practice.

As Figure 4 shows, the simulations of the soil SOC from the DNDC model were close to the corresponding field observations (nRMSE = 4.73%). A significant zero-intercept linear regression (slope = 0.98, R^2 = 0.99, p < 0.001, n = 8) was obtained to relate the simulations of the SOC to the corresponding observations. Furthermore, different trends of SOC change existed in the two treatments, which illustrated that the DNDC model could capture and distinguish management options well.

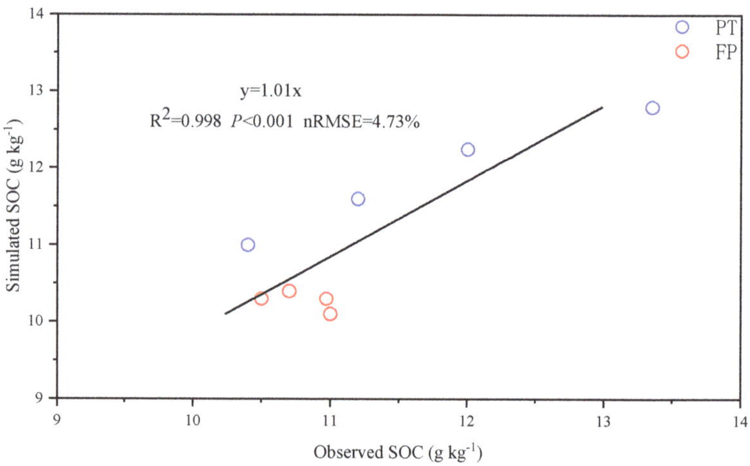

Figure 4. Comparison of simulated and observed soil SOC (0–10 cm) during 2015–2017.

3.2. *The Impact of Climate Change on Crop Yield and the Net Greenhouse Effect*

3.2.1. The Yield under Climate Change Scenarios

Crop yields under different temperature climate scenarios were basically at a relatively stable level (Figure 5), with baseline (P/T) ranging from 6305.54 to 6334.50 kg C·ha^{-1}. Crop yields had an obvious increasing (decreasing) trend over time under the increased (decreased) precipitation scenarios. The P/T−3 and P/T+3 scenarios reached the lowest yields in the 50th year, at 5615.03 and 6024.420 kg C·ha^{-1}, respectively, a reduction of 4.46% and 10.95% compared with the baseline. There was no obvious yield reduction under the remaining temperature scenarios.

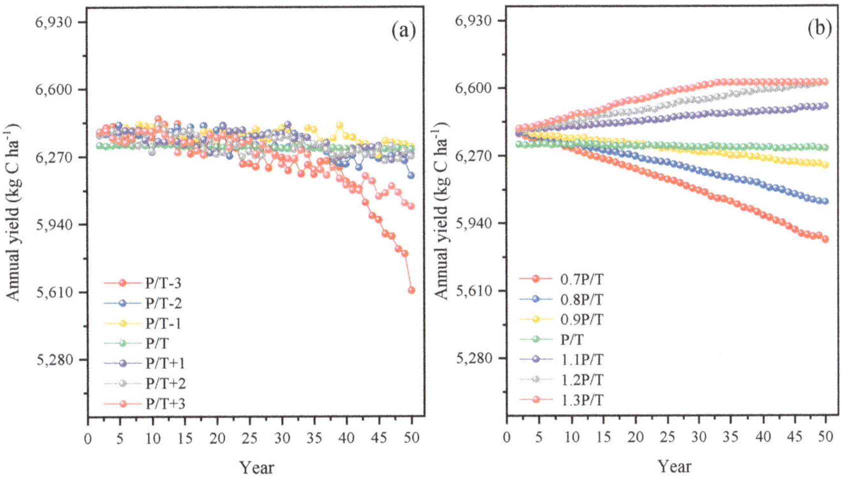

Figure 5. The 50-year effects on crop yields under temperature change scenarios (**a**) and precipitation change scenarios (**b**). P/T−3: temperature decreased by 3 °C; P/T−2, temperature decreased by 2 °C; P/T−1: temperature decreased by 1 °C; P/T: temperature unchanged; P/T+1: temperature increased by 1 °C; P/T+2: temperature increased by 2 °C; P/T+3: temperature increased by 2 °C; 0.7P/T: precipitation decreased by 30%; 0.8P/T: precipitation decreased by 20%; 0.9P/T: precipitation decreased by 10%; P: precipitation unchanged; 1.1P/T: precipitation increased by 10%; 1.2P/T: precipitation increased by 20%; 1.3P/T: precipitation increased by 30%.

3.2.2. The Net Greenhouse Effect under Climate Change Scenarios

Annual N_2O emissions showed a trend of increasing in the first ten years and then slowly decreasing over time for all climate change scenarios (Figure 6). Figure 6a,b demonstrates a positive relationship between annual N_2O emissions and temperature/ precipitation. Among the temperature/precipitation treatments, the P/T−3 and 0.7P/T scenarios consistently had the lowest annual N_2O emissions while the P/T+3 and 1.3P/T scenarios consistently had the highest annual N_2O emissions.

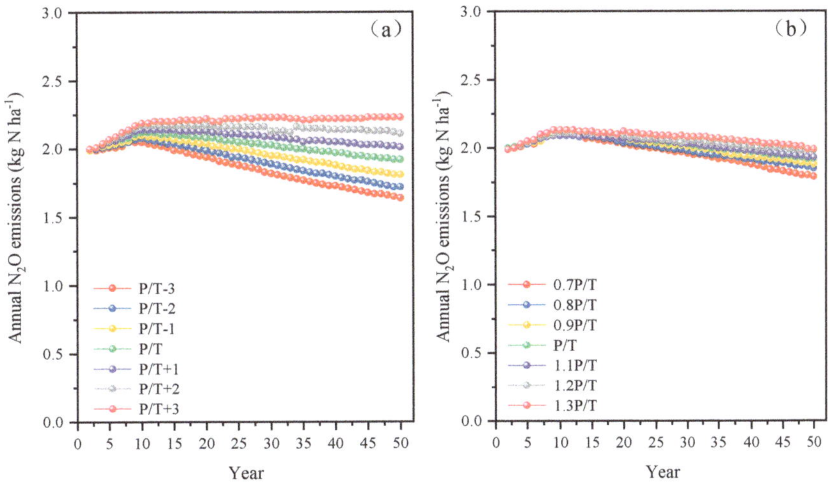

Figure 6. *Cont.*

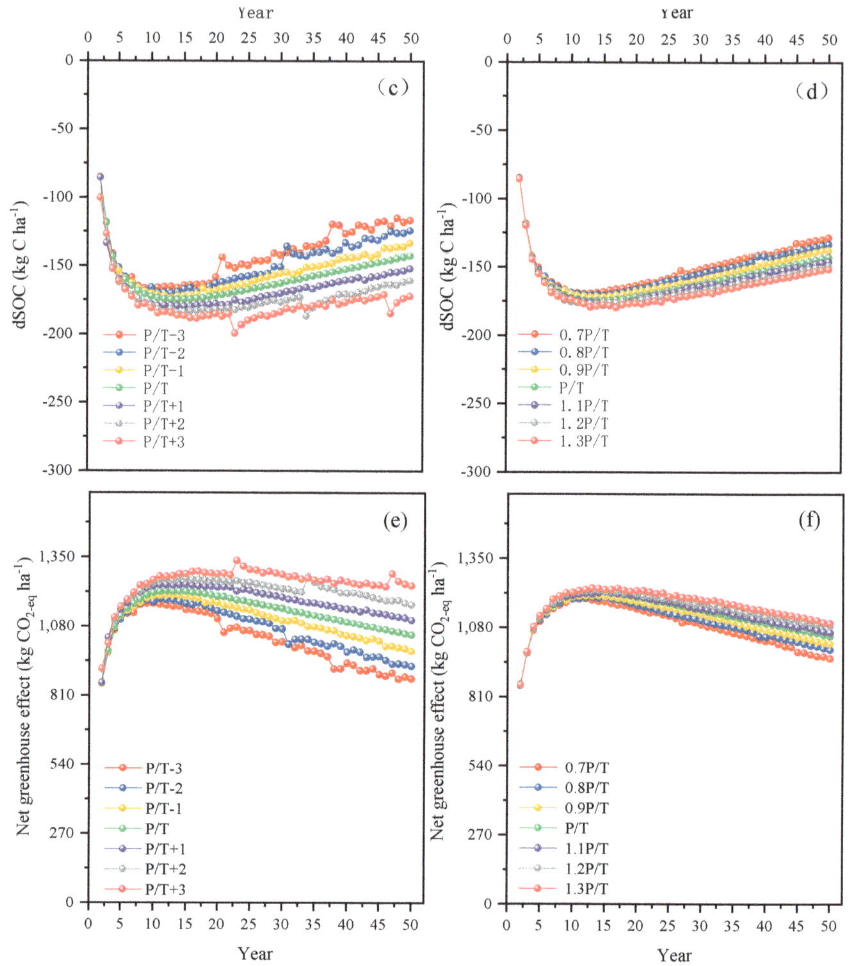

Figure 6. The 50-year effects on N_2O emissions (**a**,**b**), SOC loss (**c**,**d**), and the global warming potential (**e**,**f**) under temperature change scenarios and precipitation change scenarios. P/T−3: temperature decreased by 3 °C; P/T−2, temperature decreased by 2 °C; P/T−1: temperature decreased by 1 °C; P/T: temperature unchanged; P/T+1: temperature increased by 1 °C; P/T+2: temperature increased by 2 °C; P/T+3: temperature increased by 2 °C; 0.7P/T: precipitation decreased by 30%; 0.8P/T: precipitation decreased by 20%; 0.9P/T: precipitation decreased by 10%; P/T: precipitation unchanged; 1.1P/T: precipitation increased by 10%; 1.2P/T: precipitation increased by 20%; 1.3P/T: precipitation increased by 30%.

Overall, SOC changes in baseline ranged from −174.68 to −85.43 kg C-ha^{-1}. Figure 6c,d illustrates that increasing temperature and decreasing precipitation promoted SOC loss. For all climate change scenarios, the annual dSOC was the lowest under P/T+3 and 1.3 P/T, reaching −172.03 kg C-ha^{-1} and −128.37 kg C-ha^{-1}, respectively, in the 50th year. The highest dSOC was measured in P/T−3 and 0.7 P/T+3, but the overall loss was less than zero, which indicated that SOC was lost.

The net greenhouse effect of the baseline ranged from 853.58 to 1204.50 kg CO_2-eq-ha^{-1}. (Figure 6e,f). Along with the increase in temperature or precipitation, the greenhouse effect increased, being the lowest under P/T−3 or 0.7 P/T and the highest under P/T−3 and

1.3 P/T. The trends changed under the scenarios of temperature change and precipitation change, with a sharp increase in the first ten years but a slow decrease on a long-term basis. However, the values of the net greenhouse effect were all positive, leading to climate warming.

3.3. *Adaptive Strategies in Response to Climate Change*

3.3.1. Adaptation to the P/T−3 and P/T+3 Scenarios

In P/T−3, eight strategies reduced production by more than 9.89%, while two strategies (CP and CR) had little effect on yields (Table 4). For N_2O emissions, two strategies (0.8N and NT) resulted in the largest emission reductions of about 30%. CP, CI, IR, CR, and DR reduced N_2O emissions by 7.96% to 17.41%, while SR and FG increased N_2O emissions by 4.48% and 126.87%, respectively. For SOC changes, all strategies reduced the annual SOC loss, and SR and NT increased SOC by 1516.88 and 126.81 kg C-ha^{-1}, respectively. Considering SOC and N_2O emissions together, only FG increased the net greenhouse effect by 52.15%, while the remaining strategies reduced the net greenhouse effect, especially two strategies (SR and NT) reduced the net greenhouse effect by 562.57% and 63.60%, respectively.

Table 4. The effects on annual yield, N_2O emissions, and SOC loss under the P/T−3 and P/T+3 scenarios.

Management Strategies	P/T−3				P/T+3			
	Yield	N_2O	SOC	GWP	Yield	N_2O	SOC	GWP
0.8 N	−11.09%	−30.35%	−42.20%	−35.59%	−4.26%	1.00%	48.03%	21.81%
1.2 N	−11.06%	−1.99%	−42.12%	−19.75%	−4.28%	39.30%	47.32%	42.85%
SR	−9.89%	4.48%	−1276.94%	−562.57%	−3.02%	51.24%	−948.96%	−391.36%
NT	−11.07%	−29.35%	−106.75%	−63.60%	−4.30%	4.98%	−32.41%	−11.57%
FG	−9.89%	126.87%	−41.98%	52.15%	−2.56%	123.38%	51.41%	91.53%
CP	−0.11%	−15.92%	−39.66%	−26.42%	−5.08%	25.37%	41.25%	32.40%
CI	−11.07%	−17.41%	−42.04%	−28.31%	−4.28%	18.41%	48.25%	31.61%
IR	−11.06%	−7.96%	−42.08%	−23.06%	−4.26%	34.33%	47.51%	40.16%
CR	0.20%	−16.42%	−46.01%	−29.51%	−9.04%	21.89%	51.83%	35.14%
DR	−9.89%	−16.42%	43.18%	−28.26%	−2.56%	21.39%	49.74 %	33.94%

Note: A negative value represents a percentage decrease relative to the baseline, and a positive value represents a percentage increase relative to the baseline. 0.8N: Decrease N by 20%; 1.2N: Increase N fertilize by 20%; SR: Straw Return; NT: No Tillage; FG: Fertigation; CP: Change the crop planting time; CI: Change the time of irrigation and fertilization; IR: Increase irrigation by 30%; CR: Cold-resistant variety; DR: Drought-resistant variety.

In P/T+3, all strategies reduced the wheat–corn yields and increased the N_2O emissions (Table 4). For the annual SOC change, SR increased the soil carbon pool sequestration to 1008.48 kg C-ha^{-1}, and the remaining strategies resulted in carbon loss. Considering SOC and N_2O emissions together, only SR and NT reduced the net greenhouse effect by 391.36% and 11.57%, but the yields were reduced by 3.02% and 4.30%, respectively.

3.3.2. Adaptation to the 0.7 P/T and 1.3 P/T Scenarios

Under the 0.7 P/T scenario (Table 5), only FG and DR did not reduce the crop yields greatly, while other strategies caused a yield reduction more than 5.96%. For N_2O emissions, all of the remaining seven strategies except 1.2N, SR, and IR reduced N_2O emissions; 0.8N and NT caused a reduction of more than 20%. For the net greenhouse effect (3.11–491.03%), although SR largely reduced the net greenhouse effect, it reduced the yield by 5.96%. FG and DR were able to reduce the net greenhouse effect by 8.34% and 10.91% while stabilizing yield.

Table 5. The effects on annual yield, N_2O emissions, and SOC loss under the 0.7P/T and 1.3P/T scenarios.

Management Strategies	0.7 P/T				1.3 P/T			
	Yield	N_2O	SOC	GWP	Yield	N_2O	SOC	GWP
0.8 N	−7.64%	−22.39%	−19.93%	−21.30 %	3.87%	−16.92%	15.66%	−2.50%
1.2 N	−7.64%	7.46%	−20.18%	−4.77%	3.87%	24.88%	15.28%	20.63%
SR	−5.96%	20.90%	−1135.97%	−491.03%	3.87%	32.34%	−1084.43%	−461.85%
NT	−7.56%	−22.39%	−101.75%	−57.51%	3.87%	−6.47%	−54.69%	−27.81%
FG	2.72%	−5.47%	−11.95%	−8.34%	3.87%	86.57%	16.39%	55.51%
CP	−7.15%	−1.00%	−5.77%	−3.11%	4.15%	12.94%	4.63%	9.26%
CI	−7.58%	−7.96%	−16.93%	−11.93%	3.87%	5.97%	15.30%	10.10%
IR	−7.64%	2.99%	−20.03%	−7.20%	3.87%	17.91%	15.36%	16.78%
CR	−6.77%	−5.97%	−10.07%	−7.78%	4.86%	10.45%	20.94%	15.09%
DR	−0.32%	−7.46%	−15.25%	−10.91%	3.87%	6.97%	16.13%	11.02%

Note: A negative value represents a percentage decrease relative to the baseline, and a positive value represents a percentage increase relative to the baseline. 0.8N: Decrease N fertilizer by 20%; 1.2N: Increase N fertilize by 20%; SR: Straw Return; NT: No Tillage; FG: Fertigation; CP: Change the crop planting time; CI: Change the time of irrigation and fertilization; IR: Increase irrigation by 30%; CR: Cold-resistant variety; DR: Drought-resistant variety.

In 1.3 P/T (Table 5), all strategies increased yields by more than 3.87%. Among the ten strategies, only 0.8 N and NT reduced N_2O emissions by 16.92% and 6.47%, respectively. Similar to the P/T+3 scenario, SR increased the soil carbon pool sequestration to 1169.4 kg C-ha^{-1}, while the remaining strategies reduced the soil carbon pool, but NT still reduced the amount of SOC loss compared to the baseline. Considering SOC and N_2O emissions together, only 0.8N, SR, and NT reduced the net greenhouse effect by 2.50%, 461.85%, and 27.81%, respectively, while the rest of the strategies increased the net greenhouse effect (9.26% to 55.51%).

4. Discussion

4.1. Model Performance

In this study, adaptability of the DNDC model to a wheat–corn rotation system on the NCP was confirmed through combination with field monitoring data. There is ample evidence that soil temperature influences plant decomposition, soil organic carbon, and N_2O emissions by affecting the ratio of N_2/N_2O [8,34]. Thus, the accurate estimation of soil temperature is vital for a reliable model estimation of crop yields and the net greenhouse effect. The three indexes of soil temperature in our study confirmed that the DNDC model is capable of correcting simulated soil temperature. Figure 4 shows that the soil SOC under rotary tillage was significantly lower than that under no-tillage, which decreased soil disturbance. It showed that DNDC effectively distinguished the differences caused by management in our research. For N_2O emissions, there were uncertainties, i.e., that the simulations were not directly evaluated against observed values in 2017 due to the absence of N_2O emission data. However, the simulation results of total N_2O emissions in 2016 in Figure 3 are good. Similarly, previous studies showed that DNDC provided a good simulation of crop yield, N_2O emissions, and SOC for different management options in a wheat–corn system [27,35,36].

However, we noted that the phenomena of overestimation or underestimation of the simulated values were still present for two treatments and validated indicators. Human errors in the field experiments (sampling, index measurements, and data processing) are inevitable. In addition, some parameters used in the model were acquired from default values rather than experimental data, ignoring various external environments [24]. Furthermore, the model only includes some of the soil indicators, far fewer than the real environment. Therefore, we need to pay attention to the influence of those defects, but as a whole, the accuracy of the model simulation is unquestionable, which can provide a certain guarantee for subsequent research.

4.2. Climate Change Impact on Crop Yield and the Net Greenhouse Effect

The changed relationship between wheat–corn yields and temperature/precipitation in our study was meaningful for yield forecasting and agricultural planning. Crop yields with small temperature fluctuations (±1 °C and ±2 °C) were basically consistent with the baseline in this study due to small range fluctuations that caused relatively little stress to crop growth [37]. However, a 3 °C decrease (increase) would have a significant impact on crop yields because accumulated temperature cannot be tolerated by crops, especially wheat [38]. If we only consider climate warming, an average 2.58% of crop yield is lost for every degree of warming at the national level [39]. Among the three main crops, wheat, corn, and rice, the negative impact of rising temperature on wheat is the most evident, followed by corn. However, the response of rice yields was not significant at the subregional level [39,40]. We can see that precipitation was positively correlated with crop yields in this study. Liu et al. [9] also proved that the projected gain of crop yields increased due to increasing precipitation (+15% or 30%). The result confirmed that water shortage is a strong limiting factor for agricultural development on the NCP [41].

We researched N_2O emissions from wheat–corn rotations as a hot-spot agricultural cropping system of GHG emissions [29,42]. In our study, the total amount of N_2O emissions showed an increasing trend with rising temperature or precipitation. Factually, soil moisture status and soil temperature are influenced by climate conditions [31,38,43]. The positive feedback of soil temperature to air temperature accelerates N_2O emissions through the decomposition of soil organic matter, biological activity of denitrifying bacteria, and indirect dissolution and release of oxygen in water [44]. Precipitation changes the soil moisture and thus has a decisive influence on nitrification and denitrification processes by affecting the metabolic activity of soil microbial cells and nutrient transport [37].

SOC is an important component of soil, and we found that the effect of temperature change on SOC was greater than the effect of precipitation change. This result differs from that of Hursh et al. [45], who reported that soil moisture was the most significant predictor variable. Soil moisture was affected in our simulation because irrigation was set up in the model simulation. SOC decreased with the increase in temperature. The reason is that warming temperature can accelerate the decomposition of SOC [46] and alter the return of plant residues to the soil by affecting plant growth [47]. Although precipitation has a small impact on SOC compared to temperature, more precipitation contributed to the formation of a higher temperature–humidity environment in the soil. The activities of soil microorganisms and mineralization are accelerated under a higher temperature–humidity environment, leading to the decomposition of carbon inputs and ultimately reducing the organic carbon conversion efficiency [48]. It is worth noting that SOC was lost regardless of changes in temperature or precipitation in our study. The results stress the necessity of observing SOC changes when evaluating the impact of strategies on GHG emissions.

4.3. Adaptive Strategies in Response to Climate Change

The study on the change characteristics of crop yields under various strategies can provide theoretical support for agricultural development under climate change. Our results suggested that CP and CR could effectively counteract the yield reduction caused by decreasing temperature. Changing the planting time of crops can increase the accumulated temperature required during crop growth, and cold-resistant varieties compensate for the cumulative temperature required for crop growth by reducing the demand [9]. As temperatures rise, the water demand of crops increases due to evaporation, which will result in lower crop yields. FG satisfied the water requirement of crops by watering, and DR required less water; thus, the two strategies effectively mitigated a yield reduction.

In our study, the N_2O emission in FG was the highest under P/T−3, P/T+3, and 1.3P/T. Fertilization leads to a high risk of N_2O emissions due to pulses of excessive N and water in the soil [49]. However, some authors [50] believe that fertigation can reduce N_2O emissions under the premise of watering. One possible reason for our result was that the irrigation frequency, the amount of irrigation water, and the fertilization frequency were increased to meet the needs of crop growth during the model simulation in FG. We also found that more N_2O was emitted from 1.2N than from 0.8N under the scenarios of increasing temperature and precipitation. The amount of N inputs converted to N_2O emissions was higher when total N inputs were higher, especially under high temperature and high humidity [31]. SR indirectly affected N_2O emissions by influencing soil carbon, nitrogen availability, and soil aeration after straw application. Kravchenko et al. [42] found that straw application could increase soil N_2O emissions. Shan et al. [51] also found that straw return significantly increased N_2O emissions and nitrogen in the soil in dryland areas.

The management options that significantly increased SOC were SR and NT, which increased the input of SOC. Plant C input and microbial decomposition are two main processes for SOC storage. Straw return can directly return carbon fixed by crops to soil, thus there was a significant positive correlation between the amount of straw returned and the SOC content [52,53]. Lehtinen et ai. [54] showed that straw return increased the SOC content by 7% on average. Liao et al. [55] studied farmland in North China and demonstrated that the soil SOC content increased from 7.8 $g \cdot kg^{-1}$ to 11.0 $g \cdot kg^{-1}$ from 1981 to 2011 after straw was applied. Lenka et al. [56] also revealed that the return of straw to the field can increase the SOC content in the topsoil (0–10 cm) in a long-term positioning test. No tillage is also an effective management option to increase SOC by improving the soil structure and by accelerating the formation of soil aggregates [57,58]. Our results are similar to a previous study, i.e., that the SOC sequestration rate was higher under no tillage than under rotary tillage and conventional tillage [59].

Taking into account the yield, N_2O emissions, and SOC, we filtered out the appropriate strategies that maintained the crop yields. CP and CR reduced the net greenhouse effect and guaranteed crop yields under P/T−3. The two management options may satisfy the accumulated temperature conditions for crop growth. The N_2O emissions and the annual loss of SOC decreased with decreasing temperature, comprehensively resulting in an overall reduction of the net greenhouse effect. However, the yields under each management option did not reach the baseline level in P/T+3, although there were some differences in yield. It may be difficult to cope with the climate warming by using a single management option; hence, a combination of management options should be considered. Under the 0.7P/T scenario, the yields were maintained and the net greenhouse effect was reduced at the same time in FG and DR due to the reduction of N_2O emissions and annual SOC loss by ensuring the water requirement of the crops. For 1.3P/T, the simulation results showed that the SOC from straw return was higher and exceeded the GHG emissions brought about by straw return. Thus, the net greenhouse effect was reduced compared with traditional benchmark measures, and crop yield was increased at the same time. Therefore, straw returned to the field is a recommended management option.

4.4. Limitations

We investigated the effects of ten management practices on crop yields, NO_2 emissions, and SOC based on different climate scenarios, which can provide some reference for addressing climate change. However, our study is still limited by many factors. The climate change scenarios in our study may differ from future situations, and we need to find more reliable climate scenarios. It is essential to justify and improve the input parameters of DNDC to reduce the uncertainty of the model simulation. There were large differences in the contributions of crop yield and the N_2O emissions among the ten strategies in our study. When developing adaptive strategies for sustainable wheat–corn production, more strategies should be explored while simultaneously measuring yields, SOC changes, and GHG emissions.

5. Conclusions

The DNDC model was evaluated using field monitoring data of the wheat–corn yield, annual cumulative N_2O emissions, and SOC under two treatments with different tillage practices and N applications. The model evaluations demonstrated that the simulations were consistent with the observations and captured the SOC change under no tillage and rotatory tillage. The temperature scenarios contributed to a crop yield reduction with the greatest impact under scenarios of a 3 °C change, while the net greenhouse effect increased with increasing temperature. Although accompanied by increased crop yields, more precipitation intensified the greenhouse effect, especially in the 30% change in the precipitation scenarios. The results suggested that the application of cold-resistant varieties and a change in the planting time mitigated the decrease in yield caused by decreasing temperature. Resistant varieties minimized the negative impact of decreasing precipitation. N fertilizer reduction effectively mitigated N_2O emissions but had little effect on SOC changes under increasing temperature/precipitation scenarios. Through the consideration of yields, N_2O emissions, and SOC together, straw return could reduce the net greenhouse effect by increasing SOC while maintaining crop yields under warming temperature situations.

Author Contributions: X.Y.: Conceptualization, Formal analysis, Methodology, Investigation, Data Curation; Writing—Original Draft. N.C.: Conceptualization, Formal analysis, Methodology, Investigation, Data Curation; Writing—Original Draft. W.D.: Methodology, Writing—Review & Editing. C.X.: Methodology, Data Curation. J.Z. (Jing Zhang): Resources, Funding acquisition. J.Z. (Jianfeng Zhang): Methodology, Resources. H.L.: Funding acquisition, Data Curation, Supervision. All authors have read and agreed to the published version of the manuscript.

Funding: This research was funded by the earmarked fund for China Agriculture Research System (No. CARS-23-B18).

Institutional Review Board Statement: Not applicable.

Informed Consent Statement: Informed consent was obtained from all subjects involved in the study.

Data Availability Statement: The data presented in this study are available on request from the corresponding author.

Acknowledgments: We would like to thank our Institute of Agricultural Resources and Regional Planning, Chinese Academy of Agricultural Sciences for support in data collection, field visits, and valuable thoughts for the preparation of this manuscript.

Conflicts of Interest: The authors declare no conflict of interest.

Glossary

Abbreviations.	Explanation.
DNDC.	The Denitrification–Decomposition model.
GHG.	Greenhouse gas.
GWP.	Global warming potential.
NCP.	North China Plain.
SOC.	Soil organic carbon.
0.8N.	Reduce N fertilizer by 20%.
1.2N.	Increase N fertilizer by 20%.
SR.	Straw Return.
NT.	No Tillage.
FG.	Fertigation.
CP.	Change the crop planting time.
CI.	Change the timing of irrigation and fertilization.
IR.	Increase irrigation by 30%.
CR.	Cold-resistant varieties.
DR.	Drought-resistant varieties.

References

1. Zhang, D.D.; Lee, H.F.; Wang, C.; Li, B.S.; Pei, Q.; Zhang, J.; An, Y.L. The causality analysis of climate change and large-scale human crisis. *Proc. Natl. Acad. Sci. USA* **2011**, *108*, 17296–17301. [CrossRef] [PubMed]
2. IPCC. *Climate Change 2013: The Pysical Science Basis. Contribution of Working Group I to the Fifth Assessment Report of IPCC the Intergovernmental Panel on Climate Change*; IPCC: Geneva, Switzerland, 2013.
3. IPCC. *Global Warming of 1.5 °C. An IPCC Special Report on the Impacts of Global Warming of 1.5 °C above Pre-industrial Levels and Related Global Greenhouse Gas Emission Pathways*; IPCC: Geneva, Switzerland, 2018.
4. Contractor, S.; Donat, M.G.; Alexander, L.V. Changes in Observed Daily Precipitation over Global Land Areas since 1950. *J. Clim.* **2021**, *34*, 3–19. [CrossRef]
5. Zhao, W.J. Extreme weather and climate events in China under changing climate. *Natl. Sci. Rev.* **2020**, *7*, 938–943. [CrossRef] [PubMed]
6. Donat, M.G.; Lowry, A.L.; Alexander, L.V.; O'gorman, P.A.; Maher, N. More extreme precipitation in the world's dry and wet regions. *Nat. Clim. Change* **2016**, *6*, 508–513. [CrossRef]
7. Zydelis, R.; Weihermueller, L.; Herbst, M. Future climate change will accelerate maize phenological development and increase yield in the Nemoral climate. *Sci. Total Environ.* **2021**, *784*, 147175. [CrossRef]
8. Vargas, V.P.; Soares, J.R.; Oliveira, B.G.; Lourenco, K.S.; Martins, A.A.; Del Grosso, S.J.; Do Carmo, J.B.; Cantarella, H. Sugarcane Straw, Soil Temperature, and Nitrification Inhibitor Impact N_2O Emissions from N Fertilizer. *Bioenergy Res.* **2019**, *12*, 801–812. [CrossRef]
9. Liu, S.X.; Mo, X.G.; Lin, Z.H.; Xu, Y.Q.; Ji, J.J.; Wen, G.; Richey, J. Crop yield responses to climate change in the Huang-Huai-Hai Plain of China. *Agric. Water Manag.* **2010**, *97*, 1195–1209. [CrossRef]
10. Ming, B.; Guo, Y.Q.; Tao, H.B.; Liu, G.Z.; Li, S.K.; Wang, P. SPEIPM-based research on drought impact on maize yield in North China Plain. *J. Integr. Agric.* **2015**, *14*, 660–669. [CrossRef]
11. Streimikiene, D.; Balezentis, T.; Krisciukaitiene, I. Promoting interactions between local climate change mitigation, sustainable energy development, and rural development policies in Lithuania. *Energy Policy* **2012**, *50*, 699–710. [CrossRef]
12. Crippa, M.; Solazzo, E.; Huang, G.L.; Guizzardi, D.; Koffi, E.; Muntean, M.; Schieberle, C.; Friedrich, R.; Janssens-Maenhout, G. High resolution temporal profiles in the Emissions Database for Global Atmospheric Research. *Sci. Data* **2020**, *7*, 121. [CrossRef]
13. Shaffril HA, M.; Krauss, S.E.; Samsuddin, S.F. A systematic review on Asian's farmers' adaptation practices towards climate change. *Sci. Total Environ.* **2018**, *644*, 683–695. [CrossRef] [PubMed]
14. St. Luce, M.; Grant, C.A.; Ziadi, N.; Zebarth, B.J.; O'donovan, J.T.; Blackshaw, R.E.; Harker, K.N.; Johnson, E.N.; Gan, Y.T.; Lafond, G.P.; et al. Preceding crops and nitrogen fertilization influence soil nitrogen cycling in no-till canola and wheat cropping systems. *Field Crops Res.* **2016**, *191*, 20–32.
15. Grassini, P.; Cassman, K.G. High-yield maize with large net energy yield and small global warming intensity. *Proc. Natl. Acad. Sci. USA* **2012**, *109*, 1074–1079. [CrossRef] [PubMed]
16. Abdalla, M.; Osborne, B.; Lanigan, G.; Forristal, D.; Williams, M.; Smith, P.; Jones, M.B. Conservation tillage systems: A review of its consequences for greenhouse gas emissions. *Soil Use Manag.* **2013**, *29*, 199–209. [CrossRef]
17. Li, J.Z.; Luo, Z.K.; Wang, Y.C.; Li, H.; Xing, H.T.; Wang, L.G.; Wang, E.L.; Xu, H.; Gao, C.Y.; Ren, T.Z. Optimizing Nitrogen and Residue Management to Reduce GHG Emissions while Main-taining Crop Yield: A Case Study in a Mono-Cropping System of Northeast China. *Sustainability* **2019**, *11*, 5015. [CrossRef]
18. Waongo, M.; Laux, P.; Kunstmann, H. Adaptation to climate change: The impacts of optimized planting dates on attainable maize yields under rainfed conditions in Burkina Faso. *Agric. For. Meteorol.* **2015**, *205*, 23–39. [CrossRef]
19. Zhao, J.; Yang, X.G.; Dai, S.W.; Lv, S.; Wang, J. Increased utilization of lengthening growing season and warming temperatures by ad-justing sowing dates and cultivar selection for spring maize in Northeast China. *Eur. J. Agron.* **2015**, *67*, 12–19. [CrossRef]
20. Kahil, M.T.; Connor, J.D.; Albiac, J. Efficient water management policies for irrigation adaptation to climate change in Southern Europe. *Ecol. Econ.* **2015**, *120*, 226–233. [CrossRef]
21. Zhang, Y.Y.; Liu, J.F.; Mu, Y.J.; Pei, S.W.; Lun, X.X.; Chai, F.H. Emissions of nitrous oxide, nitrogen oxides and ammonia from a maize field in the North China Plain. *Atmos. Environ.* **2011**, *45*, 2956–2961. [CrossRef]
22. Zhou, M.H.; Zhu, B.; Wang, S.J.; Zhu, X.Y.; Vereecken, H.; Bruggemann, N. Stimulation of N_2O emission by manure application to agricultural soils may largely offset carbon benefits: A global meta-analysis. *Glob. Change Biol.* **2017**, *23*, 4068–4083. [CrossRef]
23. Li, C.S. Modeling trace gas emissions from agricultural ecosystems. *Nutr. Cycl. Agroecosyst.* **2000**, *58*, 259–276. [CrossRef]
24. Tonitto, C.; David, M.B.; Drinkwater, L.E.; Li, C.S. Application of the DNDC model to tile-drained Illinois agroecosystems: Model calibration, validation, and uncertainty analysis. *Nutr. Cycl. Agroecosyst.* **2007**, *78*, 51–63. [CrossRef]
25. Abdalla, M.; Wattenbach, M.; Smith, P.; Ambus, P.; Jones, M.; Williams, M. Application of the DNDC model to predict emissions of N_2O from Irish agri-culture. *Geoderma* **2009**, *151*, 327–337. [CrossRef]
26. Zhu, E.Y.; Deng, J.S.; Wang, H.Q.; Wang, K.; Huang, L.Y.; Zhu, G.J.; Belete, M.; Shahtahmassebi, A. Identify the optimization strategy of nitrogen fertilization level based on trade-off analysis between rice production and greenhouse gas emission. *J. Clean. Prod.* **2019**, *239*, 118060. [CrossRef]
27. Abdalla, M.; Song, X.; Ju, X.; Topp, C.F.E.; Smith, P. Calibration and validation of the DNDC model to estimate nitrous oxide emissions and crop productivity for a summer maize-winter wheat double cropping system in Hebei, China. *Environ. Pollut.* **2020**, *262*, 114199. [CrossRef]

28. Li, H.; Qiu, J.J.; Wang, L.G.; Xu, M.Y.; Liu, Z.Q.; Wang, W. Estimates of N_2O Emissions and Mitigation Potential from a Spring Maize Field Based on DNDC Model. *J. Integr. Agric.* **2012**, *11*, 2067–2078. [CrossRef]
29. Wang, M.X.; Wu, W.L.; Liu, W.N.; Bao, Y.H. Life cycle assessment of the winter wheat-summer maize production system on the North China Plain. *Int. J. Sustain. Dev. World. Ecol.* **2007**, *14*, 400–407. [CrossRef]
30. Cui, J.X.; Yan, P.; Wang, X.L.; Yang, J.; Li, Z.J.; Yang, X.L.; Sui, P.; Chen, Y.Q. Integrated assessment of economic and environmental consequences of shifting cropping system from wheat-maize to monocropped maize in the North China Plain. *J. Clean. Prod.* **2018**, *193*, 524–532. [CrossRef]
31. Zhang, J.; Li, H.; Wang, Y.C.; Deng, J.; Wang, L.G. Multiple-year nitrous oxide emissions from a greenhouse vegetable field in China: Effects of nitrogen management. *Sci. Total Environ.* **2018**, *616*, 1139–1148. [CrossRef]
32. IPCC. *Climate Change 2021: The Physical Science Basis. Contribution of Working Group I to the Sixth Assessment Report of the Intergovernmental Panel on Climate Change*; IPCC: Geneva, Switzerland, 2021.
33. Anderson, M. Climate Change 2014: Synthesis Report. *Libr. J.* **2016**, *141*, 28.
34. Wennman, P.; Katterer, T. Effects of moisture and temperature on carbon and nitrogen mineralization in mine tailings mixed with sewage sludge. *J. Environ. Qual.* **2006**, *35*, 1135–1141. [CrossRef] [PubMed]
35. Lv, F.L.; Song, J.S.; Giltrap, D.; Feng, Y.T.; Yang, X.Y.; Zhang, S.L. Crop yield and N_2O emission affected by long-term organic manure substitution fertilizer under winter wheat-summer maize cropping system. *Sci. Total Environ.* **2020**, *732*, 139321. [CrossRef] [PubMed]
36. Li, H.; Qiu, J.J.; Wang, L.G.; Tang, H.J.; Li, C.S.; Van Ranst, E. Modelling impacts of alternative farming management practices on greenhouse gas emissions from a winter wheat-maize rotation system in China. *Agric. Ecosyst. Environ.* **2010**, *135*, 24–33. [CrossRef]
37. Hu, H.W.; Macdonald, C.A.; Trivedi, P.; Holmes, B.; Bodrossy, L.; He, J.Z.; Singh, B.K. Water addition regulates the metabolic activity of ammonia oxidizers responding to environmental perturbations in dry subhumid ecosystems. *Environ. Microbiol.* **2015**, *17*, 444–461. [CrossRef]
38. Parn, J.; Verhoeven, J.T.; Butterbach-Bahl, K.; Dise, N.B.; Ullah, S.; Aasa, A.; Egorov, S.; Espenberg, M.; Jarveoja, J.; Jauhiainen, J.; et al. Nitrogen-rich organic soils under warm well-drained conditions are global nitrous oxide emission hotspots. *Nat. Commun.* **2018**, *9*, 1135. [CrossRef]
39. Liu, Y.; Li, N.; Zhang, Z.T.; Huang, C.F.; Chen, X.; Wang, F. The central trend in crop yields under climate change in China: A systematic review. *Sci. Total Environ.* **2020**, *704*, 135355. [CrossRef]
40. Zhang, J.; Zhang, Z.; Tao, F.L. Performance of Temperature-Related Weather Index for Agricultural Insurance of Three Main Crops in China. *Int. J. Disaster Risk Sci.* **2017**, *8*, 78–90. [CrossRef]
41. Piao, S.L.; Ciais, P.; Huang, Y.; Shen, Z.H.; Peng, S.S.; Li, J.S.; Zhou, L.P.; Liu, H.Y.; Ma, Y.C.; Ding, Y.H.; et al. The impacts of climate change on water resources and agriculture in China. *Nature* **2010**, *467*, 43–51. [CrossRef]
42. Kravchenko, A.N.; Toosi, E.R.; Guber, A.K.; Ostrom, N.E.; Yu, J.; Azeem, K.; Rivers, M.L.; Robertson, G.P. Hotspots of soil N_2O emission enhanced through water absorption by plant residue. *Nat. Geosci.* **2017**, *10*, 496–500. [CrossRef]
43. Nosalewicz, M.; Stepniewska, Z.; Nosalewicz, A. Effect of soil moisture and temperature on N_2O and CO_2 concentrations in soil irrigated with purified wastewater. *Int. Agrophysics* **2013**, *27*, 299–304. [CrossRef]
44. Butterbach-Bahl, K.; Dannenmann, M. Denitrification and associated soil N_2O emissions due to agricultural activities in a changing climate. *Curr. Opin. Environ. Sustain.* **2011**, *3*, 389–395. [CrossRef]
45. Hursh, A.; Ballantyne, A.; Cooper, L.; Maneta, M.; Kimball, J.; Watts, J. The sensitivity of soil respiration to soil temperature, moisture, and carbon supply at the global scale. *Glob. Change Biol.* **2017**, *23*, 2090–2103. [CrossRef] [PubMed]
46. Smith, J.; Smith, P.; Wattenbach, M.; Zaehle, S.; Hiederer, R.; Jones RJ, A.; Montanarella, L.; Rounsevell MD, A.; Reginster, I.; Ewert, F. Projected changes in mineral soil carbon of European croplands and grasslands, 1990–2080. *Glob. Change Biol.* **2005**, *11*, 2141–2152. [CrossRef] [PubMed]
47. Xu, X.; Zhou, Y.; Ruan, H.H.; Luo, Y.Q.; Wang, J.S. Temperature sensitivity increases with soil organic carbon recalcitrance along an elevational gradient in the Wuyi Mountains, China. *Soil Biol. Biochem.* **2010**, *42*, 1811–1815. [CrossRef]
48. Silver, W.L.; Miya, R.K. Global patterns in root decomposition: Comparisons of climate and litter quality effects. *Oecologia* **2001**, *129*, 407–419. [CrossRef]
49. Chai, L.L.; Hernandez-Ramirez, G.; Dyck, M.; Pauly, D.; Kryzanowski, L.; Middleton, A.; Powers, L.A.; Lohstraeter, G.; Werk, D. Can fertigation reduce nitrous oxide emissions from wheat and canola fields? *Sci. Total Environ.* **2020**, *745*, 141014. [CrossRef]
50. Tian, D.; Zhang, Y.Y.; Zhou, Y.Z.; Mu, Y.J.; Liu, J.F.; Zhang, C.L.; Liu, P.F. Effect of nitrification inhibitors on mitigating N_2O and NO emissions from an agri-cultural field under drip fertigation in the North China Plain. *Sci. Total Environ.* **2017**, *598*, 87–96. [CrossRef]
51. Shan, J.; Yan, X.Y. Effects of crop residue returning on nitrous oxide emissions in agricultural soils. *Atmos. Environ.* **2013**, *71*, 170–175. [CrossRef]
52. Li, S.; Li, Y.B.; Li, X.S.; Tian, X.H.; Zhao, A.Q.; Wang, S.J.; Wang, S.X.; Shi, J.L. Effect of straw management on carbon sequestration and grain production in a maize-wheat cropping system in Anthrosol of the Guanzhong Plain. *Soil Tillage Res.* **2016**, *157*, 43–51. [CrossRef]
53. Raffa, D.W.; Bogdanski, A.; Tittonell, P. How does crop residue removal affect soil organic carbon and yield? A hierarchical analysis of management and environmental factors. *Biomass Bioenergy* **2015**, *81*, 345–355. [CrossRef]

54. Lehtinen, T.; Schlatter, N.; Baumgarten, A.; Bechini, L.; Kruger, J.; Grignani, C.; Zavattaro, L.; Costamagna, C.; Spiegel, H. Effect of crop residue incorporation on soil organic carbon and greenhouse gas emissions in European agricultural soils. *Soil Use Manag.* **2014**, *30*, 524–538. [CrossRef]
55. Liao, Y.; Wu, W.L.; Meng, F.Q.; Smith, P.; Lal, R. Increase in soil organic carbon by agricultural intensification in northern China. *Biogeosciences* **2015**, *12*, 1403–1413. [CrossRef]
56. Lenka, N.K.; Lal, R. Soil aggregation and greenhouse gas flux after 15 years of wheat straw and fertilizer management in a no-till system. *Soil Tillage Res.* **2013**, *126*, 78–89. [CrossRef]
57. Six, J.; Elliott, E.T.; Paustian, K. Soil macroaggregate turnover and microaggregate formation: A mechanism for C sequestration under no-tillage agriculture. *Soil Biol. Biochem.* **2000**, *32*, 2099–2103. [CrossRef]
58. Zhang, J.; Yao, Y.; Lv, J.; Wang, Y.; Jin, K.; Wang, C.; Hu, B.; Li, J.; Ding, Z. Effects of Different Tillages on Soil Organic Carbon, Soil Microbial Biomass Carbon and Water Use Efficiency in the Slopping Field of Arid Areas. *J. Soil Sci.* **2007**, *38*, 882–886.
59. Zhang, M.Y.; Wang, F.J.; Chen, F.; Malemela, M.P.; Zhang, H.L. Comparison of three tillage systems in the wheat-maize system on carbon sequestration in the North China Plain. *J. Clean. Prod.* **2013**, *54*, 101–107. [CrossRef]

agriculture

MDPI

Article

Development of Data-Driven Models to Predict Biogas Production from Spent Mushroom Compost

Reza Salehi [1], Qiuyan Yuan [2] and Sumate Chaiprapat [1,3,*]

[1] Department of Civil and Environmental Engineering, Prince of Songkla University, Hat Yai 90110, Thailand; reza.salehi@polymtl.ca
[2] Department of Civil Engineering, University of Manitoba, Winnipeg, MB R3T 5V6, Canada; qiuyan.yuan@umanitoba.ca
[3] PSU Energy Systems Research Institute, Prince of Songkla University, Hat Yai 90110, Thailand
* Correspondence: sumate.ch@psu.ac.th; Tel.: +1-438-889-6591

Abstract: In this study, two types of data-driven models were proposed to predict biogas production from anaerobic digestion of spent mushroom compost supplemented with wheat straw as a nutrient source. First, a *k*-nearest neighbours (*k*-NN) model (k = 1–10) was constructed. The optimal k value was determined using the cross-validation (CV) method. Second, a support vector machine (SVM) model was developed. The linear, quadratic, cubic, and Gaussian models were examined as kernel functions. The kernel scale was set to 6.93, while the box constraint (C) was optimized using the CV method. Results demonstrated that R^2 for the *k*-NN model (k = 2) was 0.9830 at 35 °C and 0.9957 at 55 °C. The Gaussian-based SVM model (C = 1200) provided an R^2 of 0.9973 at 35 °C and 0.9989 at 55 °C, which are slightly better than those achieved by *k*-NN. The Gaussian-based SVM model produced *RMSE* of 0.598 at 35 °C and 0.4183 at 55 °C, which are 58.4% and 49.5% smaller, respectively, than those produced by the *k*-NN. These findings imply that SVM modeling can be considered a robust technique in predicting biogas production from AD processes as they can be implemented without requiring prior knowledge of biogas production kinetics.

Keywords: anaerobic digestion; biogas production; *k*-nearest neighbours; support vector machine

Citation: Salehi, R.; Yuan, Q.; Chaiprapat, S. Development of Data-Driven Models to Predict Biogas Production from Spent Mushroom Compost. *Agriculture* **2022**, *12*, 1090. https://doi.org/10.3390/agriculture12081090

Academic Editors: Dengpan Xiao and Wenjiao Shi

Received: 21 June 2022
Accepted: 20 July 2022
Published: 24 July 2022

1. Introduction

The electrical and thermal energy production processes that use non-renewable resources (i.e., fossil fuels; oil, and coal) are becoming less attractive globally. Even though such resources are rich in energy and relatively inexpensive to process, they are limited in supply and will soon be depleted. In addition, the utilization of fossil fuels emits additional greenhouse gases into the atmosphere, which has instigated climate change [1]. Hence, a large number of research bodies have aligned to overcome such an increasing universal concern. One of the most promising and attractive alternative solutions is the use of biogas derived from wastes or renewable feedstock [2,3].

Biogas, a mixture consisting chiefly of methane (CH_4) and carbon dioxide (CO_2), is the end-product of anaerobic digestion of organic matters (e.g., agricultural residues, livestock manure, food waste, sewage sludge, etc.) [4–8]. Anaerobic digestion is a complex multi-step process that is carried out by a consortium of different microbial species known as anaerobes. Uniquely, they do not need molecular oxygen for their metabolism and growth [9]. The key steps of the anaerobic digestion process, together with the possible applications of biogas, and its adverse environmental impacts are outlined in Figure 1.

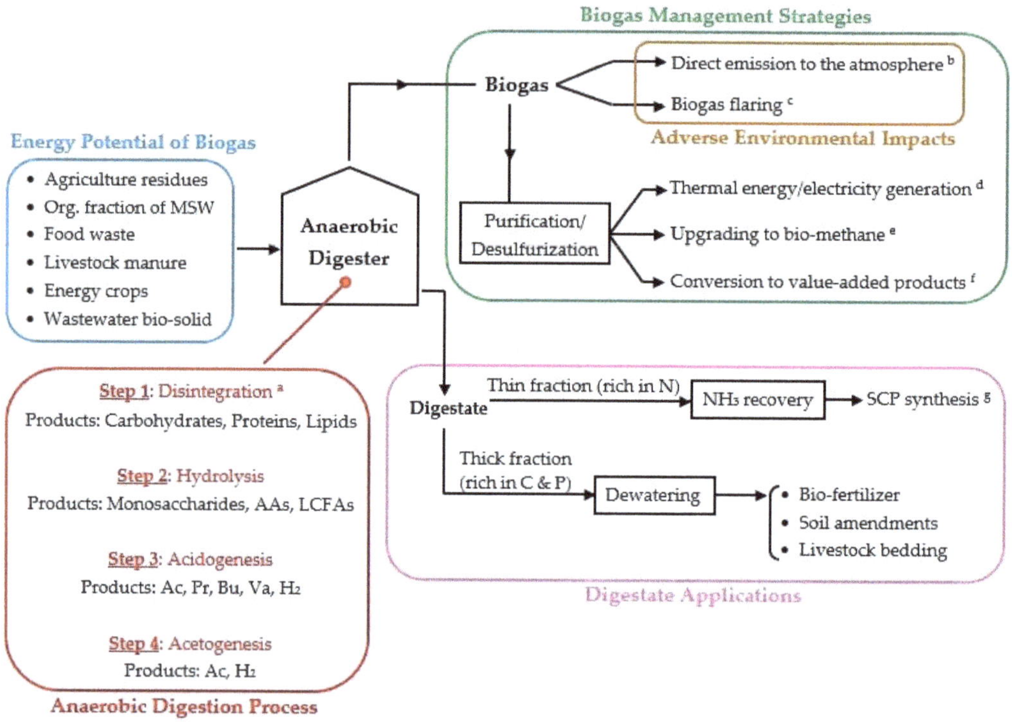

Figure 1. A schematic flowchart showing the simple representation of biogas generation during anaerobic digestion process, along with its applications and environmental impacts [10–14]. Notes: [a] A non-biological process in which the cell walls are physically or chemically broken down to release intracellular substrate. [b] Biogas release to the atmosphere should be avoided because CO_2 and CH_4, the main constituents of biogas, are contributors to global warming. [c] Biogas combustion should be avoided because it is associated with the release of pollutants (e.g., CO, SO_2, and NO_X) to the atmosphere; SO_2, and NO_X can react with moisture in the atmosphere to form sulfuric/nitric acid resulting in acid rain. [d] Solar energy and shale gas, due to being plentiful and cheap, can drive out the biogas application in electricity generation in the near future. [e] bio-methane can be used as a vehicular fuel or can be injected into the natural gas network. [f] bio-methanol, and syngas (a mixture of CO and H_2) that can be generated via reforming technology. [g] Biogas can be converted to SCP by the action of methanotrophic bacteria alone, or in combination with autotrophic hydrogen oxidizing bacteria or algae; SCP has excellent potential as an animal feed supplement. Abbreviations: AAs: amino acids; Ac: acetate; Bu: butyrate; LCFAs: long chain fatty acids; MSW: municipal solid waste; Pr: propionate; SCP: single-cell protein; Va: valerate.

The increasing global interest in biogas power plant establishment via anaerobic digestion of various organic matters has resulted in attempts to develop numerous mathematical models to predict and suggest optimal operations. Hill [15] developed a model to describe the digestion of animal wastes, assuming that the main five bacterial groups involved in the overall digestion process (acidogenic bacteria, hydrogenotrophic bacteria, homoacetogenic bacteria, acetoclastic bacteria, and H_2 utilizing methane bacteria) are inhibited by a high concentration of fatty acids (FAs). Mosey [16] proposed a model consisting of four reactions (one acidogenic reaction, one acetogenic reaction, and two methanogenic reactions), which also takes into account the role of H_2. According to this model, in case of a sudden rise in the organic loading rate, an accumulation of volatile fatty acids (VFAs) is likely to occur; this results in a decrease in pH that inhibits H_2 utilizing methanogenic bacteria.

In other words, H_2 partial pressure is increased, which leads to further accumulation of propionic/butyric acid (CH_4 generation is stopped when pH drops below 5.5). Based on Mosey's model, Pullammanappallil et al. [17] introduced a model taking into account the gas phase, and acetoclastic inhibition by undissociated FAs. Angelidaki et al. [18] presented a model considering hydrolysis, acidogenesis, acetogenesis, and methanogenesis, which is suitable to describe the behavior of anaerobic digesters fed with manures. This model was developed by incorporating some assumptions as follows: (i) methanogenesis is inhibited by free NH_3, (ii) acetogenesis is inhibited by acetic acid, (iii) acidogenesis is inhibited by total VFAs, and (iv) the degree of NH_3 ionization, the maximum specific growth rate of bacteria are pH and temperature dependent.

In all the above-mentioned models, organic material was taken into account as a whole; in other words, they are incapable of dealing with complex feed composition. In this regard, the International Water Association (IWA) task group for mathematical modeling of the anaerobic digestion process developed a model known as Anaerobic Digester Model No 1 (more often abbreviated as ADM1), that takes the complex organic substrates into account [19].

Although the kinetic-based mathematical models for describing the anaerobic digestion process can help engineers and asset managers to better plan the management of the biogas plants, it is often criticized that most of them are inherently too complex due to a large number of stoichiometric coefficients and parameters reflecting the kinetic properties of the enzymes and microorganisms that govern the physicochemical and biochemical reactions through anaerobic digestion processes [20]. In addition, these models typically involve physicochemical equilibrium expressions and differential mass balance equations for components in the liquid phase (substrates for acidogenic/acetogenic/methanogenic organisms and their corresponding microbial masses) and in the gas phase (e.g., CH_4 and CO_2). Hence, these models are often complicated to solve, and many simplifying assumptions must be made to reduce their complexity. However, incorporating simplifying assumptions into the models may not hold in practice. Fedailaine et al. [21] modeled the biokinetics of the anaerobic digestion process involving eight simplifying assumptions, which inevitably limited the application of this model to full-scale anaerobic digesters. In addition, applying assumptions to the models lowers the precision of the models; in other words, an under- or over-estimation of the response of the models will likely occur. For these reasons, developing a simple yet highly predictive model to estimate biogas production from the anaerobic digestion process is highly desired. As such, a different branch of models, called artificial intelligence (AI)-based models (more often known as easy-to-use black-box models) may be recruited. These models have advantages over complex mathematical models because they are constructed on a measured dataset (i.e., input–output data pairs for a given system) without requiring complicated kinetic relationships between the input variables and the corresponding outputs [22,23]. In addition, the AI modeling approach is proven as a robust tool with high generalization power. Holubar et al. [24] used an artificial neural network (ANN) to model an anaerobic digester fed with a mixture of primary (raw) sludge and surplus activated sludge originating from a local municipal wastewater treatment plant. The results showed that ANN is a suitable tool for modeling such a process. Cakmakci [25] applied an adaptive neuro-fuzzy inference system (ANFIS) to predict methane yield in an anaerobic digester fed with pre-thickened raw sludge. According to the findings, there was good agreement between the measured and predicted values. Kusiak and Wei [26] developed several predictive models through data mining algorithms to predict methane production from the anaerobic digesters in the Des Moines Wastewater Reclamation Facility. The results showed that the model built by the ANFIS algorithm offered excellent predictive accuracy with a coefficient of determination (R^2) of 0.99, and a percentage error of 0.08. Nair et al. [27] used ANN to evaluate the effects of the types of substrates (such as food/vegetable waste and yard trimming), and organic loading rate on CH_4 production. The training and validation R^2 values were greater than 0.88, indicating that the model's learning and generalization power were satisfactory. Dach et al. [28] reported that ANN can be considered an appropriate tool to estimate CH_4 from anaerobic digestion of slurry from animal waste and agricultural residues. Tan et al. [29] compared the performance of ANFIS and the ADM1 to predict biogas production from the anaerobic digestion of palm oil mill effluent under thermophilic conditions. The authors reported that ANFIS yielded higher predictive accuracy compared with the results obtained using the ADM1. In another study conducted by Beltramo et al. [30], an ANN model was constructed to predict the biogas production rate from a mesophilic anaerobic digester fed with a mixture of maize, grass silages, and pig/cattle manure. The authors conclude that the ANN modeling approach can be considered a promising alternative to ADM1.

This study aimed to develop, validate, and test two different predictive models based on the AI modeling approach, including *k*-nearest neighbors and support vector machine (referred to hereafter as *k*-NN and SVM, respectively) to predict biogas production from anaerobic digestion of spent mushroom compost (SMC). The independent variables involved include temperature, carbon-to-nitrogen ratio (C/N), and retention time (RT). SMC is a bulky residue from mushroom farms, and the waste generated by the mushroom processing industry. It is an ideal source of general nutrients (e.g., nitrogen and phosphorus) and is rich in organic matter that can be used for producing biogas. It is worth mentioning that the nutritional value and the content of organic matter of SMC depend on the types of cultivated mushroom species.

The predictive performance of these models was separately investigated and eventually compared with each other and with the ANN, ANFIS, and logistic models developed by Najafi and Faizollahzadeh Ardabili [31] by means of two statistical indices, including R^2, and root mean squared error (*RMSE*). To the best of the authors' knowledge, the application of *k*-NN and SVM modeling approaches to predict biogas production from *SMC* has never been exploited.

2. Materials and Methods

A schematic portrait depicting the workflow of this study is shown in Figure 2; see text for further details.

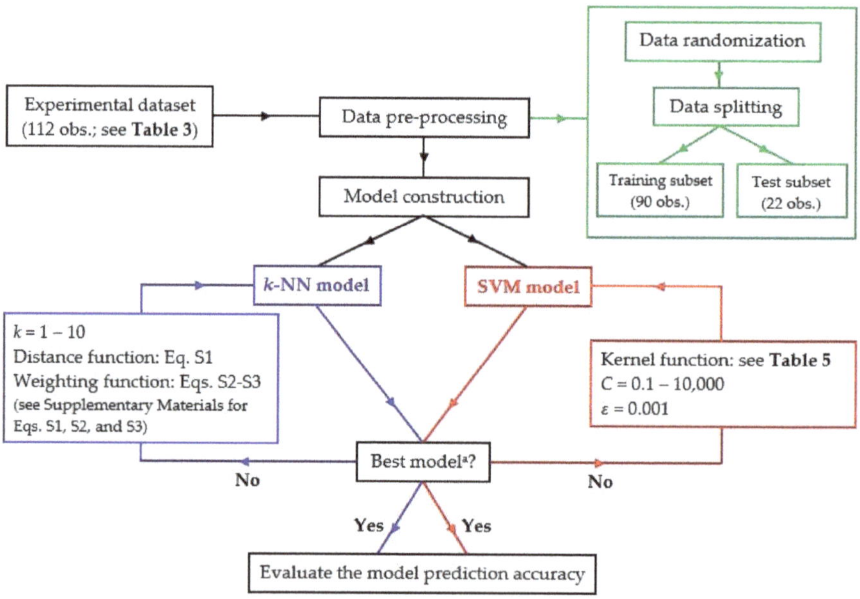

Figure 2. A schematic illustration of the workflow of this study ([a] the one that provides the least validation error); see text for further details.

2.1. Dataset

The experimental data were taken from the study of Najafi and Faizollahzadeh Ardabili [31]. Briefly, four 2.5 L batch mode anaerobic digesters, each with an effective volume of 1.5 L, were fed with a mixture of SMC and wheat straw (WS) to induce different C/N ratios of 12.2, 20, 30, and 40. The characteristics of SMC and WS are provided in Table 1.

Table 1. Characteristics of SMC and WS [31].

Variable	SMC	WS	Unit
TS	19.1	86.7	% g TS/g SMC (or g WS)
VS/TS	64.2	81.7	% g VS/g TS
Nitrogen	2.4	0.78	% g N/g SMC (or g WS)
Organic carbon	29	63	% g C/g SMC (or g WS)

SMC: spent mushroom compost; WS: wheat straw; TS: total solids; VS: volatile solids.

The authors considered the initial TS content of the substrate in the anaerobic digesters as a constant value (8%), and referring to the values of nitrogen and organic carbon for the SMC and WS (Table 1), the contents of SMC and WS in terms of g TS and g VS as a function of C/N ratio were computed as shown in Table 2.

Table 2. The content of substrate (SMC and WS) fed to the anaerobic digesters as a function of the C/N ratio examined.

C/N	SMC			WS		
	Total Mass of SMC (g)	TS (g)	VS (g)	Total Mass of WS (g)	TS (g)	VS (g)
12.2	613.11	117.10	75.18	3.21	2.78	2.27
20	222.90	42.57	27.33	89.35	77.47	63.29
30	105.95	20.24	12.99	115.05	99.75	81.50
40	59.45	11.35	7.29	125.26	108.60	88.73

SMC: spent mushroom compost; WS: wheat straw; TS: total solids; VS: volatile solids; C/N: carbon-to-nitrogen.

Each anaerobic digester was inoculated with a 10 g bovine rumen solution with a concentration of 1000 g bovine rumen per liter; the bovine rumen solution was kept at a temperature of 37 °C for five days to assist bacteria in growing more rapidly. The anaerobic digesters then were placed in hot water baths at mesophilic temperature (35 °C) and thermophilic temperature (55 °C). The biogas produced from the reactors was measured by a water displacement method for two weeks. All the tests were conducted with three replications. The produced biogas from the reactors was measured by a water displacement method for two weeks. Table 3 shows the experimental data used in this study.

Data Pre-Processing

The dataset shown in Table 3 was used to develop different predictive models compared with those presented by Najafi and Faizollahzadeh Ardabili [31]. As seen in Table 3, the dataset consists of a total number of 112 input–output data pairs (referred to hereafter as observations); the j-th observation contains a collection of 4 data points as $\{x_1^j, x_2^j, x_3^j, y^j\}$ for j = 1 to 112, where x_1, x_2, and x_3 stand for temperature, C/N ratio, and RT, respectively, while y stands for the cumulative biogas production.

Prior to utilizing the dataset to develop a predictive model, it was randomized using Excel (version 2016, Microsoft Corp., Redmond, WA, USA), and then split into two disjoint subsets, including training and testing ones. Ninety observations corresponding to 80% of the dataset were assigned to the training subset, while the remaining 20% of the dataset (i.e., 22 observations) were used as the testing subset. The training subset allowed to adjust the model parameters in order to minimize the error between the experimental data and the model predictions. Meanwhile, the testing subset was employed to evaluate the accuracy of the trained (developed) model for predicting the output. The training and testing subsets were stored in the workspace of MATLAB® (trial version, R2020a) (MathWorks Inc., Natick, MA, USA) in the form of arrays.

Table 3. Biogas production in the experimental-anaerobic digester runs [31].

T = 35 °C								T = 55 °C							
Exp. Code	**C/N**	**RT**	**CBP**	**Exp. Code**	**C/N**	**RT**	**CBP**	**Exp. Code**	**C/N**	**RT**	**CBP**	**Exp. Code**	**C/N**	**RT**	**CBP**
E1	12	1	3.77	E3	30	1	3.20	E5	12	1	3.42	E7	30	1	2.50
		2	6.87			2	5.87			2	5.51			2	4.58
		3	9.69			3	8.98			3	7.99			3	6.56
		4	12.27			4	12.45			4	11.10			4	8.83
		5	15.11			5	15.59			5	16.08			5	13.18
		6	18.92			6	20.66			6	19.52			6	18.02
		7	22.36			7	25.96			7	21.95			7	24.08
		8	25.11			8	30.07			8	24.42			8	29.52
		9	26.88			9	32.53			9	27.03			9	33.37
		10	28.15			10	34.16			10	28.56			10	36.19
		11	29.13			11	34.74			11	29.20			11	38.04
		12	29.78			12	35.24			12	29.97			12	39.99
		13	30.27			13	35.60			13	31.04			13	42.00
		14	30.52			14	36.11			14	32.26			14	43.75
E2	20	1	3.60	E4	40	1	2.75	E6	20	1	3	E8	40	1	2.14
		2	6.64			2	5.15			2	4.82			2	4.06
		3	9.30			3	7.44			3	6.98			3	5.74
		4	12.02			4	9.88			4	9.70			4	7.23
		5	16.35			5	13.52			5	13.90			5	9.87
		6	21.66			6	17.86			6	19.54			6	13.07
		7	27.45			7	22.91			7	23.89			7	17.06
		8	32.06			8	26.49			8	26.72			8	22.07
		9	35.07			9	28.05			9	30.01			9	26.84
		10	37.15			10	28.84			10	33.60			10	30.41
		11	38.51			11	29.18			11	35.99			11	32.49
		12	39.53			12	29.58			12	37.10			12	34.10
		13	40.15			13	30.20			13	38.31			13	35.90
		14	40.62			14	30.92			14	39.83			14	37.77

C/N: carbon-to-nitrogen ratio; RT: retention time (d); CBP: cumulative biogas production (mL g VS^{-1}); VS: volatile solids; anaerobic digester's volume = 2.5 L; operation mode: batch; feedstock: a mixture of spent mushroom and wheat straw.

2.2. *Modeling Approaches*

2.2.1. *k*-NN

The *k*-NN approach was initially proposed by Fix and Hodges [32] and was later expanded by Cover and Hart [33]. It is recognized as one of the top 10 influential data mining algorithms in machine learning research due to its simplicity in implementation and efficacy in terms of prediction performance [34]. The *k*-NN algorithm was initially developed with successful application in solving problems with pattern classification, and it was later utilized as a valuable tool for regression purposes. In other words, the *k*-NN algorithm can be used to predict either class labels or continuous variables. Over the past few decades, *k*-NN algorithm has attracted impressive attention and is applied in the fields of engineering, science, business, medicine, etc. When using the *k*-NN algorithm, the main challenges are associated with the determination of the number of neighbors (*k*), the distance function, and the weighting function [35]. A brief description of the determination of these hyperparameters is provided in the Supplementary Materials (Section A) [36–38].

In order to demonstrate how *k*-NN algorithm is used in developing regression models, let us suppose that Figure 3 shows a number of observations (input–output data pairs) indicated as black square points for a particular system (*X* stands for the number of observations, and Y stands for its corresponding output). Let the blue square be the query observation whose output is unknown, and suppose that the k-NN algorithm uses five nearest neighbors. The black/red solid lines connecting the query data point with other data points represent the distances, which can be computed based on a distance function specified by the user (e.g., Equation (S1)). The output of the query data point can be estimated by applying a weighting function (e.g., Equation (S4)) considering the distances between the query data point and the five nearest neighbors. The computational procedure of the k-NN algorithm is depicted in Figure 4. It consists of three steps as follows: Step 1 computes the distances between each observation in the testing subset (called query observations) and every observation in the training subset. Step 2 sorts the distances measured from the smallest to the largest, while in Step 3, an appropriate value is assigned to k. Once a weighting function is used, the target output is determined.

In this study, a k-NN model was developed based on the experimental data (shown in Table 3) using a script written in a MATLAB environment. The Euclidean distance function (Equation (S1)) was used to determine the distances between each query observation (output from the testing subset) and all observations in the training subset. Once all the distances were computed, the k neighbors (k varied from 1 to 10) with the minimum distances from the query observation were assigned a weight (Equations (S2) and (S3)). Thereafter, the output of the query observation was computed in accordance with Equation (S4).

A five-fold cross-validation (CV) approach was performed in order to obtain an optimal value for k. A brief description of an example of a q-fold CV is provided in the Supplementary Materials (Section B) [39].

After determining the optimal k value, the trained model was used to make predictions using the testing dataset, which was unseen throughout the training process.

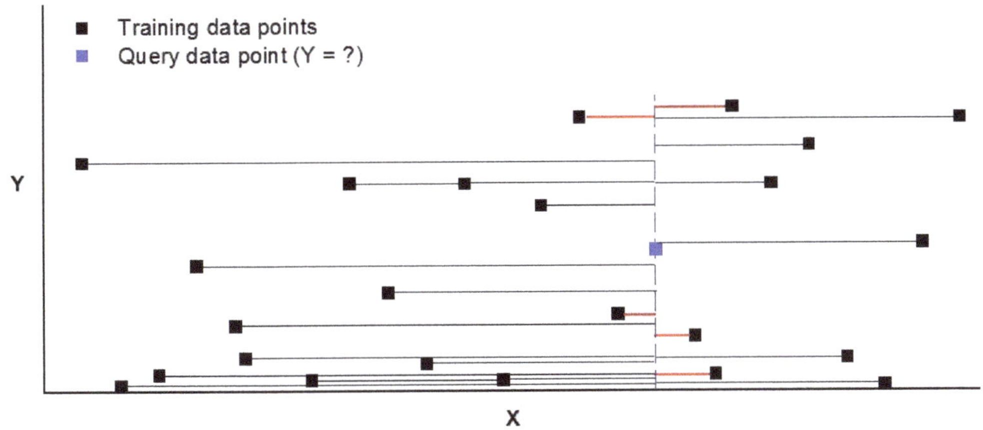

Figure 3. A basic illustration of how k-NN algorithm is used in developing regression models. Notes: The solid lines connecting the query data point with other data points represent the distances, which are computed using a distance function specified by the user. The distances from the five nearest neighbors (k assumed to be 5), shown as red solid lines, are considered herein to calculate the output of the query data point using a weighting function specified by the user (see Figure 4 for the detailed computational procedure).

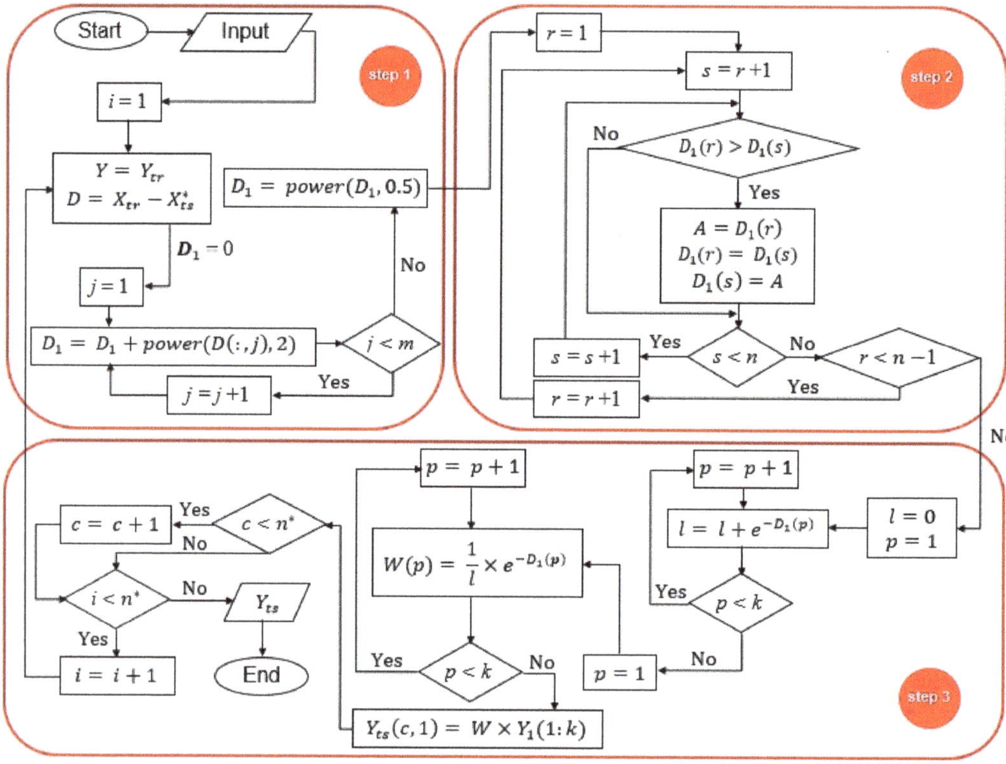

Figure 4. A graphical representation of the computational steps of *k*-NN algorithm for solving regression problems. Notes: The inputs to the algorithm are X_{tr}, Y_{tr}, X_{ts}, Y_{ts} (a column vector with n^* elements where all the elements are initially set to zero), and *k*; Y_{ts} is the target output. The reader is referred to Table 4 for the description of the symbols used in this figure.

Table 4. Description of the symbols used in Figure 4.

Symbol	Description
X	Input matrix, in which each row represents an observation that consists of the values of the input variables
$X_{ts}{}^*$	Repmat (X_{ts} (i,:), n, 1) [a]
n	Size (X_{tr},1) [b]
n^*	Size (X_{ts},1) [c]
m	Size (X_{tr}, 2) [d]
Y	A column vector whose i-th element is the output of the i-th observation
D	Distance measure (see Equation (S1))
W	Weight measure (see Equation (S3))
k	Number of the nearest neighbors (specified by the user)
i, j, r, s, p, and c	Loop control variables
D_1, l	Accumulator variable

[a] A function found in MATLAB® (trial version, R2020a) (MathWorks Inc., Natick, MA, USA) that produces a matrix consisting of n rows, each is a copy of the i-th row of matrix X_{ts}; [b] A function that returns the number of rows in matrix X_{tr}; [c] A function that returns the number of rows in matrix X_{ts}; [d] A function that returns the number of columns in matrix X_{tr}; Subscripts "*tr*", and "*ts*" stand for "training" and "testing", respectively.

The k-NN model performance was assessed by two commonly used statistical indices: R^2 and $RMSE$. R^2 represents the goodness-of-fit between the measured (actual) values and their corresponding predicted values, which is defined by Equation (1).

$$R^2 = 1 - \frac{\sum_{i=1}^{n} Y_i - Y_{pred.,i}}{\sum_{i=1}^{n} Y_i - Y_{avg.}} \tag{1}$$

$RMSE$, a measure of the average magnitude of the error, is calculated in accordance with Equation (2):

$$RMSE = \sqrt{\frac{1}{n}\sum_{i=1}^{n} (Y_i - Y_{pred.,i})^2} \tag{2}$$

where Y_i is the actual value of the output, and $Y_{pred.,I}$ is the corresponding model prediction for the i-th observation; $Y_{avg.}$ is the average value of Y_i (i = 1, 2, ... , n); and n is the total number of observations (in the training or testing subset), on which the R^2 and $RMSE$ are estimated.

It is evident from Equations (1) and (2), that the values of R^2 closer to one and $RMSE$ closer to zero demonstrate a smaller value of $(Y_i - Y_{pred.,I})$. In other words, the model perfectly fits the data when R^2 = 1 and $RMSE$ = 0.

2.2.2. SVM

SVM, a supervised learning technique within the field of computational intelligence, was originally developed at AT&T Bell Laboratories (Holmdel, NJ, USA) by Vapnik [40]. It can be used to solve data classification tasks, which is beyond the scope of this paper and can be extended to solve regression problems, which is the focus of this paper.

Suppose a certain problem is represented by a dataset $\{(x_i, y_i)\}_{i=1}^{n}$ where $x_i \in R^d$ is a vector of d input features, $y_i \in R$ is the corresponding scalar output value, and n is the total number of data patterns. The goal of SVM is to find a regression function $f(x)$ that estimates the output value whose deviation from the target (actual) value y_i, for all x_i, is at most epsilon (ε). In other words, an error larger than ε is not tolerated. In addition, $f(x)$ should be as flat as possible.

For simplicity, let us first consider the case of a linear SVM regression, which can be expressed in the following form:

$$f(x) = \langle w, x \rangle + b \tag{3}$$

where $w \in R^d$ is the weight vector, $b \in R$ is the so-called bias term, and $\langle w, x \rangle$ denotes the dot product between the weight vector w and vector x that is defined as:

$$\langle w, x \rangle = \sum_{j=1}^{d} w_j x_j \tag{4}$$

In order to ensure that $f(x)$ is as flat as possible, the Euclidean norm of w, i.e., $\|w\|$, should be minimized. This can be represented as a convex optimization problem to minimize:

$$J(w) = \frac{1}{2}\|w\|^2 \tag{5}$$

$$\text{subject to} \begin{cases} \forall i : y_i - \langle w, x_i \rangle + b \leq \varepsilon \\ \forall i : \langle w, x_i \rangle + b - y_i \leq \varepsilon \end{cases}$$

However, it is necessary to point out that such a function $f(x)$ that satisfies these constraints may not exist. Therefore, the slack variables ξ_i and $\xi_i^* \in R$ are required to be introduced. Including the slack variables, Equation (5) can be written as follows (also called the primal objective function):

$$J(w) = \frac{1}{2}\|w\|^2 + C\sum_{i=1}^{n}(\xi_i + \xi_i^*) \quad (6)$$

$$\text{subject to} \begin{cases} \forall i\ :\ y_i - \langle w, x_i \rangle - b \le \varepsilon + \xi_i \\ \forall i\ :\ \langle w, x_i \rangle + b - y_i \le \varepsilon + \xi_i^* \\ \xi_i, \xi_i^* \ge 0 \\ C > 0 \end{cases}$$

where parameter C is a user-defined constant, known as box constraint, which determines the trade-off between the flatness of $f(x)$ and the amount up to which deviations greater than ε are acceptable.

To solve Equation (3), it is possible to use the Lagrangian function and optimal constraints, to obtain a linear SVM regression [41] (see Section C in Supplementary Materials for the detailed computational procedure). In the case of a non-linear relationship between the input variables and the output, the SVM model can be simply constructed by mapping the inputs into a high-dimensional feature space, F:

$$\varphi\ :\ R^d \to F \quad (7)$$

Thus, Equation (S12) can be formulated in the following form (so-called non-linear SVM regression):

$$f(x) = \sum_{i=1}^{n}(\alpha_i - \alpha_i^*)K(x_i,\ x) + b \quad (8)$$

where the term $K(x_i,\ x)$ is defined as the kernel function:

$$K(x_i,\ x) = \langle \varphi(x_i), \varphi(x) \rangle \quad (9)$$

where $\langle \varphi(x_i), \varphi(x) \rangle$ is the dot product of the input vectors in the high-dimensional feature space, $\varphi(x_i)$ and $\varphi(x)$.

In order to develop the SVM model, the Regression Learner App in the framework of MATLAB® (trial version, R2020a) (MathWorks Inc., Natick, MA, USA) was used. On the Apps tab, in the Machine Learning and Deep Learning group, the Regression Learner was selected. The training and testing datasets were loaded from the MATLAB workspace, and then a 5-fold CV was chosen as a validation scheme to protect against overfitting.

The key to the establishment of an SVM model is to specify an appropriate kernel function. In addition, the hyperparameters, i.e., kernel scale (γ), C, and ε greatly affect the performance of the model, which are typically determined by trial-and-error method. For the system under consideration in this study, four types of kernel functions, including linear, quadratic, cubic, and fine/medium/coarse Gaussian were tested (see Table 5 for the mathematical definition of these kernel functions).

Table 5. Mathematical definition of the SVM kernel functions and their kernel scales used in this study.

Type of SVM Regression	Kernel Function	Kernel Scale (γ)
Linear	$k(x, x_i) = \gamma \times \langle x, x_i \rangle$	1
Quadratic	$k(x, x_i) = (\gamma \times \langle x, x_i \rangle + 1)^2$	1
Cubic	$k(x, x_i) = (\gamma \times \langle x, x_i \rangle + 1)^3$	1
Fine Gaussian	$k(x, x_i) = \exp\left(-\|x - x_i\|^2 / (\gamma)^2\right)$	$0.25N^{0.5}$
Medium Gaussian		$N^{0.5}$
Coarse Gaussian		$4N^{0.5}$

$\langle x, x_i \rangle$ denotes the dot product between the vectors x and x_i; $\|x - x_i\|$ denotes the Euclidean distance between the two feature vectors x and x_i; the values assigned to γ are the MATLAB default values; N is the number of predictor variables (N = 3 for the system under consideration).

The value of ε was set to 0.001 (the smallest acceptable value in MATLAB R2020a), while the value of parameter C was varied in the range of 0.1 to 10,000 (total number of data points = 23) in order to pick the best model with the least validation error (the smaller the validation error, the better the model generalization ability). Each SVM model was trained with the training subset using the SMO algorithm, considering that the model validation error was estimated by means of a 5-fold CV method (the default validation scheme in MATLAB R2020a). The SMO algorithm stopped iterating when the feasibility gap (see Equation (10)) was less than the pre-specified gap tolerance (the gap tolerance was set to 0.001).

$$\text{Feasibility gap } (\Delta) = \frac{J(w) + L(\alpha, \alpha^*)}{J(w) + 1} \tag{10}$$

where $J(w)$ and $L(\alpha, \alpha^*)$ denote the primal objective (Equation (6)) and the dual objective (Equation (S10)), respectively.

Once the algorithm met the convergence criterion, in other words, the model training process was complete, the trained model was fed as input to make a prediction using the testing dataset. The SVM model performance was assessed by means of the two aforementioned statistical indices (R^2 and *RMSE*; see Equations (1) and (2)).

3. Results and Discussion

3.1. Evaluation of k-NN Model

The optimal k value of the k-NN model was obtained with the aid of a 5-fold CV approach. The optimal k value was defined as the value that allows the k-NN model to produce the smallest *RMSE* (and the highest R^2) on the validation folds in runs 1–5. Figure 5 displays R^2 and *RMSE* values of the validation fold, as a function of the k value varying from 1 to 10, for the k-NN models 1 and 2; k-NN model 1 uses Equation (S2) as the weighting function, whereas k-NN model 2 uses Equation (S3) as the weighting function. It can be seen from Figure 5 that the optimal k value for both k-NN models 1 and 2 was found to be 2; however, model 2 performed better with validation R^2 and *RMSE* of 0.964 and 1.957, respectively, compared with the R^2 value of 0.925 and *RMSE* value 2.969 obtained using model 1. Figure 6 shows the prediction accuracy of k-NN model 2 (k = 2) against the whole dataset under mesophilic condition (35 °C) and thermophilic condition (55 °C) as a scatter plot of the measured and the model-predicted values. As seen in Figure 6, the data points on the plot are well-dispersed around the 45° line (called 100% correlation line or line 1:1) with R^2 and *RMSE* values equal to 0.983 and 1.487, respectively, in the case of mesophilic temperature, and 0.996 and 0.829 in the case of thermophilic temperature, respectively. This implies that only 0.4–1.7% of the total variability in the response cannot be explained by the developed k-NN model 2.

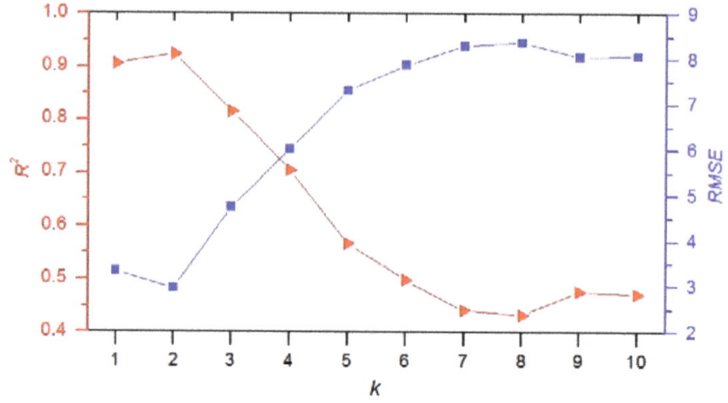

(A)

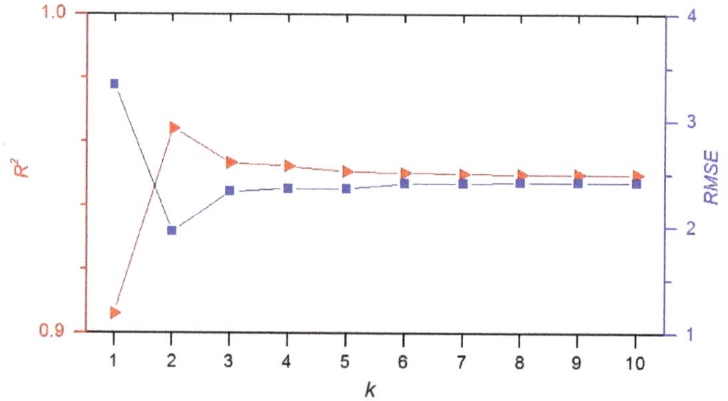

(B)

Figure 5. Validation curves for (**A**) *k*-NN model 1, and (**B**) *k*-NN model 2. *k*-NN models 1 and 2 use Equations (S2) and (S3), respectively, as a weighting function (R^2: coefficient of determination; *RMSE*: root mean squared error).

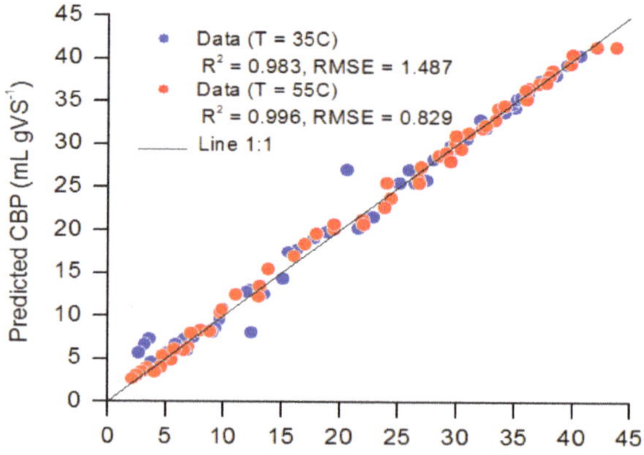

Figure 6. The measured and predicted CBP using *k*-NN model 2 at 35 °C and 55 °C. R^2: coefficient of determination; *RMSE*: root mean squared error; VS: volatile solids; CBP: cumulative biogas production; $k = 2$; weighting function: Equation (S3).

3.2. Evaluation of SVM Model

A 5-fold CV approach was applied to find an appropriate kernel function for the SVM model, and to optimize the parameters *C* and ε by means of SMO algorithm. Figure 7 illustrates the variation in validation *RMSE* as a function of the type of kernel function (linear, quadratic, cubic, fine Gaussian, medium Gaussian, and coarse Gaussian), and C value. Parameter *C* varied from 0.1 to 10,000, whereas ε was set to 0.001. The MATLAB default value was assigned to γ (1.0 for all the linear, quadratic, and cubic functions; 0.43 for the fine Gaussian, 1.73 for the medium Gaussian, and 6.93 for the coarse Gaussian function). As seen in Figure 7, among the different kernel functions that were fitted to the training subset, the coarse Gaussian kernel function yielded the least validation error (*RMSE* equals 0.932), which was obtained at a C value equal to 1200. The detailed specifications of the trained course Gaussian-based SVM model are tabulated in Table 6.

Table 6. Detailed specifications of the best trained SVM model [a].

Number of data patterns	90
Training algorithm	SMO
Convergence criterion	Feasibility gap [b]
Gap calculated	9.8398×10^{-4}
Gap tolerance	1×10^{-3}
Number of iterations	27,537
b	20.4169
w	0.01111
C	1200
ε	1×10^{-3}
γ	6.93
n_{sv}	90 [c]
α	See Supplementary Materials (Table S1)
Training runtime [d]	1.175 s

Symbols: b: bias; w: weight; C: box constraint; ε: deviation from the target output value; γ: kernel scale; n_{sv}: number of support vectors, $\alpha = \alpha_i - \alpha_i^*$ where α_i, and α_i^* are the Lagrange multipliers associated with the vector x_i whose elements are x_1 (temperature), x_2 (C/N), and x_3 (RT). Abbreviations: SVM: support vector machine; SMO: sequential minimal optimization; C/N: carbon-to-nitrogen; RT: retention time. Notes: [a] The SVM model constructed based on coarse Gaussian as a kernel function (for a solved example, refer to Section D in the Supplementary Materials). [b] See Equation (10); the SMO algorithm is converged at an iteration at which the feasibility gap is smaller than the gap tolerance (MATLAB R2020a). [c] All 90 data patterns are considered as support vectors ($\alpha \neq 0$); see Supplementary Materials (Table S1) for α values corresponding to the support vectors. [d] The model was implemented in MATLAB R2020a on a Dell laptop with Intel® Core™ i3-2330M CPU @ 2.20 GHz, and 4.00 GB RAM.

The prediction accuracy of the coarse Gaussian-based SVM model against the whole dataset under mesophilic condition (35 °C) and thermophilic condition (55 °C) is visualized in Figure 8. This figure indicates an excellent agreement between the measured and the model predicted values with R^2 and *RMSE* values equal to 0.997 and 0.598 in the case of the mesophilic condition, respectively, and 0.999 and 0.418 in the case of the thermophilic condition, respectively. This indicates that only 0.1–0.3% of the total variability in the response cannot be explained by the developed coarse Gaussian-based SVM model.

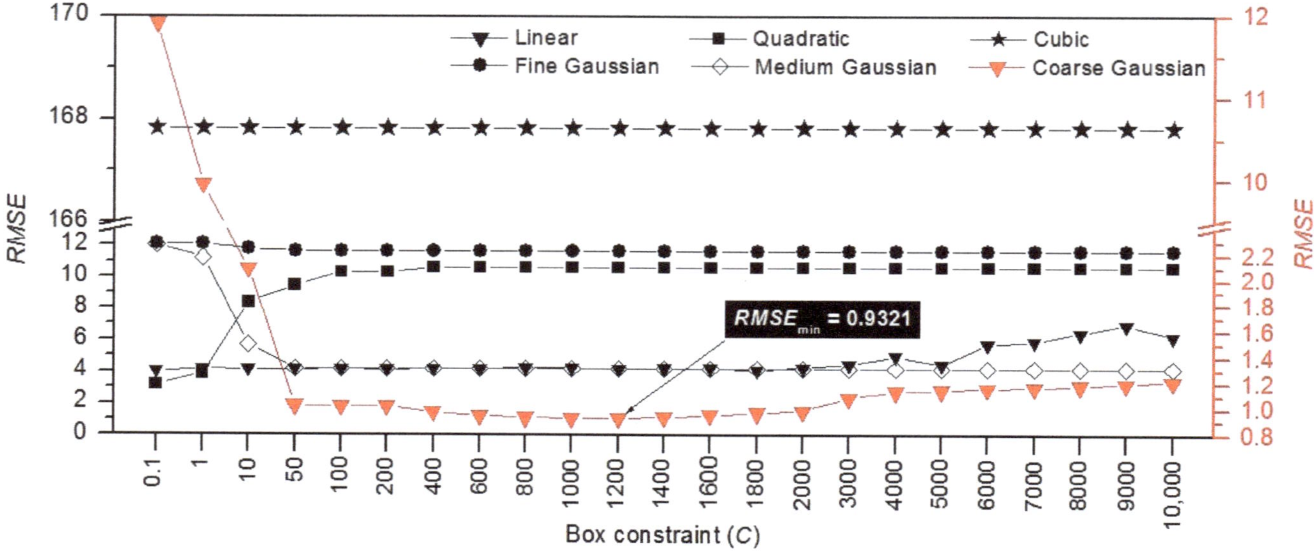

Figure 7. Validation curve for the SVM model as a function of C and type of kernel function. C: box constraint; R^2: coefficient of determination; *RMSE*: root mean squared error; ε was set to 0.001, which is the smallest acceptable value in MATLAB R2020a.

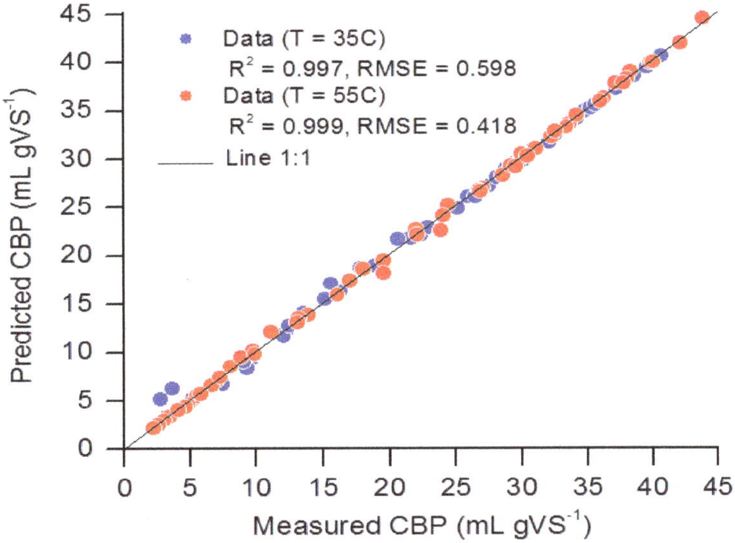

Figure 8. The measured and predicted CBP using the coarse Gaussian-based SVM model. ε = 0.001, C = 1200, and γ = 6.93 under mesophilic condition (35 °C) and thermophilic condition (55 °C); R^2: coefficient of determination; *RMSE*: root mean squared error; VS: volatile solids; CBP: cumulative biogas production.

De Clercq et al. [42] proposed *k*-NN-, SVM-, and random forest-based models to predict biogas production from "Hainan BioCNG", an industrial-scale biogas facility located in the south of China, which is capable of treating daily 750 tons of a wide range of agricultural, municipal and industrial bio-wastes, with a daily maximum production of 30,000 m^3 bio-methane vehicular fuel. Results indicated that the best performance was achieved by the k-NN model, offering a prediction accuracy of 0.86 and 0.85 on the training dataset and testing dataset, respectively. The SVM and random forest models had accuracy in the range of 0.95–0.97 on the training dataset; however, both of these models produced a testing accuracy far lower (about 0.50) than that of the training accuracy, which shows that the SVM and random forest models were noticeably overfitting the training dataset. The authors claimed that one of the possible reasons for the low testing accuracy of the SVM and random forest models was that the dataset used to tune the hyperparameters was too small. Dong and Chen [43] proposed a novel modeling method, which integrated orthogonal experimental design (OED) with SVM, to establish a relationship between the biogas produced from anaerobic digestion of corn stalk (CS) and the pretreatment process parameters, including mass of CS, ultrasonic duration time, alkali pretreatment time, and single-/dual-frequency ultrasound. The anaerobic digester, composed of a 1.0 L bottle with an effective volume of 0.8 L, operated at pH 7–8, a constant temperature of 35 °C, and at an initial TS and C/N ratio of 15 g/L and 20:1, respectively. The results of the validation experiment demonstrated that OED-SVM was an efficient method for optimizing the pretreatment process parameters and predicting biogas production from anaerobic digestion of CS. In the study performed by Yang et al. [44], two different models, including SVM and ANFIS were developed to estimate biogas production for anaerobic digestion of fruits, vegetables, and food wastes as a function of temperature, pH, VS, biomass type, reactor volume, HRT, organic loading rate, and reactor/feeding type. Findings showed that the proposed SVM model demonstrated a superior capability of predicting biogas with *RMSE* and R^2 of 0.0111 and 0.998 against 0.0683 and 0.946 for ANFIS model. Gao et al. [45] performed a multiple linear regression (MLR) analysis to estimate methane production from anaerobic co-digestion of yellow back fungus spent mushroom and different types of livestock manures (e.g., chicken, dairy, and pig manures) at a constant temperature of 35 °C. The feedstock ratio (spent mushroom-to-manure: 10–90 w/w and TS content (5–15%w)) were considered as the independent variables. From the results, a quadratic polynomial model was found to be a suitable regression model fitting the experimental data, with R^2 value greater than 0.95. The author also showed that the Modified Gompertz model could fit the cumulative methane production data with high accuracy ($R^2 > 0.98$). In another study carried out by Kumar et al. [46], two different computational tools, including a feed-forward-backpropagation neural network (FFBPNN) with logistic function, and response surface methodology (RSM) were used to optimize the performance of an electrochemical-assisted anaerobic digester of 1 L capacity fed with the spent mushroom substrate (i.e., wheat straw-based mushroom left over after cultivation of *Agaricus bisporus* mushroom). Sugar mill wastewater (SMWW), and cow dung were utilized as a supplementary nutrient source and as an inoculum, respectively. The digester temperature (30, 35, and 40 °C), direct electrical current (0, 1.5, and 3 V), and SMWW loading (0, 50, and 100% conc.) were taken as the models' input variables, whereas the biogas production was the output of the models. The modeling results demonstrated that the FFBPNN models showed an excellent ability to estimate biogas production with a prediction accuracy of 99.91%, which was slightly better than that obtained by the quadratic model of RSM (99.79%). However, from the perspective of error generated, the FFBPNN model produced a smaller *RMSE* (97.3) compared with that produced by the RSM (117.6).

3.3. Comparison of the Models

Figure 9 illustrates the measured and predicted values for the cumulative biogas production as a function of RT (1 to 14 days) while different levels of temperature (35 °C and 55 °C) and C/N ratios (12, 20, 30, and 40) were investigated. It is evident from Figure 9 that the predicted lines (generated using the developed *k*-NN and SVM models) follow the trend of experimental data points most closely.

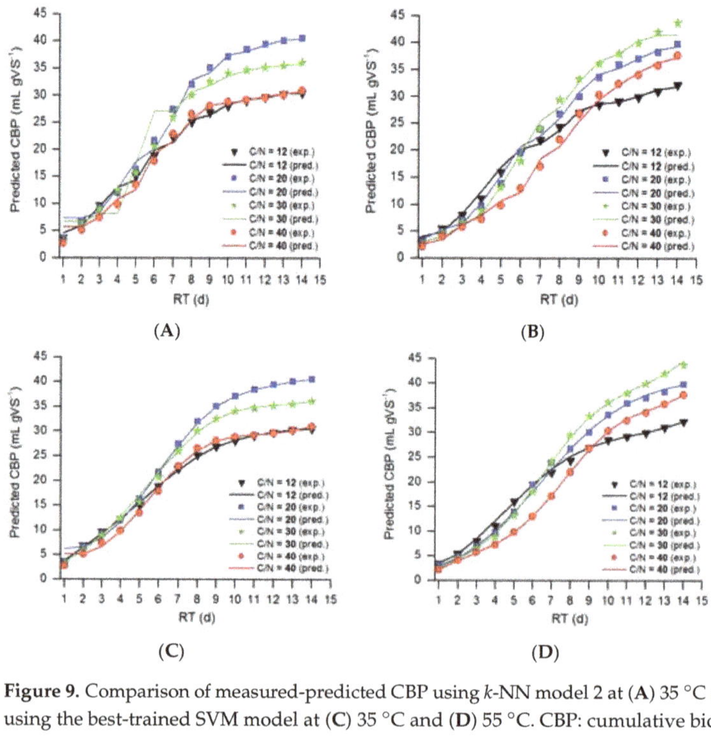

Figure 9. Comparison of measured-predicted CBP using *k*-NN model 2 at (**A**) 35 °C and (**B**) 55 °C and using the best-trained SVM model at (**C**) 35 °C and (**D**) 55 °C. CBP: cumulative biogas production.

Performance comparison of the k-NN and SVM models is tabulated in Table 7. The results of ANN, ANFIS, and logistic models developed by Najafi and Faizollahzadeh Ardabili [31] are also included in Table 7; two statistical indices (R^2 and *RMSE* between the measured and predicted values) were used in order to make the comparison. It can be observed from Table 7 that the total values of R^2 for the developed k-NN model under mesophilic digestion (35 °C) and thermophilic digestion (55 °C) were 0.9830 and 0.9957, respectively. These findings indicate that the k-NN model performs well in predicting biogas production. In addition to its high predictive performance, the k-NN model was straightforward to implement for the problem under consideration because the dataset (composed of 112 observations) and the number of features (i.e., three features) were small. However, it should be noted that in the case of problems that involve several features and a huge dataset, k-NN modeling is not a feasible technique because it is computationally expensive in terms of runtime and memory requirement. Furthermore, the k-NN algorithm calculates and stores the distance of each observation in the testing dataset from all the observations in the training dataset. The total values of R^2 for the SVM model under mesophilic digestion (35 °C) and thermophilic digestion (55 °C) were 0.9973 and 0.9989, respectively, which are slightly better than those obtained using the k-NN model (Table 7).

Table 7. Comparison of models developed in this study and those developed by Najafi and Faizollahzadeh Ardabili [31].

		Models Developed in This Study				**Models Developed by Najafi and Faizollahzadeh Ardabili [31]**					
T (°C)	**C/N**	**k-NN**		**SVM**		**ANN**		**ANFIS**		**Logistic**	
		R^2	*RMSE*	R^2	*RMSE*	R^2	*RMSE*	R^2	*RMSE*	R^2	*RMSE*
35	12	0.9958	0.5899	0.9995	0.1986	1	0.0364	0.9994	0.2346	0.9986	0.4094
	20	0.9903	1.3076	0.9967	0.7584	0.9942	1.3064	0.9998	0.2202	0.999	0.5
	30	0.9592	2.3786	0.9981	0.5166	0.9966	0.7756	0.9998	0.1475	0.9984	0.5327
	40	0.9888	1.0648	0.9946	0.7408	0.9992	0.3606	0.9998	0.1593	0.9974	0.5865
	Total	0.983	1.4374	0.9973	0.598	0.9962	0.78	0.9996	0.194	0.9984	0.5111
55	12	0.9961	0.6068	0.9981	0.422	0.9984	0.5584	0.9992	0.286	0.9972	0.5691
	20	0.9961	0.8085	0.9978	0.6023	0.9998	0.2004	0.999	0.4233	0.9984	0.5501
	30	0.9951	1.0185	0.9994	0.3554	0.9998	0.2733	0.9998	0.2512	0.9986	0.5771
	40	0.9956	0.8304	0.9998	0.1811	0.9998	0.2093	0.9998	0.2098	0.9986	0.5035
	Total	0.9957	0.829	0.9989	0.4183	0.9984	0.343	0.9994	0.3033	0.9982	0.5506

k-NN: k-nearest neighbors; SVM: support vector machine; ANN: artificial neural network; ANFIS: adaptive neuro-fuzzy inference system; R^2: coefficient of determination; *RMSE*: root mean squared error; C/N: carbon-to-nitrogen.

Comparing the performance of the k-NN and SVM models in terms of error produced, the total values of *RMSE* under mesophilic digestion for the SVM model was 0.5980, which is 58.4% smaller than that obtained using the k-NN model under the same conditions. In the case of thermophilic digestion, the total value of *RMSE* for the SVM model was 0.4183, which is 49.5% smaller than that obtained using the k-NN model under the same conditions. These results imply that the SVM model is a better choice for predicting biogas production. It is worth mentioning that the SVM modeling technique is less computationally demanding than the k-NN technique and can effectively handle any complex problems involving many features and a massive dataset with high generalization power. However, SVM is very sensitive to the input hyperparameters, and hence, caution must be taken to properly tune the hyperparameters for any given problem. Parameters that may yield an excellent prediction accuracy for problem A may yield a poor prediction accuracy for problem B.

The total values of R^2 and *RMSE* at both mesophilic and thermophilic conditions for the SVM model developed in this study were in the range of 0.9973–0.9989 and 0.4183–0.5980, respectively, which are in agreement with the results of Najafi and Faizollahzadeh Ardabili [31] who developed ANN, ANFIS, and logistic models (R^2 = 0.9962–0.9996, *RMSE* = 0.1940–0.7800). Overall, it can be concluded that the SVM can be a useful alternative tool with the capability of accurately predicting biogas production under both mesophilic and thermophilic conditions.

4. Conclusions

In this study, two data-driven modeling techniques, including *k*-nearest neighbor (*k*-NN) and support vector machine (SVM), were successfully trained, validated, and tested to estimate biogas production from anaerobic digestion of spent mushroom compost. It is evident from the results that both the developed *k*-NN and SVM models can estimate biogas production-under mesophilic and thermophilic conditions-with high prediction accuracy (R^2 = 98.3–99.9%). However, the SVM model generated a smaller error (*RMSE* = 0.418–0.598) than that of the *k*-NN model (0.829–1.437). These findings imply that the SVM model is a versatile yet more effective tool for predicting biogas production during anaerobic digestion.

Supplementary Materials: The following supporting information can be downloaded at: https://www.mdpi.com/article/10.3390/agriculture12081090/s1, Section A: a brief description of the determination of *k*-NN hyperparameters; Section B: an example of a *q*-fold cross-validation (CV) method; Section C: To derive a linear SVM regression with the use of Lagrangian function and optimal constraints; Section D: A solved example of how to use the developed SVM model in this study; Table S1: α values for the support vectors; Figure S1: Schematic illustration of *q*-fold CV approach; Equations (S1)–(S12).

Author Contributions: Conceptualization, methodology, software, formal analysis, and writing-original draft preparation, R.S.; review and editing, Q.Y. and S.C. All authors have read and agreed to the published version of the manuscript.

Funding: This research was funded by Prince of Songkla University and the Ministry of Higher Education, Science, Research and Innovation, Thailand, under the Reinventing University Project, grant number REV64061 and the APC was funded by the same grant number REV64061.

Institutional Review Board Statement: Not applicable.

Data Availability Statement: Not applicable.

Acknowledgments: The authors thank the support from the Department of Civil and Environmental Engineering, and Research and Development Office, Prince of Songkla University, Thailand. We also thank the support from Biogas and Biorefinery Laboratory at the Faculty of Engineering, and PSU Energy Systems Research Institute, Prince of Songkla University, Thailand.

Conflicts of Interest: The authors declare no conflict of interest.

References

1. Franta, B. Early oil industry disinformation on global warming. *Environ. Polit.* **2021**, *30*, 663–668. [CrossRef]
2. Kaparaju, P.; Rintala, J. Generation of heat and power from biogas for stationary applications: Boilers, gas engines and turbines, combined heat and power (CHP) plants and fuel cells. In *The Biogas Handbook: Science, Production and Applications*; Wellinger, A., Murphy, J., Baxter, D., Eds.; Woodhead Publishing: Cambridgeshire, UK, 2013; pp. 404–427.
3. Hakawati, R.; Smyth, B.M.; McCullough, G.; De Rosa, F.; Rooney, D. What is the most energy efficient route for biogas utilization: Heat, electricity or transport? *Appl. Energy* **2017**, *206*, 1076–1087. [CrossRef]
4. Nasir, I.M.; Mohd Ghazi, T.I.; Omar, R. Production of biogas from solid organic wastes through anaerobic digestion: A review. *Appl. Microbiol. Biotechnol.* **2012**, *95*, 321–329. [CrossRef] [PubMed]
5. Caruso, M.C.; Braghieri, A.; Capece, A.; Napolitano, F.; Romano, P.; Galgano, F.; Altieri, G.; Genovese, F. Recent updates on the use of agro-food waste for biogas production. *Appl. Sci.* **2019**, *9*, 1217. [CrossRef]
6. Mirmohamadsadeghi, S.; Karimi, K.; Tabatabaei, M.; Aghbashlo, M. Biogas production from food wastes: A review on recent developments and future perspectives. *Bioresour. Technol. Rep.* **2019**, *7*, 100202. [CrossRef]
7. Zhang, L.; Loha, K.C.; Zhang, J. Enhanced biogas production from anaerobic digestion of solid organic wastes: Current status and prospects. *Bioresour. Technol. Rep.* **2019**, *5*, 280–296. [CrossRef]
8. Uddin, M.M.; Wright, M.M. Anaerobic digestion fundamentals, challenges, and technological advances. *Phys. Sci. Rev.* **2022**. [CrossRef]
9. Tchobanoglous, G.; Burton, F.L.; Stensel, H.D. *Wastewater Engineering Treatment and Reuses*; McGraw-Hill: New York, NY, USA, 2003.

10. Adekunle, K.F.; Okolie, J.A. A review of biochemical process of anaerobic digestion. *Adv. Biosci. Biotechnol.* **2015**, *6*, 205–212. [CrossRef]
11. Sikora, A.; Detman, A.; Chojnacka, A.; Błaszczyk, M.K. Anaerobic digestion: I. A common process ensuring energy flow and the circulation of matter in ecosystems. II. A tool for the production of gaseous biofuels. In *Fermentation Processes*; Jozala, A.F., Ed.; Intech: Rijeka, Croatia, 2017; pp. 271–301. [CrossRef]
12. Paolini, V.; Petracchini, F.; Segreto, M.; Tomassetti, L.; Naja, N.; Cecinato, A. Environmental impact of biogas: A short review of current knowledge. *J. Environ. Sci. Health A* **2018**, *53*, 899–906. [CrossRef]
13. Verbeeck, K.; De Vrieze, J.; Pikaar, I.; Verstraete, W.; Rabaey, K. Assessing the potential for up-cycling recovered resources from anaerobic digestion through microbial protein production. *Microb. Biotechnol.* **2021**, *14*, 897–910. [CrossRef]
14. Salehi, R.; Chaiprapat, S. Conversion of biogas from anaerobic digestion to single cell protein and bio-methanol: Mechanism, microorganisms and key factors—A review. *Environ. Eng. Res.* **2022**, *27*, 210109. [CrossRef]
15. Hill, D.T. A comprehensive dynamic model for animal waste methanogenesis. *Trans. ASAF* **1982**, *25*, 2129–2143. [CrossRef]
16. Mosey, F.E. Mathematical modeling of the anaerobic digestion process: Regulatory mechanisms for the formation of short-chain volatile acids from glucose. *Water Sci. Technol.* **1983**, *15*, 209–232. [CrossRef]
17. Pullammanappallil, P.; Owens, J.M.; Svoronos, S.A.; Lyberatos, G.; Chynoweth, D.P. Dynamic model for conventionally mixed anaerobic digestion reactors. *AIChE Annu. Meet.* **1991**, *277C*, 43–53.
18. Angelidaki, I.; Ellegaard, L.; Ahring, B.K. A mathematical model for dynamic simulation of anaerobic digestion of complex substrates: Focusing on ammonia inhibition. *Biotechnol. Bioeng.* **1993**, *42*, 159–166. [CrossRef]
19. Batstone, D.J.; Keller, J.; Angelidaki, I.; Kalyuzhnyi, S.V.; Pavlostathis, S.G.; Rozzi, A.; Sanders, W.T.M.; Siegrist, H.; Vavilin, V.A. *Anaerobic Digestion Model No.1*; IWA Publishing: London, UK, 2002.
20. Siegrist, H.; Vogt, D.; Garcia-Heras, J.L.; Gujer, W. Mathematical model for meso- and thermophilic anaerobic sewage sludge digestion. *Environ. Sci. Technol.* **2002**, *36*, 1113–1123. [CrossRef]
21. Fedailaine, M.; Moussi, K.; Khitous, M.; Abada, S.; Saber, M.; Tirichine, N. Modeling of the anaerobic digestion of organic waste for biogas production. *Procedia Comput. Sci.* **2015**, *52*, 730–737. [CrossRef]
22. Enitan, A.M.; Adeyemo, J.; Swalaha, F.M.; Kumari, S.; Bux, F. Optimization of biogas generation using anaerobic digestion models and computational intelligence approaches. *Rev. Chem. Eng.* **2017**, *33*, 309–335. [CrossRef]
23. Cinar, S.; Cinar, S.O.; Wieczorek, N.; Sohoo, I.; Kuchta, K. Integration of artificial intelligence into biogas plant operation. *Processes* **2021**, *9*, 85. [CrossRef]
24. Holubar, P.; Zani, L.; Hagar, M.; Froschl, W.; Radak, Z.; Braun, R. Modeling of anaerobic digestion using self-organizing maps and artificial neural nets. *Water Sci. Technol.* **2000**, *41*, 149–156. [CrossRef]
25. Cakmakci, M. Adaptive neuro-fuzzy modelling of anaerobic digestion of primary sedimentation sludge. *Bioprocess Biosyst. Eng.* **2007**, *30*, 349–357. [CrossRef] [PubMed]
26. Kusiak, A.; Wei, X. Prediction of methane production in wastewater treatment facility: A data-mining approach. *Ann. Oper. Res.* **2014**, *216*, 71–81. [CrossRef]
27. Nair, V.V.; Dhar, H.; Kumar, S.; Thalla, A.K.; Mukherjee, S.; Wong, J.W.C. Artificial neural network based modeling to evaluate methane yield from biogas in a laboratory-scale anaerobic bioreactor. *Bioresour. Technol.* **2016**, *217*, 90–99. [CrossRef] [PubMed]
28. Dach, J.; Koszela, K.; Boniecki, P.; Zaborowicz, M.; Lewicki, A.; Czekała, W.; Skwarcz, J.; Qiao, W.; Piekarska-Boniecka, H.; Białobrzewskid, I. The use of neural modelling to estimate the methane production from slurry fermentation processes. *Renew. Sustain. Energy Rev.* **2016**, *56*, 603–610. [CrossRef]
29. Tan, H.M.; Gouwanda, D.; Poh, P.E. Adaptive neural-fuzzy inference system vs. anaerobic digestion model No.1 for performance prediction of thermophilic anaerobic digestion of palm oil mill effluent. *Process Saf. Environ. Prot.* **2018**, *117*, 92–99. [CrossRef]
30. Beltramo, T.; Klocke, M.; Hitzmann, B. Prediction of the biogas production using GA and ACO input features selection method for ANN model. *Inf. Process Agric.* **2019**, *6*, 349–356. [CrossRef]
31. Najafi, B.; Faizollahzadeh Ardabili, S. Application of ANFIS, ANN, and logistic methods in estimating biogas production from spent mushroom compost (SMC). *Resour. Conserv. Recycl.* **2018**, *133*, 169–178. [CrossRef]
32. Fix, E.; Hodges, J. *Discriminatory Analysis—Nonparametric Discrimination: Consistency Properties*; USAF School of Aviation: Randolph Field Historic District (near San Antonio), TX, USA, 1951.
33. Cover, T.; Hart, P. Nearest neighbor pattern classification. *IEEE Trans. Inf. Theory* **1967**, *13*, 21–27. [CrossRef]
34. Wu, X.; Kumar, V.; Quinlan, J.R.; Ghosh, J.; Yang, Q.; Motoda, H.; McLachlan, G.J.J.; Ng, A.; Liu, B.; Yu, P.S.; et al. Top 10 algorithms in data mining. *Knowl. Inf. Syst.* **2008**, *14*, 1–37. [CrossRef]
35. Kang, S. k-nearest neighbor learning with graph neural networks. *Mathematics* **2021**, *9*, 830. [CrossRef]
36. Abu Alfeilat, H.A.; Hassanat, A.B.A.; Lasassmeh, O.; Tarawneh, A.S.; Alhasanat, M.B.; Eyal Salman, H.S.; Surya Prasath, V.B.S. Effects of distance measure choice on k-nearest neighbor classifier performance: A review. *Big Data* **2019**, *7*, 221–248. [CrossRef] [PubMed]
37. Ehsani, R.; Drabløs, F. Robust distance measures for *k*NN classification of cancer data. *Cancer Inf.* **2020**, *19*, 1176935120965542. [CrossRef] [PubMed]
38. Imandoust, S.B.; Bolandraftar, M. Application of K-nearest neighbor (KNN) approach for predicting economic events: Theoretical background. *J. Eng. Res. Appl.* **2013**, *3*, 605–610.

39. Salehi, R.; Lestari, R.A.S. Predicting the performance of a desulfurizing bio-filter using an artificial neural network (ANN) model. *Environ. Eng. Res.* **2021**, *26*, 200462. [CrossRef]
40. Vapnik, V.N. *The Nature of Statistical Learning Theory*; Springer: New York, NY, USA, 1995. [CrossRef]
41. Platt, J.C. *Sequential Minimal Optimization: A Fast Algorithm for Training Support Vector Machines*; Microsoft Research: Washington, DC, USA, 1998.
42. De Clercq, D.; Jalota, D.; Shang, R.; Ni, K.; Zhang, Z.; Khan, A.; Wen, Z.; Caicedo, L.; Yuan, K. Machine learning powered software for accurate prediction of biogas production: A case study on industrial-scale Chinese production data. *J. Clean. Prod.* **2019**, *218*, 390–399. [CrossRef]
43. Dong, C.; Chen, J. Optimization of process parameters for anaerobic fermentation of corn stalk based on least squares support vector machine. *Bioresour. Technol.* **2019**, *271*, 174–181. [CrossRef]
44. Yang, Y.; Zheng, S.; Ai, Z.; Molla Jafari, M.M. On the prediction of biogas production from vegetables, fruits, and food wastes by ANFIS- and LSSVM-based models. *BioMed Res. Int.* **2021**, *2021*, 9202127. [CrossRef]
45. Gao, X.; Tang, X.; Zhao, K.; Balan, V.; Zhu, Q. Biogas production from anaerobic co-digestion of spent mushroom substrate with different livestock manure. *Energies* **2021**, *14*, 570. [CrossRef]
46. Kumar, P.; Kumar, V.; Singh, J.; Kumar, P. Electrokinetic assisted anaerobic digestion of spent mushroom substrate supplemented with sugar mill wastewater for enhanced biogas production. *Renew. Energy* **2021**, *179*, 418–426. [CrossRef]

agriculture

MDPI

Article

Spatiotemporal Change of Heat Stress and Its Impacts on Rice Growth in the Middle and Lower Reaches of the Yangtze River

Shuai Zhang [1,2]

[1] Key Laboratory of Land Surface Pattern and Simulation, Institute of Geographic Sciences and Natural Resources Research, Chinese Academy of Sciences, Beijing 100101, China; zhangshuai@igsnrr.ac.cn
[2] College of Resources and Environment, University of Chinese Academy of Sciences, Beijing 100049, China

Abstract: Heat stress will restrict rice yield in the middle and lower reaches of the Yangtze River. An understanding of the meteorological conditions of heat stress of rice production is important for improving the accuracy of the phenology simulation. Based on the observations of phenology and heat stress of rice agrometeorological stations in this region, as well as meteorological observations and future scenarios, this study analyzed the spatiotemporal change of heat stress and its impacts on rice growth in this region from 1990 to 2009. The results showed that the heat stress frequency of early rice increased in this region from 2000 to 2009, and that of late rice and single-season rice decreased. Moreover, rice phenology will advance under heat stress conditions. The spatiotemporal consistency of the observations and the meteorological index of heat stress shows that the change in heat stress is attributed to climate changes and extreme meteorological events. Under future climate scenarios, it is found that the frequency of heat stress will increase, which will have a serious impact on rice production. The results suggest that positive and effective measures should be taken to adapt to climate change for rice production.

Keywords: middle and lower reaches of Yangtze River; rice; climate change; heat stress

Citation: Zhang, S. Spatiotemporal Change of Heat Stress and Its Impacts on Rice Growth in the Middle and Lower Reaches of the Yangtze River. *Agriculture* **2022**, *12*, 1097. https://doi.org/10.3390/agriculture12081097

Academic Editor: Josef Eitzinger

Received: 19 June 2022
Accepted: 22 July 2022
Published: 26 July 2022

1. Introduction

In recent years, the global climate has shown a significant change characterized by warming. The global surface temperature increased by 0.69~1.08 °C from 1901 to 2012 [1]. Heat stress caused by climate warming is a risk for global food production for the regions where heat stress happens frequently [2–4], such as East Asia, Southeast Asia, and South Asia.

Rice is one of the most important main food crops in the world [5,6]. China accounts for a quarter of the world's planting area [7,8]. And rice production in China has reached one-third of the world's total rice production [9]. Rice is sensitive to temperature changes [10,11]. Heat stress is the main climate variable which has substantial impact on the rice production [12]. Studies have shown that when the temperature rises by 1 °C, the rice production is reduced by 10% [13]. As the frequency of extreme heat events increases, it will lead to an obvious decrease in rice production [14]. The impact of heat stress on rice yield based on a meta-analysis was studied. This analysis found that the mean yield of rice was reduced by 39.6% due to heat stress worldwide [15].

The responses of rice growth to heat stress is different for different developmental stages. The sensitivity of rice growth for heat stress during different rice growing periods was in the order of heading stage and flowering stage > young panicle development stage > filling stage [16]. Heat stress occurring in the reproductive growing period will cause a short pollen germination rate and decreased pollen viability, which result in rice grain spikelet sterility [17]. If it occurs in the young panicle development stage, it will inhibit the differentiation and degeneration of the spikelet. Heat stress during the grain-filling period resulted in faster translocation of photosynthates, it will shorten the length of the grain-filling period, reduce grain weight and imperfect grains. Warming stress from heading to maturity will cause spikelet sterility [18,19]. In the vegetative growing period of rice, heat stress will limit photosynthesis and reduce leaf area and tiller [20]. In addition, the reproductive growing period of rice is sensitive to heat stress, which leads to a decrease in leaf area, plant height, and harvest index and hinders the development of reproductive organs [21]. During the grain-filling period, the total grain-filling duration of rice is shortened by 21.3–37.1% when heat stress happens [20]. Studies have shown that heat stress (35 °C) occurs in the early 72 h of seed development, and the development of endosperm and embryo will be damaged [22], which seriously affects the development of rice.

Heat stress often occurs in the Yangtze River Basin in China [3,23]. This region is an important rice planting areas in China. Because this area belongs to the transition zone of subtropical and warm temperate climate, the regional climate environment is complex, and disastrous weather is prone to occur. Due to global warming, heat, heat waves, and other events, the frequency of heat stress will increase in this region. It is of great significance to study the impact of heat stress on rice growth, which is beneficial to ensure food security in the future in China.

The study aims to explore the spatiotemporal change of heat indices from 1981 to 2009, as proxy of thermal damage to rice production in China by using observations from agrometeorological stations and to project the changes of heat damages under future climate scenarios.

2. Materials and Methods

2.1. Data

The study area is the middle and lower reaches of the Yangtze River (Figure 1). The details of the agrometeorological stations in the study area were shown in Table 1. Data of rice heat stress from 1991 to 2009 were obtained. These data were maintained by the China Meteorological Administration (CMA). Daily weather data were obtained from the Chin Meteorological Administration (CMA). The future climate scenarios were constructed from the Regional Integrated Environmental Model (RIEMS) [24–26].

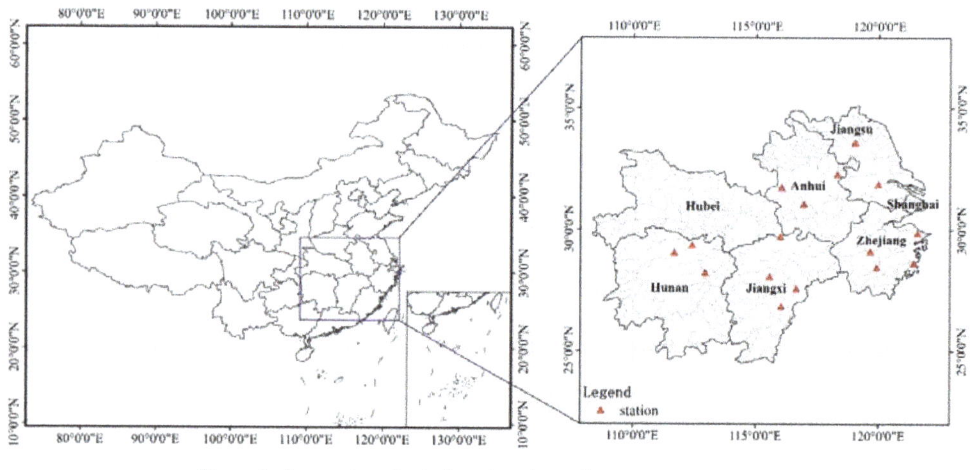

Figure 1. Agrometeorological stations in study area.

Table 1. Information of agrometeorological stations in study area.

Agrometeorological Stations	Longitude	Latitude	Annual Mean Precipitation (mm)	Annual Mean of Daily Average Temperature (°C)	Annual Mean of Daily Maximum Temperature (°C)	Annual Mean of Daily Minimum Temperature (°C)
Nanxian	112.40	29.36	336.87	18.15	21.25	14.21
Changde	111.68	29.05	368.70	17.82	21.53	14.32
Huaiyin	119.03	33.60	188.04	15.16	20.00	11.16
Chuxian	118.30	32.30	311.93	16.12	20.56	12.16
Liuan	116.50	31.75	285.85	16.42	20.80	12.64
Tongcheng	116.93	31.05	422.68	16.61	20.64	12.57
Wujin	119.93	31.77	316.33	16.36	20.62	12.97
Jiujiang	116.00	29.73	333.43	18.25	22.19	15.30
Jinhua	119.65	29.11	355.95	18.28	22.59	14.63
Yinxian	121.57	29.87	335.14	17.60	21.75	14.09
Zhangshu	115.55	28.06	277.17	18.59	22.64	15.14
Lishui	119.92	28.45	311.23	19.10	24.07	14.74
Jiaojiang	122.25	28.63	336.23	18.33	22.04	14.81
Nancheng	116.65	27.58	383.53	18.26	22.88	15.04
Guangchang	116.33	26.85	391.61	18.64	23.84	15.01

2.2. Methods

The occurrence frequency of heat events is usually used to represent heat stress for crop studies [19]. The occurrence of heat events is defined by using air temperature measurements according to "The National Agrometeorological Disasters Standard" published by Chinese Meteorological Administration [27]. The daily mean and maximum temperature are the key indicators for counting numbers of occurrence of heat events. If a daily mean temperature above 30 °C for more than three consecutive days or a daily maximum temperature above 35 °C for more than three consecutive days are detected, a heat event is counted. We count numbers of heat events in 1991–2000 and 2000–2009 respectively to explore spatiotemporal change of heat stress. Based on temperature predictions from RIEMS, we also count the occurrence of heat stress events from 2021 to 2040 by using the indices described above. Then the occurrence frequencies in 1981–2000 and 2021–2040 are compared to implicate the possible changes of heat stress in future in study area.

3. Results

3.1. Spatiotemporal Change of Heat Stress

The observations of early rice heat stress from 1990 to 2000 and from 2000 to 2009 showed that the heat stress on rice in Hunan occurred in a few stations, and the frequency of occurrence was the largest. Jiangxi Province has the largest number of stations with heat stress. Comparing the two periods of 1990–2000 and 2000–2009, the frequency of heat stress of early rice increased in most stations (Figure 2).

According to the statistics of heat stress in different growing periods of early rice, it can be found that heat stress of early rice mainly occurs from the tillering stage to the maturity stage. Heat stress occurred mainly at the booting stage, milky ripe stage, and maturity stage at the early rice stations in the study area. From 2000 to 2009, the heat stress of early rice stations in this region mainly occurred in the stage from transplanting to milky ripe, which occurred earlier than that in 1990–2000, and the number of occurrences in each stage increased compared with that in 1990–2000 (Figure 3).

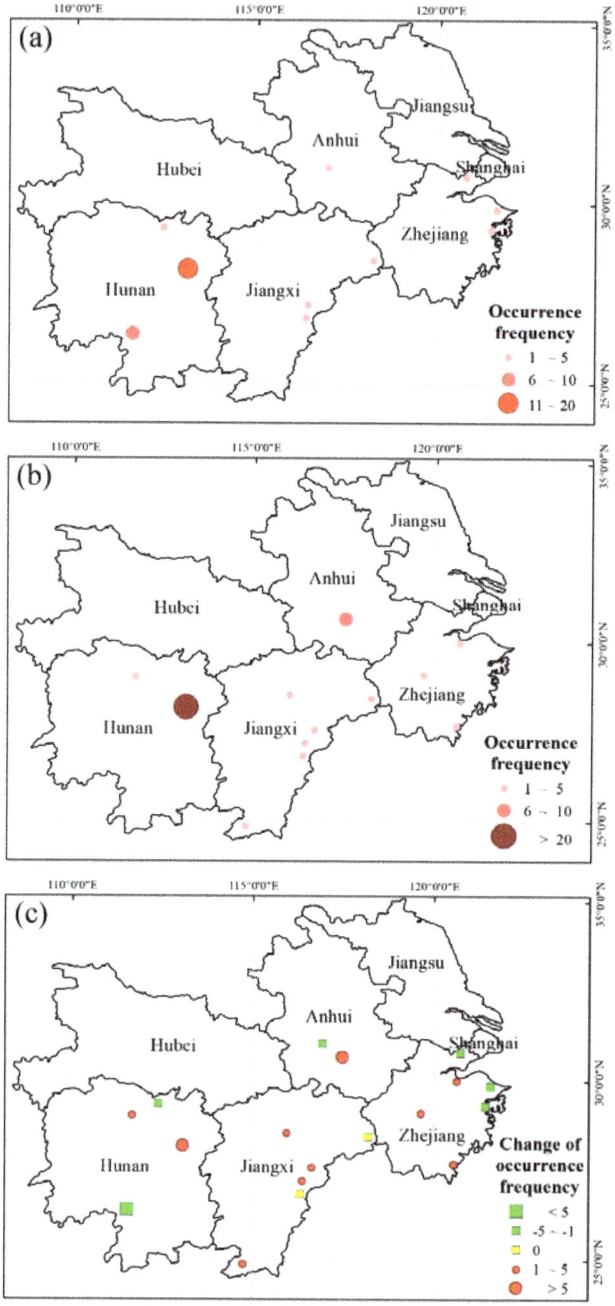

Figure 2. Occurrence frequency of heat stress to early rice in study area from 1991 to 2000 (**a**) and from 2000 to 2009 (**b**) and change of occurrence frequency (**c**).

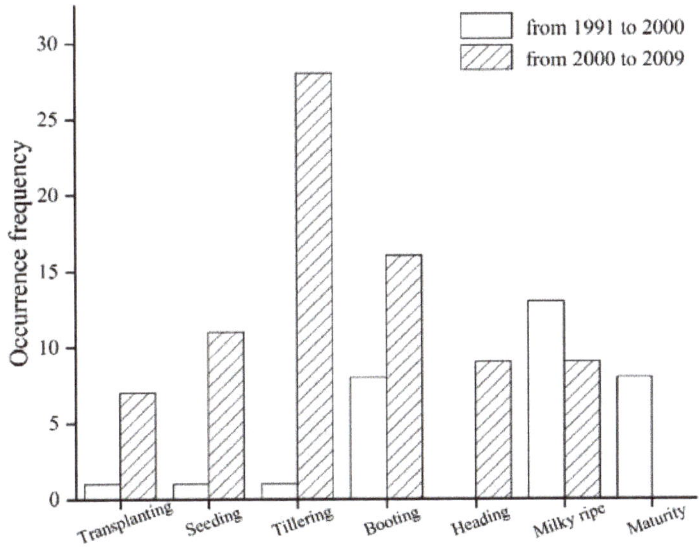

Figure 3. Occurrence frequency of heat stress in different phenological periods at early rice stations in the study area during 1991–2000 and 2000–2009.

For late rice and single-season rice, the stations with a higher frequency of heat stress were mainly concentrated in Hunan Province. Comparing the two periods of 1990–2000 and 2000–2009, the occurrence frequency of heat stress of late rice and single season decreased at most stations (Figure 4).

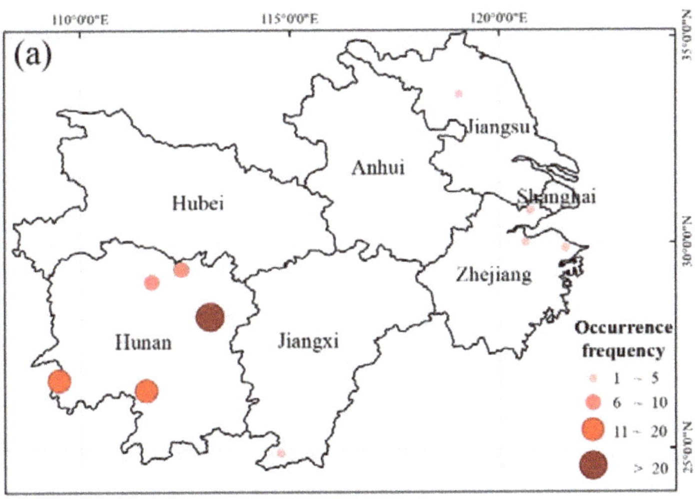

Figure 4. *Cont.*

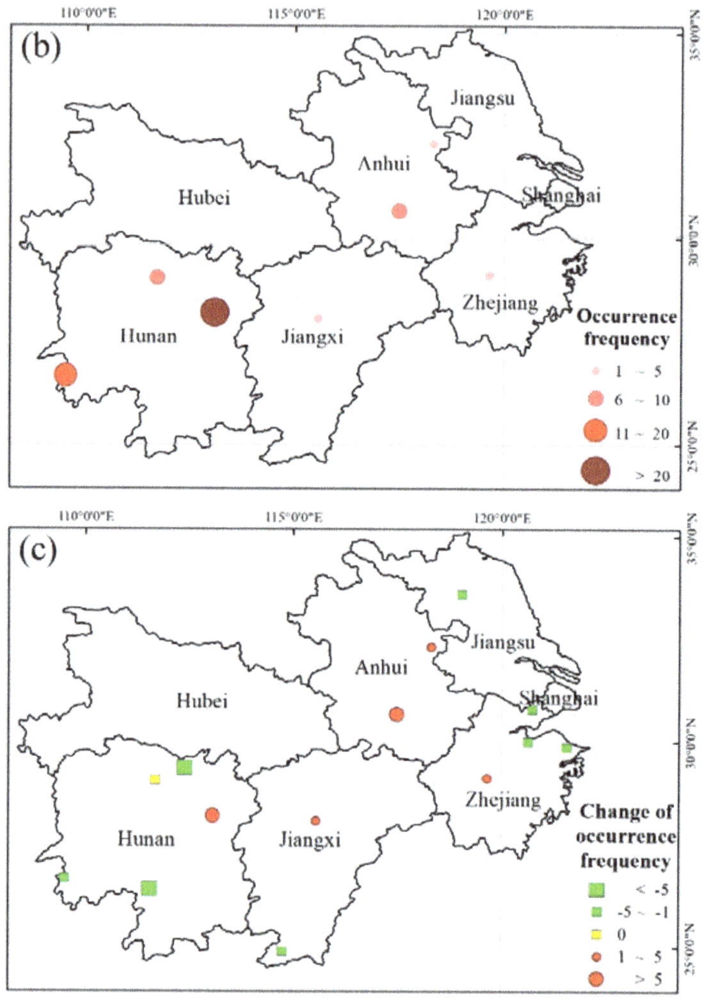

Figure 4. Occurrence frequency of heat stress to single-season rice and late rice in the study area from 1991 to 2000 (**a**) and from 2000 to 2009 (**b**) and change of occurrence frequency (**c**).

Through the statistics of the heat stress of late rice and single-season rice in different phenology stages, it can be found that the heat stress of late rice and single-season rice in the study area mainly occurred from tillering to maturity from 1990 to 2000. From 2000 to 2009, the heat stress of late rice and single-season rice in the middle and lower reaches of the Yangtze River mainly occurred in the stage from transplanting to milky ripe, which was earlier than that in 1990–2000 (Figure 5).

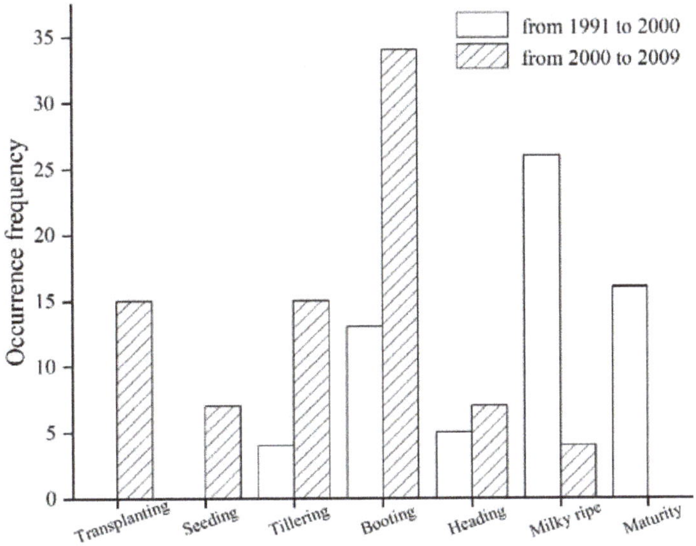

Figure 5. Occurrence frequency of heat stress in different phenological periods at late rice and single-season rice stations in study area during 1991–2000 and 2000–2009.

3.2. Simulation of Heat Stress

Due to the limited observations of national agrometeorological stations, it is difficult to investigate the occurrence of rice heat stress in the study area. Therefore, based on the standard of rice heat stress and meteorological data issued by the China Meteorological Administration, we counted the occurrence of disasters in the corresponding period (Figure 6).

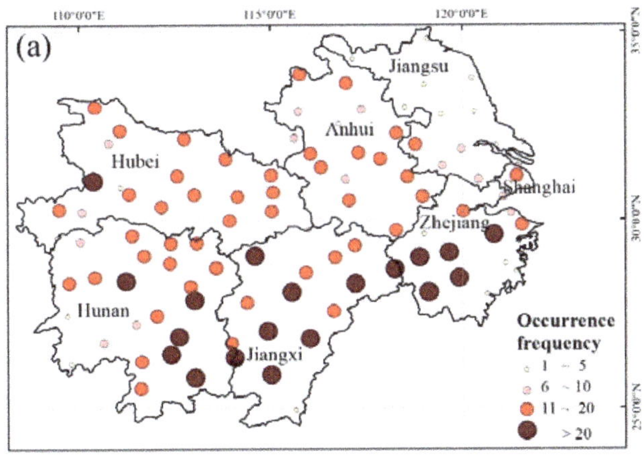

Figure 6. *Cont.*

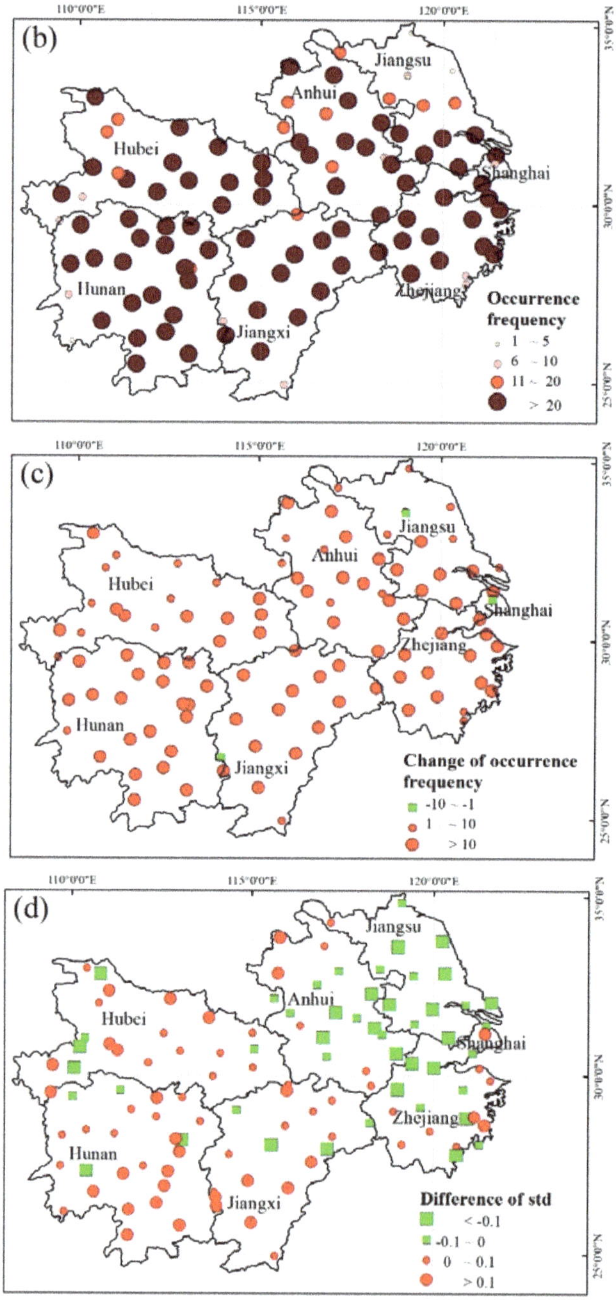

Figure 6. *Cont.*

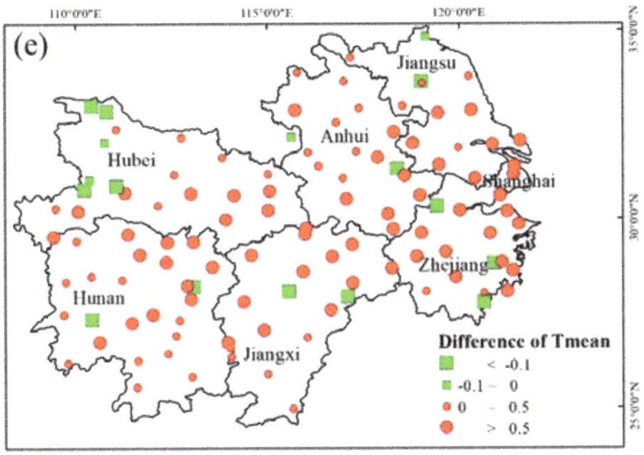

Figure 6. Occurrence frequency of rice heat stress events in study area from 1991 to 2000 (**a**) and from 2000 to 2009 (**b**), and the changes of occurrence frequency in heat stress events (**c**), mean temperature in different stations (**d**), and the standard deviation (**e**) from June to August in these two periods.

From 1991 to 2000, the frequency of heat stress events was less than 20 in the northern of the study area, but more than 20 in the southern part. From 2000 to 2009, the occurrence frequency of heat stress in six provinces in this region was more than 20. The frequency of heat stress events generally shows an increasing trend. The average daily temperature of most stations in summer increased by 0.5 °C in the late period compared with the early period. Except for some stations in eastern China, the standard deviation of mean temperature has increased, which means an increase in extreme heat events (Figure 6).

3.3. Response of Heat Stress of Rice to Climate Change

Under future climate scenarios, we studied the change of rice heat stress in the study area. Based on temperature data of RIEMS, the occurrence frequency and change of heat stress in the study area during 1981–2000 and 2021–2040 were simulated.

From 1981 to 2000, the occurrence of heat stress in this region was generally low. It was 1–3 times in the south and 4–8 times in the north. From 2021 to 2040, the occurrence frequency of heat stress will increase significantly, and the occurrence times of most stations could be more than nine times. Comparing these two time periods, however, the frequency of heat stress increased to varying degrees at all sites from 2021 to 2040 (Figure 7).

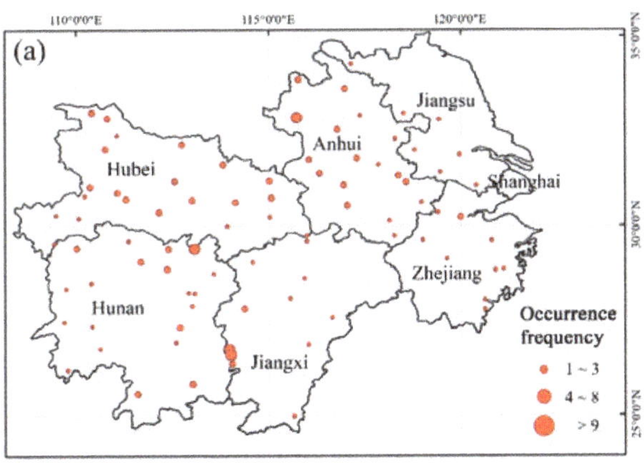

Figure 7. *Cont.*

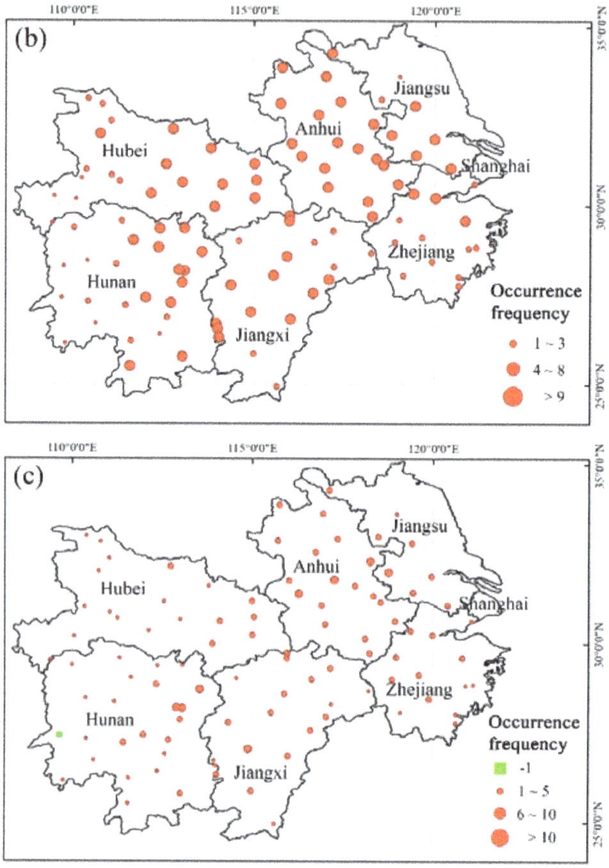

Figure 7. Simulation result of heat stress in study area during 1981–2000 (**a**) and 2021–2040 (**b**), and change of occurrence frequency (**c**).

From the above results, it can be concluded that the occurrence frequency of heat stress will increase in the future climate scenarios, which will have a serious impact on rice production. The period from June to August is the booting, flowering, and maturing stage of rice, during which the occurrence of extreme heat will seriously weaken the yield of rice. Therefore, the study suggests that the heat at the booting and heading stages of rice can be staggered as far as possible by changing varieties or sowing dates, to improve the yield of rice.

4. Discussion

In this study, the observations of phenology and heat stress of rice agrometeorological stations and the future meteorological predictions are used to investigate the spatiotemporal change of heat stress and its impact on rice growth. It suggested that heat stress occurrence frequency of early rice increased in the study area, whereas that of late rice and single-season rice decreased. Moreover, rice phenology will advance under heat stress. The conclusions are consistent with the previous studies. For example, Liu et al., suggests the heat stress of early rice increased the Yangtze River Basin [17]. They also indicate that heat stress also widely occurs in the most parts of Hubei, Anhui, Jiangsu, Zhejiang, and Hunan. For the growth of early rice, the average occurrence frequency of heat stress was 7.81 days in this region [17]. From 1980 to 2009, heat stress in a single rice planting area had an increase of 1.49 °C day, where the increases for early and late rice are about 0.35 °C per day and 1.57 °C per day, respectively [28]. These previous studies indicate that heat stress has substantial impacts on rice growth. Our study further finds that early rice and the late rice had different responses to heat stress. Many previous studies also indicate that relative humidity is an important factor for impacting rice growth [29]. In actuality, changes of relative humidity are highly dependent on temperature changes. Reducing relative humidity under heat conditions is essential for maintaining spikelet fertility [29]. These results suggest that relative humidity is a non-negligible factor for impacting rice growth. In addition, Rehmani et al., suggest that distributions of heat stress are quite different in this region [19]. Our study also finds the different distributions of occurrence frequency of heat stress events in coastline regions and inland regions (Figure 7). This could be caused by the difference of water vapor in these regions. Therefore, it is important to study the complex impacts of relative humidity temperature changes on rice growth. Our results show an increasing trend in heat stress of rice under future scenarios. These results are similar with previous studies. For example, He et al. [14] indicated the occurrence of heat stress of rice will show an increasing trend, and will have an impact on rice yield from the future predictions during 2016–2100. Their results also found that the heat events in the study area will increase by up to 185% and 319% for the RCP 4.5 scenario and the RCP 8.5 scenario, respectively, which will further affect the development of rice [14]. The flowering and maturity of rice will advance under the future climate scenario in the study area [30]. Heat stress will increase with average annual rates of 0.13% and 0.09%, respectively, during 2021–2050 and 2071–2100. Based on all these results, this region will face more heat stress in the future.

In this study, the risk assessment of heat stress is mainly based on statistical methods, meteorological indicators, and climate data under future scenarios. Nowadays, crop models and machine learning methods have also been widely used in meteorological disaster warnings and crop yield assessments. Due to the global climate warming, the response function of development rate to temperature was used to improve the prediction ability of crop models for heading and maturity of double season rice simulations in this region. The heading and maturity simulation accuracy of the improved model increased by 26.2% and 22.9% on average [31]. The improved crop models are also beneficial for improving the prediction skills of warming risks for rice caused by the heat stress. Based on the machine learning method, Li and others have explored the relationship between extreme climate and crop yield which also could be used to monitor extreme heat stress [32]. Therefore, crop models and machine learning methods could be useful tools for accurate prediction of rice yield in the future.

5. Conclusions

In this study, the spatiotemporal change of heat stress and its impacts on rice growth in the study area from 1990 to 2009 were investigated. The occurrence frequency of heat stress events generally shows an increasing trend from 1990 to 2009. For early rice, heat stress occurred mainly at the booting stage, milky ripe stage, and maturity stage at the early rice stations from 1990 to 2000, whereas the stress mainly occurred in the stage from transplanting to milky ripe from 2000 to 2009. For late rice and single rice, the heat stress mainly occurred from the tillering stage to the maturity stage from 1990 to 2000, whereas the heat stress mainly occurred in the stage from transplanting to the milky ripe from 2000 to 2009, which is earlier than that in 1990–2000. The heat stress frequency of early rice increased in the study area from 2000 to 2009, whereas that of single rice and late rice decreased. From 2021 to 2040, the occurrence frequency of heat stress could increase significantly. The frequency of most stations could increase by more than nine times when compared with the frequency from 1981 to 2000.

Funding: This study was supported by the National Key Research and Development Program of China (Project No. 2016YFD0300201) and the National Science Foundation of China (Project No. 41801078).

Institutional Review Board Statement: Not applicable.

Informed Consent Statement: Not applicable.

Data Availability Statement: The data presented in this study are available on request from the authors.

Conflicts of Interest: The author declares no conflict of interest.

References

1. IPCC. *Climate Change 2013. The Physical Science Basis*; Cambridge University Press: New York, NY, USA, 2013.
2. Bailey-Serres, J.; Parker, J.E.; Ainsworth, E.A.; Oldroyd, G.E.; Schroeder, J.I. Genetic strategies for improving crop yields. *Nature* **2019**, *575*, 109–118. [CrossRef] [PubMed]
3. Wang, P.; Hu, T.; Kong, F.; Zhang, D. Changes in the spatial pattern of rice exposure to heat stress in China over recent decades. *Clim. Chang.* **2019**, *154*, 229–240. [CrossRef]
4. Ishimaru, T.; Xaiyalath, S.; Nallathambi, J.; Sathishraj, R.; Yoshimoto, M.; Phoudalay, L.; Samson, B.; Hasegawa, T.; Hayashi, K.; Arumugam, G.; et al. Quantifying rice spikelet sterility in potential heat-vulnerable regions: Field surveys in Laos and southern India. *Field Crops Res.* **2016**, *190*, 3–9. [CrossRef]
5. Iizumi, T.; Furuya, J.; Shen, Z.; Kim, W.; Okada, M.; Fujimori, S.; Hasegawa, T.; Nishimori, M. Responses of crop yield growth to global temperature and socioeconomic changes. *Sci. Rep.* **2017**, *7*, 7800. [CrossRef]
6. Zhao, C.; Liu, B.; Piao, S.; Wang, X.; Lobell, D.B.; Huang, Y.; Huang, M.; Yao, Y.; Bassu, S.; Ciais, P.; et al. Temperature increase reduces global yields of major crops in four independent estimates. *Proc. Natl. Acad. Sci. USA* **2017**, *114*, 9326–9331. [CrossRef] [PubMed]
7. Seck, P.A.; Diagne, A.; Mohanty, S.; Wopereis, M.C. Crops that feed the world 7: Rice. *Food Secur.* **2012**, *4*, 7–24. [CrossRef]
8. Tao, F.; Zhang, Z.; Zhang, S.; Zhu, Z.; Shi, W. Response of crop yields to climate trends since 1980 in China. *Clim. Res.* **2012**, *54*, 233–247. [CrossRef]
9. Tang, H.; Pang, J.; Zhang, G.; Takigawa, M.; Liu, G.; Zhu, J.; Kobayashi, K. Mapping ozone risks for rice in China for years 2000 and 2020 with flux-based and exposure-based doses. *Atmos. Environ.* **2014**, *86*, 74–83. [CrossRef]
10. Bheemanahalli, R.; Sathishraj, R.; Manoharan, M.; Sumanth, H.N.; Muthurajan, R.; Ishimaru, T.; Krishna, J.S. Is early morning flowering an effective trait to minimize heat stress damage during flowering in rice? *Field Crop Res.* **2017**, *203*, 238–242. [CrossRef]
11. Espe, M.B.; Hill, J.E.; Hijmans, R.J.; McKenzie, K.; Mutters, R.; Espino, L.A.; Leinfelder-Miles, M.; van Kessel, C.; Linquist, B.A. Point stresses during reproductive stage rather than warming seasonal temperature determines yield in temperate rice. *Glob. Chang. Biol.* **2017**, *23*, 4386–4395. [CrossRef]
12. Ortiz-Bobea, A.; Wang, H.; Carrillo, C.M.; Ault, T.R. Unpacking the climatic drivers of US agricultural yields. *Environ. Res. Lett.* **2019**, *14*, 064003. [CrossRef]
13. Chen, W.F.; Xu, Z.J.; Tang, L. 20 years' development of super rice in China-The 20th anniversary of the super rice in China. *J. Integr. Agric.* **2017**, *16*, 981–983. [CrossRef]
14. He, L.; Cleverly, J.; Wang, B.; Jin, N.; Mi, C.; Liu, D.L.; Yu, Q. Multi-model ensemble projections of future extreme heat stress on rice across southern China. *Theor. Appl. Climatol.* **2018**, *133*, 1107–1118. [CrossRef]
15. Xiong, D.; Ling, X.; Huang, J.; Peng, S. Meta-analysis and dose-response analysis of high temperature effects on rice yield and quality. *Environ. Exp. Bot.* **2017**, *141*, 1–9. [CrossRef]
16. Xu, Y.; Chu, C.; Yao, S. The impact of high-temperature stress on rice: Challenges and solutions. *Crop J.* **2021**, *9*, 963–976. [CrossRef]
17. Liu, J.; Hasanuzzaman, M.; Wen, H.; Zhang, J.; Peng, T.; Sun, H.; Zhao, Q. High temperature and drought stress cause abscisic acid and reactive oxygen species accumulation and suppress seed germination growth in rice. *Protoplasma* **2019**, *256*, 1217–1227. [CrossRef]
18. Rehmani, M.I.; Wei, G.; Hussain, N.; Ding, C.; Li, G.; Liu, Z.; Wang, S.; Ding, Y. Yield and quality responses of two indica rice hybrids to post-anthesis asymmetric day and night open-field warming in lower reaches of Yangtze River delta. *Field Crops Res.* **2014**, *156*, 231–241. [CrossRef]
19. Rehmani, M.I.; Ding, C.; Li, G.; Ata-Ul-Karim, S.T.; Hadifa, A.; Bashir, M.A.; Hashem, M.; Alamri, S.; Al-Zubair, F.; Ding, Y. Vulnerability of rice production to temperature extremes during rice reproductive stage in Yangtze River Valley, China. *J. King Saud Univ.–Sci.* **2021**, *33*, 101599. [CrossRef]
20. Xu, J.; Henry, A.; Sreenivasulu, N. Rice yield formation under high day and night temperatures—A prerequisite to ensure future food security. *Plant Cell Environ.* **2020**, *43*, 1595–1608. [CrossRef]
21. Sita, K.; Sehgal, A.; HanumanthaRao, B.; Nair, R.M.; Vara Prasad, P.V.; Kumar, S.; Gaur, P.M.; Farooq, M.; Siddique, K.H.; Varshney, R.K.; et al. Food legumes and rising temperatures: Effects, adaptive functional mechanisms specific to reproductive growth stage and strategies to improve heat tolerance. *Front. Plant Sci.* **2017**, *8*, 1658. [CrossRef]
22. Begcy, K.; Sandhu, J.; Walia, H. Transient heat stress during early seed development primes germination and seedling establishment in rice. *Front. Plant Sci.* **2018**, *9*, 1768. [CrossRef]
23. Huang, J.; Zhang, F.; Xue, Y.; Lin, J. Recent changes of rice heat stress in Jiangxi Province, southeast China. *Int. J. Biometeorol.* **2017**, *61*, 623–633. [CrossRef]
24. Xiong, Z.; Wang, S.Y.; Zeng, Z.M.; Fu, C.B. Analysis of simulated heavy rain over the Yangtze River Valley during 11–30 June 1998 using RIEMS. *Adv. Atmos. Sci.* **2003**, *20*, 815–824.
25. Xiong, Z.; Fu, C.B.; Zhang, Q. On the ability of the regional climate model RIEMS to simulate the present climate over Asia. *Adv. Atmos. Sci.* **2006**, *23*, 784–791.162. [CrossRef]

26. Xiong, Z.; Fu, C.B.; Yan, X.D. Regional integrated environmental model system and its simulation of East Asia summer monsoon. *Chin. Sci. Bull.* **2009**, *54*, 4253–4261.161. [CrossRef]
27. Lv, H.; Zhang, Y.; Feng, M.; Li, B.; Liu, A.; Mao, F.; Zhuang, L.; Li, Y.; Wu, M. *Assessment of Heating Harm for Rice*; National Standard of People's Republic of China: Beijing, China, 2011.
28. Zhang, J.; Zhang, Z.; Tao, F. Performance of temperature-related weather index for agricultural insurance of three main crops in China. *Int. J. Disaster Risk Sci.* **2017**, *8*, 78–90. [CrossRef]
29. Yan, C.; Ding, Y.; Wang, Q.; Liu, Z.; Li, G.; Muhammad, I.; Wang, S. The impact of relative humidity, genotypes and fertilizer application rates on panicle, leaf temperature, fertility and seed setting of rice. *J. Agric. Sci.* **2010**, *148*, 329–339. [CrossRef]
30. Zhang, Z.; Chen, Y.; Wang, C.; Wang, P.; Tao, F. Future extreme temperature and its impact on rice yield in China. *Int. J. Climatol.* **2017**, *37*, 4814–4827. [CrossRef]
31. Zhang, S.; Tao, F. Improving rice development and phenology prediction across contrasting climate zones of China. *Agric. For. Meteorol.* **2019**, *268*, 224–233. [CrossRef]
32. Zhang, Z.; Li, Z.; Chen, Y.; Zhang, L.; Tao, F. Improving regional wheat yields estimations by multi-step-assimilating of a crop model with multi-source data. *Agric. For. Meteorol.* **2020**, *290*, 107993. [CrossRef]

agriculture

MDPI

Article

A Gas Diffusion Analysis Method for Simulating Surface Nitrous Oxide Emissions in Soil Gas Concentrations Measurement

K. M. T. S. Bandara [1,2,*], Kazuhito Sakai [1,3,*], Tamotsu Nakandakari [1,3] and Kozue Yuge [1,4]

1 United Graduate School of Agricultural Sciences, Kagoshima University, 1-21-24 Korimoto, Kagoshima 890-0065, Japan; zhunai@agr.u-ryukyu.ac.jp (T.N.); yuge@cc.saga-u.ac.jp (K.Y.)
2 Department of Agricultural Engineering, Faculty of Agriculture, University of Ruhuna, Kamburupitiya 81100, Sri Lanka
3 Faculty of Agriculture, University of the Ryukyus, 1 Senbaru, Okinawa 903-0213, Japan
4 Faculty of Agriculture, Saga University, 1 Honjo-machi, Saga 840-8502, Japan
* Correspondence: thusihta@gmail.com (K.M.T.S.B.); ksakai@agr.u-ryukyu.ac.jp (K.S.)

Abstract: The detection of low gas concentrations from the soil surface demands expensive high-precision devices to estimate nitrous oxide (N_2O) flux. As the prevalence of N_2O concentration in the soil atmosphere is higher than its surface, the present study aimed to simulate N_2O surface flux (CF) from soil gas measured in a soil-interred silicone diffusion cell using a low-cost device. The methodological steps included the determination of the diffusion coefficient of silicone membrane (D_{slcn}), the measurement of the temporal variations in the N_2O gas in the soil (C_{si}) and on the surface (MF), and the development of a simulation process for predicting CF. Two experiments varying the procedure and periods of soil moisture saturation in each fertilized soil sample were conducted to detect C_{si} and MF. Using D_{slcn} and C_{si}, the variations in the soil gas (C_{soil}) were predicted by solving the diffusion equation using the implicit finite difference analysis method. Similarly, using six soil gas diffusivity models, the CF values were simulated from C_{soil}. For both experiments, statistical tests confirmed the good agreement of CF with MF for soil gas diffusivity models 4 and 5. We suggest that the tested simulation method is appropriate for predicting N_2O surface emissions.

Keywords: nitrous oxide; soil gas flux; silicone diffusion cell; soil gas diffusivity; passive gas sampling; soil gas diffusion coefficient; soil gas flux simulation

Citation: Bandara, K.M.T.S.; Sakai, K.; Nakandakari, T.; Yuge, K. A Gas Diffusion Analysis Method for Simulating Surface Nitrous Oxide Emissions in Soil Gas Concentrations Measurement. *Agriculture* **2022**, *12*, 1098. https://doi.org/10.3390/agriculture12081098

Academic Editors: Dengpan Xiao and Wenjiao Shi

Received: 19 June 2022
Accepted: 23 July 2022
Published: 26 July 2022

1. Introduction

Accelerated crop production ensures food security for the global population. High-yielding crop varieties that consume significant amounts of synthetic fertilizers are being cultivated to secure food production needs [1]. Nitrogen-based fertilizers are essential for plants during their growth stages [2]. In addition to synthetic fertilizers, the use of organic manure in the forms of crop residues, animal waste, and biological N-fixing plants is common in plant nutrient supply chains [3,4]. By 2019, the global average atmospheric nitrous oxide (N_2O) concentration had increased to 333.2 ppbv [5]. The use of synthetic fertilizers is one of the major causes of these changes [6]. The contribution of N_2O to atmospheric warming is 298 times greater than that of carbon dioxide, and it significantly contributes to the depletion of the stratospheric ozone layer [7,8]. Therefore, control of fertilizer application levels is highly needed to limit the N_2O emissions from the crop fields. Soil can be considered as a large bioreactor that produces various materials as output [9,10], especially greenhouse gases (GHGs) [11]. Among the GHGs, methane (CH_4), carbon dioxide (CO_2), and nitrous oxide (N_2O), the nitrogen-based fertilizers highly contribute to the N_2O production via the microbial denitrification and nitrification processes in the soil [12]. The results of various experiments convey the impact of environmental and agronomic factors on the N_2O emissions from uplands [13–16]. Globally, collective strategies are being implemented to achieve the sustainable development goals (SDGs) by 2030 [17]. The assessment of N_2O emissions is vital for implementing climate-responsive actions, such as control measures to limit GHG emissions from cropping fields, controlling the use of synthetic fertilizers, and testing alternative soil nutrient enrichment methods.

According to Denmead 2008, the main techniques that are used for measuring methane and nitrous oxide fluxes in inland ecosystems and atmospheres are known as enclosure-based and micrometeorological measurements [18]. Butterbach-Bahl et al. have mentioned that the selection of the appropriate technique is mainly based on investment capacity, the demand for the findings, and the type of research question to be analyzed [19]. As mentioned in the work of Maljanen et al. [20], the chamber technique is commonly used to estimate N_2O fluxes [21–23]. According to Butterbach-Bahl et al., homogenous fields, including those with trees, slopes, and that are building-free, and the mono-crop condition are required for micrometeorological measurements, and the related measurement instruments should be high-precision fast-response devices. When using this technique, the expense for conducting the onsite measurement in one station has been indicated to be USD 60,000–80,000 for CO_2, and this price can be increased by USD 30,000 to USD 40,000 for each additional gas (CH_4, N_2O, etc.) if needed. In 2014, Chikowo et al. also mentioned the inappropriateness of micrometeorological techniques in small-scale farming systems that insist on intercropping, a soil fertility gradient, and complicated land use patterns [24]. The chamber method can be identified as a system that directly samples gas from the soil surface through the use of permanently installed enclosed chambers. In the chamber method, suites for small-scale cultivated areas are created by establishing small, simple chambers on the soil surface. The chamber method can operate manually or automatically when taking gas samples for gas monitoring [19,25]. Since gas sampling is performed within the chamber, a small area of the field is represented when calculating the emissions, and the installation of the chamber disturbs the natural environmental conditions of the soil gas diffusion process. According to the air circulation schedule, there are two types of chambers called static and dynamic chambers, in which the chamber gas is replaced rapidly while measurements continue and while scheduled ventilation takes place after the gas has been collected in the chamber for a period of time [26,27]. Low-level N_2O gas concentrations are more prevalent in the chamber than in the soil atmosphere, since gas diffuses from the soil surface to the atmosphere. Therefore, chamber methods also require high-precision gas-monitoring devices and related accessories, increasing the measurement cost. Conversely, the operating costs of the gas-monitoring process limit the expansion of estimation activities for GHG emissions. It is necessary to develop techniques that use low-cost gas-monitoring devices together with an accurate gas-sampling system to better estimate N_2O gas emissions.

Compared to surface emissions, a higher N_2O gas concertation level is prevalent in the soil atmosphere [28]. This supports the development of a low-cost gas-monitoring instrument to measure the N_2O gas releasing activities in the soil atmosphere if gas samples are collected effectively. The soil gas sampling can be carried out indirectly by inserting a perforated pipe or silicone tube into the soil, and the process is known as the passive gas sampling method. The soil gas then surrounds the tube and equilibria with the inner gas level via a gas diffusion process [29–31]. Because of its molecular structure, silicone has special characteristics that make it suitable for soil gas sampling, such as water repellence, releasability, cold resistance, and high gas diffusivity. In various studies, these silicone tubes have been tested to determine their gas sampling ability [32–34]. The results of previous studies have demonstrated the suitability of silicone tubes [35] and perforated stainless-steel pipes [36] for soil gas sampling in continuous automated sampling methods and manual sampling methods. Nondispersive infrared (NDIR) technology has been successfully applied to develop GHG monitoring devices for CO_2, CH_4 [37–40], and N_2O [28]. In the upper-level detection limits (upper ppm levels) shown by low-cost devices compared to high-precision devices (QCL, CRL laser-based, photoacoustic, and FTIR devices), where the detection limit shows the sub-ppm level, higher soil gas levels are more prevalent, meaning that low-cost devices can be used for gas measurements with soil gas diffusion cells.

In the process of numerical solving of soil gas flux, the gradient method has gained the most attention. In porous soil, gas flux can be estimated based on the measured gas concentrations at each level of the soil profile and the gas diffusivity of the soil by assuming that gas diffusion is the dominant gas transport mechanism in the soil [41]. Since there are high levels of uncertainty when testing the soil gas diffusivity, some research works used soil gas diffusivity model-based approximation approaches. Since flux is more sensitive when its own calculation steps are used, the use of implicit assumptions was suggested [31,42,43]. Therefore, the soil gas diffusivity model-based implicit finite difference simulation method can be used to effectively estimate the N_2O flux from measured soil gas levels.

In summary, N_2O flux assessment activities in the agricultural fields need to be accelerated to control GHG emissions. The operational costs induced by high-precision gas-monitoring devices and the related accessories that are necessary for existing gas-monitoring methodologies are major barriers to achieving the targeted gas emission estimations. Mutually, low-cost N_2O gas-monitoring devices use passive sampling methods to sample the gases in the soil (where there are higher gas levels than on the surface), and numerical simulation approaches can be used to predict gas fluxes in cost-efficient ways. Therefore, this study demonstrated a simulation approach for estimating N_2O flux on the soil surface according to the soil gas concentration measured using a low-cost measuring device connected to a silicone soil gas sampler.

2. Materials and Methods

The conceptual framework of the experiment was mainly based on the soil N_2O flux simulation process of the recorded soil N_2O levels determined by a low-cost NDIR device with a diffusion cell (silicone tube) entombed in the soil region, where there are higher gas concentration levels than in the atmosphere (Figure 1). Together with the diffusion coefficient, the inner N_2O gas concentration of the diffusion cell changes, along with the temporal variations in the soil gas concentration. From the gas concentrations recorded in the silicone diffusion cell, the predicted values on temporal changes in the soil N_2O gas concentration, hereafter C_{soil}, were simulated by solving the diffusion equation (Equation (4)) using the implicit finite difference method.

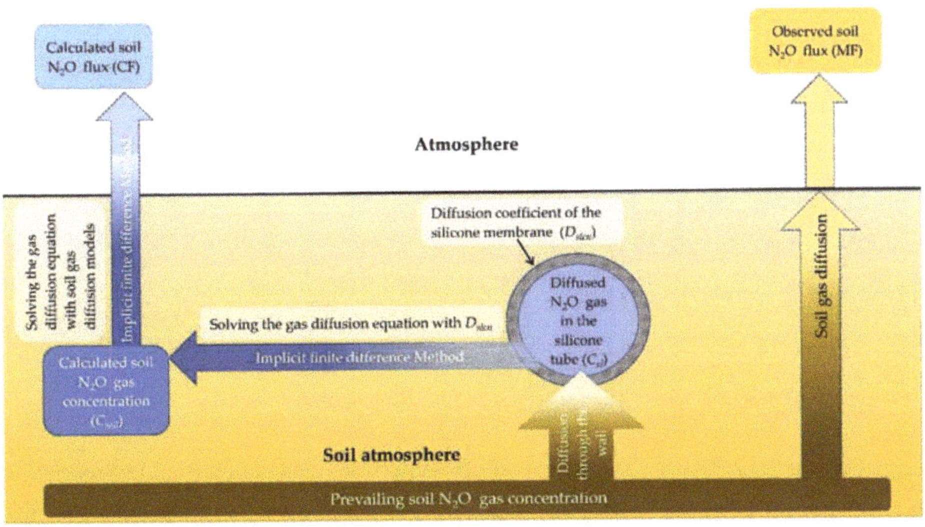

Figure 1. The main simulation steps for predicting the N_2O flux from measured soil gas levels in the silicone diffusion cell.

Secondly, the values calculated for N_2O flux on the soil surface (CF) were simulated by solving the diffusion equation (implicit finite difference method) using the six soil gas diffusivity models. To compare the simulated values, the observed surface flux of N_2O gas (MF) was calculated from monitored soil surface emission levels using a high-resolution gas-monitoring device followed by the chamber method. Accordingly, the overall experiment was based on three main steps: (1) the determination of the diffusion coefficient of the silicone membrane; (2) the measurement of the concurrent N_2O gas concentrations in the soil gas and soil surface; and (3) the development of a simulation process to determine the CF values. The simulation process is described below.

2.1. Determination of the Diffusion Coefficient of the Silicone Membrane

The observed N_2O concentration in the silicone diffusion cell was the result of the diffused soil gas travelling through the wall of the silicone tube. Therefore, the N_2O concentration in soil gas should be estimated by solving the diffusion equation according to the observed N_2O concentration in the silicone tube. The diffusion coefficient of the silicone membrane (D_{slcn}) was determined, since it was required to solve the diffusion equation during the simulation process. In the experimental setup, two silicone tubes (length 59 cm, internal diameter 6 mm, and external diameter 8 mm for each) were placed in an enclosed chamber (8000 mL), and the two edges of each tube were serially connected with an air circulation pump, FTIR device (system 1; Perkin Elmer—Spectrum Two FT-IR spectrometer with a long path gas cell system, 7 m optical path length, volume 500 mL, Infrared Analysis, Inc., Anaheim, USA, model 7.2-V), and a new NDIR device [28] (system 2; volume 320.3 mL) to create two separate diffused gas measurement systems (Figure 2). After adding a known N_2O gas (purity level 99.5%) volume (50 mL) into the gas chamber, the accumulated gas level in the silicone tube was measured by each device at 30-min intervals for 12 h. From the data recorded by each device, the gas concentration in the silicone tubes (system1 and system 2) and the balanced gas concentration in the gas chamber were approximately determined at each successive period by applying the difference equations below. The gas concentrations of system 1 and system 2 were calculated by solving Equation (3), and D_{slcn} was optimized using an R package.

$$q_1 = D_s \frac{C_c - C_1}{\Delta x} A = V_1 \frac{C_1 - C_1^o}{\Delta t} \quad (1)$$

$$q_2 = D_s \frac{C_c - C_2}{\Delta x} A = V_2 \frac{C_2 - C_2^o}{\Delta t} \quad (2)$$

$$V_c \frac{C_c - C_c^o}{\Delta t} = -(q_1 + q_2) \quad (3)$$

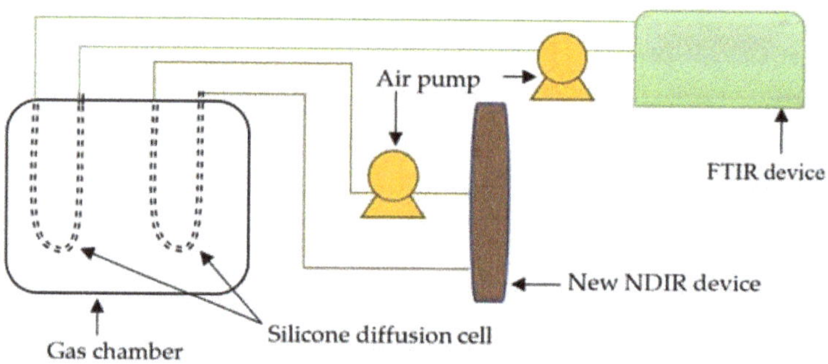

Figure 2. Experimental setup for determining the diffusion coefficient of the silicone diffusion cell.

q_1: rate of the diffused gas volume from container to system 1 (10^{-6} cm^3 s^{-1}); q_2: rate of the diffused gas volume from container to system 2 (10^{-6} cm^3 s^{-1}); D_s: gas diffusion rate in silicone membrane ($cm^2\ s^{-1}$); C_c: gas concentration in the container ($10^{-6}\ cm^3/cm^3$); C_1: gas concentration in system 1 ($10^{-6}\ cm^3/cm^3$); C_2: gas concentration in system 2 ($10^{-6}\ cm^3/cm^3$); C_c^o: gas concentration of previous time step in the container ($10^{-6}\ cm^3/cm^3$); C_1^o gas concentration of previous time step in system 1 ($10^{-6}\ cm^3/cm^3$); C_2^o: gas concentration of previous time step in system 2 ($10^{-6}\ cm^3/cm^3$); V_c: volume of the container (cm^3); V_1: volume of system 1 (cm^3); V_2: volume of system 2 (cm^3); Δx: thickness of the silicone tube (cm); Δt: time step (s); A: surface area of the silicone tube (cm^2).

2.2. The Experimental Setup for Monitoring the N_2O Flux and Soil Atmospheric N_2O Gas Concentrations

To simulate the surface flux (CF) and compare it with the measured flux (MF), temporal variation in the soil and atmospheric N_2O concentrations should be monitored. Therefore, a laboratory test was conducted to monitor the gas concentrations in both regions (Figure 3). A soil sample (*shimajiry maji*) weight of 4 kg taken from the research field of the University of the Ryukyus was sieved with a 2 mm sieve (particle density: 2.685 g/cm^3, bulk density: 0.958 g/cm^3) and was repacked in the testing chamber (c). As an ammonium-based nitrogen source, 3 g of $(NH_4)_2SO_4$ was mixed with the soil. As shown in Figure 3, the experimental setup mainly consisted of serially interconnected air-drying sections (g,i) in each gas-monitoring device (h,i) to monitor the gas concentrations in the soil and atmospheric regions. The soil chamber (c) had a hole underneath and was connected to the water drainage system (f) to ensure drainage after the completion of the saturation events. There were two water supply systems that were used in the soil saturation events: the top water supply system (a) applied water to the top surface of the soil, and the bottom water supply unit (Mariotte cell) (b) provided a controlled water supply from the bottom side of the soil layer.

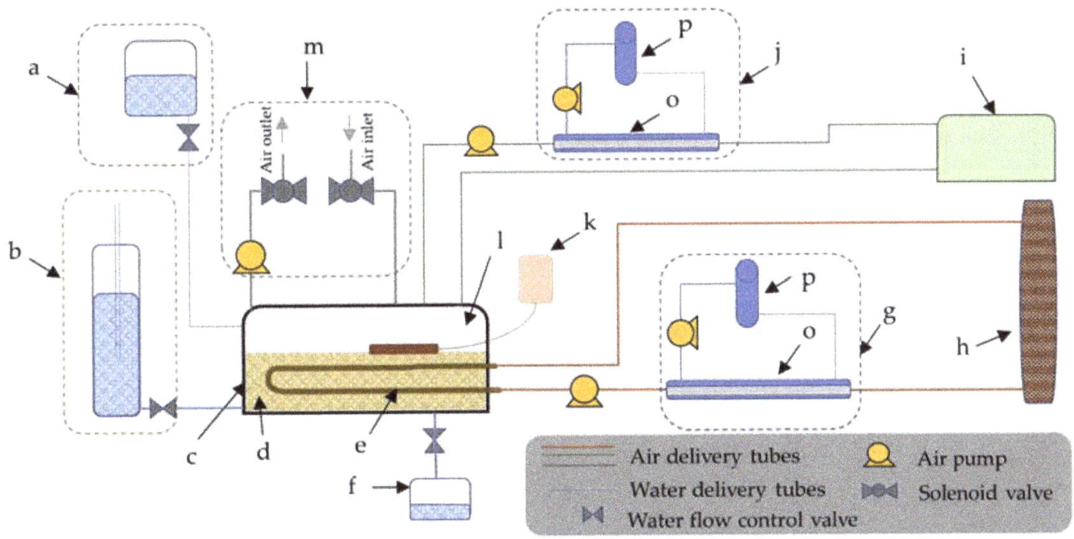

Figure 3. A diagram of the laboratory experimental setup for monitoring variations in the headspace and soil atmospheric N_2O gas concentrations: (**a**) top-side water supply unit, (**b**) bottom-side water supply unit (Mariotte cell), (**c**) soil chamber, (**d**) soil region, (**e**) diffusion cell (silicone tube), (**f**) water drainage system, (**g**,**j**) membrane air-drying system, (**h**) low-cost NDIR gas-monitoring device, (**i**) FTIR device, (**k**) data logger connected to soil moisture and temperature sensors, (**l**) headspace of the soil chamber, and (**m**) ventilation system for chamber head space.

A silicone tube diffusion cell (length: 59 cm, internal diameter: 0.6 cm, and wall thickness: 0.1 mm) (e) was buried in the soil 3.5 cm from the top so that the diffused soil N_2O gas could accumulate into it through the wall, and the two ends of the tube were serially connected with a low-cost gas-monitoring device (h). Data were recorded in 30-min intervals. The diffused air was circulated within the system using an air pump (AS ONE-EAP-01). The circulated air was sent through a drying section to avoid water accumulation in the system when it was operating over long periods of time. The drying sections were developed by using membrane-type dryers (Suncep SWG-A01-03) (o) consisting of two eccentric tubes. The outward tube for dry gas circulation was connected to a silicone moisture absorber (p), and the middle tube was for soil gas circulation.

To monitor the headspace gas levels, an FTIR spectrometer (i) was used. The measurement schedule for the headspace gas was every 1 min over a 30-min period in a closed chamber and over 1 h of chamber ventilation (by ventilation system (m)) period. The gas flux was calculated from the slope according to the gas concentration in the last seven minutes during the closed-chamber period. A sensor connected to a data logger (moisture meter embedded with a thermometer) (k) was placed in the soil container, and data were recorded at 30-min intervals.

To test the gas diffusion at different volumetric water levels, the water supply and drainage events were carried out under different conditions for each experiment. Using the different water supply methods, two experiments were conducted in which saturation took place over different periods of time. The first experiment was based on the bottom-side water supply, in which a Mariotte cell (b) was used to control the water supply drop by drop. The saturation and drainage periods were both two days long. During the saturation period, the bottom edge of the Mariotte cell's air intake tube was maintained at the same level as the soil in the chamber. In the second experiment, the top-side water supply unit (a) was used to control the water supply (rate: 0.05 mL S^{-1}) over a two-day saturation period, and the soil was allowed to drain with the help of gravity over the course of three days. The drainage steps in each experiment included open gravity drainage (experiment 1) and flow-controlled gravity drainage (rate: 0.05 mL S^{-1}) (experiment 2). For both experiments, concurrent soil gas concentration was measured by each device according to the abovementioned measurement schedules.

2.3. Simulation Steps for Predicted N_2O Surface Flux (CF)

The two simulation steps were followed as indicated in Figure 1. The first step was to simulate C_{soil} according to the measured gas concentrations in a silicone diffusion cell (C_{si}). The second step was to estimate the values for the predicted N_2O flux from the soil surface (CF). In the second step, the parameters of the soil diffusivity models were optimized on a trial-and-error basis until the CF values fit with the measured N_2O flux (MF) from the soil surface.

2.3.1. Simulation Steps for Predicting the Soil N_2O Gas Level from the Measured Gas in Silicone Diffusion Cell

Theoretically, the soil N_2O gas can be estimated from the inverse analysis of the diffusion equation on C_{si}. However, the results of the calculation did not converge satisfactorily. Therefore, we estimated the concentration of the N_2O gas in the soil (C_{soil}) according to the following steps: (1) C_{soil} was assumed to be the boundary condition of the soil side for solving the diffusion equation (Equation (4)). In the first equation, the observed C_{si} was used by shifting the assumed C_{soil} a few hours forward. For the first calculation, the observed C_{si} was shifted the same number of hours forward as the assumed C_{soil}. (2) The assumed C_{soil} was used to simulate the gas concentration in the silicone tube (C_{ssi}) by solving the diffusion equation according to the implicit finite difference method. (3) The simulated values of the C_{ssi} *time series* were compared with those of the C_{si}. Steps 1 to 4 were repeated until the C_{ssi} matched the C_{si} on a trial-and-error basis (Figure 4). An assumption was made about the completion of the gas measurements soon after the gas mixture entered the silicone diffusion cell.

$$\frac{\partial C_{si}}{\partial t} = D_{slcn}\frac{\partial^2 C_{si}}{\partial x^2} \tag{4}$$

where:

C_{si}: N_2O concentration in the silicone diffusion cell (g gas m^{-3} diffusion cell air);
D_{slcn}: gas diffusion coefficient of the silicone membrane (m^3 gas m^{-1} silicone s^{-1});
Δx: thickness of the silicone tube (m); Δt: time step (s).

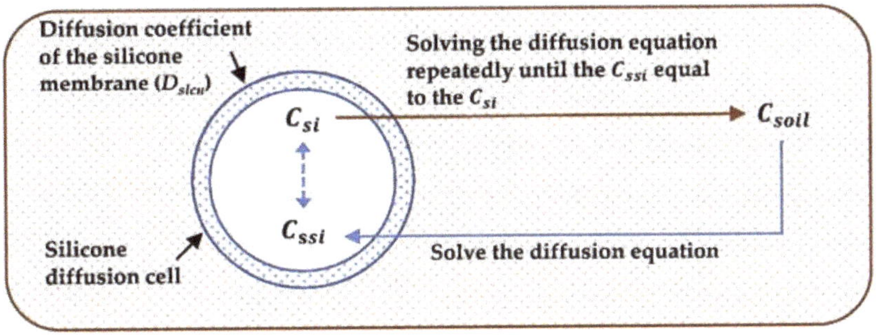

Figure 4. The repetitive simulation steps for obtaining the predicted N_2O gas levels in soil.

2.3.2. Steps for Simulating the Predicted N_2O Flux from the Soil Surface

Using the gas diffusion coefficients for soil, the predicted N_2O flux from the soil surface (CF) was simulated by solving the diffusion equation (Equation (5)) according to the implicit finite difference method. The boundary conditions of the simulation were considered to be as follows: the soil side used simulated C_{soil} values, and the atmospheric side was set to the N_2O gas concentration of 0 ppm. Since the gas diffusion coefficient was the main factor influencing the simulation results, the six models that were used to calculate the gas diffusion coefficient in soil (D_{soil}) were used to test the best-fitting cumulative values of CF with the observed flux (MF). The cumulative values of MF were calculated from the measured values of the N_2O concentration in the upper chamber using an FTIR device.

$$\frac{\partial C_{soil}}{\partial t} = \frac{D_{soil}}{\varepsilon}\frac{\partial^2 C_{soil}}{\partial x^2} \tag{5}$$

where:

C_{soil}: predicted soil N_2O gas concentration (g gas m^{-3} soil air);
D_{soil}: N_2O gas diffusion coefficient in soil (m^3 soil air m^{-1} soil s^{-1});
ε: air-filled porosity (m^3 soil-air m^{-3} soil);
Δx: thickness of the soil layer above the silicone tube (m);
Δt: time step (s).

Six soil gas diffusivity models based on diffusive transport gas movement (soil-type-independent and soil-water characteristic-based models) were used to determine the gas diffusion coefficients of the soil. The models were defined as the relative diffusion coefficient (diffusion coefficient in the soil air (D_{soil})/diffusion coefficient in free air (D_O)) as a function of air-filled porosity (ε) (m^3 soil-air m^{-3} soil). According to the first model, the model suggested by Buckingham (1904) and known as a power function of (ε), Equation (6) was applied [44]. According to the soil type: sand, loam, or clay, the power function n varies from 1.7 to 2.3. In the current experiment, we assigned the value of n to be 2. The models suggested by Millington and Quirk (1960, 1961), indicated in Equations (7) and (8), where the ratio of the power function of air-filled porosity (ε) to *Sat* is known as soil total porosity (m^3 pore space m^{-3} soil), were used as models 2 and 3 [45,46]. As large tortuosity develops, the presence of water can affect gas diffusion in soil, and some soil gas diffusivity models have been developed for wet soils. Model 4 (WLR–Marshall model) (Equation (9)), which was developed by Moldrup et al., 2000a, according to the Marshall (1959) model and assumes water-induced linear reduction (WLR), was used [47]. A model based on the gas diffusivities at the soil water potential (−100 cm H_2O) and the corresponding air-filled porosities (macroporosity) developed by Moldrup et al., 2000b, was applied as the fifth model (Equation (10)) in the simulation [48]. The sixth model (Equation (11)), which was based on the gas diffusivity in unsaturated soil as suggested by Moldrup et al., 1996, was applied [49]. The required soil gas diffusivity for the gas flux simulation process was determined by each model using recorded temporal soil moisture data (volumetric water content-VWC).

$$\text{Model 1 } \frac{D_{soil}}{D_0} = f(\varepsilon)f(\varepsilon) = \varepsilon^n \tag{6}$$

$$\text{Model 2 } f(\varepsilon) = \varepsilon^2 / Sat^{2/3} \tag{7}$$

$$\text{Model 3 } f(\varepsilon) = \varepsilon^{10/3} / Sat^2 \tag{8}$$

$$\text{Model 4 } f(\varepsilon) = \varepsilon^{2.5} / Sat \tag{9}$$

$$\text{Model 5 } f(\varepsilon) = 2 \times \varepsilon^3 + 0.04 \times \varepsilon \tag{10}$$

$$\text{Model 6 } f(\varepsilon) = 0.1\left\{2\left(\frac{\varepsilon}{Sat}\right)^3 + 0.04\left(\frac{\varepsilon}{Sat}\right)\right\} \tag{11}$$

From the CF values for each model and the MF values, the cumulative flux variation in the CF and MF (hereinafter *CFcu* and *MFcu*) was calculated for the ease of comparing the simulated and predicted values at each time step. From the simulation output, there were six cumulative flux (*CFcu*) lines for all of the models. Therefore, to test the statistical accuracy of the developed *CFcu* lines with the *MFcu* of soil gas diffusivity models, the symmetric mean absolute percentage error (*SMAPE*) (Equation (12)) was used. The *SMAPE* expresses the averaged percentages of the sums of ratios between the absolute differences in *CFcu* and *MFcu*, and half the sum of the absolute values of the *CFcu* and *MFcu* at each time point (n).

$$SMAPE = \frac{100\%}{n} \sum_{t=1}^{n} \frac{|CFcu_t - MFcu_t|}{\frac{(|MFcu_t| + |CFcu_t|)}{2}} \tag{12}$$

where:

$SMAPE$: symmetric mean absolute percentage error (%);
$CFcu_t$: cumulative calculated flux at time t;
$MFcu_t$: cumulative measured flux at time t;
n: number of measurement points in the time series.

To test the model fitness of the *CFcu* to the *Mfcu*, *"Willmott's agreement index (d)"* was applied. *"d"* is a dimensionless measurement of model accuracy developed by Willmott (1981) to standardize (range between 0 and 1) the measured degree of the model prediction error. A standardized value of 1 indicates a perfect match between *CFcu* and *MFcu*, and 0 that there is no agreement between them. *"d"* is described by Equation (13). To compare the results of both accuracy tests, heatmaps were developed for *SMAPE*, as well as the *d* values of both experiments 1 and 2.

$$d = 1 - \frac{\sum_{t=1}^{n} (CFcu_t - MFcu_t)^2}{\sum_{t=1}^{n} \left(|CFcu_t - \overline{MFcu}| + |MFcu_t - \overline{MFcu}|\right)^2} \tag{13}$$

where:

d: Willmott's agreement index (goodness of fit);
$CFcu_t$: cumulative calculated flux at time t;
$MFcu_t$: cumulative measured flux at time t;
n: number of measurement time points t.

3. Results and Discussion

3.1. Diffusion Coefficient of the Silicone Membrane

The results of the gas diffusion test and simulation process are shown in Figure 5. Depending on the variations in the concentration gradient between the gas chamber and silicone tubes, the graphs of the observed N_2O gas accumulation rate in the silicone tubes changed from being steep to moderate. Because of the different internal volumes of the two gas-monitoring devices (system 1: 500 mL and system 2: 320.3 mL), the steepness of the two curves demonstrates steeper N_2O gas accumulation in system 1 than in system 2. In the simulation output, the temporal variation in calculated gas concentrations in both systems match with the measured values. For the N_2O gas, the diffusion coefficient of the silicone membrane (D_{slcn}) was 1.1×10^{-8} cm^2/s according to the steps resolved in the difference equations (Equation (3)).

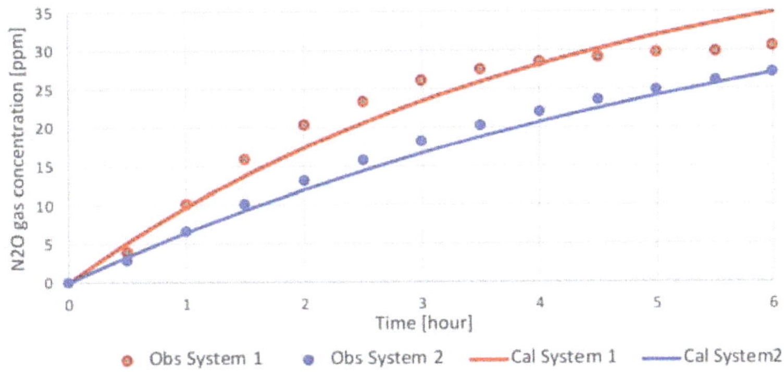

Figure 5. Temporal variations in measured and calculated N_2O gas concentrations in the silicone tubes of systems 1 and 2.

3.2. Results of the Variations in N_2O Gas Concentration in the Silicone diffusion cell and N_2O Gas flux of the Headspace of Experiment 1 and 2

Considering experiments 1 and 2, the variations in the gas flux from soil surface and the gas concentrations in the silicone diffusion cell show similar patterns during the irrigation and drainage events (Figures 6 and 7).

During the saturation events, higher gas production was observed in both experiments than during the drainage period. As discussed in previous studies [50–53], during saturation events, microbes produce more N_2O gas via nitrification and denitrification processes, since the soil moisture regulates the oxygen availability in the soil pores. Zheng et al. [54] also demonstrated a similar N_2O gas variation pattern in in situ measurements testing the impacts of soil moisture on gas emissions on crop land. However, in this study, during the drainage period, considerably elevated gas levels were observed for a short period of time when the water filling the space in the soil pores was replaced when the soil gas changed its diffusion direction temporarily to be downwards. This condition, which was determined to be a function of soil gas diffusion under low water-filled pore spaces, disappeared soon after the drainage completion and was clearly shown in both experiments. In experiment 1, the drainage and saturation periods were both two days long, and water was allowed to drain with the help of gravity (faster draining process than experiment 2). A very sharp elevated gas level was observed at the beginning of the drainage period compared to that in experiment 2, where the drainage process was carried out at a lower rate. The graphs showing the volumetric water content in Figures 6 and 7 also demonstrate the two drainage flow rates in experiments 1 and 2.

In both experiments, at peak levels, the variations in the N_2O gas concentration levels that were measured in the silicone diffusion cell were approximately 10 times higher than they were in the measurements for the headspace concentration. This characteristic explains the possibility of using low-cost and less precise gas-monitoring devices to monitor soil gas levels compared with the costly surface chamber methods [25] associated with high-precision gas measurement events for determining trace gas levels. During the two experiments, the average soil temperatures were 22.6 ± 0.41 °C and 26.52 ± 1.01 °C for experiments 1 and 2, respectively.

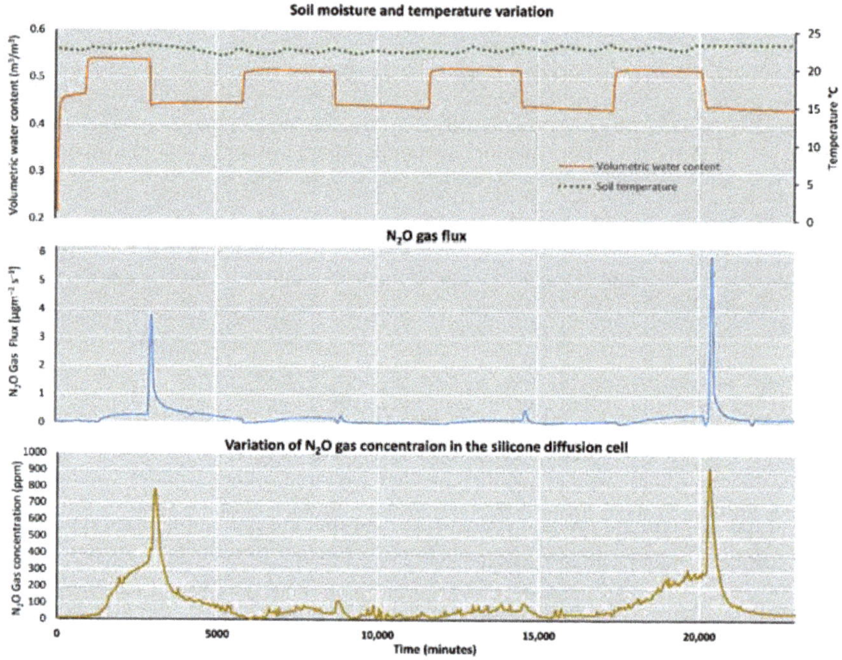

Figure 6. Temporal variations in the N_2O gas concentrations in the silicone diffusion cell and N_2O gas flux from the headspace, soil temperature, and moisture levels in experiment 1.

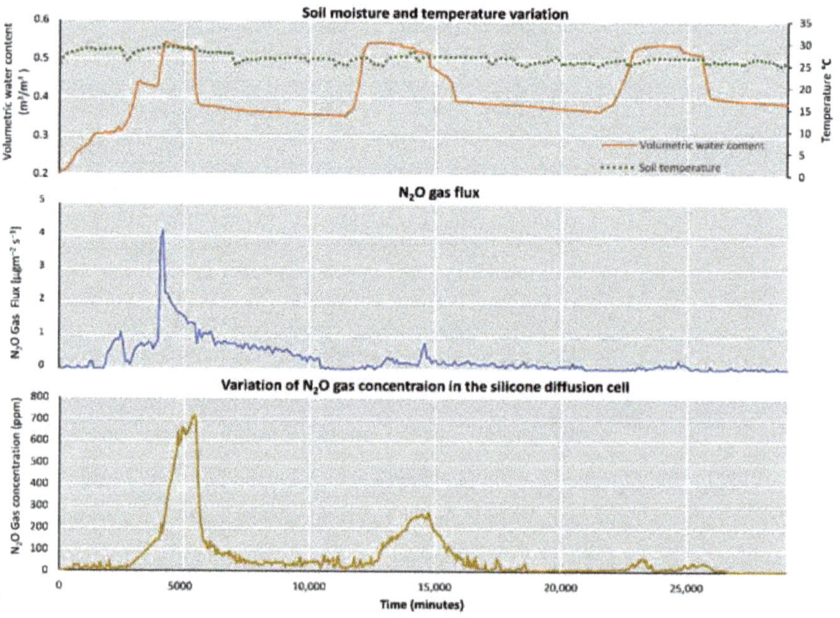

Figure 7. Temporal variations in the N_2O gas concentrations in the silicone diffusion cell and N_2O gas flux from the headspace, soil temperature, and moisture levels in experiment 2.

3.3. Results of the Simulation Steps of the Soil N_2O Gas Level Prediction

For both experiments 1 and 2, the series of C_{soil} were shifted to an earlier position (time lag) in the time series compared to the C_{si} series (Figures 8 and 9). Because of the diffusion coefficient of the silicone diffusion cell, theoretically, the simulated values of soil N_2O should appear earlier on in the time series than the time points of the corresponding gas levels of C_{si}, and this time lag confirms the condition. The time lag was the result of the repeated simulation steps that were conducted to achieve the overlapping C_{ssi} and C_{si} series. Since C_{ssi} is simulated from C_{soil}, the accuracy of the simulation process is confirmed by the clearly overlapping C_{si} series on the C_{ssi} series in both experiments. In the case of experiment 1, the high C_{ssi} peaks that occurred soon after drainage began were smaller than those that were observed for C_{si}. We considered that these peaks could be caused by gas diffusion, as well as by rapid changes in the soil gas pressure under the forcible drainage. Therefore, these peaks could not be calculated by solving the diffusion equation.

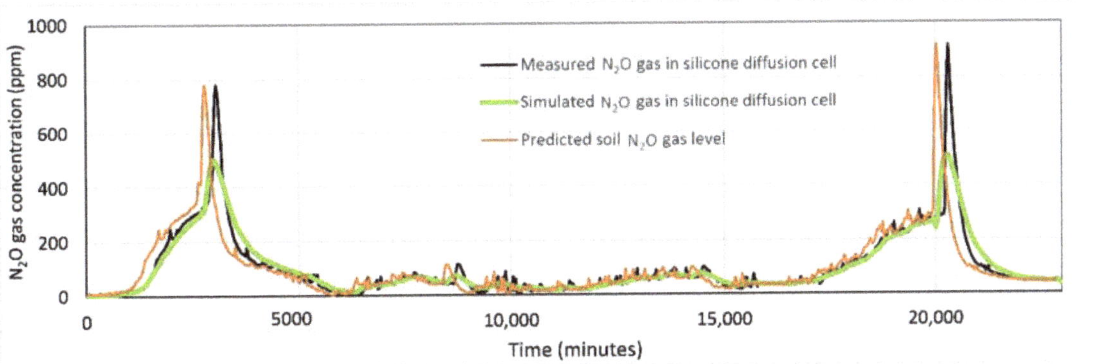

Figure 8. Temporal variations in measured and simulated N_2O gas levels in the silicone diffusion cell with predicted soil gas levels for experiment 1.

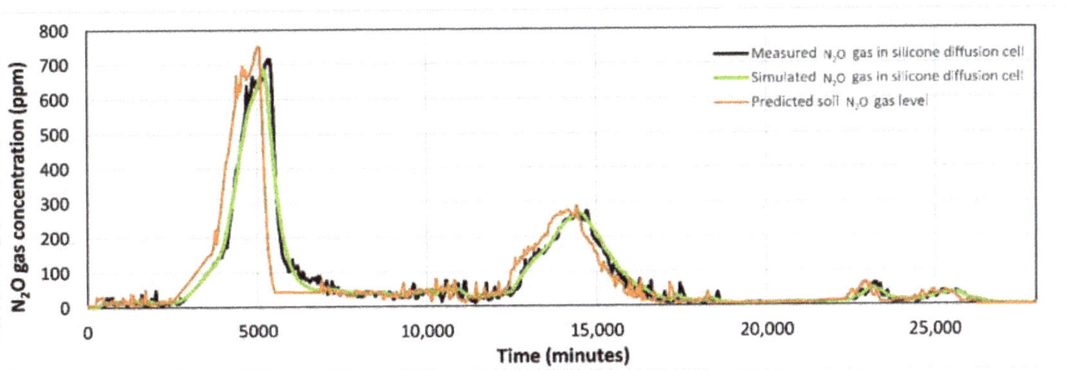

Figure 9. Temporal variations in measured and simulated N_2O gas levels in the silicone diffusion cell with predicted soil gas levels for experiment 2.

According to the graphical descriptions of Figures 10 and 11 for experiments 1 and 2, the highest agreement level between *CFcu* and *MFcu* is shown by soil gas diffusivity models 4 and 5. The results (Table 1) of the accuracy tests (*SMAPE:* symmetric mean absolute percentage error (%), *d:* Willmott's agreement index) for model fitting with *MFcu* also confirm the output of said graphical explanation. Among the tested characteristic-based soil gas diffusivity models for soil water, models 4 and 5 demonstrate lower *SMAPE* (%) values (experiment 1: 8.18%, 10.18%; experiment 2: 10.73%, 8.02%) and higher *d* values (experiment 1: 0.9996, 0.9994; experiment 2: 0.9992, 0.9997). The independent models for the different soil types (models 1, 2, and 3) showed higher SMAPE and lower *d* values for both experiments. Therefore, according to the selected soil category, by considering the common simulation approach for predicting soil surface N_2O flux from the soil gas levels, soil gas diffusivity models 4 and 5 are the most appropriate regardless of whether or not the water supply method is controlled.

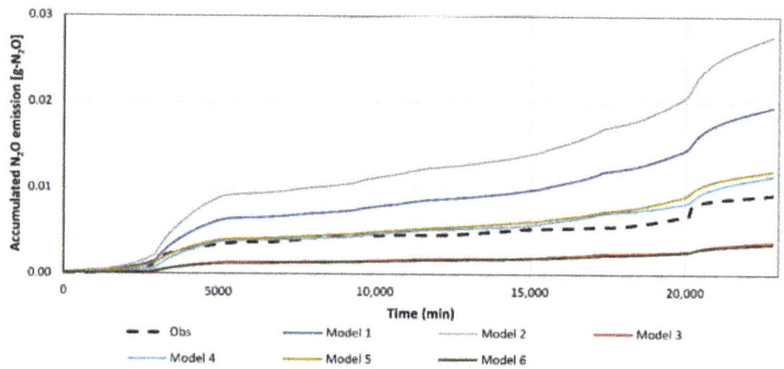

Figure 10. Experiment 1: temporal progression of the simulated cumulative N_2O emissions in each of the soil gas diffusivity models with the observed emissions.

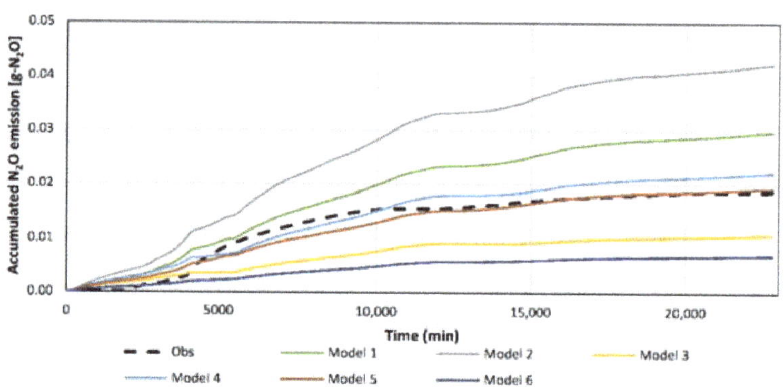

Figure 11. Experiment 2: temporal progression of the simulated cumulative N_2O emissions for each of the soil gas diffusivity models with the observed emissions.

Considering the results of the two experiments, a simulation approach for predicting N_2O soil surface flux according to measured soil N_2O gas data was successfully carried out at the laboratory level. However, additional experiments need to be conducted on cultivated land before this method is adopted at the field level.

The tested method, including the arrangement of its hardware arrangement together with the simulation steps, makes it easier to estimate soil N_2O emissions once the gas diffusion coefficients of the silicone membrane and the soil have been determined. Compared to closed-chamber methods, the tested method requires less accessories for gas sampling and circulating and uses low-cost measurement devices. As a passive gas sampling unit, the special characteristics (water repellency, structural stability of the wall, and higher gas permeability) of the silicone diffusion cell allow it to be layered under the soil. This keeps the upper surface free of gas-sampling devices and enables natural soil gas diffusion to the air. Moreover, the gas sampler does not require operating power or intensive maintenance, which allows it to be used for long-term monitoring.

Table 1. Results of the accuracy tests conducted for *CFcu* with all models with *MFcu*.

Experiment	Accuracy Test	Model					
		1	2	3	4	5	6
1	*SMAPE* (%)	16.45	31.85	20.02	8.18	10.18	19.72
	d	0.9954	0.9905	0.9965	0.9996	0.9994	0.9964
2	*SMAPE* (%)	12.63	27.89	18.18	10.73	8.02	20.65
	d	0.9947	0.9844	0.9924	0.9992	0.9997	0.9893

Color scale of SMAPE*: Symmetric mean absolute percentage error (%).*

0	5	10	15	20	25	30	35

Color scale of d*: Willmott's agreement index (goodness of fit (scale 0–1) of the model's output (CF) to the measured cumulative N_2O flux (MF)).*

0.9	0.92	0.94	0.96	0.98	1

The gas sampling technique of the proposed simulation method has demonstrated the function of fewer gas sampling structures above the ground, which allows the natural gas diffusion process from the soil surface. Reflecting the current greenhouse gas sampling and analyzing techniques, compared to the precise eddy covariance technique [55,56], the chamber enclosure method was greatly applied because of the relative inexpensiveness and simplicity of the measurement mechanism [57,58]. However, the artifacts of established gas sampling structures (chamber artifacts) form errors and higher uncertainty in the ultimate results [59] via the altered diffusion gradient [60] and pressure artifacts [61,62]. The long covering period of the transparent chamber results in higher internal temperatures, and the opaque chamber method was proposed for blocking the impact of the light and tested for CO_2 [63] and CH_4 [64] gases. For the N_2O flux, the DN-based opaque static chamber measurement method has been tested to further improvement of accuracy in the opaque chamber [65]. Therefore, the tested method including soil-entombed silicone diffusion cell can minimize the artifacts and biases made by above-ground gas sampling structures of the chamber methods.

In the simulation steps, there are only two stages of analysis in the implicit finite difference method, which are based on the volumetric water content of the soil and the diffusion coefficients of the silicone membrane and the soil. Moreover, special models and solution methods are not required, and only a simple solution for diffusion equation analysis is enough for estimating the flux. Therefore, said parameters for the soil and gas sampler are found, and the analysis of the replicated data from various sampling points can be continued effortlessly. As a low-cost method, it is more likely that this method will be adopted at the field level, and this method increases the number of measurement points per field, accelerating the estimation of N_2O gas emission activities in agricultural fields.

4. Conclusions

We tested a simulation method for estimating the soil surface flux of N_2O gas from measured soil atmospheric gas levels. The methodology consisted of three steps: determining the diffusion coefficient of the silicone membrane, monitoring N_2O gas variations in the soil and gas flux from soil surface, and carrying out the simulation steps for predicting the surface emissions from the measured soil gas levels. The diffusion coefficient of the silicone membrane for the N_2O gas was determined to be 1.1×10^{-8} cm^2/s by solving a diffusion equation. In the first stage of simulation, D_{slcn} was used to predict the variation in the soil N_2O gas level according to the measured N_2O gas concentrations in a silicone diffusion cell by solving the gas diffusion equation via the implicit finite difference analysis method. In the second stage, using soil gas diffusion coefficients from six soil gas diffusivity models, the N_2O gas flux from the soil surface emissions was predicted from the predicted soil gas levels. At the laboratory level, we successfully simulated the cumulative values of the predicted soil surface N_2O flux and confirmed the good agreement with the measured cumulative flux graphically and statistically in both experiments. From the tested six soil gas diffusivity models, models 4 and 5 demonstrated lower *SMAPE* (%) and higher *d* values for *CFcu* and *Mfcu* in both experiments. The results of the two experiments using varying soil saturation methods and durations in the fertilized soil samples confirmed that the simulation methodology was acceptable for predicting the surface flux from measured soil gas levels. The overlapping *CFcu* and *Mfcu* curves and the results of the statistical tests (*SMAPE* (%) and *d*) demonstrate how expensive conventional N_2O flux estimation methods can be replaced with the use of low-cost gas-monitoring devices for soil gas measurements with gas flux simulation steps. Further field-level studies are needed before the simulation method is adopted for use in cultivated cropland.

Author Contributions: Conceptualization, methodology, and formal analysis, K.M.T.S.B., K.S., T.N., and K.Y.; investigation and writing—original draft preparation, K.M.T.S.B. and K.S.; writing—review and editing, K.M.T.S.B. and K.S.; supervision, K.S., T.N., and K.Y. All authors have read and agreed to the published version of the manuscript.

Funding: This research was supported by the JSPS KAKENHI Grant-in-Aid for Scientific Research(B), number 21H02307.

Institutional Review Board Statement: Not applicable.

Informed Consent Statement: Not applicable.

Data Availability Statement: The data are contained within the article.

Conflicts of Interest: The authors declare no conflict of interest.

References

1. Skiba, U.; Rees, B. Nitrous oxide, climate change and agriculture. *CAB Rev. Perspect. Agric. Vet. Sci. Nutr. Nat. Resour.* **2014**, *9*, 1–7. [CrossRef]
2. Mosier, A.R.; Syers, J.K.; Freney, J.R. Nitrogen fertilizer: An essential component of increased food, feed, and fiber production. In *Agriculture and the Nitrogen Cycle: Assessing the Impacts of Fertilizer Use on Food Production and the Environment*; Mosier, A.R., Syers, J.K., Freney, J.R., Eds.; SCOPE 65 Island Press: Washington, DC, USA, 2004; pp. 3–15.
3. Kaur, R.; Kaur, S. Biological alternates to synthetic fertilizers: Efficiency and future scopes. *Indian J. Agric. Res.* **2018**, *52*, 587–595. [CrossRef]
4. Gollner, G.; Starz, W.; Friedel, J.K. Crop performance, biological N fixation and pre-crop effect of pea ideotypes in an organic farming system. *Nutr. Cycl. Agroecosyst.* **2019**, *115*, 391–405. [CrossRef]
5. USGCRP (U.S. Global Change Research Program). *Climate Science Special Report: Fourth National Climate Assessment*; Wuebbles, D.J., Fahey, D.W., Hibbard, K.A., Dokken, D.J., Stewart, T.K., Maycock, B.C., Eds.; U.S. Global Change Research Program: Washington, DC, USA, 2017; Volume I. [CrossRef]
6. Global Annual Surface Man Abundances (2020) and Trends of Key Greenhouse Gases from the GAW In-Situ Observational Network for GHG. The Averaging Method is Described in GAW Report No. 184 9. Available online: https://gaw.kishou.go.jp/static/publications/bulletin/Bulletin2019/ghg-bulletin-16.pdf (accessed on 20 April 2022).
7. Ravishankara, A.R.; Daniel, J.; Portmann, R. Nitrous Oxide (N_2O): The Dominant Ozone-Depleting Substance Emitted in the 21st Century. *Science* **2009**, *326*, 123–125. [CrossRef]
8. Zschornack, T.; Rosa, C.; Reis, C.; Pedroso, G.; Camargo, E.; Santos, D.; Boeni, M.; Bayer, C. Soil CH_4 and N_2O Emissions from Rice Paddy Fields in Southern Brazil as Affected by Crop Management Levels: A Three-Year Field Study. *Rev. Bras. Ciênc. Solo* **2018**. [CrossRef]
9. Wang, B.; Brewer, P.E.; Shugart, H.H.; Lerdau, M.T.; Allison, S.D. Soil aggregates as biogeochemical reactors and implications for soil-atmosphere exchange of greenhouse gases—A concept. *Glob. Chang. Biol.* **2019**, *25*, 373–385. [CrossRef] [PubMed]
10. Chesworth, W. *Encyclopedia of Soil Science*; Springer: Dorderecht, The Netherlands, 2009; p. 902.
11. Smith, K.A.; Ball, T.; Conen, F.; Dobbie, K.E.; Rey, A. Exchange of greenhouse gases between soil and atmosphere: Interactions of soil physical factors and biological processes. *Eur. J. Soil Sci.* **2018**, *69*, 2–4. [CrossRef]

12. Barnard, R.; Leadley, P.W.; Hungate, B.A. Global change, nitrification, and denitrification: A review. *Glob. Biogeochem. Cycles.* **2005**, *19*, GB1007. [CrossRef]
13. Van Groenigen, K.J.; Osenberg, C.W.; Hungate, B.A. Increased soil emissions of potent greenhouse gases under increased atmospheric CO_2. *Nature* **2011**, *475*, 214–216. [CrossRef]
14. Wolf, B.; Chen, W.; Brüggemann, N.; Zheng, X.; Pumpanen, J.; Butterbach-Bahl, K. Applicability of the soil gradient method for estimating soil-atmosphere CO_2, CH_4, and N_2O fluxes for steppe soils in Inner Mongolia. *J. Plant Nutr. Soil Sci.* **2011**, *174*, 359–372. [CrossRef]
15. Dijkstra, F.A.; Prior, S.A.; Runion, G.B.; Torbert, H.A.; Tian, H.Q.; Lu, C.Q.; Venterea, R.T. Effects of elevated carbon dioxide and increased temperature on methane and nitrous oxide fluxes: Evidence from field experiments. *Front. Ecol. Environ.* **2012**, *10*, 520–527. [CrossRef]
16. Ma, F.; Li, M.; Wei, N.; Dong, L.; Zhang, X.; Han, X.; Li, K.; Guo, L. Impacts of Elevated Atmospheric CO_2 and N Fertilization on N_2O Emissions and Dynamics of Associated Soil Labile C Components and Mineral N in a Maize Field in the North China Plain. *Agronomy* **2022**, *12*, 432. [CrossRef]
17. Sustainable Development Goals. Available online: https://www.un.org/sustainabledevelopment/climate-change/ (accessed on 15 April 2022).
18. Denmead, O.T. Approaches to measuring fluxes of methane and nitrous oxide between landscapes and the atmosphere. *Plant Soil.* **2008**, *309*, 5–24. [CrossRef]
19. Butterbach-Bahl, K.; Sander, B.O.; Pelster, D.; Díaz-Pinés, E. Quantifying Greenhouse Gas Emissions from Managed and Natural Soils. In *Methods for Measuring Greenhouse Gas Balances and Evaluating Mitigation Options in Smallholder Agriculture*; Rosenstock, T., Rufino, M., Butterbach-Bahl, K., Wollenberg, L., Richards, M., Eds.; Springer: Cham, Switzerland, 2016. [CrossRef]
20. Maljanen, M.; Liikanen, A.; Silvola, J.; Martikainen, P.J. Measuring N_2O emissions from organic soils by closed chamber or soil/snow N_2O gradient methods. *Eur. J. Soil Sci.* **2003**, *54*, 625–631. [CrossRef]
21. Martikainen, P.J.; Nykänen, H.; Crill, P.; Silvola, J. Effect of a lowered water table on nitrous oxide fluxes from northern peatlands. *Nature* **1993**, *366*, 51–53. [CrossRef]
22. Nykänen, H.; Alm, J.; Lång, K.; Silvola, J.; Martikainen, P.J. Emissions of CH_4, N_2O and CO_2 from a virgin fen and a fen drained for grassland in Finland. *J. Biogeogr.* **1995**, *22*, 351–357. [CrossRef]
23. Flessa, H.; Wild, U.; Klemisch, M.; Pfadenhauer, J. Nitrous oxide and methane fluxes from organic soils under agriculture. *Eur. J. Soil Sci.* **1998**, *49*, 327–335.
24. Chikowo, R.; Zingore, S.; Snapp, S.; Johnston, A. Farm typologies, soil fertility variability and nutrient management in smallholder farming in Sub-Saharan Africa. *Nutr. Cycl. Agroecosyst.* **2014**, *100*, 1–18. [CrossRef]
25. Breuer, L.; Papen, H.; Butterbach-Bahl, K. N_2O-emission from tropical forest soils of Australia. *J. Geophys. Res.* **2000**, *105*, 26353–26378. [CrossRef]
26. Butterbach-Bahl, K.; Papen, H.; Rennenberg, H. Impact of gas transport through rice cultivars on methane emission from rice paddy fields. *Plant Cell Environ.* **1997**, *20*, 1175–1183. [CrossRef]
27. Butterbach-Bahl, K.; Gasche, R.; Breuer, L.; Papen, H. Fluxes of NO and N_2O from temperate forest soils: Impact of forest type, N deposition and of liming on the NO and N_2O emissions. *Nutr. Cycl. Agroecosyst.* **1997**, *48*, 79–90. [CrossRef]
28. Bandara, K.M.T.S.; Sakai, K.; Nakandakari, T.; Yuge, K. A Low-Cost NDIR-Based N_2O Gas Detection Device for Agricultural Soils: Assembly, Calibration Model Validation, and Laboratory Testing. *Sensors* **2021**, *21*, 1189. [CrossRef]
29. Pihlatie, M.; Pumpanen, J.; Rinne, J.; Ilvesniemi, H.; Simojoki, A.; Hari, P.; Vesala, T. Gas concentration driven fluxes of nitrous oxide and carbon dioxide in boreal forest soil. *Tellus B Chem. Phys. Meteorol.* **2007**, *59*, 458–469. [CrossRef]
30. Goffin, S.; Longdoz, B.; Maier, M.; Schack-Kirchner, H.; Aubinet, M. Soil Respiration in forest Ecosystems: Combination of a multilayer Approach and an Isotopic Signal Analysis. *Commun. Agric. Appl. Biol. Sci.* **2012**, *72*, 139–144.
31. Maier, M.; Longdoz, B.; Laemmel, T.; Schack-Kirchner, H.; Lang, F. 2D profiles of CO_2, CH_4, N_2O and gas diffusivity in a well aerated soil: Measurement and Finite Element Modeling. *Agric. Met.* **2017**, *247*, 21–33. [CrossRef]
32. Jacinthe, P.-A.; Dick, W.A. Use of silicone tubing to sample nitrous oxide in the soil atmosphere. *Soil Biol. Biochem.* **1996**, *28*, 721–726. [CrossRef]
33. Jacinthe, P.-A.; Groffman, P.M. Silicone rubber sampler to measure dissolved gases in saturated soils and waters. *Soil Biol. Biochem.* **2001**, *33*, 907–912. [CrossRef]
34. Kammann, C.; Grünhage, L.; Jäger, H.J. A new sampling technique to monitor concentrations of CH_4, N_2O and CO_2 in air at well-defined depths in soils with varied water potential. *Eur. J. Soil Sci.* **2001**, *52*, 297–303. [CrossRef]
35. Maier, M.; Machacova, K.; Lang, F.; Svobodova, K.; Urban, O. Combining soil and tree-stem flux measurements and soil gas profiles to understand CH_4 pathways in *Fagus sylvatica* forests. *J. Plant Nutr. Soil Sci.* **2018**, *181*, 31–35. [CrossRef]
36. Maier, M.; Gartiser, V.; Schengel, A.; Lang, V. Long Term Soil Gas Monitoring as Tool to Understand Soil Processes. *Appl. Sci.* **2020**, *10*, 8653. [CrossRef]
37. Marinov, M.; Nikolov, G.; Gieva, E.; Ganev, B. Improvement of NDIR carbon dioxide sensor accuracy. In Proceedings of the 38th International Spring Seminar on Electronics Technology (ISSE), Eger, Hungary, 6–10 May 2015; pp. 466–471.
38. Yasuda, T.; Yonemura, S.; Tani, A. Comparison of the characteristics of small commercial NDIR CO_2 sensor models and development of a portable CO_2 measurement device. *Sensors* **2012**, *12*, 3641–3655. [CrossRef] [PubMed]

39. Honeycutt, W.T.; Ley, M.T.; Materer, N.F. Precision and Limits of Detection for Selected Commercially Available, Low-Cost Carbon Dioxide and Methane Gas Sensors. *Sensors* **2019**, *19*, 3157. [CrossRef]
40. Hodgkinson, J.; Smith, R.; Wo, H.; Saffell, J.R.; Tatam, R.P. A low cost, optically efficient carbon dioxide sensor based on nondispersive infra-red (NDIR) measurement at 4.2 μm. In *Optical Sensing and Detection II*; International Society for Optics and Photonics: Bellingham, WA, USA, 2012; Volume 8439, p. 843919.
41. Maier, M.; Schack-Kirchner, H. Using the gradient method to determine soil gas flux: A review. *Agric. For. Meteorol.* **2014**, *192–193*, 78–95. [CrossRef]
42. Atsuhiro, T.; Tetsuya, H.; Hiroshi, A.T.; Yoshihiro, F. Analytical estimation of the vertical distribution of CO2 production within soil: Application to a Japanese temperate forest. *Agric. For. Meteorol.* **2004**, *126*, 223–235. [CrossRef]
43. Schack-Kirchner, H.; Kublin, E.; Hildebrand, E.E. Finite-Element Regression to Estimate Production Profiles of Greenhouse Gases in Soils. *Vadose Zone J.* **2011**, *10*, 169–183. [CrossRef]
44. Buckingham, E. *Contributions to Our Knowledge of the Aeration of Soils*; USDA Burr Soil Bulletin, 25; US Government Printing Office: Washington, DC, USA, 1904.
45. Millington, R.J.; Quirk, J.M. Transport in porous media. In *Transactions of the 7th International Congress of Soil Science*; Van Beren, F.A., Ed.; Elsevier: Amsterdam, The Netherlands, 1960; pp. 97–106.
46. Millington, R.J.; Quirk, J.M. Permeability of porous solids. *Trans. Faraday Soc.* **1961**, *57*, 1200–1207. [CrossRef]
47. Moldrup, P.T.; Olesen, J.; Gamst, P.; Schjønning, P.; Yamaguchi, T.; Rolston, D.E. Predicting the gas diffusion coefficient in repacked soil: Water-induced linear reduction model. *Soil Sci. Soc. Am. J.* **2000**, *64*, 1588–1594. [CrossRef]
48. Moldrup, P.T.; Olesen, P.; Schjønning, P.; Yamaguchi, T.; Rolston, D.E. Predicting the gas diffusion coefficient in undisturbed soil from soil water characteristics. *Soil Sci. Soc. Am. J.* **2000**, *64*, 94–100. [CrossRef]
49. Moldrup, P.; Kruse, C.W.; Rolston, D.E.; Yamaguchi, T. Modelling diffusion and reaction in soils: III. Predicting gas diffusivity from the Campbell soil water retention model. *Soil Sci.* **1996**, *161*, 366–375. [CrossRef]
50. Williams, E.J.; Hutchinson, G.L.; Fehsenfeld, F.C. NOx and N_2O emissions from soil. *Glob. Biogeochem. Cycles* **1992**, *4*, 351–388. [CrossRef]
51. Poth, M.; Focht, D.D. 15N kinetic analysis of N_2O production by Nitrosomonas europea: An examination of nitrifier denitrification. *Appl. Environ. Microbiol.* **1985**, *49*, 1134–1141. [CrossRef]
52. Davidson, E.A. Sources of nitric oxide and nitrous oxide following wetting of dry soil. *Soil Sci. Soc. Am. J.* **1992**, *56*, 95–102. [CrossRef]
53. Davidson, E.A. Soil water content and the ratio of nitrous oxide to nitric oxide emitted from soil. In *The Biogeochemistry of Global Change: Radiatively Active Trace Gases*; Oremland, R.S., Ed.; Chapman & Hall: New York, NY, USA, 1992.
54. Xunhua, Z.; Mingxing, W.; Yuesi, W.; Renxing, S.; Gou, J.; Jing, L.; Jisheng, J.; Laotu, L. Impacts of soil moisture on nitrous oxide emission from croplands: A case study on the rice-based agro-ecosystem in Southeast China. *Chemosphere—Glob. Chang. Sci.* **2000**, *2*, 207–224.
55. Wang, J.M.; Murphy, J.G.; Geddes, J.A.; Winsborough, C.L.; Basiliko, N.; Thomas, S.C. Methane fluxes measured by eddy covariance and static chamber techniques at a temperate forest in central Ontario, Canada. *Biogeosciences* **2013**, *10*, 4371–4382. [CrossRef]
56. Kroon, P.S.; Hensen, A.; Jonker, H.J.J.; Ouwersloot, H.G.; Vermeulen, A.T.; Bosveld, F.C. Uncertainties in eddy covariance flux measurements assessed from CH4 and N_2O observations. *Agric. For. Meteorol.* **2010**, *150*, 806–816. [CrossRef]
57. Cao, G.; Xu, X.; Long, R.; Wang, Q.; Wang, C.; Du, Y.; Zhao, X. Methane emissions by alpine plant communities in the Qinghai-Tibet Plateau. *Biol. Lett.* **2008**, *4*, 681–684. [CrossRef]
58. Davidson, E.A.; Nepstad, D.C.; Ishida, F.Y.; Brando, P.M. Effects of an experimental drought and recovery on soil emissions of carbon dioxide, methane, nitrous oxide, and nitric oxide in a moist tropical forest. *Glob. Chang. Biol.* **2008**, *14*, 2582–2590. [CrossRef]
59. Davidson, E.A.; Savage, K.; Verchot, L.V.; Navarro, R. Minimizing artifacts and biases in chamber-based measurements of soil respiration. *Agric. For. Meteorol.* **2002**, *113*, 21–37. [CrossRef]
60. Conen, F.; Smith, K.A. An explanation of linear increases in gas concentration under closed chambers used to measure gas exchange between soil and the atmosphere. *Eur. J. Soil Sci.* **2000**, *51*, 111–117. [CrossRef]
61. Longdoz, B.; Yernaux, M.; Aubinet, M. Soil CO_2 efflux measurements in a mixed forest: Impact of chamber disturbances, spatial variability and seasonal evolution. *Glob. Chang. Biol.* **2000**, *6*, 907–917. [CrossRef]
62. Lund, C.P.; Riley, W.J.; Pierce, L.L.; Field, C.B. The effects of chamber pressurization on soil-surface CO_2 flux and the implications for NEE measurements under elevated CO_2. *Glob. Chang. Biol.* **1999**, *5*, 269–281. [CrossRef]
63. Zou, J.; Huang, Y.; Zheng, X. Static opaque chamber-based technique for determination of net exchange of CO_2 between terrestrial ecosystem and atmosphere. *Chin. Sci. Bull.* **2004**, *49*, 381. [CrossRef]
64. Junwei, L.; Jianghua, W. Gross photosynthesis explains the 'artificial bias' of methane fluxes by static chamber (opaque versus transparent) at the hummocks in a boreal peatland. *Environ. Res. Lett.* **2014**, *9*, 105005.
65. Lei, M.; Wei, Z.; Xunhua, Z.; Zhisheng, Y.; Han, Z.; Rui, W.; Bo, Z.; Kai, W.; Chunyan, L.; Guangmin, C.; et al. Attempt to correct grassland N_2O fluxes biased by the DN-based opaque static chamber measurement. *Atmos. Environ.* **2021**, *264*, 118687. [CrossRef]

agriculture

MDPI

Article

Cropland Expansion Mitigates the Supply and Demand Deficit for Carbon Sequestration Service under Different Scenarios in the Future—The Case of Xinjiang

Mingjie Shi [1,2,3], Hongqi Wu [1,2,*], Pingan Jiang [1,2,*], Wenjiao Shi [3,4], Mo Zhang [3,4], Lina Zhang [1,2], Haoyu Zhang [5], Xin Fan [6,7], Zhuo Liu [1,2], Kai Zheng [1,2], Tong Dong [8] and Muhammad Fahad Baqa [4,9]

1 College of Resources and Environment, Xinjiang Agricultural University, Urumqi 830052, China
2 Xinjiang Key Laboratory of Soil and Plant Ecological Processes, Xinjiang Agricultural University, Urumqi 830052, China
3 Key Laboratory of Land Surface Pattern and Simulation, Institute of Geographic Sciences and Natural Resources Research, Chinese Academy of Sciences, Beijing 100101, China
4 College of Resources and Environment, University of Chinese Academy of Sciences, Beijing 100049, China
5 College of Geography and Remote Sensing Sciences, Xinjiang University, Urumqi 830046, China
6 School of Public Administration, China University of Geosciences (Wuhan), Wuhan 430074, China
7 Center for Turkmenistan Studies, China University of Geosciences, Wuhan 430074, China
8 Key Laboratory of Coastal Science and Integrated Management, First Institute of Oceanography, Ministry of Natural Resources, Qingdao 266061, China
9 Key Laboratory of Digital Earth Sciences, Aerospace Information Research Institute, Chinese Academy of Sciences, Beijing 100094, China
* Correspondence: whq@xjau.edu.cn (H.W.); xjaudbxb@xjau.edu.cn (P.J.)

Citation: Shi, M.; Wu, H.; Jiang, P.; Shi, W.; Zhang, M.; Zhang, L.; Zhang, H.; Fan, X.; Liu, Z.; Zheng, K.; et al. Cropland Expansion Mitigates the Supply and Demand Deficit for Carbon Sequestration Service under Different Scenarios in the Future—The Case of Xinjiang. *Agriculture* **2022**, *12*, 1182. https://doi.org/10.3390/agriculture12081182

Academic Editor: María Martínez-Mena

Received: 16 June 2022
Accepted: 4 August 2022
Published: 9 August 2022

Abstract: China's double carbon initiative faces huge challenges, and understanding the carbon sequestration service of terrestrial ecosystems under future interannual regional land use change is important to respond to China's carbon policy effectively. Previous studies have recognized the important impact of land use/land cover (LULC) planning on carbon sequestration in terrestrial ecosystem services (ESs). However, exploring trends in carbon sequestration under sustainable development scenarios that combine economic and ecological development, particularly the mechanisms that balance the supply and demand of carbon sequestration, still requires in-depth exploration in different geographical contexts. In this study, we present the LULC simulation framework from 2000 to 2030 for four different development scenarios in the Xinjiang region, located in an important Belt and Road region, including business as usual (BAU), rapid economic development (RED), ecological land protection (ELP), and sustainable development with both economic and ecological development (SD). Our results suggest that both the supply and demand of carbon stock in Xinjiang will increase in 2025 and 2030, with the demand exceeding the supply. However, our scenario planning mitigates the supply and demand deficit situation for carbon sequestration in the context of future cropland expansion in different scenarios. In summary, our study's findings will enrich the study of carbon sequestration under future scenarios in the Belt and Road region. Xinjiang should pay more attention to the dynamic changes in landscape type structure and its carbon storage supply and demand caused by cultivated land expansion. Among the four scenarios, the spatial difference between carbon storage supply and demand based on the SD scenario is the smallest, which is more in line with the high-quality development of regional ecological security in Xinjiang.

Keywords: carbon sequestration; different scenarios; land use; sustainable development; Xinjiang

1. Introduction

Along with the Chinese government's goal of achieving peak carbon by 2030 and carbon neutrality by 2060, the timely assessment of the terrestrial ecosystem carbon sequestration service has become one of the most important issues in response to the current carbon neutrality policy [1]. As the paramount indicator of ecosystem carbon stock services, terrestrial ecosystem carbon sequestration is critical to the carbon cycle [2–4]. Land use/land cover (LULC) change is one of the major factors influencing carbon sequestration in terrestrial ecosystems, as land use changes affect the material cycling and energy flow of soils and vegetation carbon sequestration by altering the structure and function of the original ecosystem [5]. Most studies only consider the supply of ecosystem services, ignoring the human demand for ecosystem services [6,7]. Therefore, exploring the coupling of supply and demand in terrestrial ecosystem carbon sequestration is crucial to deepening future human, economic, and social knowledge of carbon source sinks.

The methods currently used to assess carbon sequestration at national and regional scales fall into three broad categories [4,8,9]. The first is the field survey method, which is primarily an area-weighted average method based on soil profiles [3]. However, this type of study may cause some multi-scale variation in results due to differences in soil profile size, location, methods, and sampling periods [10]. The second approach is empirical biogeochemical modeling [11,12]. This approach creates much uncertainty in assessing carbon sequestration, mainly due to differences in the mechanisms or structures of different models [10]. Third are remote sensing methods for calculating net primary production (NPP) which are often used to estimate carbon stocks, but they produce very large errors in some arid and semi-arid regions [13,14]; further, using spatial scales smaller than NPP, typically <1 km resolution, does not provide a true per-pixel NPP output [15]. Currently, the combination of land use and terrestrial ecosystem carbon stock models is widely used in studies to estimate carbon sequestration and their future spatial variability, and the application of such methods is an important trend concerning the development of dynamic carbon stock assessments for the future [12,16]. Among the many models quantifying the carbon sequestration of ecosystem services, machine learning is considered a feasible and reliable method for assessing carbon sequestration, and it has been widely used in carbon stock assessments at national and regional scales to balance overexploitation and environmental protection [8,17]. However, there are still some limitations to the abovementioned research methods. First, they fail to analyze carbon sequestration under different future scenarios, and only assess current carbon sequestration in a single way. Secondly, it has not been possible to explore the coupling between the supply and demand of carbon sequestration in terrestrial ecosystems. Third, they fail to address the deficit in the supply and demand of carbon sinks resulting from the expansion of cropland under different future scenarios. Therefore, it is important to explore the coupling between the supply and demand of carbon sequestration under different future LULC policy scenarios for planning and analyzing the surplus/deficit of carbon sequestration in the context of cropland expansion to provide a balance of supply and demand for a sustainable landscape pattern.

While there is growing recognition of the impacts of rapid LULC change due to urbanization on ecosystem services (ESs), the LULC landscape continues to be transformed in an unsustainable manner [18]. Land management is one of the most important factors influencing land cover, either directly or indirectly, with policy and environmental planning decisions having a significant impact on how land is managed [19]. Moreover, at the landscape level, the current main challenge is to identify alternative best management scenarios for different LULC change scenarios [9]. Numerous studies have shown that the environmental impact could be improved by changing LULC dynamics [20–23]. For example, a study conducted in Hawaii, USA, examined various LULC scenarios, with an increase in the carbon sequestration service of 3458 tons of carbon in each specific scenario [20]. Research in the Willamette Basin of Oregon has shown that different scenarios of LULC can influence the spatial pattern of the carbon sequestration service and that optimized scenarios can increase carbon sequestration in terrestrial ecosystems [21]. Furthermore, a study in Beijing–Tianjin–Hebei, China, planned four different scenarios to explore the maximum area of ESs loss, thus ensuring that the critical ESs are not affected [22]. However, while previous studies have explored carbon sequestration from the perspective of maximizing economic or ecological benefits [21], there is still a paucity of studies that have examined the targeting of sustainable development goals (SDGs) for assessing carbon sequestration under sustainability scenarios that combine economic and ecological development.

The UN Sustainable Development Goals (SDGs) focus on regional development and ecological security. In the context of the SDGs, it is important to understand regional sustainable development planning and to assess local ecological security [24]. To fill the above research gaps, this paper takes the Xinjiang Uyghur Autonomous Region (hereinafter referred to as Xinjiang) as the study area, because this core area of the Silk Road along the Belt and Road can better reveal the spatial distribution characteristics and evolutionary patterns of mountain ecosystems in a temperate arid zone [25,26]. The study uses the gray multi-objective optimization–patch generation land use simulation (GMOP-PLUS) model to simulate the variation in land use landscape patterns under various scenarios and propose a sustainable development scenario that balances economic and ecological development. The study further applies a random forest model to quantify the carbon sequestration of terrestrial ecosystems in Xinjiang under different scenarios from 2000 to 2030 and to explore the coupling between the supply and demand of carbon sequestration. The main objectives of this study are three-fold: (i) to predict spatial–temporal patterns of land use in Xinjiang from 2020 to 2030 by the PLUS model under the business as usual, rapid economic development, ecological land protection, and sustainable development scenario; (ii) to quantify the spatial and temporal variation characteristics of terrestrial ecosystem carbon sequestration under different scenarios in Xinjiang during 2020–2030 using random forest models; and (iii) to elucidate the relationship between the supply and demand of carbon sequestration in Xinjiang, and explore the difference between the supply and demand of LULC on carbon sequestration under different scenarios.

2. Materials and Methods

2.1. Study Area

Xinjiang is located inland in northwestern China, with a geographical location bounded by (73°40′~96°18′ E, 34°25′~48°10′ N), spanning 2000 km from east to west and 1650 km from north to south, with an area of about 1.66 $\times$ 10^6 km^2, this accounting for about one-sixth of China's land area (Figure 1). The average annual temperature in Xinjiang is 10.5 °C, and there is ca. 2600 h of sunshine per year. The average annual rainfall is 145.5 mm, and the average annual evaporation is 1000–4500 mm.

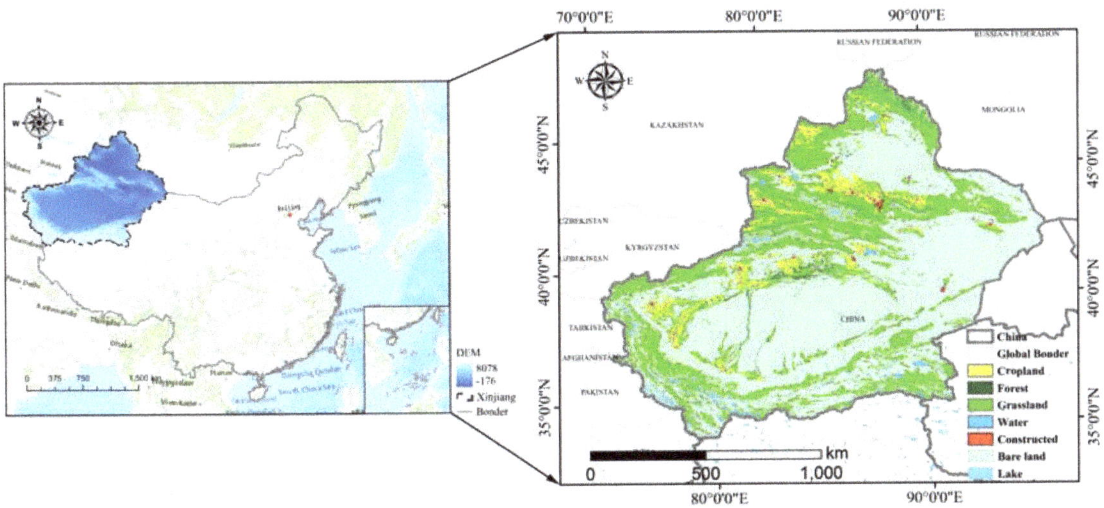

Figure 1. Digital elevation model of the study area.

As the core region of the overland Silk Road Economic Belt, Xinjiang is an important link for political, economic, and cultural exchanges between China and other Belt and Road countries. The Xinjiang government has historically attached importance to the multiple roles played by ecological and environmental protection, enacting and implementing several master land use plans in conjunction with an ecologically sustainable development agenda. Quantifying green spatial patterns and exploring trends in green spatial change in Xinjiang are essential for assessing and mapping the mismatch between supply and demand for ESs and providing guidance for future landscape and urban planning [25].

2.2. Data and Processing

The LULC dataset used in the study mainly includes: (1) Five periods of land use data with a spatial resolution of 30 m for 2000, 2005, 2010, 2015, and 2020 from the CAS Data Centre for Resource and Environmental Sciences (http://www.resdc.cn, (accessed on: 15 June 2022)). All these data were combined with field surveys, visual interpretation, and confusion matrix judgment, allowing for the total accuracy of the interpretation to reach 94.3% and the total accuracy of the 25 sub-categories to reach 91.2% [27]. According to the national land use category 1 classification system, there are six types of land: cropland, forest land, grassland, construction land, bare land, and water. (2) The annual average temperature and precipitation data used to discern suitability conditions for different land types were obtained from the CAS Data Centre (http://www.resdc.cn (accessed on: 15 June 2022)). For the latest year of meteorological data, we obtained raster data at a 250 m resolution by spatial interpolating the annual average data for 2020 from meteorological stations. (3) Digital elevation model (DEM) data, used to drive the LULC simulations for natural environmental factors, were obtained from the Geospatial Data Cloud (http://www.gscloud.cn (accessed on: 15 June 2022)) at a spatial resolution of 30 m. Soil type raster data came from the FAO dataset of the Food and Agriculture Organization of the United Nations (https://www.fao.org, (accessed on: 15 June 2022)). (4) Socioeconomic data, mainly containing the spatial distribution of population and gross domestic product (GDP) 1 km gridded data, came from the CAS Data Centre (http://www.resdc.cn (accessed on: 15 June 2022)). Vector datasets for assessing the distance to major roads and the distance to secondary roads came from Open Street Map (http://www.openstreetmap.org (accessed on: 15 June 2022)), and the vector data for river systems came from the National Geographic Information Resource Service (http://www.webmap.cn (accessed on 15 June 2022)). Urban night lighting data were obtained from the China Research Data Service Platform (https://www.cnrds.com (accessed on: 15 June 2022)). (5) The carbon density data of China's terrestrial ecosystems were taken from papers published between 2004 and 2014 and coupled with relevant experimental data from the same time period to generate a complete, systematic database of China's vegetation and soil organic carbon density [28]. In addition, all raster data were resampled to a spatial resolution of 250 m.

2.3. The GMOP-PLUS Model

2.3.1. PLUS

To better understand, assess, and predict future land use changes, research scholars have developed numerous land use simulation models. However, such models are usually linear and numerically based and cannot simulate all land use change processes [29]. However, the PLUS model can make use of the rule mining framework of the land expansion analysis strategy (LEAS) to yield a higher simulation accuracy than other models and better portray the landscape patterns of different future scenarios [25,29].

Under the influence of human social activities and regional socio-economic development, both the natural environment and policy factors can promote certain land use. Natural environmental factors include temperature and precipitation, among others. The process by which they drive such changes is complex and relatively stable, and the ensuing change is often small in magnitude over a short period. Policy factors that affect land use changes include GDP and population. In this paper, 12 driving factors affecting land change are used to reflect the changes of regional ecological environment and provide guidance and reference for the future planning of local land use [9,25,29].

To simulate the patch evolution of different scenarios of land use types, a multi-type random patch seeding mechanism based on threshold descent was used in the PLUS model:

$$OP_{i,k}^{1,t} = \begin{cases} P_{i,k}^{1} \times (r \times \mu_k) \times D_k^t & if\ \Omega_{i,k}^t = 0\ and\ r < P_{i,k}^1 \\ P_{i,k}^{1} \times \Omega_{i,k}^t \times D_k^t & all\ others \end{cases}, \quad (1)$$

where r is a random value in the range 0–1 and μ_k is the threshold value for generating new land-use patches of land use type k. The land use type k can be used to generate new land use patches. The number of decision trees is 50, the sampling rate is 0.01, and the number of features used to train the random forest is 12 (i.e., the same as the number of drivers) [29].

2.3.2. Gray Multi-Objective Optimization (GMOP)

GMOP is a dynamic multi-objective planning method that searches for ways to optimize the use of land given a variety of constraints imposed by different scenarios. It also takes into account the uncertainty of those constraints [30]. Accordingly, it is better able to make accurate models of how land use is spread out in space. The goal of this study was to find a sustainable way to use land with GMOP by using the objective optimization functions, constraints, and parameters that have been suggested by other studies [29,31].

2.3.3. GMOP-PLUS

Having been developed from the GM model and gray theory combined with multiple objectives, GMOP can consider the uncertainty of future LULC occurrence and solve the optimization problem of LULC by handling multiple constraints [32]. Previous studies have shown that the GMOP coupled PLUS model can play a comprehensive and decisive role in directing policy concerning the spatial allocation of land use [29,31]. Hence, the sustainable development scenario projected in this paper goes a step further than those used in previous studies [31,32]. In addition, we used Lingo 12.0 software to predict the spatial quantitative changes to the SD scenario in 2025 and 2030.

In our study, the land use structure of the SD scenario is assumed to maximize all three objectives simultaneously (Table 2) [30]. That is, with (1) $max_{E_{d(x)}}$ to maximize economic benefits, and (2) $max_{E_{p(x)}}$ to maximize ecological benefits, GMOP's optimization objectives are as follows:

$$E_{d(x)} = \sum_{i=1}^{n} d_i \cdot x_i, \quad (2)$$

$$E_{p(x)} = \sum_{i=1}^{n} p_i \cdot x_i\ . \quad (3)$$

where $E_{d(x)}$ and $E_{p(x)}$ denote economic and ecological benefits, respectively; x_i denotes the i category of land variable (i = 1, 2, … , 6); and d_i and p_i are the coefficients of economic and ecological benefits of the land category per unit area, respectively.

Table 1. Constraints on the objective function for the 2025 SD scenario (and likewise for 2030).

Subject to (Unit: Pixel Number)	Description
$x_1 + x_2 + x_3 + x_4 + x_5 + x_6 = 26,214,249$	The sum of the total area of all land use types generally remained constant before and after the simulation.
$1,384,334 \leq x_1 \leq 1,507,440$	To guarantee regional food security, the cropland area should not be lower than the 2020 level and less than the maximum number of pixels in the three scenarios (BAU, RED, and ELP).

Table 2. Constraints on the objective function for the 2025 SD scenario (and likewise for 2030).

Subject to (Unit: Pixel Number)	Description
$436,000 \leq x_2 \leq 448,145$	Forest is the ecological barrier of Xinjiang and should not be less than the 2020 level and less than the maximum number of pixels in the three scenarios (BAU, RED, and ELP).
$7,685,687 \leq x_3 \leq 7,851,922$	Grassland can contribute to livestock development, soil and water conservation, and ecological balance and should not be less than the 2020 level, and less than the maximum number of pixels in the three scenarios (BAU, RED, and ELP).
$539,509 \leq x_4 \leq 5,664,23$	The water area should be at least 90% of the 2020 level and less than the maximum number of pixels in the three scenarios (BAU, RED, and ELP).
$149,029 \leq x_5 \leq 153,558$	With the steady development of Xinjiang, which is bound to attract more people, the constructed area should be no less than the 2020 level and less than the maximum number of pixels in the three scenarios (BAU, RED, and ELP).
$15,761,239 \leq x_6 \leq 15,841,522$	We set the area of bare land to be greater than the 2020 level and below the maximum number of pixels in the three scenarios (BAU, RED, and ELP).

Achieving an optimal land use structure requires maximizing both objectives:

$$max\{E_{d(x)}, E_{p(x)}\} \tag{4}$$

2.4. Scenario Setting and Land Use Requirements

2.4.1. Scenario Setting

The research can be broadly divided into the following three steps. First, data on the LULC and the various drivers were prepared, and transformation rules for the LULC were developed. Second, spatial optimization of future LULC was carried out based on the PLUS model and Markov chain, and four different development scenarios were planned and simulated. Third, we explored the supply and demand balance relationships of the carbon sequestration service in Xinjiang terrestrial ecosystems under the different scenarios (Figure 2).

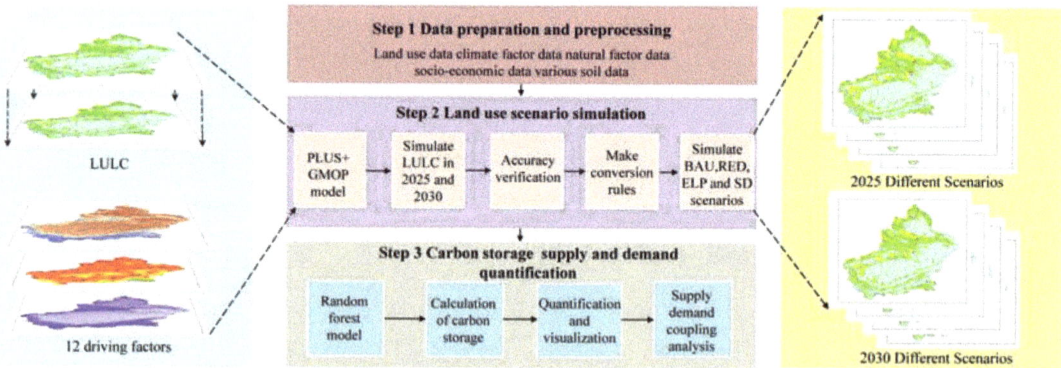

Figure 2. Science–policy framework linking institutional and ecological information.

Four alternative potential land use change scenarios are presented in this study, namely the business as usual (BAU), rapid economic development (RED), ecological land protection (ELP), and sustainable development (SD) scenarios. The principles and objectives of the design scenarios are as follows:

1. BAU scenario: This scenario assumes that land use change trends from the past will continue and that land demand for the BAU year of 2025 will be calculated based on the transition probability of shifts in the Markov chain for the 2015–2020 period and the 2030 BAU year based on the transition probability of shifts in the Markov chain for 2020–2025 BAU [9,25].
2. RED scenario: This scenario is based on the policy of rapid development of urban construction land in the region of the General Land Use Plan of the Xinjiang Uygur Autonomous Region [25]. The RED scenario prioritizes rapid economic development, leading to more demand for urban space. Based on the BAU scenario, and through a combination of thresholds set by previous studies, expert opinions, etc., we assume that the RED scenario accelerates the rate of conversion of grassland, construction land, and bare land to cropland by 50% and that the rate of conversion of cropland, grassland, and water to built-up land increases by the same 50% [33].
3. ELP scenario: This scenario is based on the Grain for Green Project, the Three-North Shelter Forest Program (TNSFP), and the 14th Five-Year Plan for Ecological Protection and represents the strengthening of the local government's commitment to forestry. This scenario represents the execution of the local government's policy of strengthening the protection of forests, grasslands, and water sources, strictly controlling the growth of cropland and construction land, and encouraging the return of farmland to forests, grass, and lakes. In this scenario, we modify the development potential of the cropland layer to convert farmland with a slope between 6° and 25° into grassland, and farmland with a slope greater than 25° into woodland. In addition, a buffer zone of 100 m near river waters was established as a woodland–grassland buffer zone [34].
4. SD scenario: The first three scenarios are more extreme, but the future development of Xinjiang cannot necessarily be modeled using a single scenario, and a trade-off between the three scenarios is needed to find the most appropriate development model for the region. To this end, this study proposes a sustainable development (SD) scenario, which provides a perspective on the trade-offs between the three scenarios. SDGs 15.3.1 represents the proportion of total land area that is degraded, which is a combination of three sub-indicators: land use change, land productivity, and carbon sequestration above and below ground. Given the data availability in our study, we have simplified the SDG 15.3.1 scenario by using only the land use scenario [35]. Although the SDG 15.3.1 scenario calculated here may not sufficiently reflect future realities, using GMOP-PLUS results to characterize the SDGs may provide a new perspective for planning SDGs under future land use change scenarios. Most importantly, specific implementation data for the SDGs model are not yet available for individual countries [36].

2.4.2. SD Scenario Setting

This study uses the GMOP-PLUS model to simplify the SD scenario not only to protect, restore, and promote the sustainability of terrestrial ecosystems, but also to account for rapid economic development. We first set up the land use economic value indicators to parameterize the individual land categories in the land use data. Here, x_1 = cropland; x_2 = forest; x_3 = grassland; x_4 = water; x_5 = construction; and x_6 = bare land. The average land economic value (RMB/hm^2) of each land category can be obtained from the Xinjiang Government Work Report and the Xinjiang 2020 Statistical Yearbook [30,37], and finally the economic value indicator formula was obtained as follows:

$$E_{red(x)} = 2.8x_1 + 0.22x_2 + 0.16x_3 + 0.08x_4 + 85.52x_5 + 0x_6 \quad (5)$$

Setting the ecological value index of land use, through the Xinjiang government work report and previous research results [31,37,38], the ecological value per unit area of land use (RMB/hm^2) was obtained, and the ecological value index formula was obtained as follows:

$$E_{elp(x)} = 1.31x_1 + 7.83x_2 + 2.57x_3 + 35.80x_4 + 0.0082x_5 + 0.016x_6 \quad (6)$$

To achieve the optimal sustainability scenario, the land use structure needs to maximize both of these indicators so that $E_{sdgs(x)}$ reaches the optimal ratio:

$$E_{sdgs(x)} = max\{E_{d(x)}, E_{p(x)}\} = \alpha E_{red(x)} + \beta E_{elp(x)} \quad (7)$$

The optimal adjustment of the land use structure should be designed according to the actual development of the region with a variety of structural optimization and adjustment options to be considered for positioning Xinjiang's development in the next five years with the simultaneous enhancement of economic and ecological benefits.

2.5. Carbon Sequestration Service Supply and Demand

2.5.1. Carbon Sequestration Service Supply

In this study, we chose the random forest (RF) model to estimate the spatial and temporal dynamics of carbon sequestration in Xinjiang [17,39]. RF is an ensemble of decision tree predictors that uses bootstrap resampling methods to build decision trees for each sample [39,40]. For the construction of each tree, samples were selected independently; however, the distribution of all decision trees in the forest is the same, which guarantees the robustness of the model. The advantage of random forest is that it can prevent the overfitting of data, and it is favored for its relatively high overall accuracy and Kappa coefficient, interpretability of results, and accuracy of spatial display results for soil carbon sequestration prediction. The RF model is available in the Random Forest R 4.1.2 package. In this study, we divided the carbon sequestration of terrestrial ecosystems into three carbon pools: aboveground biomass carbon pool, underground biomass carbon pool, and soil carbon pool (0–20 cm). RF was used to model these three data parts separately and then add them together.

The carbon density data for this study were obtained from an open access database of the Chinese Academy of Sciences [28,41]. This is a publicly available carbon density dataset that includes 3026 soil samples taken from the soil surface layer across China through 2014. These samples were obtained from 1036 published papers and field survey data. The number of points in Xinjiang is 231, which covers its six major land use types. Thus, this database could provide new insights for future carbon sequestration strategies in Xinjiang. Because the data for dead organic carbon is relatively complex and difficult to observe and obtain, only carbon stored aboveground, belowground, and in the soil was considered in this study [28]. The model was calculated as follows:

$$C_{cs} = C_{i-above} + C_{i-below} + C_{i-soil} \quad (8)$$

$$C_{stocks} = C_{0-20} \times AREA_i \quad (9)$$

where C_{CS} is the carbon density; $C_{i-above}$ is the carbon density of the aboveground plant biomass, kg/m^2; $C_{i-below}$ is the carbon density of belowground biomass of plant roots, kg/m^2; and C_{i-soil} refers to the density of soil organic carbon in the soil layer, kg/m^2. C_{stocks} is the total carbon sequestration and $AREA_i$ denotes the area of different land use types or soil types.

2.5.2. Carbon Sequestration Service Demand

The demand for carbon sequestration service was estimated as the difference between the actual carbon emissions and the allowable carbon dioxide emissions set by local governments, as per Equation (10), consistent with previous research [27]. For spatial mapping, the amount of carbon emissions from industry was split evenly between construction, grassland, woodland, and cropland. The demand for carbon sinks from personal energy was split evenly across construction land:

$$D_{CS} = E_{industry} + E_{transportation} + E_{living} \tag{10}$$

where D_{CS} is the carbon sequestration demand; $E_{industry}$, $E_{transportation}$, and E_{living} are the carbon emission data of industry, transportation, and personal energy, respectively; $E_{industry}$ is the amount of carbon dioxide released by industrial production, whose value comes from the Xinjiang Statistical Yearbook; and $E_{transportation}$ is the carbon emitted by transportation. Each car uses about 1564.9 kg of gasoline per year, and one vehicle in the Xinjiang Uyghur Autonomous Region generates 4.67 tons of carbon per year [42]. The number of vehicles can be found in the Xinjiang Statistical Yearbook; E_{living} is the carbon emissions caused by each person's energy use. In the Xinjiang Uyghur Autonomous Region, one person is responsible for emitting about 5.84 tons of carbon per year [43]. Based on industrial output, vehicles, and population data from 2000 to 2020 (at 5-year intervals), linear regression was used to calculate industrial output, vehicles, and population in 2025 and 2030.

2.6. LULC Accuracy Verification

We compared the actual LULC data for 2015 and 2020 in the study area with the LULC data for the same years simulated based on the PLUS model, and then calculated the Kappa coefficient and overall accuracy (OA). The closer these two values are to 1, the higher the accuracy of the simulation; values greater than 0.8 indicate that the statistical accuracy of the model is satisfactory [25,29]. In this study, the Kappa coefficients of the simulated LULC for 2015 and 2020 were 0.931 and 0.905, respectively, and the overall accuracy was 0.964 and 0.949, respectively, indicating a high degree of confidence in the simulation results.

3. Results and Analysis

3.1. LULC Simulation under Multi-Scenarios

We applied the PLUS model to simulate the spatial distribution of land use in the Xinjiang region under different scenarios in 2025 and 2030, respectively, and calculated the dynamic rate of land change under four different scenarios for the two periods (Tables 3 and 4). The land use types in the Xinjiang region are dominated by bare land, this accounting for about 60.55% of the total study area. The LULC of the region also shows different trends in future scenarios, with the BAU scenario continuing the trend of urbanization in Xinjiang (Figure 3), with a dynamic land use index of 0.0045 and 0.0043 for construction in 2025 and 2030, respectively; this indicated land use change under this scenario is characterized by a slow, naturally expanding trend of construction. In the RED scenario, Xinjiang's construction expanded further, with land use dynamics of 0.0089 and 0.0145, respectively, in 2025 and 2030, corresponding to about 1.85% and 1.67% of other land types being converted to construction land (Figure 3); this indicates a more pronounced expansion of construction in the 2025–2030 period. Under this scenario, cropland also expands further, with land use dynamics of 0.0087 and 0.0095 in 2025 and 2030, respectively. Under the ELP scenario, the area of forest and grassland increases somewhat as a result of reforestation and ecological engineering policies, with about 7.28% and 0.57% of other land types converted to grassland in 2025 and 2030, respectively (Figure 3). In the SD scenario, we consider both the rapid economic development and the implementation of ecological projects to optimize the economic and ecological benefits. In this case, forest land increases by 554 km^2 and 2089 km^2 in 2025 and 2030, respectively, compared with 2020. There is a similar trend of construction land expansion, with an increase of 413.3 km^2 and 609.8 km^2 in the 2025 and 2030 SD scenarios, respectively, compared with 2020.

Table 3. LULC and its dynamic index K (%) in Xinjiang for each of the 2020–2025 scenarios.

LULC Type	Areal Coverage (km^2)					LULC Dynamic Index K (%)			
	2020	2025 BAU	2025 RED	2025 ELP	2025 SD	2020–2025 BAU	2020–2025 RED	2020–2025 ELP	2020–2025 SD
Cropland	90,255.7	93,594.9	94,215	86,520.5	90,956.1	0.007399	0.008774	−0.008277	0.001552
Forest	27,454.1	27,321.6	27,250	28,004.6	28,008.1	−0.00097	−0.001487	0.004010	0.004036
Grassland	48,4605	487,602	480,355	490,659	486,124	0.001237	−0.001754	0.002499	0.000627
Water	34,784.8	35,401.4	34,900.2	33,694.9	33,503.6	0.003545	0.000664	−0.006267	−0.007366
Constructed	9185.8	9393.3	9597.38	9314.3	9599.1	0.004518	0.008961	0.002798	0.008999
Bare land	992,105	985,077	992,073	989,940	989,942	−0.00142	−0.000006	−0.000436	−0.000436

Table 4. LULC and its dynamic index K (%) in Xinjiang for each of the 2020–2030 scenarios.

LULC Type	Areal Coverage (km^2)					LULC Dynamic Index K (%)			
	2025 BAU	2030 BAU	2030 RED	2030 ELP	2030 SD	2025 BAU–2030 BAU	2025 BAU–2030 RED	2025 BAU–2030 ELP	2025 BAU–2030 SD
Cropland	93,594.9	96,911.3	98,051.9	83,327.4	91,802	0.007087	0.009524	−0.021940	−0.003831
Forest	27,321.6	27,189.9	26,917.4	28,120.4	29,410.6	−0.000964	−0.002959	0.005847	0.015292
Grassland	487,602	490,577	476,350	496,092	489,906	0.001220	−0.004615	0.003482	0.000945
Water	35,401.4	36,013.8	35,011.7	34,974.5	34,784.3	0.003460	−0.002202	−0.002412	−0.003486
Constructed	9393.3	9599.2	10,074.6	9398	10,003.1	0.004384	0.014506	0.000100	0.012984
Bare land	985,077	978,099	991,985	986,479	982,485	−0.001417	0.001403	0.000285	−0.000526

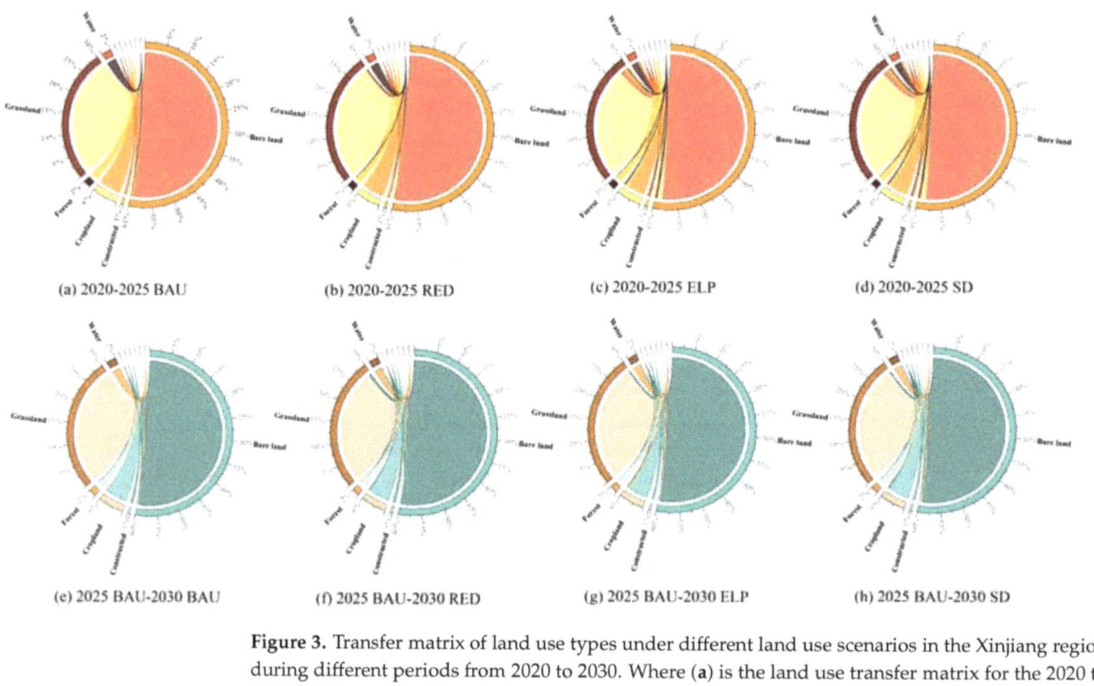

Figure 3. Transfer matrix of land use types under different land use scenarios in the Xinjiang region during different periods from 2020 to 2030. Where (**a**) is the land use transfer matrix for the 2020 to 2025 BAU scenario; (**b**) is the land use transfer matrix for the 2020 to 2025 RED scenario; (**c**) is the land use transfer matrix for the 2020 to 2025 ELP scenario; (**d**) is the land use transfer matrix for the 2020 to 2025 SD scenario; (**e**) is the land use transfer matrix for the 2025 BAU to 2030 BAU scenario; (**f**–**h**) and so on.

To explore the spatial and temporal characteristics of different land use types in Xinjiang under four future scenarios, we calculated the area of land use types during 2020–2030. Figure 4 shows the changes in the spatial patterns of cropland, forest land, grassland, and construction land in Xinjiang between 2020 and 2030 under the BAU, RED, ELP, and SD scenarios. Cropland significantly increased under the BAU, RED, and SD scenarios in 2025, increasing by 3339.2 km^2, 3959.3 km^2, and 700.4 km^2, respectively. These locations were mainly concentrated near the urban expansion zone along the northern slopes of the Tianshan Mountains, the Yili River Valley, the Aksu region, and the urban–rural farming belt in the Hotan region. In contrast, forest area decreased by 132.5 km^2 and 204.1 km^2 in 2025 under the BAU and RED scenarios, respectively, mainly in the Altai Mountains, the Yili River Valley, and the valley buffer zone near the Kunlun Mountains. By 2030, grassland under the RED scenario degraded extensively, with a decline of about 11,252 km^2, mainly in the Altai Mountains in the north, the Tianshan Mountains in the center, and near the Kunlun Mountains in the south of the study area, probably due to rapid urbanization at the expense of some forest and grassland. In addition, construction land shows an increasing trend in all four scenarios in 2030; the only difference is the magnitude of the increase, with the largest increases evidently occurring under the RED scenario, where the area increased by about 681.3 km^2, mainly in the urban agglomeration on the northern slopes of the Tianshan Mountains, the Aksu region, and the Kashgar region.

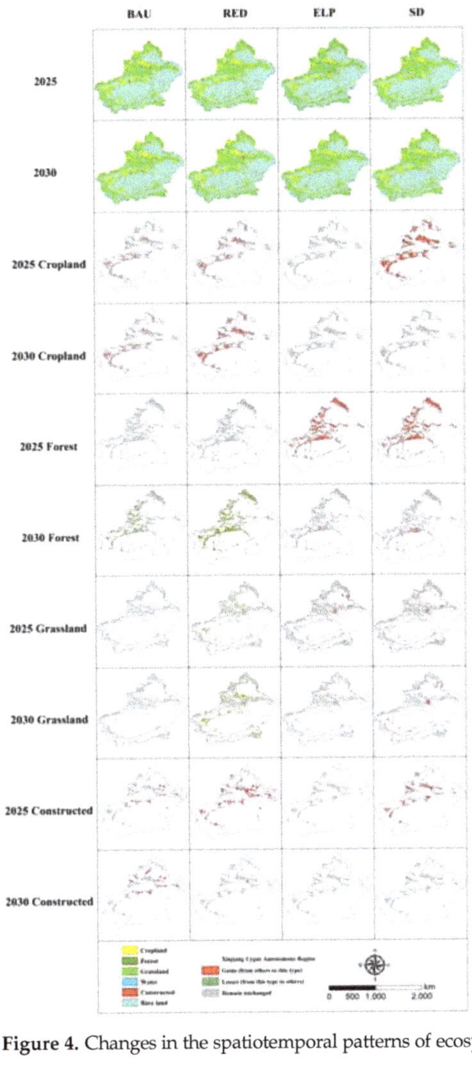

Figure 4. Changes in the spatiotemporal patterns of ecosystem types in each scenario from 2020 to 2030.

3.2. Spatial and Temporal Changes in the Supply of Carbon Sequestration under Different Scenarios

We used a random forest technique that incorporates environmental factors in our approach to assess changes in the landscape pattern of LULC-induced carbon sequestration service in terrestrial ecosystems in Xinjiang under different scenarios from 2020 to 2030 (Figure 5). The results show a clear spatial and temporal divergence in carbon sequestration under different scenarios. Under the BAU scenario in 2025, carbon sequestration shows a small annual increase compared to 2020 (interval of five years), of about 540 Tg. Under the RED scenario in 2025, carbon sequestration shows a decreasing trend compared to the BAU scenario, with an overall decrease of about 30 Tg, likely due to the continued expansion of construction driven by greater land use, resulting in the production of carbon from terrestrial ecosystems. In the 2025 SD scenario, carbon sequestration increased by another 370 Tg compared to the BAU scenario. This is because the SD scenario combines ecological and economic development, so the increase in forest and grassland areas leads to an increase in total carbon sequestration. In 2030, both the ELP and RED scenarios show an increase in carbon sequestration compared to the BAU scenario, by 20 Tg and 60 Tg, respectively. In the 2030 SD scenario, carbon sequestration increases significantly due to the pronounced profound expansion of forested grassland and the high carbon sequestration service capacity of forested land, making this scenario a carbon sink.

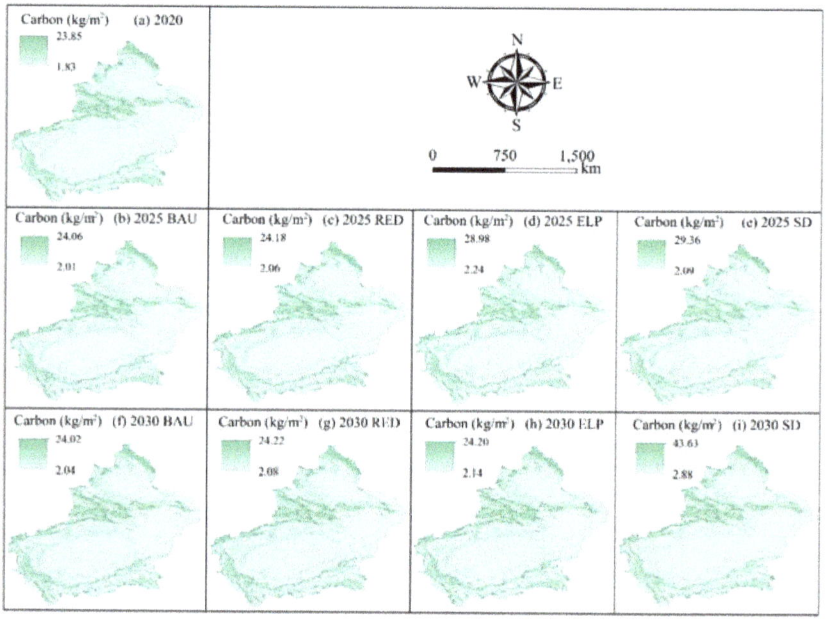

Figure 5. Spatial distribution characteristics of carbon sequestration service in each scenario during 2020–2030.

3.3. Analysis of the Supply and Demand for Carbon Sequestration under Different Future Scenarios

Xinjiang's carbon supply under different scenarios during 2020–2030 can hardly meet the current demand for carbon emissions (Figure 6), and the impact of land use on carbon supplies under different scenarios is also significant. In particular, the carbon supply in Xinjiang changes from 9.26 Pg in 2020 to 14.26 Pg under the SD scenario in 2030, while the carbon demand increases from 147.93 Pg in 2020 to 195.79 Pg, equivalent to an increase of about 32.35%. Considering the land use patterns under the different scenarios, the high-value areas of carbon stock are mainly distributed in the Altai Mountains, Tianshan Mountains, and Yili River Valley in the northern part of the study area due to the spatial distribution of forests and grasslands (Figure 6). Areas with low values of carbon sequestration service are distributed around bare land, construction land, and cultivated areas near river valley plains. Because of the high population density and industrialization of construction land, the carbon demand in this area is high, and the high-value areas of carbon demand are all concentrated around construction areas. In 2025 and 2030, in all the different scenarios for carbon sequestration in Xinjiang, the demand for a carbon sequestration service is exacerbated by the increasing expansion of construction areas, but the SD sustainability scenarios planned for this study can partly mitigate the deficit levels of carbon sequestration supply and demand.

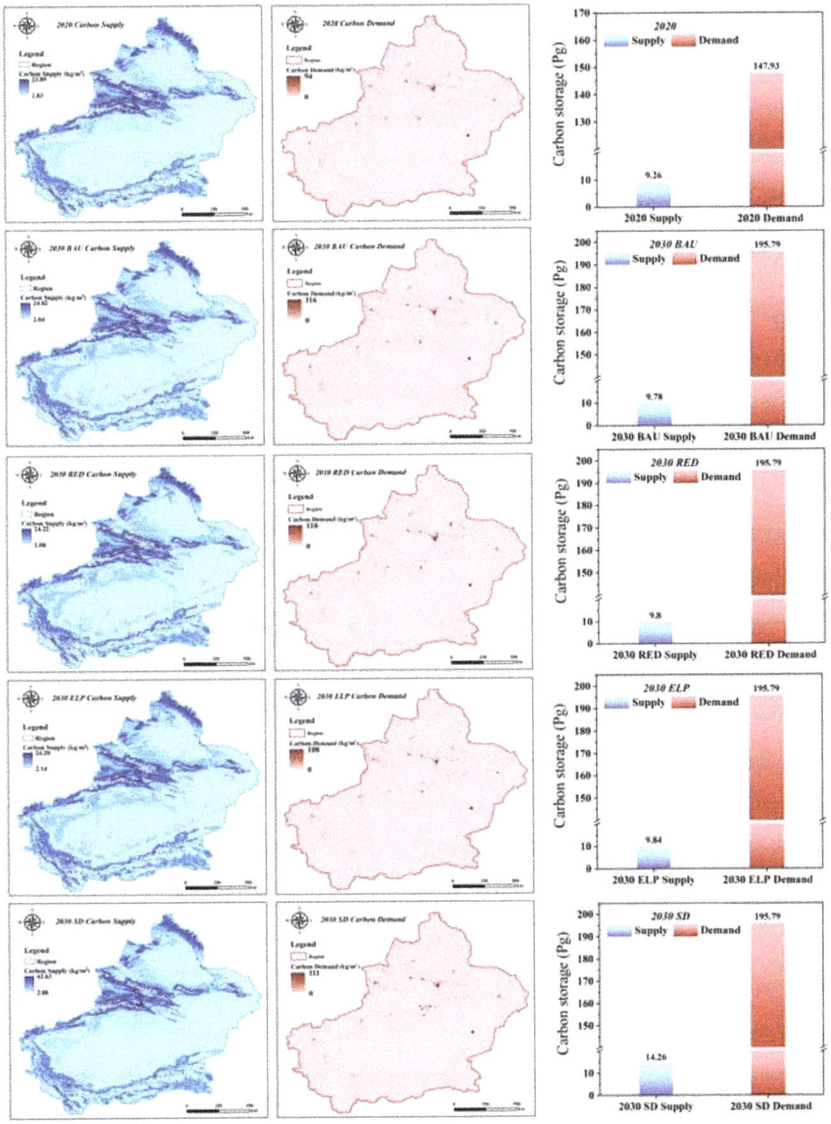

Figure 6. Spatial distribution of carbon stock supply and demand in Xinjiang under four different scenarios during 2020–2030.

4. Discussion

4.1. Analyses of Future Land Use Change under Different Scenarios

Based on the concept of sustainable development and the perspective of system evolution, this study proposes a land use evaluation framework for regional sustainable development that is oriented towards future development dynamics [44]. This study also attempts to apply global sustainable development goals at the local scale, which can effectively place regional development at the provincial level in the context of global sustainable development assessment and provide a basis for making decisions to help integrate regional development into the process of globalization. In addition, the United Nations Environment Program recognizes the key role of terrestrial ecosystem services in the SDGs, and in our study, we focus on terrestrial ecosystems, which echoes SDG 15.3 to combat desertification and restore degraded land and soil [35]. Land use planning is important for achieving SDG 15.3, and we should work towards a land-degradation-neutral world via better land use scenario planning. In addition, ecosystem services are central to achieving SDG 15.9 and should be integrated into national and local developmental processes [24,35,36].

To ensure the study's accuracy, we simulated the LULC data for 2015 and 2020 using the PLUS model, with Kappa coefficients of 0.931 and 0.905 and overall accuracies of 0.964 and 0.949, respectively, indicating a high degree of confidence in the simulation results. In addition, the PLUS model was used to simulate land use patterns in Xinjiang from 2020 to 2030 under four different scenarios, with a slight magnitude growth trend for cropland and construction land in the BAU scenario and a sharp expansion pattern in the RED scenario, both consistent with the findings of Fu et al. [33] in Xinjiang. In the ELP scenario, the large expansion of forests and grasslands was concentrated in the alpine forest–grassland and Yili River valley regions of the study area, which is also consistent with the findings of Shi et al. in the same valley [25]. We also found that the SD sustainability scenario accounts for the urbanization process while paying more attention to ecological protection, limiting the uncontrolled growth of urban space, and slowing down or even reversing the rising trend of constructed patches in some areas [35].

4.2. Analysis of the Impact of LULC on Carbon

This study completes the first high-resolution mapping of terrestrial ecosystem carbon sequestration in Xinjiang under different future scenarios. This fine-scale mapping shows that we can combine the contributions of nature and the needs of people [45]. Moreover, the RF approach applied in this study is superior to other methods for estimating carbon sequestration, and our carbon density raster data are spatially continuous rather than using the same carbon density value fixed for each land use type [20,21].

In addition, to ensure the accuracy of the study, we calibrated the output result parameters of Xinjiang carbon sequestration simulated by the RF model with the results of other studies on Xinjiang stocks. Via soil profiling, Yan et al. estimated the soil carbon stock in Xinjiang residing in a 0 to 100 cm depth to be about 19.56 Pg [46]. However, one study did find that the organic carbon in the top 20 cm of soils in Xinjiang accounts for 37.9% of the percentage of organic carbon in a 1 m deep soil layer [47]. Therefore, Yan et al. estimated the carbon stock in the topsoil layer of Xinjiang to be about 7.41 Pg, whereas this study estimated a higher carbon stock of terrestrial ecosystems in Xinjiang, at about 9.26 Pg. This discrepancy may be because we modeled not only the soil carbon sequestration service but also the aboveground and belowground biomass carbon sequestration, this being a plausible explanation for the relatively high results of our study.

4.3. Analysis of the Supply and Demand of the Carbon Sequestration Service in Different Scenarios

Studies have shown that LULC is considered a key anthropogenic driver of ecosystem service change at the regional scale [48,49]. Our study used measured datasets of aboveground biomass carbon density, belowground biomass carbon density, and soil carbon density, combined with a random forest model and spatial mapping of raster layers of different future land uses, as factorial environmental variables [17]. In this way, we proposed a new scheme for the spatial simulation of the terrestrial ecosystem carbon sequestration service with the GMOP-PLUS model combined with the random forest method.

According to our findings, the ELP and SD scenarios of this study simulated the implementation of ecological projects, whose primary aim is restoring forests and protecting grasslands, mainly to prevent increased desertification and soil erosion in Xinjiang, and to increase the productivity of vegetation for more carbon sequestration service, objectives closely related to dozens of ecological projects carried out in China in the last half-century [9]. Furthermore, our study shows that the growth of one land use patch comes at the cost of the decline of another land use type. For example, in the RED scenario, the growth of built-upon land comes at the expense of forested grassland [23,31,50]. We expect the supply of carbon storage in Xinjiang to increase by about 5.72% in 2020, but, at the same time, the demand will rise by 21.80%. In 2030, this supply and demand servicing will intensify the deficit. Xinjiang's supply of carbon stock in 2030 will increase by approximately 5.61% compared to 2020, yet the demand for it will increase by 32.35% in the meantime. Still, in the SD scenario, our projections can serve to mitigate some of the carbon sources through the implementation of ecological engineering projects, and the different scenarios are set up to help clarify the relationship between different LULC structures and ESs carbon stock. For example, in the 2025 RED scenario, the increase in carbon sequestration services from cropland expansion is about 12.61 Mg, while in 2030 this trend is reversed and carbon sourcing occurs, with a cumulative net release of 2.42 Mg. Alternatively, in the 2025 SD scenario, cropland expansion is expected to generate a carbon sink of 63.26 Mg, while in 2030 this trend is slowed down, with a projected net carbon sequestration of 11.24 Mg. This result is likely attributed to the expectation that cropland will reach an expansion saturation in 2030 in the scenario simulation setting so that cropland expansion is eventually slowed down. Our findings suggest that different scenarios can help clarify the relationship between different LULC structures and ESs carbon sequestration.

4.4. Limitations and Perspectives

In this study, future land use patterns under different scenarios were generated through the PLUS model, and carbon stock supply and demand services were assessed for various scenarios. However, there are still some uncertainties and limitations. For example, this study only portrayed four different future scenarios of LULC through policy guidelines; the four alternative scenarios do not represent all possible LULC realities, and more comprehensive scenarios should be explored in subsequent studies. For example, the impact of future climate change on LULC could be considered, among others, to address multi-stakeholder needs for optimal land use policies [25,51].

In addition, with the development of low carbon technologies and the policy direction of the national dual carbon targets, whether future carbon demand will still develop in line with the original demand trend to address the need to achieve China's peak carbon policy by 2030 could potentially impact the results of the carbon stock demand component. Therefore, future studies should plan the LULC scenarios more rationally and truly consider the future carbon stock demand in the context of China's policy. This will help provide better scientific references for future regional decision-making and sustainable development planning.

5. Conclusions

In this study, we coupled the gray multi-objective optimization (GMOP) and patch generation land use simulation (PLUS) models and proposed a new SD sustainability scenario framework for optimizing the structure of future land use in Xinjiang by using the GMOP-PLUS model. This work also explores the carbon sequestration services of terrestrial ecosystems in key regions from the perspective of land use change, and addresses the disparities arising between the supply and demand of carbon sequestration services in

Xinjiang in the future years of 2025 and 2030. The starkest findings to emerge from this study are as follows: (1) the future expansion of arable land in Xinjiang will occur at the expense of some forest and grassland areas, which are particularly prominent in the interlocking zones of river valleys and plains, especially in the Ili Valley, the Altay Mountains, etc.; (2) the supply and demand of carbon stock in Xinjiang will increase in 2025, but the demand is much greater than the supply, and in 2030 this supply and demand imbalance is exacerbated; and (3) Xinjiang, in the context of future cropland expansion, could alleviate the supply and demand deficit situation threatening Xinjiang's carbon stock; the occurrence of this mitigation is most likely under in the SD scenario. Nonetheless, some of the carbon sources can be mitigated by the implementation of ecological engineering in our planned SD scenario, and the analysis of the SD scenario and other scenarios can help to clarify the relationship between different LULC structures and carbon sequestration. Therefore, local governments can increase their efforts to protect ecosystem carbon sequestration services through policies such as returning farmland to forest, reasonable ecological land regulation, and appropriate afforestation activities, in addition to sequestering carbon belowground, while minimizing the loss of ecosystem service functions, to achieve the sustainable development of agroecology in key areas along the Belt and Road.

Author Contributions: H.W. and P.J. were responsible for the research design, analysis, and the manuscript's design and its review. M.S. drafted the manuscript and was responsible for data preparation, experiments, and analyses. W.S. was responsible for the research design and reviewing the manuscript. Resources and funding were procured by H.W. and P.J. M.Z., Z.L., K.Z. have performed the data processing work. L.Z., H.Z., X.F. performed the manuscript editing. T.D., M.F.B. performed the manuscript proofreading and retouching. All authors have read and agreed to the published version of the manuscript.

Funding: This research was financially supported by Major Science and Technology Special Projects in Xinjiang Uygur Autonomous Region, China (Integrated demonstration of high quality, high yield and high efficiency standardized production technology for cotton, No. 2020A01002).

Institutional Review Board Statement: Not applicable.

Informed Consent Statement: Written informed consent has been obtained from the patient(s) to publish this paper.

Data Availability Statement: The data presented in this study are available on request from the corresponding author.

Acknowledgments: Special thanks go to the College of Resources and Environment, University of Chinese Academy of Sciences, for supporting the implementation of this study. We thank Xiaozhen Wang at the Northwest A&F University for her constructive suggestions. We also thank the anonymous reviewers for their constructive comments on the manuscript.

Conflicts of Interest: The authors declare no conflict of interest.

References

1. Wang, Y.; Wang, X.; Wang, K.; Chevallier, F.; Zhu, D.; Lian, J.; He, Y.; Tian, H.; Li, J.; Zhu, J. The size of the land carbon sink in China. *Nature* **2022**, *603*, E7–E9. [CrossRef] [PubMed]
2. Liu, C.; Liang, Y.; Zhao, Y.; Liu, S.; Huang, C. Simulation and Analysis of the Effects of Land Use and Climate Change on Carbon Dynamics in the Wuhan City Circle Area. *Int. J. Environ. Res. Public Health* **2021**, *18*, 11617. [CrossRef] [PubMed]
3. Tang, X.; Zhao, X.; Bai, Y.; Tang, Z.; Wang, W.; Zhao, Y.; Wan, H.; Xie, Z.; Shi, X.; Wu, B. Carbon pools in China's terrestrial ecosystems: New estimates based on an intensive field survey. *Proc. Natl. Acad. Sci. USA* **2018**, *115*, 4021–4026. [CrossRef] [PubMed]
4. Eggleston, H.; Buendia, L.; Miwa, K.; Ngara, T.; Tanabe, K. *2006 IPCC Guidelines for National Greenhouse Gas Inventories*; U.S. Department of Energy Office of Scientific and Technical Information: Oak Ridge, TN, USA, 2006.
5. Houghton, R. The annual net flux of carbon to the atmosphere from changes in land use 1850–1990. *Tellus B* **1999**, *51*, 298–313. [CrossRef]
6. Jiang, H.; Peng, J.; Dong, J.; Zhang, Z.; Xu, Z.; Meersmans, J. Linking ecological background and demand to identify ecological security patterns across the Guangdong-Hong Kong-Macao Greater Bay Area in China. *Landsc. Ecol.* **2021**, *36*, 2135–2150. [CrossRef]

7. Schwartz, C.; Shaaban, M.; Bellingrath-Kimura, S.D.; Piorr, A. Participatory Mapping of Demand for Ecosystem Services in Agricultural Landscapes. *Agriculture* **2021**, *11*, 1193. [CrossRef]
8. Li, H.; Wu, Y.; Liu, S.; Xiao, J.; Zhao, W.; Chen, J.; Alexandrov, G.; Cao, Y. Decipher soil organic carbon dynamics and driving forces across China using machine learning. *Glob. Chang. Biol.* **2022**, *28*, 3394–3410. [CrossRef] [PubMed]
9. Zhao, M.; He, Z.; Du, J.; Chen, L.; Lin, P.; Fang, S. Assessing the effects of ecological engineering on carbon storage by linking the CA-Markov and InVEST models. *Ecol. Indic.* **2019**, *98*, 29–38. [CrossRef]
10. Stockmann, U.; Adams, M.A.; Crawford, J.W.; Field, D.J.; Henakaarchchi, N.; Jenkins, M.; Minasny, B.; McBratney, A.B.; De Courcelles, V.d.R.; Singh, K. The knowns, known unknowns and unknowns of sequestration of soil organic carbon. *Agric. Ecosyst. Environ.* **2013**, *164*, 80–99. [CrossRef]
11. Wieder, W.R.; Bonan, G.B.; Allison, S.D. Global soil carbon projections are improved by modelling microbial processes. *Nat. Clim. Chang.* **2013**, *3*, 909–912. [CrossRef]
12. Nayak, A.K.; Rahman, M.M.; Naidu, R.; Dhal, B.; Swain, C.K.; Nayak, A.D.; Tripathi, R.; Shahid, M.; Islam, M.R.; Pathak, H. Current and emerging methodologies for estimating carbon sequestration in agricultural soils: A review. *Sci. Total Environ.* **2019**, *665*, 890–912. [CrossRef] [PubMed]
13. Fensholt, R.; Sandholt, I.; Rasmussen, M.S.; Stisen, S.; Diouf, A. Evaluation of satellite based primary production modelling in the semi-arid Sahel. *Remote Sens. Environ.* **2006**, *105*, 173–188. [CrossRef]
14. Zhao, M.; Running, S.W.; Nemani, R.R. Sensitivity of Moderate Resolution Imaging Spectroradiometer (MODIS) terrestrial primary production to the accuracy of meteorological reanalyses. *J. Geophys. Res. Biogeosci.* **2006**, *111*, G1. [CrossRef]
15. Rahman, A.; Sims, D.A.; Cordova, V.D.; El-Masri, B.Z. Potential of MODIS EVI and surface temperature for directly estimating per-pixel ecosystem C fluxes. *Geophys. Res. Lett.* **2005**, *32*, 19. [CrossRef]
16. Scharlemann, J.P.; Tanner, E.V.; Hiederer, R.; Kapos, V. Global soil carbon: Understanding and managing the largest terrestrial carbon pool. *Carbon Manag.* **2014**, *5*, 81–91. [CrossRef]
17. Zhang, M.; Shi, W.; Xu, Z. Systematic comparison of five machine-learning models in classification and interpolation of soil particle size fractions using different transformed data. *Hydrol. Earth Syst. Sci.* **2020**, *24*, 2505–2526. [CrossRef]
18. Euliss, N.H., Jr.; Smith, L.M.; Liu, S.; Feng, M.; Mushet, D.M.; Auch, R.F.; Loveland, T.R. The need for simultaneous evaluation of ecosystem services and land use change. *Environ. Sci. Technol.* **2010**, *44*, 7761–7763. [CrossRef]
19. Maring, L.; Blauw, M. Asset management to support urban land and subsurface management. *Sci. Total Environ.* **2018**, *615*, 390–397. [CrossRef] [PubMed]
20. Goldstein, J.H.; Caldarone, G.; Duarte, T.K.; Ennaanay, D.; Hannahs, N.; Mendoza, G.; Polasky, S.; Wolny, S.; Daily, G.C. Integrating ecosystem-service tradeoffs into land-use decisions. *Proc. Natl. Acad. Sci. USA* **2012**, *109*, 7565–7570. [CrossRef] [PubMed]
21. Nelson, E.; Mendoza, G.; Regetz, J.; Polasky, S.; Tallis, H.; Cameron, D.; Chan, K.M.; Daily, G.C.; Goldstein, J.; Kareiva, P.M. Modeling multiple ecosystem services, biodiversity conservation, commodity production, and tradeoffs at landscape scales. *Front. Ecol. Environ.* **2009**, *7*, 4–11. [CrossRef]
22. Zhang, D.; Huang, Q.; He, C.; Yin, D.; Liu, Z. Planning urban landscape to maintain key ecosystem services in a rapidly urbanizing area: A scenario analysis in the Beijing-Tianjin-Hebei urban agglomeration, China. *Ecol. Indic.* **2019**, *96*, 559–571. [CrossRef]
23. Li, Z.; Cheng, X.; Han, H. Future impacts of land use change on ecosystem services under different scenarios in the ecological conservation area, Beijing, China. *Forests* **2020**, *11*, 584. [CrossRef]
24. Wu, X.; Fu, B.; Wang, S.; Song, S.; Li, Y.; Xu, Z.; Wei, Y.; Liu, J. Decoupling of SDGs followed by re-coupling as sustainable development progresses. *Nat. Sustain.* **2022**, *5*, 452–459. [CrossRef]
25. Shi, M.; Wu, H.; Fan, X.; Jia, H.; Dong, T.; He, P.; Baqa, M.F.; Jiang, P. Trade-offs and synergies of multiple ecosystem services for different land use scenarios in the yili river valley, China. *Sustainability* **2021**, *13*, 1577. [CrossRef]
26. Wang, X.; Dong, X.; Liu, H.; Wei, H.; Fan, W.; Lu, N.; Xu, Z.; Ren, J.; Xing, K. Linking land use change, ecosystem services and human well-being: A case study of the Manas River Basin of Xinjiang, China. *Ecosyst. Serv.* **2017**, *27*, 113–123. [CrossRef]
27. Chen, J.; Jiang, B.; Bai, Y.; Xu, X.; Alatalo, J.M. Quantifying ecosystem services supply and demand shortfalls and mismatches for management optimisation. *Sci. Total Environ.* **2019**, *650*, 1426–1439. [CrossRef]
28. Xu, L.; He, N.; Yu, G. A dataset of carbon density in Chinese terrestrial ecosystems (2010s). *China Sci. Data* **2019**, *4*, 49–54.
29. Liang, X.; Guan, Q.; Clarke, K.C.; Liu, S.; Wang, B.; Yao, Y. Understanding the drivers of sustainable land expansion using a patch-generating land use simulation (PLUS) model: A case study in Wuhan, China. *Comput. Environ. Urban Syst.* **2021**, *85*, 101569. [CrossRef]
30. Wang, Y.; Li, X.; Zhang, Q.; Li, J.; Zhou, X. Projections of future land use changes: Multiple scenarios-based impacts analysis on ecosystem services for Wuhan city, China. *Ecol. Indic.* **2018**, *94*, 430–445. [CrossRef]
31. Li, C.; Wu, Y.; Gao, B.; Zheng, K.; Wu, Y.; Li, C. Multi-scenario simulation of ecosystem service value for optimization of land use in the Sichuan-Yunnan ecological barrier, China. *Ecol. Indic.* **2021**, *132*, 108328. [CrossRef]
32. Zhang, H.-B.; Zhang, X.-H. Land use structural optimization of Lilin based on GMOP-ESV. *Trans. Nonferrous Met. Soc. China* **2011**, *21*, s738–s742. [CrossRef]
33. Fu, Q.; Hou, Y.; Wang, B.; Bi, X.; Li, B.; Zhang, X. Scenario analysis of ecosystem service changes and interactions in a mountain-oasis-desert system: A case study in Altay Prefecture, China. *Sci. Rep.* **2018**, *8*, 12939. [CrossRef] [PubMed]

34. Liang, J.; Li, S.; Li, X.; Li, X.; Liu, Q.; Meng, Q.; Lin, A.; Li, J. Trade-off analyses and optimization of water-related ecosystem services (WRESs) based on land use change in a typical agricultural watershed, southern China. *J. Clean. Prod.* **2021**, *279*, 123851. [CrossRef]
35. Peng, K.; Jiang, W.; Ling, Z.; Hou, P.; Deng, Y. Evaluating the potential impacts of land use changes on ecosystem service value under multiple scenarios in support of SDG reporting: A case study of the Wuhan urban agglomeration. *J. Clean. Prod.* **2021**, *307*, 127321. [CrossRef]
36. Giuliani, G.; Mazzetti, P.; Santoro, M.; Nativi, S.; Van Bemmelen, J.; Colangeli, G.; Lehmann, A. Knowledge generation using satellite earth observations to support sustainable development goals (SDG): A use case on Land degradation. *Int. J. Appl. Earth Obs. Geoinf.* **2020**, *88*, 102068. [CrossRef]
37. Xinjiang Statistical Yearbook. 2020. Available online: http://www.stats.gov.cn/tjsj./ndsj/ (accessed on 1 May 2022).
38. Xie, G.; Zhang, C.; Zhang, C.; Xiao, Y.; Lu, C. The value of ecosystem services in China. *Resour. Sci.* **2015**, *37*, 1740–1746.
39. Breiman, L. Bagging predictors. *Mach. Learn.* **1996**, *24*, 123–140. [CrossRef]
40. Sun, Y.; Zhang, S.; Tao, F.; Aboelenein, R.; Amer, A. Improving Winter Wheat Yield Forecasting Based on Multi-Source Data and Machine Learning. *Agriculture* **2022**, *12*, 571. [CrossRef]
41. Xu, L.; Yu, G.; He, N.; Wang, Q.; Gao, Y.; Wen, D.; Li, S.; Niu, S.; Ge, J. Carbon storage in China's terrestrial ecosystems: A synthesis. *Sci. Rep.* **2018**, *8*, 2806. [CrossRef]
42. Li, Y.; Qian, Y. An analysis of operation cost of Chinese family car. *Market Res.* **2008**, *1*, 52–54.
43. Cai, H. *Factors of Northern and Southern China per Capita CO_2 Emission in China's Household*; Jinan University: Guangzhou, China, 2016. (In Chinese)
44. Eustachio, J.H.P.P.; Caldana, A.C.F.; Liboni, L.B.; Martinelli, D.P. Systemic indicator of sustainable development: Proposal and application of a framework. *J. Clean. Prod.* **2019**, *241*, 118383. [CrossRef]
45. Chaplin-Kramer, R.; Sharp, R.P.; Weil, C.; Bennett, E.M.; Pascual, U.; Arkema, K.K.; Brauman, K.A.; Bryant, B.P.; Guerry, A.D.; Haddad, N.M. Global modeling of nature's contributions to people. *Science* **2019**, *366*, 255–258. [CrossRef] [PubMed]
46. Yan, A. *Spatial Distribution and Storages Estimation of Soil Organic Carbon and Soil Inorganic Carbon in Xinjiang, China*; China Agricultural University: Beijing, China, 2015. (In Chinese)
47. Wang, X.; Yang, D.; Xiong, H. Characteristics of soil organic carbon under different vegetation types in Xinjiang. *Arid. Zone Res.* **2017**, *34*, 782–788.
48. Bai, Y.; Wong, C.P.; Jiang, B.; Hughes, A.C.; Wang, M.; Wang, Q. Developing China's Ecological Redline Policy using ecosystem services assessments for land use planning. *Nat. Commun.* **2018**, *9*, 3034. [CrossRef]
49. Fu, B.; Zhang, L.; Xu, Z.; Zhao, Y.; Wei, Y.; Skinner, D. Ecosystem services in changing land use. *J. Soils Sediments* **2015**, *15*, 833–843. [CrossRef]
50. Chen, Y.; Zhang, L.; He, L.; Men, M. Multi-scenario simulation of land use structure based on dual combined models. *Acta Ecol. Sin.* **2016**, *36*, 5391–5400.
51. Wang, Y.; Huang, M.; Wang, X. Impacts of land use and climate change on agricultural productivity in Shanghai. *Acta Sci. Circumstantiae* **2010**, *30*, 641–648.

agriculture

Article

Spatial and Temporal Variability of ET_0 in Xinjiang Autonomous Region of China during 1957–2017

Yanhui Jia [1], Xiaojun Shen [2,3,*], Ruochen Yi [3] and Ni Song [1,*]

1 Farmland Irrigation Research Institute, Chinese Academy of Agricultural Sciences, Xinxiang 453002, China
2 School of Agricultural and Wine Sciences, Charles Sturt University, Wagga Wagga, NSW 2650, Australia
3 College of Water Conservancy Engineering, Tianjin Agricultural University, Tianjin 300392, China
* Correspondence: shenxiaojun@tjau.edu.cn (X.S.); songni@caas.cn (N.S.); Tel.: +86-22-2386-8236 (X.S.); +86-373-339-3384 (N.S.)

Abstract: This article scientifically studies the direct impact of climate problems on the time transition of reference crop evapotranspiration in the Xinjiang Autonomous Region of China from 1957 to 2017, which is conducive to formulating irrigation scheduling and adaptive capacity countermeasures. The objective of this study is to investigate the impacts of climate change on ET_0 for the cotton growing seasons. The meteorological data were collected from 48 meteorological stations in the region and analyzed using the Mann–Kendall test and linear trend. The results show the following points: (1) the ET_0 decreases from low to high elevations, and with the increase in northern latitude. (2) The annual mean ET_0 and average values of ET_0 during the growing seasons for cotton exhibited two abrupt changes in the period 1957–2017, with the first abrupt change in 1995 to 1999 and the second abrupt change in 2006 to 2011. (3) The ET_0 in Xinjiang of China demonstrates a decreasing trend during 1957–1996; a significant decreasing trend during 1997–2008; and a significant increasing trend during 2009–2017.

Keywords: climate change; ET_0; yield response factor; irrigation water requirement; cotton; Xinjiang

Citation: Jia, Y.; Shen, X.; Yi, R.; Song, N. Spatial and Temporal Variability of ET_0 in Xinjiang Autonomous Region of China during 1957–2017. *Agriculture* **2022**, *12*, 1380. https://doi.org/10.3390/agriculture12091380

Academic Editor: Rabin Bhattarai

Received: 17 July 2022
Accepted: 30 August 2022
Published: 2 September 2022

1. Introduction

Much of the evidence from the past half century show that the increase in the global average temperature is mainly due to the increase in greenhouse gas emissions [1–3]. Global warming is a significant manifestation of the present observable and directly felt future climate change [4,5]. The period 2011–2015 was the hottest based on the previous records, about 0.57 °C higher than the reference temperature for the period of 1961–1990 [6]. Climate change is already affecting many aspects of life for people [7–9] and it will affect crop water demand [10,11], and the spatial and temporal redistribution of water resources in the future [12]. Reference crop water requirement (ET_0) is an important parameter for the calculation of crop water requirement [13–15]. In order to improve the efficiency of the irrigation water resources, it is important to study the spatiotemporal variation in ET_0 under climate change.

In the past half century, China has produced many studies on regional temperature and rainfall [16,17]. However, there are relatively few studies on Northwest China against the background of continuous climate warming, and these are mainly focused on precipitation [18,19], air temperature [20], and small pan evaporation [21,22]. These studies showed the change feature of precipitation in Xinjiang as it changed from dry to wet in the middle and late 1980s, and the evapotranspiration potential showed a downward trend.

This study investigated the spatial transformation of ET_0 in mountainous areas, oasis and desert areas of Xinjiang from 1957 to 2017. This scientific research plays an important role in revealing the fluctuation regularity of ET_0, mastering the evolution of drought and improving the utilization rate of irrigation water sources.

2. Materials and Methods

2.1. Description of the Xinjiang in the Northwest of China

Xinjiang is located in the most inland core of the Eurasian continent, with a longitude ranging from 73.40° E to 96.18° E and north latitude ranging from 34.25° N to 48.10° N and elevation of ranging from −155 m to 8611 m. It has a vast territory, with a total area of about 1.66 million hectares, accounting for more than 17% of China's land area. Tianshan, Altay and other mountains surround the Tarim Basin, Junggar Basin and other wide desert basins. The drought-stricken intermountain basins and humid and cold mountain areas are sensitive to climate issues [23]. Because it is far away from the surrounding deep sea, the arid climate is a very common inland continental climate, which is mainly affected by the inland drought norms, while the temperate monsoon climate in Asia is less harmful [24]. The annual average temperature is about 8 °C, and the annual average rainfall is less than 150 mm, slowly decreasing from north to south.

2.2. Data Processing

Continuous and long-time series of the observed daily maximum/minimum air temperature (T_{max}/T_{min}), relative humidity (Hr), wind speed at 10 m height (U_{10}) and global solar radiation (Rs) during 1957–2017 were collected from typical weather stations in Xinjiang (Figure 1). The FAO Penman–Monteith method used to calculate ET_o and can be shown in the following equation [25]:

$$ET_o = \frac{0.408\Delta(R_n - G) + \gamma \frac{900}{T_a+273} u_2 (e_s - e_a)}{\Delta + \gamma(1 + 0.34u_2)} \quad (1)$$

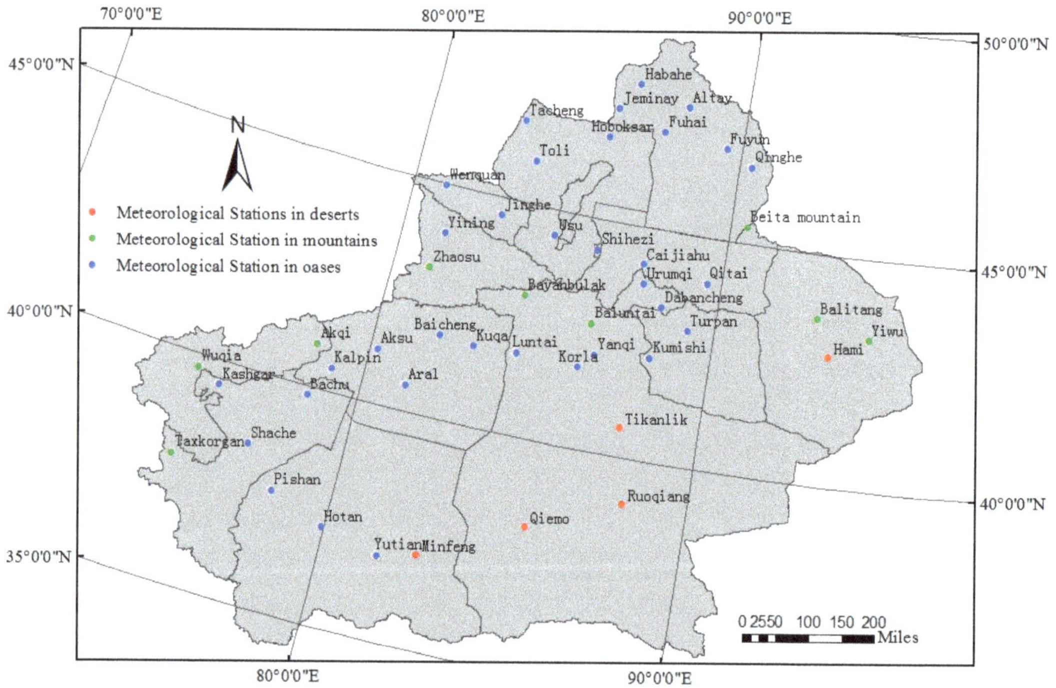

Figure 1. Xinjiang meteorological observation station is located in Northwest China.

In the above equation, ET_o is the reference evapotranspiration of crops (mm), R_n is the net radioactive material (MJ·m^{-2}·day^{-1}), G is the thermal diffusion coefficient of the soil layer ($MJ \cdot m^{-2} \cdot day^{-1}$), γ is the air relative humidity constant ($kPa \cdot {}^{\circ}C^{-1}$), e_s is the saturated vapor pressure (kPa), e_a is the specific saturated vapor pressure (kPa), Δ is the slope of the saturated vapor pressure temperature curve ($kPa \cdot {}^{\circ}C^{-1}$), T_a is the daily mean temperature (°C) and u_2 is the daily mean wind speed at 2 m ($m \cdot s^{-1}$).

The original daily temperature, air relative humidity, sunshine duration and wind speed data information are given by the National Climate Center of the China Meteorological Administration (NCC-CMA). The 48 meteorological stations selected in this study (Figure 1) have been maintained in accordance with the standards of the National Meteorological Administration of China. The standard requires strict quality control of the whole process before releasing the data information, including extreme inspections and regular inspections of duration consistency. For the oases, we chose large, medium and small urban meteorological stations located in densely populated areas to reflect the negative impact of human activities.

2.3. Methods

2.3.1. Mann–Kendall Method for Nonparametric Test

The nonparametric Mann–Kendall method used in climate trend analyses and gene mutation was created by Mann [26] and was changed by Kendall [27]. The Mann–Kendall method evaluates the development trend in the climate independent variable time series model, and has been widely used in the change trend, because it does not use the unique spread of data information samples [28].

In that way, H_o represents the distribution range of random variables, and H_1 represents the probability of double transformation. The test statistic S is obtained from the following formula:

$$S = \sum_{i=1}^{n-1} \sum_{k=i+1}^{n} \operatorname{sgn}(x_k - x_i) \tag{2}$$

where x_k and x_i are the sequential data values, n is the length of the data set, and

$$\operatorname{sgn}(\theta) = \begin{cases} +1, & \theta > 0 \\ 0, & \theta = 0 \\ -1, & \theta < 0 \end{cases} \tag{3}$$

If the sample size exceeds 10, the statistic S is basically normal, that is, the statistic is a normal random variable, and its expected value and standard deviation are as follows:

$$Z_C = \begin{cases} \frac{S-1}{\sqrt{var(S)}}, & S > 0; \\ 0, & S = 0; \\ \frac{S+1}{\sqrt{var(S)}}, & S < 0. \end{cases} \tag{4}$$

$$E(S) = 0 \tag{5}$$

$$var(S) = [n(n-1)(2n+5)]/18 \tag{6}$$

t represents the type of all given connections and Σ represents the number of all connections. In the Mann–Kendall test, another very useful technical index is the Kendall straight-line slope, which is the compressive strength with a monotonic development trend, which is calculated by the following formula:

$$\beta = Median\left(\frac{x_i - x_j}{i - j}\right), \quad j < i \tag{7}$$

In the above equation, $1 < j < i < n$. It indicates an "upward" trend, that is, it increases over time, and a negative value indicates a "downward" trend, that is, it decreases for a long time at any time.

2.3.2. Mann–Kendall's Method of Genetic Variation Test

Under zero assumption with the out trend, the time series of the variables do not change, and the time series can be regarded as $x_1, x_2, \ldots \ldots, x_n$. For each new item, m_i is calculated as the number of subsequent new items whose value in the coding sequence exceeds x_i. The test statistical analysis and calculation are as follows:

$$d_k = \sum_{i=1}^{k} r_k \quad (2 \le k \le N) \tag{8}$$

$$r_i = \begin{cases} 1\ , x_i > x_j \\ 0\ , x_i \le x_j \end{cases} \tag{9}$$

Assuming that the sequence is arbitrary and independent, the variance in the expected values $E(d_k)$ and d_k can be calculated as follows:

$$\begin{cases} E(d_k) = k(k-1)/4 \\ var(d_k) = \frac{k(k-1)(2k+S)}{72} (2 \le k \le N) \end{cases} \tag{10}$$

We can, therefore, obtain the statistic $u(d_k)$ with the following equation:

$$u(d_k) = (d_k - E(d_k))/\sqrt{var(d_k)} \tag{11}$$

The terms of the $u(d_k)$ $(1 \le k \le n)$ constitute curve C_1. If the normalized normal probability Pr(|u|) < |u(d_k)| > a, the null hypothesis of no trend will be rejected at the confidence level a. The coding sequence of annual total precipitation applies the typical 95% confidence level. If this method is applied to the inverse level, it can obtain $\bar{u}$ (d_k), as shown in the following equation:

$$\begin{cases} \overline{u}(d_i) = -u(d_i) \\ i = n+1-i \end{cases} \cdots (i, i = 1,\ 2,\ 3,\ \cdots,\ n) \tag{12}$$

The terms of the $\bar{u}$ (d_k) $(1 \le k \le n)$ constitute another curve C_2. If C_1 exceeds the confidence line, it indicates that there is an important upward or downward trend in the coding sequence. If the intersection point of the C_1 and C_2 is between the two lines, we can conclude that climate gene mutations should be produced in this way [24].

2.3.3. Pre-Whitening Mann–Kendall Method

Pre-whitening [29] was proposed by von Storch (1995), and it was used to reduce the influence of serial correlation on the MK test. The formula is shown below.

$$Y_t = X_t - r \times X_{t-1} \tag{13}$$

where X_t represents the original sequential data values, Y_t represents the sequential data values without autocorrelation and r is the autocorrelation coefficient of the sequence data.

2.3.4. Trend-Free Pre-Whitening Mann–Kendall Method

By considering the influence of the dominant trend of the data series on the autocorrelation coefficient estimation, the method of removing the preset white trend is adopted to eliminate the influence of the trend on the autocorrelation coefficient estimation, and the MK test of the data series is more accurate. The procedure is as follows [29]:

$$X'_t = X_t - \beta \times t \tag{14}$$

$$Y'_t = X'_t - r \times X'_{t-1} \tag{15}$$

$$Y_t = Y'_t + \beta \times t \tag{16}$$

where β is the Kendall straight-line slope, which can be calculated by Formula (7), X_t' is the detrended series, Y_t' is the detrended series without autocorrelation and the Y_t series preserved the true trend without autocorrelation.

2.3.5. Two-Phase Linear Regression Model

The simple linear regression model with a change point is given as follows [30]:

$$X_i = \begin{cases} \alpha_1 + \beta_1 \times t_i + \varepsilon_1, & t_{min} \leq t_i \leq t_C \\ \alpha_2 + \beta_2 \times t_i + \varepsilon_2, & t_C \leq t_i \leq t_{Max} \end{cases} \quad (17)$$

where (X_i,t_i) are the observations that correspond to the dependent and the independent variables, α_1, α_2, β_1, β_2 are the regression coefficients with the usual interpretations, ε_1, ε_2 are the random error terms for each line of the regression model and the point t_C is the unknown change point.

C ($t_C \leq C < t_C + 1$) and the other parameter values can be obtained by fitting, and the fitting formula is as follows:

$$X_i = \alpha_1 + \beta_1 \times t_i + \beta'(1 - C) \times INDc(t_i) + \varepsilon \quad (18)$$

$$\beta' = \beta_2 - \beta_1 \quad (19)$$

$$INDc(t_i) = \begin{cases} 0, & t_i \leq C \\ 1, & t_i > C \end{cases} \quad (20)$$

Each regression coefficient is obtained by the least square method, and the sum of the squares of the residuals of the equation (S) is obtained; then, the significance test is carried out for the mutation trend. The statistic U calculation formula is as follows:

$$U = \frac{(S_0 - S)/3}{S/(n-4)} \quad (21)$$

In this equation, S_0 is the residual sum of squares of the linear regression equation without catastrophe and S is the residual sum of squares corresponding to the optimal mutation point. The mutation point was accepted at the significance level of 0.05, otherwise the mutation is not considered obvious. U is the F(3, $n-4$) distribution.

2.3.6. Dominant Analysis

Dominant analysis, also known as the advantage analysis method, was proposed by Budescu [31]. In order to analyze the relative importance of different influencing factors, firstly, it is necessary to carry out regression analysis on the dependent variables by various indicators (influencing factors) and different combinations of these indicators. Calculate the determination coefficient R^2 of the regression equation containing these indexes and various combinations of indexes, compare these R^2, and then analyze the improvement of R^2 after adding one index or a combination of indexes into the regression equation. The most improved indicators or combination of indicators are better than others. This method has the following two advantages: (1) it has model independence, that is, the relative importance of the prediction index is kept constant in each sub-model; (2) according to the relative importance of each prediction index, the total prediction variance in the regression model is decomposed and expressed as a percentage. Accordingly, the order of the influence of meteorological elements on ET_0 can be realized.

3. Results and Discussion

3.1. Spatial Patterns of Annual Mean ET_0

The annual mean ET_0 during 1957–2017 ranges from 660.94 mm to 1532.54 mm in Xinjiang, the ET_0 ranges from 1532.54 mm to 660.64 mm with a latitude from 36.85° N to 45.93° N. It ranges from 660.64 mm to 1094.00 mm with a latitude from 43.03° N to 48.05° N (Figure 2a). The average value of ET_0 for the cotton growing season during 1957–2017 ranges from 1108.15 mm to 519.01 mm with a latitude from 36.85° N to 44.96° N; the average value of ET_0 ranges 519.01 mm to 870.30 mm, with a latitude from 43.03° N to 48.05° N (Figure 2a). The average value of ET_0 for the cotton seedling stage during 1957–2017 ranges from 340.67 mm to 153.49 mm in Xinjiang; the average value of ET_0 for the cotton squaring stage ranges from 224.82 mm to 105.15 mm, with a latitude from 36.85 to 48.05 degrees north latitude; the average value of ET_0 for the cotton flowing-boll stage ranges from 431.79 mm to 209.29 mm, with a latitude from 36.85 to 48.05 degrees north latitude; the average value of ET_0 for the cotton boll opening stage ranges from 110.87 mm to 50.92 mm, with a latitude from 36.85 to 48.05 degrees north latitude (Figure 2b). The value of the annual mean ET_0 decreases with the values of the degrees for north latitude (Figure 2a,b); the value of annual mean ET_0 for all years and the growing season of cotton increases with the values of the degree of elevation from 34.5 m to 935.0 m, and decreases with the degree of elevation from 935.0 m to 3090.1 m (Figure 2c,d).

High values of the annual mean ET_0 in Xinjiang are found in deserts and the low values in mountainous areas (Figure 3). The average value of ET_0 is found in different growth periods of the cotton area (Figure 3). The average value of ET_0 in south Xinjiang is greater than that in north Xinjiang (Figure 3).

3.2. ET_0 Change with Altitude and Elevation

The relationship between ET_0 and relative height used to reveal the layout of ET_0 conversion space is shown in Figure 2. The results show that ET_0 is inversely proportional to latitude. The annual average ET_0 decreases by 27.86 mm with the increase of 1° of the latitude, and the average ET_0 in the cotton wadding growth season decreases by 8.21 mm with the increase of 1° of the latitude (Figure 2g,h). The results show that ET_0 has vertical zonality; in other words, with the rise in altitude in Xinjiang, China, ET_0 decreased significantly (Figure 2). The results show that ET_0 is inversely proportional to altitude. The annual average ET_0 decreases by 0.07 mm with the increase in altitude by 1 m, and the average ET_0 in the cotton boll growing season decreases by 0.09 mm with the increase in altitude by 1 m (Figure 2i,j).

(a) (b)

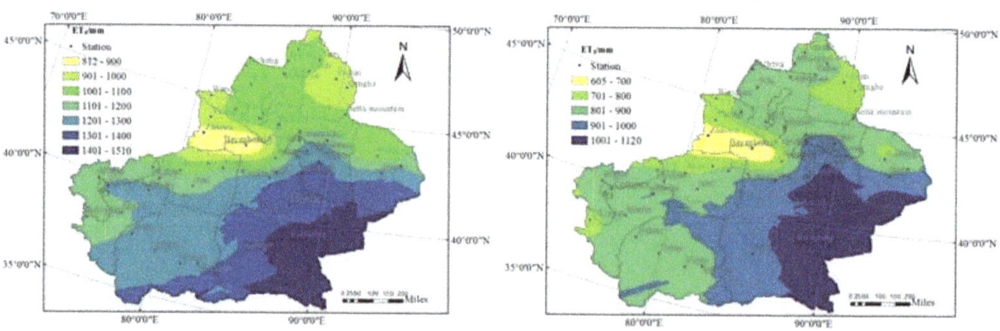

Figure 2. *Cont.*

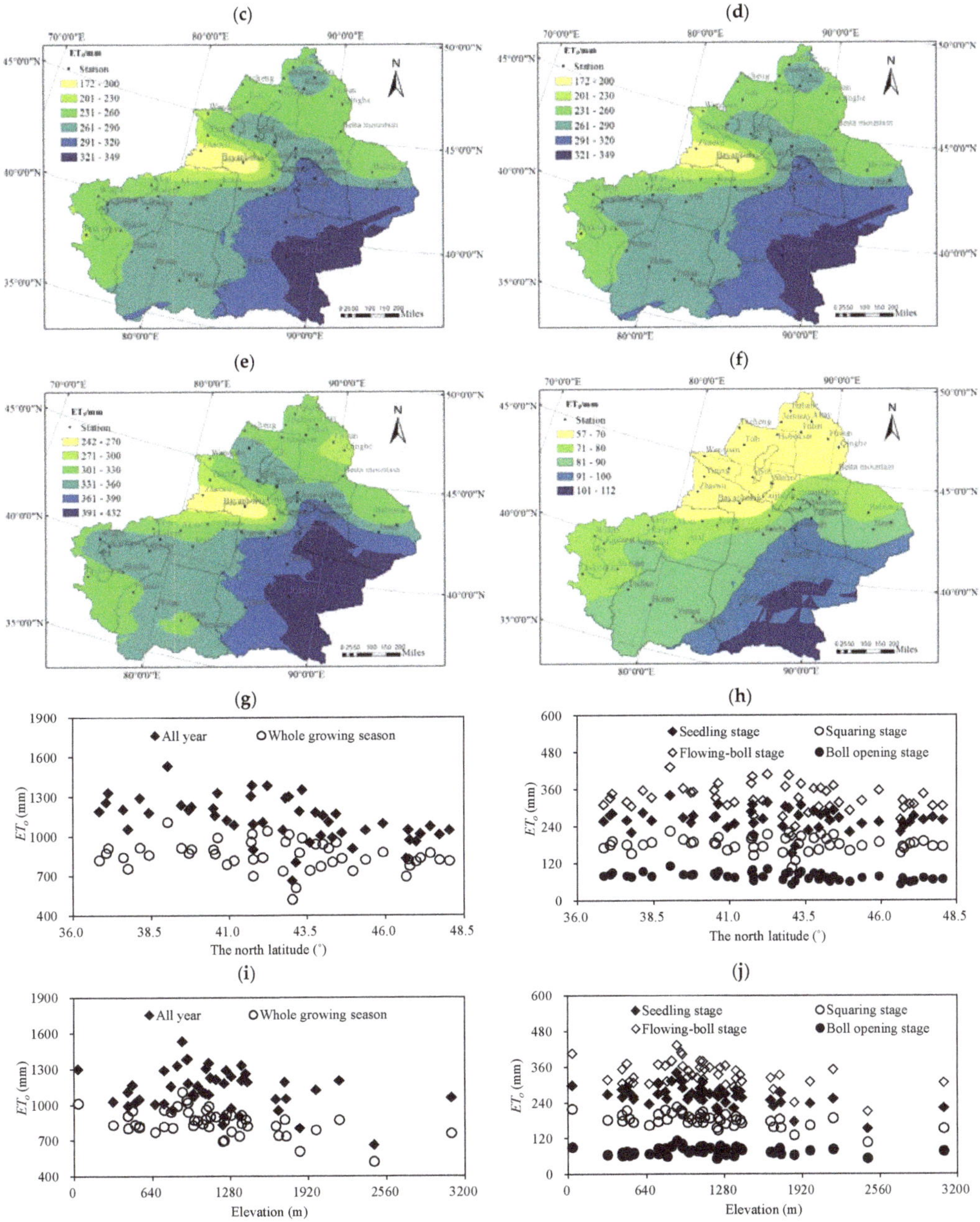

Figure 2. ET_o varies with space. (**a**): Annual average ET_o of Xinjiang; (**b**): the average ET_o value of the entire cotton growing season in Xinjiang; (**c**): the average ET_o value of cotton seedling stage in Xinjiang; (**d**): the average ET_o value of cotton squaring stage in Xinjiang; (**e**): the average ET_o value of cotton flowing-boll stage in Xinjiang; (**f**): the average ET_o value of cotton boll opening stage in Xinjiang; (**g**,**h**): relationship between the ET_o and latitude in Xinjiang; (**i**,**j**): relationship between the ET_o and elevation in Xinjiang of Northwest China.

3.3. ET_0 Change with Precipitation

The relationship between the ET_0 and the precipitation in Xinjiang is based on the long-term (1957–2017) meteorological data series during the cotton growing season (Figure 4). The results show that ET_0 decreased with the increase in precipitation in the growing season for cotton in Xinjiang, the average ET_0 decreases by 1.40 mm with the precipitation increase of 1 mm during the cotton growing season in Xinjiang (a), the average ET_0 decreases by 1.57 mm with the precipitation increase of 1 mm during the cotton growing season in North Xinjiang (b), the average ET_0 decreases by 1.42 mm with the precipitation increase of 1 mm during the cotton growing season in South Xinjiang (c), the average ET_0 decreases by 1.24 mm with the precipitation increase of 1 mm during the cotton growing season in East Xinjiang and (d) the values of the correlation coefficient between rainfall and ET_0 are −0.7245, −0.8241, −0.6065 and −0.5433. Figure 4 shows that the reference crop evapotranspiration rate in northern Xinjiang was highly affected by precipitation during the cotton growing seasons.

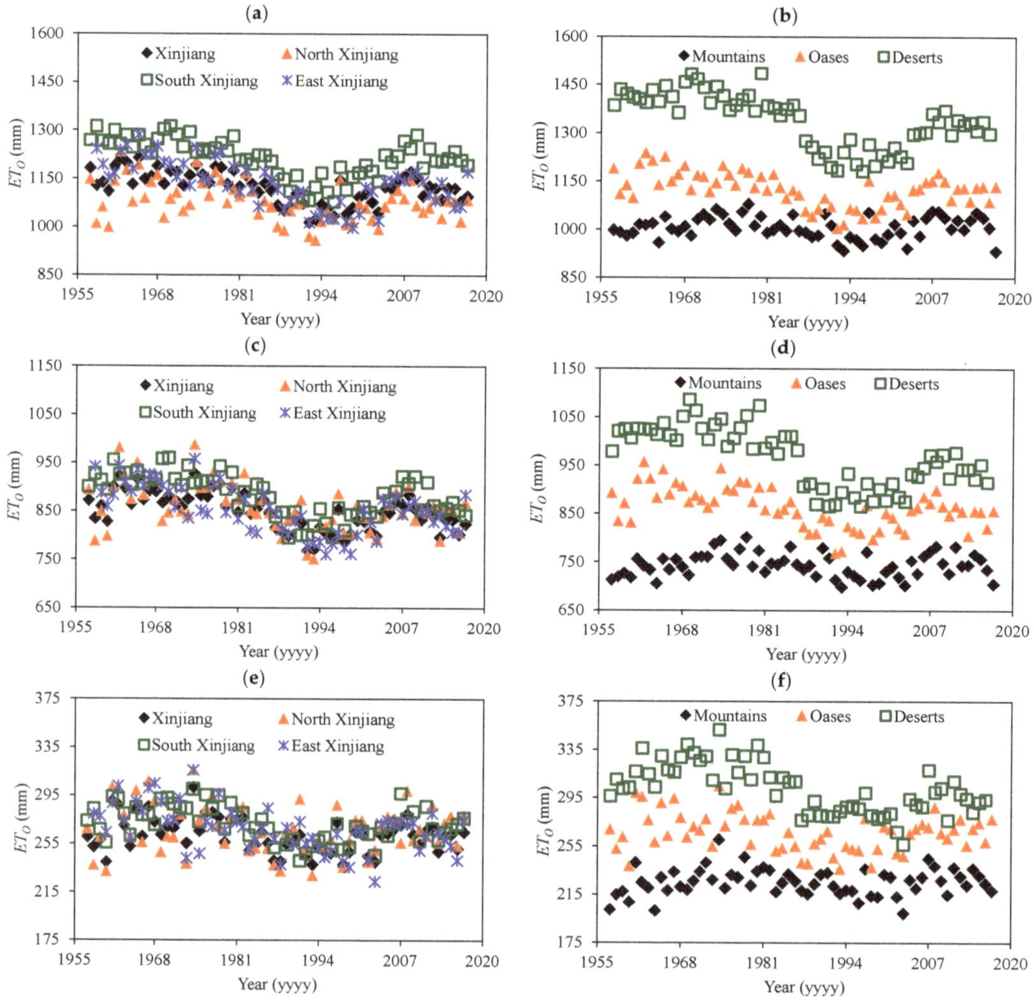

Figure 3. *Cont.*

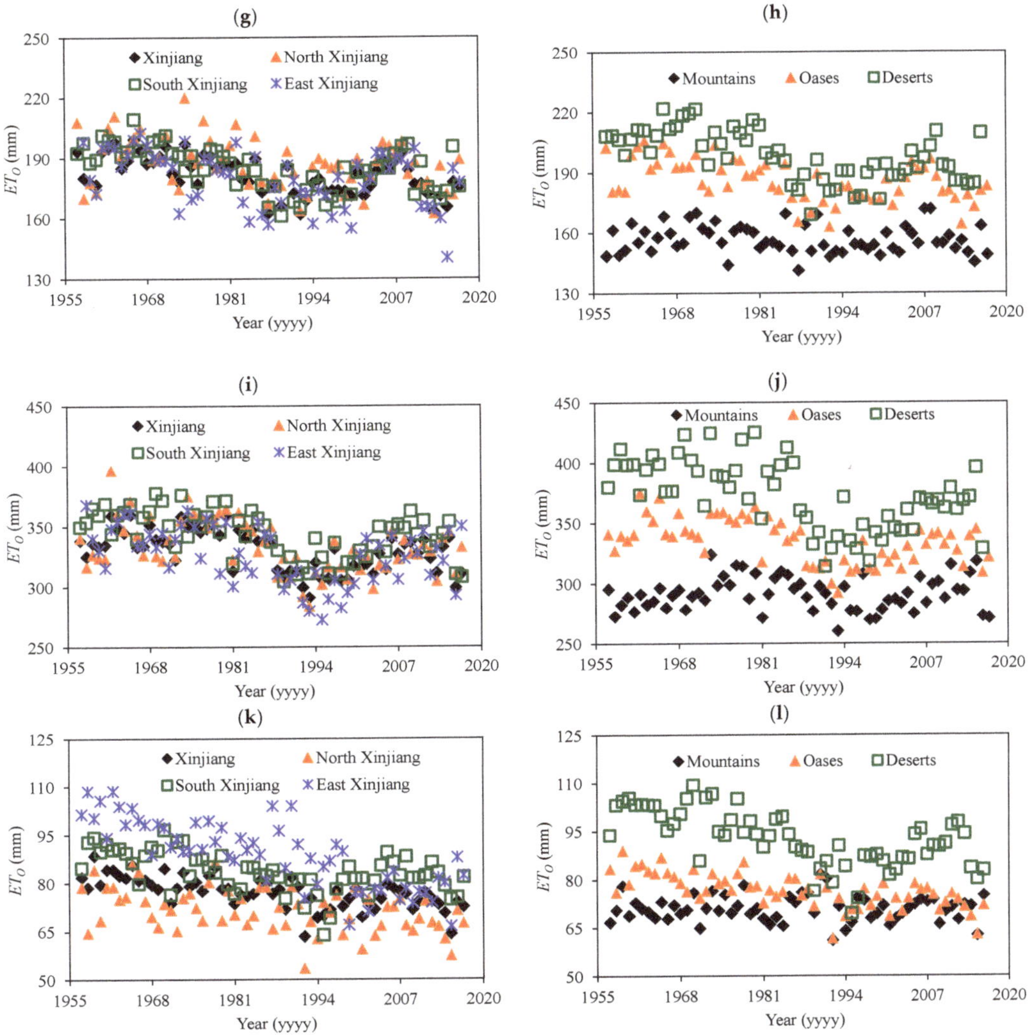

Figure 3. The change process of the ET_o in Xinjiang of Northwest China during 1957–2017. (**a**,**b**): Annual average ET_o over the entire year; (**c**,**d**): average ET_o value of cotton growing season; (**e**,**f**): average ET_o value of cotton seedling stage; (**g**,**h**): average ET_o value of cotton squaring stage; (**i**,**j**): average ET_o value of cotton flowing-boll stage; (**k**,**l**): average ET_o value of cotton boll opening stage.

The values of water consumption for cotton were directly affected by the change in ET_o and precipitation during the cotton growing season under the climate changes; furthermore, the irrigation water demand for cotton in Xinjiang was affected. If one considers the Shihezi region as an example, the value of K_c (crop coefficient) is 0.6 for cotton at Shihezi in northern Xinjiang [32], the values of water consumption for cotton were 124.65–554.58 mm during the growing seasons from 1957–2017, and the values of precipitation ranged from 50.5 mm to 235.4 mm, the values of irrigation water requirement for cotton ranged from 234.27 mm to 504.08 mm during the 1957–2017, and the irrigation water requirement decreased by 3.8 mm/10a during the 1950s–1990s, and increased by 1.84 mm/a during the 1990s–2010s.

Against the background of climate change, the contradiction between supply and demand of irrigation water for cotton in Xinjiang has become more serious.

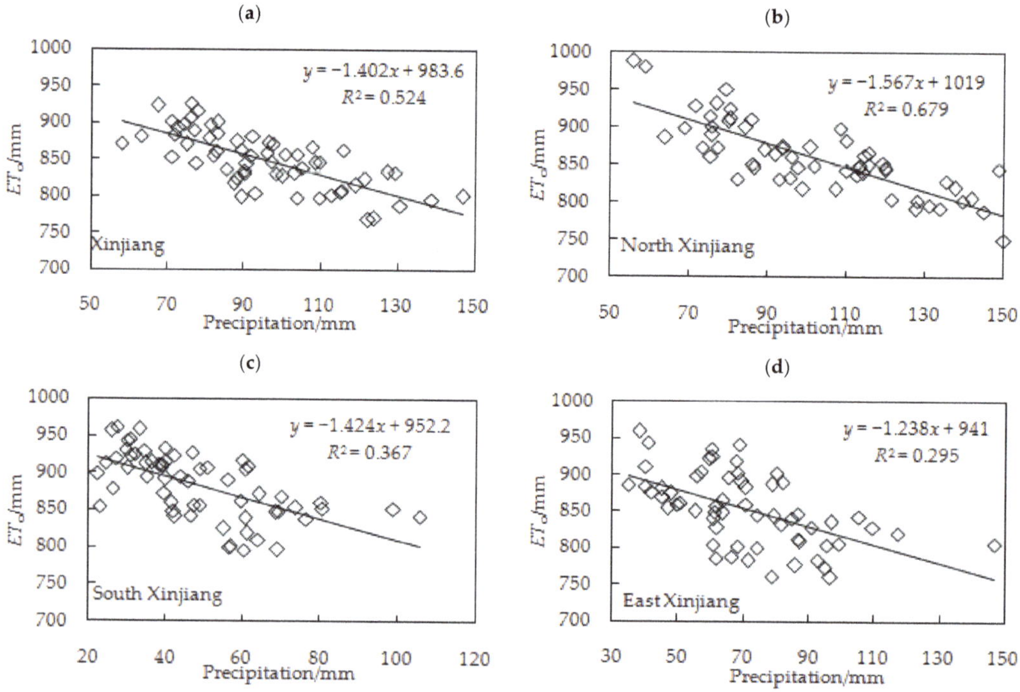

Figure 4. The relationship of the ET_0 and the precipitation during the cotton growing seasons in Xinjiang. (**a**): Average value of Xinjiang; (**b**): Average value of North Xinjiang; (**c**): Average value of South Xinjiang; (**d**): Average value of East Xinjiang.

3.4. ET_0 Change Trend

Through the autocorrelation test, it was found that almost all the ET_0 series in Xinjiang show strong autocorrelation. Pre-whitening and trend-free pre-whitening are used to pre-process all the ET_0 sequences. The typical treatment results are shown in Table 1. The results show that the two methods can effectively reduce the autocorrelation of the ET_0 series, and the PW method is better than TFPW. However, the trend of the sequences processed by the PW method is seriously reduced. Therefore, the TFPW method is chosen to preprocess the sequences, and then the Mann–Kendall (m–k) nonparametric statistics is carried out. The Mann–Kendall (m–k) nonparametric statistical method showed that ET_0 demonstrated a downward trend in general ($p < 0.01$).

The results of two-phase linear regression showed that the annual average ET_0 in Xinjiang had a downward trend from 1957 to 1996 (Figure 5a), with a decline rate of 33.96 mm/10a; there was a significant growth trend from 1997 to 2008 (Figure 5a), with an annual growth rate of 71.41 mm/10a. From 2009 to 2017, there was an obvious downward trend (Figure 5b), with a decline rate of 66.64 mm/10a. Such growth trends are consistent with global and Chinese climate change [23,33].

Table 1. Mann–Kendall value and first-order coefficient of autocorrelation for original and two pre-whitening ET_o series.

Series	Average Value of ET_o during the Growing Seasons for Cotton				Annual Mean of ET_o			
	Xinjiang	North Xinjiang	South Xinjiang	East Xinjiang	Xinjiang	Mountains	Oases	Deserts
MK-original ET_o series	−4.3179	−2.7106	−4.4327	−4.4837	−3.8969	−0.1084	−3.0359	−5.3766
r1-Original ET_o series	0.6493	0.4124	0.7069	0.5870	0.6310	0.2817	0.5933	0.8346
MK-pre-whitening	−1.7539	−1.3968	−2.0473	−1.9708	−1.7157	−0.2487	−1.3840	−1.8560
r1-Pre-whitening	−0.1721	−0.0541	−0.3380	−0.1727	−0.1759	−0.0017	−0.1488	−0.1911
MK-trend-free pre-whitening	−4.6240	−2.9402	−5.0449	−4.5730	−4.3689	−0.3253	−3.0168	−6.7670
r1-Trend-free pre-whitening	0.3190	0.1580	0.2366	0.3092	0.2213	−0.0010	0.1514	0.5218

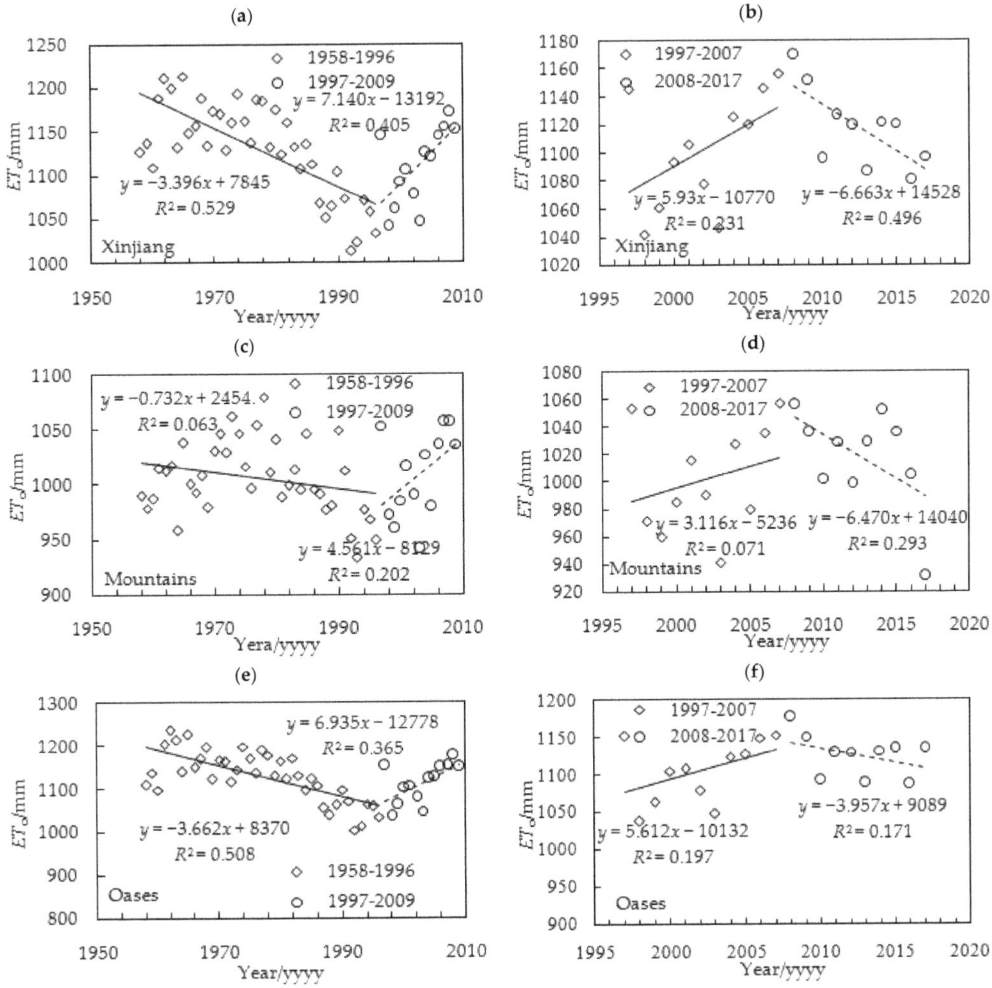

Figure 5. *Cont.*

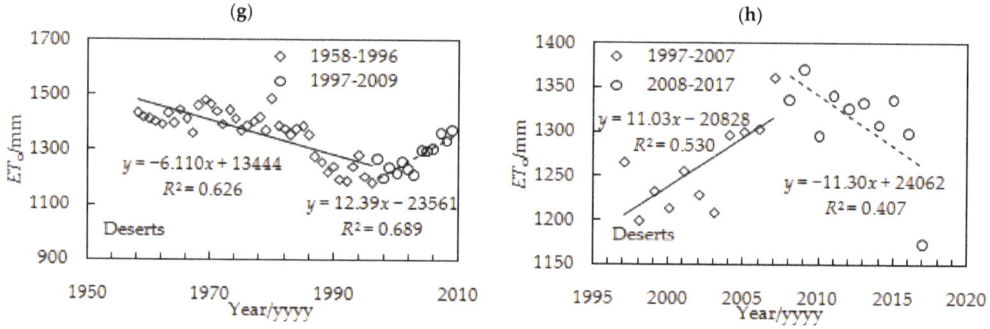

Figure 5. Abrupt change in annual mean ET_0 in Xinjiang during 1957–2017. (**a**,**b**): Annual mean of ET_0 in Xinjiang; (**c**,**d**): annual mean of ET_0 in mountains; (**e**,**f**): annual mean of ET_0 in oases; (**g**,**h**): annual mean of ET_0 in deserts.

Figure 5 shows that the annual average ET_0 mutation point occurred in Xinjiang in 1996 and 2008 (Figure 5a,b); for the oasis, it occurred in 1995 and 2008 (Figure 5e,f), followed by the mountain region in 1998 and 2010 (Figure 5c,d), and desert in 1999 and 2009 (Figure 5g,h); this is related to the unique geographical location and climatic condition garden landscapes. Because the mutation analysis shows that all garden landscapes and three specific garden landscapes have relative mutations around 1997 and 2009, we divide 60 years into the following three periods: before 1997, 1997–2009 and after 2009, and carry out a more detailed analysis of each link.

Among the three landscapes here, the mountainous area shows decreasing trends from 1957 to 1998 (Figure 5c), with a downward rate of 7.33 mm/10a, an upward trend from 1999 to 2010 (Figure 5c) with a rate of 45.62 mm/10a, and a downward trend from 2011 to 2017 (Figure 5d), with a downward rate of 64.71 mm/10a; the oases showed a downward trend from 1957 to 1995 (Figure 5e), with a speed of 3.63 mm/10a, an upward trend from 1996 to 2008 (Figure 5e) with a speed of 69.35 mm/10a, and a downward trend from 2009 to 2017 (Figure 5f), with a speed of 39.57 mm/10a; the desert showed a downward trend from 1957 to 1999 (Figure 5g), with a downward rate of 61.11 mm/10a, an upward trend from 1997 to 2009 (Figure 5g) with a rate of 123.99 mm/10a, and a downward trend from 2010 to 2017 (Figure 5h), with a downward rate of 113.05 mm/10a. The reasons for these trends mainly lie in the following two aspects: first, the ecosystem in the desert area has low stability and the mountain ecosystem has high stability; the other reason is that the Xinjiang has experienced rapid population growth, along with the fast expansion of urbanization, industrialization, and tourism. Human activity might have caused the differences among the different landscapes against the background of global warming and regional climate change.

Figure 6 shows that the ET_0 gene mutation point of the cotton growing seasons in Xinjiang occurred in 1997 and 2008 (Figure 6a,b); the mutation point occurred in eastern Xinjiang in 1997 and 2006 (Figure 6g,h), in the south of Xinjiang in 1997 and 2007 (Figure 6e,f), and in the north of Xinjiang in 1997 and 2009 (Figure 6c,d); this is related to the unique geographical location and climatic conditions in different landscapes. Because the mutation analysis shows that all landscapes and three specific landscapes have relative gene mutations around 1997 and 2008, we divide 60 years into the following three periods: before 1997, 1997–2009 and after 2009, and carry out a more detailed analysis of each link.

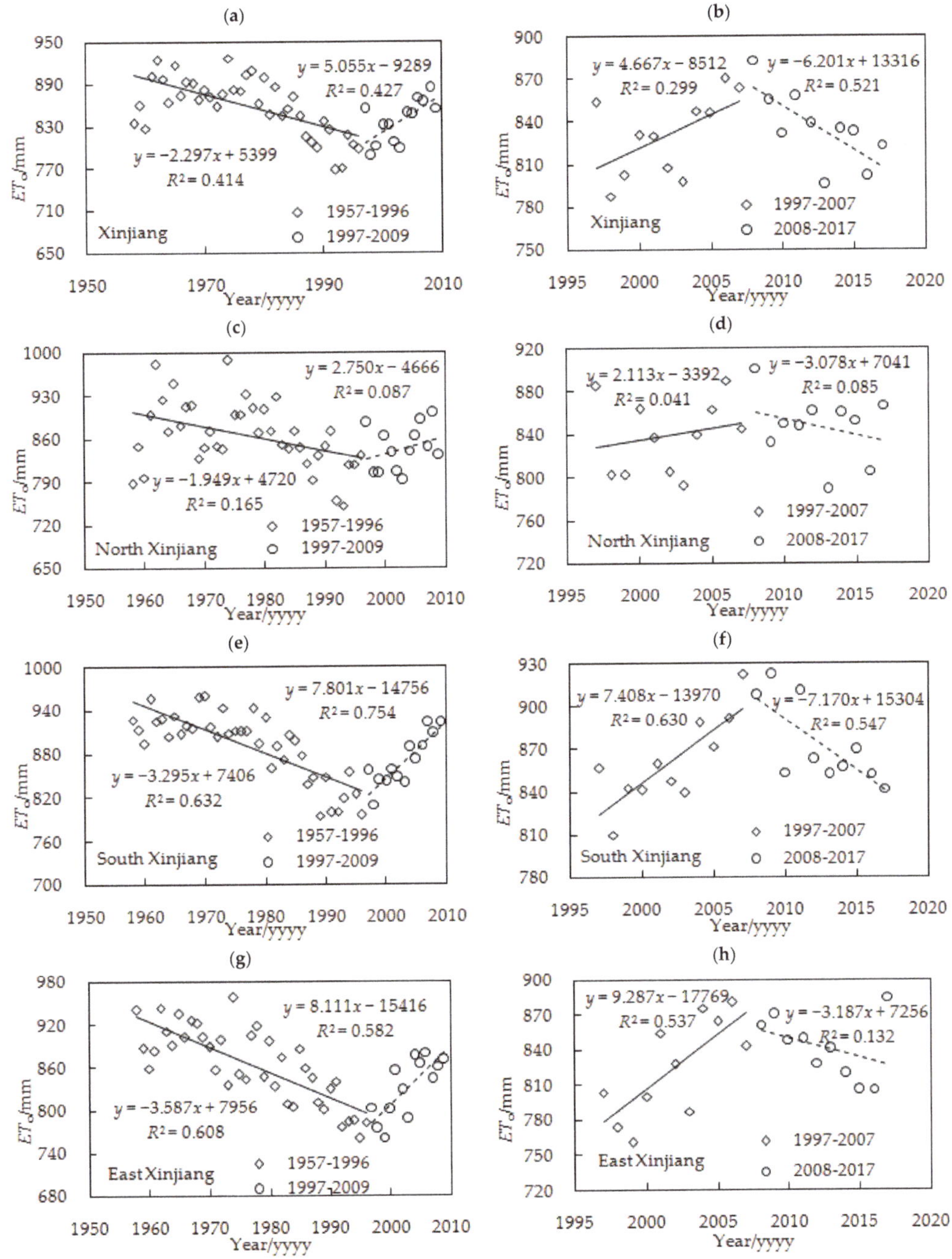

Figure 6. Abrupt change in average values of ET_o during the growing seasons for cotton in Xinjiang. (**a**,**b**): average value of ET_o during the growing seasons for cotton in Xinjiang; (**c**,**d**): average value of ET_o during the growing seasons for cotton in north Xinjiang; (**e**,**f**): average value of ET_o during the growing seasons for cotton in south Xinjiang; (**g**,**h**): average value of ET_o during the growing seasons for cotton in east Xinjiang.

The average values of ET_0 during the growing season for cotton in Xinjiang showed a significant downward trend during 1957–1997 (Figure 6a), with a rate of 22.97 mm/10a, a significant upward trend during 1998–2008 (Figure 6a), with a rate of 50.53 mm/10a and a significant downward trend during 2009–2017 (Figure 6b), with a rate of 62.02 mm/10a. The average values of ET_0 during the growing season for cotton in north Xinjiang showed a significant downward trend during 1957–1997 (Figure 6c), with a rate of 19.50 mm/10a, a significant upward trend during 1997–2009 (Figure 6c), with a rate of 27.51 mm/10a and a significant downward trend during 2010–2017 (Figure 6d), with a rate of 30.78 mm/10a. The average values of ET_0 during the growing season for cotton in south Xinjiang showed a significant downward trend during 1957–1997 (Figure 6e), with a rate of 32.95 mm/10a, a significant upward trend during 1998–2007 (Figure 6e), with a rate of 78.01 mm/10a and a downward trend during 2008–2017 (Figure 6f), with a rate of 71.71 mm/10a. The average values of ET_0 during the growing season for cotton in east Xinjiang showed a significant decreasing trend during 1957–1997 (Figure 6c), with a rate of 35.87 mm/10a, a significant increasing trend during 1998–2006 (Figure 6g), with a rate of 81.12 mm/10a and a decreasing trend during 2007–2017 (Figure 6c), with a rate of 31.88 mm/10a.

The cause of the ET_0 change trend can be reduced to two main reasons, the first reason is the increase in irrigation and water conservancy after the founding of the People's Republic of China, which included increasing the irrigated area in Xinjiang and the regional air humidity; the second reason is the planting of windbreaks in agricultural areas that reduced the regional wind speed, finally leading to the continuous decline in the water requirement of reference crops from the 1950s to 1990s. Under the conditions of climate change, the temperature has been increasing since the 1990s, which explains why the values of ET_0 during the cotton growing seasons showed an increasing trend from 1990s to 2010s.

3.5. Abrupt Change in ET_0

Two-phase linear regression analysis was used to analyze the abrupt trend of the ET_0 time series, and the abrupt trend of each characteristic time point passed the significance level test of 0.05 ($F_{0.05}(3,48) = 2.80$, $F_{0.05}(3,16) = 3.24$). The Alar station (81.27° E, 40.55° N) was selected as an example to demonstrate the abrupt change in the time series of ET_0 during the growing seasons for cotton (Figure 7). The results showed that the mutation occurred in 2001 and 2011. Using this method of detection, the abrupt changes in the annual average ET_0 for the entirety of Xinjiang, southern Xinjiang, eastern Xinjiang, oasis and desert areas show obvious trends. For the average ET_0 of the cotton growing season, there is a clear trend for Xinjiang, northern Xinjiang, southern Xinjiang, eastern Xinjiang and the oasis and desert areas.

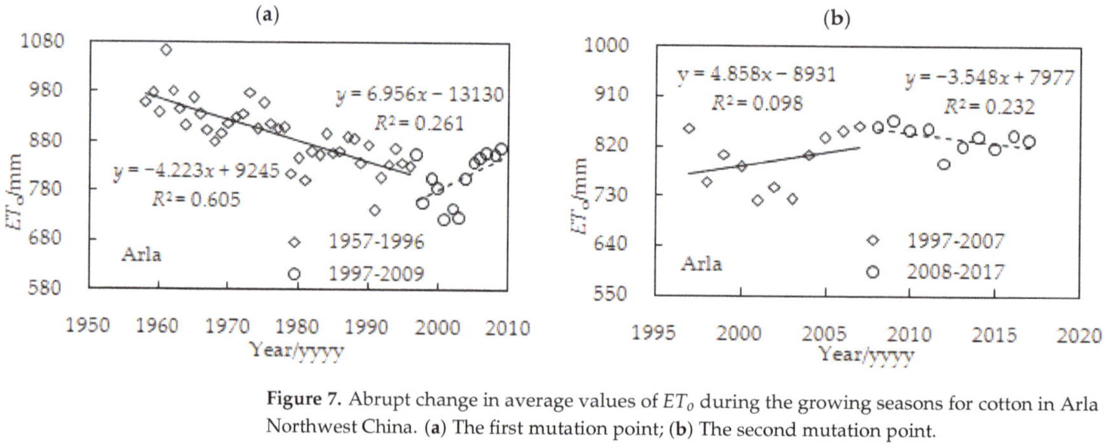

Figure 7. Abrupt change in average values of ET_0 during the growing seasons for cotton in Arla of Northwest China. (**a**) The first mutation point; (**b**) The second mutation point.

The mutation age detected by the m-k method is shown in Figures 5–7. The results showed that in the past 60 years, the first mutation occurred in 1995–1999, and the second mutation occurred in 2006–2011 from 1957 to 2017. In this abrupt change year, due to climate change and human activities, vegetation cover underwent a great change. In 1987, the climate in Xinjiang changed from warm and dry to warm and humid, and the vegetation coverage changed [23]. In order to more effectively study the interaction between climate change, soil resource cover and human activities, we should understand the reasons for the sudden change in transpiration rate in Xinjiang.

4. Conclusions

The time specificity of ET_0 in Xinjiang, China, from 1957 to 2017 was revealed by using the M-K nonparametric development trend and mutation detection, and two-phase linear regression model. The conclusions are as follows.

1. The annual average ET_0 of the Xinjiang region demonstrates vertical zonality and longitude and latitude zonality. Due to the rise in altitude, it gradually decreases from high to low, and the ET_0 changes significantly from southern Xinjiang to northern Xinjiang. For different garden landscapes, the low value of ET_0 occurs in desert areas, and the low value consumption occurs in mountainous areas. The large-scale atmospheric circulation, location and altitude constitute complex standards that endanger the spatial distribution of the annual average transpiration rate in Xinjiang.

2. The results show that during the period from 1957 to 2017, ET_0 showed a downward trend. The first mutation occurred in 1995–1999 and the second mutation occurred in 2006–2011. Among the three garden landscapes, ET_0 decreases faster in desert areas, followed by oasis areas and the least in mountainous areas.

3. The results of the m-k analysis and two-phase linear regression show that 48 sites have experienced ET_0 mutations, and the annual average of ET_0 in Xinjiang, China, shows a downward trend from 1957 to 1996. There was also an obvious growth trend from 1997 to 2009 and a significant downward trend in the period from 2008 to 2017.

4. The annual average ET_0 of Xinjiang demonstrates a sudden change, and Xinjiang, southern Xinjiang, eastern Xinjiang and the oases and desert areas show an obvious development trend. In terms of the average ET_0 in the cotton growing season, there is an obvious trend for Xinjiang, northern Xinjiang, southern Xinjiang, eastern Xinjiang, the oases and desert areas. Because of the negative impact of human activities, the land resource cover in this area has changed greatly, so we must further study the relationship between land resource use and cover and ET_0 change.

Author Contributions: Conceptualization, Y.J., X.S. and N.S.; methodology, X.S., Y.J. and N.S.; investigation, Y.J., X.S. and N.S.; data curation, Y.J. and N.S.; formal analysis, X.S., R.Y., Y.J. and N.S.; writing—original draft preparation, Y.J., X.S. and R.Y.; writing—editing, N.S. and X.S.; supervision, X.S. and N.S.; project administration, Y.J.; funding acquisition, Y.J. All authors have read and agreed to the published version of the manuscript.

Funding: This research was jointly supported by the National Natural Science Foundation of China (51790534), and the China Scholarship Council (No. 201703250034).

Institutional Review Board Statement: This study not involving humans or animals, so this research do not applicable.

Informed Consent Statement: This study involving humans, not applicable.

Data Availability Statement: Data involving meteorological data openly available in a public repository (the National Meteorological Information Center of China Meteorological Administration).

Acknowledgments: Thanks to the National Meteorological Information Center of China Meteorological Administration for offering the meteorological data. This study was supported by the Agricultural Science and Technology Innovation Program (ASTIP) of Chinese Academy of Agricultural Sciences. Special thanks are given to the anonymous reviewers for their constructive comments that improved this manuscript.

Conflicts of Interest: The authors declare no conflict of interest.

References

1. Solomon, S.D.; Qin, D.; Manning, M.; Chen, Z.; Miller, H.L. *Climate Change 2007-the Physical Science Basis: Working Group I Contribution to the Fourth Assessment Report of the IPCC*; Cambridge University Press: Cambridge, MA, USA, 2007.
2. Yoro, K.O.; Daramola, M.O. CO_2 emission sources, greenhouse gases, and the global warming effect. *Adv. Carbon Capture* **2020**, 3–28. [CrossRef]
3. Anderson, T.R.; Hawkins, E.; Jones, P.D. CO_2, the greenhouse effect and global warming: From the pioneering work of Arrhenius and Callendar to today's Earth System Models. *Endeavour* **2016**, *40*, 178–187. [CrossRef]
4. Thomas, J.; Brunette, M.; Leblois, A. The determinants of adapting forest management practices to climate change: Lessons from a survey of French private forest owners. *For. Policy Econ.* **2022**, *135*, 102662. [CrossRef]
5. Oertel, C.; Matschullat, J.; Zurba, K.; Zimmermann, F.; Erasmi, S. Greenhouse gas emissions from soils—A review. *Geochemistry* **2016**, *76*, 327–352. [CrossRef]
6. Kong, D.; Gu, X.; Li, J.; Ren, G.; Liu, J. Contributions of global warming and urbanization to the intensification of human-perceived heat waves over China. *J. Geophys. Res. Atmos.* **2020**, *125*, e2019JD032175. [CrossRef]
7. Lannelongue, L.; Grealey, J.; Inouye, M. Green algorithms: Quantifying the carbon footprint of computation. *Adv. Sci.* **2021**, *8*, 2100707. [CrossRef]
8. Lorenzoni, I.; Pidgeon, N.F. Public views on climate change: European and USA perspectives. *Clim. Chang.* **2006**, *77*, 73–95. [CrossRef]
9. Mcmichael, A.J.; Woodruff, R.E.; Hales, S. Climate change and human health: Present and future risks. *Lancet* **2006**, *367*, 859–869. [CrossRef]
10. Karimi, V.; Karami, E.; Keshavarz, M. Climate change and agriculture: Impacts and adaptive responses in Iran. *J. Integr. Agric.* **2018**, *17*, 1–15. [CrossRef]
11. Li, M.; Fu, Q.; Singh, V.P.; Ji, Y.; Liu, D.; Zhang, C.; Li, T. An optimal modeling approach for managing agricultural water-energy-food nexus under uncertainty. *Sci. Total Environ.* **2019**, *651*, 1416–1434. [CrossRef]
12. Luo, Q.; O'Leary, G.; Cleverly, J.; Eamus, D. Effectiveness of time of sowing and cultivar choice for managing climate change: Wheat crop phenology and water use efficiency. *Int. J. Biometeorol.* **2018**, *62*, 1049–1061. [CrossRef] [PubMed]
13. Ramanathan, K.C.; Saravanan, S.; Krishna, K.M.A.; Srinivas, T.; Selokar, A. Reference evapotranspiration assessment techniques for estimating crop water requirement. *Int. J. Eng. Technol.* **2019**, *8*, 1094–1100.
14. Djaman, K.; ONeill, M.; Diop, L.; Bodian, A.; Allen, S.; Koudahe, K.; Lombard, K. Evaluation of the Penman-Monteith and other 34 reference evapotranspiration equations under limited data in a semiarid dry climate. *Theor. Appl. Climatol.* **2019**, *137*, 729–743. [CrossRef]
15. Mirshekarnezhad, B.; Paknejad, F.; Amiri, E.; Ardakani, M.R.; Ilkaee, M.N. Functional strategies for certain growth stages of corn in response to environmental factors: Irrigation and planting date management. *Appl. Ecol. Environ. Res.* **2018**, *16*, 6169–6180. [CrossRef]
16. Qin, Y. *The Spatiotemporal Characteristics of Temperature Lapse Rate in China*; China University of Geosciences: Wuhan, China, 2021.
17. Xie, W.; Wang, S.; Yan, X. Evaluation on CMIP6 Global Climate Model Simulation of the Annual Mean Daily Maximum and Minimum Air Temperature in China. *Clim. Environ. Res.* **2022**, *27*, 63–78.
18. Tang, S. *Spatiotemporal Variations of the Temperature and Precipitation in Xinjiang during the Past 40 Years*; Si-chuan Normal University: Chengdu, China, 2021.
19. Xue, Y.; Han, P.; Feng, G. Change trend of the precipitation and air temperature in Xinjiang since recent 50 years. *Arid Zone Res.* **2003**, *20*, 127–130.
20. Yu, H.; Liu, J. Trend analysis of average temperature variations and runoff response in last 40 years in Xinjiang. *Water Power* **2007**, *22*, 169–177.
21. Hu, A.; Dong, X. A study on water surface evaporation in southern Xinjiang. *J. Arid Land Resour. Environ.* **2006**, *20*, 22–24.
22. Hu, A.; Guo, X. A model of water surface evaporation forecasting for Xinjiang plain area. *J. China Hydrol.* **2006**, *26*, 24–27+41.
23. Shi, Y.; Shen, Y.; Kang, E.; Li, D.; Ding, Y.; Zhang, G.; Hu, R. Recent and future climate change in northwest China. *Clim. Chang.* **2007**, *80*, 379–393. [CrossRef]
24. Liu, Q.; Yang, Z.; Cui, B. Spatial and temporal variability of annual precipitation during 1961–2006 in Yellow River Basin, China. *J. Hydrol.* **2008**, *361*, 330–338. [CrossRef]
25. Allen, R.; Pereira, L.; Raes, D.; Smith, M.; Allen, R.G.; Pereira, L.S.; Martin, S.; Allen, R.; Pereira, L.; Raes, D.; et al. *Crop Evapotranspiration: Guidelines for Computing Crop Water Requirements*; FAO Irrigation and Drainage Paper 56; Food and Agriculture Organization of the United Nations: Rome, Italy, 1998.
26. Mann, H.B. Non-parametric test against trend. *Econometrica* **1945**, *13*, 245–259. [CrossRef]
27. Kendall, M.G. *Rank Correlation Methods*; Hafner: New York, NY, USA, 1948.
28. Kahya, E.; Kalayc, S. Trend analysis of stream flow in Turkey. *J. Hydrol.* **2004**, *289*, 128–144. [CrossRef]
29. Sheng, Y.; Pilon, P.; Phinney, B.; Cavadias, G. The influence of autocorrelation on the ability to detect trend in hydrological series. *Hydrol. Processes* **2002**, *16*, 1807–1829.

30. Acitas, S.; Senoglu, B. Robust change point estimation in two-phase linear regression models: An application to metabolic pathway data. *J. Comput. Appl. Math.* **2019**, *363*, 337–349. [CrossRef]
31. Budescu, D.V. Dominance analysis: A new approach to the problem of relative importance of predictors in multiple regression. *Psychol. Bull.* **1993**, *114*, 542–551. [CrossRef]
32. Shen, X. *Water-Saving Mechanism and Optimal Irrigation Pattern for Effective Water Use and High-Quality of Cotton under Drip Irrigation*; Chinese Academy of Agricultural Sciences: Beijing, China, 2012.
33. Ren, G.; Xu, M.; Chu, Z.; Guo, J.; Ying, W. Changes of surface air temperature in China during 1951–2004. *Clim. Environ. Res.* **2005**, *10*, 717–727.

agriculture

MDPI

Article

Climate Change Affects the Utilization of Light and Heat Resources in Paddy Field on the Songnen Plain, China

Ennan Zheng [1,†], Mengting Qin [1,†], Peng Chen [2], Tianyu Xu [1,*] and Zhongxue Zhang [3,*]

[1] School of Hydraulic and Electric Power, Heilongjiang University, Harbin 150080, China
[2] College of Agricultural Science and Engineering, Hohai University, Nanjing 210098, China
[3] School of Water Conservancy and Civil Engineering, Northeast Agricultural University, Harbin 150030, China
[*] Correspondence: 2021007@hlju.edu.cn (T.X.); zhangzhongxue@163.com (Z.Z.)
[†] These authors contributed equally to this work.

Abstract: Efficient utilization of light and heat resources is an important part of cleaner production. However, exploring the changes in light and heat resources utilization potential in paddy under future climate change is essential to make full use of the potential of rice varieties and ensure high-efficient, high-yield, and high-quality rice production, which has been seldom conducted. In our study, a process-based crop model (CERES-Rice) was calibrated and validated based on experiment data from the Songnen Plain of China, and then driven by multiple global climate models (GCMs) from the coupled model inter-comparison project (CMIP6) to predict rice growth period, yield, and light and heat resources utilization efficiency under future climate change conditions. The results indicated that the rice growth period would be shortened, especially in the high emission scenario (SSP585), while rice yield would increase slightly under the low and medium emission scenarios (SSP126 and SSP245), it decreased significantly under the high emission scenario (SSP585) in the long term (the 2080s) relative to the baseline of 2000–2019. The light and temperature resources utilization (E_{RT}), light utilization efficiency (E_R), and heat utilization efficiency (HUE) were selected as the light and heat resources utilization evaluation indexes. Compared with the base period, the mean E_{RT} in the 2040s, 2060s, and 2080s were −6.46%, −6.01%, and −6.03% under SSP126, respectively. Under SSP245, the mean E_{RT} were −7.89%, −8.41%, and −8.27%, respectively. Under SSP585, the mean E_{RT} were −6.88%, −13.69%, and −28.84%, respectively. The E_R would increase slightly, except for the 2080s under the high emission scenario. Moreover, the HUE would reduce as compared with the base period. The results of the analysis showed that the most significant meteorological factor affecting rice growth was temperature. Furthermore, under future climate conditions, optimizing the sowing date could make full use of climate resources to improve rice yield and light and heat resource utilization indexes, which is of great significance for agricultural cleaner production in the future.

Keywords: GCMs; CERES-rice model; climate change; rice yield; light and heat resource utilization

Citation: Zheng, E.; Qin, M.; Chen, P.; Xu, T.; Zhang, Z. Climate Change Affects the Utilization of Light and Heat Resources in Paddy Field on the Songnen Plain, China. *Agriculture* **2022**, *12*, 1648. https://doi.org/10.3390/agriculture12101648

Academic Editors: Dengpan Xiao and Wenjiao Shi

Received: 10 August 2022
Accepted: 4 October 2022
Published: 9 October 2022

1. Introduction

Global climate change, which affects agricultural production and human health, is a central issue of constant concern [1]. It has been estimated that global warming will reach 1.5 °C in the near-term [2]. Agriculture is very sensitive to climate change and is also one of the industries most affected by climate change [3,4]. The United Nations Intergovernmental Panel on Climate Change (IPCC) Sixth Assessment Report (AR6) on agriculture extends from crop production systems to food supply systems, and evidence of the adverse effects of climate change on crop production is strengthening [2]. Greenhouse gas emissions have a significant impact on climate warming [5–7]. Anthropogenic warming hampers crop yield increase. Moreover, crop yield is decreased with the increase in surface O_3 concentrations and CH_4 emissions exacerbate these adverse effects [8]. Climate change will increase pressure on food production, especially in vulnerable areas [2]. Therefore, for ensuring global food security, assessing the impacts of future climate change on crop growth is one of the most important issues in the 21st century [9].

Rice is a staple food for the global population, and its production is key to global food security. Stable growth of rice production has been an issue in achieving food security, especially in developing countries [10]. A slight decline in rice productivity will have a significant impact on global food security [11]. The Songnen Plain is an important rice producer in China, with its vast area and large inter-annual variability of heat and precipitation. Climate change changes the conditions of light, temperature, and water during the growth and development of crops, affecting the allocation of their heat, radiation, and water resources [12]. The light and heat resources utilization of rice is a decisive factor affecting its yield, and the light and heat resources utilization in a certain time and space range determines the production potential of the agricultural system [13]. Efficient utilization of local light and heat resources is of great importance to provide full play to the potential of crop varieties and ensure high-efficient, high-yield, and high-quality rice production. However, most of the previous studies only focused on crop phenology and yield changes and rarely in terms of light and heat resources utilization. Therefore, it is necessary to analyze the response of light and heat resource utilization efficiency of rice to future climate change on the Songnen Plain and ensure the rice supply.

At this stage, field experiments cannot simulate climate change, such as temperature, precipitation, and solar radiation very well. The decision support system for agrotechnology transfer (DSSAT) can simulate the process of crop growth and yield formation, soil and crop water balance, nutrient and carbon dynamics, etc. Research and application on DSSAT have focused on yield prediction, crop breeding, land use, water and fertility management, climate change, and other areas. It is one of the most widely used crop modeling systems at present [14]. In the context of climate change, the application of crop growth models to study the impacts of climate conditions on crop production status and yield in historical periods has become quite extensive and mature [15]. The use of global climate models (GCMs) or regional climate models (RCMs) to construct future climate change scenarios and then coupled with crop growth models has developed into an important tool to assess the impacts of future climate change on agricultural production [16–18]. Kang et al. used an enhanced soil and water assessment tool (SWAT) model in combination with five global climate models to assess the potential impacts of climate change on the yield of two major crops (i.e., potato and barley) in Western Canada [18]. Rosenzweig et al. used seven global grid crop models (GGCMs) combined with five global climate models to analyze crop responses to climate change, showing strong negative effects of climate change, particularly at higher levels of warming and lower latitudes [17]. Crop model yield projections derived from averaging multiple GCM ensembles and different shared socioeconomic pathway and representative concentration (SSP) scenarios can provide a more reliable assessment of climate change impacts.

In our study, we used a calibrated and validated crop model, the CERES-Rice model, driven by future climate data from six GCMs for phase six of the coupled model intercomparison project (CMIP6) under the scenario of three different shared socioeconomic pathway and representative concentration pathway (SSP), including SSP126, SSP245, and SSP585, to project impacts of future climate change on rice production on the Songnen Plain of China. The objectives of our study were to (1) evaluate the performance of the CERES-Rice model in simulating the rice growth on the Songnen Plain of China, (2) explore the effects of future climate change on rice phenology, yield, and the light and heat resources utilization efficiency in the study area, and (3) investigate the ability to optimize sowing dates to cope with the effects of climate change on rice growth.

2. Materials and Methods

2.1. Study Site

The Songnen Plain is located in the central and western part of Northeast China (Figure 1). We selected the representative site, Qing'an National Irrigation Experimentation

Key Station (46°52′41″ N, 127°30′04″ E) on the Songnen Plain for our study. The study site is a typical cold black soil distribution area, that belongs to the cold temperate continental monsoon climate, the annual frost-free period is 128 days, the average annual temperature is 2.5 °C, the average annual rainfall is 566 mm, the rainfall is mostly concentrated in July and August, the geographical environment and natural resources conditions in the area are superior, favorable to the growth of crops, and the crop growing period is 156~171 days.

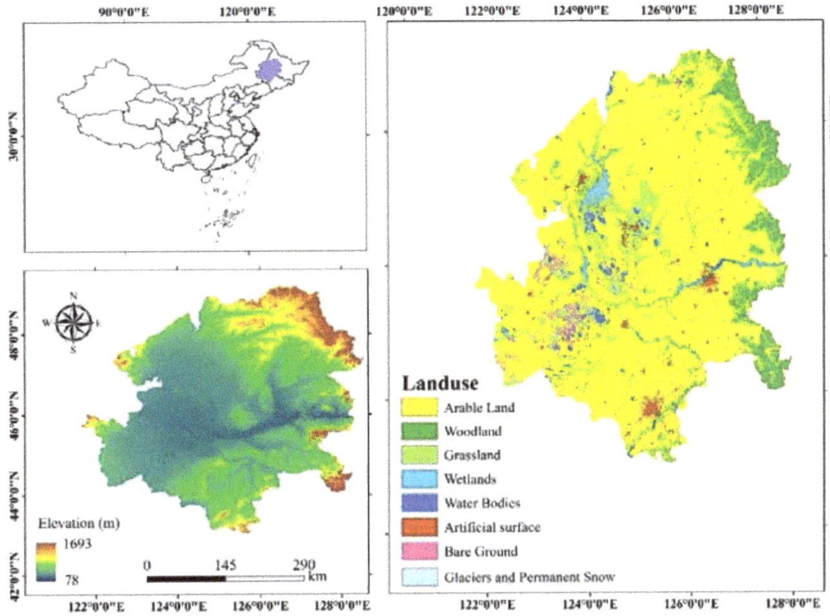

Figure 1. Location site.

2.2. Field Experimental Data

The field trials were conducted for 2 years with the rice cultivar "Suijing No.18". The rice was sown on 17 April, at a planting density of 24 plants/m^2, row spacing was 30 cm, the area of the test plot was 100 m^2 (10 × 10 m), and each plot was applied with N (110 kg/hm^2), P_2O_5 (45 kg/hm^2), and K_2O (80 kg/hm^2) in the ratio of base fertilizer/tillering fertilizer/heading fertilizer (5:3:2). The three different irrigation methods, namely control irrigation (CI), wet irrigation (WI), and flood irrigation (FI) were designed [19,20]. The observation data included phenology date and yield, which were used to calibrate and validate the CERES-Rice model. The soil parameters of the study site were obtained from the 1:1 million soil data provided by the Nanjing Soil Institute of the Second National Land Survey in the Harmonized World Soil Database (HWSD) and the empirical data automatically generated by the CERES-Rice model, including soil profile characteristics, soil physical and chemical properties, i.e., soil color, hydrology group, bulk density, organic carbon, soil texture of each layer (clay, silt, and stones), soil nitrogen content, pH in water, etc.

2.3. Climate Data

The historical observed meteorological data of the study site were obtained from the National Meteorological Information Center-China Meteorological Data Network (http://data.cma.cn/, accessed on 1 March 2022), including four meteorological indicators of daily maximum and minimum temperature (°C), precipitation (mm), and solar radiation (calculated based on the sunshine hours using the Ångström-Prescott formula, MJ/m^2).

Based on the results of the existing study evaluating the CMIP6 model, six GCMs (Table 1) were selected that were more effective in simulating the study site [21]. Compared with CMIP5, CMIP6 improved projections of climate features, such as extreme temperatures and precipitation [22,23]. Future climate data were derived from monthly-scale meteorological data output from multi-modal ensemble averaging (MME) under three shared socioeconomic pathways and typical concentration pathway combination scenarios SSP126, SSP245, and SSP585 of the CMIP6 (https://esgf-node.llnl.gov/projects/cmip6/, accessed on 17 April 2022), including four meteorological indicators: Daily maximum and minimum temperature, precipitation, and solar radiation for three future periods: 2031–2050 (2040s), 2051–2070 (2060s), and 2071–2090 (2080s). The monthly climate projections from GCMs were down-scaled to the study site using an inverse distance-weighted (IDW) interpolation method, then bias correction was conducted by transferring the resulting monthly site data using functions obtained from analyzing the observed and GCM data for the period of 1961–2000. Daily climate variables were downscaled through the WGEN stochastic weather generator based on the spatially down-scaled monthly model [24].

Table 1. Information on the six global climate models (GCMs) selected in our study.

No.	Model Name	Institute and Country	Atmospheric Resolution (lon × lat: Number of Grids, L: Vertical Levels)
1	ACCESS-CM2	Commonwealth Scientific and Industrial Research Organization, Australian	192 × 144, L85
2	BCC-CSM2-MR	Beijing Climate Center, China	320 × 160, L46
3	EC-Earth3	EC-Earth Consortium, Europe	512 × 256, L91
4	GFDL-ESM4	National Oceanic and Atmospheric Administration, Geophysical Fluid Dynamics Laboratory, USA	288 × 180, L49
5	MPI-ESM1-2-HR	Max Planck Institute for Meteorology, Germany	384 × 192, L95
6	MRI-ESM2-0	Meteorological Research Institute, Japan	320 × 160, L80

Statistical down-scaling technique has been widely used for providing daily climate data to drive crop models in the assessment of the impacts of future climate change on agricultural systems in different countries and regions [25–28]. We used the statistical down-scaling model NWAL-WG provided by Liu and Zuo (2012). The main advantage of this statistical down-scaling approach, especially compared with dynamic down-scaling, is that it can be easily applied to any location where long-term daily historical climate records are available [29].

2.4. Model Simulations

2.4.1. CERES-Rice Model

The CERES-Rice model used in this study is included in decision support system for agrotechnology transfer (DSSAT) version 4.7.5 [30]. The CERES-Rice model is a process-based crop model driven by daily climate data (daily maximum and minimum temperature, precipitation, solar radiation). It can simulate rice growth, development, leaf area index, and dry matter content. Moreover, it can simulate soil water balance and nitrogen balance. Minimum input data include weather, soil, and crop variety genetic parameters. The CERES-Rice model has been widely tested and applied in many countries, including China [31–33]. In our study, the CERES-Rice model from DSSAT version 4.7.5 (https://get.dssat.net/, accessed on 21 December 2021) was used to simulate rice growth and development.

2.4.2. Model Calibration and Validation

During model calibration and validation, the varietal genetic parameters of the crop were adjusted using the trial-and-error method and the generalized likelihood uncertainty estimation (CLUE) tuning tool that comes with the DSSAT system. The observed dates of anthesis, maturity, and yield were made close to the CERES-Rice model simulation results.

The parameters were calibrated using trial data for anthesis, maturity, and yield in 1 year. The parameters were validated using trial data in another year.

2.4.3. Simulation Scenarios

Rice growth was simulated by the CERES-Rice model, the irrigation was set to automatic when required (i.e., when the moisture content was <50% of water capacity at 30 cm depth, the rice was irrigated with 10 mm of water), the harvest time was set at maturity, and other field management methods were the same as in the field experiment. In our study, rice growth was also simulated under different future sowing dates, including 27 March, 3 April, 10 April, 17 April, 24 April, 1 May, and 8 May. The transplanting dates were 35 days after the sowing date.

2.5. *Indicator Calculation Method*

2.5.1. Statistical Indices for Model Evaluation

To evaluate the CERES-Rice model performance in simulating rice growth on the Songnen Plain, we used the following three evaluation metrics: (1) Root mean squared error (*RMSE*), (2) normalized root mean squared error (NRMSE), and (3) determination coefficient (R^2):

$$RMSE = \sqrt{\frac{\sum_{i=1}^{n}(S_i - O_i)^2}{n}} \tag{1}$$

$$\text{NRMSE}(\%) = \frac{RMSE}{\overline{O}} \times 100 \tag{2}$$

$$R^2 = \frac{\sum_{i=1}^{n}(S_i - \overline{S})^2}{\sum_{i=1}^{n}(O_i - \overline{O})^2} \tag{3}$$

where S_i is the simulated value; O_i is the observed value; $\overline{S}$ and $\overline{O}$ are the average values of the simulated and observed values, respectively; n is the number of samples.

2.5.2. Light and Heat Resources Utilization Evaluation Index

In our study, we computed the following indices to evaluate the light and heat resources utilization efficiency:

(1) Light and temperature resources utilization

$$\text{E}_{RT} = \frac{Y}{Y_P} \times 100\% \tag{4}$$

$$Y_P = Y_1 \cdot f(t) \tag{5}$$

$$Y_1 = \varepsilon(1-\alpha)(1-\beta)(1-\gamma)(1-\rho)(1-\omega)f(L)E\varphi(1-\lambda)^{-1}(1-\chi)^{-1}H^{-1}\sum Q \times 10{,}000 \tag{6}$$

$$f(t) = \begin{cases} 0(t < t_{min}, t > t_{max}) \\ \frac{t-t_{min}}{t_{opt}-t_{min}}(t_{min} \le t < t_s) \\ \frac{t_{max}-t}{t_{max}-t_{opt}}(t_s \le t \le t_{max}) \end{cases} \tag{7}$$

where E_{RT} is the light and temperature resources utilization, Y is the crop yield in kg/hm^2, Y_P is light and temperature potential productivity in kg/hm^2, i.e., the upper limit of yield per unit area per unit time determined by local solar radiation and temperature at optimum conditions. Y_1 is photosynthetic potential productivity in kg/hm^2, i.e., the upper limit of yield per unit area per unit time determined solely by local solar radiation at optimum conditions. $\sum Q$ is the total solar radiation projected onto the unit area in MJ/m^2, H is the dry weight calorific value of the crop and rice taking the value of 16.9 MJ/kg, ε is the ratio of photosynthetically active radiant energy to total radiant energy taking the value of 0.49. α, β, γ, ρ, ω, φ, λ, χ, f(L), E are photosynthetic production potential parameters; α is the plant population reflectance, taking a value of 0.06, β is the plant luxuriant population light transmission, taking a value of 0.08, γ is the proportion of light above the light saturation point, taking a value of 0.05, ρ is the proportion of radiation intercepted by non-photosynthetic organs of the crop, taking a value of 0.10, ω is the depletion rate of respiration, taking a value of 0.33, φ is the quantum efficiency of photosynthesis, taking a value of 0.22, and λ is the inorganic nutrient content of the plant body, taking a value of 0.08, χ is the water content and takes the value of 0.14, f(L) is the revised positive value of crop leaf area dynamics and takes the value of 0.56, and E is the economic coefficient and takes the value of 0.45.

(2) Light utilization efficiency

$$\mathrm{E}_R = \frac{H \times Y}{\sum_{i=SD}^{MD} PAR} \times 100\% \tag{8}$$

where E_R is light utilization efficiency, i.e., the ratio of the energy stored in the crop harvest per unit of land area during the plant's reproductive life to the photosynthetically active radiation projected onto that unit area during the same period. $\sum\limits_{i=SD}^{MD} PAR$ is photosynthetically active radiation during crop growth (seeding to maturity), in $\mathrm{MJ/m^2}$, i.e., the radiant energy in the solar radiation spectrum that can be used by green plants in the photosynthetic band, it accounts for 49% of the total solar radiation. *SD* is the sowing date, and *MD* is the harvest date. The total solar radiation can be obtained by Equations (9)–(14):

$$R_s = \left(a_s + b_s \frac{n}{N}\right) R_a \tag{9}$$

$$R_a = 24 \frac{60}{\pi} G_{sc} d_r [\omega_s \sin(\varphi) \sin(\delta) + \cos(\varphi) \cos(\delta) \sin(\omega_s)] \tag{10}$$

$$d_r = 1 + 0.033 \cos\left(\frac{2\pi}{365} J\right) \tag{11}$$

$$\delta = 0.409 \sin\left(\frac{2\pi}{365} J - 1.39\right) \tag{12}$$

$$\omega_s = \arccos[-\tan(\varphi) \tan(\delta)] \tag{13}$$

$$N = \frac{24}{\pi} \omega_s \tag{14}$$

where R_s is the solar radiation ($\mathrm{MJ/m^2/d}$), a_s and b_s are empirical parameters, where a_s indicates the fraction of astronomical radiation reaching the Earth's surface on cloudy days, and b_s indicates the transport properties (aerosol density) of the cloud-free atmosphere, a_s takes 0.19 and b_s takes 0.54; n is the actual insolation duration (h); N is the maximum possible insolation duration (h); n/N is the relative insolation duration, also called the insolation percentage; R_a is the astronomical radiation ($\mathrm{MJ/m^2/d}$); π is the circumference; G_{sc} is the solar constant, taking the value of 0.082; d_r is the reverse solar-terrestrial relative distance; ω_s is the sunset time angle; φ is the latitude (rad); δ is the solar declination; J is the number of days.

(3) Heat utilization efficiency

$$\mathrm{HUE} = \frac{Y}{\sum_{i=SD}^{MD} (T_{i,mean} - t_{min})} \tag{15}$$

where HUE is heat utilization efficiency, in kg/(hm^2·°C·d); $T_{i,mean}$ is the daily average temperature of the day i; $\sum_{i=SD}^{MD} (T_{i,mean} - t_{min})$ is the effective cumulative temperature during crop growth that is greater than the biological lower limit temperature t, in °C·d, the t_{min} of rice is 10 °C.

2.6. Data Analysis

A stepwise multiple linear regression model and correlation analysis were used to quantify the effects of climate factors, including mean temperature (average of maximum and minimum temperatures), precipitation, and solar radiation on future rice yield, the light and temperature resources utilization, light utilization efficiency, and heat utilization efficiency. The Pearson correlation coefficient was used for the correlation analysis.

The framework of our study is shown in Figure 2.

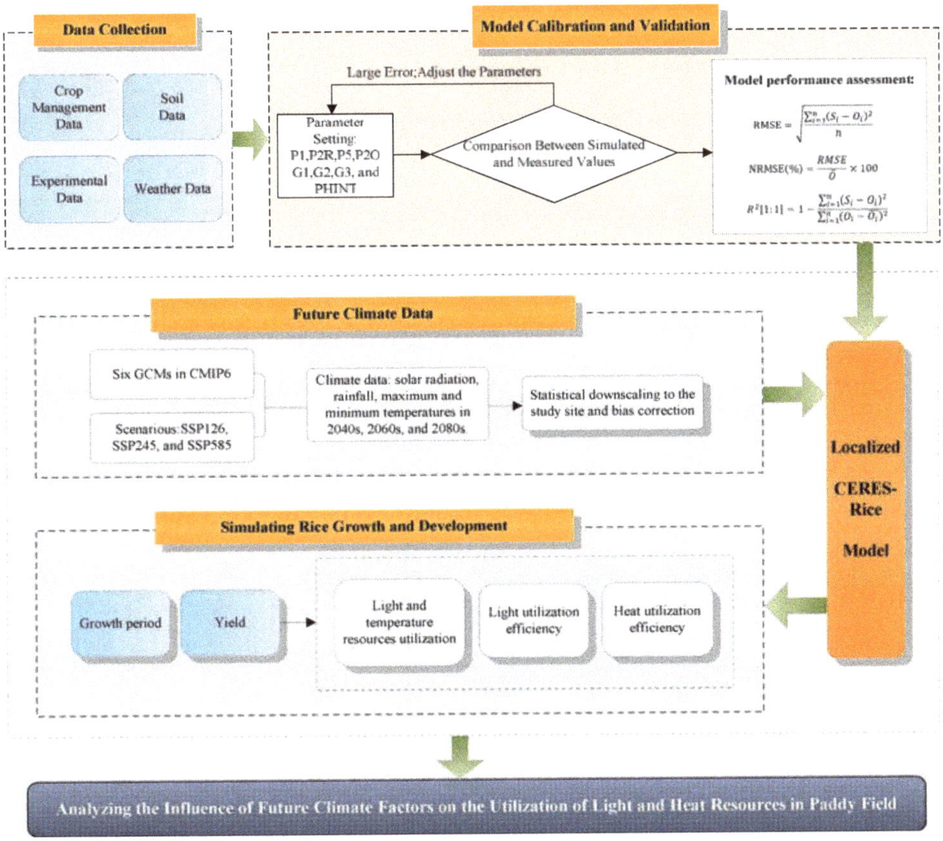

Figure 2. The framework of analysis.

3. Results

3.1. Calibration and Validation of the CERES-Rice Model

The calibrated CERES-Rice model was able to simulate rice phenology and yield reasonably well in the study area. The CERES-Rice model was calibrated and validated using 2 years of field trial data, including three irrigation methods. The calibrated parameters for "Suijing 18" are shown in Table 2, including phenology and growth parameters.

Table 2. The calibration parameters for "Suijing 18".

Cultivar Parameter		Physical Significance	Range	Value
Phenology Parameters	P1	The amount of heat hours required during the basic vegetative phase of the plant (°C·d)	150~800	338.4
	P2R	The extent to which phase development leading to panicle initiation is delayed (°C·d)	5~300	57.56
	P5	The amount of heat hours required from the beginning of grout to physiological maturity (°C·d)	150~850	328.3
	P2O	Critical photoperiod or the longest day length at which the development occurs at a maximum rate (h)	11~13	12.73
Growth Parameters	G1	Potential spikelet number coefficient	50~75	74.3
	G2	Single grain weight under ideal growing conditions (g)	0.015~0.03	0.019
	G3	Tillering coefficient relative to IR64 cultivar under ideal conditions	0.7~1.3	0.747
	PHINT	Phyllochron interval (°C·d)	55~90	74.01

The results of the validation have shown that the simulated dates of anthesis and maturity were consistent with the observed dates (Figure 3a). The error of simulating anthesis and maturity was generally within 5 days. The normalized root means square error (NRMSE) of anthesis and maturity dates between simulated and observed values was 3%, and the R^2 value was 0.968. The simulated and observed yields were also in general agreement (Figure 3b), with NRMSE = 3.4% and R^2 close to 1.0. Therefore, the validated results indicated that the CERES-Rice model could effectively simulate rice growth and development on the Songnen Plain of China.

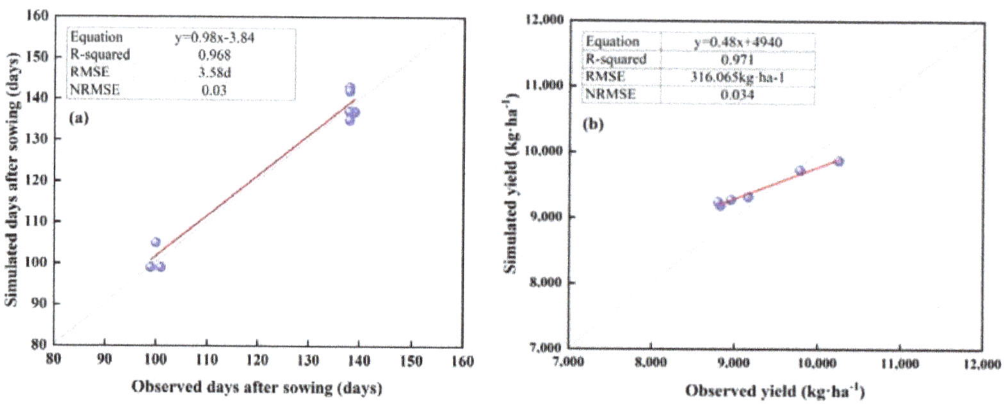

Figure 3. Verification results for the simulated and observed phenology and yield in the study field. The dots are simulated and observed values, the red line is the fitted line.

3.2. Projected Climate Change in Rice Growth Period in the Future

The ensemble average of six GCMs under SSP126, SSP245, and SSP585 conditions were selected to reduce the uncertainty of future climate change projections. Compared with the baseline of 2000–2019, the ensemble-mean daily maximum temperature increased 1.06, 1.26, and 1.25 °C under SSP126, respectively, increased 1.31, 1.64, and 1.92 °C under SSP245, respectively, and increased 1.29, 2.36, and 3.54 °C under SSP585 in the 2040s, 2060s, and 2080s (Figure 4a). The ensemble-mean daily minimum temperature increased 1.81, 1.28, and 1.28 °C under SSP126, respectively, increased 1.40, 1.93, and 2.21 °C under SSP245, respectively, and increased 1.55, 2.95, and 4.38 °C under SSP585 in the 2040s, 2060s, and 2080s relative to the baseline (Figure 4b). In general, the ensemble-mean daily temperature showed an increasing trend during the rice growth period in the future. The largest increase in temperature was observed in the SSP585 scenario and the lower increase in temperature in the SSP126 scenario.

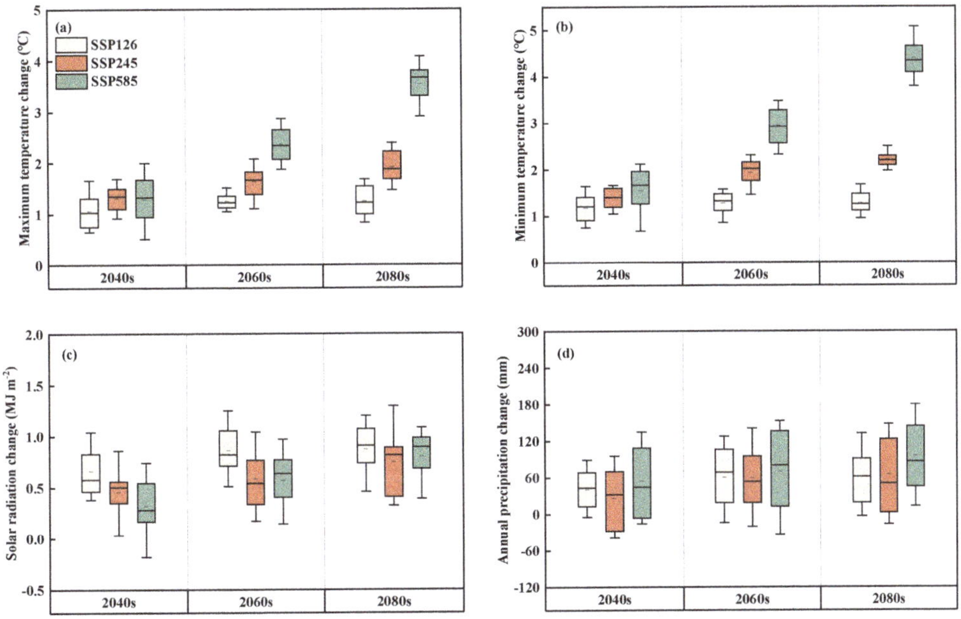

Figure 4. Projected changes in mean maximum and minimum temperature (°C), solar radiation (MJ m^{-2}), and annual precipitation (mm) during the rice growth period (April to September) for the 2040s, 2060s, and 2080s compared with 2000–2019 under the SSP126, SSP245, and SSP585 scenarios using six GCMs for the study station on the Songnen Plain of China. The box boundaries indicate the 25th and 75th percentiles; the black line and short line within the box mark the median and mean, respectively; and whiskers below and above the box indicate the 10th and 90th percentiles, respectively.

The daily solar radiation was also projected to increase relative to the baseline in the future under both SSPs (Figure 4c). The ensemble-mean solar radiation increased 0.66, 0.86, and 0.87 MJ/m^2 in the 2040s, 2060s, and 2080s under SSP126, respectively, increased 0.46, 0.56, and 0.74 MJ/m^2 in the future period under SSP245, respectively, and increased 0.32, 0.57, 0.80 MJ/m^2 in the future period under SSP585, respectively. Solar radiation increased more in the SSP126 scenario than in the SSP245 and SSP585 scenarios. The multi-model mean changes in annual precipitation were similar to other climatic factors (Figure 4d). In the 2040s, 2060s, and 2080s, there could be an increase in 40.71, 60.10, and 61.60 mm under SSP126, respectively, increase in 26.46, 58.94, and 64.87 mm under SSP245, respectively, and increase in 54.61, 69.73, and 95.35 mm under SSP585 relative to the baseline. Overall, the multi-model mean annual precipitation was also increased in the future period, especially in the SSP585 scenario, where the increase was more pronounced.

3.3. Impacts of Climate Change on Rice Phenology and Yield

Climate change could directly affect the rice growth period, including anthesis and maturity, which in turn could have an impact on rice yield. Our simulation results indicated that the rice growth period would be shortened in the future (Figure 5). In the 2040s, 2060s, and 2080s, the multi-model mean changes in anthesis date were −5.2, −4.2, and −5.5 days under SSP126, respectively, and −6.55, −7.4, and −7.15 days under SSP245, respectively, and −7.15, −11.85, and −19.45 days under SSSP585, respectively.

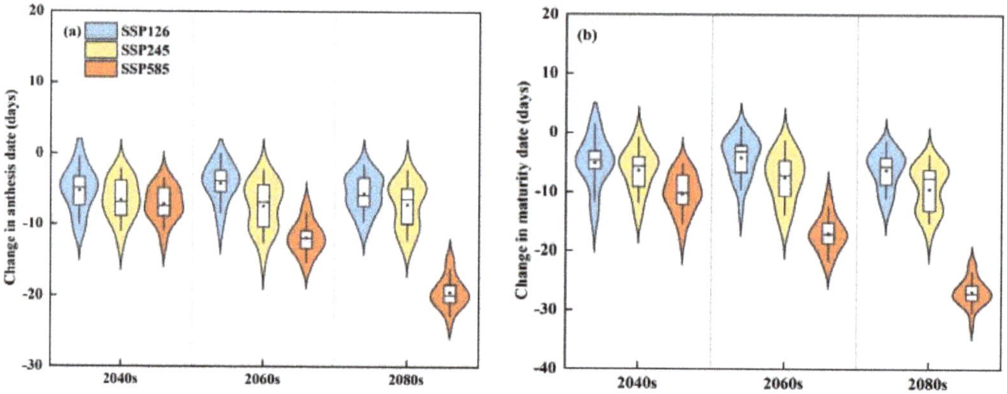

Figure 5. Simulated change in a rice growth period in the 2040s, 2060s, and 2080s under SSP126, SSP245, and SSP585 scenarios relative to 2000–2019 using six GCMs for Qing'an station on the Songnen Plain of China. The box boundaries indicate the 25th and 75th percentiles; the black line and dot within the box mark the median and mean, respectively; and whiskers below and above the box indicate mean +1.5 SD and mean −1.5 SD, respectively.

The maturity date would be changed to −5.15, −4.2, and −6.25 days under SSP126, respectively, and −6.35, −7.55, and −9.4 days under SSP245, respectively, and −10.35, −17.05, and −26.8 days under SSP585, respectively. The results showed that the greatest variation in rice growth period under the future climate scenario was observed for SSP585, followed by SSP245, and the least significant for SSP126, which was more similar to the multi-model mean changes in temperature. Correlation analysis showed a significant negative correlation between rice maturity and temperature, with correlation coefficients reaching above 0.9 (Figure 6).

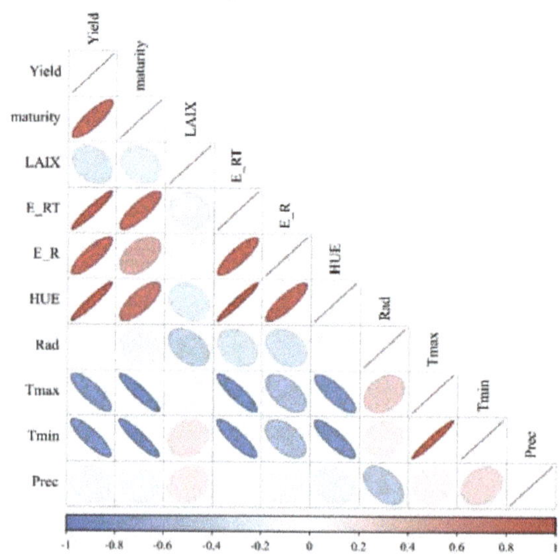

Figure 6. Correlations between the projected rice yield, maturity, maximum leaf area index (LAIX), the light and heat resource utilization indices (including E_{RT}, E_R, and HUE), and climate indices (including Rad, T_{max}, T_{min}, and Prec). The gradient of legend color is the function of strength of the correlation; the color and the size of ellipse indicates the strength of the correlation.

Compared with the baseline of 2000–2019, in the 2040s, 2060s, and 2080s, the multi-model mean changes in rice yield were +1.90%, +4.68%, and +2.46% under SSP126, respectively, and +0.25%, +0.48%, and +0.80% under SSP245, respectively, and −1.60%, −3.80%, and −24.89% under SSSP585, respectively (Figure 7). In general, rice yield varied inconsistently under different future scenarios, with a slight increase under SSP126 and SSP245 and a significant downward trend under SSP585. The result showed the greatest reduction in rice yield in the 2080s under SSP585. We used stepwise multiple linear regression analysis to reflect the relationship between rice yield and climatic variables change (including mean temperature, precipitation, and solar radiation). The value of the determination coefficient (R^2) was 0.717. As the regression coefficient was shown in Table 3, rice yield was significantly positively correlated with solar radiation and negatively correlated with mean temperature, while rice yield had a slight positive correlation with precipitation. In addition, temperature, precipitation, and solar radiation could explain 71.7% of the rice yield change.

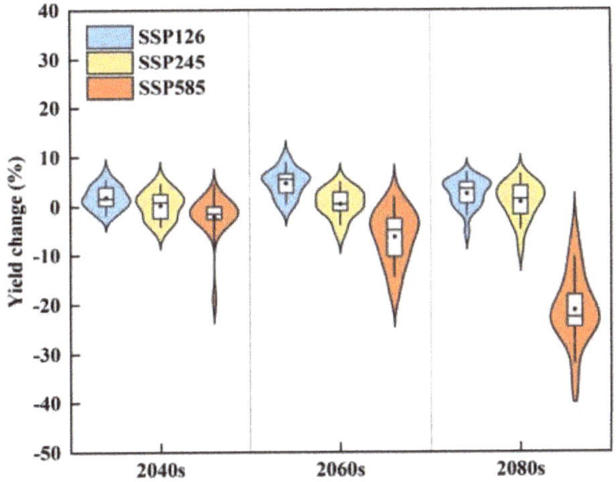

Figure 7. Simulated change in rice yield in the 2040s, 2060s, and 2080s under SSP126, SSP245, and SSP585 scenarios relative to 2000–2019 using six GCMs for Qing'an station on the Songnen Plain. The box boundaries indicate the 25th and 75th percentiles; the black line and dot within the box mark the median and mean, respectively; and whiskers below and above the box indicate mean + 1.5 SD and mean −1.5 SD, respectively.

Table 3. Coefficients of regression analysis of the impacts of climate change in rice growth period (April to September) on rice yield (Y), the light and temperature resources utilization (E_{RT}), light utilization efficiency (E_R), and heat utilization efficiency (HUE) change. Shown in the table are the ΔY (kg·ha^{-1}), ΔE_{RT} (%), ΔE_R (%), and Δ HUE (kg/(hm^2·°C·d) as a function of change in mean temperature (ΔT_{mean}, °C), precipitation ($\Delta Prec$, mm), and solar radiation (ΔRad, MJ·m^{-2}).

Item	a	b	c	R^2
$\Delta Y = a\Delta T_{mean} + b\Delta Prec + c\Delta Rad$	−7.972 ***	2.904 *	4.569 ***	0.717
$\Delta E_{RT} = a\Delta T_{mean} + b\Delta Prec + c\Delta Rad$	−7.440 ***	1.905	−1.365	0.779
$\Delta E_R = a\Delta T_{mean} + b\Delta Prec + c\Delta Rad$	−0.044 ***	0.024	−0.035 *	0.331
$\Delta HUE = a\Delta T_{mean} + b\Delta Prec + c\Delta Rad$	−0.619 ***	0.152	0.252 **	0.697

Note: *, **, and *** indicate significance at $p < 0.05$, $p < 0.01$, and $p < 0.001$, respectively.

3.4. Impacts of Climate Change on the Light and Heat Resources Utilization

Within a certain range, the light and heat resources utilization in rice determines the production potential of an agricultural system, and research on the light and heat resources utilization was particularly important but has been less studied in the past. Therefore, we focused on the study of this issue. In the baseline period (2000–2019), the calculated light and temperature resources utilization (E_{RT}) reached 89.74%. In the 2040s, 2060s, and 2080s, the calculated E_{RT} changes were −6.46%, −6.01%, and −6.03% under SSP126, respectively, and −7.89%, −8.41%, and −8.27% under SSP245, respectively, and −6.88%, −13.69%, and −28.84% under SSP585, respectively (Figure 8a). In general, the calculated E_{RT} showed a decreasing trend in the future, especially in the SSP585 scenario, and the reduction was more apparent. The light utilization efficiency (E_R) was 1.39% in the baseline period and decreased by 0.13% under SSP585 in the 2080s, while slightly increased in other scenarios and periods (Figure 8b). Compared with the baseline of 2000–2019, in the 2040s, 2060s, and 2080s, the calculated heat utilization efficiency (HUE) changes were −0.17, −0.08, and −0.01 kg/(hm^2·°C·d) under SSP126, respectively, and −0.37, −0.40, and −0.39 kg/(hm^2·°C·d) under SSP245, respectively, and −0.26, −0.74, and −2.00 kg/(hm^2·°C·d) under SSP585, respectively (Figure 8c). In this case, the changes in HUE were similar to the E_{RT}.

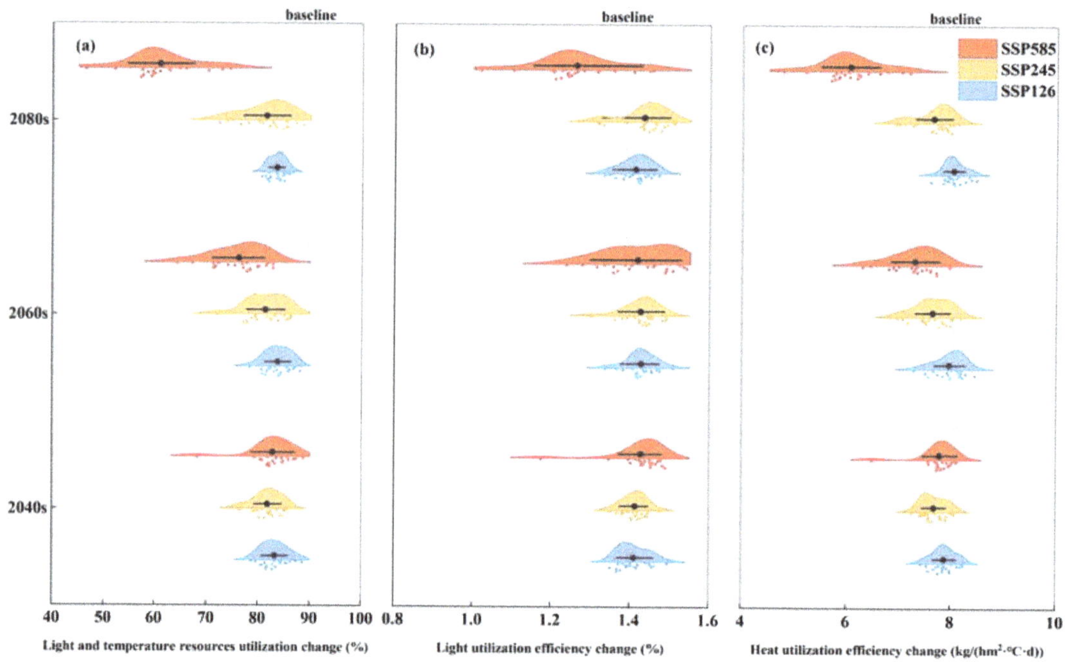

Figure 8. Simulated change in the light and temperature resources utilization, the light utilization efficiency, and heat utilization efficiency in the 2040s, 2060s, and 2080s under SSP126, SSP245, and SSP585 scenarios relative to 2000–2019 (baseline) using six GCMs for Qing'an station in the Songnen Plain.

As shown in Table 3, the results of multiple linear regression analysis showed the relationship between the indexes of rice, light, and heat resources utilization (including E_{RT}, E_R, and HUE) and changes in climatic variables (including mean temperature, precipitation, and solar radiation). The values of the determination coefficient (R^2) were 0.779, 0.331, and 0.697, respectively. All photothermal resource utilization indicators showed a significant negative correlation with mean temperature, while there was no significant correlation with precipitation. The E_R had a slight negative correlation with solar radiation and the HUE had a slight positive correlation with solar radiation.

3.5. The Impact of Different Sowing Dates on Rice Yield and Light and Heat Resources Utilization

To adapt to the impact of future climate change on rice growth, we have adjusted the sowing dates of rice to change the efficiency of light and heat resources utilization in rice. We simulated the light and temperature resources utilization (E_{RT}), light utilization efficiency (E_R), and heat utilization efficiency (HUE) for different sowing periods under two typical scenarios, SSP245 and SSP585 (Figure 9). The results showed that early sowing was beneficial for improving the E_{RT} under all emission scenarios, while light energy utilization was different and appropriately delayed sowing was beneficial for improving light energy utilization. Early sowing under the SSP126 and SSP245 scenarios increased the HUE, while the effect of the sowing date on the HUE was different in different periods under the SSP585 scenario.

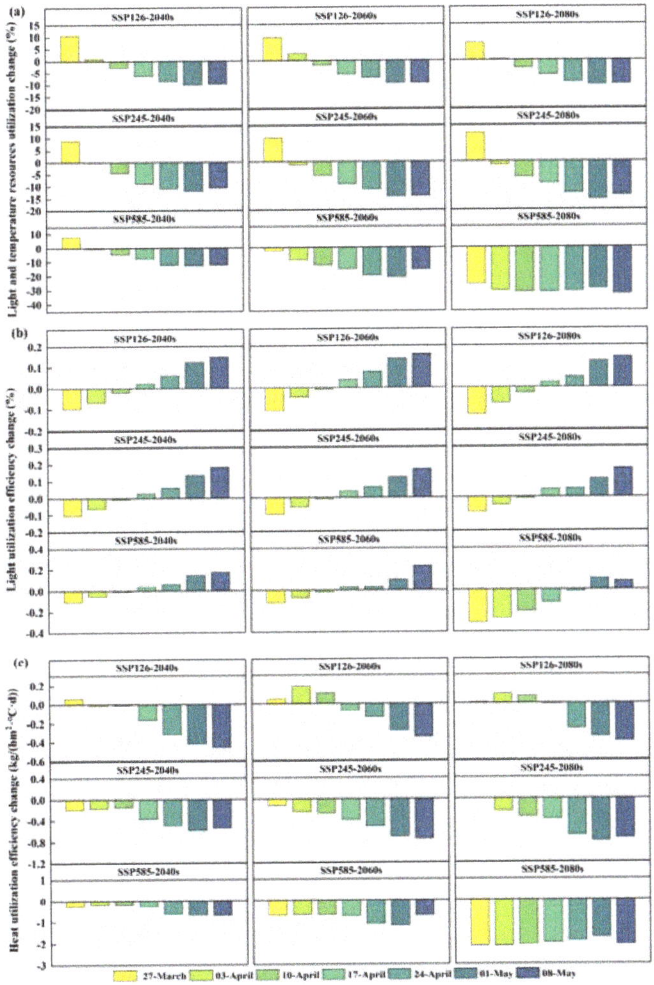

Figure 9. Simulated change in the light and heat resource utilization indices under different sowing dates in the 2040s, 2060s, and 2080s under SSP126, SSP245, and SSP585 scenarios based on the six GCMs compared with sowing on 17 April 2000–2019 for Qing'an station in the Songnen Plain.

In the context of climate change, by adjusting the sowing date, rice could use light and heat resources more efficiently during the growth period, thus contributing to rice yield. According to the rice yield changes under different sowing dates in the 2040s, 2060s, and 2080s relative to 2000–2019, we found that advancing or delaying the sowing period in the future period could mitigate the negative effects of climate change to some extent, but could not fully offset them, especially under the SSP585 scenario (Figure 10). Under the SSP126 and SSP245 scenarios, earlier sowing dates would increase rice yield, regardless of the period, while delayed sowing would result in lower rice yield. Our study showed that in the 2040s, the optimum sowing date for rice was before 27 March and would increase yield by 3.35% to 4.76%, while in the 2060s and 2080s, the optimum sowing date was around 3 April and would increase yield by 1.30% to 2.50% compared with sowing on 17 April. Under the SSP585 scenario, in the 2040s and 2060s, early sowing could increase yield by 3.36% to 3.76% compared with the baseline sowing date of 17 April. Whereas, in the 2080s, an appropriate delay in sowing would have a positive impact on rice yield, sowing on 1 May would increase yield by 8.37% compared with sowing on 17 April.

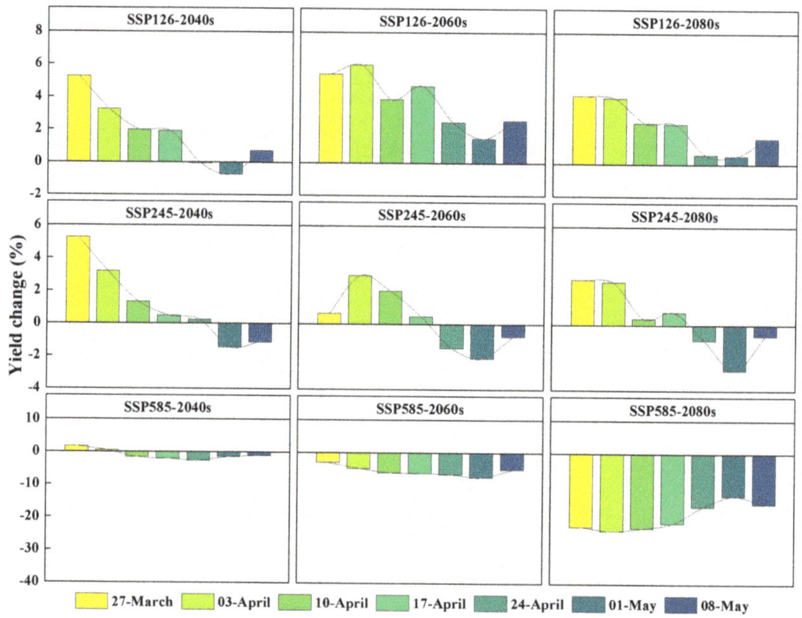

Figure 10. Simulated change in rice yield under different sowing dates in the 2040s, 2060s, and 2080s under SSP126, SSP245, and SSP585 scenarios based on the six GCMs compared with sowing on 17 April 2000–2019 for Qing'an station in the Songnen Plain.

4. Discussion

4.1. Performance of the CERES-Rice Model

The performance of the crop model needs to be validated before it can be applied to future simulations. The results of our calibration indicated that the DSSAT-Rice model could simulate rice growth well in our study area (Figure 3). The simulated phenological periods and yield were in agreement with our experimental observations. Similar findings had been obtained in previous scholarly studies [34–36]. For example, Boonwichai et al. [34] validated the DSSAT-Rice model in the Songkhram River Basin, Thailand, with calibrated and validated R^2 values of 0.84 and 0.78, respectively, and simulated rice yield similar to the observed yield. Kontgis et al. [36] calibrated the DSSAT-Rice and DSSAT-Maize models using experimental data and observed a close agreement between the observed and simulated values for anthesis, maturity, yield, biomass, and N uptake in both crops, with d-values of 0.89 to 0.99, indicating an acceptable performance. Overall, our results showed that the calibrated and validated DSSAT-Rice model had an acceptable error rate and could better simulate the effects of climate change on rice growth on the Songnen Plain of China.

4.2. Impacts of Future Climate Change on Rice Growth

In our study, the future climate variables were averaged from the set of six GCM outputs. We found that under different future scenarios, the daily maximum temperature increased by 1.06–3.54 °C, the daily minimum temperature increased by 1.28–4.38 °C, the average daily solar radiation increased by 0.32–0.87 MJ m^{-2}, and the average annual precipitation increased by 26.26–95.35 mm (Figure 4). This prediction was consistent with many previous studies [37–41]. Arunrat et al. [37] used GCMs to predict future climate change in Thailand, with an increase in precipitation, as well as maximum and minimum temperatures for the SSP245 and SSP585 scenarios compared with the historical period. The maximum temperature increase for the SSP245 and SSP585 scenarios were 0.7–2.2 °C and 0.7–3.9 °C, respectively, and the minimum temperatures were 0.7–2.1°C and 0.8–3.8 °C, respectively. Precipitation increased by 2.2–3.9% and 1.8–5.8%, respectively. Tan et al. [38] also showed an increasing trend in both temperature and solar radiation in the future climate scenario. Multi-model ensemble averaging methods are widely used in climate modeling to reduce the large bias and uncertainty introduced by model parameterization errors, model structure, assumptions, and input variables [42].

The predicted future climate change resulted in varying degrees of shortening of rice maturity, with the largest magnitude during the 2080s under the SSP585 scenario, with an average shortening of 26.8 days (Figure 5). Many studies indicated that future climate change would shorten the crop growing period [43–45]. Our simulations projected that the mean rice yield increased by 1.9–4.68% under SSP126 and by 0.25–0.80% under SSP245 in the future period; however, under the SSP585 scenario, rice yield decreased by 1.6–24.89% (Figure 7). A relevant study showed that a gradual increase in 3.0–4.3% in rice yield in irrigated areas was predicted under the SSP245 scenario in Thailand, while a decrease in 6.0–17.7% in rice yield in irrigated areas in the medium and long term was predicted under the SSP585 scenario [8]. The same findings were found in wheat on the Sichuan basin: Wheat yield was significantly higher under the SSP245 scenario and negatively affected under the SSP585 scenario [46]. Temperature and solar radiation were considered to be the main environmental factors affecting the yield potential of rice [47,48]. Moreover, our results demonstrated that rice yield showed a significant positive correlation with solar radiation and a significant negative correlation with temperature (Table 3).

Furthermore, we simulated and calculated the E_{RT}, E_R, and HUE for different scenarios in the future period (Figure 8). The E_{RT} was not common in previous studies, which takes the light and temperature potential productivity as the light-temperature resources available in a region, where the light and temperature potential productivity was calculated using the step-by-step revision method. E_R and HUE are the more common evaluation metrics in solar thermal resource utilization assessment [49]. Our results showed that there was a relatively significant downward trend in E_{RT} and HUE under future climate scenarios. The main reason for this is the change in light-temperature production potential and effective cumulative temperature during the rice reproductive period due to future temperature and solar radiation increases. The E_R increased in the low and medium emission scenarios and decreased more significantly in the high emission scenario in the 2080s, which was mainly due to the alteration of temperature and photosynthetically active radiation during the reproductive period of rice.

4.3. Adaptive Strategies for Rice Production in Response to Climate Change

In rice production, a suitable sowing date could make the rice fertility process coincide with the local suitable water temperature and climate conditions; adjusting the sowing period is considered an effective strategy to slow down the development of rice and avoid heat damage [50–52]. In this study, climate resources could be fully utilized to increase rice yield by adjusting the sowing date, but the negative impact of climate change on rice yield could not be fully offset in the high emission scenario (Figure 9). Therefore, there is a need for more adaptation strategies, and some relevant studies have shown that with advances in technologies, such as breeding and crop management, more diverse rice varieties could provide a greater buffer for rice production to better adapt to environmental anomalies, such as climate change [53].

4.4. Uncertainty and Limitations of This Study

In this study, we evaluated the changes in rice yield and simulated the utilization of light and heat resources by the DSSAT-Rice model in future climate change under SSP126, SSP245, and SSP585 scenarios based on the six GCMs. Different climate change projections may result in different results. To reduce the uncertainty of climate projections, we selected six GCMs with better projection results in the study area for ensemble averaging. Nevertheless, there were still some uncertainties. First, climate change scenarios and subsequently estimated impacts may have some uncertainties due to the inherent systematic biases of GCMs. Second, although the DSSAT-Rice model has been widely used in global climate change assessment studies, it still has uncertainties, and a single crop model may be overconfident and ignore the uncertainty of crop model structure and parameters on climate change. Based on the existing crop models that use different functional relationships to simulate crop growth processes, multi-model ensembles are often considered superior to single-model simulations, and future studies are recommended to use multiple models for simulations. Third, there may be some uncertainty in the timing of varietal adaptation, since we assumed no future adaptation while keeping other agronomic management, including planting density and cropping pattern constant in the future. In addition, we evaluated the performance of only one commonly used variety in the study site, without explicitly considering the impact of other factors that could harm rice yields, such as pests and diseases, extreme climatic events (droughts and floods), irrigation water availability, and socioeconomic conditions.

5. Conclusions

In this study, we used six GCMs in CMIP6 in combination with the CERES-Rice crop model to assess the impact of climate change on the utilization potential of light and heat resources of rice on the Songnen Plain. The main results could be summarized as follows: (1) Under the SSPs, the maximum and minimum temperatures, precipitation, and solar radiation increased in the future period relative to the baseline, (2) the rice growth period showed different degrees of shortening under the future climate scenario, with a slight increase in rice yield under the SSP126 and SSP245 scenarios, while under the SSP585 scenario, rice yield was significantly lower than the baseline. The E_{RT} and HUE would decrease in the future, while E_R would slightly increase in the rest of the scenarios except under SSP585 in the 2080s, which would decrease. The results of data analysis showed that both rice yield and light and heat resource utilization indicators showed a significant negative correlation with temperature, which was the most dominant meteorological factor, and (3) under future climate conditions, optimizing the sowing date could make full use of climate resources to improve rice yield and light and heat resource utilization indexes.

Author Contributions: E.Z. contributed to the design of the experiment, revision of the paper, and evaluated the obtained results; M.Q. contributed to the simulation of the model, the analysis of the data, and the writing of the paper; P.C. revised the paper; T.X. and Z.Z. evaluated the obtained results. All authors have read and agreed to the published version of the manuscript.

Funding: This work was jointly supported by the Basic Scientific Research Fund of Heilongjiang Provincial Universities (2020-KYYWF-1042). We are grateful to the staff of the National Key Irrigation Experimental Station for their technical assistance.

Institutional Review Board Statement: Not applicable.

Informed Consent Statement: Not applicable.

Data Availability Statement: Not applicable.

Conflicts of Interest: The authors declare no conflict of interest.

References

1. IPCC. *Climate Change 2021: The Physical Science Basis*; Cambridge University Press: Cambridge, UK, 2021.
2. IPCC. *Climate Change 2022: Impacts, Adaptation, and Vulnerability*; Cambridge University Press: Cambridge, UK, 2022.
3. Asseng, S.; Ewert, F.; Martre, P.; Rotter, R.P.; Lobell, D.B.; Cammarano, D.; Kimball, B.A.; Ottman, M.J.; Wall, G.W.; White, J.W.; et al. Rising temperatures reduce global wheat production. *Nat. Clim. Change* **2014**, *5*, 143–147. [CrossRef]
4. Wang, J.-X.; Huang, J.-K.; Yang, J. Overview of impacts of climate change and adaptation in china's agriculture. *J. Integr. Agric.* **2014**, *13*, 1–17. [CrossRef]
5. Gao, M.; Yang, H.; Xiao, Q.; Goh, M. A novel method for carbon emission forecasting based on gompertz's law and fractional grey model: Evidence from american industrial sector. *Renew. Energy* **2022**, *181*, 803–819. [CrossRef]
6. Gao, M.; Yang, H.; Xiao, Q.; Goh, M. COVID-19 lockdowns and air quality: Evidence from grey spatiotemporal forecasts. *Socio-Econ. Plan. Sci.* **2022**, *83*, 101228. [CrossRef]
7. Gao, M.; Yang, H.; Xiao, Q.; Goh, M. A novel fractional grey Riccati model for carbon emission prediction. *J. Clean. Prod.* **2021**, *282*, 124471. [CrossRef]
8. Ahmad, M.J.; Iqbal, M.A.; Choi, K.S. Climate-driven constraints in sustaining future wheat yield and water productivity. *Agric. Water Manag.* **2021**, *231*, 105991. [CrossRef]
9. Cui, Z.; Zhang, H.; Chen, X.; Zhang, C.; Ma, W.; Huang, C.; Zhang, W.; Mi, G.; Miao, Y.; Li, X.; et al. Pursuing sustainable productivity with millions of smallholder farmers. *Nature* **2018**, *555*, 363–366. [CrossRef]
10. Bandumula, N. Rice Production in Asia: Key to Global Food Security. *Proc. Natl. Acad. Sci. India Sect. B Biol. Sci.* **2017**, *88*, 1323–1328. [CrossRef]
11. Zhang, L.; Zhang, Z.; Tao, F.; Luo, Y.; Zhang, J.; Cao, J. Adapting to climate change precisely through cultivars renewal for rice production across China: When, where, and what cultivars will be required? *Agric. For. Meteorol.* **2022**, *316*, 108856. [CrossRef]
12. Barrios, S.; Ouattara, B.; Strobl, E. The impact of climatic change on agricultural production: Is it different for africa? *Food Policy* **2008**, *33*, 287–298. [CrossRef]
13. He, H.; Hu, Q.; Li, R.; Pan, X.; Huang, B.; He, Q. Regional gap in maize production, climate and resource utilization in China. *Field Crop. Res.* **2020**, *254*, 107830. [CrossRef]
14. Qu, C.-H.; Li, X.-X.; Ju, H.; Liu, Q. The impacts of climate change on wheat yield in the Huang-Huai-Hai Plain of China using DSSAT-CERES-Wheat model under different climate scenarios. *J. Integr. Agric.* **2019**, *18*, 1379–1391. [CrossRef]
15. Thorburn, P.J.; Boote, K.J.; Nendel, C.; Rötter, R.P.; Ewert, F. Recent advances in crop modelling to support sustainable agricultural production and food security under global change. *Eur. J. Agron.* **2018**, *100*, 1–3. [CrossRef]
16. Wang, W.; Yu, Z.; Zhang, W.; Shao, Q.; Luo, Y.; Jiao, X.; Xu, J. Responses of rice yield, irrigation water requirement and water use efficiency to climate change in China: Historical simulation and future projections. *Agric. Water Manag.* **2014**, *146*, 249–261. [CrossRef]
17. Rosenzweig, C.; Elliott, J.; Deryng, D.; Ruane, A.C.; Müller, C.; Arneth, A.; Boote, K.J.; Folberth, C.; Glotter, M.; Khabarov, N.; et al. Assessing agricultural risks of climate change in the 21st century in a global gridded crop model intercomparison. *Proc. Natl. Acad. Sci. USA* **2014**, *111*, 3268–3273. [CrossRef]
18. Kang, X.; Qi, J.; Li, S.; Meng, F.-R. A watershed-scale assessment of climate change impacts on crop yields in Atlantic Canada. *Agric. Water Manag.* **2022**, *269*, 107680. [CrossRef]
19. Zheng, E.; Zhang, C.; Qi, Z.; Zhang, Z. Experimental study on radiation utilization efficiency and soil temperature in paddy field with different irrigation methods in Northeast China. *Paddy Water Environ.* **2019**, *18*, 179–191. [CrossRef]
20. Chen, P.; Xu, J.; Zhang, Z.; Wang, K.; Li, T.; Wei, Q.; Li, Y. Carbon pathways in aggregates and density fractions in Mollisols under water and straw management: Evidence from 13C natural abundance. *Soil Biol. Biochem.* **2022**, *169*, 108684. [CrossRef]
21. Yang, X.; Zhou, B.; Xu, Y.; Han, Z. CMIP6 Evaluation and Projection of Temperature and Precipitation over China. *Adv. Atmos. Sci.* **2021**, *38*, 817–830. [CrossRef]
22. Xin, X.; Wu, T.; Zhang, J.; Yao, J.; Fang, Y. Comparison of CMIP6 and CMIP5 simulations of precipitation in China and the East Asian summer monsoon. *Int. J. Clim.* **2020**, *40*, 6423–6440. [CrossRef]
23. Chen, H.; Sun, J.; Lin, W.; Xu, H. Comparison of CMIP6 and CMIP5 models in simulating climate extremes. *Sci. Bull.* **2020**, *65*, 1415–1418. [CrossRef]
24. Liu, D.L.; Zuo, H. Statistical downscaling of daily climate variables for climate change impact assessment over New South Wales, Australia. *Clim. Chang.* **2012**, *115*, 629–666. [CrossRef]
25. Wang, B.; Liu, D.L.; O'Leary, G.J.; Asseng, S.; Macadam, I.; Lines-Kelly, R.; Yang, X.; Clark, A.; Crean, J.; Sides, T.; et al. Australian wheat production expected to decrease by the late 21st century. *Glob. Chang. Biol.* **2018**, *24*, 2403–2415. [CrossRef]

26. Jiang, T.; Wang, B.; Xu, X.; Cao, Y.; Liu, D.L.; He, L.; Jin, N.; Ma, H.; Chen, S.; Zhao, K.; et al. Identifying sources of uncertainty in wheat production projections with consideration of crop climatic suitability under future climate. *Agric. For. Meteorol.* **2022**, *319*, 108933. [CrossRef]
27. Huang, M.; Wang, J.; Wang, B.; Liu, D.L.; Feng, P.; Yu, Q.; Pan, X.; Li, S.; Jiang, T. Dominant sources of uncertainty in simulating maize adaptation under future climate scenarios in China. *Agric. Syst.* **2022**, *199*, 103411. [CrossRef]
28. Wang, B.; Feng, P.; Liu, D.L.; O'Leary, G.J.; Macadam, I.; Waters, C.; Asseng, S.; Cowie, A.; Jiang, T.; Xiao, D.; et al. Sources of uncertainty for wheat yield projections under future climate are site-specific. *Nat. Food* **2020**, *1*, 720–728. [CrossRef]
29. Wang, B.; Liu, D.L.; Macadam, I.; Alexander, L.V.; Abramowitz, G.; Yu, Q. Multi-model ensemble projections of future extreme temperature change using a statistical downscaling method in south eastern Australia. *Clim. Chang.* **2016**, *138*, 85–98. [CrossRef]
30. Jones, J.W.; Hoogenboom, G.; Porter, C.H.; Boote, K.J.; Batchelor, W.D.; Hunt, L.A.; Wilkens, P.W.; Singh, U.; Gijsman, A.J.; Ritchie, J.T. The dssat cropping system model. *Eur. J. Agron.* **2003**, *18*, 235–265. [CrossRef]
31. Gao, F.; Luan, X.; Yin, Y.; Sun, S.; Li, Y.; Mo, F.; Wang, J. Exploring long-term impacts of different crop rotation systems on sustainable use of groundwater resources using DSSAT model. *J. Clean. Prod.* **2022**, *336*, 130377. [CrossRef]
32. Tooley, B.E.; Mallory, E.B.; Porter, G.A.; Hoogenboom, G. Predicting the response of a potato-grain production system to climate change for a humid continental climate using DSSAT. *Agric. For. Meteorol.* **2021**, *307*, 108452. [CrossRef]
33. Chen, S.; Jiang, T.; Ma, H.; He, C.; Xu, F.; Malone, R.W.; Feng, H.; Yu, Q.; Siddique, K.H.; Dong, Q.; et al. Dynamic within-season irrigation scheduling for maize production in Northwest China: A Method Based on Weather Data Fusion and yield prediction by DSSAT. *Agric. For. Meteorol.* **2020**, *285–286*, 107928. [CrossRef]
34. Boonwichai, S.; Shrestha, S.; Babel, M.S.; Weesakul, S.; Datta, A. Evaluation of climate change impacts and adaptation strategies on rainfed rice production in Songkhram River Basin, Thailand. *Sci. Total Environ.* **2019**, *652*, 189–201. [CrossRef]
35. Kontgis, C.; Schneider, A.; Ozdogan, M.; Kucharik, C.; Tri, V.P.D.; Duc, N.H.; Schatz, J. Climate change impacts on rice productivity in the Mekong River Delta. *Appl. Geogr.* **2019**, *102*, 71–83. [CrossRef]
36. Kadiyala, M.; Jones, J.; Mylavarapu, R.; Li, Y.; Reddy, M. Identifying irrigation and nitrogen best management practices for aerobic rice–maize cropping system for semi-arid tropics using CERES-rice and maize models. *Agric. Water Manag.* **2015**, *149*, 23–32. [CrossRef]
37. Arunrat, N.; Sereenonchai, S.; Chaowiwat, W.; Wang, C. Climate change impact on major crop yield and water footprint under CMIP6 climate projections in repeated drought and flood areas in Thailand. *Sci. Total Environ.* **2022**, *807*, 150741. [CrossRef]
38. Tan, L.; Feng, P.; Li, B.; Huang, F.; Liu, D.L.; Ren, P.; Liu, H.; Srinivasan, R.; Chen, Y. Climate change impacts on crop water productivity and net groundwater use under a double-cropping system with intensive irrigation in the Haihe River Basin, China. *Agric. Water Manag.* **2022**, *266*, 107560. [CrossRef]
39. Yue, Y.; Yan, D.; Yue, Q.; Ji, G.; Wang, Z. Future changes in precipitation and temperature over the Yangtze River Basin in China based on CMIP6 GCMs. *Atmos. Res.* **2021**, *264*, 105828. [CrossRef]
40. Osman, R.; Ata-Ul-Karim, S.T.; Tahir, M.N.; Ishaque, W.; Xu, M. Multi-model ensembles for assessing the impact of future climate change on rainfed wheat productivity under various cultivars and nitrogen levels. *Eur. J. Agron.* **2022**, *139*, 105828. [CrossRef]
41. Chen, X.; Wang, L.; Niu, Z.; Zhang, M.; Li, C.; Li, J. The effects of projected climate change and extreme climate on maize and rice in the Yangtze River Basin, China. *Agric. For. Meteorol.* **2020**, *282*, 107867. [CrossRef]
42. Ahmed, K.; Sachindra, D.A.; Shahid, S.; Demirel, M.C.; Chung, E.-S. Selection of multi-model ensemble of general circulation models for the simulation of precipitation and maximum and minimum temperature based on spatial assessment metrics. *Hydrol. Earth Syst. Sci.* **2019**, *23*, 4803–4824. [CrossRef]
43. Funes, I.; Savé, R.; de Herralde, F.; Biel, C.; Pla, E.; Pascual, D.; Zabalza, J.; Cantos, G.; Borràs, G.; Vayreda, J.; et al. Modeling impacts of climate change on the water needs and growing cycle of crops in three Mediterranean basins. *Agric. Water Manag.* **2021**, *249*, 106797. [CrossRef]
44. Li, N.; Li, Y.; Biswas, A.; Wang, J.; Dong, H.; Chen, J.; Liu, C.; Fan, X. Impact of climate change and crop management on cotton phenology based on statistical analysis in the main-cotton-planting areas of China. *J. Clean. Prod.* **2021**, *298*, 126750. [CrossRef]
45. Liu, Y.; Dai, L. Modelling the impacts of climate change and crop management measures on soybean phenology in China. *J. Clean. Prod.* **2020**, *262*, 121271. [CrossRef]
46. Li, M.; Tang, Y.; Li, C.; Wu, X.; Tao, X.; Liu, M. Climate warming causes changes in wheat phenological development that benefit yield in the Sichuan Basin of China. *Eur. J. Agron.* **2022**, *139*, 105828. [CrossRef]
47. Zhou, N.-B.; Zhang, J.; Fang, S.-L.; Wei, H.-Y.; Zhang, H.-C. Effects of temperature and solar radiation on yield of good eating-quality rice in the lower reaches of the Huai River Basin, China. *J. Integr. Agric.* **2021**, *20*, 1762–1774. [CrossRef]
48. Rao, G.-S.; Ashraf, U.; Kong, L.-L.; Mo, Z.-W.; Xiao, L.-Z.; Zhong, K.-Y.; Rasul, F.; Tang, X.-R. Low soil temperature and drought stress conditions at flowering stage affect physiology and pollen traits of rice. *J. Integr. Agric.* **2019**, *18*, 1859–1870. [CrossRef]
49. Mi, N.; Zhang, F.; Zhao, Y.; Zhang, Y.; Ji, R.; Yu, W.; Chen, N.; Jiang, Y.; Qi, H.; Wu, J. Evaluation on the utilization efficiency of light and heat resources in maize interval tillage straw strip returning technology. *J. Agric. Resour. Environ.* **2021**, *38*, 475–483.
50. Ding, Y.; Wang, W.; Zhuang, Q.; Luo, Y. Adaptation of paddy rice in China to climate change: The effects of shifting sowing date on yield and irrigation water requirement. *Agric. Water Manag.* **2020**, *228*, 105890. [CrossRef]

51. Tu, D.; Jiang, Y.; Liu, M.; Zhang, L.; Chen, L.; Cai, M.; Ling, X.; Zhan, M.; Li, C.; Wang, J.; et al. Improvement and stabilization of rice production by delaying sowing date in irrigated rice system in central China. *J. Sci. Food Agric.* **2020**, *100*, 595–606. [CrossRef]
52. Deng, F.; Zhang, C.; He, L.; Liao, S.; Li, Q.; Li, B.; Zhu, S.; Gao, Y.; Tao, Y.; Zhou, W.; et al. Delayed sowing date improves the quality of mechanically transplanted rice by optimizing temperature conditions during growth season. *Field Crop. Res.* **2022**, *281*, 108493. [CrossRef]
53. Wang, X.; Jing, Z.-H.; He, C.; Liu, Q.-Y.; Jia, H.; Qi, J.-Y.; Zhang, H.-L. Breeding rice varieties provides an effective approach to improve productivity and yield sensitivity to climate resources. *Eur. J. Agron.* **2021**, *124*, 126239. [CrossRef]

agriculture

Article

Characteristics of Potential Evapotranspiration Changes and Its Climatic Causes in Heilongjiang Province from 1960 to 2019

Tangzhe Nie [1], Rong Yuan [1], Sihan Liao [1], Zhongxue Zhang [2,3,*], Zhenping Gong [4], Xi Zhao [5], Peng Chen [6], Tiecheng Li [2,3], Yanyu Lin [7], Chong Du [1], Changlei Dai [1] and Hao Jiang [8]

1 School of Water Conservancy and Electric Power, Heilongjiang University, Harbin 150006, China
2 School of Water Conservancy and Civil Engineering, Northeast Agricultural University, Harbin 150030, China
3 Key Laboratory of Efficient Use of Agricultural Water Resources, Ministry of Agriculture and Rural Affairs of the People's Republic of China, Northeast Agricultural University, Harbin 150030, China
4 College of Agriculture, Northeast Agricultural University, Harbin 150030, China
5 Department of Science and Technology, Heilongjiang University, Harbin 150006, China
6 College of Agricultural Science and Engineering, Hohai University, Nanjing 210024, China
7 College of Civil Engineering and Water Conservancy, Heilongjiang Bayi Agricultural University, Daqing 163319, China
8 Xingyang Water Conservancy Bureau, Zhengzhou 450000, China
* Correspondence: zhangzhongxue@neau.edu.cn; Tel.: +86-138-4517-1015

Abstract: Climate change refers to the statistically significant changes in the mean and dispersion values of meteorological factors. Characterizing potential evapotranspiration (ET_0) and its climatic causes will contribute to the estimation of the atmospheric water cycle under climate change. In this study, based on daily meteorological data from 26 meteorological stations in Heilongjiang Province from 1960 to 2019, ET_0 was calculated by the Penman–Monteith formula, linear regression method and the Mann–Kendall trend test were used to reveal the seasonal and inter-annual changing trend of ET_0. The sensitivity-contribution rate method was used to clarify the climatic factors affecting ET_0. The results showed that: (1) From 1960 to 2019, the maximum temperature (T_{max}), minimum temperature (T_{min}) and average temperature (T_{mean}) showed an increasing trend, with climate tendency rate of 0.22 °C per decade (10a), 0.49 °C/(10a), 0.36 °C/(10a), respectively. The relative humidity (RH), wind speed (U) and net radiation (Rn) showed a decreasing trend, with a climate tendency rate of −0.42%/(10a), −0.18 m/s/(10a), −0.08 MJ/m^2/(10a), respectively. (2) ET_0 showed a decreasing trend on seasonal and inter-annual scales. Inter-annually, the average climate tendency rate of ET_0 was −8.69 mm/(10a). seasonally, the lowest climate tendency rate was −6.33 mm/(10a) in spring. (3) ET_0 was negatively sensitive to T_{min}, and RH, while positively sensitive to T_{max}, T_{mean} U and Rn, its sensitivity coefficient of U was the highest, which was 1.22. (4) The contribution rate of U to ET_0 was the highest on an inter-annual scale as well as in spring and autumn, which were −8.96%, −9.79% and −13.14%, respectively, and the highest contribution rate to ET_0 were Rn and T_{min} in summer and winter, whose contribution rates were −4.37% and −11.46%, respectively. This study provides an understanding on the response of evapotranspiration to climatic change and further provides support on the optimal allocation of regional water resource and agricultural water management under climate change.

Keywords: potential evapotranspiration (ET_0); climate change; Penman–Monteith; sensitivity analysis; contribution rate

Citation: Nie, T.; Yuan, R.; Liao, S.; Zhang, Z.; Gong, Z.; Zhao, X.; Chen, P.; Li, T.; Lin, Y.; Du, C.; et al. Characteristics of Potential Evapotranspiration Changes and Its Climatic Causes in Heilongjiang Province from 1960 to 2019. *Agriculture* **2022**, *12*, 2017. https://doi.org/10.3390/agriculture12122017

Academic Editors: Dengpan Xiao and Wenjiao Shi

Received: 17 October 2022
Accepted: 24 November 2022
Published: 26 November 2022

1. Introduction

Over the past 100 years (1906–2005), the global average surface temperature has increased by 0.74 °C and is expected to increase by at least 4 °C in 2100 if carbon dioxide emissions are not reduced [1]. Climate change will inevitably lead to changes in the global hydrological cycle [2]. Apart from precipitation and runoff, actual evapotranspiration (ET)

is the most important climatic factor and a main component in the hydrological cycle [3]. Evapotranspiration is affected by a variety of meteorological factors, such as temperature, relative humidity (*RH*), wind speed (*U*), and net radiation (*Rn*). Characteristics of *ET* changes and its climatic causes under climate change will reveal the water cycle process and its driving mechanism.

Due to the complexity of the *ET* process, it is difficult to directly measure the *ET*: however, with the deepening of vegetation *ET* research, obtaining regional continuous *ET* data by calculating the *ET* becomes particularly vital. Potential evapotranspiration (ET_0), as the basis for calculating the *ET* [4], represents the maximum *ET* that can be achieved on a fixed underlying surface with unlimited water supply under certain meteorological conditions. ET_0 is widely used in the analysis of climate dry and wet conditions, rational use and evaluation of water resources, crop water demand and agricultural production management, and ecological environment research [5]. There are many ways to calculate ET_0, such as temperature and the solar radiation-based Hargreaves method, the equilibrium evaporation-based Priestley-Taylor model and the FAO Blaney–Criddle formula [6]. However, many scholars choose the Penman-Monteith method recommended by the FAO56 to calculate ET_0. The advantage of the Penman–Monteith method is that standard meteorological data are easily obtained or obtained through routine observation, and all calculation procedures can be standardized by the calculation of available meteorological data and time scales [7].

ET_0 was heterogeneous in temporal and spatial distribution. Jung et al. [8] pointed out that the global average annual *ET* increased at a rate of 7.1 ± 1.0 mm/10a from 1982 to 1997, but since 1998, the global average *ET* decreased by -7.9 mm/10a. Roderick et al. [9] found that in the past 50 years, the ET_0 in the northern hemisphere was decreasing at a faster rate of 2 to 4 mm/a. ET_0 also showed an increasing trend in some areas, such as West Africa [10] and Turkey [11]. Wu et al. [12] pointed out that the national annual average ET_0 decreased at a rate of 0.52 mm/a based on the daily meteorological data of 552 meteorological stations in China from 1961 to 2015, and the ET_0 of most stations in arid and humid regions showed a significant decreasing trend; however, neither the increasing nor decreasing trend of ET_0 is significant in semi-arid and semi-humid regions. The decrease or increase of ET_0 may affect the hydrological cycle differently. For example, when the ET_0 decreases, the transport of water vapor in the atmosphere is reduced, resulting in corresponding changes in precipitation patterns. For agriculture, the water use efficiency of crops may be improved by reducing the adverse effects of drought on crops. However, the increase of ET_0 will increase water consumption, resulting in increased land water loss and drought, and the atmospheric circulation may be affected, which may lead to strong rainfall, thus changing the distribution of water resources.

The spatio–temporal heterogeneity of ET_0 was attributed to the common effect of different climatic factors, including maximum temperature ($T_{\max}$), minimum temperature ($T_{\min}$), average temperature (T_{mean}), *RH*, *U* and sunshine hours, while the dominant factors affecting ET_0 varied in different regions. For example, Bandyopadhyay et al. [13] concluded that the main reason for the decrease of ET_0 in India was the significant increase in *RH* and decrease of *U* by the non-parametric method of Sen's slope. Roderick et al. [14] believed that the decrease in solar radiation caused by the increase in cloud cover and aerosol concentration in the southern hemisphere was the main reason for the decrease of ET_0 in New Zealand. Hossein et al. [15] believed that *U* has the greatest impact on ET_0 in Iranian region under arid climate condition by sensitivity analysis. Guo et al. [16] used the Sobol global sensitivity analysis method to analyze the sensitivity of ET_0 in Australia calculated by the Penman–Monteith and Priestley–Taylor models and obtained that air temperature was the most sensitive factor to ET_0. China has seven geographic regions, with complex and diverse climatic characteristics. ET_0 has shown a fluctuating downward trend in recent decades, such as Henan Province in Central China [17], Anhui Province in East China [18], Sichuan Basin [19] and Huaihe basin [20] in Southwest China; the results of the sensitivity-contribution rate method showed that the contribution of *U* to ET_0 was higher than that of sunshine hours, indicating that U was the dominant meteorological factor in the above area. On the contrary, in North China, the results of the sensitivity coefficient method showed that sunshine hours were the dominant factor, followed by U [21]. Zhou et al. [22] analyzed the partial correlation between ET_0 and meteorological factors by discussing climate attribution and concluded that the increase of the T_{max} in the Three River Headwaters region located in Northwest China was the main reason for the increasing trend of ET_0, however, the contribution of U was the smallest among all meteorological factors, which was different from the meteorological factors affecting the trend of ET_0 in other regions.

Heilongjiang Province, located in one of the four major black soil belts in the world, has an existing land area of 454,600 km^2 and an area of 23,900 km^2 of arable land. It is the largest province in terms of grain production in the country. In recent years, Heilongjiang Province has become one of the regions with the largest temperature rise in the country and one of the regions with more serious climate disasters, threatening the safe production of food crops. The impact of floods and droughts on the agricultural economy in Heilongjiang Province accounted for about 89.4% of the total impact of various natural disasters [23]; the reduction of grain production due to floods and droughts accounted for about 12% of the total grain output in the same period [24]. Nie et al. [25] speculated that the climate in Heilongjiang Province would continue to warm in the next 30 and 50 years, and water resources would be scarcer. The probability of drought is increasing. ET_0, as a key factor to characterize the regional dry and wet conditions, is of great significance for guiding water resource management and agricultural production management in Heilongjiang Province. Jiang et al. [26] expounded on the various characteristics of ET_0 in Heilongjiang Province from the scale of the crop growing season and pointed out that ET_0 was the most sensitive to RH, which provided an important reference index for formulating a reasonable irrigation system for the crop growing season in Heilongjiang Province. However, their study was limited to the growing season of crops, which could not fully reflect the changes of ET_0 in Heilongjiang Province between inter-annual and different seasons; moreover, contributing rates of different climate factors were not analyzed. Su et al. [27] pointed out that the contribution rate of water vapor pressure and U to the ET_0 change in Heilongjiang Province was the largest by multiple regression analysis, and it has a certain significance for understanding the influencing factors on the inter-annual change of ET_0. However, they failed to study the impact of more meteorological factors on a shorter time scale. The weather of Heilongjiang Province changes significantly in four seasons, and ET_0 may be affected by different climatic factors in seasonal scales.

This study used a linear regression equation and the Mann–Kendall method to analyze the trend of ET_0 in Heilongjiang Province in the past 60 years on the inter-annual and seasonal scales. The sensitivity coefficient and contribution of different climate factors to ET_0 were comprehensively analyzed by the sensitivity-contribution rate method, aiming to provide an important basis for the rational and efficient use of water resources and agricultural water management in different regions of Heilongjiang Province under climate change.

2. Materials and Methods

2.1. Overview of the Study Area and Data Sources

Heilongjiang Province, located in the east of Eurasia, is the northernmost province with the highest latitude in China, starting at 121°11′ in the west, 135°05′ in the east, 43°26′ in the south, and 53°33′ in the north, spanning 14 longitudes from east to west and 10 latitudes from north to south, with an average altitude of 481 m. It has a temperate continental monsoon climate. It spans the four major river systems of the Heilongjiang River, Wusuli River, Songhua River and Suifenhe River. The average annual temperature of Heilongjiang Province is between −5 and 5 °C, gradually increasing from the north to the south. The annual precipitation is between 400 and 650 mm, with more in the central mountainous areas, followed by the east, and less in the west and north. The annual sunshine hours in the province are mostly between 2400 and 2800 h, where the west is more than the east.

The meteorological data in this study were from the daily meteorological data of 26 meteorological stations (Figure 1) in Heilongjiang Province from 1960 to 2019 recorded by the China Meteorological Data Network, including T_{max}, T_{min}, T_{mean}, RH, U, and sunshine hours; Rn was calculated according to the method recommended by the Food and Agriculture Organization of the United Nations (FAO) [7]. According to the meteorological division method, March to May is divided into spring, June to August is summer, September to November is autumn, and December to February is winter [28].

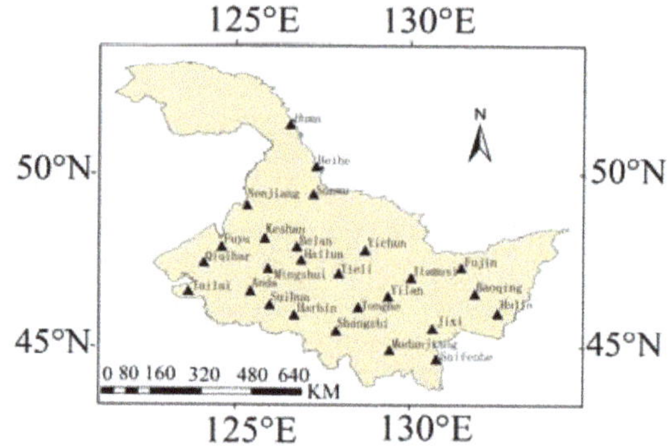

Figure 1. Map of the study area and distribution of meteorological stations in Heilongjiang Province.

2.2. Mann–Kendall Trend and Mutation Test

For the time series X, (containing n samples), an order column is constructed:

$$S_k = \sum_{i=1}^{k} r_i \ (k = 2, 3, \ldots, n) \tag{1}$$

where

$$r_i = \begin{cases} +1 & x_i > x_j \\ 0 & or \end{cases} (\text{j} = 1, 2, \ldots, i) \tag{2}$$

The order column S_k is the sum of the number of values when the value of the i moment is greater than the j moment.

Under the assumption that the time series is random, define the statistic:

$$UF_k = \frac{[S_k - E(S_k)]}{\sqrt{Var(S_k)}} \ (\text{k} = 1, 2, \ldots, n) \tag{3}$$

where $UF_1 = 0$, $E(S_k)$ and $Var(S_k)$ are the mean and variance of S_k, respectively, and when $X_1, X_2, \ldots, X_n$ are independent of one other, they have the same continuous distribution, which can be deduced from the following equation:

$$E(S_k) = \frac{n(n+1)}{4},\ Var(S_k) = \frac{n(n-1)(2n+5)}{72} (2 \leq \text{k} \leq n) \tag{4}$$

UF_k is a standard normal distribution, it is a sequence of statistics calculated in the order of time series X ($X_1, X_2, \ldots, X_n$); given the significance level α by checking the normal distribution table, if $UF_i > U_\alpha$, it indicates that there is a significant trend change in the sequence. Then repeat the above process in the reverse order of time series X ($X_n, X_{n-1}, \ldots, X_1$) and make $UB_k = UF_k$ ($k = n, n-1, \ldots, 1$), $UB_1 = 0$.

Generally, if the significance level $\alpha = 0.05$, then the critical value $Z = \pm 1.96$. The curves of the two statistical sequences of UF, UB and the two straight lines of ± 1.96 are plotted on one plot. If the values of UF and UB are greater than 0, it indicates an upward trend in the series, and if the values of UF and UB are less than 0, it indicates a downtrend. When UF and UB exceed the critical line, it indicates a significant upward or downward trend, and the range above the critical line is determined as the time zone in which the mutation occurred. If the two curves of *UF* and *UB* intersect and the intersection point is between the critical lines, then the moment corresponding to the intersection point is the time when the mutation begins [29].

2.3. Climate Tendency Rate

The climate tendency rate was calculated by the least squares method, and the unary linear regression equation of Y_i and X_i was established:

$$Y_i = AX_i + B \quad (5)$$

where A and B are regression constants, and A × 10 is the climate tendency rate, representing the changing rate of each climatic factor every 10 years (10a). A positive value indicates an increasing trend of climate in the corresponding time series, while a negative value indicates a decreasing trend.

2.4. Calculation of ET_0

The daily ET_0 of 26 sites in Heilongjiang Province was calculated using the Penman–Monteith model recommended by the FAO. The formula is as follows:

$$ET_0 = \frac{0.408\Delta(R_n - G) + \gamma\frac{900}{T+273}U_2(e_s - e_a)}{\Delta + \gamma(1 + 0.34U_2)} \quad (6)$$

where ET_0 is the potential evapotranspiration, mm; Δ is the slope of the temperature change with the saturated water vapor pressure, kPa/°C; *Rn* is the surface net radiation, MJ/(m^2·d); *G* is the soil heat flux, MJ/(m^2·d); γ is the hygrometer constant, kPa/°C; U_2 is the *U* at a height of 2 m above the ground, m/s; e_s is the saturated vapor pressure, kPa; e_a is the actual vapor pressure, kPa; *T* is the temperature, °C.

2.5. Sensitivity-Contribution Rate Method Based on Partial Derivatives

Sensitivity analysis makes each input variable change within the corresponding value range, and studies and predicts the influence degree of the changes of these input variables on the output value. The influence degree is called the sensitivity coefficient [30] and is used to judge the interference degree of the relative changes of climatic factors to the changes of ET_0. This paper mainly analyzed the sensitivity and contribution rate of T_{max}, T_{min}, T_{mean}, *RH*, *U* and *Rn* to ET_0. The sensitivity coefficient Sv_i of ET_0 to each climatic factor was calculated by the following formula:

$$Sv_i = \lim_{\Delta\to 0}\left(\frac{\Delta ET_0}{\Delta v_i}\cdot\frac{ET_0}{v_i}\right) = \frac{\partial ET_0}{\partial v_i}\cdot\frac{v_i}{ET_0} \quad (7)$$

where v_i is the climatic factor; Sv_i is the sensitivity coefficient of the climatic factor. The positive and negative values of Sv_i reflect the correlation between ET_0 and climatic factor. A negative sensitivity coefficient indicates that ET_0 decreases with the decrease of climatic factors, and vice versa. The absolute value reflects the impact of climatic factors on ET_0. The greater the absolute value, the greater the impact, and vice versa.

The contribution rate Gv_i is equal to Sv_i multiplied by the annual change rate of meteorological variables (Rv_i), which is:

$$R_{v_i} = \frac{n \times Trend_{v_i}}{|av_{v_i}|} \times 100\% \quad (8)$$

$$Gv_i = Rv_i \cdot Sv_i \quad (9)$$

where Rv_i is the annual relative change rate of v_i; n is the total number of years; $Trend_{vi}$ is the annual climate tendency rate of v_i, which is the slope of the univariate linear regression equation between v_i and n; av_{vi} is the annual average value of v_i; Gv_i is the contribution rate, the magnitude of the absolute value of Gv_i reflects the contribution of the relative change of v_i to the change of ET_0.

2.6. Data Processing

The CROPWAT 8.0 (FAO, Rome, Italy) software was used to calculate the daily ET_0 by the Penman–Monteith formula, Matlab 2021a (MathWorks, Natick, MA, USA) was used to calculate the rate of change of ET_0, T_{max}, T_{min}, T_{mean}, *RH*, *U* and *Rn*, and perform a Mann–Kendall mutation test. The Mann–Kendall test was used to analyze the long-term change trend and mutation of ET_0. The ArcMap 10.4 (ESRI, Redlands, CA, USA) toolkit was used for spatial analysis to perform spatial interpolation and mapping for each meteorological variable, the spatial interpolation method was Inverse Distance Weighting (IDW) and the spatial resolution was 500 dpi.

3. Results

3.1. Spatial Distribution of the Mean Values of Meteorological Factors

Spatial distribution of average T_{max}, T_{min}, T_{mean}, *RH*, *U*, *Rn* from 1960 to 2019 is shown in Figure 2. On the inter-annual scale, the T_{max}, T_{min} and T_{mean} showed an increasing trend from north to south, their inter-annual ranges were 5.5~11.5 °C, 0.5~7.0 °C and 1.0~5.5 °C, respectively. A higher *RH* was mainly distributed in the central region, while the *RH* in east and west region was relatively lower. A higher *U* was mainly distributed in the east and west regions. Spatially, *Rn* increased from north to south.

On the seasonal scale, the T_{max}, T_{min} and T_{mean} also showed an increasing trend from north to south. The *RH* was higher in summer and lower in spring. However, the *U* was higher in spring and lower in summer, which were 3.71 m/s and 2.67 m/s, respectively. The *Rn* is larger in summer and smaller in winter; the ranges were 11.5~12.8 MJ/m^2 and 0.30~1.90 MJ/m^2, respectively.

3.2. Temporal and Spatial Variation of the Climate Tendency Rate of Meteorological Factors

At time series on inter-annual and seasonal scales, the T_{max}, T_{min}, T_{mean} showed a significant upward trend ($p < 0.05$) for T_{max} in winter (Table 1). Whereas, the *RH*, *U* and *Rn* showed a significant downward trend ($p < 0.05$), except for *RH* in spring and summer, *U* in autumn and *Rn* in autumn and winter. The inter-annual climate tendency rate for the T_{max}, T_{min}, T_{mean}, *RH*, *U* and *Rn* was 0.22 °C/10a, 0.49 °C/10a, 0.36 °C/10a, −0.42 (%/10a), −0.18 (m/s/10a), and −0.04 (MJ/m^2/10a), respectively.

Table 1. Seasonal and inter-annual climate tendency rate of meteorological factors in Heilongjiang Province.

Climatic Factors	Spring	Summer	Autumn	Winter	Inter-Annual
T_{max} (°C/(10a))	0.25 *	0.16 *	0.19 *	0.28	0.22 *
T_{min} (°C/(10a))	0.57 *	0.39 *	0.43 *	0.65 *	0.49 *
T_{mean} (°C/(10a))	0.41 *	0.26 *	0.31 *	0.47 *	0.36 *
RH (%/(10a))	−0.13	−0.28	−0.51 *	−0.78 *	−0.42 *
U (m/s/(10a))	−0.24 *	−0.14 *	−0.18	−0.15 *	−0.18 *
Rn (MJ/m^2/(10a))	−0.05 *	−0.10 *	−0.01	−0.01	−0.04 *

Note: * indicates a significant level of 0.05.

In terms of spatial distribution on inter-annual scales, the T_{max}, T_{min}, and T_{mean} showed an increasing trend, higher climate tendency rate of them were mainly distributed in the northern region (Figure 3). The *RH* showed a decreasing climate tendency rate except for some central and southern regions. The *U* was shown a decreasing trend except Jixi, and the climate tendency rate was lower in western and higher in central region. The *Rn* showed a decreasing climate tendency rate except for Hulin and Tonghe. On seasonal scales, The T_{max}, T_{min}, and T_{mean} showed an increasing trend except for T_{max} in Suifenhe and T_{mean} in Hulin in summer. The climate tendency rate of T_{max}, T_{min}, T_{mean} were greatest in winter compared with the other 3 seasons (Figure 3). The *RH* showed a decreasing climate tendency rate except in the central regions in spring, summer, and autumn. The climate tendency rate of *U* was lower in the western and eastern regions, and the climate tendency rate of *Rn* showed a decreasing trend except for some central and eastern regions.

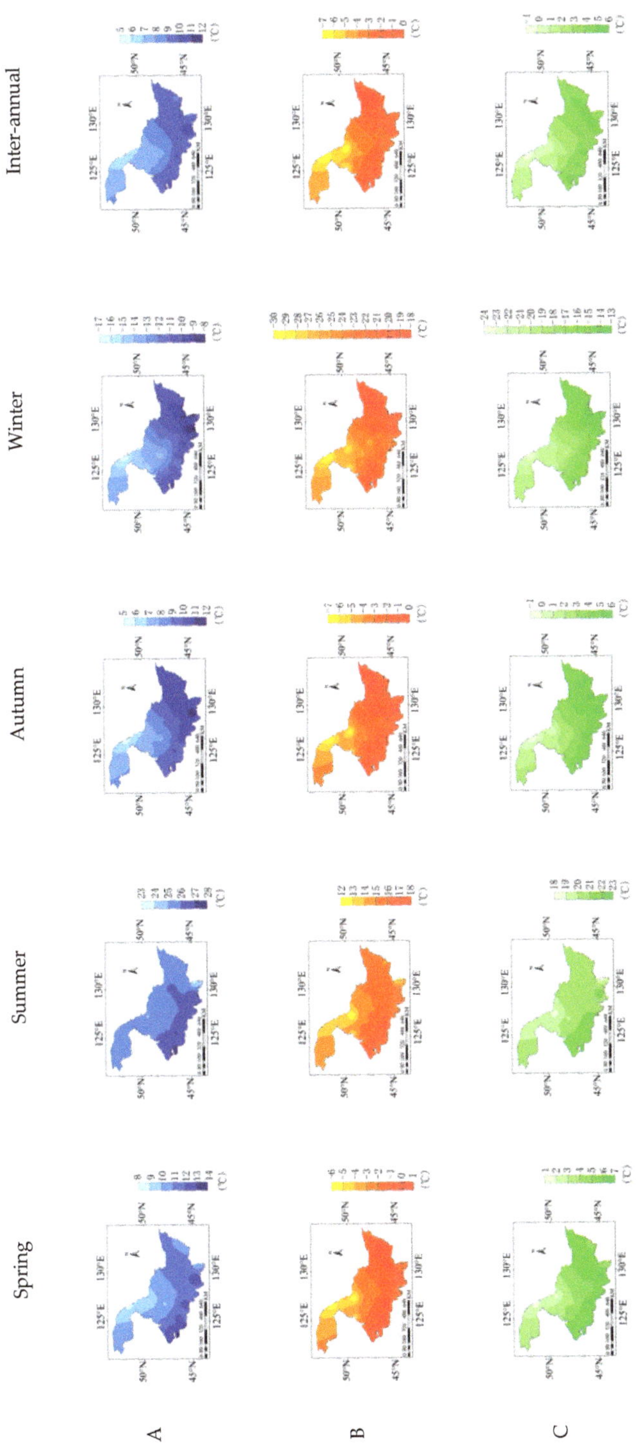

Figure 2. *Cont.*

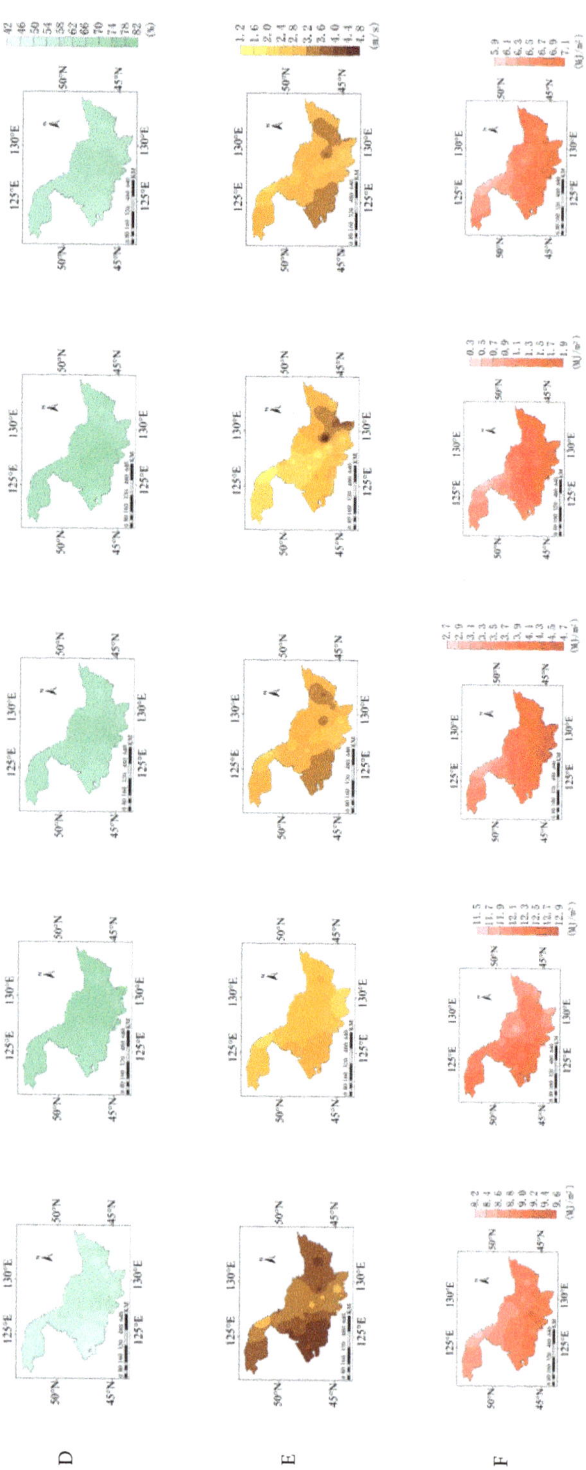

Figure 2. Spatial distribution of average (**Row A**) T_{max}, (**Row B**) T_{min}, (**Row C**) T_{mean}, (**Row D**) *RH*, (**Row E**) *U*, (**Row F**) *Rn* from 1960 to 2019.

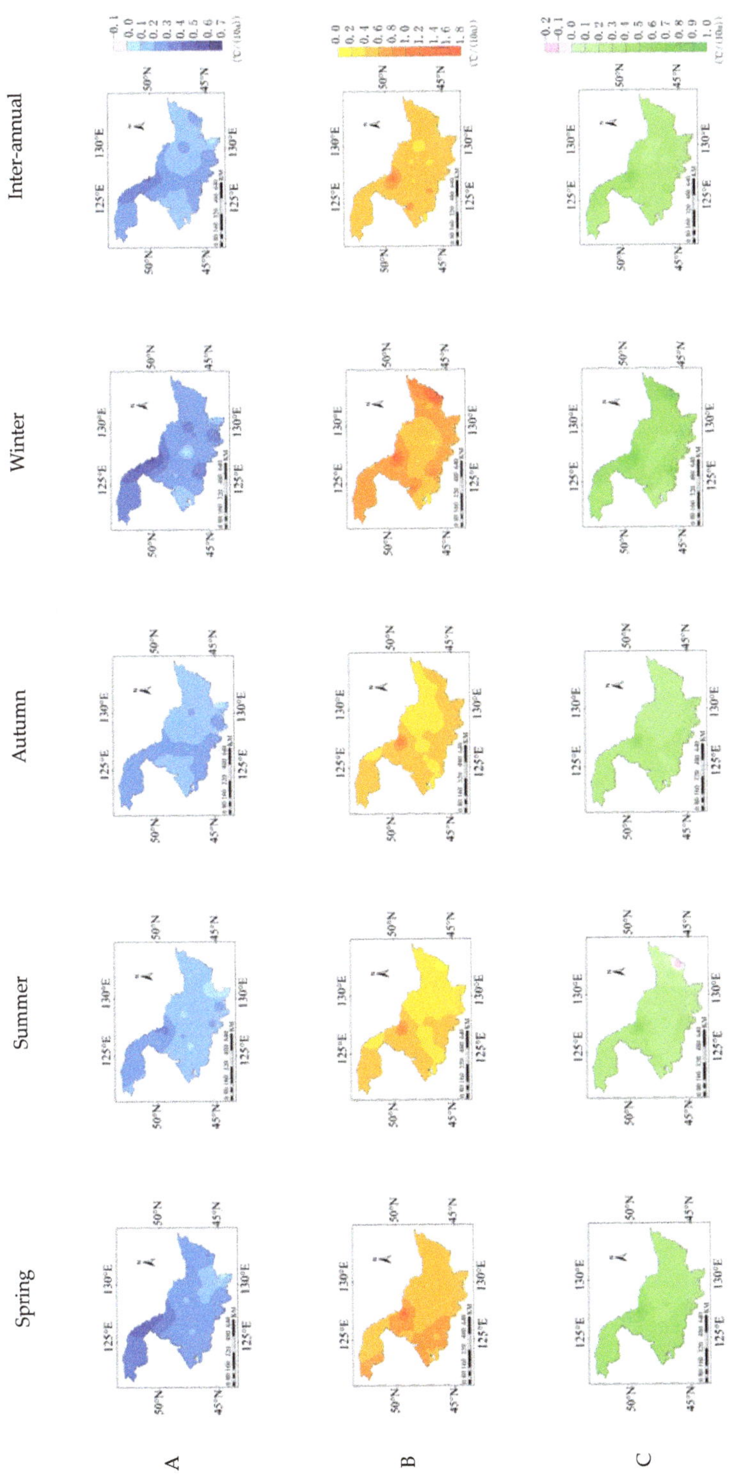

Figure 3. *Cont.*

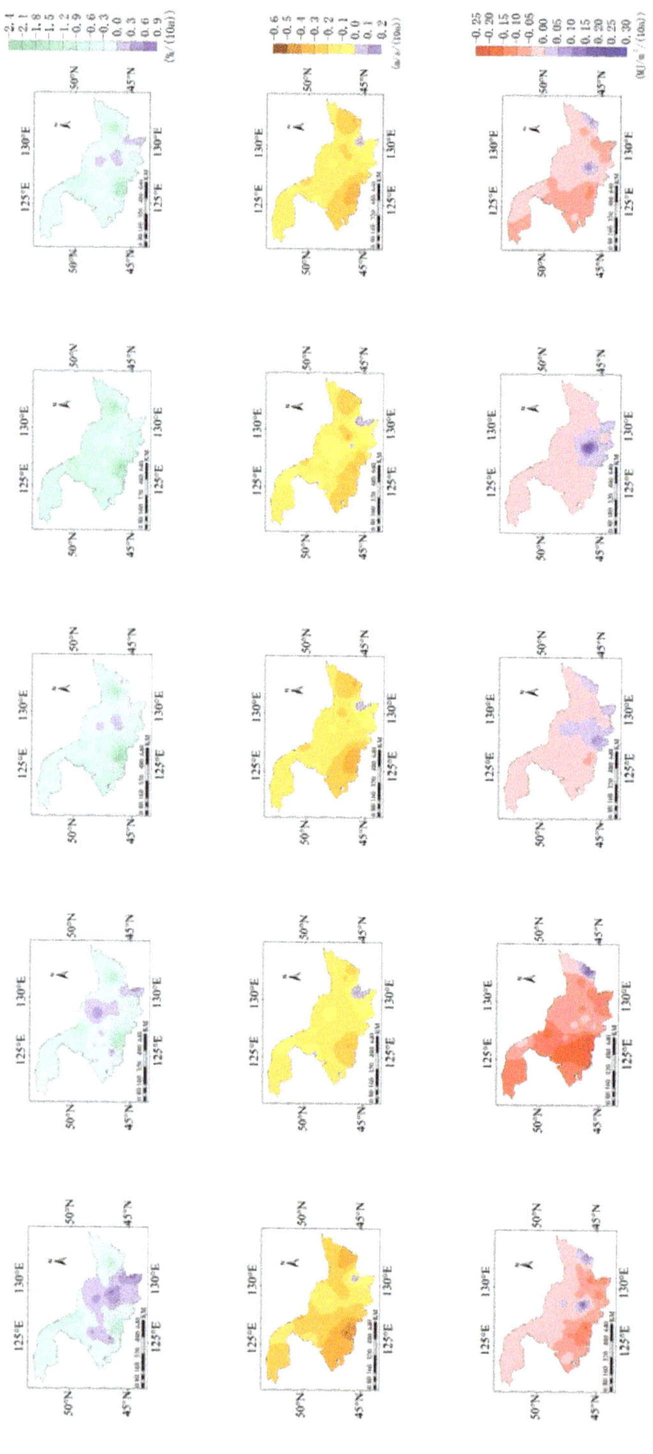

Figure 3. Climatic tendency rate of (**Row A**) T_{max}, (**Row B**) T_{min}, (**Row C**) T_{mean}, (**Row D**) *RH*, (**Row E**) *U*, (**Row F**) *Rn* from 1960 to 2019.

3.3. Temporal and Spatial Variation of ET_0

ET_0 was the highest in summer and lowest in winter, with daily averages of 4.11 mm and 0.35 mm, respectively (Figure 4). ET_0 showed decreasing trends in seasons and inter-annual; the higher ET_0 values were mainly distributed in the southwest region. The average climate tendency rate of ET_0 in spring, summer, autumn, winter and inter-annual was −6.33 mm/(10a), −2.72 mm/(10a), −2.58 mm/(10a), −0.65 mm/(10a), and −8.69 mm/(10a), respectively (Figure 4). In the western region, the seasonal lower ET_0 climate tendency rate, especially in spring and summer, led to a more rapid decrease trend of inter-annual ET_0.

The inter-annual ET_0 in Heilongjiang Province showed a significant increasing trend from 1977 to 1983 ($Z > 1.96$), while it showed a significant decreasing trend in 2017–2019 ($Z < -1.96$). The mutation point of ET_0 appeared in 2011, and the changing trend of ET_0 changed from increase to decrease (Figure 5e). On a seasonal scale, the mutation years range from 2006 to 2015 (Figure 5a–d).

3.4. Sensitivity of ET_0 to Meteorological Factors

The sensitivity coefficients of ET_0 to U, Rn, RH, T_{max}, T_{min} and T_{mean} are shown in Table 2. On the inter-annual scale, in the case that other climatic factors remain unchanged, when the U, Rn, T_{max}, T_{mean}, RH and T_{min} increase by 10%, ET_0 will increase by 12.2%, 4.0%, 4.2%, 1.4% or decrease by 11.5% and 1.4%, respectively. The sensitivity order of ET_0 change to each climatic factor was $U > RH > T_{max} > Rn > T_{min} = T_{mean}$.

Table 2. Sensitivity coefficients of seasonal and inter-annual ET_0 change to climate factors.

Season	U	Rn	RH	T_{max}	T_{min}	T_{mean}
Spring	1.35	0.36	−0.12	0.44	−0.10	0.25
Summer	0.40	0.65	−0.91	0.69	0.41	0.55
Autumn	1.40	0.28	−1.34	0.39	−0.14	0.15
Winter	0.53	0.10	−1.67	−1.64	−1.76	−1.32
Inter-annual	1.22	0.40	−1.15	0.42	−0.14	0.14

On the seasonal scale (Table 2), the changes of ET_0 in spring, summer and autumn were all negatively sensitive to RH, positively sensitive to U, Rn, T_{max}, and T_{mean}, while the changes of ET_0 to RH, T_{max}, T_{min} and T_{mean} were negatively sensitive in winter. ET_0 was most sensitive to U in spring and autumn, and most sensitive to RH and T_{min} in summer and winter, with sensitivity coefficients of 1.35, 1.40, −0.91 and −1.76, respectively.

Figure 6 shows the spatial distribution of sensitivity coefficients of ET_0. On the inter-annual scale, The sensitivity coefficients of ET_0 to T_{max}, T_{min} and T_{mean} gradually increased from north to south. The sensitivity coefficients of ET_0 to RH decreased from west to east. A higher sensitivity coefficient to U was mainly distributed in the eastern and western regions, while the sensitivity coefficient to Rn was mainly distributed in the central region.

On the seasonal scales, the sensitivity of ET_0 to T_{max}, T_{min}, T_{mean} was higher in the western region in spring and summer, and higher in the southeast and south in autumn and winter. ET_0 showed lower sensitivity to RH in the central region throughout the 4 seasons, the sensitivity coefficient of ET_0 to U was higher in spring and autumn, while the sensitivity coefficient of ET_0 to Rn was higher in summer.

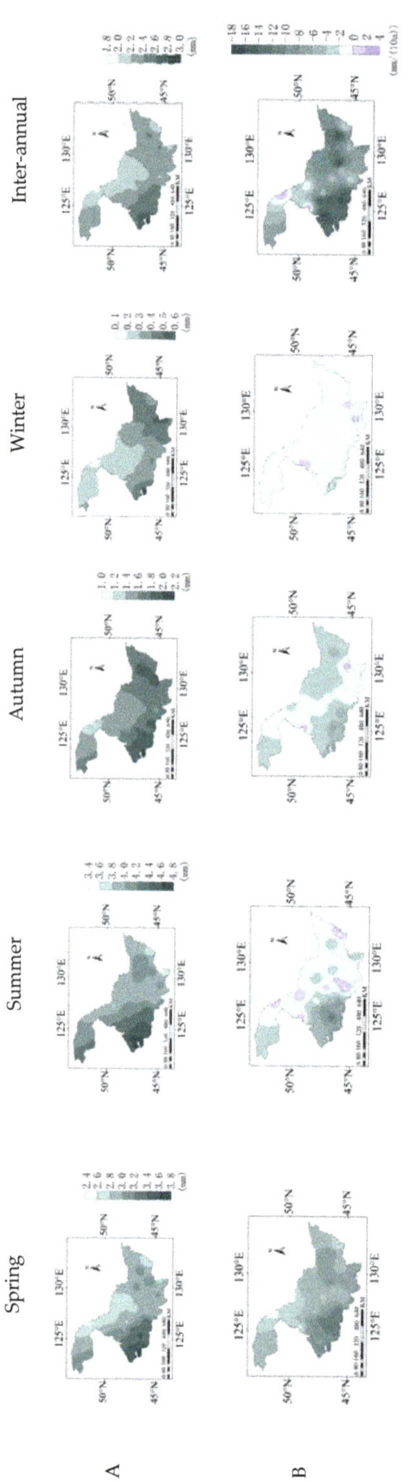

Figure 4. **(Row A)** spatial distribution and **(Row B)** climate tendency of ET_0 from 1960 to 2019.

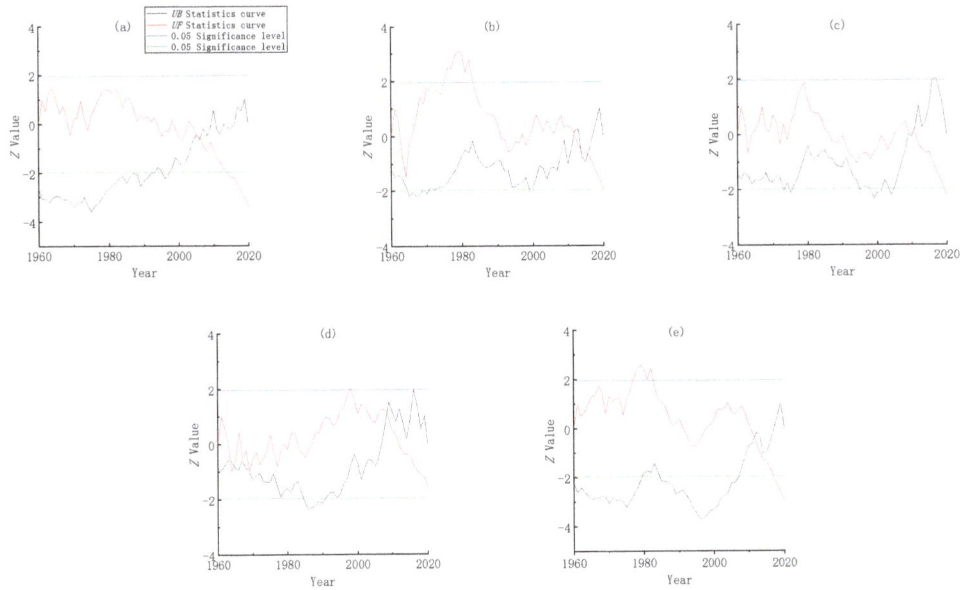

Figure 5. Mann–Kendall (MK) analysis of ET_0 in (**a**) spring, (**b**) summer, (**c**) autumn, (**d**) winter and (**e**) inter-annually from 1960 to 2019.

3.5. Dominant Climatic Factors for ET_0 Change

On the inter-annual scale, U was the dominant climatic factor for the change of ET_0, followed by T_{max}, T_{mean}, RH, Rn and T_{min}, with contribution rates of 6.15%, 5.03%, 4.34%, −1.82%, and −1.41%, respectively (Table 3). The positive contribution rates of RH, T_{max} and T_{mean} to ET_0 have not been able to offset the negative contribution rates of U, Rn and T_{min}; therefore, the ET_0 showed a decreasing trend from 1960 to 2019.

Table 3. Contribution of seasonal and inter-annual climate factors to ET_0 changes from 1960 to 2019.

Time Scales	Contribution Rates/%						Total Contribution/%	Dominant Factor
	U	Rn	RH	T_{max}	T_{min}	T_{mean}		
Spring	−9.79	−1.56	0.18	6.04	−5.82	3.14	−7.81	U
Summer	−3.63	−4.37	2.02	2.57	2.1	2.09	0.78	Rn
Autumn	−13.14	−0.3	7.33	4.7	−4.5	3.78	−2.13	U
Winter	−10.86	−0.3	3.09	−10.1	−11.46	−9.33	−38.96	T_{min}
Inter-annual	−8.96	−2.82	3.34	5.15	−1.41	4.03	−0.67	U

On the seasonal scale, U was the dominant factor for the decrease of ET_0 in spring and autumn, followed by T_{max} and RH; the largest contribution rates to the change of ET_0 were −9.79% and −13.14%, respectively. In summer, Rn was the dominant factor for ET_0 change, the contribution rates of T_{max}, T_{min}, and T_{mean} to ET_0 were less different, which were 2.57%, 2.1% and 2.09%,respectively, while in winter, Rn contributes the least to ET_0, and T_{min} became the dominant factor.

The spatial distribution of dominant meteorological factors is shown in Figure 7. In spring, the dominant factor for ET_0 was U in the study area except for Fujin (Figure 7a), In summer, Rn was the dominant factor in most regions of the study area, T_{min} and RH were the dominant factors in the partial northern and eastern regions. In autumn, U was the dominant factor for 85% of the total 26 sites (Figure 7c); in winter, T_{min} was the dominant factor in the northern and eastern regions, while U was the dominant factor in the western region. On the inter-annual scale, the dominant factor of all 26 sites was U (Figure 7e).

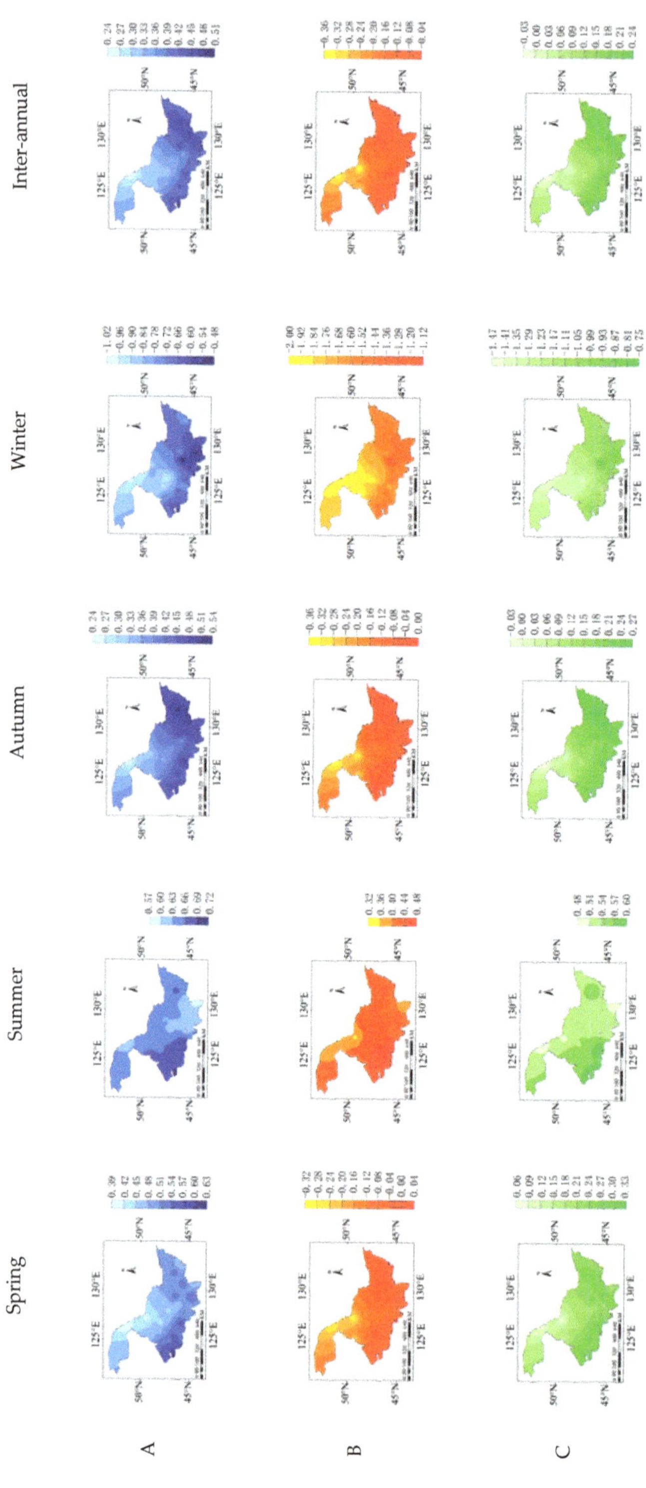

Figure 6. *Cont.*

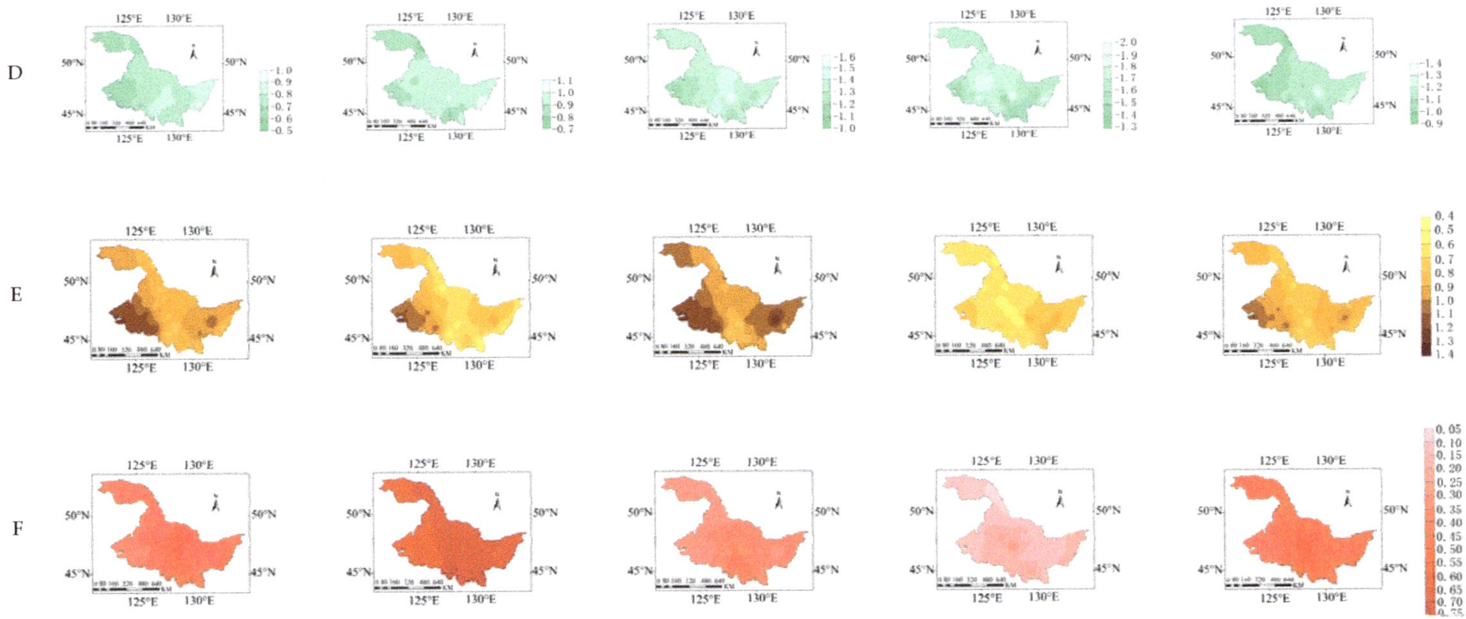

Figure 6. Spatial distribution of sensitivity coefficients of ET_0 to **(Row A)** *T*max, **(Row B)** *T*min, **(Row C)** *T*mean, **(Row D)** *RH*, **(Row E)** *U*, **(Row F)** *Rn* from 1960 to 2019.

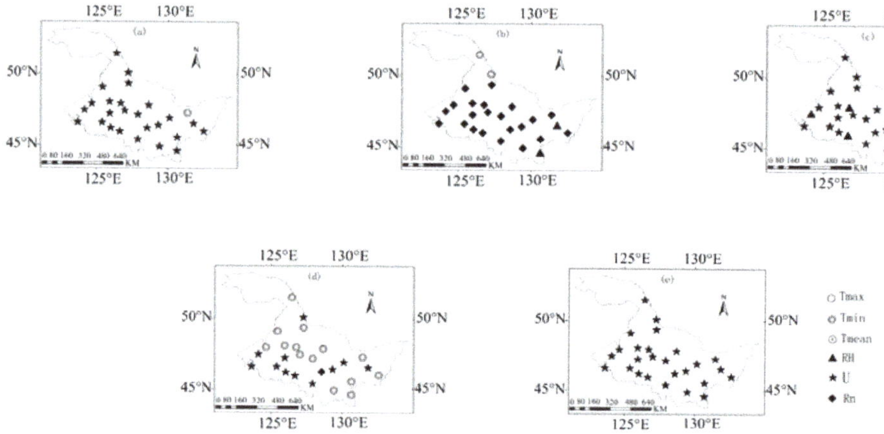

Figure 7. Spatial distribution of dominant climatic factors in (**a**) spring, (**b**) summer, (**c**) autumn, (**d**) winter and (**e**) inter-annually in the study area from 1960 to 2019.

4. Discussion

Under the general warming of the global climate, ET_0 was on the rise in most areas, such as the annual ET_0 for India, which, as a whole, has increased in the latter half of the 20th century [31]; Awash River basin, Ethiopia [32], and South Korea over the recent 100 years [33], which is consistent with the generally accepted trend that ET_0 increases with temperature. However, the phenomenon of the "evaporation paradox" appeared in many areas around the world [34], that is, with the continuous increase of temperature, the ET_0 showed a decreasing trend, such as the Canadian prairie region, the northern region of South America, Thailand, New Zealand [9,35–37], and the northwest of India [38], the Lijiang watershed [39], as well as Alor Setar, Malaysia [40]. In this study, ET_0 in Heilongjiang Province from 1960 to 2019 showed a decrease trend inter-annually and seasonally with the increasing temperature, proving that the "evaporation paradox" phenomenon also existed in our study area, which was consistent with the conclusion of Li et al. [41]. At present, there is still no clear conclusion about the mechanism of the "evaporation paradox"; the existing related research studies are mostly qualitative analysis, and the reasons need to be further explored.

In China, the sensitivity coefficient and contribution rate of ET_0 to meteorological factors varied in different climatic regions. Even in the same climatic region, the sensitivity coefficient and contribution rate of ET_0 were different spatially. For example, in the subtropical monsoon climate zone, ET_0 in the Yangtze River basin [42], Sichuan basin [43], and Yunnan-Guizhou plateau [44] were most sensitive to the T_{max}, RH, and sunshine hours, respectively. In the temperate continental climate zone, ET_0 in the middle and upper reaches of the Yellow River [45] and the Ebinur Lake basin in Xinjiang [46] were most sensitive to the actual water vapor pressure and U, respectively, with the highest contribution rate of U for both regions. In the plateau mountainous climate region, the meteorological factor with the largest sensitivity coefficient in the Qinghai Tibet Plateau was the actual water vapor pressure, and the sunshine hours had the largest contribution rate to ET_0 [47]. Our study area is located in the temperate monsoon climate zone, where ET_0 was the most sensitive to U and had the largest contribution rate to ET_0. However, in the Loess Plateau basin, which is also located in the same climate zone, ET_0 was most sensitive to actual water vapor pressure and T_{mean} had the highest contribution rate [48]. In the North China Plain of the temperate monsoon climate zone, sunshine hours had the largest contribution rate to ET_0 [21]. In summary, the plateau area has strong solar radiation, long sunshine time, and low temperature, which may cause the ET_0 to be most affected by the sunshine hours; most of the basin areas have a dry climate, less rainfall, and lack of water resources, causing vegetation to be more sensitive to the water condition; therefore, *RH* might have a larger sensitivity coefficient to ET_0. Heilongjiang Province is located in plain and mountainous areas, and affected by southeast monsoon in summer and controlled by northwest monsoon in winter [49]. This might be the reason why *U* contributed the most to ET_0 change.

In this study, ET_0 in Heilongjiang Province showed a decrease trend on the inter-annual scale from 1960 to 2019, with *U* as the dominant factor, which was consistent with the trend of ET_0 in Jilin Province analyzed by Liu et al. [50] and in Liaoning Province analyzed by Cao et al. [51]. This was probably because the above three provinces are geographically contiguous and have a similar type of climate characteristics. A research established by Xu et al. in 2003 predicted that the temperature in Heilongjiang Province would increase significantly by 2030 and 2050 with a higher increase in winter and a lower increase in summer by the CCCma (Canadian Center for Climate Modelling and analysis), CCSR (Center for Climate System Research), CSIRO (Commonwealth Scientific and Industrial Research Organization), GFDL (Geophysical Fluid Dynamics Laboratory) and Hadley climate models [52]. Zhang et al. [53] pointed out that the climate of Heilongjiang Province would tend to be warm and humid in the next 41 years by the Hadley model. Wang et al. [54] believed that ET_0 was not only affected by various meteorological factors, but interactions between meteorological factors would also interfere with ET_0 changes. Moreover, ET_0 may also be affected by human activities including aerosol emissions, air pollution [55] and rice area expansion [56]. Therefore, further exploration in these aspects will be needed for contribution analysis of ET_0 changes in Heilongjiang Province in the future, as well as in other regions of the world.

5. Conclusions

In this study, ET_0 was calculated by the Penman–Monteith formula, the sensitivity-contribution rate method was used to clarify the climatic factors affecting seasonal and inter-annual changing of ET_0., The T_{max}, T_{min}, T_{mean} showed an increasing trend and *RH*, *U*, *Rn* showed a decreasing trend from 1960 to 2019. ET_0 showed a decreasing trend with an average climate tendency rate of −8.69 mm/(10a). The annual average ET_0 change was negatively sensitive to T_{min} and *RH*, and positively sensitive to T_{max}, T_{mean}, *U* and *Rn*. *U* was the dominant factor for the ET_0 decrease in spring and autumn, while *RH* and T_{min} were the dominant factors in summer and winter, respectively. On the inter-annual scale, the sensitivity order of ET_0 change to each climatic factor was $U > RH > T_{max} > Rn > T_{min} = T_{mean}$. These results indicated that the dominant factors for ET_0 changes were *U* in Heilongjiang Province. Moreover, the dominant factors for ET_0 change varies under different climatic and geographical conditions; impacts of human activities on ET_0 should also be considered in future studies.

Author Contributions: Conceptualization, T.N.; methodology, T.N. and R.Y.; software, R.Y.; validation, R.Y., S.L. X.Z. and Y.L.; formal analysis, T.N. and R.Y.; data curation, P.C. and T.L.; writing—original draft preparation, T.N. and R.Y.; writing—review and editing, Z.G., C.D. (Chong Du), C.D. (Changlei Dai) and H.J.; visualization, R.Y.; supervision, T.N., Z.Z. and Z.G.; funding acquisition, T.N. and Z.Z. All authors have read and agreed to the published version of the manuscript.

Funding: This work was fund by the Opening Project of Key Laboratory of Efficient Use of Agricultural Water Resources, Ministry of Agriculture and Rural Affairs of the People's Republic of China (number: AWR2021002), the Basic Scientific Research Fund of Heilongjiang Provincial Universities (number: 2021-KYYWF-0019), the National Key Research and Development Program of China (2021YFD1500802) and the National Natural Science Foundation Project of China (numbers: 51779046 & 52079028).

Institutional Review Board Statement: Not applicable.

Informed Consent Statement: Not applicable.

Data Availability Statement: Not applicable.

Acknowledgments: We thank the anonymous reviewers and the editors for their suggestions which substantially improved the manuscript.

Conflicts of Interest: The authors declare no conflict of interest.

References

1. IPCC. *Climate Change 2007: The Physical Science Basis*; Cambridge University Press: Cambridge, UK, 2007.
2. Silva, C. Importance of modelling and its feasibility in the future climate in Sri Lanke. *Symp. Water Prof. Day* **2013**, *18*, 15–28.
3. Anabalón, A.; Sharma, A. On the divergence of potential and actual evapotranspiration trends: An assessment across alternate global datasets. *Earth's Future* **2017**, *9*, 905–917. [CrossRef]
4. Ding, Y.H. *Introduction to Climate Change Science in China*; China Meteorological Press: Beijing, China, 2008; Volume 128.
5. Abedi-Koupai, J.; Dorafshan, M.M.; Javadi, A. Estimating potential reference evapotranspiration using time series models (case study: Synoptic station of Tabriz in northwesternIran). *Appl. Water Sci.* **2022**, *12*, 212. [CrossRef]
6. Li, Y.L.; Cui, J.Y.; Zhang, T.H. A comparative study on the calculation methods of reference crop evapotranspiration. *China Desert* **2002**, *04*, 65–69.
7. Pereira, L.S.; Allen, R.G.; Smith, M. Crop evapotranspiration estimation with FAO56: Past and future. *Agric. Water Manag.* **2015**, *147*, 4–20. [CrossRef]
8. Jung, M.; Reichstein, M.; Ciais, P.; Seneviratne, S.I.; Goulden, M.L.; Bonan, G.; Cescatti, A.; Chen, J.Q.; De, J. Recent decline in the global land evapotranspiration trend due to limited moisture supply. *Nature* **2010**, *467*, 951–954. [CrossRef] [PubMed]
9. Roderick, M.L.; Farquhar, G.D. Changes in New Zealand pan evaporation since the 1970s. *Int. J. Climatol.* **2005**, *25*, 2031–2039. [CrossRef]
10. Onyutha, C. Statistical analyses of potential evapotranspiration changes over the period 1930–2012 in the Nile River riparian countries. *Agric. For. Meteorol.* **2016**, *226–227*, 80–95. [CrossRef]
11. Dadaser-Celik, F.; Cengiz, E.; Guzel, O. Trends in reference evapotranspiration in Turkey: 1975–2006. *Int. J. Climatol.* **2016**, *36*, 1733–1743. [CrossRef]
12. Wu, X.; Wang, P.J.; Huo, Z.G.; Bai, Y.M. Spatio-temporal distribution characteristics of potential evapotranspiration and impact factors in China from 1961 to 2015. *Resour. Sci.* **2017**, *39*, 964–977.
13. Bandyopadhyay, A.; Bhadra, A.; Raghuwanshi, N.S.; Singh, R. Temporal trends in estimates of reference evapotranspiration over India. *J. Hydrol. Eng.* **2014**, *14*, 508–515. [CrossRef]
14. Roderick, M.L.; Farquhar, G.D. Changes in Australian pan evaporation from 1970 to 2002. *Int. J. Climatol.* **2004**, *24*, 1077–1090. [CrossRef]
15. Tabari, H.; Talaee, P.H. Sensitivity of evapotranspiration to climatic change in different climates. *Glob. Planet. Chang.* **2014**, *115*, 16–23. [CrossRef]
16. Guo, D.L.; Westra, S.; Holger, R.M. Sensitivity of potential evapotranspiration to changes in climate variables for different Australian climatic zones. *Hydro. Earth Syst. Sci.* **2017**, *21*, 2107–2126. [CrossRef]
17. Zhang, Z.G.; Zheng, M.J.; Cai, M.T.; Yin, J.Y.; Li, H.H.; Wang, Y.W. Spatio-temporal evolution and genesis analysis of reference crop evapotr-anspiration in Henan Province from 1960 to 2019. *Water Sav. Irrig.* **2021**, *3*, 44–50. (In Chinese)
18. Wu, W.Y.; Kong, Q.Q.; Wang, X.D.; Ma, X.Q.; He, B.F.; Zhang, H.Q. Sensitivity analysis of reference crop evapotranspiration in Anhui Province in the past 40 years. *J. Ecol. Environ.* **2013**, *22*, 1160–1166.
19. Zhao, L.; Liang, C. Study on the causes of changes in potential evapotranspiration in Sichuan Province in the past 50 years. *Soil Water Conserv. Res.* **2014**, *21*, 26–30. (In Chinese)
20. Wang, X.D.; Ma, X.Q.; Xu, Y.; Lui, R.N.; Cao, W.; Zhu, H. The variation characteristics of reference crop evapotranspiration and the contribution of main meteorological factors in the Huaihe River Basin. *China Agric. Meteorol.* **2013**, *34*, 661–667. (In Chinese)
21. Wang, P.T.; Yan, J.P.; Jiang, C.; Liu, X.F. Spatial-temporal variation of reference crop evapotranspiration and its influencing factors in North China Plain. *Acta Ecol. Sin.* **2014**, *34*, 5589–5599.
22. Zhou, B.R.; Li, F.X.; Xiao, H.B.; Hu, A.J.; Yan, L.D. Temporal and spatial differentiation characteristics of potential evapotranspiration and climate attribution in the source region of the Three Rivers. *J. Nat. Resour.* **2014**, *29*, 2068–2077.
23. Wang, Q.J.; Lv, J.J.; Li, X.F.; Wang, P.; Ma, G.Z. Distribution characteristics of agricultural meteorological disasters in Heilongjiang Province and their impact on agricultural production. *Heilongjiang Water Conserv. Sci. Technol.* **2016**, *44*, 57–61. (In Chinese)
24. Zhang, P. Analysis on the Causes of Agricultural Natural Disasters in Heilongjiang Province. *Res. Agric. Mech.* **2011**, *33*, 249–252.
25. Nie, T.Z.; Tang, Y.; Jiao, Y.; Li, N.; Wang, T.Y.; Du, C.; Zhang, Z.X.; Chen, P.; Li, T.C.; Sun, Z.Y.; et al. Effects of Irrigation Schedules on Maize Yield and Water Use Efficiency under Future Climate Scenarios in Heilongjiang Province Based on the AquaCrop Model. *Agronomy* **2022**, *12*, 810. [CrossRef]

26. Jiang, Y.; Du, C.; Sun, H.N.; Zhong, Y. Spatio-temporal variation characteristics and sensitivity analysis of potential evapotranspiration during crop growing season in Heilongjiang Province under climate change. *Hydropower Energy Sci.* **2018**, *36*, 6–9. (In Chinese)
27. Su, J.W.; Zhang, X.L.; Shen, B. Spatio-temporal variation characteristics and influencing factors of potential evapotranspiration in Heilongjiang. *Heilongjiang Water Conserv. Sci. Technol.* **2021**, *49*, 1–8. (In Chinese)
28. Tu, A.G.; Li, Y.; Nie, X.F.; Mo, M.H. Changes of reference crop evapotranspiration in Poyang Lake Basin and its attribution analysis. *Ecol. Environ. Sci.* **2017**, *26*, 211–218.
29. Xing, L.T.; Huang, L.X.; Chi, G.Y.; Yang, L.Z.; Li, C.S.; Hou, X.Y. A Dynamic Study of a Karst Spring Based on Wavelet Analysis and the Mann-Kendall Trend Test. *Water* **2018**, *10*, 698. [CrossRef]
30. Xing, H.H.; Chen, H.D.; Lin, H.Y. Sensitivity analysis of hierarchical hybrid fuzzy-neural network based on input perturbation. *Software* **2013**, *34*, 52–55.
31. Sarma, A.; Bhaskar, V.V.; Sastry, C.M. Potential Evapotranspiration over India—An estimate of Green Water flow. *Mausam* **2014**, *65*, 365–378. [CrossRef]
32. Yadeta, D.; Kebede, A.; Tessema, N. Potential evapotranspiration models evaluation, modelling, and projection under climate scenarios, Kesem sub-basin, Awash River basin, Ethiopia. *Model. Earth Syst. Environ.* **2020**, *6*, 2165–2176. [CrossRef]
33. Jeon, M.G.; Nam, W.H.; Mun, Y.S.; Yoon, D.H.; Yang, M.H.; Lee, H.J.; Shin, J.H.; Hong, E.M.; Zhang, X. Climate change impacts on reference evapotranspiration in South Korea over the recent 100 years. *Theor. Appl. Climatol.* **2022**, *150*, 309–326. [CrossRef]
34. Roderick, M.L.; Farquhar, G.D. The cause of decreased pan evaporation over the past 50 years. *Science* **2002**, *298*, 1410–1411. [CrossRef] [PubMed]
35. Burn, D.H.; Hesch, N.M. Trends in evaporation for the Canadian Prairies. *J. Hydrol.* **2007**, *336*, 61–73. [CrossRef]
36. Quintana-Gomez, R.A. Changes in evaporation patterns detected in northernmost South America: Homogeneity testing. In Proceedings of the 7th International Meeting on Statistical Climatology, Whisler, BC, Canada, 25–29 May 1998.
37. Tebakari, T.C.; Yoshitani, J.C.; Suvanpimol, C.C. Time-space trend analysis in pan evaporation over Kingdom of Thailand. *J. Hydrol. Eng.* **2005**, *10*, 205–215. [CrossRef]
38. Das, P.K.; Midya, S.K.; Das, D.K.; Rao, G.S.; Raj, U. Characterizing Indian meteorological moisture anomaly condition using long-term (1901–2013) gridded data: A multivariate moisture anomaly index approach. *Int. J. Climatol.* **2018**, *38*, E144–E159. [CrossRef]
39. Jiao, L.; Wang, D.M. Climate Change, the Evaporation Paradox, and Their Effects on Streamflow in Lijiang Watershed. *Pol. J. Environ. Stud.* **2018**, *27*, 2585–2591. [CrossRef]
40. Ahmad, A.A.; Yusof, F.; Mispan, M.R.; Kamaruddin, H. Rainfall, evapotranspiration and rainfall deficit trend in Alor Setar, Malaysia. *Malays. J. Fundam. Appl. Sci.* **2017**, *13*, 400–404. [CrossRef]
41. Li, X.F.; Jiang, L.X.; Li, X.F.; Zhao, F.; Zhu, H.X.; Wang, P.; Gong, L.J.; Zhao, H.Y. Evolution characteristics of evaporation and its relationship with climatic factors in Heilongjiang Province from 1961 to 2017. *Meteorology* **2021**, *47*, 755–766. (In Chinese)
42. Liu, C.M.; Zhang, D.; Liu, X.M.; Zhao, C.S. Spatial and temporal change in the potential evapotranspiration sensitivity to meteorological factors in China (1960–2007). *J. Geogr. Sci.* **2012**, *22*, 3–14. (In Chinese) [CrossRef]
43. Chen, D.D.; Wang, X.D.; Wang, S.; Su, X.W. Variation of potential evapotranspiration and its climatic influencing factors in Sichuan Province. *Chin. J. Agrometeorol.* **2017**, *38*, 548–557. (In Chinese)
44. Yu, F.; Gu, X.P.; Xiong, H. Spatio-temporal variation characteristics of potential evapotranspiration in Guizhou and sensitivity to meteorological factors. *Agric. Sci. Technol.* **2015**, *16*, 2845–2848. (In Chinese)
45. Wang, Y.J.; Li, J.; Lin, Z.H.; Tong, X.J.; Xing, X.M. Estimation of the potential evapotranspiration effects of climate change on the middle and upper reaches of the Yellow River. *Chin. J. Soil Water Conserv.* **2013**, *11*, 48–56. (In Chinese)
46. Xu, J.J.; Jin, X.Y.; Qiang, H.F.; Dai, H.; Liang, C. Characteristics and causes of potential evapotranspiration change in The Ebi Lake Basin in Xinjiang. *J. Irrig. Drain.* **2018**, *37*, 89–94.
47. Shi, Y. *Climate Change on the Tibetan Plateau and Its Impact on Potential Evapotran-Spiration*; Beijing Forestry University: Beijing, China, 2019.
48. Ning, T.T. *Spatial-Temporal Variation of Evapotranspiration in the Loess Plateau under Budyko Framework and Its Attribution Analysis*; Research Center for Soil and Water Conservation and Eco-Environment of the Ministry of Education; Chinese Academy of Sciences: Beijing, China, 2017. (In Chinese)
49. Chen, P.; Xu, J.; Zhang, Z.; Wang, K.; Li, T.; Wei, Q.; Li, Y. Carbon pathways in aggregates and density fractions in Mollisols under water and straw management: Evidence from ^{13}C natural abundance. *Soil. Biol. Biochem.* **2022**, *169*, 108684. [CrossRef]
50. Liu, Y.X.; Ren, J.Q.; Wang, D.N.; Mu, J.; Cui, J.L.; Chen, C.S.; Chen, X.; Guo, C.M. Spatio-temporal distribution and genesis analysis of reference crop evapotranspiration in Jilin Province. *J. Ecol. Environ.* **2019**, *28*, 2208–2215.
51. Cao, Y.Q.; Qi, J.W.; Wang, F.; Li, L.H.; Lu, J. Evolution law and attribution analysis of potential evapotranspiration in Liaoning Province. *Acta Ecol. Sin.* **2020**, *40*, 3519–3525.
52. Xu, N.P.; Pan, H.S.; Xu, Y.; Zhang, G.H. Climate change prediction of Heilongjiang Province in the next 30 and 50 years. *J. Nat. Disasters* **2004**, *1*, 146–150.
53. Zhang, J.T.; Feng, L.P.; Pan, Z.H. Climate change trend and mutation analysis in Heilongjiang Province in the next 41 years. *Meteorol. Environ.* **2014**, *37*, 60–66.

54. Wang, Q.; Zhang, M.J.; Pan, S.K.; Ma, X.N.; Li, F.; Liu, W.L. Spatial-temporal variation characteristics of potential evapotranspiration in the Yangtze River Basin. *Chin. J. Ecol.* **2013**, *32*, 1292–1302. (In Chinese)
55. Ning, T.; Li, Z.; Liu, W.; Han, X. Evolution of potential evapotranspiration in the northern Loess Plateau of China: Recent trends and climatic drivers. *Int. J. Climatol.* **2016**, *36*, 4019–4028. [CrossRef]
56. Hu, X.; Chen, M.; Liu, D.; Li, D.; Jin, L.; Liu, S.; Cui, Y.; Dong, B.; Khan, S.; Luo, Y. Reference evapotranspiration change in Heilongjiang Province, China from 1951 to 2018: The role of climate change and rice area expansion. *Agric. Water Manag.* **2021**, *253*, 106912. [CrossRef]

agriculture

MDPI

Article

Optimal Irrigation under the Constraint of Water Resources for Winter Wheat in the North China Plain

Xiaoli Shi [1,2,3,*], Wenjiao Shi [2,4,*], Na Dai [1,5] and Minglei Wang [2,4]

1 Hebei Key Laboratory of Environmental Change and Ecological Construction, School of Geographical Sciences, Hebei Normal University, Shijiazhuang 050024, China
2 Key Laboratory of Land Surface Pattern and Simulation, Institute of Geographic Sciences and Natural Resources Research, Chinese Academy of Sciences, Beijing 100101, China
3 Hebei Technology Innovation Center for Remote Sensing Identification of Environmental Change, Geocomputation and Planning Center, Hebei Normal University, Shijiazhuang 050024, China
4 College of Resources and Environment, University of Chinese Academy of Sciences, Beijing 100049, China
5 Bayannaoer City No. 1 Middle School, Bayannaoer 015000, China
* Correspondence: shixl@mail.hebtu.edu.cn (X.S.); shiwj@lreis.ac.cn (W.S.)

Abstract: The North China Plain (NCP) has the largest groundwater depletion in the world, and it is also the major production area of winter wheat in China. For sustainable food production and sustainable use of irrigated groundwater, it is necessary to optimize the irrigation amount for winter wheat in the NCP. Previous studies on the optimal irrigation amount have less consideration of the groundwater constraint, which may result in the theoretical amount of optimal-irrigation exceeding the amount of regional irrigation availability. Based on the meteorological data, soil data, crop variety data, and field management data from field experimental stations of Tangshan, Huanghua, Luancheng, Huimin, Nangong, Ganyu, Shangqiu, Zhumadian and Shouxian, we simulated the variation of yield and water use efficiency (WUE) under different irrigation levels by using the CERES-Wheat model, and investigated the optimal irrigation amount for high yield (OI_y), water saving (OI_{WUE}), and the trade-off between high yield and water saving (OI_t) of winter wheat in the NCP. Based on the water balance theory, we then calculated the irrigation availability, which was taken as the constraint to explore the optimal irrigation amount for winter wheat in the NCP. The results indicated that the OI_y ranged from 80 mm to 240 mm, and the OI_{WUE} was 17% to 67% less than OI_y, ranging from 0 mm to 200 mm. The OI_t was between 80 mm and 240 mm, realizing the co-benefits of high yield and water saving. Finally, we determined the optimal irrigation amount (62–240 mm) by the constraint of irrigation availability. Our results can provide a realistic and scientific reference for the security of both grain production and groundwater use in the NCP.

Keywords: optimal irrigation; sustainable irrigation; yield; water use efficiency; North China Plain

Citation: Shi, X.; Shi, W.; Dai, N.; Wang, M. Optimal Irrigation under the Constraint of Water Resources for Winter Wheat in the North China Plain. *Agriculture* **2022**, *12*, 2057. https://doi.org/10.3390/agriculture12122057

Academic Editor: David Maxwell Freebairn

Received: 4 November 2022
Accepted: 28 November 2022
Published: 30 November 2022

1. Introduction

As one of the major production areas, the North China Plain (NCP) produces 60–80% of the winter wheat in China [1]. For precipitation only meets 30–45% of the water requirement of winter wheat during the growing season, the increased yield has largely relied on unsustainable overuse of water resources in the NCP, especially groundwater [2]. Local farmers irrigate the winter wheat several times, resulting in the largest groundwater depletion in the world. Consequently, a series of eco-environmental problems occurred, such as land surface subsidence, seawater intrusion, streamflow depletion, wetlands, and ecological damages, which seriously affected the material basis of food production in the NCP [3]. Therefore, for sustainable food production and to relieve the pressure of groundwater exploitation, it is necessary to optimize the irrigation amount for winter wheat in the NCP.

The main indicators of optimal irrigation amount are currently yield and water use efficiency (WUE). High yield is the direct objective of agricultural production and the most used indicator when evaluating the optimal irrigation amount. For example, Boumaza et al. [4] studied the optimal irrigation according to the maximum biomass production. Some studies applied the yield as the only indicator to assess the optimal irrigation amount, without considering the water utilization during crop growth, which caused high water consumption with high yield. Thus, some scholars combined yield and WUE to explore the optimal irrigation amount of crops. For example, based on the yield and WUE changes with irrigation, some researchers investigated the optimal irrigation amount for winter wheat in the Guanzhong Plain and the NCP respectively [5]. Compared with yield, the relationship between input water and output yield is considered in the WUE, which emphasizes the productivity of water resources, so the WUE is recommended as a common index to characterize water saving. However, most of the previous studies determined the optimal irrigation amount based on the variation patterns of yield and WUE with different irrigation levels, lacking a comparison with the water availability. Thus, the recommended optimal irrigation amount may be unsustainable if it exceeds the irrigation availability. In general, in addition to yield and WUE, the optimal irrigation amount should also focus on the water balance and the natural carrying capacity of the available water to truly realize water saving and sustainable food production.

Field experiments and model simulations are the main methods to investigate the optimal irrigation amount for crops. For example, based on a 16-year field experiment using seven irrigation schedules in the winter wheat-summer maize double cropping system, the optimal irrigation was determined in the Luancheng station of the NCP [6]. Field experiments can accurately evaluate the optimal irrigation; However, it is difficult to apply on a large spatial and temporal scale because of the differences in climate and soil conditions among regions [7]. In recent years, an increasing number of models have been employed in the evaluation of irrigation optimization [5,8]. The crop model can simulate the growth process of crops by setting different environmental parameters, which is a useful complement to the traditional field experiment and has good performance in regional applications [9]. Nevertheless, most of the current studies still simulated the relationship between indicators (yield, WUE) and irrigation level, lacking comparison with available irrigation water. In fact, the determination of optimal irrigation amount needs to account for both simulation results and constraints of water availability.

The objectives of the present analysis are: (1) to explore the optimal irrigation amount for high yield, water saving, and the trade-off between high yield and water saving of winter wheat in the NCP. (2) To calculate the irrigation availability in the NCP based on the water balance theory, and take it as the constraint to investigate the optimal irrigation amount for winter wheat in the NCP.

2. Materials and Methods

2.1. Study Area

The NCP is located in the east of China (114°–121° E, 32°–40° N), with a total area of 30×10^6 ha. It covers the Beijing, Tianjin Municipalities, Shandong Province, most of Hebei and Henan Province, and northern Anhui and Jiangsu Province (Figure 1). Prevailing soils in NCP are formed of fluvial materials from the Yellow River, which is fertile and favorable for cultivation. The climate of the NCP is temperate and monsoonal, with rainy hot summers and dry cold winters. The average annual temperature varies from 14 °C to 15 °C, and the average annual precipitation decreases from the southwest to the northeast, varying from 500 mm to 900 mm. In addition, 70% of the precipitation is concentered in the growing season of maize, while only 30% of the precipitation is in the growing season of winter wheat [10]. Water resources are merely 456 $m^3\ yr^{-1}$ per capita in NCP, which is below 1/7 of the national average and 1/24 of the world average [11]. As one of the most important agricultural areas of China, grain production, especially winter wheat, heavily relies on groundwater irrigation. With the continuous exploitation, the groundwater is experiencing rapid depletion, with a rate of 1.66–2.76 cm yr^{-1} during 2003–2020 in the NCP [12], a series of shallow groundwater depletion were formed in the pre-mountain plains, and relatively independent deep groundwater depletion appeared in the central-coastal plains. In this study, we chose nine representative experimental stations (Tangshan, Huanghua, Luancheng, Huimin, Nangong, Ganyu, Shangqiu, Zhumadian and Shouxian) from the NCP to explore the optimal irrigation.

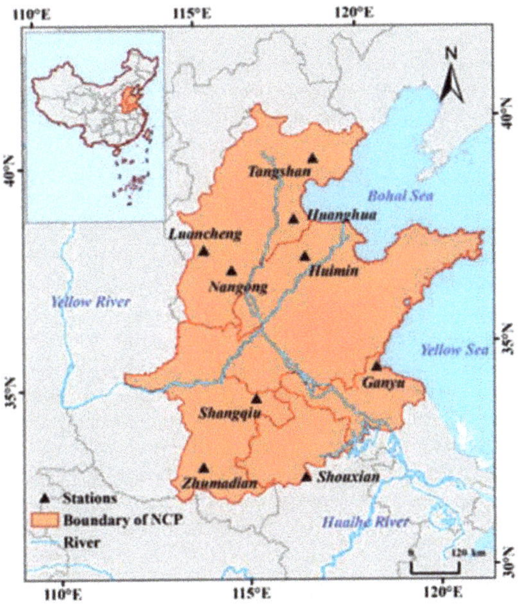

Figure 1. Location of the North China Plain.

2.2. Data

Meteorological data were obtained from the Chinese Meteorological Data Network (http://data.cma.cn/ accessed on 1 May 2020), including daily maximum temperature, daily minimum temperature, precipitation and sunlight hours from 1981 to 2017 (Figure 1). Based on the empirical formula developed by Angstrom [13], daily solar radiation was estimated by sunlight hours.

Soil data were obtained from the Chinese Soil Science Database (http://vdb3.soil.csdb.cn/ accessed on 1 May 2020), including soil type, color, soil depth, organic carbon content, soil texture, total nitrogen content, bulk density, and pH (Table 1), where organic carbon content was obtained by multiplying soil organic matter by a conversion factor of 0.58 [14].

Table 1. Soil parameters in the North China Plain.

Soil Parameters	Range
Bulk density (g cm^{-3})	0.04–1.83
Field capacity (cm^3 cm^{-3})	0.11–0.56
Wilting point (cm^3 cm^{-3})	0.04–0.48
Soil organic carbon (%)	0.01–30.64
pH	4.0–10.1
Total nitrogen content (%)	0.00–1.10

Field management data mainly include observation of phenology and yield components of winter wheat at each station, as well as the field management measures, such as varieties, irrigation, and fertilization. The field management data for 1982–2017 were from experiments conducted at the national agrometeorological stations, which are maintained by the Chinese Meteorological Administration. For the phenology, the dates of sowing (BBCH 00), emergence (BBCH 10), dormancy (start of dormancy), green up (end of dormancy), anthesis (BBCH 61) and maturity (BBCH 89) were recorded [15]. For the yield, the spikelet number, the infertile spikelet rate, the grain number, the thousand grain weight, and the plot grain yield were recorded. For irrigation and fertilization, the dates and quantities were recorded.

Groundwater data were obtained from the Atlas of Groundwater Resources and Environment in China [16], surface water resource availability and precipitation data were obtained from the Water Resources Reports of provinces and cities in the NCP, as well as the Water Resources Reports of the Haihe River, the Yellow River, and the Huaihe River from 2011 to 2014.

2.3. *Methods*

2.3.1. DSSAT Model

DSSAT is one of the widely used crop models, and CERES-Wheat is the sub-model developed specifically for wheat [17]. The model runs on DSSAT-CSM (cropping system model) public platform as a module, and uses the meteorological and soil database and soil moisture, nitrogen, and carbon balance module to simulate wheat growth and development, yield formation, nitrogen-carbon-water balance process [17]. The simulation process involves light interception and photosynthesis, nutrient absorption and root activity, dry matter distribution, water absorption and transpiration, growth and respiration, leaf area growth, development and organ formation, senescence, field management measures, etc. There are four parts of data needed for the operation of the CERES–Wheat model, which are meteorological data, soil data, crop variety data, and field management data. This model has a complete consideration of soil water balance processes and physical mechanisms, it has been validated for the assessment of wheat production [18].

After the data were prepared, we calibrated the model by using field experimental data for the nine stations. After calibration, we validated the model by comparing the observed data on phenology and yield during the other years (Table 2).

Table 2. Experimental dataset for CERES-Wheat calibration and validation at nine stations.

Station	Calibration Dataset	Validation Dataset
Tangshan	1997–2000	2004–2007
Huanghua	1995–1999	2004–2009
Luancheng	2001	2002, 2003
Huimin	2001–2004	2005–2009
Nangong	2002, 2003	2004–2005
Ganyu	2005, 2006	2007–2008
Shangqiu	1997	1998
Zhumadian	1987, 1989, 1990	1993–1995
Shouxian	2006	2007

Then we employed the Normalized Root Mean Square Error (NRMSE) and Consistency Index (D value) to check the agreement between observed and simulated values. The formulas are as follows:

$$\mathrm{NRMSE} = \frac{\sqrt{\sum_{i=1}^{n}(S_i - R_i)^2/n}}{\overline{R}} \times 100\% \tag{1}$$

$$D = 1 - \left[\frac{\sum_{i=1}^{n}(S_i - R_i)^2}{\sum_{i=1}^{n}\left(|S_i + \overline{R}| + |R_i - \overline{R}|\right)^2}\right] \tag{2}$$

where S_i is the simulated value; R_i is the observed value; $\overline{R}$ is the average of observed values; n is the sample size. The simulation is excellent if the NRMSE is less than 10%; the simulation is generally accurate if the NRMSE is between 20% and 30%; the simulation is poor if the NRMSE is greater than 30% [19]. The D value ranges from 0 to 1, a value closer to 1 indicates perfect agreement between the observed value and simulated value, while a value closer to 0 indicates poor predictability [20].

After validation, we used the CERES-Wheat model to simulate the yield and WUE under different irrigation levels (from 40 mm to 320 mm, with an interval gradient of 40 mm) for 36 years (1982–2017) at nine representative stations in the NCP, to explore the optimal amount of irrigation for high yield and water saving of the winter wheat. For each station, the irrigation dates and frequency were set according to the observed irrigation management from experiments conducted at the nine agrometeorological stations, meanwhile, based on the constraint of the total amount of given irrigation levels, the irrigation amount was set according to the percentage of the observed irrigation management from experiments on the nine agrometeorological stations.

2.3.2. Optimal Irrigation Amount for High Yield of the Winter Wheat (OI_y)

The yield was directly simulated by the CERES-Wheat model, which was set to be affected by irrigation only, the nutrients, pests, and other factors were not taken into account. The optimal irrigation amount for high yield was the irrigation value corresponding to the maximum yield of the winter wheat.

2.3.3. Optimal Irrigation Amount for Water Saving of the Winter Wheat (OI_{WUE})

WUE was derived from the ratio of yield to evapotranspiration. It was used to evaluate optimum irrigation management to ensure the most efficient use of water resources. The formula is as follows:

$$WUE = \frac{Y}{ET} \tag{3}$$

where WUE is the water use efficiency; Y is the yield of winter wheat; ET is the evapotranspiration. The yield and evapotranspiration were based on the simulation from the CERES-Wheat model. The optimal irrigation amount for water saving was the irrigation value corresponding to the maximum WUE of winter wheat.

2.3.4. Optimal Irrigation Amount for the Trade-Off between High Yield and Water Saving of the Winter Wheat (OI_t)

The optimal amount of irrigation for high yield was not necessarily equal to that value for water saving. The trade-off between high yield and water saving should be considered if we want to complete the two objectives simultaneously. Using the method of Zheng et al. [21], yield and WUE at different gradients of irrigation levels at each station were standardized, and the irrigation amount that achieved the maximum value of normalized yield and WUE was determined as the optimal irrigation amount for the trade-off between the high yield and water saving. The formula is as follows:

$$\lim_{I \to OI_t} \left(\frac{Y}{Y_{max}} + \frac{WUE}{WUE_{max}} \right) = 2 \tag{4}$$

where Y and WUE are the yield and WUE simulated by CERES-Wheat model under different irrigation gradients; Y_{max} and WUE_{max} are the maximum values of yield and WUE over the growing season at each station, respectively; OI_t is the optimal irrigation amount for the trade-off between high yield and water saving of the winter wheat.

2.3.5. Optimal Irrigation Amount Constrained by the Irrigation Availability (OI)

The calculation of irrigation availability was referred to Lei et al. [22], with the following equations:

$$W_i = W_s + W_g \times k - W_d \tag{5}$$

$$W_d = \rho \times W_i \quad (6)$$

where W_i is the available irrigation amount; W_s is the available surface water; W_g is the amount of exploitable coefficient groundwater; k is the proportion of agricultural water to groundwater; W_d is the duplication of surface water and groundwater; ρ is the exploitable coefficient, which is the ratio of exploitable groundwater to the total amount of groundwater; W_i is the infiltration. Since deep groundwater has few recharge sources, a long recovery period, and a very slow renewal rate, it can be regarded as a non-renewable resource. Therefore, the deep groundwater is not suitable for irrigation and not included in the calculation.

For the purpose of water saving and sustainable agricultural production, irrigation of winter wheat should use only the available irrigation without over-exploitation of groundwater. Taking irrigation availability as a constraint, the final optimal irrigation amount was determined by comparing the OI_t with the available irrigation amount at the corresponding station. If the available irrigation amount is greater than or equal to the OI_t, which indicates that the water resources can support the OI_t to complete high yield and water saving simultaneously, then take the OI_t as the final optimal irrigation amount. If the available irrigation amount is less than the OI_t, then the production of winter wheat is irrigation constrained, the final optimal irrigation amount should be determined according to the yield and WUE changes with irrigation levels, as well as the available irrigation amount.

3. Results

3.1. Validation of CERES-Wheat Model

For the anthesis of winter wheat in the NCP, the simulated anthesis (from 172 days to 227 days after sowing) was very close to the observed values (from day 176 days to 223 days after sowing), the simulated anthesis agreed well with the observed anthesis with NRMSE of 1.51% and D value of 0.98 (Table 3). Similarly, simulated maturity (from 209 days to 257 days after sowing) was well matched with the observations (from 210 days to 257 days after sowing), with NRMSE of 0.95% and D value of 0.99, respectively. Although the NRMSE (14.89%) and D value (0.96) of the yield were not as high as the values of anthesis and maturity, the observed yield ranged from 1650 kg ha^{-1} to 7395 kg ha^{-1}, and the simulated yield ranged from 1030 kg ha^{-1} to 7577 kg ha^{-1} (Table 3), the model also exhibited the agreement between the simulation and observation. Overall, the stability and accuracy of the calibrated model were confirmed by the above evaluation, especially for the anthesis and maturity. The calibrated model can be applied to simulate the yield and WUE of winter wheat in response to irrigation management in the NCP.

Table 3. Validation of simulation on anthesis, maturity, and yield of winter wheat in the North China Plain.

Item	Observed Value	Simulated Value	NRMSE	D Value
Anthesis (BBCH 61)	176–223 d after sowing	172–227 d after sowing	1.51%	0.98
Maturity (BBCH 89)	210–257 d after sowing	209–257 d after sowing	0.95%	0.99
Yield	1650–7395 kg ha^{-1}	1030–7577 kg ha^{-1}	14.89%	0.96

Notes: The NRMSE is the Normalized Root Mean Square Error between the observed values and simulated values; D value is the Consistency Index between the observed values and simulated values; BBCH is the Biologische Bundesanstalt, Bundessortenamt, Chemische Industrie. This code is recommended for phenological observations.

3.2. Optimal Irrigation Amount for High Yield of Winter Wheat (OI_y)

Low yield occurred in all stations with rainfed management, the range of no irrigation yields was from 2411 kg ha^{-1} to 7679 kg ha^{-1} (Figure 2). For each station in the NCP, yield exhibited an increasing trend with the increase in irrigation. However, when irrigation reached a certain level, the yield change leveled off, indicating that the high yield of winter wheat greatly relied on irrigation in the NCP (Figure 2). Conversely, excessive water was not beneficial to yield increase due to the Law of Diminishing Returns. For the maximum yield of winter wheat, Nangong and Tangshan stations required more irrigation (240 mm) than other stations, the maximum yield could reach 10,917 kg ha^{-1} and 9346 kg ha^{-1}, respectively. Luancheng, Huimin and Huanghua stations required an equal amount of irrigation (160 mm), the maximum yields at the three stations were 8669 kg ha^{-1}, 8334 kg ha^{-1} and 5856 kg ha^{-1}, respectively. Whereas the amounts of irrigation required at Ganyu and Zhumadian stations were relatively small (120 mm), and the highest yields were 8851 kg ha^{-1} and 7333 kg ha^{-1}, respectively. Shouxian and Shangqiu stations had the lowest optimal irrigation (80 mm) among the nine stations, their maximum yields were 8260 kg ha^{-1} and 6912 kg ha^{-1}, respectively.

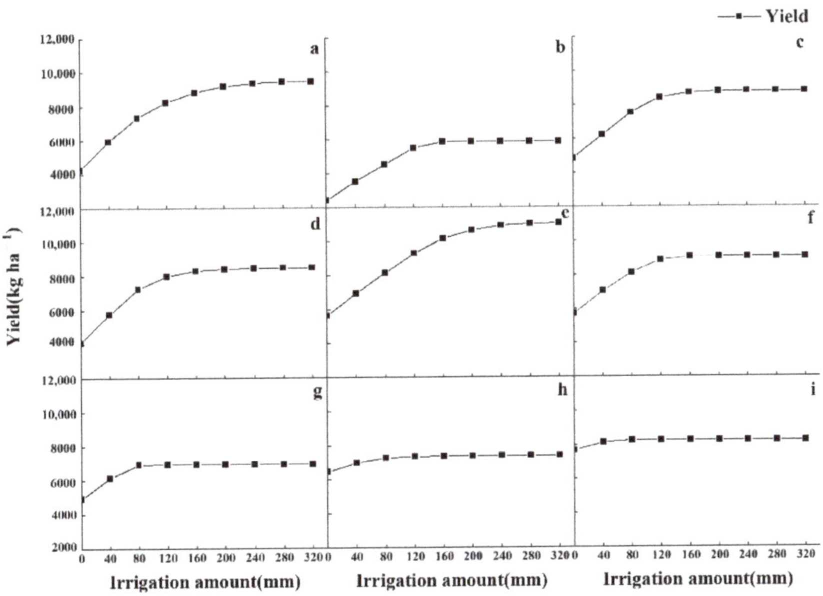

Figure 2. Changes in winter wheat yield with irrigation levels in the North China Plain. (**a**) Tangshan, (**b**) Huanghua, (**c**) Luancheng, (**d**) Huimin, (**e**) Nangong, (**f**) Ganyu, (**g**) Shangqiu, (**h**) Zhumadian, (**i**) Shouxian.

Spatially, taking the Yellow River as the dividing line, the optimal irrigation amounts in the northern five stations (Tangshan, Huanghua, Luancheng, Huimin, and Nangong stations) were all greater than 160 mm (Figure 2). Nevertheless, the range of optimal irrigation amounts in the southern four stations (Ganyu, Shangqiu, Zhumadian, and Shouxian stations) was from 80 mm to 120 mm, the difference between the southern and northern stations probably due to the spatial variability of precipitation, which drove the variation in water demand for irrigation.

3.3. Optimal Irrigation Amount for Water Saving of the Winter Wheat (OI_{WUE})

For each station in the NCP, similarly to the trend of yield change, the WUE increased gradually with the irrigation level, and then leveled off, stopped rising, or even started to decline after reaching a certain threshold (Figure 3). WUE decreased with more irrigation because the increase in irrigation was greater as compared to the increase in yield. For the maximum WUE of winter wheat, Nangong and Tangshan stations demanded the highest irrigation amount in the region (200 mm), and their maximum WUEs were 2.4 kg m^{-3} and 2.2 kg m^{-3}, respectively. When the winter wheat acquired 160 mm irrigation in Huanghua station, its WUE reached the maximum (1.6 kg m^{-3}). In addition, the irrigation amount was required for 120 mm in Luancheng, Huimin and Ganyu stations for the optimal WUE, the maximum WUE of winter wheat was 2.3 kg m^{-3}, 2.3 kg m^{-3} and 1.9 kg m^{-3}, respectively. While the maximum WUE of winter wheat was 2.0 kg m^{-3} and 1.9 kg m^{-3} at Shangqiu and Zhumadian stations, which required relatively less irrigation, 80 mm and 40 mm, respectively.

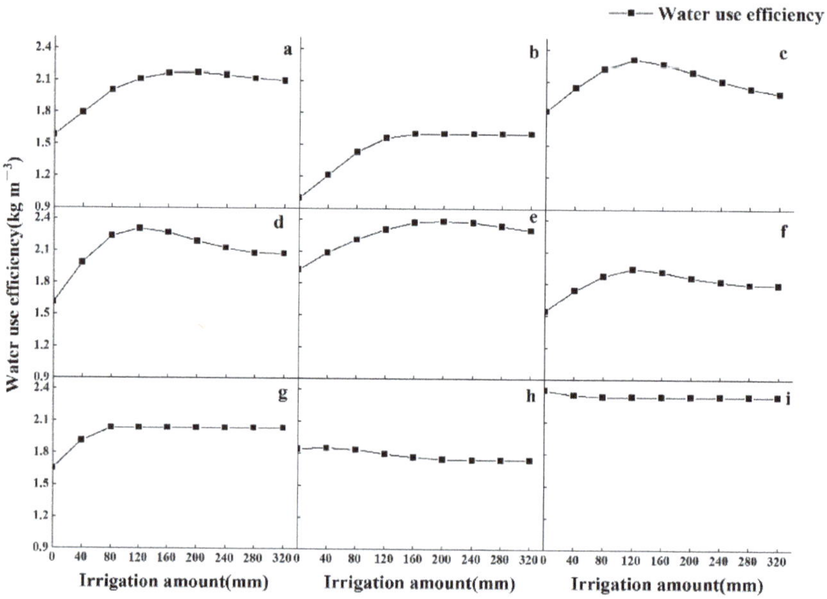

Figure 3. Changes in water use efficiency of winter wheat with irrigation levels in the North China Plain. (**a**) Tangshan, (**b**) Huanghua, (**c**) Luancheng, (**d**) Huimin, (**e**) Nangong, (**f**) Ganyu, (**g**) Shangqiu, (**h**) Zhumadian, (**i**) Shouxian.

The optimal irrigation amount for water saving (OI_{WUE}) had a similar spatial pattern to the optimal irrigation amount for high yield (OI_y), OI_{WUE} in the northern stations varied from 120 mm to 200 mm, while the OI_{WUE} range in southern stations was between 0 mm and 120 mm. For the southern station, the abundant precipitation can satisfy the water demand of winter wheat, so the increase in irrigation amount significantly decreased the WUE.

3.4. Optimal Irrigation Amount for the Trade-Off between High Yield and Water Saving (OI_t)

In terms of the relationship among OI_y, OI_{WUE}, and OI_t in each station, values of the three types of optimal irrigation amount were coincidentally equal in Huanghua, Ganyu, and Shangqiu stations, thus, the OI_t in Huanghua, Ganyu, and Shangqiu stations were 160 mm, 120 mm and 80 mm, respectively. With this irrigation management, these stations can complete the perfect combination of high yield and water saving (Figure 4). For the remaining six stations, the OI_y values were all greater than the OI_{WUE} values. After calculation, the OI_t values of Tangshan station (240 mm), Luancheng station (160 mm), Huimin station (160 mm), Nangong station (240 mm), and Shouxian station (80 mm) were all equal to their own OI_y. Thus, the trade-offs in these stations were at the expense of falling WUE to some extent. Nevertheless, Zhumadian station was a special case, for the OI_t (80 mm) was between the OI_y (120 mm) and the OI_{WUE} (40 mm), its trade-off was the optimal configuration with losses of the maximum yield and WUE (Figure 4).

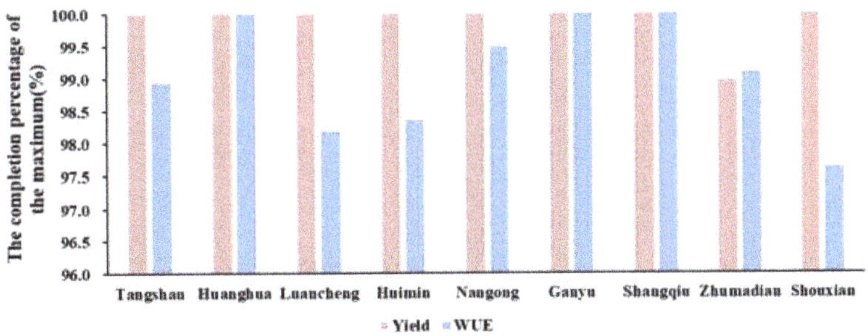

Figure 4. The completion percentage of the maximum yield and water use efficiency (WUE) based on the OI_t.

In terms of the spatial pattern, the OI_t values in the four southern stations (Ganyu, Shangqiu, Zhumadian and Shouxian stations) were all no greater than 120 mm, while the OI_t values of the five northern stations (Tangshan, Huanghua, Luancheng, Huimin and Nangong stations) were all greater than these in the southern stations, varied from 160 mm to 240 mm (Figure 5). The OI_t showed a similar spatial pattern to the OI_y and OI_{WUE}.

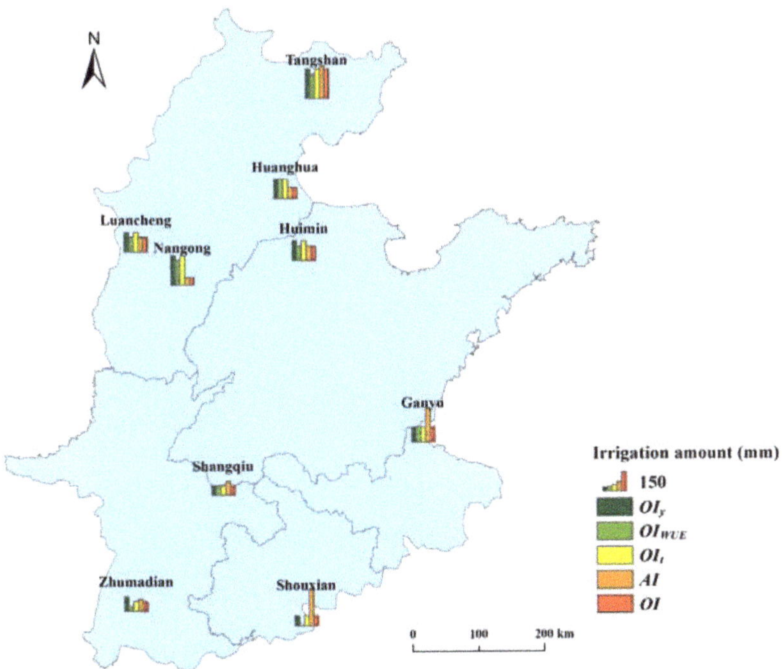

Figure 5. The optimal irrigation amount and the available irrigation amount of winter wheat in the North China Plain. Notes: OI_y is the optimal irrigation amount for a high yield of winter wheat; OI_{WUE} is the optimal irrigation amount for water saving of winter wheat; OI_t is the optimal irrigation amount for the trade-off between high yield and water saving; AI is the available irrigation amount; OI is the final optimal irrigation amount constrained by the available irrigation amount.

Under the irrigation management of OI_t, eight stations can maintain 100% of the maximum yield except for Zhumadian station, the loss percentage of yield in Zhumadian station was only 1.04%. In addition to the 100% of the maximum yield, Huanghua, Ganyu, and Shangqiu stations can simultaneously complete 100% of the maximum WUE. Besides, the completion percentages of the maximum WUE in the Nangong and Zhumadian stations were greater than 99%, and the values in Tangshan, Luancheng, and Huimin stations were all higher than 98%. Although Shouxian station had the highest loss percentage of the maximum WUE, it was still less than 2.5%. In conclusion, the OI_t contributed to the best yield and WUE of winter wheat for all the nine stations in the NCP, the OI_t can maximize the co-benefits of high yield and water saving to the greatest extent possible, it is a suitable choice for considering both yield and efficiency.

3.5. Optimal Irrigation Amount Constrained by the Irrigation Availability (OI)

Spatially, the irrigation availability in the NCP was generally high in the south and low in the north. Among these stations, Nangong and Huanghua stations had the least available irrigation amount (62 mm and 91 mm), because they are in the Heilonggang area, which has the largest groundwater depletion in the world. Tangshan station (267 mm) in the coastal plain of eastern Hebei Province, Ganyu station (273 mm) in the low plain of Xu-Huai of Jiangsu Province, and Shouxian station (300 mm) in the plain of northern Anhui Province had enough available water resources, they were all greater than 200 mm. The available irrigation amounts at the remaining stations (Zhumadian, Shangqiu, Luancheng, and Huimin stations) varied from 100 mm to 121 mm (Figure 6).

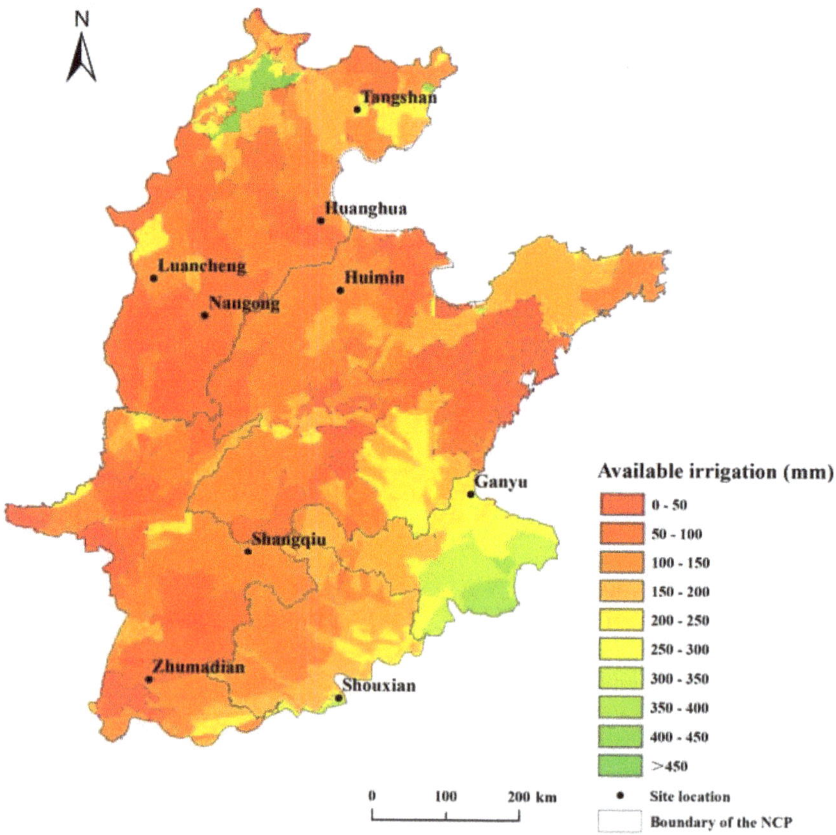

Figure 6. The irrigation availability of winter wheat in the North China Plain.

Compared with the OI_t, the irrigation availability was abundant for the southern stations (Ganyu, Shangqiu, Zhumadian, and Shouxian stations) and Tangshan station, so the final optimal irrigation amount (OI) was equal to the OI_t in the five stations (Figure 5), that is, the OI of the Tangshan and Ganyu stations was 240 mm and 120 mm, respectively, Shangqiu, Zhumadian, and Shouxian stations had the consistent value of OI (80 mm). Nevertheless, four stations in the northern region (Huanghua, Luancheng, Huimin, and Nangong stations) are now in the area of serious groundwater over-exploitation, the available irrigation amount was less than their own OI_t, that is, the water resources cannot supply sufficient irrigation for winter wheat to complete the trade-off between high yield and water saving, so the final optimal irrigation should have binding constraints on water availability. After calculation, the Nangong and Huanghua stations had the lowest OI, 62 mm and 91 mm, respectively; Luancheng (120 mm) and Huimin (121 mm) stations had close values in final optimal irrigation. Due to the differences in groundwater constraints, there was no obvious spatial pattern of optimal irrigation amount.

With the final optimal irrigation amount, for the yield, Tangshan, Ganyu, Shangqiu, and Shouxian stations can complete 100% of the maximum yield; the completion percentages of Luancheng, Huimin, and Zhumadian stations were all greater than 96%; while the Nangong and Huanghua stations had the highest loss of yield, they can only complete 69.72% and 81.87% of the maximum yield, respectively (Figure 7). For the WUE, Luancheng, Huimin, Ganyu, and Shangqiu stations can obtain 100% of the maximum WUE, Nangong and Huanghua stations still had the lowest completion percentage (90.46% and 91.58%), completion percentages of the rest of stations were all greater than 97.5%. In general, with this irrigation management, Ganyu and Shangqiu stations can simultaneously complete 100% of yield and WUE optimization; Tangshan, Luancheng, Huimin, Shouxian, and Zhumadian stations had a slight loss in the yield or WUE. The above seven stations did not lose much yield or efficiency within the irrigation constraints. Nevertheless, the yield and WUE at Nangong and Huanghua stations were greatly limited by water availability. Consequently, appropriate fallowing or adjustment of cropping systems is recommended to protect the local water ecosystem.

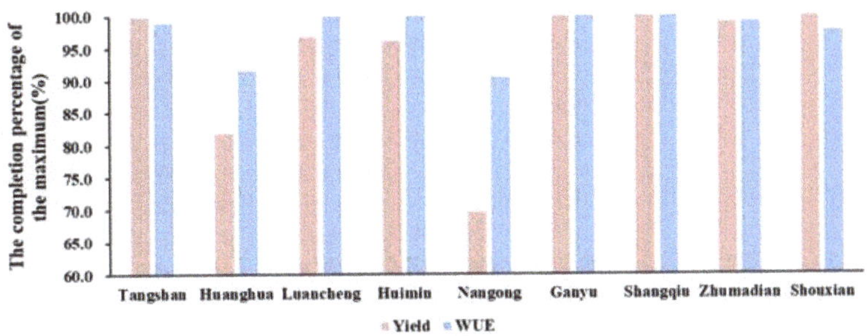

Figure 7. The completion percentage of the maximum yield and water use efficiency (WUE) based on the OI.

4. Discussion

In the NCP, irrigation is of paramount importance to increasing productivity for winter wheat, investigation of the optimal irrigation amount should consider the natural carrying capacity of water resources. In this study, we simulated the variation of yield and WUE with different irrigation levels by using the CERES-Wheat model at nine representative stations in the NCP. Then we determined the optimal irrigation amount for high yield, water saving, and the trade-off between high yield and water saving of winter wheat. Subsequently, based on the equilibrium relationship between irrigation demand and the natural carrying capacity of water resources, we investigated the optimal irrigation amount under the constraint of irrigation availability. The optimization of irrigation strategies is beneficial for wheat production and water conservation in the NCP.

Water is a dominant driver affecting yield, and it has been demonstrated that winter wheat yield had a non-linear relationship with irrigation amount, excessive irrigation did not contribute to a continuous increase in yield [23], and our results also supported that. The yield growth may be inhibited by the decrease in soil permeability due to excess water and the lack of oxygen in the root system of winter wheat. Aiming at a high yield, the optimal irrigation amount for winter wheat at nine stations in the NCP ranged from 80 mm to 240 mm. This was consistent with the field experimental results in Xinxiang station, which revealed the 175–180 mm optimal irrigation for a high yield of winter wheat in the NCP [24].

In addition to yield, we chose WUE to investigate the optimal irrigation amount for winter wheat. The WUE showed an increasing trend and then decreased with the irrigation increase. When the winter wheat is stressed by the water, WUE increases with irrigation increase, while irrigation exceeds a certain threshold, winter wheat evapotranspiration no longer changes significantly, and excessive irrigation will leach into the ground or form surface runoff. Taking WUE as the criterion, Zhang et al. [25] concluded that the optimal irrigation amount of winter wheat in the NCP ranged from 60 mm to 140 mm. Wang et al. [26] suggested that the optimal irrigation amount at Beijing station was between 192 mm and 245 mm. In the study of Ma et al. [27], the optimal irrigation amount for winter wheat in the NCP was between 60 mm and 300 mm, being 70–210 mm at the Luancheng station. Our finding also suggested that the optimal irrigation amount for winter wheat varied from 0 mm to 200 mm, with 120 mm at the Luancheng station.

The OI_t aimed at the unification and optimization of maximum yield and WUE. It has been shown that the maximum reached by yield and WUE with irrigation level did not overlap [28]. In this study, we also concluded that WUE was maximized before yield. Compared to the OI_y, the OI_{WUE} was reduced by 17% to 67%, indicating the water-saving trend. After the trade-off, the OI_t contributed to the best yield and WUE for winter wheat for all nine stations in the NCP, varying from 80 mm to 240 mm. Similar conclusions can be found in previous studies. Based on the APSIM model, Zheng et al. [21] investigated that the OI_t ranged from 3 mm to 286 mm. Moreover, the OI_t of the Luancheng station in another study (202 mm) also had a close value to our study (160 mm) [29].

We considered the sustainable use of groundwater as an upper bound for winter wheat irrigation in this study. Results showed that the available water resources for irrigation exhibited a distribution trend of high in the south and low in the north, revealing a consistent spatial pattern to Lei et al. [22]. With the rapid depletion of groundwater in the NCP, agricultural production has entered a dilemma of food security and water security. This study showed that the irrigation-constrained areas were mainly concentrated in the groundwater over-exploitation area (especially the Heilonggang area). For the regions suffering from water limitations, we should reduce the agricultural intensification, and carry out appropriate fallowing or adjustment of cropping systems to alleviate the groundwater crisis. Meanwhile, for the regions with abundant water resources for irrigation, which were mainly located in the northern Anhui Plain and the low Xu-Huai Plain in the southern part of the NCP, the potential production needs to be maximized to ameliorate the yield losses from the water-scarce regions [30].

Adjustment of the cropping systems is an effective measure to save valuable groundwater for sustainability in over-exploited aquifers. A study conducted by Davis et al. [31] showed that the optimization of cropping structure would save 12–14% of irrigation water at a global scale. Thus, the Chinese government launched the Seasonal Land Fallowing Policy in the NCP in 2014, designed to mitigate serious groundwater over-exploitation. Analysis showed that the policy reduced groundwater consumption and contributed to real water saving [32]. Besides, similar irrigation strategies had been adopted to mitigate groundwater depletion, for example, deficit irrigation [8], drip irrigation systems under the plastic mulch [33], micro-sprinkling [34], and adjustment of planting density [35]. A study concluded that combining all agricultural management could reduce groundwater exploitation intensity by around 74.58% to 96.95%, resulting that groundwater could recover to the original health level nearly in the NCP [11].

In addition to promoting water conservation technologies and implementing the relevant water use policies from the demand side, on the supply side, it encourages water users to replace groundwater with surface water delivered by the central South-to-North Water Diversion, to alleviate the water stress and groundwater storage deficit in the NCP. A study conducted by Long et al. [36] found, within the context of climate variability and policy implications, water diverted to Beijing reduces cumulative groundwater depletion by 3.6 km^3, accounting for 40% of total groundwater storage restoration between 2006 and 2018, meanwhile, increased precipitation and policy-induced contributed about 2.7 km^3 (~30%) and 2.8 km^3 (~30%) to the groundwater storage recovery [36]. It proves the important role of South-to-North Water Diversion in groundwater restoration. Overall, more efforts need to be explored to save valuable groundwater for the sustainability of irrigated agriculture in the NCP.

Our analysis was subject to considerable uncertainties. Except for the irrigation management, the rest of the cultivation conditions were assumed as optimal in our study. In general, climate, soil, cultivation practices, pests, and diseases all have influences on the optimal irrigation amount of winter wheat. Thus, the optimal irrigation of winter wheat can be further analyzed based on specific considerations of regional differences in cultivation. Furthermore, the results of the optimal irrigation should be applied to the field for supporting evidence, in order to better guide practice for local farmers. In addition, the optimal irrigation amount can be determined by considering the greenhouse gas emissions [37], soil moisture, and carbon footprint [38], in combination with the precipitation pattern, to make more targeted recommendations in the future.

5. Conclusions

For the NCP, the OI_y ranged from 80 mm to 240 mm, and the OI_{WUE} varied from 0 mm to 200 mm. After the trade-off between yield and WUE, the OI_t was between 80 mm and 240 mm, which maximized the co-benefit of high yield and water saving.

The available irrigation amount varied from 62 mm to 300 mm in the NCP. Generally, southern stations had higher water availability than northern stations, and stations located in the Heilonggang area (Nangong and Huanghua stations) had the lowest irrigation availability. As the region has the most severe groundwater depletion in the world, the optimal irrigation amount needs to be constrained by water availability. Based on the equilibrium relationship between irrigation and the natural carrying capacity of water resources, we determined the final optimal irrigation amount (62–240 mm) by the irrigation availability constraint. Yield and WUE were greatly affected in Huanghua and Nangong stations with the final optimal irrigation amount; however, the rest stations can maintain more than 96% of the maximum yield and 97.5% of the maximum WUE and complete the optimization of both high productivity and water saving.

For the water-scarce regions, the irrigation availability cannot support the optimization of yield and WUE, and it is recommended to moderately fallow, deficit irrigation, or implement appropriate cropping system adjustments, as well as use alternative water resources to ensure water security. Meanwhile, we should strengthen the production potential of the southern part of the NCP and appropriately consider the global wheat trade market for capacity substitution to maintain the sustainable use of regional water resources and grain production.

Author Contributions: Conceptualization, X.S. and W.S.; methodology, X.S. and W.S.; software, N.D. and M.W.; validation, N.D. and M.W.; visualization, N.D. and M.W.; writing—original draft preparation, X.S.; writing—review and editing, W.S.; supervision, X.S. and W.S.; funding acquisition, W.S. All authors have read and agreed to the published version of the manuscript.

Funding: This research was funded by the National Key Research and Development Program of China (2022YFB3903504), the Strategic Priority Research Program of the Chinese Academy of Sciences (Grant No. XDA20010202 and No. XDA23100202), and the Science Fund for Creative Research Groups of the National Natural Science Foundation of China (Grant No. 72221002).

Institutional Review Board Statement: Not applicable.

Data Availability Statement: Not applicable.

Conflicts of Interest: The authors declare no conflict of interest.

References

1. Lei, M.; Zhang, Y.Q.; Dang, Y.X.; Kong, X.B.; Yao, J.T. Does crop rotation enhance groundwater health? A review of the Winter Wheat Fallow Policy in the North China Plain. *Water* **2019**, *11*, 2416. [CrossRef]
2. Liu, H.J.; Yu, L.P.; Luo, Y.; Wang, X.P.; Huang, G.H. Responses of winter wheat evapotranspiration and yield to sprinkler irrigation regimes. *Agric. Water Manag.* **2011**, *98*, 483–492. [CrossRef]
3. Sun, H.Y.; Shen, Y.J.; Yu, Q.; Flerchinger, G.N.; Zhang, Y.Q.; Liu, C.M.; Zhang, X.Y. Effect of precipitation change on water balance and WUE of the winter wheat-summer maize rotation in the North China Plain. *Agric. Water Manag.* **2010**, *97*, 1139–1145. [CrossRef]
4. Boumaza, K.; Kalboussi, N.; Rapaport, A.; Roux, S.; Sinfort, C. Optimal control of a crop irrigation model under water scarcity. *Optim. Contr. Appl. Met.* **2021**, *42*, 1612–1631. [CrossRef]
5. Zhang, C.; Xie, Z.A.; Wang, Q.J.; Tang, M.; Feng, S.Y.; Cai, H.J. AquaCrop modeling to explore optimal irrigation of winter wheat for improving grain yield and water productivity. *Agric. Water Manag.* **2022**, *266*, 107580. [CrossRef]
6. Yang, X.L.; Wang, G.Y.; Chen, Y.Q.; Sui, P.; Pacenka, S.; Steenhuis, T.S.; Siddique, K.H.M. Reduced groundwater use and increased grain production by optimized irrigation scheduling in winter wheat-summer maize double cropping system-A 16-year field study in North China Plain. *Field Crop. Res.* **2022**, *275*, 108364. [CrossRef]
7. Xu, X.X.; Zhang, M.; Li, J.P.; Liu, Z.Q.; Zhao, Z.G.; Zhang, Y.H.; Zhou, S.L.; Wang, Z.M. Improving water use efficiency and grain yield of winter wheat by optimizing irrigations in the North China Plain. *Field Crop. Res.* **2018**, *221*, 219–227. [CrossRef]
8. Kheir, A.M.S.; Alrajhi, A.A.; Ghoneim, A.M.; Ali, E.F.; Magrashi, A.; Zoghdan, M.G.; Sedhom, A.M.; Abdelkhalik, F.; Ahmed, E.; Fahmy, G.; et al. Modeling deficit irrigation-based evapotranspiration optimizes wheat yield and water productivity in arid regions. *Agric. Water Manag.* **2021**, *256*, 107122. [CrossRef]
9. Aluoch, S.O.; Li, Z.T.; Li, X.X.; Hu, C.S.; Mburu, D.M.; Yang, J.Y.; Xu, Q.S.; Yang, Y.Z.; Su, H.X. Effect of mineral N fertilizer and organic input on maize yield and soil water content for assessing optimal N and irrigation rates in Central Kenya. *Field Crop. Res.* **2022**, *277*, 108420. [CrossRef]
10. Wang, Y.Q.; Zhang, Y.H.; Zhang, R.; Li, J.P.; Zhang, M.; Zhou, S.L.; Wang, Z.M. Reduced irrigation increases the water use efficiency and productivity of winter wheat-summer maize rotation on the North China Plain. *Sci. Total Environ.* **2018**, *618*, 112–120. [CrossRef]
11. Chen, X.Z.; Wang, P.X.; Muhammad, T.; Xu, Z.C.; Li, Y.K. Subsystem-level groundwater footprint assessment in North China Plain–The world's largest groundwater depression cone. *Ecol. Indic.* **2020**, *117*, 106662. [CrossRef]
12. Liu, R.L.; Zhong, B.; Li, X.P.; Zheng, K.Y.; Liang, H.; Cao, J.M.; Yan, X.; Lyu, H.X. Analysis of groundwater changes (2003–2020) in the North China Plain using geodetic measurements. *J. Hydrol.-Reg. Stud.* **2022**, *41*, 101085. [CrossRef]
13. Van, L.P.E.; Sanchez, G.A. Mapping PAR using MODIS atmosphere products. *Remote Sens. Environ.* **2005**, *94*, 554–563.
14. Zhang, H.; Tao, F.L.; Zhou, G.S. Potential yields, yield gaps, and optimal agronomic management practices for rice production systems in different regions of China. *Agric. Syst.* **2019**, *171*, 100–112. [CrossRef]
15. Xiao, D.P.; Tao, F.L.; Liu, Y.J.; Shi, W.J.; Wang, M.; Liu, F.S.; Zhang, S.; Zhu, Z. Observed changes in winter wheat phenology in the North China Plain for 1981–2009. *Int. J. Biometeorol.* **2013**, *57*, 275–285. [CrossRef] [PubMed]
16. Zhang, Z.H.; Li, L.R. *Atlas of Groundwater Resources and Environment in China*; China Map Press: Beijing, China, 2004.
17. Jones, J.W.; Hoogenboom, G.; Porter, C.H.; Boote, K.J.; Batchelor, W.D.; Hunt, L.A.; Wilkens, P.W.; Singh, U.; Gijsman, A.J.; Ritchie, J.T. The DSSAT cropping system model. *Eur. J. Agron.* **2003**, *18*, 235–265. [CrossRef]
18. Zhou, L.L.; Liao, S.H.; Wang, Z.M.; Wang, P.; Zhang, Y.H.; Yan, H.J.; Gao, Z.; Shen, S.; Liang, X.G.; Wang, J.H.; et al. A simulation of winter wheat crop responses to irrigation management using CERES-Wheat model in the North China Plain. *J. Integr. Agric.* **2018**, *17*, 1181–1193. [CrossRef]
19. Bannayan, M.; Hoogenboom, G. Using pattern recognition for estimating cultivar coefficients of a crop simulation model. *Field Crop. Res.* **2009**, *111*, 290–302. [CrossRef]
20. Willmott, C.J. Some comments on the evaluation of model performance. *Bull. Am. Meteorol. Soc.* **1982**, *63*, 1309–1313. [CrossRef]
21. Zheng, J.Q.; Wang, J.; Ren, W.; Tang, J.Z.; He, D.; Huang, M.X.; Bai, H.Q.; Wu, B.J. Modeling the impacts of climate, soil, and cultivar on optimal irrigation amount of winter wheat in the North China Plain. *Agron. J.* **2020**, *112*, 1176–1189. [CrossRef]
22. Lei, M.; Kong, X.B.; Wang, J.N. Estimation of sustainable grain productivity for arable land under water balance in the Huang-Huai-Hai Plain. *Acta Geogr. Sin.* **2018**, *73*, 535–549.

23. Dar, E.A.; Brar, A.S.; Mishra, S.K.; Singh, K.B. Simulating response of wheat to timing and depth of irrigation water in drip irrigation system using CERES-Wheat model. *Field Crop. Res.* **2017**, *214*, 149–163. [CrossRef]
24. Jha, S.K.; Ramatshaba, T.S.; Wang, G.; Liang, Y.; Liu, H.; Gao, Y.; Duan, A.W. Response of growth, yield and water use efficiency of winter wheat to different irrigation methods and scheduling in North China Plain. *Agric. Water Manag.* **2019**, *217*, 292–302.
25. Zhang, X.Y.; Chen, S.Y.; Sun, H.Y.; Pei, D.; Wang, Y.M. Dry matter, harvest index, grain yield and water use efficiency as affected by water supply in winter wheat. *Irrig. Sci.* **2008**, *27*, 1–10. [CrossRef]
26. Wang, X.P.; Huang, G.H.; Yang, J.S.; Huang, Q.Z.; Liu, H.J.; Yu, L.P. An assessment of irrigation practices: Sprinkler irrigation of winter wheat in the North China Plain. *Agric. Water Manag.* **2015**, *159*, 197–208. [CrossRef]
27. Ma, Y.; Feng, S.Y.; Song, X.F. Evaluation of optimal irrigation scheduling and groundwater recharge at representative sites in the North China Plain with SWAP model and field experiments. *Comput. Electron. Agric.* **2015**, *116*, 125–136. [CrossRef]
28. Liu, Y.; Guan, X.K.; Yang, M.D.; Ding, C.M.; Cui, J.Y.; Huang, J.; Mei, F.J.; Wang, H.P.; Wang, T.C. Optimizing the summer maize irrigation schedule in North Henan Province based on the DSSAT model. *Acta Ecol. Sin.* **2019**, *39*, 5348–5358.
29. Wang, H.X.; Zhang, L.; Dawes, W.R.; Liu, C.M. Improving water use efficiency of irrigated crops in the North China Plain–Measurements and modelling. *Agric. Water Manag.* **2001**, *48*, 151–167. [CrossRef]
30. Hu, Y.; Moiwo, J.P.; Yang, Y.H.; Han, S.M.; Yang, Y.M. Agricultural water-saving and sustainable groundwater management in Shijiazhuang Irrigation District, North China Plain. *J. Hydrol.* **2010**, *393*, 219–232. [CrossRef]
31. Davis, K.F.; Rulli, M.C.; Seveso, A.; D'Odorico, P. Increased food production and reduced water use through optimized crop distribution. *Nat. Geosci.* **2017**, *10*, 919–924. [CrossRef]
32. Deng, H.B.; Guan, B.Z.; Wang, J.X.; Zuo, A.; Wang, Z.L.; Sun, T.H. Seasonal Land Fallowing Policy in response to groundwater overdraft in the North China Plain. *Water Altern.* **2021**, *14*, 371–394.
33. Zhang, G.Q.; Liu, C.W.; Xiao, C.H.; Xie, R.Z.; Ming, B.; Hou, P.; Liu, G.Z.; Xu, W.J.; Shen, D.P.; Wang, K.R.; et al. Optimizing water use efficiency and economic return of super high yield spring maize under drip irrigation and plastic mulching in arid areas of China. *Field Crop. Res.* **2017**, *211*, 137–146. [CrossRef]
34. Li, J.P.; Zhang, Z.; Liu, Y.; Yao, C.S.; Song, W.Y.; Xu, X.X.; Zhang, M.; Zhou, A.N.; Gao, Y.M.; Wang, Z.M.; et al. Effects of micro-sprinkling with different irrigation amount on grain yield and water use efficiency of winter wheat in the North China Plain. *Agric. Water Manag.* **2019**, *224*, 105736. [CrossRef]
35. Guo, Q.; Huang, G.M.; Guo, Y.L.; Zhang, M.C.; Zhou, Y.Y.; Duan, L.S. Optimizing irrigation and planting density of spring maize under mulch drip irrigation system in the arid region of Northwest China. *Field Crop. Res.* **2021**, *266*, 108141. [CrossRef]
36. Long, D.; Yang, W.T.; Scanlon, B.R.; Zhao, J.S.; Liu, D.G.; Burek, P.; Pan, Y.; You, L.Z.; Wada, Y. South-to-North Water Diversion stabilizing Beijing's groundwater levels. *Nat. Commun.* **2020**, *11*, 3665. [CrossRef]
37. Zhang, X.; Xiao, G.M.; Li, H.; Wang, L.G.; Wu, S.X.; Wu, W.L.; Meng, F.Q. Mitigation of greenhouse gas emissions through optimized irrigation and nitrogen fertilization in intensively managed wheat-maize production. *Sci. Rep.* **2020**, *10*, 5907. [CrossRef]
38. Gao, J.; Xu, C.C.; Luo, N.; Liu, X.W.; Huang, S.B.; Wang, P. Mitigating global warming potential while coordinating economic benefits by optimizing irrigation managements in maize production. *J. Environ. Manag.* **2021**, *298*, 113474. [CrossRef]

agriculture

MDPI

Article

Assessing Drought, Flood, and High Temperature Disasters during Sugarcane Growth Stages in Southern China

Pei Yao [1,†], Long Qian [2,*,†], Zhaolin Wang [1], Huayue Meng [1] and Xueliang Ju [3,4]

1 School of Water Resources and Hydropower Engineering, Wuhan University, Wuhan 430072, China
2 School of Civil Engineering, Sun Yat-sen University, Guangzhou 510275, China
3 State Grid Shaanxi Electric Power Research Institute, Xi'an 710100, China
4 State Grid (Xi'an) Environmental Protection Technology Center Co., Ltd., Xi'an 710100, China
* Correspondence: qianlong@mail.sysu.edu.cn; Tel.: +86-15626003798
† These authors contributed equally to this work.

Abstract: As a globally important sugarcane-producing region, Southern China (SC) is severely affected by various agrometeorological disasters. This study aimed to comprehensively assess multiple sugarcane agrometeorological disasters with regards to sugarcane yield in SC. The standardized precipitation evapotranspiration index and the heat degree-days were employed to characterize drought, flood, and high temperature (HT) during sugarcane growth stages in three provinces in SC in the period 1970–2020. Moreover, the relationships between sugarcane climatic yield and disaster intensities were investigated. The results indicated that the most recent decade witnessed the most intensive sugarcane agrometeorological disasters; sugarcane drought and HT intensities significantly ($p < 0.05$) increased in one and two provinces, respectively. Central and western SC was most drought-prone, while eastern SC was most flood-prone; sugarcane HT was concentrated in southwestern SC. The mature stage exhibited the greatest monthly intensities of drought and flood; the most HT-prone growth stage varied with provinces. The relationships between drought/flood intensity and sugarcane climatic yield were significant in seven districts; the yield-reducing effect of sugarcane flood was more obvious than that of drought. In conclusion, this study provides references for agrometeorological disaster risk reduction for sugarcane in SC.

Keywords: sugarcane; heat stress; SPEI; waterlogging; climate change; growth stage; climatic yield

Citation: Yao, P.; Qian, L.; Wang, Z.; Meng, H.; Ju, X. Assessing Drought, Flood, and High Temperature Disasters during Sugarcane Growth Stages in Southern China. *Agriculture* **2022**, *12*, 2117. https://doi.org/10.3390/agriculture12122117

Academic Editors: Dengpan Xiao and Wenjiao Shi

Received: 21 October 2022
Accepted: 7 December 2022
Published: 9 December 2022

1. Introduction

Agriculture is sensitive to climatic environments, and the increasing levels of climate change are having profound effects on agricultural crops from various aspects. Sugarcane is one of the most important economic crops in the world; to date, its response and adaptation to climate change have been extensively investigated in different regions around the world [1,2]. Among the effects of climate change on sugarcane production, some are considered positive, such as elevated CO_2 concentration and increased air temperature, which could benefit sugarcane yield [3,4]; however, some are negative and can severely restrict sugarcane production. Such negative effects mainly refer to extreme weather events, e.g., drought, flood, and high temperatures [2,5–7]. Due to the long life cycle of sugarcane [8], its growth and yields are objectively affected by drought and flooding stresses [9–12], as well as high temperatures [13,14]. Given these concepts, it is considered that the most challenging problems induced by climate change for sugarcane production are extreme meteorological disasters [2]. Hence, from the perspective of the sustainability of sugarcane production, it is meaningful to reveal the characteristics of agrometeorological disasters occurring during sugarcane growth stages.

China is the third largest sugarcane-producing country in the world, following Brazil and India [14,15]. In China, sugarcane-producing areas are concentrated in southern China (SC for short), since the warm climate in SC is suitable for the growth and development of such tropical plants. In SC, the growing seasons of sugarcane are generally from March to December; during this long life-cycle, typical meteorological disasters in SC—including drought, flood, and high temperature—frequently occur during sugarcane growing seasons. Drought and flood are extensively distributed abiotic stresses for most crops; for sugarcane crops, a series of sugarcane field experiments have demonstrated that soil water deficit and excessive water (induced by drought and flood) can significantly reduce the growth and yield of sugarcane [16,17]. In addition to flood and drought, HT is also common in summer in SC. Although sugarcane crops are relatively tolerant to HT, the growth and yield of sugarcane is affected when air temperatures are above the upper limit of its optimal temperature range [18–20]. In conclusion, for sugarcane in SC, it is necessary to study the impacts of drought, flood, and HT during sugarcane growth stages [21].

Currently, at the regional scale, meteorological indices are powerful tools for evaluating the impacts of agrometeorological disasters on crops. Moreover, many studies employing these tools have further examined the relationships between meteorological disaster intensities and crop yield fluctuations. For agricultural drought and flood, various precipitation-based indices, e.g., the renowned SPI and SPEI, have been employed to quantify the water conditions during crop growing seasons; moreover, the drought and flood intensities characterized by meteorological indices have been found to be significantly related to the climatic yield of various crops, including rice, corn, wheat, and cotton in different regions [22–27]. For HT, many temperature-based indices, such as heat degree-days, have been applied to study the impacts of HT on the yields of various crops, e.g., maize, rice, and wheat [28–30]. However, relevant exploration regarding sugarcane crops has been insufficiently performed. For sugarcane in SC, as drought is the most noticeable agrometeorological disaster, a previous study revealed the spatial-temporal variations of drought during sugarcane growth stages in Guangxi province (the largest sugarcane-producing province in SC) by using the SPEI [25]. However, in addition to drought, many other disasters—such as flood and HT—also severely threaten sugarcane yield in SC; so far, comprehensive assessment research accounting for multiple agrometeorological disasters during sugarcane growth stages in SC is still lacking. More importantly, the potential effects of drought, flood, and HT on sugarcane yield fluctuations have not been explored on a regional scale.

Given these concepts, the primary aims of the present work were to reveal the spatial-temporal characteristics of drought, flood, and HT during sugarcane growth stages in SC, and further, to investigate the relations between the disaster intensities and sugarcane yield fluctuations. The obtained results can provide guidance for guaranteeing a high yield of sugarcane in SC in future climates.

2. Materials and Methods

2.1. Study Region

The basic information on sugarcane production in China is illustrated in Figure 1. It is apparent that Southern China (SC) is the dominating sugarcane-producing region in the country. In particular, Guangxi, Guangdong, and Yunnan provinces, i.e., three major provinces in SC lying between 20–29° N and 97–117° E (Figure 2), produce approximately 90% of the total sugarcane yield in China [31]; in recent years, they have produced over one hundred million tons of sugarcane per annum. Thus, this study took these three provinces as the study regions. SC has tropical and subtropical climate characteristics; local water and heat resources are abundant, providing ideal conditions for the growth of tropical crops, such as sugarcane.

2.2. Data Collection

The study period was 1970–2020. The employed meteorological data, mainly including daily air temperature, precipitation, wind speed, and relative humidity from 1970 to 2020, were collected from the national meteorological state of China (http://data.cma.cn, accessed on 5 August 2021). Up to 81 national level meteorological stations distributed in SC that had consistent meteorological data were employed.

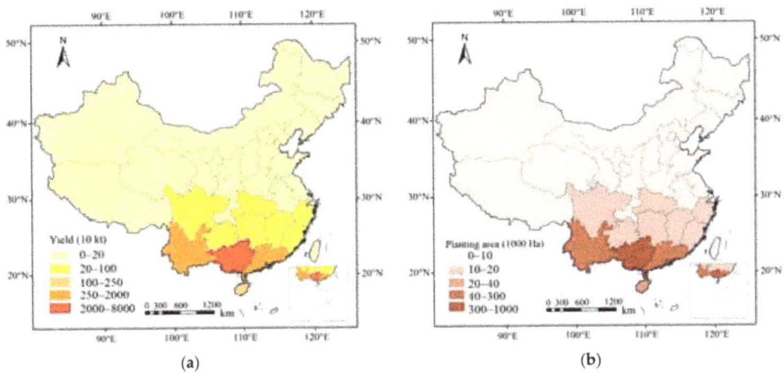

Figure 1. Basic information on Chinese sugarcane production over the last decade. (**a**,**b**) refer to sugarcane yield (10 kt) and planting areas (1000 ha), respectively.

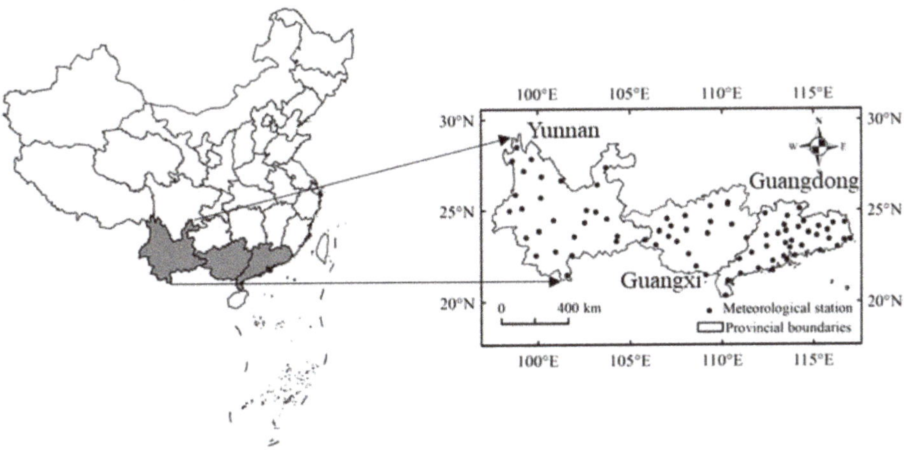

Figure 2. Description of the study region. The colored area on the left-side subgraph indicates southern China, including Yunnan, Guangxi, and Guangdong provinces. The points on the right-side subgraph indicate the national level meteorological stations that were employed in this work.

The whole growing season of sugarcane can be generally divided into four growth stages, i.e., seedling stage, tillering stage, stem elongation stage, and mature stage [25]. Since monthly indices were calculated to describe the drought and flood conditions at different sugarcane growth stages, the specific months for each growth stage were first determined. Depending on a series of existing reports concerning sugarcane in SC [25,32,33], the months of each sugarcane growth stage were determined in each province before computing the corresponding monthly SPEI. Resultingly, in Guangxi and Guangdong provinces, the seedling stage lasts from March to April, the tillering stage lasts from May to June, the stem elongation stage lasts from June to October, and the mature stage is from November to December. For Yunnan province, the seedling stage is from March to May, the tillering stage is from June to July, the stem elongation stage is from July to November, and the mature stage is in December.

The annual observed yield of sugarcane (kg/ha) in every district in SC was obtained from the provincial statistical materials that were accessible to us, including Guangdong rural year books (1992–2020), Guangxi year books (2001–2020), and Yunnan year books (1991–2020).

2.3. Study Method

The methodological diagram of this study is shown in Figure 3. First, basic meteorological data were employed to compute the drought and flood index (i.e., SPEI) and the heat index (i.e., HDD) in our study regions. Accordingly, the intensities of drought, flood, and high temperature disasters during four major sugarcane growth stages were quantified, and the spatial and temporal characteristics of these sugarcane agro-meteorological disasters were revealed. In addition, annual sugarcane yield data at different areas were collected; then, they were detrended, and the climatic yields of sugarcane were obtained. Afterwards, the variations in the actual sugarcane yield and the climatic yield were analyzed, respectively. Finally, to explore the potential impacts of agrometeorological disasters on sugarcane yield, the relationships between sugarcane climatic yield and the intensities of sugarcane drought, flood, and high temperature were examined.

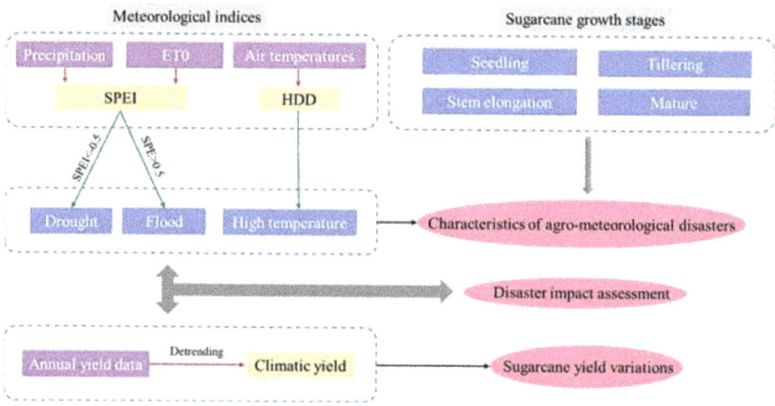

Figure 3. Methodological diagram.

2.3.1. Meteorological Indices for Drought, Flood, and HT

SPEI is a highly recognized index appropriate for agriculture drought and flood monitoring [22,23,26]. In essence, the SPEI describes water conditions based on the difference between precipitation and potential evapotranspiration. In accordance with actual requirements, the SPEI can be computed at month-scale, season-scale, and year-scale. In the present work, monthly SPEI was chosen to quantify the water conditions during each sugarcane growth stage.

First, the difference (D) between monthly precipitation P and monthly PET was computed:

$$D_i = P_i - PET_i \quad (1)$$

where P_i and PET_i were precipitation (mm) and PET (mm) in month i, respectively. In this study, PET was calculated using the Penman–Monteith method, which was recommended by FAO due to its solid physical bases.

Afterwards, the D_i series over the study period was fitted by the probability density distribution of three-parameter log-logistic function F(x). Finally, the monthly SPEI values were obtained from the standardized values of F(x). Detailed instructions for the theories and calculating processes of SPEI can be found in original documents [34]. According to a series of previous studies using the SPEI to identify agricultural drought and flood conditions [26,27,35–37], SPEI < −0.5 and SPEI > 0.5 were set as the criterion for identifying the drought and flood months, respectively. Since sugarcane growth stages last for several months, a growth stage often witnesses both sugarcane drought and flood conditions. To individually characterize sugarcane drought and flood intensities during the growth stages, we employed an accumulative index derived from a simple and commonly used waterlogging index named SEW_{30} (sum of excess water table within 30 cm soil profile) [38]. SEW_{30}

accumulates the parts of excessive water tables over a crop growth period; accordingly, our employed accumulative index, called SESPEI (sum of excessive SPEI), accumulates the parts of excessive SPEI (relative to the drought and flood threshold) over a crop growth period. SESPEI was designed for evaluating regional drought and flood intensities during the crop growth stages [22,37].

(1) For sugarcane drought:

$$SESPEI_{DR} = \begin{cases} \sum_{i=1}^{n} (|SPEI_i| - 0.5) & SPEI_i < -0.5 \\ 0 & SPEI_i \geq -0.5 \end{cases} \quad (2)$$

(2) For sugarcane flood:

$$SESPEI_{FL} = \begin{cases} \sum_{i=1}^{n} (SPEI_i - 0.5) & SPEI_i > 0.5 \\ 0 & SPEI_i \leq 0.5 \end{cases} \quad (3)$$

where $SESPEI_{DR}$ and $SESPEI_{FL}$ indicate drought and flood intensities over the calculation stage, respectively. n is the number of months across the calculation stage. $SPEI_i$ indicates the SPEI value in month i. The −0.5 and 0.5 refer to the thresholds value for drought and flood in the SPEI, respectively.

To quantify sugarcane HT, we employed the heat degree-days index (HDD), which is a simple heat index accounting for both heat duration and heat intensity over a given period [29]. In describing the overall impacts of disasters over a given period, the calculation considerations of the HDD are similar to that of the abovementioned SEW_x; thus, the HDD was calculated as:

$$HDD = \begin{cases} \sum_{i=1}^{n} (T_{max,i} - T_h) & T_{maxi,i} > T_h \\ 0 & T_{maxi,i} \leq T_h \end{cases} \quad (4)$$

where $T_{max,i}$ is the daily maximum temperature on day i, T_h is the threshold temperature for crop HT stress. n is the number of days of the calculation stage. Since sugarcane is a tropical crop, the upper temperature threshold for its normal growth is higher than that for many other crops, e.g., 30~35 °C [27,39]. Resultingly, in this paper, the T_h of sugarcane was set to 38 °C based on previous research regarding sugarcane heat stress [18,40].

It Is noted that the length of the growth stages influenced the accumulative intensities of drought/flood/HT over the given growth stages. Hence, when comparing the disaster intensities at different growth stages in this study, the accumulative disaster intensity of each growth stage was divided by the months of this growth stage. In this way, the disaster intensities (i.e., monthly intensity of disasters) between different growth stages were compared more fairly.

2.3.2. Spatial-Temporal Characteristics of Sugarcane Agrometeorological Disasters

The linear trend method was performed to detect the changing tendency of the intensity of different agrometeorological disasters:

$$y = kx + b \quad (5)$$

where y indicates the years of calculation period, x is the examined disaster intensity index, e.g., SESPEI. K is the regression coefficient which represents the climate inclination rate of the disaster. K > 0 and k < 0 indicate the upward and downward trends, respectively. A significant regression result ($p < 0.05$) indicated that the disaster intensity significantly changed over the years.

For the spatial characteristic analysis, the ArcGIS software (version 10.2; ESRI, Redlands, CA, USA) was applied to illustrate the spatial distribution of the intensities of sugarcane drought, flood, and HT during different sugarcane growth stages. The disaster intensity indices, including SESPEIDR, SESPEIFL, and HDD, were first calculated yearly for each station and then averaged over the study period (1970–2020); afterwards, these results were spatially interpolated for the whole study region for the spatial analysis. The kriging method was used for interpolation, and the spatial resolution of computations was 4.2 km × 4.2 km.

2.3.3. Sugarcane Climatic Yield and Its Relations to Meteorological Disaster Intensities

The time series of crop yield can be primarily divided into two parts, i.e., trend yield and detrended yield (also known as the crop-restricting climatic yield—climatic yield for short in this study). The trend yield was determined by non-climatic factors, such as advances in agricultural technology and improvements in field management. Currently, there are various methods for detrending crop yield, but it should be noted that these methods cannot fully remove the influence of external factors. For the present work, detrending yield is not a research issue. Hence, we selected the quadratic polynomial [22,23,26] to detrend sugarcane yield, since this commonly used method can capture the non-linear trend of the time series of a crop yield. On the other hand, the climatic yield was determined by climatic factors, such as precipitation (relevant to drought and flood disasters) and air temperatures (relevant to high temperature disasters). Hence, the climatic yield of sugarcane (Y_{cl}, kg/ha) was calculated as the difference between the actual sugarcane yield (Y_{act}, kg/ha, which refers to the abovementioned annual observed yield derived from provincial year books), and the trend yield (Y_{tr}, kg/ha)

$$Y_{cl} = Y_{act} - Y_{tr} \tag{6}$$

When the sugarcane climatic yield was obtained, its relations to the intensities of sugarcane drought, flood, and HT were investigated by performing a Pearson correlation analysis. Considering that the final sugarcane yield was affected by the combined impacts from the agrometeorological disasters occurring during various growth stages, we related the sugarcane climatic yield to the accumulative $SESPEI_{DR}$/$SESPEI_{FL}$/HDD over the whole sugarcane growth period. It should be noted that the calculation data were preprocessed to be more representative before the correlation analysis was performed. In particular, both sugarcane drought and flood disasters are common in SC; hence, when investigating the relationships between drought intensity (or flood intensity) and sugarcane climatic yield, it is meaningful to minimize the influence of flood (or drought). According to a previous drought-relevant paper [23], we considered calculating the years with the middle 40 percent of $SESPEI_{DR}$/$SESPEI_{FL}$/HDD values as "near-normal conditions"; as a result, the years with the lowest 30 percent of the index values were excluded to reduce the influence of other influential factors. Taking drought as an example, the years with the top 30 percent and the middle 40 percent of $SESPEI_{DR}$ values referred to "dry conditions" and "near normal conditions", respectively; thus, the remaining years (with the lowest 30 percent of $SESPEI_{DR}$) were excluded from correlation analysis because they were relatively flood-prone.

3. Results

3.1. *Spatial-Temporal Characteristics of Sugarcane Drought, Flood, and HT*

3.1.1. Temporal Trends

As displayed in Figure 4a, over the past five decades, the intensities of sugarcane drought have increased in Yunnan and Guangdong but decreased in Guangxi; more importantly, a significant ($p < 0.05$) increasing trend of drought intensity was detected in Yunnan (the $SESPEI_{DR}$ increased by 0.132 per decade). For sugarcane flood (Figure 4b), the intensities did not show significant trends in any province; additionally, the flood intensity in Guangdong was slightly higher than that in the other provinces. Compared with sugarcane drought and flood, sugarcane HT's intensity exhibited a more obviously increasing trend (Figure 4c). HT intensities in Yunnan and Guangdong significantly increased; in particular, Yunnan saw an obvious increase in sugarcane HT (the HDD increased by 0.466 per decade). In general, the temporal trends of sugarcane drought and flood were relatively unobvious, except a significant increasing trend of drought intensity in Yunnan.

In contrast, the sugarcane HT showed obvious increasing trends, and it was the most enhanced agrometeorological disaster over the past five decades in SC.

3.1.2. Interdecadal Analysis

The intensities of sugarcane drought, flood, and HT during different decades are displayed in Figure 5. Regarding sugarcane drought, the most recent two decades, i.e., the 2000s and 2010s, witnessed more intensive droughts than before. In the 2000s, sugarcane drought was intensive in Guangdong and Guangxi; additionally, in the 2010s, drought intensity in Yunnan reached a historic high. Generally, the most sugarcane drought-prone decade in SC was the 2000s. For sugarcane flood, the most recent decade (2010s) was apparently the most flood-prone decade for Guangdong and Guangxi. However, in the 2010s, the sugarcane flood intensity in Yunnan was at a historic low; therefore, for Yunnan, the most recent decade was severely affected by drought (Figure 5a) and slightly affected by flood (Figure 5b). Finally, for sugarcane HT, the most recent decade was the most HT-prone decade, which was in accordance with previous temporal trend results (Figure 5c). In particular, Yunnan showed a dramatically increasing tendency of sugarcane HT. In conclusion, taking drought, flood, and HT into overall consideration, the most recent decade was most affected by agrometeorological disasters for sugarcane in SC.

3.1.3. Spatial Characteristic Analysis

As depicted in Figure 6, at the seedling stage, drought-prone areas were concentrated in southwestern SC, i.e., Yunnan. Afterwards, during the following stage (i.e., tillering stage), the high-prone areas gradually expanded to all the three provinces, including western SC (Yunnan), central SC (western Guangxi) and eastern SC (eastern Guangdong). During the stem elongation stage, the high-prone areas were central SC (western Guangxi) and eastern SC (central Guangdong). Finally, at the mature stage, the high-prone area became northwestern SC (northern Yunnan). Hence, the drought-prone areas varied greatly with the growth stages. For sugarcane flood (Figure 6b1–b4), the high-prone areas at the initial stage were in eastern SC (Yunnan), which was generally similar to the drought-prone areas during this period. Nevertheless, during the remaining three growth stages, the distribution of flood-prone areas (Figure 6b2–b4) was totally different from that of drought-prone areas (Figure 6a2–a4). As for sugarcane HT (Figure 6c1–c4), during the first two growth stages, the high-prone areas were concentrated in southern Yunnan (Figure 6c1,c2). Then, at the stem elongation stage (Figure 6c3), the intensity of sugarcane HT reduced but the HT-prone areas became extensive, covering most parts of Yunnan and Guangxi, and northern Guangdong. At the mature growth stage, which corresponded to local winter seasons, HT did not occur.

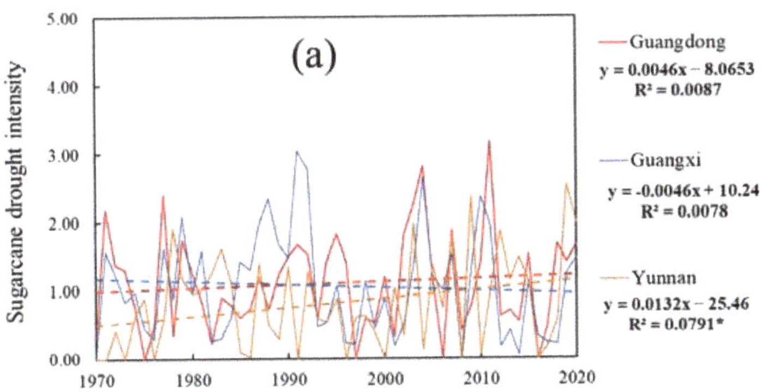

Figure 4. *Cont.*

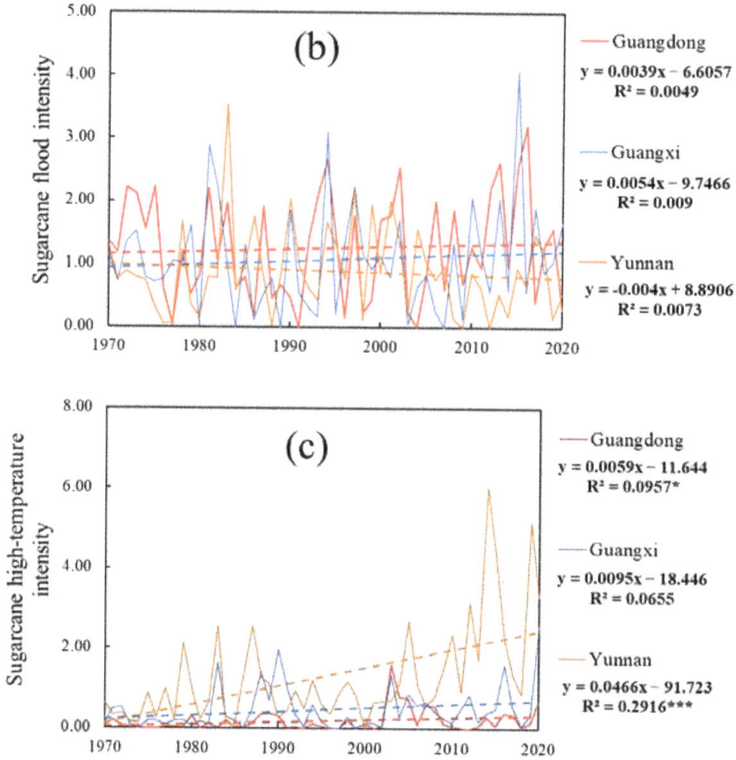

Figure 4. Temporal trends of drought (**a**), flood (**b**), and high temperatures (**c**) during sugarcane growth stages in Guangdong, Guangxi, and Yunnan provinces. The calculation sample size of every regression model is 51 (from 1971 to 2020). * and *** indicate $p < 0.05$ and $p < 0.001$. The regression model in Guangxi is significant at $p < 0.10$.

In summary, the spatial distribution of the drought-prone and flood-prone areas varied greatly with the sugarcane growth stages; during critical growth stages, the drought-prone and flood-prone areas were quite different. Additionally, HT-prone areas were consistently concentrated in southern Yunnan during different growth stages.

Figure 7 illustrates the spatial distribution of sugarcane drought, flood, and HT over the entire sugarcane growing season. The spatial distributions of sugarcane drought (Figure 7a) and flood conditions (Figure 7b) were quite different; this sharp difference was also found in Figure 6. The former was concentrated in western and central SC, including Yunnan and western Guangxi; the latter was concentrated in eastern SC, including eastern Guangdong. For sugarcane HT, the most affected areas were concentrated in Yunnan.

3.1.4. Inter-Growth-Stage Distribution of Sugarcane Drought, Flood, and HT

The comparison of the monthly intensities of disasters at different growth stages is illustrated in Figure 8. It was found that the mature stage was the period most affected by drought and flood (Figure 8a,b) in Guangdong and Guangxi. Meanwhile, in these two provinces, the tillering stage had the lowest monthly intensities of drought and flood. Finally, for sugarcane HT, the greatest monthly intensity in Guangdong and Guangxi was found at the stem elongation stage. In comparison, sugarcane HT was most intensive at the seedling stage in Yunnan (Figure 8c) which was mainly because a few HT-intensive places in Yunnan (i.e., Huaping station and Yuanjiang station) suffered severe HT during the seedling stage.

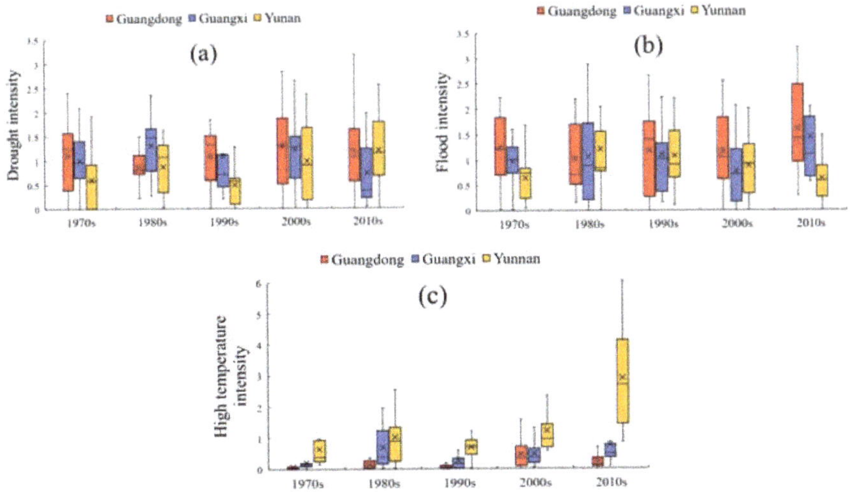

Figure 5. The intensities of drought (**a**), flood (**b**), and high temperature (**c**) during sugarcane growth stages in Guangdong, Guangxi, and Yunnan provinces. × and — in the boxes represent the mean and median values, respectively.

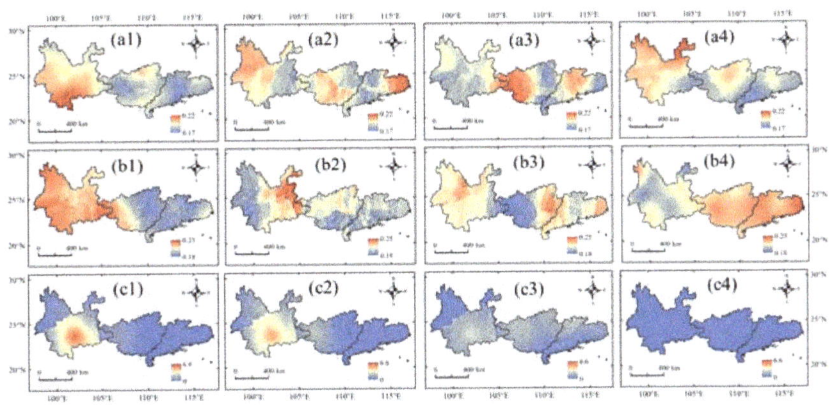

Figure 6. Spatial distribution of the monthly intensities of drought (**a1–a4**), flood (**b1–b4**), and high temperatures (**c1–c4**) during the seedling (**a1**,**b1**,**c1**), tillering (**a2**,**b2**,**c2**), stem elongation (**a3**,**b3**,**c3**), and mature stages (**a4**,**b4**,**c4**) of sugarcane in southern China. The intensities were averaged over 1970–2020 and displayed here for spatial analysis.

3.2. Sugarcane Yield Variations over the Past Few Decades

Figure 9a displays the variations in sugarcane yield (kg/ha) in the three provinces in SC. It is obvious that over the past few decades, sugarcane yield in SC has increased significantly ($p < 0.001$). In particular, Guangdong and Guangxi have witnessed a highly increasing rate of sugarcane yield (increased 801.94 and 867.27 kg/ha per year). These results suggest that sugarcane yield in SC maintained a sustainable increase. The obtained sugarcane climatic yield after detrending the sugarcane yield is displayed in Figure 9b. As illustrated by the negative values of the sugarcane climatic yield, the periods witnessing the most severe sugarcane yield losses in the three provinces were different; most severe sugarcane yield losses in Guangdong, Guangxi, and Yunnan occurred around the years 2017, 2011, and 2006, respectively. In terms of SC, the years around 2000 were the only period during which all three provinces simultaneously suffered severe losses in sugarcane yield. According to Figure 5b, around 2000, sugarcane flood intensity in SC reached a high level, contributing to severe sugarcane yield losses during that period. During the most recent decade, sugarcane climatic yield varied greatly among the three provinces. In Yunnan, sugarcane climatic yield was always near zero, indicating that the sugarcane yield was slightly influenced by agrometeorological disasters. In Guangxi, the sugarcane climate yield reached a historic low in the 2010s and then continued to increase. However, an opposite trend was found in Guangdong; its sugarcane climatic yield was high during 2010–2015, but then decreased sharply and continued to maintain low levels.

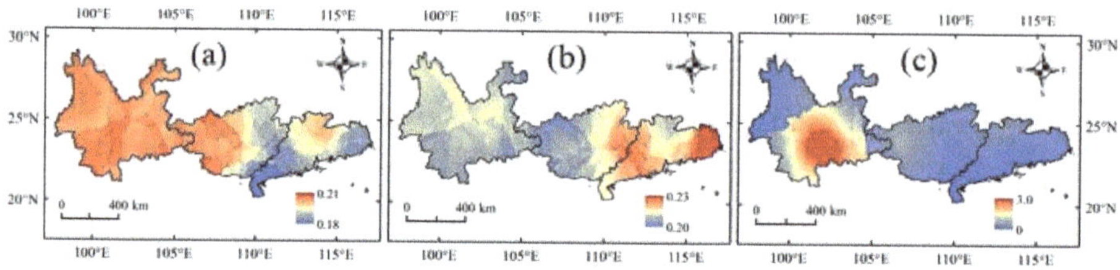

Figure 7. Spatial distribution of the monthly intensities of drought (**a**), flood (**b**), and high temperatures (**c**) during the entire sugarcane growing season in southern China. The intensities were averaged over 1970–2020 and displayed here for spatial analysis.

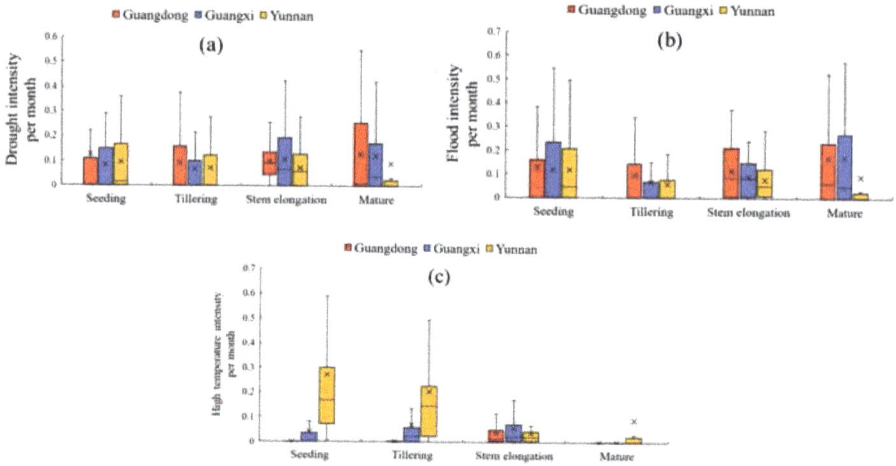

Figure 8. The monthly intensities of drought (**a**), flood (**b**), and HT (**c**) during different sugarcane growth stages in southern China. × and — in the boxes represent the mean and median values, respectively.

To analyze the spatial characteristic of sugarcane climatic yield in SC, the negative parts of sugarcane climatic yield, which describe yield losses, were calculated during the 1990s, 2000s, and 2010s (Figure 10a–d). It was noted that northwestern SC (including northern Yunnan) and eastern SC (including eastern Guangdong) were consistently low-climatic-yielding regions in different decades; thus, sugarcane yields in these regions were greatly affected by agrometeorological disasters (Figure 10d). In addition, the differences of climatic yield between districts were greater in Guangdong than in the other two provinces. We also mapped the spatial distribution of the actual yield of sugarcane per year (kg/ha/year), and the result is shown in Figure 10e. It was found that high-yielding areas were concentrated in eastern SC, mainly including eastern Guangxi and western Guangdong. Moreover, in most high-yielding areas, the corresponding climatic yield was also generally high, implying strong risk-resistant abilities.

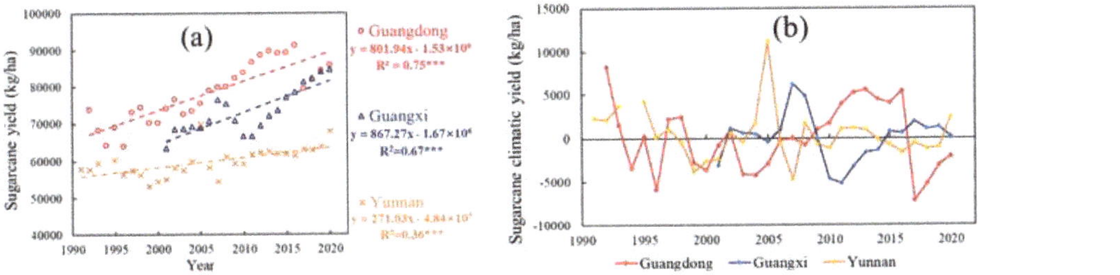

Figure 9. The actual yield (**a**) and climatic yield (**b**) of sugarcane in Guangdong, Guangxi, and Yunnan provinces. *** indicates $p < 0.001$.

3.3. The Relationships between Sugarcane Climatic Yield and Agrometeorological Disaster Intensity

As shown in Figure 11a, for sugarcane drought, the relationships between drought intensity and sugarcane climatic yield were found to be significant in four districts (i.e., Meizhou and Dongguan in Guangdong, and Beihai and Laibin in Yunnan). Two of them (in Dongguan and Laibin) were negative and reached the significance of $p < 0.01$, indicating a yield-reducing effect of sugarcane drought. However, the remaining two significant results were positive (in Meizhou and Beihai), implying positive relationships between sugar drought intensity and climatic yield. In Guangxi and Yunnan, the number of negative correlations was more common than positive correlations; however, more positive results existed in Guangdong. Therefore, the effects of drought on sugarcane yield were not simply negative at the reginal scale. Differing from sugarcane drought, sugarcane flood intensity was majorly negatively related to sugarcane climatic yield in all the three provinces (Figure 11b). Moreover, there were three significant results (Huizhou, Puer, and Beihai; Beihai was significant at $p < 0.01$)—all of them negative—which demonstrated the obvious yield-reducing effect of sugarcane flood. For sugarcane HT (Figure 11c), the relationships of HT intensity vs. sugarcane climatic yield were generally weak, and no significant relationships were found.

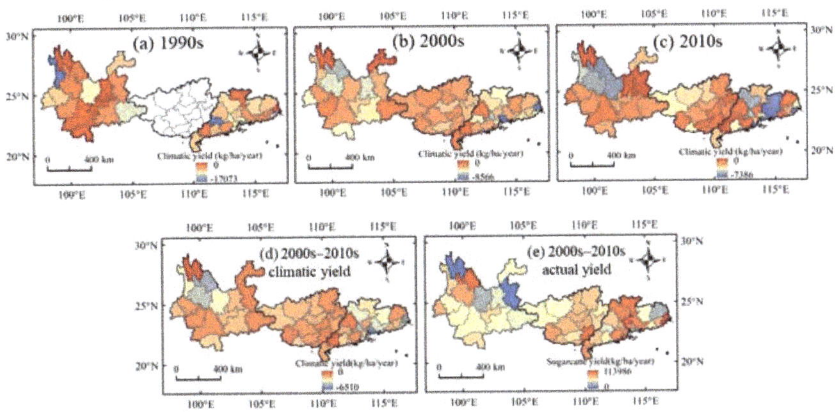

Figure 10. Sugarcane climatic yield (kg/hg/year) during 1990s (**a**), 2000s (**b**), 2010s (**c**), and 2000s–2010s (**d**), as well as actual sugarcane yield during 2000s–2010s (**e**) in southern China. Negative values of sugarcane climatic yield were used here to describe sugarcane yield losses. Guangxi has no yield data in the 1990s, so only Yunnan and Guangdong were included in Figure 10a.

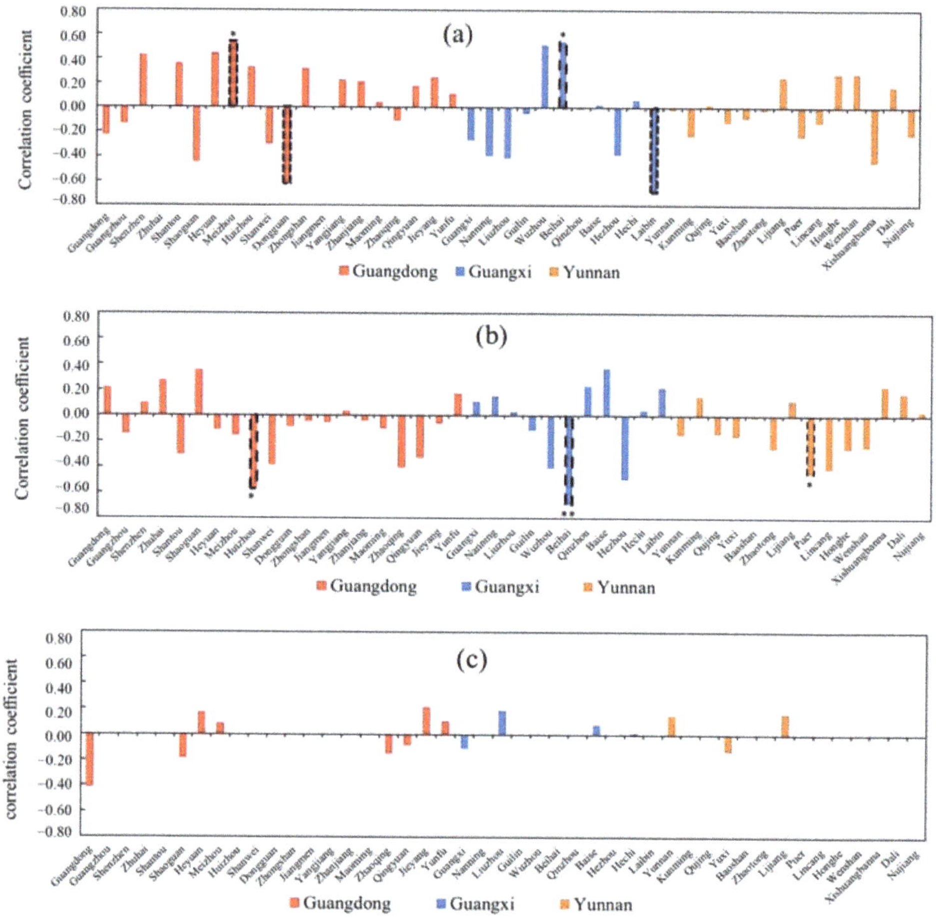

Figure 11. Correlation coefficients between sugarcane climatic yield and the intensities of drought (**a**), flood (**b**), and HT (**c**) in southern China. The black dotted boxes indicate $p < 0.05$. * and ** indicate $p < 0.05$ and $p < 0.01$, respectively.

4. Discussion

4.1. Spatial-Temporal Characteristics of Sugarcane Drought and Flood in SC

Drought and flood are widely distributed agrometeorological disasters with severe impacts on agricultural crops. Our results (Figure 4a) indicated that over the past five decades in SC, the only significant trend of sugarcane drought and flood was an upward trend in Yunnan. This result is consistent with a previous report [41], in which a noticeable drying trend was found in Yunnan over the past six decades. In addition, our results (Figure 7a) also pointed out that Yunnan was obviously the most drought-prone region in SC. Following Yunnan, western and central Guangxi were also at a relatively high risk of sugarcane drought. These spatial characteristics generally matched with a drought risk map constructed in a previous drought-relevant report in SC [42].

From the perspective of interdecadal difference, our results (Figure 5a) suggested that the 2000s witnessed more severe sugarcane drought than the other decades in SC. This finding is consistent with a previous study concerning agricultural drought in SC [43]; in that work, agricultural drought was found to be more intensive in the 2000s than in the other decades. Moreover, our results of sugarcane climatic yield (Figure 9b) showed that the 2000s was the only historic period during which all the provinces in SC suffered severe sugarcane yield losses (as demonstrated by low climatic yield), which can be explained by intensive flooding (Figure 5b) during this decade.

In terms of the sugarcane growth stages, the stem elongation stage was found to be the period during which sugarcane drought and flood were most intensive in a year. A similar finding was observed in another study on soil droughts in Guangxi [44]; that report indicated that soil droughts occurred frequently during autumn and winter seasons in Guangxi, which mainly corresponded to the stem elongation stage. However, when the influences of growth stage length were considered, the mature stage had the greatest probability of drought and flood occurrence per month; by contrast, the tillering stage had the minimum probability, which has been reported in a relevant study that focused on sugarcane drought in Guangxi [25].

Since drought and flood were quantified by the same standardized index (i.e., SPEI) in this paper, the SPEI-based drought intensity (SPEI < -0.5) and flood intensity (SPEI > 0.5) for sugarcane were comparable, and the results are displayed in Figure 12. Sugarcane flood intensity was found to be greater than drought intensity during the sugarcane growth stages in Guangdong and Yunnan. In a previous investigation on drought and flood disasters in Yunnan [45], it was concluded that flood frequency was slightly higher than drought frequency in Yunnan over the past 620 years. Furthermore, Guangdong was usually considered as having a very humid climate, but Zhang et al. [46] suggested that the impacts of drought disasters in Guangdong increased and became non-negligible.

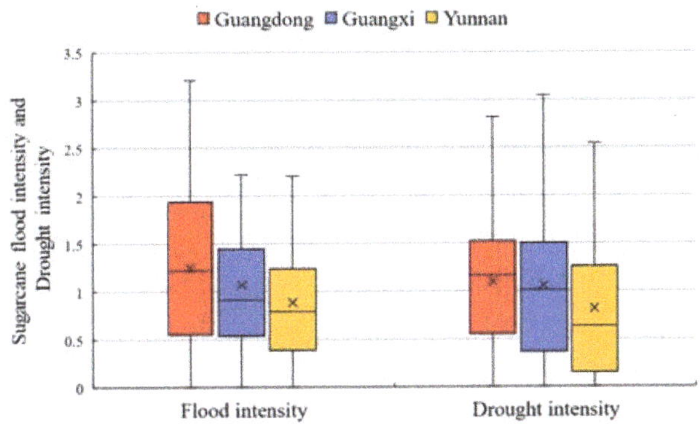

Figure 12. The intensities of drought and flood during sugarcane growth stages in SC from 1970 to 2020. × and — in the boxes represent the mean and median values, respectively.

Finally, it should be noted that in the present work, only the months during the sugarcane growth stages (from March to December) were considered in computing sugarcane drought and flood. However, as Wang and Yan [47] pointed out in an investigation concerning drought in SC, although drought events occurred throughout the entire year, January and February were among the most drought-prone months. Therefore, the difference in the calculation months may result in a few of our findings differing from previous investigations regarding drought and flood in SC.

4.2. Spatial-Temporal Characteristics of Sugarcane HT in SC

In addition to drought and flood, HT is also an extensively distributed agrometeorological disaster. Although sugarcane growth requires warm conditions, HT is regarded as a threat to sugarcane crop in future climates due to global warming [13]. As concluded by Zhang et al. [42], air temperatures in SC exhibited significant increasing trends, and annual air temperatures were expected to increase in the future (2020–2050). Correspondingly, in our results (Figure 4c), the intensity of sugarcane HT increased in all three provinces in SC; moreover, the increasing trend was highly significant ($p < 0.01$) in Yunnan and significant ($p < 0.05$) in Guangdong. We found that the increasing rate of HT intensity was greater in

Yunnan than in Guangdong and Guangxi (Figure 4c). This finding was fairly consistent with a relevant study by Zu et al. [21]; they found that the temperature during the sugarcane growing seasons obviously increased during 1970–2014, and Yunnan exhibited the greatest increase. Moreover, this finding was also in accordance with previous literature concerning extreme weather in Yunnan [41], in which the extreme temperatures were found to increase over the past six decades. Finally, in our results (Figure 4c), Guangdong, following Yunnan, also witnessed a significant increase in sugarcane HT intensity; this finding was in line with a conclusion drawn by Yuan et al. [48]: all cities in Guangdong had significant warming trends during 1958–2016.

4.3. Relationships between Sugarcane Climatic Yield and Agrometeorological Disasters

To date, the relationships between the intensity of drought/flood/HT and the yields of various crops have been widely investigated. These relationships can be used to assess the impacts of agrometeorological disasters on crop yield fluctuation; for many crops, such as rice, cotton, corn, and wheat, the significant relationships of disaster intensity vs. yield fluctuation have been established in a wide range of reports [23,26,28,29,49,50]. To the best of our knowledge, such relations concerning sugarcane have not yet been investigated. Accordingly, in the present work, we examined the relationships between sugarcane climatic yield and the intensities of drought, flood, and HT in SC; however, no significant relationships were obtained at the provincial scale (Figure 11) and only a few districts had significant relationships of drought/flood intensity vs. sugarcane climatic yield. A probable reason is the relatively strong tolerance of sugarcane to waterlogging and drought stresses. For example, short-term waterlogging treatments can significantly reduce cotton and wheat yields [22,51]; however, in most waterlogging stress experiments using sugarcane, long-term waterlogging was performed, and some sugarcane clones can well adapt to short-term flooding [17]. Nevertheless, the present findings provide evidence for the regional yield-reducing effects of sugarcane drought and flood (i.e., significantly negative correlation analysis results in a few districts), as well as a reference for future investigations into the impacts of drought and flood on sugarcane yield. In addition, a previous study [52] took Guangxi as the study area and found that the impact on crops from drought disasters was weaker than that from flood disasters. Consistent with that conclusion, we found that sugarcane flood impacts were more obvious than drought impacts in terms of yield-reducing effects (Figure 11a,b).

The relationships between HT and sugarcane climatic were not significant in any district (Figure 11c). Considering that sugarcane is a tropical plant, we adopted 38 °C, rather than the commonly-used 30 °C or 35 °C for other crops [27,39], as the threshold temperature of sugarcane HT; however, these air temperatures are not common in SC, resulting in a low level of HDD. In fact, the significantly negative impacts of HT on sugarcane yield have been detected in Australia [5] and northeastern Brazil [6]. Even so, we detected a significant increase in HT intensity (Figure 5c). This is a vigilant finding because the increasing sugarcane HT threats in SC probably induce unpredictable consequences. Similar to our findings, an increasing trend of air temperatures has also been detected in the largest sugarcane-producing country, i.e., Brazil [6]. Hence, under the context of global warming, sugarcane-growing regions may be confronted with higher air temperatures and more heat waves.

4.4. Implications for Future Sugarcane Irrigation and Drainage in SC

Irrigation and drainage are basic means for reducing the impacts of agricultural drought and flood disasters. Figure 7a,b displays the high-prone regions of sugarcane drought and flood in SC; in these regions, including the whole Yunnan and western Guangxi, timely irrigation is required to prevent sugarcane crops from drought stress. In comparison, sugarcane flood was concentrated in eastern Guangdong, where timely field drainage is important for eliminating sugarcane waterlogging stress. More importantly, according to our district-level results on the relationships between the intensities of drought/flood and sugarcane climatic yield (Figure 11), five districts (including Dongguan and Meizhou in Guangdong, Beihai and Laibin in Guangxi, and Puer in Yunnan) deserve special attention to be paid to irrigation and drainage for sugarcane, because sugarcane yield in these places was significantly and negatively affected by flood/drought over the past few decades. Furthermore, the impacts of sugarcane drought and flood were different. As discussed above, the yield-reducing effects of sugarcane flood were stronger than sugarcane drought (Figure 11). More importantly, in terms of near-term disaster characteristics, the interdecadal analysis results (Figure 5b) showed that sugarcane flood intensity in SC reached a historic high in the most recent decade. Therefore, high-efficient drainage in sugarcane fields is of great importance for improving sugarcane yield in SC; additionally, differing from sugarcane flood, sugarcane drought seems to be efficiently relieved by irrigation.

Although HT has no direct associations with irrigation and drainage, it can affect the consequences of sugarcane drought and flood. According to recent reports, the coupling of drought and HT, or of flood and HT, can result in more impacts on crop yields [27,39]. Hence, the occurrence of sugarcane HT will have indirect influences on sugarcane irrigation and drainage when HT becomes more intensive. Our results (Figure 4c) demonstrated that sugarcane HT intensity in SC has increased significantly over the past decades; increasing HT provides breeding grounds for the coupling events of drought/flood and HT. Similarly, Xu et al. [39] deduced that future sugarcane HT would be more severe and threaten sugarcane production in China. In conclusion, when making irrigation and drainage schedules for sugarcane in SC, we should be cautious with the increasing influences from HT.

4.5. Future Perspective and Research Limitations

The effect of climate change on sugarcane production in future climates is a complex and important issue. On the positive side, it is expected that due to increased air temperature and CO_2 concentration, sugarcane yield may increase in future scenarios [3]. However, on the negative side, the most challenging problems arising from the increasing risk of extreme meteorological disasters [2] can dramatically restrict sugarcane yield. For SC, many studies have consistently demonstrated that the air temperatures and heat waves in SC will increase in the coming decades [21,41,53]. Although elevated air temperatures may benefit sugarcane growth, according to previous investigations in Brazil and Australia, sugarcane yield can be obviously reduced by HT [5,6]. Hence, HT disasters in SC will probably become more harmful and nonnegligible. In addition, being consistent with HT, drought intensity in SC is also likely to increase in the future [41]. Considering that water deficit stress can indubitably affect sugarcane yield, we should pay special attention to sugarcane drought disaster reduction in SC.

The present work attempted to make contributions to agrometeorological disaster assessment, but it also faces some limitations which are expected to be overcome in future work. First, the division of sugarcane growth stages in this study was spatially and temporally coarse. It is considered that the phenology stages of crops likely have strong variations on small geographic scales (e.g., districts). Hence, to obtain more reasonable disaster intensities, it is crucial to include more detailed and precise databases for the crop phenology stage in different districts; phenology simulation models may provide efficient support. In addition, it is crude to determine growth stages on a monthly scale, which renders the drought and flood results insufficiently specific for sugarcane crops. Therefore, in future investigations, more specific timings of growth stages are expected to be included; accordingly, daily scale indices are preferable to monthly scale indices, such as SPEI. Another limitation of this work lies in computing the special results of disaster intensities. We employed a traditional method to obtain spatial results, i.e., calculating SPEI based on station-specific weather data and then performing spatial interpolation. However, the results of this approach can be affected by some factors, such as the distribution of weather stations and the employed interpolation method. In comparison, if high-quality data in uniform grid cells are available, one can simply attach them to the targeted areas and obtains more accurate and stable spatial results.

5. Conclusions

SC is the dominating sugarcane-producing region in China, also playing an important role in global sugarcane industry. However, sugarcane crops in SC are severely affected by agrometeorological disasters, mainly including drought, flood, and HT. This work employed commonly used meteorological indices, i.e., SPEI and HDD, to characterize drought, flood, and HT during sugarcane growth stages in SC over the past five decades. Moreover, relationships between sugarcane climatic yield and disaster intensities were also examined. The main results are as following:

(1) During 1970–2020, the most recent decade witnessed the most severe agrometeorological disasters during sugarcane growth stages in SC. Sugarcane drought intensity significantly increased in Yunnan; in addition, sugarcane HT exhibited increasing trends in all three provinces, and the trends in Yunnan and Guangdong were significant ($p < 0.05$). In addition, in terms of the comparison between sugarcane drought and flood, flood intensity was considered slightly greater than drought intensity in Yunnan and Guangdong.

(2) In terms of the monthly intensity of sugarcane drought and flood disasters (i.e., disaster occurrence probability per month), the mature stage was more affected than the other growth stages. Additionally, the stem elongation was the most HT-prone period in Guangdong and Guangxi. However, for Yunnan, the seedling stage was the period most affected by sugarcane HT.

(3) The most drought-prone and flood-prone regions for sugarcane were western SC (i.e., Yunnan and western Guangxi) and eastern SC (i.e., eastern Guangdong), respectively. In addition, the high-prone regions of sugarcane HT were concentrated in southern Yunnan.

(4) Sugarcane yield in northwestern SC (i.e., northern Yunnan) and eastern SC (i.e., eastern Guangdong) was most affected by agrometeorological disasters, resulting in a lower climatic yield than other regions during different decades.

(5) The relationships between flood intensity and sugarcane climatic yield were majorly negative, and significant relationships were found in three districts, which demonstrates the yield-reducing effect of sugarcane flood. In comparison, sugarcane drought intensity had significant relations to climatic yield in four districts, but two of them were positive. In summary, for sugarcane in SC, the yield-reducing effect of flood was more obvious than drought. No significant effect of sugarcane HT on sugarcane climatic yield was detected.

Author Contributions: Conceptualization, L.Q.; methodology, L.Q. and P.Y.; software, P.Y. and Z.W.; validation, P.Y. and H.M.; formal analysis, P.Y.; investigation, P.Y., Z.W. and H.M.; resources, L.Q. and X.J.; data curation, L.Q.; writing—original draft preparation, P.Y. and L.Q. (equal contributions to draft preparation); writing—review and editing, L.Q. and X.J.; visualization, P.Y. and H.M.; supervision, L.Q.; project administration, L.Q.; funding acquisition, L.Q. All authors have read and agreed to the published version of the manuscript.

Funding: This research was funded by the National Natural Science Foundation of China, grant number 51909286; the Fundamental Research Funds for the Central Universities, grant number 2021qntd15. The APC was funded by the National Natural Science Foundation of China, grant number 51909286.

Data Availability Statement: Not applicable.

Conflicts of Interest: The authors declare no conflict of interest. The funders had no role in the design of the study; in the collection, analyses, or interpretation of data; in the writing of the manuscript; or in the decision to publish the results.

References

1. Linnenluecke, M.K.; Nucifora, N.; Thompson, N. Implications of climate change for the sugarcane industry. *WIREs Clim. Chang.* **2018**, *9*, e498. [CrossRef]
2. Zhao, D.; Li, Y.R. Climate change and sugarcane production: Potential impact and mitigation strategies. *Int. J. Agon.* **2015**, *2015*, 547386. [CrossRef]
3. Santos, D.L.; Sentelhas, P.C. Climate change scenarios and their impact on water Balance and Sugarcane Yield in Southern Brazil. *Sugar Tech* **2014**, *16*, 356–365. [CrossRef]
4. Gouvêa, J.R.F.; Sentelhas, P.C.; Gazzola, S.T.; Santos, M.C. Climate changes and technological advances: Impacts on sugarcane productivity in tropical southern Braizl. *Sci. Agric.* **2009**, *66*, 593–605. [CrossRef]
5. Linnenluecke, M.K.; Zhou, C.; Smith, T.; Thompson, N.; Nucifora, N. The impact of climate change on the Australian sugarcane industry. *J. Clean. Prod.* **2020**, *246*, 118974. [CrossRef]
6. Carvalho, A.L.; Menezes, R.S.C.; Nobrega, R.S.; Pinto, A.S.; Ometto, J.P.H.; Randow, C.; Giarolla, A. Impact of climate changes on potential sugarcane yield in Pernambuco, northeastern region of Brazil. *Renew. Energy* **2015**, *78*, 26–34. [CrossRef]
7. Dias, H.B.; Sentelhas, P.C. Sugarcane yield gap analysis in Brazil—A multi-model approach for determining magnitudes and causes. *Sci. Total Environ.* **2018**, *637–638*, 1127–1136. [CrossRef]
8. Kumar, D.; Malik, N.; Sengar, R.S. Physio-biochemical insights into sugarcane genotypes under water stress. *Biol. Rhythm Res.* **2021**, *52*, 92–115. [CrossRef]
9. Berning, C.; Viljoen, M.; Du Plessis, L. Loss functions for sugar-cane: Depth and duration of inundation as determinants of extent of flood damage. *Water SA* **2000**, *26*, 527–530. [CrossRef]
10. Misra, V.; Solomon, S.; Mall, A.K.; Prajapati, C.P.; Hashem, A.; Abd_Allah, E.F.; Ansari, M.I. Morphological assessment of water stressed sugarcane: A comparison of waterlogged and drought affected crop. *Saudi J. Biol. Sci.* **2020**, *27*, 1228–1236. [CrossRef]
11. Bamrungrai, J.; Tubana, B.; Tre-loges, V.; Promkhambut, A.; Polthanee, A. Effects of water stress and auxin application on growth and yield of two sugarcane cultivars under greenhouse conditions. *Agriculture* **2021**, *11*, 613. [CrossRef]
12. Jaiphong, T.; Tominaga, J.; Watanabe, K.; Nakabaru, M.; Takaragawa, H.; Suwa, R.; Ueno, M.; Kawamitsu, Y. Effects of duration and combination of drought and flood conditions on leaf photosynthesis, growth and sugar content in sugarcane. *Plant Prod. Sci.* **2016**, *19*, 427–437. [CrossRef]
13. Castro-Nava, S.; Lopez-Rubio, E. Thermotolerance and physiological traits as fast tools to heat tolerance selection in experimental sugarcane genotypes. *Agriculture* **2019**, *9*, 251. [CrossRef]
14. Moreira, B.R.D.; Viana, R.D.; de Figueiredo, P.A.M.; Lisboa, L.A.M.; Miasaki, C.T.; Magahaes, A.C.; Ramos, S.B.; Almeida Viana, C.R.D.; Trindade, V.D.R.; May, A. Glyphosate plus carboxylic compounds boost activity of free radical-scavenging enzymes in sugarcane. *Agriculture* **2020**, *10*, 106. [CrossRef]
15. Li, Y.R.; Yang, L.T. Sugarcane agriculture and sugar industry in China. *Sugar Tech* **2015**, *17*, 1–8. [CrossRef]
16. Ferreira, T.H.S.; Tsunada, M.S.; Bassi, D.; Araújo, P.; Mattiello, L.; Guidelli, G.V.; Righetto, G.L.; Gonçalves, V.R.; Lakshmanan, P.; Menossi, M. Sugarcane water stress tolerance mechanisms and its implications on developing biotechnology solutions. *Front. Plant Sci.* **2017**, *8*, 1077. [CrossRef]
17. Gomathi, R.; Gururaja Rao, P.N.; Chandran, K.; Selvi, A. Adaptive responses of sugarcane to waterlogging stress: An over view. *Sugar Tech* **2015**, *17*, 325–338. [CrossRef]
18. Nawae, W.; Shearman, J.R.; Tangphatsornruang, S.; Punpee, P.; Yoocha, T.; Sangsrakru, D.; Naktang, C.; Sonthirod, C.; Wirojsirasak, W.; Ukoskit, K.; et al. Differential expression between drought-tolerant and drought-sensitive sugarcane under mild and moderate water stress as revealed by a comparative analysis of leaf transcriptome. *PeerJ* **2020**, *8*, e9608. [CrossRef]
19. Xu, C.H.; Li, C.J.; Su, H.S.; Lu, X.; Li, X.J.; Liu, H.B.; Lin, X.Q.; Mao, Y.; Zi, Y.Q.; Liu, X.L. Process in the studies on abiotic stress resistance of sugarcane (*Saccharum* spp.). *J. Plant Genet. Resour.* **2017**, *18*, 483–493. [CrossRef]
20. Ebrahim, M.K.; Zingsheim, O.; El-Shourbagy, M.N.; Moor, P.H.; Komor, E. Growth and sugar storage in sugarcane grown at temperatures below and above optimum. *J. Plant Physiol.* **1998**, *153*, 593–602. [CrossRef]
21. Zu, Q.; Mi, C.R.; Liu, D.L.; He, L.; Kuang, Z.M.; Fang, Q.X.; Ramp, D.; Li, L.; Wang, B.; Chen, Y.L.; et al. Spatio-temporal distribution of sugarcane potential yields and yield gaps in Southern China. *Eur. J. Agron.* **2018**, *92*, 72–83. [CrossRef]
22. Qian, L.; Chen, X.H.; Gao, Y.W.; Deng, K.N.; Wang, X.G.; Zeng, W.Z.; Luo, Y.Y. Quantifying the impacts of waterlogging on cotton at different growth stages: A case study in Hubei Province, China. *Agron. J.* **2021**, *113*, 1831–1851. [CrossRef]
23. Tian, L.Y.; Yuan, S.S.; Quiring, S.M. Evaluation of six indices for monitoring agricultural drought in the southcentral United States. *Agric. For. Meteorol.* **2018**, *249*, 107–119. [CrossRef]
24. Raja, R.; Nayak, A.K.; Panda, B.B.; Lal, B.; Tripathi, R.; Shahid, M.; Kumar, A.; Mohanty, S.; Samal, P.; Gautam, P.; et al. Monitoring of meteorological drought and its impact on rice (*Oryza sativa* L.) productivity in Odisha using standardized precipitation index. *Arch. Agron. Soil Sci.* **2014**, *60*, 1701–1715. [CrossRef]
25. Chen, Y.L.; Meng, L.L.; Huang, X.H.; Mo, J.F.; Feng, L.P. Spatial and temporal evolution characteristics of drought in Guangxi during sugarcane growth period based on SPEI. *Trans. Chin. Soc. Agric. Eng.* **2019**, *35*, 149–158. [CrossRef]
26. Gao, C.; Yin, Z.X.; Xu, Y. Space-time characteristics of drought and flood in main growing periods of winter wheat in Huaihe River Basin and its impact on yield. *Trans. Chin. Soc. Agric. Eng.* **2017**, *33*, 101–111. [CrossRef]
27. Zhu, X.F.; Liu, T.T.; Xu, K.; Chen, C.X. The impacts of high temperature and drought stress on the yield of major staple crops in northern China. *J. Environ. Manag.* **2022**, *314*, 115092. [CrossRef]

28. Sun, X.S.; Long, Z.W.; Song, G.S.; Chen, C.Q. High-temperature episodes with spatial-temporal variation impacted middle-season rice yield in China. *Agron. J.* **2018**, *110*, 961–969. [CrossRef]
29. Liu, B.; Liu, L.L.; Tian, L.Y.; Cao, W.X.; Zhu, Y.; Asseng, S. Post-heading heat stress and yield impact in winter wheat of China. *Glob. Chang. Biol.* **2014**, *20*, 372–381. [CrossRef]
30. Lobell, D.B.; Hammer, G.L.; McLean, G.; Messina, C.; Roberts, M.J.; Schlenker, W. The critical role of extreme heat for maize production in the United States. *Nat. Clim. Chang.* **2013**, *3*, 497–501. [CrossRef]
31. Li, Y.R.; Song, X.P.; Wu, J.M.; Li, C.N.; Liang, Q.; Liu, X.H.; Wang, W.Z.; Tan, H.W.; Yang, L.T. Sugar Industry and Improved Sugarcane Farming Technologies in China. *Sugar Tech* **2016**, *18*, 603–611. [CrossRef]
32. Zhong, C.; Zhou, C.; Xu, M.Y. Forecasting growth and development stage of sugarcane based on physiological development time. *Sugar Crops China* **2012**, *2*, 49–51.
33. Sun, S.; Liu, L.L.; Hu, Z.W.; Yu, X. Inversion method of sugarcane plant height based on sentinel 1A dual-polarization time series data. *Trans. Chin. Soc. Agric. Mach.* **2022**, *53*, 186–194. [CrossRef]
34. Vicente-Serrano, S.M.; Beguería, S.; López-Moreno, J.I. A multiscalar drought index sensitive to global warming: The standardized precipitation evapotranspiration index. *J. Clim.* **2010**, *23*, 1696–1718. [CrossRef]
35. Yao, N.; Li, Y.; Liu, Q.Z.; Zhang, S.Y.; Chen, X.G.; Ji, Y.D.; Liu, F.G.; Pulatov, A.; Feng, P.Y. Response of wheat and maize growth-yields to meteorological and agricultural droughts based on standardized precipitation evapotranspiration indexes and soil moisture deficit indexe. *Agric. Water Manag.* **2022**, *266*, 107566. [CrossRef]
36. Geng, G.P.; Yang, R.; Liu, L.Z. Downscaled solar-induced chlorophyll fluorescence has great potential for monitoring the response of vegetation to drought in the Yellow River Basin, China: Insights from an extreme event. *Ecol. Indic.* **2022**, *138*, 108801. [CrossRef]
37. Meng, H.Y.; Wang, Z.L.; Yao, P.; Qian, L.; Chen, C.; Luo, Y.Y.; Ju, X.L. Comparative study of different meteorological yield separation methods in waterlogging disaster assessment. *Chin. J. Eco-Agric.* **2022**, *30*, 976–989. [CrossRef]
38. Shaw, R.E.; Meyer, W.S. Improved empirical representation of plant responses to waterlogging for simulating crop yield. *Agron. J.* **2015**, *107*, 1711–1723. [CrossRef]
39. Qian, L.; Wang, X.G.; Luo, Y.Y.; Sun, H.W.; Luo, W.B. Responses of cotton at different growth stages to aeration stress under the influence of high temperature. *Crop Sci.* **2018**, *58*, 342–353. [CrossRef]
40. Selvarajan, D.; Mohan, C.; Dhandapan, V.; Nerkar, G.; Jayanarayanan, A.; Mohanan, M.V.; Murugan, N.; Kaur, L.; Chennappa, M.; Kumar, P.; et al. Differential gene expression profiling through transcriptome approach of *Saccharum spontaneum* L. under low temperature stress reveals genes potentially involved in cold acclimation. *3 Biotech* **2018**, *8*, 195. [CrossRef]
41. Yan, W.B.; He, Y.L.; Cai, Y.; Cui, X.L.; Qu, X.X. Analysis of spatiotemporal variability in extreme climate and potential driving factors on the Yunnan Plateau (Southwest China) during 1960–2019. *Atmosphere* **2021**, *12*, 1136. [CrossRef]
42. Zhang, Q.; Yao, Y.B.; Wang, Y.; Wang, S.P.; Wang, J.S.; Yang, J.H.; Wang, J.; Li, Y.P.; Shang, J.L.; Li, W.J. Characteristics of drought in Southern China under climatic warming, the risk, and countermeasures for prevention and control. *Theor. Appl. Climatol.* **2019**, *136*, 1157–1173. [CrossRef]
43. Wei, S.G.; Zhang, R.Q.; Li, L.; Zhang, S.L.; Zhang, Y.; Huang, F.N.; Li, J.D.; Liu, W. Assessment of agricultural drought based on reanalysis soil moisture in southern China. *Land* **2022**, *11*, 502. [CrossRef]
44. Wu, W.X.; Li, R.H.; Shao, J.H. Assessment of regional spatiotemporal variations in drought from the perspective of soil moisture in Guangxi, China. *Water* **2022**, *14*, 289. [CrossRef]
45. Niu, J.; Wang, L.M.; Yang, K.W.; Liu, S.Z.; Yin, J.Q.; Liang, Q.L.; Zhang, W.X. Spatial–temporal characteristics of drought and flood disasters in Yunnan Province on the margin of the Tibetan Plateau over the past 620 years. *Int. J. Climatol.* **2022**, *42*, 1388–1398. [CrossRef]
46. Zhang, Q.; Zhang, W.; Chen, Y.D.; Jiang, T. Flood, drought and typhoon disasters during the last half-century in the Guangdong province, China. *Nat. Hazards* **2011**, *57*, 267–278. [CrossRef]
47. Wang, Y.J.; Yan, F. Spatial and temporal variations of regional drought events in southern China. *Meteorol. Appl.* **2017**, *24*, 481–490. [CrossRef]
48. Yuan, Z.J.; Liang, C.; Li, D.Q. Urban stormwater management based on an analysis of climate change: A case study of the Hebei and Guangdong provinces. *Landsc. Urban Plan.* **2018**, *177*, 217–226. [CrossRef]
49. Chen, Y.Y.; Huang, J.F.; Song, X.D.; Gao, P.; Wan, S.Q.; Shi, L.; Wang, X.Z. Spatiotemporal characteristics of winter wheat waterlogging in the Middle and Lower Reaches of the Yangtze River, China. *Adv. Meteorol.* **2018**, *2018*, 3542103. [CrossRef]
50. Zhang, Y.; Deng, K.N.; Gao, Y.W.; Qian, L.; Chen, C.; Huang, T.X. Using accumulative humidity index to estimate drought and flooding hazards in cotton growth season in Hubei Province. *J. Irrig. Drain.* **2021**, *40*, 57–65. [CrossRef]
51. Malik, A.I.; Colmer, T.D.; Lambers, H.; Setter, T.L.; Schortemeyer, M. Short-term waterlogging has long-term effects on the growth and physiology of wheat. *New Phytol.* **2002**, *153*, 225–236. [CrossRef]
52. Qin, N.X.; Wang, J.N.; Gao, L.; Hong, Y.; Huang, J.L.; Liu, Q.Q. Observed trends of different rainfall intensities and the associated spatiotemporal variations during 1958–2016 in Guangxi, China. *Int. J. Climatol.* **2021**, *41*, 2880–2895. [CrossRef]
53. Luo, M.; Lau, N.C. Heat waves in Southern China: Synoptic behavior, long-term change, and urbanization effects. *J. Clim.* **2017**, *30*, 703–720. [CrossRef]

agriculture

MDPI

Article

The Prediction of Wheat Yield in the North China Plain by Coupling Crop Model with Machine Learning Algorithms

Yanxi Zhao [1,2,3], Dengpan Xiao [1,2,3,*], Huizi Bai [1,*], Jianzhao Tang [1], De Li Liu [4,5], Yongqing Qi [6] and Yanjun Shen [6,7]

1 Hebei Technology Innovation Center for Geographic Information Application, Institute of Geographical Sciences, Hebei Academy of Sciences, Shijiazhuang 050011, China
2 College of Geography Science, Hebei Normal University, Shijiazhuang 050024, China
3 Hebei Laboratory of Environmental Evolution and Ecological Construction, Shijiazhuang 050024, China
4 NSW Department of Primary Industries, Wagga Wagga Agricultural Institute, Wagga, NSW 2650, Australia
5 Climate Change Research Centre, University of New South Wale, Sydney, NSW 2052, Australia
6 Key Laboratory for Agricultural Water Resources, Hebei Key Laboratory for Agricultural Water Saving, Center for Agricultural Resources Research, Institute of Genetics and Developmental Biology, Chinese Academy of Sciences, Shijiazhuang 050021, China
7 School of Advanced Agricultural Sciences, University of the Chinese Academy of Sciences, Beijing 100049, China
* Correspondence: xiaodp@sjziam.ac.cn (D.X.); baihz272@126.com (H.B.)

Abstract: The accuracy prediction for the crop yield is conducive to the food security in regions and/or nations. To some extent, the prediction model for crop yields combining the crop mechanism model with statistical regression model (SRM) can improve the timeliness and robustness of the final yield prediction. In this study, the accumulated biomass (AB) simulated by the Agricultural Production Systems sIMulator (APSIM) model and multiple climate indices (e.g., climate suitability indices and extreme climate indices) were incorporated into SRM to predict the wheat yield in the North China Plain (NCP). The results showed that the prediction model based on the random forest (RF) algorithm outperformed the prediction models using other regression algorithms. The prediction for the wheat yield at SM (the period from the start of grain filling to the milky stage) based on RF can obtain a higher accuracy (r = 0.86, RMSE = 683 kg ha^{-1} and MAE = 498 kg ha^{-1}). With the progression of wheat growth, the performances of yield prediction models improved gradually. The prediction of yield at FS (the period from flowering to the start of grain filling) can achieve higher precision and a longer lead time, which can be viewed as the optimum period providing the decent performance of the yield prediction and about one month's lead time. In addition, the precision of the predicted yield for the irrigated sites was higher than that for the rainfed sites. The APSIM-simulated AB had an importance of above 30% for the last three prediction events, including FIF event (the period from floral initiation to flowering), FS event (the period from flowering to the start of grain filling) and SM event (the period from the start of grain filling to the milky stage), which ranked first in the prediction model. The climate suitability indices, with a higher rank for every prediction event, played an important role in the prediction model. The winter wheat yield in the NCP was seriously affected by the low temperature events before flowering, the high temperature events after flowering and water stress. We hope that the prediction model can be used to develop adaptation strategies to mitigate the negative effects of climate change on crop productivity and provide the data support for food security.

Keywords: yield prediction; machine learning; APSIM model; climate indices; North China Plain

Citation: Zhao, Y.; Xiao, D.; Bai, H.; Tang, J.; Liu, D.L.; Qi, Y.; Shen, Y. The Prediction of Wheat Yield in the North China Plain by Coupling Crop Model with Machine Learning Algorithms. *Agriculture* **2023**, *13*, 99. https://doi.org/10.3390/agriculture13010099

Academic Editor: Yanqun Zhang

Received: 24 November 2022
Revised: 19 December 2022
Accepted: 26 December 2022
Published: 29 December 2022

1. Introduction

Food security is related to a series of major issues such as national food security, social stability and sustainable development of the national economy, which is highly concerned by the country [1–3]. Increasing food productivity is an important measure to ensure food security. However, the trend of global warming became more severe throughout the 20th century [4,5]. Generally, climate warming can shorten the crop growth period, which negatively influences the formation of a crop yield and, ultimately, causes crop failure [6,7]. Predicting the yield can provide data support to take appropriate management for farmers. Wheat is one of the three major grain crops in China, with a wide planting range, large planting area and high yield [8]. Therefore, the studies on yield prediction in wheat are conducive to the timely and accurate grasp of the grain production status and scientific formulation of policies for the government [9,10].

The statistical regression model (SRM) directly developed statistical models based on the relationship between selected predictors and target variables to achieve the goal [11–14]. Guan et al. [15] used partial least-square regression (PLSR) to estimate the relationships between crop yield and the predictor variables. In general, the models combined with statistical regression algorithms are easy to understand and require fewer parameters, so the methods are commonly used in yield predictions worldwide [16–18]. However, with the increasing volume and dimension of observation data, it is a great challenge to fully explore the information of datasets for effective analysis and utilization. The most current SRM based on linear regression have some problems in application due to the complexity of the crop production system. For example, crop yields exhibit nonlinear responses to extreme climate events, while previous linear regression models may not perform well under frequent extreme climate conditions [19,20]. Compared with the linear regression analysis, the machine learning algorithm (MLA) is an advanced method for yield estimation that can capture nonlinear relations between the dependent and independent variables [21–24]. The MLA can explore the information of the training data, obtain a higher generalization level, and enhance the robustness and universality of the prediction model [15]. For example, Cai et al. [25] developed the prediction model for wheat in Australia by using some machine learning, while the support vector machine (SVM) algorithm performs better than other statistical regression algorithms. Hunt et al. [26] used the random forest (RF) algorithm to evaluate the crop yield and achieved a good performance. Nevertheless, MLA are not mechanistic and can not fully consider the dynamic process of crop growth.

The crop mechanism model (CM) has good mechanical properties, which is a simulation program that can dynamically describe the process of crop growth and yield formation under various environmental conditions by importing weather data, a variety of parameters, soil data and so on [19]. With the development of CM, the studies on the estimation of the crop yield increased gradually. For example, Huang et al. [27], Xiao et al. [28] and Zhang et al. [29], based on CMs, estimated the yields of maize, wheat and rice in China, respectively. However, the results of the most related studies were end-of-season yield predictions. The greatest limitation of within-season predictions is the lack of meteorological data from the prediction date to the maturity date [30]. Some studies achieved the prediction results by coupling the CM with seasonal weather forecasts. Pagani et al. [31] developed a high-resolution integrated prediction system for rice yield at the district level based on the combination of the WARM model, weather forecasts and remote sensing images. However, the real weather conditions may deviate from the weather forecast data, thus increasing the uncertainty of the prediction model [32].

We can reduce unnecessary errors by combining MLA and CM for yield prediction. Feng et al. [32] used the integration of the MLA and APSIM model to predict the yield of wheat under rainfed conditions in Southeastern Australia, and the hybrid model obtained a decent yield prediction at one month leading time before harvest. Nevertheless, there are few studies on using a hybrid model to predict crop yields under irrigated conditions. Furthermore, CMs can simulate the effects of complicated climate conditions on crop growth to a certain extent but are not sufficient. The quantity variation of key climatic factors (e.g., temperature and precipitation) can be transformed into the climate suitability of crop growth based on the membership function method in fuzzy mathematics [33,34]. Meanwhile, the extreme climate indices (ECIs) can quantify the destructive effects of extreme climate events on crop growth [35,36]. The climate suitability and ECIs can be included in the hybrid model as predictive indicators to further explore the information reflected by the climate factors and improve the robustness of the hybrid model. However, there were few studies on using the combination of climate suitability and ECIs as predictive variables.

The North China Plain (NCP) is an important grain production base and occupies an important position in the national grain production in China [8]. In this study, we investigated the yield prediction of wheat in the NCP by using the CM and SRM. The main objectives of the study were (1) to develop the yield prediction model of wheat based on the combination of the multiple growth period-specific variables and SRM, (2) to identify the optimal lead time before maturity of yield prediction with acceptable accuracy, and (3) to evaluate the relative importance of input variables during different growth stages in the yield prediction model.

2. Data and Methods

2.1. Study Area

The NCP is delimitated in the east by the sea, the west by the Taihang Mountains, the south by the main stream of the Huaihe River, and the north by the Yan Mountains (Figure 1) [37]. The region has a warm temperate monsoon climate with plenty of light and heat resources [37]. The annual precipitation is not evenly distributed, with over 70% of precipitation appearing in July through September. The main soil type in the NCP is the loam of Aeolian origin, a soil type deposited by rivers over geological periods. The NCP is an important grain production region in China, where the main cropping system is the double cropping systems of winter wheat–summer maize [38]. Winter wheat is usually planted in early or middle October and harvested in early June. We selected 20 agro-meteorological sites distributed across the NCP (Figure 1). Table 1 presents basic information for the 20 study sites, including location, irrigation condition, and wheat phenology and yield.

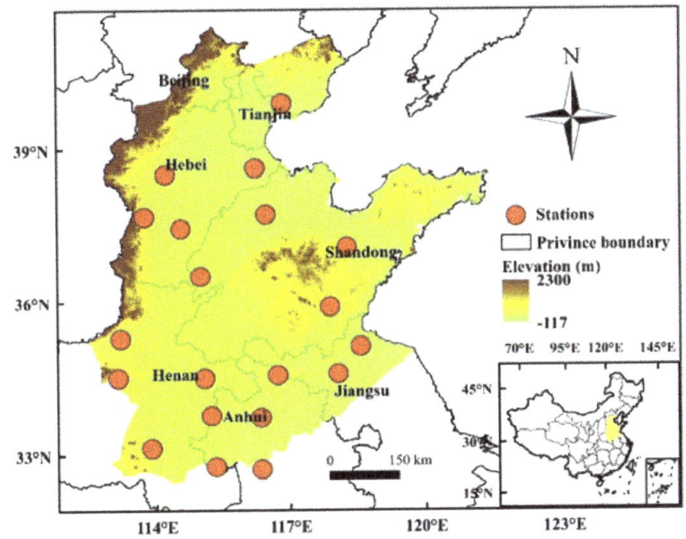

Figure 1. The spatial distribution of 20 agro-meteorological sites across the North China Plain.

Table 1. Related information about the 20 investigated sites in the study.

Site	Longitude (°E)	Latitude (°N)	Harvest Years	Irrigation	FDm (DOY)	MDm (DOY)	WYm (kg/ha)
Bozhou	115.8	33.9	2000–2010	no	114	150	5213
Dingzhou	115.0	38.3	2000–2010	yes	129	162	5728
Fuyang	115.8	32.9	2000–2010	no	112	150	6233
Ganyu	119.1	34.5	2000–2010	yes	123	159	6559
Huanghua	117.2	38.2	2004–2010	no	130	157	2924
Huimin	117.4	37.3	2000–2010	yes	128	160	6591
Juxian	118.8	35.6	2000–2010	yes	127	162	7120
Liaocheng	116.0	36.4	2000–2010	yes	124	158	5908
Luancheng	114.6	37.9	2000–2010	yes	127	162	6845
Nangong	115.3	37.3	2000–2010	yes	124	157	5580
Shangqiu	115.7	34.5	2000–2010	yes	115	150	5099
Shouxian	116.8	32.6	2000–2010	no	112	146	5316
Shuyang	118.8	34.1	2000–2010	no	123	158	6069
Suxian	116.6	33.4	2000–2010	no	116	151	6173
Tangshan	118.1	39.4	2000–2010	yes	132	166	6123
Weifang	119.2	36.8	2000–2010	yes	127	158	6017
Xinxiang	114.0	35.3	2000–2010	yes	119	151	6016
Xuzhou	117.4	34.3	2000–2010	no	118	154	7406
Zhengzhou	113.4	34.4	2000–2010	yes	112	148	5033
Zhumadian	114.1	33.0	2000–2010	no	107	143	5667

Notes: FDm, MDm and WYm denote the mean flowering date, the mean maturity date, and the mean yield for wheat during the investigated period, respectively. DOY is day of year.

2.2. Climate, Soil and Crop Data

The historical records about daily climate data, including mean temperature (Tmean), maximum temperature (Tmax), minimum temperature (Tmin), precipitation (Prec), and sunshine hours (Sh) during 2000 to 2010 for 20 agro-meteorological sites across the NCP, were obtained from China's Meteorological Administration (CMA). Soil profile data of all the sites were obtained from the 1:1 million scale soil map of China included in the Harmonized World Soil Database (HWSD) version 1.2 [39]. The climate and soil data were used to run the APSIM model.

Detailed field experimental data records included the phenology (sowing date (SD), flowering date (FD), and maturity date (MD)); grain yield; and management data at the agro-meteorological experiment sites for 2000–2010 were also obtained from CMA. The phenology data was observed by experimenters in the specific fields at the agro-meteorological experiment sites, while the grain yield was the weight of the harvested crop in the specific fields. We used the experimental crop data to calibrate and validate the crop parameters in the APSIM model.

2.3. Methodology

2.3.1. Agricultural Production Systems SIMulator (APSIM) Simulations

The APSIM model is a comprehensive model developed to simulate biophysical processes in agricultural production systems [40,41]. The APSIM model can provide an acceptable prediction accuracy of crop productivity under the combined influences of climate change, soil condition, and management measures [42,43]. In this study, the APSIM model was implemented to simulate crop phenology, biomass, and grain yield during 2000–2010 at the 20 selected sites.

2.3.2. Climate Indices

In the study, we took account of four main growth periods, including the period from end of the juvenile stage to floral initiation (JF), the period from floral initiation to flowering (FIF), the period from flowering to the start of grain filling (FS), and the period from the start of grain filling to the milky stage (SM). We assessed the impacts of 10 extreme climate indices (ECIs) [44,45] and 3 for climate suitability (CS) [46] during different growth periods for wheat (Table 2). The calculation methods of the ECIs were shown in Table 2. The CS can further explore the information of the mean climate variables. We can develop the climate suitability model according to related studies [46].

Table 2. The information about the thirteen climate indices (CIs) used in the study.

Index	Name	Definition	Growth Stage
TS	Temperature suitability	The indicator of measurement when temperature is less or greater than physiological temperature requirement	JF, FIF, FS, SM
SS	Sunshine suitability	The indicator of measurement when sunshine is less or greater than physiological sunshine requirement	JF, FIF, FS, SM
PS	Precipitation suitability	The indicator of measurement when precipitation is less or greater than physiological water requirement	JF, FIF, FS, SM
HD	Hot days	The number of days with Tmax ≥ 30 °C	FS, SM
HCD	Consecutive hot days	The number of days with three or more continuous days of Tmax ≥ 30 °C	FS, SM
WD	Warm days	The number of days with Tmax > 22 °C	JF, FIF, FS, SM
WCD	Consecutive warm days	The number of days with three or more continuous days of Tmax ≥ 22 °C	JF, FIF, FS, SM
FD	Frost days	The number of days with Tmin < 2 °C	JF, FIF
FCD	Consecutive cold days	The number of days with three or more continuous days of Tmin < 2 °C	JF, FIF
R10	Heavy precipitation days	The number of days when precipitation ≥ 10 mm	JF, FIF, FS, SM
CDD	Consecutive dry days	The number of days with three or more continuous days of daily precipitation < 1 mm	JF, FIF, FS, SM
CWD	Consecutive wet days	The number of days with three or more continuous days of daily precipitation ≥ 1 mm	JF, FIF, FS, SM
SDII	Simple daily intensity index	The ratio of total precipitation to the number of wet days (≥ 1 mm)	JF, FIF, FS, SM

Note: JF, FIF, FS, and SM denote the periods from end of the juvenile stage to floral initiation, from floral initiation to flowering, from flowering to the start of grain filling, and from the start of grain filling to the milky stage, respectively.

The sunshine suitability (SS) of wheat was calculated as follows [47–49]:

$$SS = \begin{cases} e^{-[(S_i - S_0)/b]^2} & S_i < S_0 \\ 1 & S_i \geq S_0 \end{cases} \quad (1)$$

where S_0 is the daily sunshine hours when the percentage of the daily sunshine hours reaches 70%, S_i is the daily sunshine hours (h), and b is a constant that can be determined according to the climatic conditions across the NCP and relevant studies [49,50]. The values for b at different growth periods are shown in Table 3. The arithmetic mean of the daily SS for a specific growth period is the SS for the corresponding period.

Table 3. Values of related parameters for calculating the sunshine suitability (SS), temperature suitability (TS), and precipitation suitability (PS) at four growth periods of wheat.

Parameters	JF	FIF	FS	SM
b	4.26	4.5	4.61	4.96
T_1	−2	3	8	12
T_0	5	10	16	20
T_2	10	20	27	30
Kc	0.55	0.8	1.05	1.0

Note: JF, FIF, FS, and SM denote the periods from the end of the juvenile stage to floral initiation, from floral initiation to flowering, from flowering to the start of grain filling, and from start of the grain filling to the milky stage, respectively.

The temperature suitability (TS) of wheat was calculated as follows [47–49]:

$$TS = \frac{\left[(T_i - T_1)(T_2 - T_i)^B\right]}{\left[(T_0 - T_1)(T_2 - T_0)^B\right]} \tag{2}$$

$$\text{Among}\left\{B = \frac{(T_2 - T_0)}{(T_0 - T_1)}\right\} \tag{3}$$

where T_i is the daily mean temperature (°C), T_0 is the optimal temperature (°C) at different growth periods, T_1 is the lower limit temperature (°C) at different growth periods, and T_2 is the upper limit temperature (°C) at different growth periods. The specific values of T_0, T_1, and T_2 refer to the climatic conditions across the NCP and relevant studies [49,51]. The values for T_0, T_1, and T_2 at different growth periods are listed in Table 3. The arithmetic mean of the daily TS for a specific growth period is the TS for the corresponding period.

The precipitation suitability (PS) of wheat was calculated as follows [52]:

$$PS = \begin{cases} P/P_0 & P_i < P_0 \\ P_0/P & P_i \geq P_0 \end{cases} \tag{4}$$

where P is precipitation (mm), and P_0 is the physiological water requirement of crops, which can be calculated as follows:

$$P_0 = Kc \bullet ET_0 \tag{5}$$

where Kc is the crop coefficient, and ET_0 is the reference crop evapotranspiration (mm). The Kc values of wheat at different growth stages listed in Table 3 are determined according to the relevant studies [53,54]. The ET_0 values of wheat are calculated based on the Penman–Monteith formula [54].

2.3.3. Regression Models

Two machine learning algorithms, i.e., random forest (RF) and light gradient boosting machine (LGB), were selected to predict the wheat yield. RF is an ensemble learning algorithm [26,55], which creates multiple decision trees in a random way and applies them in training samples. Among all the current algorithms, RF has high accuracy and stability, which can effectively process input samples with large data volumes and high-dimensional features. LGB is an implementation of the gradient boosting decision tree, which is essentially based on decision tree training integration to gain the optimal model [56,57]. The LGB model uses the histogram algorithm to find the best branching point, which greatly improves the training speed of the model. At the same time, LGB optimizes the growth strategy of the decision tree and uses the leaf-wise algorithm with depth limitation to create the decision tree, which can reduce the unnecessary amount of computation. In addition, multiple linear regression (MLR) was selected as the benchmark model in this study to compare with the above two machine learning models.

2.3.4. The Framework for the Procedures

The diagram for the procedures in this study is shown in Figure 2. We developed a yield-predicting system based on multi-source environmental data using the APSIM model and regression models (MLR and RF). Firstly, the APSIM model was calibrated and validated based on observed phenology data and grain yield data at the selected sites. Then, we ran the implemented model to obtain the biomass and main growth stages, including the end of the juvenile stage, floral initiation, flowering, start of grain filling, and the milky stage. The main growth stages were used to calculate the 13 CIs. We aggregated the APSIM-accumulated biomass (AB) and climate variables into four groups by different growth periods. In the study, four prediction events (JF, FIF, FS, and SM) were triggered successively, while the predictive indicators were added with crop growth progression. Therefore, the number of predictive indicators would increase with progressing phases from JF to SM. Furthermore, we conducted "leave-one-year-out" experiments [25,58] for 2000–2010 to test the performances of the yield prediction models. Finally, the importance values for the input characteristic variables were analyzed based on the RF model and LGB model.

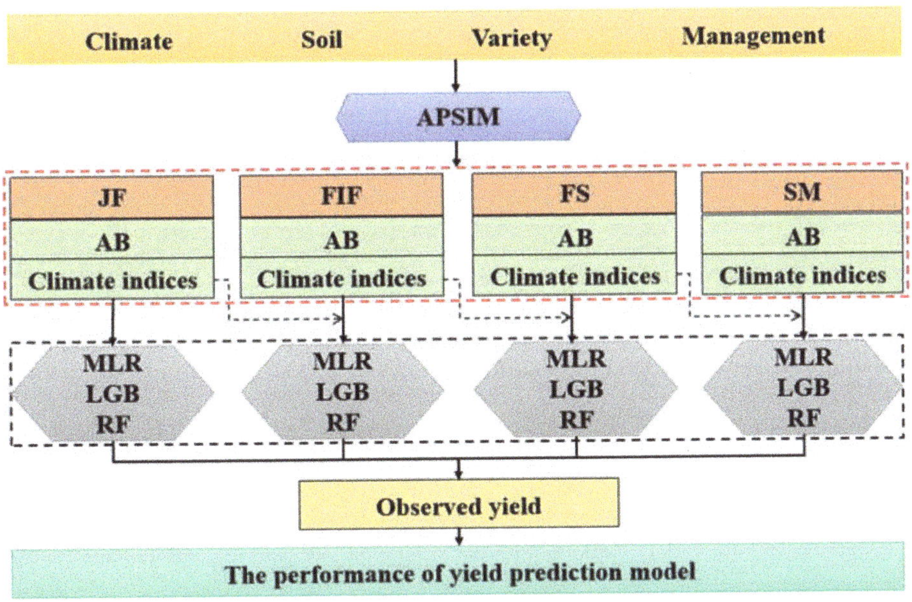

Figure 2. The diagram for the procedures used in this study, where JF, FIF, FS, and SM were the periods from the end of the juvenile stage to floral initiation, from floral initiation to flowering, from flowering to the start of grain filling, and from the start of grain filling to the milky stage, respectively. AB was accumulated biomass; MLR, LGB, and RF the multiple linear regression, light gradient boosting machine, and random forest, respectively.

2.3.5. Model Performance Assessment

The performance of the yield prediction model was validated by calculating the root mean square error (RMSE), Pearson's correlation coefficient (r), and mean absolute error

(MAE) between the estimated data and the observed data. The calculation formulas were as follows:

$$r = \frac{\sum_{i=1}^{n}(O_i - \overline{O})(S_i - \overline{S})}{\sqrt{\sum_{i=1}^{n}(O_i - \overline{O})^2}\sqrt{\sum_{i=1}^{n}(S_i - \overline{S})^2}} \quad (6)$$

$$RMSE = \sqrt{\frac{1}{n}\sum_{i=1}^{n}(O_i - S_i)^2} \quad (7)$$

$$MAE = \frac{\sum_{i=1}^{n}|O_i - S_i|}{n} \quad (8)$$

where O_i, S_i, $\overline{O}$, $\overline{S}$, and n represent the observed data, estimated data, mean value of the observed data, mean value of the estimated data, and the number of samples, respectively.

3. Results

3.1. Validation of the APSIM Model

The comparison of the observed and APSIM-simulated values of the flowering date (FD), maturity date (MD), and yield from 2000 to 2010 at the 20 sites is shown in Figure 3. The simulated FD and MD were in good agreement with observed values. The r values for the simulated and observed values of FD and MD were 0.78 and 0.82, respectively. The RMSE values between the simulated and observed values of FD and MD were 5.46 d and 4.94 d, respectively. On the other hand, the simulated grain yield was consistent with the observed yield, with r of 0.81 and RMSE of 792 kg ha^{-1}. Overall, the APSIM model can provide an acceptable assessment for the phenology and grain yield of wheat. Therefore, the simulation results from the APSIM model for wheat phenology and grain yield were reliable, and we could use the simulations to develop a hybrid model for predicting the wheat yield.

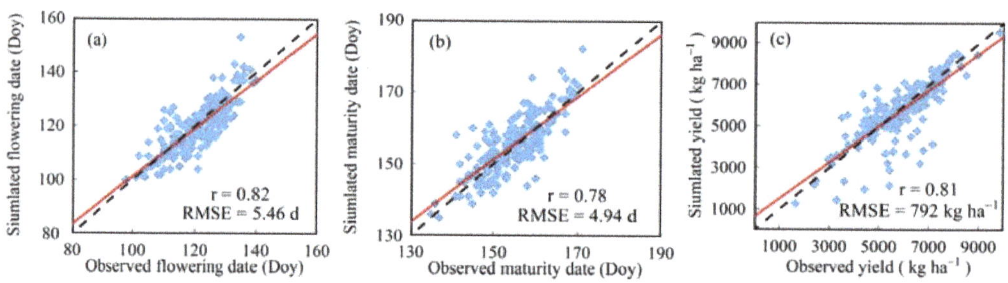

Figure 3. Comparison of observed and APSIM-simulated values of the flowering date (**a**), maturity date (**b**), and yield (**c**) from 2000 to 2010 at the 20 sites across the North China Plain. Red lines are the linear regression fit. Dashed lines represent the 1:1 lines.

3.2. The Model Performance and Optimum Leading Time for Yield Prediction

We developed a hybrid model to predict wheat yield based on the APSIM-simulated AB, climate indices at different growth stages and regression algorithms. The performances of three regression models are shown in Figures 4 and 5. In the early stage, the yield prediction accuracy of the three regression models was generally lower, with RMSE values of above 1000 kg ha^{-1} and MAE values of more than 700 kg ha^{-1} (Figures 4a,e,i and 5a,c,e). With the progression of wheat growth period, the input variables also increased, and the performances of the prediction models improved further. From JF to SM, the prediction accuracy increased significantly for the three regression models. For the MLR model, r increased from 0.22 to 0.79, RMSE decreased from 1237 kg ha^{-1} to 778 kg ha^{-1}, and MAE decreased from 957 kg ha^{-1} to 619 kg ha^{-1} (Figures 4a–d and 5b). Compared with the machine learning model, MLR was less effective in predicting the wheat yield. The machine learning model can capture the nonlinear relationship between the characteristic variables and the yield, and the overall performance of the machine learning models was good, especially the RF model. For the RF model, r increased from 0.66 to 0.86, RMSE decreased from 1026 kg ha^{-1} to 683 kg ha^{-1}, and MAE decreased from 756 kg ha^{-1} to 498 kg ha^{-1} (Figures 4i–l and 5f). The performances of the yield prediction models improved gradually with the development of crop growth. However, the tradeoff between the accuracy and leading time needs to be taken into account. The yield prediction at JF will achieve the target of the prediction approximate with three months leading time before the maturity but with a poor performance (r < 0.66). The yield prediction at SM outperformed the yield prediction at other growth periods, while the leading time of the prediction decreased to below 15 d. A higher precision and longer lead time were taken into consideration for the prediction of the yield at FS. Therefore, FS can be regarded as the optimal period, providing the best performance of yield prediction and about one month of lead time.

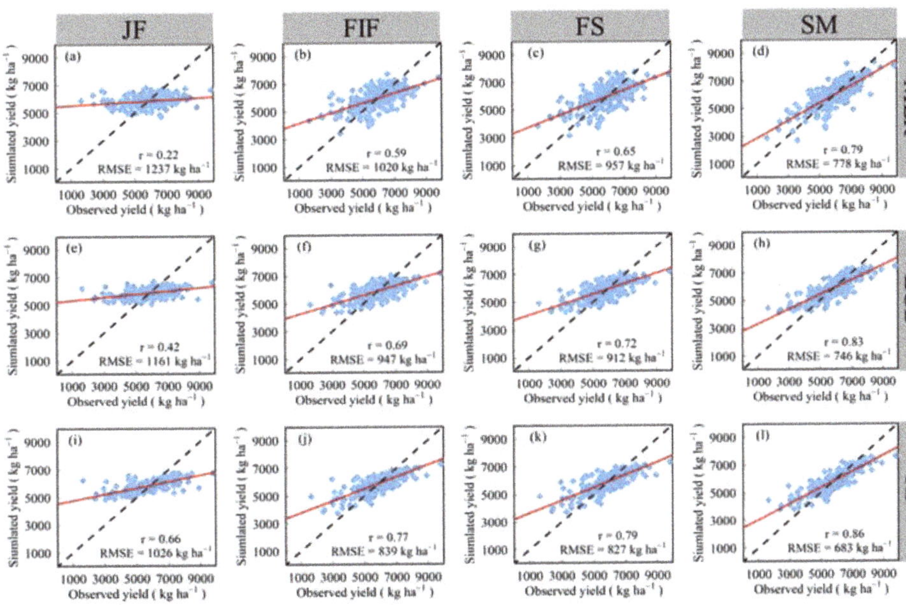

Figure 4. Comparison of the observed and predicted wheat yields for the period from the end of the juvenile stage to floral initiation (JF) (**a**,**e**,**i**), from floral initiation to flowering (FIF) (**b**,**f**,**j**), from the start of grain filling to the milky stage (FS) (**c**,**g**,**k**), and from flowering to the start of grain filling (SM) (**d**,**h**,**l**) from multiple linear regression (MLR) (**a**–**d**), light gradient boosting machine (LGB) (**e**–**h**), and random forest (RF) (**i**–**l**). Red lines are the linear regression fit. Dashed lines represent the 1:1 lines.

We compared the performance of the predicted yield across the study sites under irrigated conditions with the performance of the predicted yield across the study sites under rainfed conditions (Figures 6 and 7). The errors of the predicted yield from three regression models at all growth periods for the study sites under irrigated conditions (MAE ranged from 419 kg ha^{-1} to 789 kg ha^{-1}) were lower than those for the rainfed sites (MAE ranged from 624 kg ha^{-1} to 1130 kg ha^{-1}) (Figures 6 and 7). The accuracy of the predicted yield for the irrigated sites was higher than that for the rainfed sites. The water shortage caused by drought limited photosynthesis and carbon allocation, which was not conducive to the formation of the crop yield and affected the prediction accuracy [59,60]. However, the impacts of water stress on the crop yield was reduced by irrigation, which improved the accuracy of the yield prediction under irrigated conditions [61–64]. Nevertheless, the predicted yield for the study sites under irrigated conditions were underestimated compared to the observed yield, while the predicted yield for the rainfed sites were overestimated compared to the observed yield (Figures 6a,c,e and 7a,c,e).

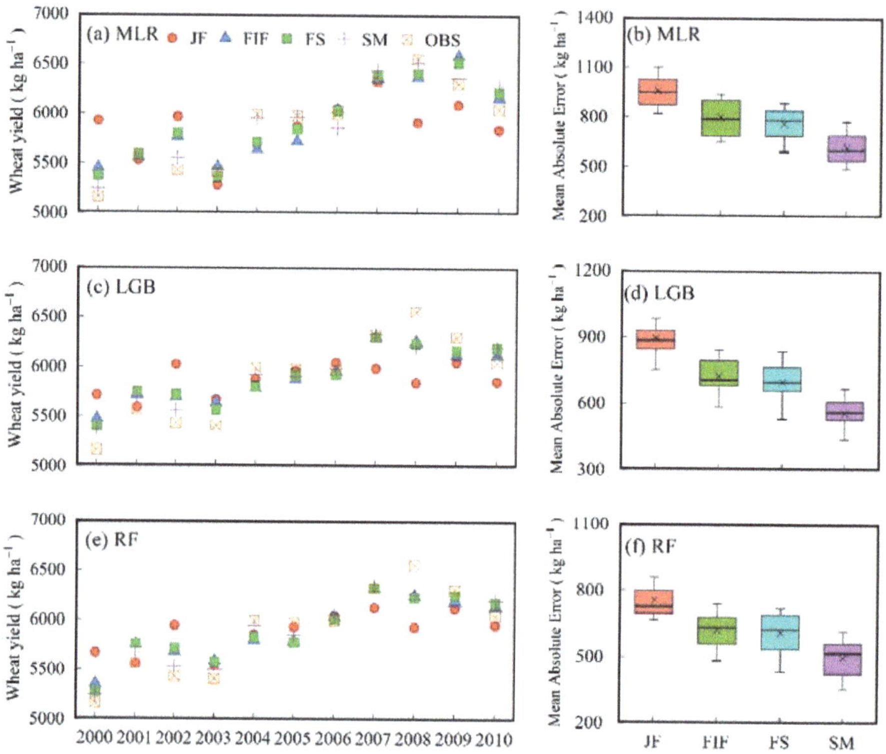

Figure 5. Time series of observed and predicted wheat yields across the 20 investigated sites based on the four prediction events from multiple linear regression (MLR) (**a**,**b**), light gradient boosting machine (LGB) (**c**,**d**), and random forest (RF) (**e**,**f**). Wheat yields for each year were averaged across the 20 investigated sites. Data were generated from the "leave-one-year-out" cross-validation procedure from the three regression models. JF, FIF, FS, and SM were the periods from the end of the juvenile stage to floral initiation, from floral initiation to flowering, from flowering to the start of grain filling, and from the start of grain filling to the milky stage, respectively.

3.3. *Relative Importance of Selected Predictors at Different Growth Stages*

The RF model and LGB model were used to assess the importance of the input characteristic variables in the yield prediction model. The relative importance of the input predictors as determined from the average of the LGB model and RF model for each prediction event is shown in Figure 8. With the crop growth and progression, the importance of the APSIM-simulated accumulated biomass (AB) increased rapidly, while the importance of AB at the last three prediction events was over 30% (Figure 8). For the CIs, the climate suitability indices were most important for the yield prediction at the early prediction event, such as TS and SS (Figure 8a). The roles of the climate suitability indices in the prediction model should not be ignored, though some extreme climate indices had higher importance than the climate suitability indices in the last three prediction events (Figure 8b–d). In the middle prediction events (FIF and FS), SDII and FD at FIF generally ranked high in the climate indices, which may be because the wheat yield was very sensitive to low-temperature stress and water stress before flowering (Figure 8b,c). However, SDII and HCD at SM ranked first in the late prediction events, suggesting that the impact of heat stress and water stress after flowering on the wheat yield was more significant than low-temperature stress and water stress before flowering (Figure 8b–d).

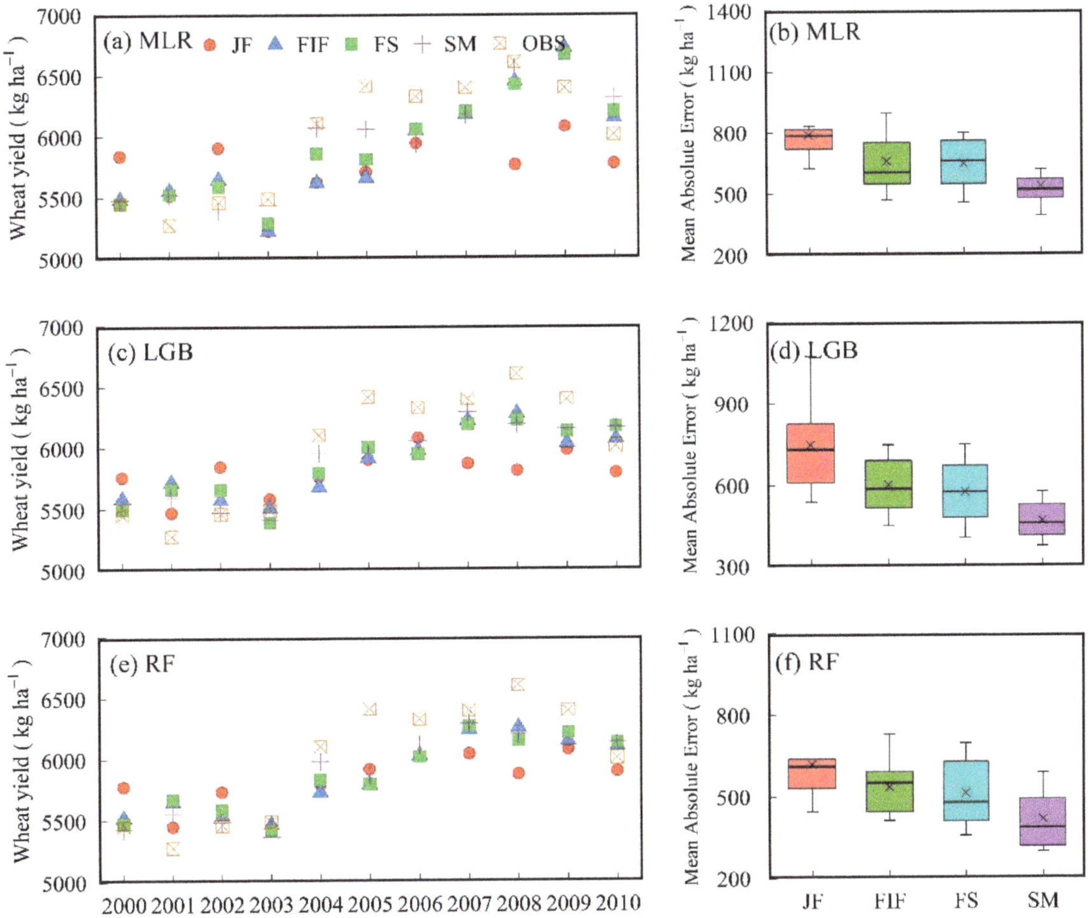

Figure 6. Time series of observed and predicted wheat yields across the investigated sites under irrigated conditions based on the four prediction events from multiple linear regression (MLR) (**a**,**b**), light gradient boosting machine (LGB) (**c**,**d**), and random forest (RF) (**e**,**f**). Wheat yields for each year were averaged across the investigated sites under irrigated conditions. Data were generated from the "leave-one-year-out" cross-validation procedure from the three regression models. JF, FIF, FS, and SM were the periods from the end of the juvenile stage to floral initiation, from floral initiation to flowering, from flowering to the start of grain filling, and from the start of grain filling to the milky stage, respectively.

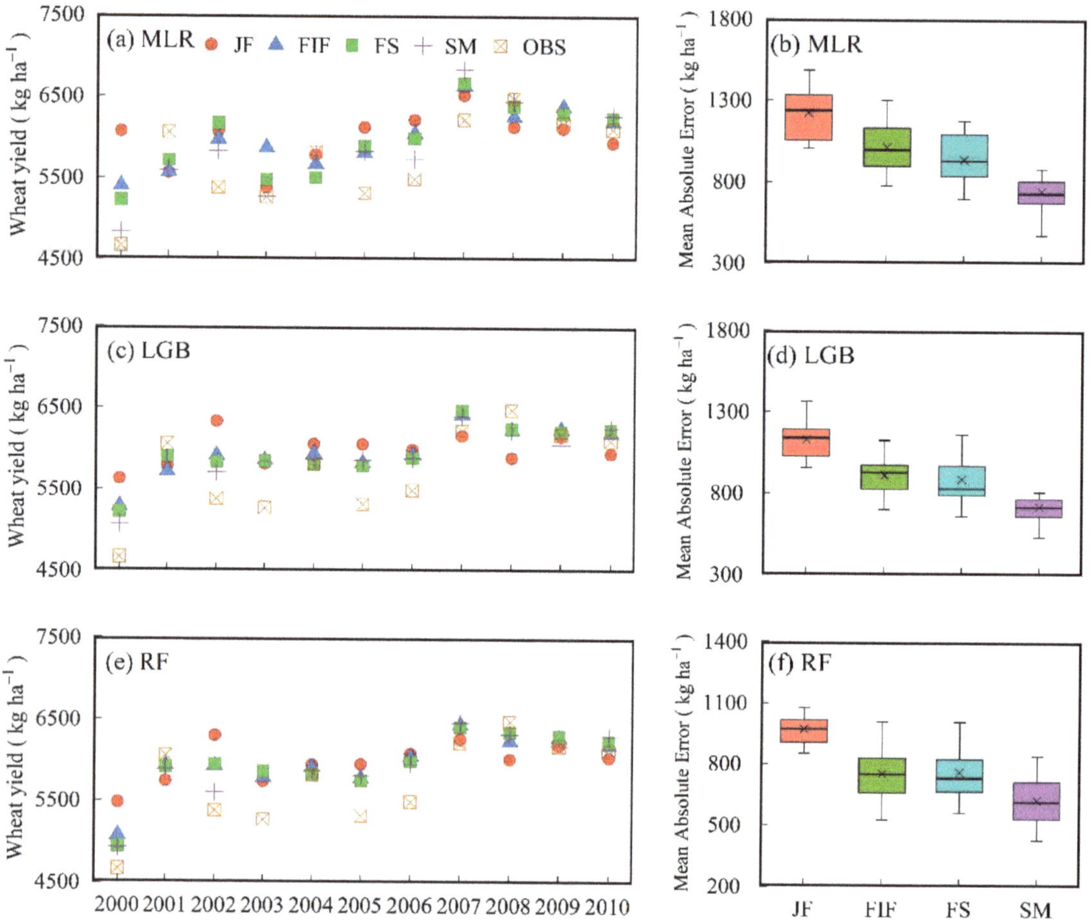

Figure 7. Time series of observed and predicted wheat yields across the investigated sites under rainfed conditions based on the four prediction events from multiple linear regression (MLR) (**a**,**b**), light gradient boosting machine (LGB) (**c**,**d**), and random forest (RF) (**e**,**f**). Wheat yields for each year were averaged across the investigated sites under rainfed conditions. Data were generated from the "leave-one-year-out" cross-validation procedure from the three regression models. JF, FIF, FS, and SM were the periods from the end of the juvenile stage to floral initiation, from floral initiation to flowering, from flowering to the start of grain filling, and from the start of grain filling to the milky stage, respectively.

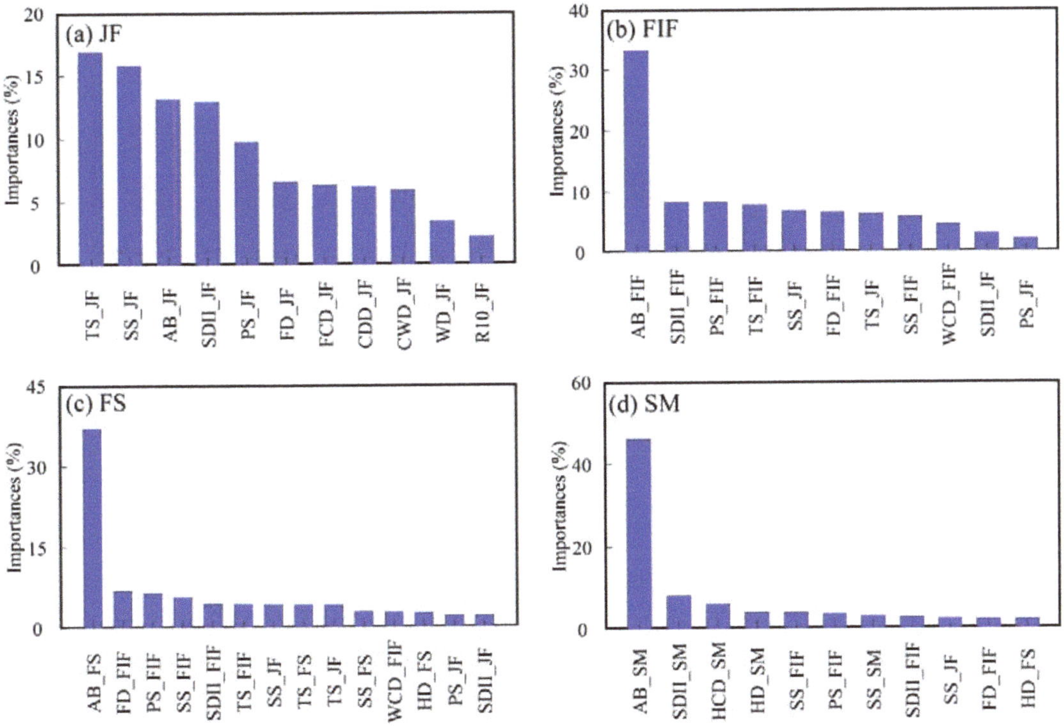

Figure 8. Relative importance of the input predictors as determined from the average of LGB (light gradient boosting machine) model and RF (random forest) model for the period from the end of the juvenile stage to floral initiation (JF) (**a**), from floral initiation to flowering (FIF) (**b**), from flowering to the start of grain filling (FS) (**c**), and from the start of grain filling to the milky stage (SM) (**d**). The results are normalized to sum 100% and shown in decreasing order in the figure (The input predictors lower than 2% were not shown in the figure).

4. Discussion

A crop model can dynamically describe the process of crop growth and development under various environmental conditions [65]. A growing body of studies have investigated the effects of climate change during the past few decades on crop phenology and yield using various crop models to develop adaptive measures (such as adjustment of the sowing date and renewal of crop variety) for reducing the yield loss [66–68]. However, there were fewer studies on the yield prediction using the crop model due to the limitation of the meteorological data. Some studies used the combination of statistical regression models and crop models to estimate the crop yield. For example, Everingham et al. [69] built one prediction model for sugarcane yield by incorporating the biomass simulated by the crop model and several climate indices into the RF algorithm and obtained a high accuracy. Similarly, Feng et al. [32] conducted the study on the yield prediction for wheat in South-Eastern Australia through combining the APSIM model and RF model, obtaining a high accuracy (r = 0.87, RMSE = 640 kg ha^{-1}). In this study, we developed a hybrid model for the yield prediction of the wheat coupling crop model and several statistical regression models. The yield prediction model based on the crop model and RF algorithm outperformed the yield prediction model based on the crop model and other regression algorithms (MLR and LGB), with r of 0.86 and RMSE of 683 kgha^{-1}. The precision of the study was similar to the related study [32]. This may be because that RF algorithm has a strong ability of data processing, which improved the accuracy and robustness of the yield prediction based on the RF algorithm [70–72].

Global climate change has a significant impact on social economy and the natural environment, especially on agricultural production [73–80]. Different crops have different demands for climate resources, and more or less, climate resources are not conducive to crop growth and development [81,82]. As compared with extreme climate events, mean climate conditions generally made more contributions to the variations of wheat growth in the NCP [83]. The climate suitability can be used to estimate the sensitivity of crops to climate factors, such as mean temperature, precipitation, and sunshine, and there is a certain correlation between climate suitability and climatic yield [84,85]. Climate suitability can further explore the information reflected by the mean climate conditions, though the crop model can simulate the effects of the mean climate conditions on crop growth to some extent. In this study, the climate suitability indices (TS, SS, and PS) played an important role in predicting the final yield of wheat, which generally ranked high in the models for every prediction event (Figure 8). The roles of climate suitability indices in the prediction model should not be ignored.

Extreme temperature events have a negative influence on crop growth and yield formation, which could cause crop yield loss [2,86–88]. The low-temperature events before flowering and high-temperature events after flowering are two major extreme temperature events affecting winter wheat [89,90]. Xiao et al. [91] found that there was the greatest frost duration and intensity in the NCP, which suffered the largest yield losses due to spring frost events. Warm temperatures can improve the growth of crops before the temperature reaches the threshold, but yields will abruptly diminish subsequently [92,93]. Around flowering or the grain filling period, extreme high temperature could affect pollination, reduce male fertility, and the efficiency of grain yield, and a large yield loss would be caused by continuous heat stress [94–97]. Bai et al. [98] found that heat stress after flowering significantly negative impacts wheat production in the NCP, while the wheat yield might have a higher frequency of exposure to extreme high-temperature stress in the future. The findings of the study showed that FD at FIF ranked high for the middle prediction events (FIF and FS), while HCD at SM ranked first in the climate indices for the late prediction events (Figure 8). Low-temperature events before flowering and heat stress after flowering are the main natural disasters affecting wheat growth in the NCP [90,99]. It is of great significance to take appropriate measures to alleviate the negative effects of these disasters on crops.

Drought is also closely correlated to agricultural production [100–103]. In this study, the rank of SDII related to water stress was consistently high for all prediction events, indicating that water stress has a significant impact on wheat yield in the North China Plain. Water stress can affect the coupling mechanism of environmental driving factors and crop yield, while it is difficult to achieve the acceptable yield prediction in the rainfed system [61]. The predicted yield for the sites under rainfed conditions would be overestimated due to the water stress, while irrigation can effectively reduce the effect of drought on the crop yield and increase the accuracy of the crop yield prediction [62,83]. In the study, the predicted yield for the sites in the rainfed system was overestimated, while the MAE of the predicted yield for the sites under the irrigated condition was significantly lower than that of the sites in the rainfed system (Figures 6 and 7). More predictive variables may need to be incorporated into the hybrid model to improve the performance of the model under irrigated conditions.

There are still some uncertainties and limitations in our study. The RF model is more dependent on data. Sufficient data samples are conducive to improving the accuracy and robustness of the model, while the lack of training samples may lead to overfitting and increase the uncertainty of the model [104]. The data processing ability of machine learning algorithms can fully function by obtaining more yield samples, while the performance of the model can be improved further. Furthermore, the model developed in this study is limited to the yield prediction at the site scale, which is difficult to be applied in a large-scale region. Lobell et al. [105] developed a scalable satellite-based crop yield mapper (SCYM) based on satellite images and crop models, which successfully explained 35% of the maize yield variation and 32% of the soybean yield variation in the study area. In the future, we can incorporate the SCYM model into the hybrid model to predict the crop yield at a large-scale region. This is a study direction with great development potential.

5. Conclusions

Based on the APSIM-simulated AB, climate indices at different growth stages, and statistical regression algorithms, we developed a hybrid model to predict wheat yields in the NCP. The results showed that the prediction model based on machine learning algorithms outperformed the prediction models using MLR regression, especially the RF algorithm. The performances of the yield prediction models improved gradually with the development of crop growth. A higher precision and longer lead time were taken into consideration for the prediction of the yield at FS. The FS can be regarded as the optimal period, providing the acceptable performance of yield prediction and about one month lead time. Moreover, the accuracy of the predicted yield for the irrigated sites was higher than that for the rainfed sites. The APSIM-simulated AB dominated the last three prediction events, with the importance above 30%. The climate suitability indices played an important role in predicting the wheat yield, with high rankings for every prediction event. Among extreme climate events, the low temperature events before flowering, high temperature events after flowering, and water stress were major extreme climate events affecting the winter wheat yield.

In general, the hybrid model can be used to predict the wheat yield under both rainfed and irrigated conditions in the NCP. This model is helpful in developing adaptation strategies to alleviate the negative effects of climate change on crop productivity and improve agricultural risk management. Nevertheless, the hybrid model is dependent on the quantity and quality of the data samples. Furthermore, the model developed in this study is limited in the yield prediction at the large-scale region. In the next study, we can incorporate the SCYM model into the hybrid model to predict crop yield at large-scale regions.

Author Contributions: Conceptualization, D.X.; methodology and data analysis, Y.Z.; writing—original draft preparation, Y.Z.; and writing—review and editing, D.X., H.B., J.T., D.L.L., Y.Q. and Y.S. All authors have read and agreed to the published version of the manuscript.

Funding: This research and the APC were funded by a grant from the Hebei Provincial Science Foundation for Distinguished Young Scholars (No. D2022205010), the National Natural Science Foundation of China (No. 41901128), the High-level Talents Training and Subsidy Project of Hebei Academy of Science (2022G04), and the Technology Program of Hebei Academy of Sciences (22102).

Institutional Review Board Statement: Not applicable.

Data Availability Statement: Not applicable.

Conflicts of Interest: The authors declare no conflict of interest.

References

1. Godfray, H.C.J.; Beddington, J.R.; Crute, I.R.; Haddad, L.; Lawrence, D.; Muir, J.F.; Pretty, J.; Robinson, S.; Thomas, S.M.; Toulmin, C. Food Security: The Challenge of Feeding 9 billion People. *Science* **2010**, *327*, 812–818. [CrossRef] [PubMed]
2. Bailey-Serres, J.; Parker, J.E.; Ainsworth, E.A.; Oldroyd, G.E.D.; Schroeder, J.I. Genetic strategies for improving crop yields. *Nature* **2019**, *575*, 109–118. [CrossRef] [PubMed]
3. Kotz, M.; Levermann, A.; Wenz, L. The effect of rainfall changes on economic production. *Nature* **2022**, *601*, 223–227. [CrossRef] [PubMed]
4. IPCC. *Climate Change 2021: The Physical Science Basis. Contribution of Working Group I to the Sixth Assessment Report of the Intergovernmental Panel on Climate Change*; Masson-Delmotte, V., Zhai, P., Pirani, A., Connors, S.L., Péan, C., Berger, S., Caud, N., Chen, Y., Goldfarb, L., Gomis, M.I., et al., Eds.; Cambridge University Press: Cambridge, UK, 2021.
5. Piao, S.; Ciais, P.; Huang, Y.; Shen, Z.; Peng, S.; Li, J.; Zhou, L.; Liu, H.; Ma, Y.; Ding, Y.; et al. The impacts of climate change on water resources and agriculture in China. *Nature* **2010**, *467*, 43–51. [CrossRef] [PubMed]

6. Estrella, N.; Sparks, T.H.; Menzel, A. Trends and temperature response in the phenology of crops in Germany. *Glob. Chang. Biol.* **2007**, *13*, 1737–1747. [CrossRef]
7. Li, E.; Zhao, J.; Pullens, J.W.M.; Yang, X. The compound effects of drought and high temperature stresses will be the main constraints on maize yield in Northeast China. *Sci. Total Environ.* **2022**, *812*, 152461. [CrossRef]
8. National Bureau of Statistics of China. *China Rural Statistical Yearbook*; China Statistics Press: Beijing, China, 2020.
9. Becker-Reshef, I.; Vermote, E.; Lindeman, M.; Justice, C. A generalized regression-based model for forecasting winter wheat yields in Kansas and Ukraine using MODIS data. *Remote Sens. Environ.* **2010**, *114*, 1312–1323. [CrossRef]
10. Tao, F.; Zhang, L.; Zhang, Z.; Chen, Y. Designing wheat cultivar adaptation to future climate change across China by coupling biophysical modelling and machine learning. *Eur. J. Agron.* **2022**, *136*, 126500. [CrossRef]
11. Bolton, D.K.; Friedl, M.A. Forecasting crop yield using remotely sensed vegetation indices and crop phenology metrics. *Agric. Forest Meteorol.* **2013**, *173*, 74–84. [CrossRef]
12. Franch, B.; Vermote, E.F.; Becker-Reshef, I.; Claverie, M.; Huang, J.; Zhang, J.; Justice, C.; Sobrino, J.A. Improving the timeliness of winter wheat production forecast in the United States of America, Ukraine and China using MODIS data and NCAR Growing Degree Day information. *Remote Sens. Environ.* **2015**, *161*, 131–148. [CrossRef]
13. Erfanian, S.; Ziaullah, M.; Tahir, M.A.; Ma, D. How does justice matter in developing supply chain trust and improving information sharing-an empirical study in Pakistan. *Int. J. Manuf. Technol. Manag.* **2021**, *35*, 354–368. [CrossRef]
14. Razzaq, A.; Xiao, M.; Zhou, Y.; Anwar, M.; Liu, H.; Luo, F. Towards Sustainable Water Use: Factors Influencing Farmers' Participation in the Informal Groundwater Markets in Pakistan. *Front. Environ. Sci.* **2022**, *10*, 944156. [CrossRef]
15. Guan, K.; Wu, J.; Kimball, J.S.; Anderson, M.C.; Frolking, S.; Li, B.; Hain, C.R.; Lobell, D.B. The shared and unique values of optical, fluorescence, thermal and microwave satellite data for estimating large-scale crop yields. *Remote Sens. Environ.* **2017**, *199*, 333–349. [CrossRef]
16. Mkhabela, M.S.; Bullock, P.; Raj, S.; Wang, S.; Yang, Y. Crop yield forecasting on the Canadian Prairies using MODIS NDVI data. *Agric. Forest Meteorol.* **2011**, *151*, 385–393. [CrossRef]
17. Michel, L.; Makowski, D. Comparison of statistical models for analyzing wheat yield time series. *PLoS ONE* **2013**, *8*, e78615. [CrossRef]
18. Satir, O.; Berberoglu, S. Crop yield prediction under soil salinity using satellite derived vegetation indices. *Field Crops Res.* **2016**, *192*, 134–143. [CrossRef]
19. Feng, P.; Wang, B.; Liu, D.L.; Waters, C.; Yu, Q. Incorporating machine learning with biophysical model can improve the evaluation of climate extremes impacts on wheat yield in south-eastern Australia. *Agric. Forest Meteorol.* **2019**, *275*, 100–113. [CrossRef]
20. Li, Y.; Guan, K.; Yu, A.; Peng, B.; Zhao, L.; Li, B.; Peng, J. Toward building a transparent statistical model for improving crop yield prediction: Modeling rainfed corn in the U.S. *Field Crops Res.* **2019**, *234*, 55–65. [CrossRef]
21. Kamir, E.; Waldner, F.; Hochman, Z. Estimating wheat yields in Australia using climate records, satellite image time series and machine learning methods. *ISPRS J. Photogramm.* **2020**, *160*, 124–135. [CrossRef]
22. Oliveira, R.A.; Näsi, R.; Niemeläinen, O.; Nyholm, L.; Honkavaara, E. Machine learning estimators for the quantity and quality of grass swards used for silage production using drone-based imaging spectrometry and photogrammetry. *Remote Sens. Environ.* **2020**, *246*, 111830. [CrossRef]
23. Wan, L.; Cen, H.; Zhu, J.; Zhang, J.; Zhu, Y.; Sun, D.; Du, X.; Zhai, L.; Weng, H.; Li, Y.; et al. Grain yield prediction of rice using multi-temporal UAV-based RGB and multispectral images and model transfer—A case study of small farmlands in the South of China. *Agric. Forest Meteorol.* **2020**, *291*, 108096. [CrossRef]
24. Erfanian, S.; Zhou, Y.; Razzaq, A.; Abbas, A.; Safeer, A.A.; Li, T. Predicting Bitcoin (BTC) Price in the Context of Economic Theories: A Machine Learning Approach. *Entropy* **2022**, *24*, 1487. [CrossRef]
25. Cai, Y.; Guan, K.; Lobell, D.; Potgieter, A.B.; Wang, S.; Peng, J.; Xu, T.; Asseng, S.; Zhang, Y.; You, L.; et al. Integrating satellite and climate data to predict wheat yield in Australia using machine learning approaches. *Agric. Forest Meteorol.* **2019**, *274*, 144–159. [CrossRef]
26. Hunt, M.L.; Blackburn, G.A.; Carrasco, L.; Redhead, J.W.; Rowland, C.S. High resolution wheat yield mapping using Sentinel-2. *Remote Sens. Environ.* **2019**, *233*, 111410. [CrossRef]
27. Huang, S.; Lv, L.; Zhu, J.; Li, Y.; Tao, H.; Wang, P. Extending growing period is limited to offsetting negative effects of climate changes on maize yield in the North China Plain. *Field Crops Res.* **2018**, *215*, 66–73. [CrossRef]
28. Xiao, D.; Shen, Y.; Zhang, H.; Moiwo, J.P.; Qi, Y.; Wang, R.; Pei, H.; Zhang, Y.; Shen, H. Comparison of winter wheat yield sensitivity to climate variables under irrigated and rain-fed conditions. *Front. Earth Sci.* **2015**, *10*, 444–454. [CrossRef]
29. Zhang, T.; Huang, Y.; Yang, X. Climate warming over the past three decades has shortened rice growth duration in China and cultivar shifts have further accelerated the process for late rice. *Glob. Chang. Biol.* **2013**, *19*, 563–570. [CrossRef]
30. Basso, B.; Liu, L. Chapter Four—Seasonal crop yield forecast: Methods, applications, and accuracies. In *Advances in Agronomy*; Sparks, D.L., Ed.; Academic Press: Cambridge, MA, USA, 2019; Volume 154, pp. 201–255.
31. Pagani, V.; Guarneri, T.; Busetto, L.; Ranghetti, L.; Boschetti, M.; Movedi, E.; Campos-Taberner, M.; Garcia-Haro, F.J.; Katsantonis, D.; Stavrakoudis, D.; et al. A high-resolution, integrated system for rice yield forecasting at district level. *Agric. Syst.* **2019**, *168*, 181–190. [CrossRef]
32. Feng, P.; Wang, B.; Liu, D.L.; Waters, C.; Xiao, D.; Shi, L.; Yu, Q. Dynamic wheat yield forecasts are improved by a hybrid approach using a biophysical model and machine learning technique. *Agric. For. Meteorol.* **2020**, *285*, 107922. [CrossRef]

33. Ma, S.Q. The climatic and ecological suitability of central Jilin Province for developing maize zones. *J. Ecol.* **1990**, *54*, 40–45.
34. Gong, L.J.; Yang-Hui, J.I.; Wang, P.; Zhu, H.X.; Wang, L.L.; Wang, Q.J. Variation of Climate Suitability of Maize in Northeast of China. *J. Maize Sci.* **2013**, *21*, 140–146.
35. Lobell, D.B.; Hammer, G.L.; McLean, G.; Messina, C.; Roberts, M.J.; Schlenker, W. The critical role of extreme heat for maize production in the United States. *Nat. Clim. Chang.* **2013**, *3*, 497–501. [CrossRef]
36. Wu, J.; Han, Z.; Xu, Y.; Zhou, B.; Gao, X. Changes in Extreme Climate Events in China under 1.5–4 °C Global Warming Targets: Projections Using an Ensemble of Regional Climate Model Simulations. *J. Geophys. Res. Atmos.* **2020**, *125*, e2019JD031057. [CrossRef]
37. Xiao, D.; Liu, D.L.; Wang, B.; Feng, P.; Waters, C. Designing high-yielding maize ideotypes to adapt changing climate in the North China Plain. *Agric. Syst.* **2020**, *181*, 102805. [CrossRef]
38. Xiao, D.; Tao, F. Contributions of cultivar shift, management practice and climate change to maize yield in North China Plain in 1981–2009. *Int. J. Biometeorol.* **2016**, *60*, 1111–1122. [CrossRef]
39. Shi, X.; Yu, D.; Warner, E.; Pan, X.; Petersen, G.; Gong, Z.; Weindorf, D. Soil database of 1:1,000,000 digital soil survey and reference system of the Chinese genetic soil classification system. *Soil Surv. Horiz.* **2004**, *45*, 129–136. [CrossRef]
40. Asseng, S.; Keating, B.A.; Fillery, I.R.P.; Gregory, P.J.; Abrecht, D.G. Performance of the APSIM-wheat model in Western Australia. *Field Crops Res.* **1998**, *57*, 163–179. [CrossRef]
41. Asseng, S.; Van Keulen, H.; Stol, W. Performance and application of the APSIM Nwheat model in the Netherlands. *Eur. J. Agron.* **2000**, *12*, 37–54. [CrossRef]
42. Keating, B.A.; Carberry, P.S.; Hammer, G.L.; Probert, M.E.; Robertson, M.J.; Holzworth, D.; Huth, N.I.; Hargreaves, J.N.G.; Meinke, H.; Hochman, Z. An overview of APSIM, a model designed for farming systems simulation. *Eur. J. Agron.* **2003**, *18*, 267–288. [CrossRef]
43. Arshad, A.; Raza, M.A.; Zhang, Y.; Zhang, L.; Wang, X.; Ahmed, M.; Habib-ur-Rehman, M. Impact of Climate Warming on Cotton Growth and Yields in China and Pakistan: A Regional Perspective. *Agriculture* **2021**, *11*, 97. [CrossRef]
44. Bai, H.; Xiao, D.; Wang, B.; Liu, D.L.; Feng, P.; Tang, J. Multi-model ensemble of CMIP6 projections for future extreme climate stress on wheat in the North China plain. *Int. J. Climatol.* **2020**, *41*, E171–E186. [CrossRef]
45. Xiao, D.; Bai, H.; Liu, D.L.; Tang, J.; Wang, B.; Shen, Y.; Cao, J.; Feng, P. Projecting future changes in extreme climate for maize production in the North China Plain and the role of adjusting the sowing date. *Mitig. Adapt. Strateg. Glob. Chang.* **2022**, *27*, 1–21. [CrossRef]
46. Zhao, Y.; Xiao, D.; Bai, H.; Tang, J.; Liu, D. Future Projection for Climate Suitability of Summer Maize in the North China Plain. *Agriculture* **2022**, *12*, 348. [CrossRef]
47. Zhao, F.; Qian, H.S.; Jiao, S.X. The climatic suitability model of crop: A case study of winter wheat in Henan province. *Resour. Sci.* **2003**, *25*, 77–82.
48. Zhao, J.; Guo, J.; Xu, Y.; Mu, J. Effects of climate change on cultivation patterns of spring maize and its climatic suitability in Northeast China. *Agric. Ecosyst. Environ.* **2015**, *202*, 178–187. [CrossRef]
49. Zhao, Y.; Xiao, D.; Bai, H.; Tang, J.Z. Climatic suitability degrees of winter wheat and summer maize in the North China Plain. *Chin. J. Ecol.* **2020**, *39*, 1–12.
50. Huang, H. A study on the climatic ecology adaptability of the crop production in the red and yellow soils region of China. *J. Nat. Resour.* **1996**, *11*, 340–346.
51. Pu, J.Y.; Yao, X.Y.; Yao, R.X. Variations of summer and autumn grain crops' climatic suitability in the areas east of Yellow River in Gansu in recent 40 years. *Agric. Res. Arid. Areas* **2011**, *29*, 253–258.
52. Hou, Y.Y.; Zhang, Y.H.; Wang, L.Y.; Hou-Quan, L.; Song, Y.B. Climatic suitability model for spring maize in Northeast China. *Chin. J. Appl. Ecol.* **2013**, *24*, 3207–3212.
53. Wang, E.; Han, X. The productivity evaluation and its application of winter wheat and summer maize in Huang-Huai-Hai region. *Chin. J. Agrometeorol.* **1990**, *11*, 41–46.
54. Allen, R.; Pereira, L.; Dirk, R.; Smith, M. *Crop Evapotranspiration Guidelines for Computing Crop Water Requirements—FAO Irrigation and Drainage Paper No 56*; FAO: Rome, Italy, 1998; Volume 300, pp. 53–62.
55. Breiman, L. Random Forests. *Mach. Learn.* **2001**, *45*, 5–32. [CrossRef]
56. Ke, G.; Meng, Q.; Finley, T.; Wang, T.; Chen, W.; Ma, W.; Ye, Q.; Liu, T.-Y. LightGBM: A highly efficient gradient boosting decision tree. *Adv. Neural Inf. Process. Syst.* **2017**, *30*, 3146–3154.
57. Sun, X.; Liu, M.; Sima, Z. A novel cryptocurrency price trend forecasting model based on LightGBM. *Financ. Res. Lett.* **2020**, *32*, 101084. [CrossRef]
58. Cao, J.; Zhang, Z.; Luo, Y.; Zhang, L.; Zhang, J.; Li, Z.; Tao, F. Wheat yield predictions at a county and field scale with deep learning, machine learning, and google earth engine. *Eur. J. Agron.* **2021**, *123*, 126204. [CrossRef]
59. Chaves, M.M.; Pereira, J.S.; Maroco, J.; Rodrigues, M.L.; Ricardo, C.P.; Osorio, M.L.; Carvalho, I.; Faria, T.; Pinheiro, C. How plants cope with water stress in the field. Photosynthesis and growth. *Ann. Bot.* **2002**, *89*, 907–916. [CrossRef]
60. Sun, H.; Zhang, X.; Liu, X.; Liu, X.; Shao, L.; Chen, S.; Wang, J.; Dong, X. Impact of different cropping systems and irrigation schedules on evapotranspiration, grain yield and groundwater level in the North China Plain. *Agric. Water Manag.* **2019**, *211*, 202–209. [CrossRef]

61. Grassini, P.; Yang, H.; Cassman, K.G. Limits to maize productivity in Western Corn-Belt: A simulation analysis for fully irrigated and rainfed conditions. *Agric. Forest Meteorol.* **2009**, *149*, 1254–1265. [CrossRef]
62. Sibley, A.M.; Grassini, P.; Thomas, N.E.; Cassman, K.G.; Lobell, D.B. Testing Remote Sensing Approaches for Assessing Yield Variability among Maize Fields. *Agron. J.* **2014**, *106*, 24–32. [CrossRef]
63. Razzaq, A.; Qing, P.; Naseer, M.; Abid, M.; Anwar, M.; Javed, I. Can the informal groundwater markets improve water use efficiency and equity? Evidence from a semi-arid region of Pakistan. *Sci. Total Environ.* **2019**, *666*, 849–857. [CrossRef]
64. Razzaq, A.; Xiao, M.; Zhou, Y.; Liu, H.; Abbas, A.; Liang, W.; Naseer, M.A.U.R. Impact of Participation in Groundwater Market on Farmland, Income, and Water Access: Evidence from Pakistan. *Water* **2022**, *14*, 1832. [CrossRef]
65. De Wit, C.T. *Photosynthesis of Leaf Canopies*; Pudoc: Wageningen, The Netherlands, 1965.
66. Xiao, D.; Shen, Y.; Qi, Y.; Moiwo, J.P.; Min, L.; Zhang, Y.; Guo, Y.; Pei, H. Impact of alternative cropping systems on groundwater use and grain yields in the North China Plain Region. *Agric. Syst.* **2017**, *153*, 109–117. [CrossRef]
67. Yan, Z.; Zhang, X.; Rashid, M.A.; Li, H.; Jing, H.; Hochman, Z. Assessment of the sustainability of different cropping systems under three irrigation strategies in the North China Plain under climate change. *Agric. Syst.* **2020**, *178*, 102745. [CrossRef]
68. Zhu, G.; Liu, Z.; Qiao, S.; Zhang, Z.; Huang, Q.; Su, Z.; Yang, X. How could observed sowing dates contribute to maize potential yield under climate change in Northeast China based on APSIM model. *Eur. J. Agron.* **2022**, *136*, 126511. [CrossRef]
69. Everingham, Y.; Sexton, J.; Skocaj, D.; Inman-Bamber, G. Accurate prediction of sugarcane yield using a random forest algorithm. *Agron. Sustain. Dev.* **2016**, *36*, 1–9. [CrossRef]
70. Wang, L.A.; Zhou, X.; Zhu, X.; Dong, Z.; Guo, W. Estimation of biomass in wheat using random forest regression algorithm and remote sensing data. *Crop J.* **2016**, *4*, 212–219. [CrossRef]
71. Zhang, J.; Okin, G.S.; Zhou, B. Assimilating optical satellite remote sensing images and field data to predict surface indicators in the Western U.S.: Assessing error in satellite predictions based on large geographical datasets with the use of machine learning. *Remote Sens. Environ.* **2019**, *233*, 111382. [CrossRef]
72. Sakamoto, T. Incorporating environmental variables into a MODIS-based crop yield estimation method for United States corn and soybeans through the use of a random forest regression algorithm. *ISPRS. J. Photogramm.* **2020**, *160*, 208–228. [CrossRef]
73. Lobell, D.B.; Asner, G.P. Climate and management contributions to recent trends in U.S. agricultural yields. *Science* **2003**, *299*, 1032. [CrossRef]
74. Lobell, D.B.; Burke, M.B.; Tebaldi, C.; Mastrandrea, M.D.; Falcon, W.P.; Naylor, R.L. Prioritizing climate change adaptation needs for food security in 2030. *Science* **2008**, *319*, 607–610. [CrossRef]
75. Tao, F.; Zhang, Z.; Shi, W.; Liu, Y.; Xiao, D.; Zhang, S.; Zhu, Z.; Wang, M.; Liu, F. Single rice growth period was prolonged by cultivars shifts, but yield was damaged by climate change during 1981–2009 in China, and late rice was just opposite. *Glob. Change Biol.* **2013**, *19*, 3200–3209. [CrossRef]
76. Tao, F.; Zhang, S.; Zhang, Z.; Rotter, R.P. Maize growing duration was prolonged across China in the past three decades under the combined effects of temperature, agronomic management, and cultivar shift. *Glob. Change Biol.* **2014**, *20*, 3686–3699. [CrossRef]
77. Liu, B.; Asseng, S.; Müller, C.; Ewert, F.; Elliott, J.; David, B.L.; Martre, P.; Ruane, A.C.; Wallach, D.; Jones, J.W.; et al. Similar estimates of temperature impacts on global wheat yield by three independent methods. *Nat. Clim. Chang.* **2016**, *6*, 1130–1136. [CrossRef]
78. Fletcher, A.L.; Chen, C.; Ota, N.; Lawes, R.A.; Oliver, Y.M. Has historic climate change affected the spatial distribution of water-limited wheat yield across Western Australia? *Clim. Chang.* **2020**, *159*, 347–364. [CrossRef]
79. Ye, Z.; Qiu, X.; Chen, J.; Cammarano, D.; Ge, Z.; Ruane, A.C.; Liu, L.; Tang, L.; Cao, W.; Liu, B.; et al. Impacts of 1.5 °C and 2.0 °C global warming above pre-industrial on potential winter wheat production of China. *Eur. J. Agron.* **2020**, *120*, 126149. [CrossRef]
80. Salman, S.A.; Shahid, S.; Sharafati, A.; Salem, G.S.A.; Bakar, A.A.; Farooque, A.A.; Chung, E.-S.; Ahmed, Y.A.; Mikhail, B.; Yaseen, Z.M. Projection of Agricultural Water Stress for Climate Change Scenarios: A Regional Case Study of Iraq. *Agriculture* **2021**, *11*, 1288. [CrossRef]
81. Mall, R.K.; Lal, M.; Bhatia, V.S.; Rathore, L.S.; Singh, R. Mitigating climate change impact on soybean productivity in India: A simulation study. *Agric. Forest Meteorol.* **2004**, *121*, 113–125. [CrossRef]
82. Wang, P.; Wu, D.; Yang, J.; Ma, Y.; Feng, R.; Huo, Z. Summer maize growth under different precipitation years in the Huang-Huai-Hai Plain of China. *Agric. Forest Meteorol.* **2020**, *285*, 107927. [CrossRef]
83. Li, J.; Lei, H. Impacts of climate change on winter wheat and summer maize dual-cropping system in the North China Plain. *Environ. Res. Commun.* **2022**, *4*, 075014. [CrossRef]
84. Cao, Y.; Qi, J.; Wang, F.; Li, L.; Lu, J. Analysis of climate suitability of spring maize in Liaoning Province based on modulus and mathematics. *Sci. Geogr. Sin.* **2020**, *40*, 1210–1220.
85. Liu, B.; Liu, L.; Asseng, S.; Zhang, D.; Ma, W.; Tang, L.; Cao, W.; Zhu, Y. Modelling the effects of post-heading heat stress on biomass partitioning, and grain number and weight of wheat. *J. Exp. Bot.* **2020**, *71*, 6015–6031. [CrossRef]
86. Song, X.; Zhang, Z.; Chen, Y.; Wang, P.; Xiang, M.; Shi, P.; Tao, F. Spatiotemporal changes of global extreme temperature events (ETEs) since 1981 and the meteorological causes. *Nat. Hazards* **2013**, *70*, 975–994. [CrossRef]
87. Zhu, X.; Liu, T.; Xu, K.; Chen, C. The impact of high temperature and drought stress on the yield of major staple crops in northern China. *J. Environ. Manag.* **2022**, *314*, 115092. [CrossRef] [PubMed]
88. Chen, R.; Wang, J.; Li, Y.; Song, Y.; Huang, M.; Feng, P.; Qu, Z.; Liu, L. Quantifying the impact of frost damage during flowering on apple yield in Shaanxi province, China. *Eur. J. Agron.* **2023**, *142*, 126642. [CrossRef]

89. Liu, B.; Liu, L.; Tian, L.; Cao, W.; Zhu, Y.; Asseng, S. Post-heading heat stress and yield impact in winter wheat of China. *Glob. Chang. Biol.* **2014**, *20*, 372–381. [CrossRef] [PubMed]
90. Xiao, L.; Liu, B.; Zhang, H.; Gu, J.; Fu, T.; Asseng, S.; Liu, L.; Tang, L.; Cao, W.; Zhu, Y. Modeling the response of winter wheat phenology to low temperature stress at elongation and booting stages. *Agric. Forest Meteorol.* **2021**, *303*, 108376. [CrossRef]
91. Xiao, L.; Liu, L.; Asseng, S.; Xia, Y.; Tang, L.; Liu, B.; Cao, W.; Zhu, Y. Estimating spring frost and its impact on yield across winter wheat in China. *Agric. Forest Meteorol.* **2018**, *260*, 154–164. [CrossRef]
92. Lobell, D.B.; Roberts, M.J.; Schlenker, W.; Braun, N.; Little, B.B.; Rejesus, R.M.; Hammer, G.L. Greater sensitivity to drought accompanies maize yield increase in the U.S. Midwest. *Science* **2014**, *344*, 516–519. [CrossRef]
93. Zhao, C.; Liu, B.; Piao, S.; Wang, X.; Lobell, D.B.; Huang, Y.; Huang, M.; Yao, Y.; Bassu, S.; Ciais, P.; et al. Temperature increase reduces global yields of major crops in four independent estimates. *Proc. Natl. Acad. Sci. USA* **2017**, *114*, 9326–9331. [CrossRef]
94. Bita, C.E.; Gerats, T. Plant tolerance to high temperature in a changing environment: Scientific fundamentals and production of heat stress-tolerant crops. *Front. Plant Sci.* **2013**, *4*, 273. [CrossRef]
95. Wang, N.; Wang, E.; Wang, J.; Zhang, J.; Zheng, B.; Huang, Y.; Tan, M. Modelling maize phenology, biomass growth and yield under contrasting temperature conditions. *Agric. Forest Meteorol.* **2018**, *250*, 319–329. [CrossRef]
96. Siddik, M.A.; Zhang, J.; Chen, J.; Qian, H.; Jiang, Y.; Raheem, A.K.; Deng, A.; Song, Z.; Zheng, C.; Zhang, W. Responses of indica rice yield and quality to extreme high and low temperatures during the reproductive period. *Eur. J. Agron.* **2019**, *106*, 30–38. [CrossRef]
97. You, L.; Rosegrant, M.W.; Wood, S.; Sun, D. Impact of growing season temperature on wheat productivity in China. *Agric. Forest Meteorol.* **2009**, *149*, 1009–1014. [CrossRef]
98. Bai, H.; Xiao, D.; Wang, B.; Liu, L.; Tang, J. Simulation of Wheat Response to Future Climate Change Based on Coupled Model Inter-Comparison Project Phase 6 Multi-Model Ensemble Projections in the North China Plain. *Front. Plant Sci.* **2022**, *13*, 829580. [CrossRef]
99. Liu, B.; Asseng, S.; Wang, A.; Wang, S.; Tang, L.; Cao, W.; Zhu, Y.; Liu, L. Modelling the effects of post-heading heat stress on biomass growth of winter wheat. *Agric. Forest Meteorol.* **2017**, *247*, 476–490. [CrossRef]
100. Glotter, M.; Elliott, J. Simulating US agriculture in a modern Dust Bowl drought. *Nat. Plants* **2016**, *3*, 16193. [CrossRef]
101. Xu, C.; McDowell, N.G.; Fisher, R.A.; Wei, L.; Sevanto, S.; Christoffersen, B.O.; Weng, E.; Middleton, R.S. Increasing impacts of extreme droughts on vegetation productivity under climate change. *Nat. Clim. Chang.* **2019**, *9*, 948–953. [CrossRef]
102. Chiang, F.; Mazdiyasni, O.; AghaKouchak, A. Evidence of anthropogenic impacts on global drought frequency, duration, and intensity. *Nat. Commun.* **2021**, *12*, 2754. [CrossRef]
103. Kreibich, H.; Van Loon, A.F.; Schroter, K.; Ward, P.J.; Mazzoleni, M.; Sairam, N.; Abeshu, G.W.; Agafonova, S.; AghaKouchak, A.; Aksoy, H.; et al. The challenge of unprecedented floods and droughts in risk management. *Nature* **2022**, *608*, 80–86. [CrossRef]
104. Ma, Y.; Zhang, Z.; Kang, Y.; Özdoğan, M. Corn yield prediction and uncertainty analysis based on remotely sensed variables using a Bayesian neural network approach. *Remote Sens. Environ.* **2021**, *259*, 112408. [CrossRef]
105. Lobell, D.B.; Thau, D.; Seifert, C.; Engle, E.; Little, B. A scalable satellite-based crop yield mapper. *Remote Sens. Environ.* **2015**, *164*, 324–333. [CrossRef]

agriculture

Article

Modelling the Geographical Distribution Pattern of Apple Trees on the Loess Plateau, China

Wei Xu [1,2], Yuqi Miao [3], Shuaimeng Zhu [4], Jimin Cheng [2,5] and Jingwei Jin [5,*]

1 School of Humanities and Social Sciences, Jiangsu University of Science and Technology, Zhenjiang 212000, China
2 Institute of Soil and Water Conservation, Chinese Academy of Sciences and Ministry of Water Resources, Xianyang 712100, China
3 College of Grassland Agriculture, Northwest Agriculture and Forestry University, Xianyang 712100, China
4 School of Surveying and Land Information Engineering, Henan Polytechnic University, Jiaozuo 454000, China
5 Institute of Soil and Water Conservation, Northwest Agriculture and Forestry University, Xianyang 712100, China
* Correspondence: jjw@ms.iswc.ac.cn

Abstract: The Loess Plateau, known for its fragile ecosystems, is one of the traditional apple-producing regions in China. Although some management measures are needed to enhance sustainable agriculture in response to the rising pressure of climate change, the geographic distribution of apple trees considering multiple variables has not been considered. In this study, we used three software (the maximum entropy model, IDRISI, and ArcGIS) to simulate the potential distribution of suitable habitats and range shifts of apple trees in the near present and near future (i.e., the 2030s and the 2050s) under two climate scenarios (the Shared Socioeconomic Pathways (SSP)1-26 and SSP5-85), while taking a variety of environmental factors into account (e.g., temperature, precipitation, and terrain). After optimization, the class unsuitable habitat (CUH) changed the potential distribution pattern of apple trees on the Loess Plateau. Currently, the areas of lowly suitable habitat (LSH), moderately suitable habitat (MSH), highly suitable habitat (HSH), and CUH were 7.66×10^4, 2.80×10^4, 0.23×10^4, and 18.05×10^4 km^2, respectively. Compared to the centroid estimated under the climate of 1970–2000, the suitability range of apple trees was displaced to the northwest in both the 2030s and the 2050s in SSP5-85 (i.e., 63.88~81.30 km), causing a larger displacement in distance than SSP1-26 (i.e., 40.05~50.32 km). This study demonstrates the possible changes in the spatial distribution of apple trees on the Loess Plateau in the near future and may provide a strong basis for future policy making.

Keywords: suitable habitat; climate scenario; range shift; ArcGIS; MaxEnt; apple trees

Citation: Xu, W.; Miao, Y.; Zhu, S.; Cheng, J.; Jin, J. Modelling the Geographical Distribution Pattern of Apple Trees on the Loess Plateau, China. *Agriculture* **2023**, *13*, 291. https://doi.org/10.3390/agriculture13020291

Academic Editors: Dengpan Xiao and Wenjiao Shi

Received: 23 October 2022
Revised: 18 January 2023
Accepted: 23 January 2023
Published: 25 January 2023

1. Introduction

The Loess Plateau is located in the arid and semi-arid regions of Northwest China. Its world-famous loess deposition, soil erosion, and huge spatial heterogeneity in precipitation have resulted in a unique and fragile plateau ecosystem [1]. In the 20th century, soil degradation and dust storms were further aggravated by the unsustainable land use practices of local farmers and herdsmen (overgrazing and farmland reclamation, etc.) as a result of their intense struggle to survive [2–4]. Additionally, since the 1970s, after the implementation of several ecological improvement projects (e.g., the Grain for Green Project, the Natural Forest Project [5]), soil erosion in this region has been significantly reduced [6], and a large amount of farmland has been transformed into forests and grasslands [7]. Thus, the ecological environment has been significantly improved, and the economic forestry and fruit industries have rapidly developed [8,9]. The primary responsibilities of the Chinese government in this century are gradually turning to maintaining the work of soil and water conservation [10], increasing ecosystem biodiversity [11], raising incomes of local inhabitants [12], and developing sustainable ecological agriculture [6]. Global apple consumption is increasing annually [13], and the Loess Plateau has become the largest apple production area in China and even in the world [14]. As the main income-generating economic fruit in the region, the development of the apple industry is of great significance to reduce poverty [14,15]. The rational planning of cultivation patterns, as the cornerstone of realizing the healthy and stable development of the apple industry [16], plays an important role in achieving the long-term goal of sustainable development in the Loess Plateau. However, few studies have focused on the suitable habitats (SH) of this economic tree in this region, particularly in the context of climate change [17].

The continuous pressure of climate change and rapid social development affect the structure and function of global ecosystems [18], and are changing the distribution range of species to a great extent [19]. Throughout the past 100 years, the global average temperature has risen by about 0.85 °C [20], and is expected to rise by 1.5–2.1 °C by 2050 [21]. Until now, CO_2 concentrations have increased from 280 parts per million (ppm, the pre-industrial level) to around 408 ppm, and may reach 560 ppm (double pre-industrial levels) by 2060 without actions to reduce emissions [22]. Due to the adaption of the natural environment to human activities, the land use and cover change (LUCC) has become the most obvious alteration in natural ecological environments [23,24], especially in fragile ecosystems close to the range of human activities [25,26]. However, the LUCC may exacerbate climate change, limit human activities [18], and threaten global biodiversity [21].

Plant pathogens are generally ignored in the research and planning of the SH of economic trees [27] despite the fact that they have a powerful effect on the distribution of their host plants [28]. *Valsa mali*, a necrotrophic fungus belonging to ascomycete [29], causes the apple valsa canker (AVC). AVC is a serious disease affecting the quality and yields of apples [30,31], and seriously restricts the sustainable development of the apple industry in the Loess Plateau [32,33]. At the same time, relevant policies [34] and other biological factors [29] also have important impacts on the apple cultivation areas [18,19]. However, to the best of our knowledge, studies mainly focus on the geographic distribution of apple trees relying on environmental factors only (e.g., temperature, precipitation, terrain), while no attempts have been performed for coupling this with land use types and plant pathogens. Without considering these multiple related factors, the persuasiveness and accuracy of simulation results may be seriously affected.

The development of computer technology has promoted a variety of species distribution models (SDMs) [35,36]. These SDMs mainly model and calculate the distribution of species with georeferenced presence/absence data and their interrelated environmental layers (e.g., meteorological data, terrain data, and social data) [37,38]. The maximum entropy model (MaxEnt), a program that relies on continuous/classified environmental variables and associated occurrence data, produces highly precise predictions [38–42]. It should be noted that many environmental factors that are difficult to collect and quantify (e.g., impacts from the related biological species and policies such as land-use planning) are not easily taken into account by MaxEnt [29]. Hence, in order to better mimic the distribution of species, it required a technique to discern the effects of interactions between species and variables that are difficult to quantify.

In this study, we simulate the SH of apple trees while taking relevant policies and plant pathogens into account in the context of climate change. To reach this goal, we: (1) independently simulated the distribution pattern of an economically important plant species (i.e., *Malus pumila* Mill.) and its pathogen (i.e., *V. mali*); (2) evaluated and selected the limits of ranges of abiotic factors (i.e., LUCC) and biographic factors (i.e., *V. mali*) that influence the distribution pattern of an economically important plant species; (3) optimized the SH of apple trees by integrating the effects of those abiotic and biotic factors. The purpose of this study was to better understand how apple trees on the Loess Plateau will respond to climate change in the near future, and to offer a theoretical foundation for apple cultivation, structural adjustment, and policy-making in the connected businesses in this region.

2. Materials and Methods

2.1. Collection of Occurrence Data of Species and Environmental Variables

In this study, the occurrence data of apple trees and its plant pathogen (*V. mali*) in the northern area of the Yangtze River (China) were collected with field surveys, published articles, and online databases (for details see: [17,29]). In total, we collected 260 georeferenced present-only records on apple trees and 211 georeferenced present-only records on *V. mali*.

In the Coupled Model Intercomparison Projects phase 6 (CMIP6), 49 different modelling groups from different countries contributed around 100 unique climate models to represent the change in the future climate. In this initiative, the Representative Concentration Pathways used in phase 5 have been replaced by the new Shared Socioeconomic Pathways (SSP) that have approaches to different radiative forcing levels that depend on the emissions of greenhouse gases (SSP1-26: 2.6 wm^{-2}, SSP2-45: 4.5 wm^{-2}, SSP3-70: 7.0 wm^{-2}, and SSP5-85: 8.5 wm^{-2}), which in turn lead to increasing warming [43,44]. In terms of intergovernmental energy conservation and emissions reduction, SSP1-26 offers the most optimistic scenario for achieving the goal of limiting the temperature rise to 2 °C by 2100 [43], whereas SSP5-85 represents the worst case. In this study, two extreme SSPs (i.e., SSP1-26 and SSP5-85) were chosen to depict the future distribution pattern of *M. pumila* on the Loess Plateau. We chose the climate system model of the Beijing Climate Center (BCC-CSM2-MR) as the source data for this study as it has been widely utilized in previous studies [45,46] in East Asia. We downloaded nineteen bioclimate layers (i.e., bio1-bio19) with a spatial resolution of 2.5 arc-min from the WorldClim (www.worldclim.org/, accessed on 18 May 2020) [47]. These geodatabases include the climatic conditions of the near to present period (period 1970–2000) and climatic conditions estimated for the near future (results of simulations for period 2021–2040 and period 2041–2060). In this article, we use the 2030s and the 2050s for referring to the time periods 2021–2040 and 2041–2060 respectively. Moreover, we downloaded one elevation datum (1 km) from the RESDC (http://www.resdc.cn/, accessed on 19 May 2020), three soil texture data (clay, sand, and silt, 1 km), and one soil type datum (1 km) from the FAO (www.fao.org/soils-portal/, accessed on 11 May 2020). Including the bioclimatic layers, we prepared a total of 27 variables for MaxEnt: bioclimatic layers (19: bio1–bio19), terrain data (4: one elevation datum and its three derived terrain variables: aspect, curvature, and slope), and soil data (4: three soil texture data and one soil type datum) to MaxEnt (Table 1).

Table 1. The environmental factors used in the corresponding simulation (variables with labels of "+" for *Valsa mali*, "−" for *Malus pumila* Mill., and "*" for land use).

Factor	Variables	Description	Unit	Resolution	Labels	Sources
Bioclimatic layers	bio1	Annual Mean Temperature	°C	2.5 arc-min	±	www.worldclim.org/, accessed on 18 May 2020
	bio5	Max Temperature of Warmest Month	°C	2.5 arc-min	−	
	bio6	Min Temperature of Coldest Month	°C	2.5 arc-min	±	
	bio11	Mean Temperature of Coldest Quarter	°C	2.5 arc-min	+	
	bio12	Annual Mean Precipitation	mm	2.5 arc-min	±	
	bio15	Precipitation Seasonality		2.5 arc-min	+	
	bio16	Precipitation of Wettest Quarter	mm	2.5 arc-min	−	
	bio17	Precipitation of Driest Quarter	mm	2.5 arc-min	−	

Table 1. *Cont.*

Factor	Variables	Description	Unit	Resolution	Labels	Sources
Terrain data	aspect	Aspect			±	www.resdc.cn/, accessed on 19 May 2020
	curvature	Curvature			±	
	elevation	Elevation	m	1 km	±	
	slope	Slope	°		±	
Soil data	sand	Soil Texture		1 km	±	www.fao.org/soils-portal/, accessed on 11 May 2020
	soil	Soil Type		1 km	±	
Land use data	land use	Land Use and Cover Change		300 m	*	www.climate.copernicus.eu/, accessed on 8 August 2020

2.2. The Screening and Pre-Processing of Data

Model overfitting can be decreased by variable screening [48,49]. We first removed duplicated records [21] from the occurrence data of the species. The occurrence data were then evaluated in compliance with the requirements of the subsequent study simulations. Considering the geographic location of the Loess Plateau, we transformed the environmental layers with a resolution of 2.5 arc-min into the Asia North Albers Equal Area Conic (ANAEAC) with the resolution of ~4857 m. We then filtered the species occurrence data with a boundary distance of 5000 m to ensure that each grid involved in the model simulation at most covers a single species occurrence point. We obtained 158 points for *V. mali* and 107 points for apple trees (Figure 1) after filtering species occurrence data using the SDMs toolbox (version 2.4) of ArcGIS 10.2 (ESRI, Redlands, CA, USA) (more details see [17,29]). By using the toolbox of ArcGIS to further analyse the elevation layer, we obtained three more terrain factors: aspect, curvature, and slope. To avoid multicollinearity [50,51], we conducted a correlation analysis [48] on 27 variables of bioclimatic layers, terrain data, and soil data. We retained the Annual Mean Temperature (bio1) and Annual Mean Precipitation (bio12), and eliminated other bioclimatic variables with a correlation coefficient value greater than 0.8 (Table 1, see details in [29]). Based on the physiological growth requirements of apple trees [50] and the incidence rate trends of *V. mali* [32], we added six extra bioclimatic variables and finally obtained eight bioclimatic variables for MaxEnt (Table 1). This study also made the assumption that these terrain variables will not change in the near future due to the long-term stability of the terrain [48]. We then resampled the environmental variables (i.e., terrain data and soil data) into a spatial resolution of 2.5 arc-min and converted all the layers used in this study into WGS1984 (the geographic coordinate system) and ANAEAC (the projection coordinate system) to ensure that all software did not need to consider coordinate system transformation.

2.3. Model Processing

2.3.1. Processing with the Binary Maps of MaxEnt and Land Use Data

The self-evaluation capability facilites of MaxEnt [18,52] were used to assess the accuracy of the resulting models, including receiver operating characteristic (ROC) curves and the area under these curves (AUC). The AUC values range between 0 and 1, with higher AUC values indicating more accurate simulation results [53,54]. When the AUC value is more than 0.8, the result is good; when it is higher than 0.9, the result is excellent [38]. In this study, we selected the automatic mode, setting 10,000 as the maximum number of background points, and choosing a random seed for MaxEnt simulation. For the occurrence data of *V. mali* and *M. pumila*, we randomly selected 30% of them as test data to assess the accuracy of the model, while the remaining 70% were used to calibrate it. Five bootstrap replications (exported in ASCII) were performed and the simulation results were exported in ArcGIS for further analysis.

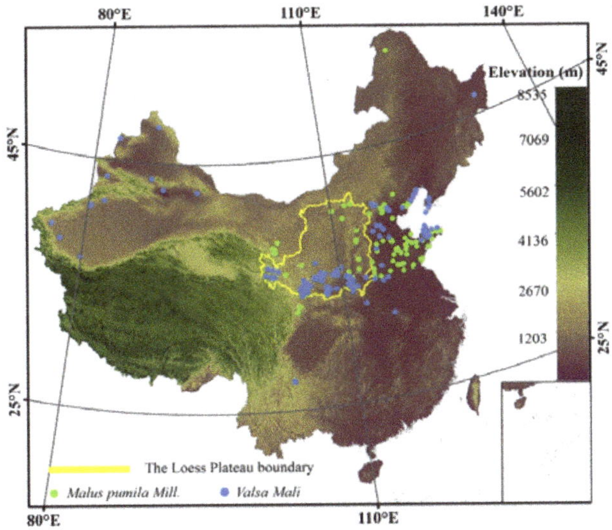

Figure 1. The occurrence data of *Malus pumila Mill.* and *Valsa mali* in China.

The selection of threshold values could improve the stability of the MaxEnt model [55]. In this study, we initially averaged the floating-point values of the acquired simulation results in order to discern between the presence and absence of species in the distribution maps. Then, the species floating matrix was divided into two types: the unsuitable part and the suitable part. For apple trees, *M. pumila*, the maximized sensitivity and specificity value (maxss, 0.2385) was set as the threshold [16,20,56]. The part with a floating value greater than the threshold was the suitable portion, and the remainder was the unsuitable portion. The suitable portion was then divided into three classes by occurrence probability values of 0.4 and 0.6 in lowly suitable, moderately suitable, and highly suitable habitats. For *V. mali*, the major plant pathogen of the AVC, the threshold was changed from its maxss value (0.160) to a new threshold (i.e., 80% of the floating value of the grid map) in order to establish the high-risk habitat of pathogen (HRHP). In other words, the HRHP only included values larger than 80% of the floating value. Additionally, the centroid of habitats was used as an essential metric to measure the range variations of the SH [57]. Hence, in this study, we also measured the centroids of apple trees in different climate conditions and drew their distribution maps with the help of the ArcGIS toolbox.

Land use data from various sources often have varying resolutions and translation criteria. Based on the land resources categorization method of the Chinese Academy of Sciences, this study reclassified the land use data from the European Aeronautics and Space Administration (given in Table S1) into seven categories (i.e., cropland, forest, grassland, mosaic area, bare area, built-up area, and water area). To create the plausible land use distribution maps for the near future, we first estimated the land use transfer matrix and transfer probability using the Markov model of IDRISI 17.0 (Clark Labs, Clark University, Worcester, MA, USA) between 2000 and 2010. The CA-Markov model of IDRISI was then utilized to complete the prediction of land use in 2030 and 2050 with a 10-year intermittent iteration, using the land use map of the starting period (years 2000 and 2010) and the newly established land use transfer and probability matrix.

2.3.2. Build the Mask Layer and Optimize the Suitable Habitats of Apple Trees

Unreasonable fruit tree management (e.g., pruning, [58]) and precipitation are major transmission pathways of *V. mali* [29]. In light of the size of orchards and the effect distance of *V. mali*, experts recommended establishing a buffer space for the HRHP with a distance of 300 m in order to avoid the AVC. With the help of the ArcGIS toolbox, we resampled the floating-point maps into about 323 m and screened the HRHP. We then added a high-risk buffer for the HRHP by setting the grids adjacent to the HRHP to be HRHP. We reclassified all values inside the high-risk buffer range in the ArcGIS toolbox, regardless of whether their attributes had previously been classified as high-risk (Figure S2).

In order to optimize the layout of urban and rural structures and promote the verification and rectification of permanent basic farmland (PBF), the Chinese government conducted their third national land survey (from 2017 to 2021) and designated 1.28×10^6 km^2 of national cultivated land (including 1.03×10^6 km^2 of actual PBF; see http://www.gov.cn/, accessed on 27 March 2021). Lacking current accurate digital distribution maps of PBF, this study initially used the land use map of 2015 as its research object and made the assumption that all farmland on the Loess Plateau was PBF. After that, the PBF of 2015 was transformed into a mask layer with a permanent transfer barrier. We finally produced the digital PBF maps for 2030 and 2050 (Figure S3) by overlapping the mask map with the predicted land use maps of the near future. Due to their unsuitability for the large-scale development of apple orchards, in this study, the built-up areas, water areas, and PBF were all reclassified as unsuitable land use types (ULUT, Figure S3).

To optimize the distribution pattern of apple trees under the two near future climate scenarios, we overlaid the ULUT mask, the HRHP mask, and the distribution map of apple trees. If the region matched at least one ULUT and HRHP, we defined it as an unsuitable habitat (CUH) for the cultivation of apple trees. Thereafter, the optimized maps were divided into the following five classes: unsuitable habitat (USH), lowly suitable habitat (LSH), moderately suitable habitat (MSH), highly suitable habitat (HSH), and CUH. Despite the fact that neither the USH nor the CUH were suitable for growing apple trees, there were important differences: the USHs were divided based on the outputs of the original model simulation, while the CUHs were split according to the distribution of the ULUT and HRHP.

3. Results

3.1. Model Robustness and the Independent Distribution Patterns of Apple Trees, ULUT and HRHP

Throughout the simulations, MaxEnt provided excellent predictions, with average AUC values of the apple trees (i.e., *M. pumila*) and their vital pathogen (i.e., *V. mali*) of 0.946 ± 0.02 and 0.965 ± 0.013 (mean $\pm$ SD), respectively. Without considering the effects of the ULUT and HRHP, the spatial distribution patterns of apple trees indicated that the MSH and HSH were mainly distributed in the south and southeast of the Loess Plateau, and the HSH increased in the west under both SSP1-26 and SSP5-85 in the 2030s and the 2050s (Figure S1). In all time periods (i.e., 1970–2000, the 2030s, and the 2050s), *V. mali* was mostly located in the central and southern Loess Plateau (Figure S2). The most noticeable changes in the modelling of future land use were the decrease in forests and the increase in built-up areas from the 1970–2000 period to the 2030s and the 2050s (Figure S3a–c). For apple trees, the ULUT was mainly dispersed in the south and southeast of the Loess Plateau in the three periods, and increased in the north from the 1970–2000 period to both the 2030s and the 2050s (Figure S3d,e).

3.2. Suitable Habitats under the Effects of Multiple Environmental Factors

The optimized results showed that, in the 1970–2000 period, the areas of USH, LSH, MSH, HSH, and CUH were 36.15×10^4, 7.66×10^4, 2.80×10^4, 0.23×10^4, and 18.05×10^4 km^2, respectively. In the south and southeast of the Loess Plateau, the CUH led to a significant decline in MSH (-5.14×10^4 km^2, ~−64.74%) and HSH (-0.63×10^4 km^2, ~−74.12%) (Figure 2). Under SSP1-26 and SSP5-85, the inclusion of the CUH resulted in an area decline of the SH in various degrees in the 2030s and the 2050s (Figure 2). In the near future, under SSP1-26, the USH, LSH, MSH, and HSH decreased by 8.75~10.58×10^4, 6.20~7.28×10^4, 3.06~3.40×10^4, and 0.04~0.25×10^4 km^2, respectively (Figure 2a,b). In the 2030s and the 2050s, the reduction in the MSH was 50.00% and 45.76%, respectively, whereas it was 50.00% and 59.52% for the HSH. Compared to the MSH and HSH, the LSH was less affected by the CUH, as the proportion of habitat decline in the near future was 39.61% (the 2030s) and 40.68% (the 2050s). Under SSP5-85, the USH, LSH, MSH, and HSH decreased by 8.65~11.07 × 10^4, 5.73~6.18 × 10^4 (−34.30%~−34.71%), 3.37~4.11 × 10^4 (−32.31%~−42.02%) and 0.21~0.26 × 10^4 km^2 (−36.62%~−53.85%), respectively (Figures 2d,e and 3).

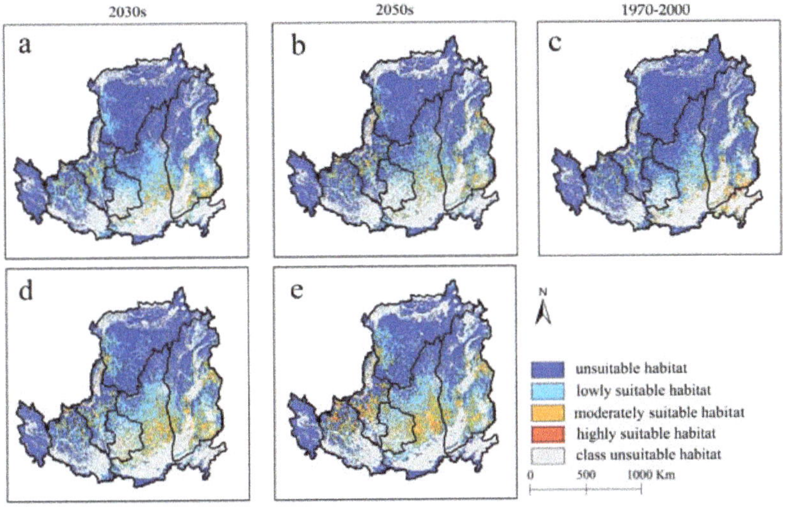

Figure 2. The spatial patterns of apple trees in the 1970–2000 (**c**) and near future under two scenarios ((**a**,**b**) SSP1-26, (**d**,**e**) SSP5-85).

According to the optimized results of the distribution of apple trees, the CUH mainly spread in the east, south, southeast, southwest, west, north, and northwest of the Loess Plateau (Figures 2 and S3). Currently, the CUH, within 1.81 × 10^5 km^2 (Figure 3), is predominantly contiguously distributed in the south, southeast, and southwest of the Loess Plateau, is sporadic in the central area, and is patchy in the west, northwest, north, and northeast; the MSH and HSH were mainly scattered in the southeast, and the LSH was mainly located in the central area (Figure 2). Compared with the 1970–2000 period, the area of the CUH expanded in different degrees in the 2030s and the 2050s under both SSP1-26 (1.07~2.37 × 10^4 km^2) and SSP5-85 (1.10~2.38 × 10^4 km^2, Figure 3). The CUH mostly extended in the east, north, and northeast of the Loess Plateau, while the changes in the south, southeast, southwest, and west were relatively slight (Figure S3). In contrast, under the two SSPs, the distribution range of the LSH and MSH were expanded to different degrees in the south, central, and southwest areas in the near future (Figure 2, Table S2). The HSH area change trend showed a V-shaped curve from the 1970–2000 period to the 2030s and the 2050s, while its area under SSP1-26 was still lower than it was under near present climatic conditions (Table S2).

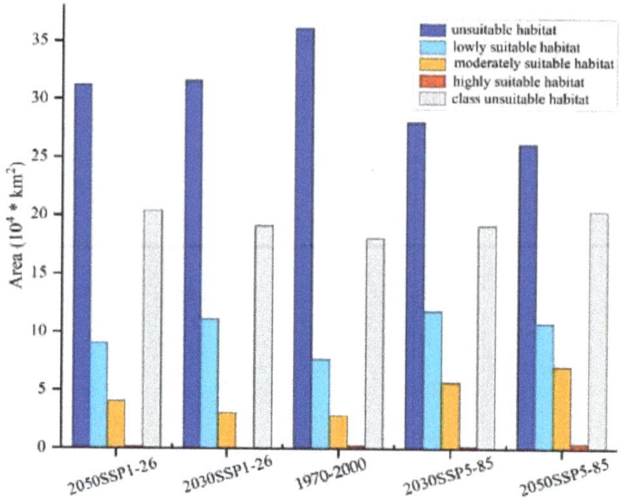

Figure 3. The habitat area of apple trees in three periods (1970–2000, the 2030s, and the 2050s) under two climate scenarios (SSP1-26 and SSP5-85).

3.3. Shifts of Centroids in the near Future under Two Climate Scenarios

The changes in the distribution patterns of SHs resulted in shifts of their centroids. Currently, the suitability centroid of apple trees is located at 109°22′11.79″ E, 36°21′11.92″ N (Figure 4). Under SSP1-26, the shift distances for the suitability centroid in the 2030s and the 2050s were 40.05 km and 50.32 km, respectively. Under SSP5-85, the shift distances were 63.88 km and 81.30 km, respectively (Figure 4). In particular, we noticed that, under both SSP1-26 and SSP5-85, all suitability centroids of apple trees displaced towards the northwest.

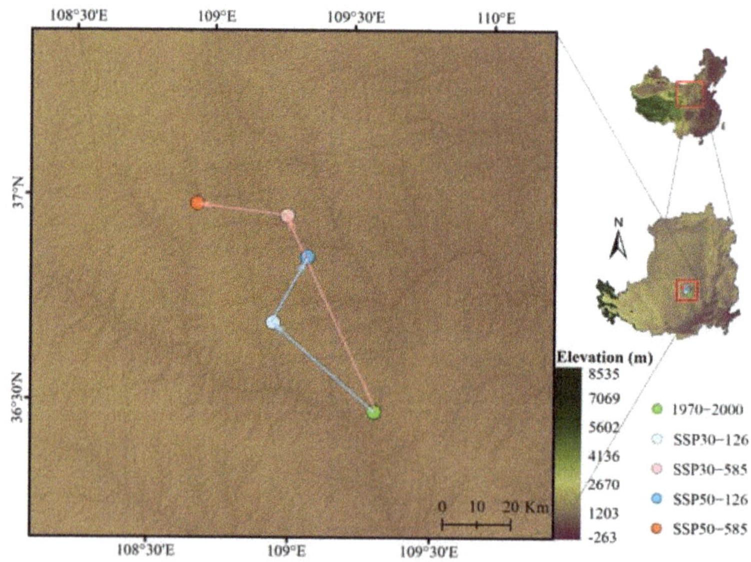

Figure 4. The suitability centroids of apple trees on the Loess Plateau.

4. Discussion

4.1. Effects of the USH and Range Shifts in Suitable Habitats of Apple Trees

In this study, the geographic distribution patterns of one economically important tree (i.e., *M. pumila*) and one of its vital pathogens (i.e., *V. mali*) under SSP1-26 and SSP5-85, as well as the corresponding land use patterns, were simulated using MaxEnt and the CA–Markov model. Based on this, more detailed maps of the geographic distribution of apple trees were produced. After considering the CUH (i.e., ULUT and HRHP), the distribution patterns of apple trees on the Loess Plateau became more fragmented as a result of the increase in the classification criteria from four SHs to five. Compared with its 1970–2000 range, the SH (i.e., the LSH, MSH, and part of the HSH in the 2050s period under SSP5-85) expanded to different degrees in the south and southwest of the Loess Plateau. This is consistent with the impact of the southeast monsoon on the direction of precipitation inside the Loess Plateau range [59]. Moreover, the geographical distribution of the Qinling Mountains (on the south and spanning from west to east, [1]) may have some impact on this. Furthermore, while not negligible, the influence of regional topography on regional climate change (i.e., temperature and precipitation, [59]) is often difficult to measure precisely.

Global climate change has brought challenges for the cultivation of apple trees on the Loess Plateau. From the 1970–2000 period to the 2030s and the 2050s, the area of the SH under SSP5-85 increased more than under SSP1-26. Under SSP1-26 and SSP5-85, the ideal cultivation habitat of apple trees on the Loess Plateau will shift northwest in the near future compared to their near present distribution. In addition, because the Loess Plateau is one of the most important apple-producing regions in China [14], a series of adaptation measures will be required to maintain the size and yield of the apple industry. In recent decades, the average elevation of apple orchards in Northern India has displaced upward by about 800 m [15], and it has moved northward and westward in China [17]. Across the period from the 2030s to the 2050s, the shift distances of suitability centroids increased between SSP1-26 and SSP5-85. We hypothesize that this may be related to the temperature influence on the growth and development of apple trees: similar to how temperature thresholds influence the development activities of particular pest species [60], an appropriate temperature increase will enhance the distribution of apple trees on the Loess Plateau, but when the temperature change exceeds a certain threshold, the promoting effect of temperature increase is likely to shift from the positive to negative. This is contrary to the practical experience of economically important forest trees seeking the most environmentally similar habitats when facing climate change [15,45].

4.2. Effects of Abiotic and Biological Factors on the CUH

Compared with the distribution patterns of the CUH in the 1970–2000 period, it will expand in different degrees in the near future. Furthermore, its expansion trends were mainly concentrated in the east, northeast, and north of the Loess Plateau, while the changes occurring in the south, southeast, southwest, west, and northwest directions were minimal. However, considering the delineation of the CUH, there were two potential causes for uncertainty in the simulations. On the one hand, the simulation uncertainty of the ULST may lead to the uncertainty in the CUH. Both natural controlling factors (i.e., temperature, precipitation, terrain, etc.) and socio-economic driving factors have impacts on the LUCC [61], and their effects vary depending on the land use type [62]. This is partly because some land use types (e.g., orchards) have more economic benefits than those with more ecological functions (e.g., forests and grasslands). This study assumed that the terrain factors would remain stable in the near future, and on this basis, the climatic factors, such as temperature and precipitation, could be the dominant natural factors impacting the Loess Plateau ecosystem. Meanwhile, given the significant impacts of socio-economic factors (i.e., policies, regulations, and systems, [63]) on the LUCC, it cannot be ignored. National ecological projects have significantly changed the land use patterns in the north, northwest, and northeast of China during the last 50 years [7], as well as the creation of nature reserves [64]. These are significant examples of how socio-economic factors have influenced the LUCC [61,63]. Hence, limited by land use policy, predicting future land use changes based on current regulations, although essential and indispensable at this point, cannot overlook the tremendous uncertainty caused by itself. On the other hand, the simulation uncertainty of the HRHP may lead to extra uncertainty in the range of the CUH. In the 2030s and the 2050s, the CUH range differs slightly between SSP1-26 and SSP5-85 (Figure 2). To lessen the computational burden of the simulation, the impact of the two SSPs on land use change was not considered separately in this study (i.e., the same land use simulations were used in both SSP1-26 and SSP5-85). Therefore, the difference in the range of the CUH under similar SSPs was caused by their HRHP range (Figure S2). For the plant pathogen *V. mali*, details about its host (i.e., apple trees, plum trees, etc.), biological factors (such as insect behaviour), and its potential transmission routes (such as seeds and stocks with the pathogen [58]) need to be considered in future studies to obtain more convincing simulation results. In addition, further research is required on the reaction of *V. mali* to temperature and precipitation. An essential factor in the prevention and treatment of the AVC is setting an adequate buffer distance for *V. mali*. Hence, future research should also pay more attention to how many relevant environmental factors such as land types, terrain, and other geographic barriers (e.g., seasonal wind, river, etc.) affect the buffer distance of *V. mali*.

4.3. Strategies to Improve the Accuracy of Simulations

This study produced an excellent simulation of the potential geographic distribution of apple trees in the near present and near future on the Loess Plateau. However, to increase its accuracy, three uncertainties should be overcome in follow-up studies. First, there are uncertainties within the MaxEnt model itself. Species occurrence data and environmental variables are the fundamental inputs of MaxEnt. Before building the model, we screened the species occurrence data (based on environmental data resolution) and environmental data (i.e., principal component analysis and correlation coefficient) separately [65]. However, it is still necessary to make sure that this is the best strategy to utilize these occurrence data. Additionally, merely considering the correlation and overlap between environmental factors may neglect crucial factors [57] that might have potential impacts on the distribution patterns of species. In order to reasonably filter environmental variables and enhance the stability and precision of the simulation of species distribution, future studies may need more extensive assessment strategies (e.g., evaluate the weight of environmental variables in the model, [66,67]). Second, there are uncertainties in dividing SHs. Though maxss has been commonly applied in SDM studies to distinguish between the potential of presence and absence of species [56], its stability in various hydrothermal environments remains uncertain. Future studies should focus more on improving and enhancing the indicators used to assess the presence and absence situation of species. Third, there are uncertainties in the mask maps. Some natural, inevitable uncertainties existed in the simulation of climate scenarios by the climate prediction organisations, resulting in the uncertainty in the mask maps of the HRHP. Human activities, policies, and climate change all have an impact on the LUCC [19,61]. However, their effects vary from ecosystem to ecosystem, especially in those that are regularly impacted by human activities [26]. The Loess Plateau ecosystem is particularly sensitive due to its peculiar climatic conditions [6,12]. It should be noted that the modelling estimates of land use in the near future in this study did not sufficiently account for relevant environmental factors. In order to increase the accuracy of modelling in this region, future studies may need to perform more in-depth related research on the driving forces [62]. At the moment, the LUCC predictions of the Loess Plateau under different climate change scenarios are still lacking, and hence their potential impacts are overlooked in this work.

5. Conclusions

In this study, we developed a model that took multiple environmental factors into account, such as temperature, precipitation, terrain, soil, climate change, and human activities, to simulate the distribution pattern of apple trees on the Loess Plateau in the near present and near future under two Shared Socioeconomic Pathways (i.e., SSP1-26 and SSP5-85). The increase in the SH in SSP5-85 was larger than in SSP1-26. In the near future, the CUH increased in different degrees in the east, northeast, and north of the Loess Plateau. The LSH, MSH, and HSH shrunk to varied degrees after optimization, taking the CUH into consideration, and their decrease in percentage was larger than that of the USH. Under the two SSPs, all suitability centroids shifted to the northwest in the near future relative to the 1970–2000 centroid. As the pressures of climate change increased from SSP1-26 to SSP5-85, the shift distances of centroids increased in both the 2030s and the 2050s. Under the same climate change pressures, the shift distance increased more in the 2050s than it did in the 2030s.

Supplementary Materials: The following supporting information can be downloaded at: https://www.mdpi.com/article/10.3390/agriculture13020291/s1. Figure S1. The spatial distribution patterns of suitable habitats of apple trees on the Loess Plateau in three periods under two climate scenarios (SSP1-26: (a,b); SSP5-85: (d,e)); Figure S2. The spatial patterns of high-risk habitat of Valsa Mali in the 1970–2000 and near future (the 2030s and 2050s) under two climate scenarios (SSP1-26: (a,b); SSP5-85: (d,e)); Figure S3. The land use patterns (a–c) and the unsuitable land use type for the cultivation of apple trees (d–f) on the Loess Plateau in the 1970–2000 and near future (the 2030s and 2050s); Table S1: The correspondence of the reclassification of the land use and cover changes; Table S2: The area of suitable habitats (before and after optimization) of apple trees on the Loess Plateau (104 × km2). The CUH represents the class unsuitable habitat.

Author Contributions: W.X. designed the study, collected occurrence data, drew the map, tables, and figures, and wrote the main body of the manuscript; Y.M. and S.Z. improved the manuscript in many parts and calculated the corresponding area of suitable habitats; J.J. and J.C. acquired the funding and revised the manuscript. All authors have read and agreed to the published version of the manuscript.

Funding: This study was financially supported by the Key Research and Development Program of Shaanxi Province (2021NY-006), Natural Science Foundation of China (31601987), China Agriculture Research System (CARS-34), Deployment Program of the Chinese Academy of Sciences (KJZD-EW-TZ-G10) and Doctoral Fund of Henan Polytechnic University (B2019-4).

Institutional Review Board Statement: Not application.

Acknowledgments: We thank the corresponding organisations that provided data and software to support this work. Importantly, we would like to thank Runzhi Mao for his spiritual support of our life and this work. We also thank the anonymous reviewers for their suggestions and contributions in improving this manuscript.

Conflicts of Interest: The authors declare no conflict of interest.

References

1. Fu, B.; Wang, S.; Liu, Y.; Liu, J.; Liang, W.; Miao, C. Hydrogeomorphic Ecosystem Responses to Natural and Anthropogenic Changes in the Loess Plateau of China. *Annu. Rev. Earth Planet. Sci.* **2017**, *45*, 223–243. [CrossRef]
2. Akiyama, T.; Kawamura, K. Grassland degradation in China: Methods of monitoring, management and restoration. *Grassl. Sci.* **2010**, *53*, 1–17. [CrossRef]
3. Zhou, W.; Gang, C.; Chen, Y.; Mu, S.; Sun, Z.; Li, J. Grassland coverage inter-annual variation and its coupling relation with hydrothermal factors in China during 1982–2010. *J. Geogr. Sci.* **2014**, *24*, 593–611. [CrossRef]
4. He, X.; Zhou, J.; Zhang, X.; Tang, K. Soil erosion response to climatic change and human activity during the Quaternary on the Loess Plateau, China. *Reg. Environ. Change* **2006**, *6*, 62–70. [CrossRef]
5. Yin, R.; Yin, G.; Li, L. Assessing China's Ecological Restoration Programs: What's Been Done and What Remains to Be Done? *Environ. Manag.* **2010**, *45*, 442–453. [CrossRef]
6. Liu, G.B. Soil conservation and sustainable agriculture on the Loess Plateau: Challenges and prospects. *Ambio* **1999**, *28*, 663–668.

7. Liu, Y.Y.; Wang, Q.; Zhang, Z.Y.; Tong, L.J.; Wang, Z.Q.; Li, J.L. Grassland dynamics in responses to climate variation and human activities in China from 2000 to 2013. *Sci. Total Environ.* **2019**, *690*, 27–39. [CrossRef]
8. Chen, H.; Marter-Kenyon, J.; Lopez-Carr, D.; Liang, X.-Y. Land cover and landscape changes in Shaanxi Province during China's Grain for Green Program (2000–2010). *Environ. Monit. Assess.* **2015**, *187*, 644. [CrossRef]
9. Treacy, P.; Jagger, P.; Song, C.; Zhang, Q.; Bilsborrow, R.E. Impacts of China's Grain for Green Program on Migration and Household Income. *Environ. Manag.* **2018**, *62*, 489–499. [CrossRef]
10. Yang, M.; Wang, S.F.; Zhao, X.N.; Gao, X.D.; Liu, S. Soil properties of apple orchards on China's Loess Plateau. *Sci. Total Environ.* **2020**, *723*, 15. [CrossRef]
11. Xu, Z.; Peng, J.; Qiu, S.; Liu, Y.; Dong, J.; Zhang, H. Responses of spatial relationships between ecosystem services and the Sustainable Development Goals to urbanization. *Sci. Total Environ.* **2022**, *850*, 157868. [CrossRef]
12. Zhao, H.F.; He, H.M.; Wang, J.J.; Bai, C.Y.; Zhang, C.J. Vegetation Restoration and Its Environmental Effects on the Loess Plateau. *Sustainability* **2018**, *10*, 4676. [CrossRef]
13. Li, Z.; Liu, G.; Zhang, X. Review of reaearches on the Valsa canker of apple trees. *North. Fruits* **2013**, *4*, 13.
14. Peng, X.X.; Guo, Z.; Zhang, Y.J.; Li, J. Simulation of Long-term Yield and Soil Water Consumption in Apple Orchards on the Loess Plateau, China, in Response to Fertilization. *Sci. Rep.* **2017**, *7*, 11. [CrossRef]
15. Sahu, N.; Saini, A.; Behera, S.K.; Sayama, T.; Sahu, L.; Van-Thanh-Van, N.; Takara, K. Why apple orchards are shifting to the higher altitudes of the Himalayas? *PLoS ONE* **2020**, *15*, e0235041. [CrossRef]
16. Liu, C.; White, M.; Newell, G. Selecting thresholds for the prediction of species occurrence with presence-only data. *J. Biogeogr.* **2013**, *40*, 778–789. [CrossRef]
17. Xu, W.; Jin, J.; Cheng, J. Predicting the Potential Geographic Distribution and Habitat Suitability of Two Economic Forest Trees on the Loess Plateau, China. *Forests* **2021**, *12*, 747. [CrossRef]
18. Li, J.J.; Fan, G.; He, Y. Predicting the current and future distribution of three Coptis herbs in China under climate change conditions, using the MaxEnt model and chemical analysis. *Sci. Total Environ.* **2020**, *698*, 8. [CrossRef]
19. Zhang, Y.H.; Liu, X.P.; Chen, G.L.; Hu, G.H. Simulation of urban expansion based on cellular automata and maximum entropy model. *Sci. China-Earth Sci.* **2020**, *63*, 701–712. [CrossRef]
20. Xu, X.; Zhang, H.; Yue, J.; Xie, T.; Xu, Y.; Tian, Y. Predicting Shifts in the Suitable Climatic Distribution of Walnut (*Juglans regia* L.) in China: Maximum Entropy Model Paves the Way to Forest Management. *Forests* **2018**, *9*, 103. [CrossRef]
21. Naeem, M.; Liu, M.; Huang, J.; Ding, G.; Potapov, G.; Jung, C.; An, J. Vulnerability of East Asian bumblebee species to future climate and land cover changes. *Agric. Ecosyst. Environ.* **2019**, *277*, 11–20. [CrossRef]
22. Lang-Yona, N.; Levin, Y.; Dannemiller, K.C.; Yarden, O.; Peccia, J.; Rudich, Y. Changes in atmospheric CO_2 influence the allergenicity of *Aspergillus fumigatus*. *Glob. Change Biol.* **2013**, *19*, 2381–2388. [CrossRef]
23. Grimm, N.B.; Faeth, S.H.; Golubiewski, N.E.; Redman, C.L.; Wu, J.G.; Bai, X.M.; Briggs, J.M. Global change and the ecology of cities. *Science* **2008**, *319*, 756–760. [CrossRef]
24. Li, J.; Wang, Z.L.; Lai, C.G.; Wu, X.Q.; Zeng, Z.Y.; Chen, X.H.; Lian, Y.Q. Response of net primary production to land use and land cover change in mainland China since the late 1980s. *Sci. Total Environ.* **2018**, *639*, 237–247. [CrossRef]
25. Tong, S.Q.; Bao, G.; Rong, A.; Huang, X.J.; Bao, Y.B.; Bao, Y.H. Comparison of the Spatiotemporal Dynamics of Land Use Changes in Four Municipalities of China Based on Intensity Analysis. *Sustainability* **2020**, *12*, 3687. [CrossRef]
26. Zhou, W.; Yang, H.; Huang, L.; Chen, C.; Lin, X.S.; Hu, Z.J.; Li, J.L. Grassland degradation remote sensing monitoring and driving factors quantitative assessment in China from 1982 to 2010. *Ecol. Indic.* **2017**, *83*, 303–313. [CrossRef]
27. Wang, C.; Wang, X.; Jin, Z.; Müller, C.; Pugh, T.A.M.; Chen, A.; Wang, T.; Huang, L.; Zhang, Y.; Li, X.Z.; et al. Occurrence of crop pests and diseases has largely increased in China since 1970. *Nat. Food* **2021**, *3*, 57–65. [CrossRef]
28. Narouei-Khandan, H.A.; Worner, S.P.; Viljanen, S.L.H.; van Bruggen, A.H.C.; Jones, E.E. Projecting the suitability of global and local habitats for myrtle rust (*Austropuccinia psidii*) using model consensus. *Plant Pathol.* **2020**, *69*, 17–27. [CrossRef]
29. Xu, W.; Sun, H.; Jin, J.; Cheng, J. Predicting the Potential Distribution of Apple Canker Pathogen (*Valsa mali*) in China under Climate Change. *Forests* **2020**, *11*, 1126. [CrossRef]
30. Bessho, H.; Tsuchiya, S.; Soejima, J. Screening methode of apple-trees for resistance to Valsa canker. *Euphytica* **1994**, *77*, 15–18. [CrossRef]
31. Wang, C.; Guan, X.; Wang, H.; Li, G.; Dong, X.; Wang, G.; Li, B. Agrobacterium tumefaciens-Mediated Transformation of *Valsa mali*: An Efficient Tool for Random Insertion Mutagenesis. *Sci. World J.* **2013**, *2013*, 968432. [CrossRef]
32. Chen, C.; Li, B.-H.; Dong, X.-L.; Wang, C.-X.; Lian, S.; Liang, W.-X. Effects of Temperature, Humidity, and Wound Age on Valsa mali Infection of Apple Shoot Pruning Wounds. *Plant Dis.* **2016**, *100*, 2394–2401. [CrossRef]
33. Wang, S.T.; Hu, T.L.; Wang, Y.A.; Luo, Y.; Michailides, T.J.; Cao, K.Q. New understanding on infection processes of Valsa canker of apple in China. *Eur. J. Plant Pathol.* **2016**, *146*, 531–540. [CrossRef]
34. Li, K.M.; Feng, M.M.; Biswas, A.; Su, H.H.; Niu, Y.L.; Cao, J.J. Driving Factors and Future Prediction of Land Use and Cover Change Based on Satellite Remote Sensing Data by the LCM Model: A Case Study from Gansu Province, China. *Sensors* **2020**, *20*, 2757. [CrossRef]
35. Halvorsen, R.; Mazzoni, S.; Dirksen, J.W.; Næsset, E.; Gobakken, T.; Ohlson, M. How important are choice of model selection method and spatial autocorrelation of presence data for distribution modelling by MaxEnt? *Ecol. Model.* **2016**, *328*, 108–118. [CrossRef]

36. Zhang, K.; Yao, L.; Meng, J.; Tao, J. Maxent modeling for predicting the potential geographical distribution of two peony species under climate change. *Sci. Total Environ.* **2018**, *634*, 1326–1334. [CrossRef]
37. Abolmaali, S.M.-R.; Tarkesh, M.; Bashari, H. MaxEnt modeling for predicting suitable habitats and identifying the effects of climate change on a threatened species, *Daphne mucronata*, in central Iran. *Ecol. Inform.* **2018**, *43*, 116–123. [CrossRef]
38. Phillips, S.J.; Anderson, R.P.; Schapire, R.E. Maximum entropy modeling of species geographic distributions. *Ecol. Model.* **2006**, *190*, 231–259. [CrossRef]
39. Araujo, M.B.; Pearson, R.G.; Thuiller, W.; Erhard, M. Validation of species-climate impact models under climate change. *Glob. Change Biol.* **2005**, *11*, 1504–1513. [CrossRef]
40. Peterson, A.T. Uses and Requirements of Ecological Niche Models and Related Distributional Models. *Biodivers. Inform.* **2006**, *3*, 59–72. [CrossRef]
41. Guisan, A.; Zimmermann, N.E. Predictive habitat distribution models in ecology. *Ecol. Model.* **2000**, *135*, 147–186. [CrossRef]
42. Elith, J.; Graham, C.H.; Anderson, R.P.; Dudik, M.; Ferrier, S.; Guisan, A.; Hijmans, R.J.; Huettmann, F.; Leathwick, J.R.; Lehmann, A.; et al. Novel methods improve prediction of species' distributions from occurrence data. *Ecography* **2006**, *29*, 129–151. [CrossRef]
43. Phillips, S.J.; Dudik, M. Modeling of species distributions with Maxent: New extensions and a comprehensive evaluation. *Ecography* **2008**, *31*, 161–175. [CrossRef]
44. Riahi, K.; van Vuuren, D.P.; Kriegler, E.; Edmonds, J.; O'Neill, B.C.; Fujimori, S.; Bauer, N.; Calvin, K.; Dellink, R.; Fricko, O.; et al. The Shared Socioeconomic Pathways and their energy, land use, and greenhouse gas emissions implications: An overview. *Glob. Environ. Change-Hum. Policy Dimens.* **2017**, *42*, 153–168. [CrossRef]
45. O'Neill, B.C.; Tebaldi, C.; van Vuuren, D.P.; Eyring, V.; Friedlingstein, P.; Hurtt, G.; Knutti, R.; Kriegler, E.; Lamarque, J.F.; Lowe, J.; et al. The Scenario Model Intercomparison Project (ScenarioMIP) for CMIP6. *Geosci. Model Dev.* **2016**, *9*, 3461–3482. [CrossRef]
46. Rana, S.K.; Rana, H.K.; Ghimire, S.K.; Shrestha, K.K.; Ranjitkar, S. Predicting the impact of climate change on the distribution of two threatened Himalayan medicinal plants of Liliaceae in Nepal. *J. Mt. Sci.* **2017**, *14*, 558–570. [CrossRef]
47. Xin, X.G.; Zhang, L.; Zhang, J.; Wu, T.W.; Fang, Y.J. Climate Change Projections over East Asia with BCC_CSM1.1 Climate Model under RCP Scenarios. *J. Meteorol. Soc. Jpn.* **2013**, *91*, 413–429. [CrossRef]
48. Fick, S.E.; Hijmans, R.J. WorldClim 2: New 1-km spatial resolution climate surfaces for global land areas. *Int. J. Climatol.* **2017**, *37*, 4302–4315. [CrossRef]
49. Zhang, Q.; Wei, H.; Zhao, Z.; Liu, J.; Ran, Q.; Yu, J.; Gu, W. Optimization of the Fuzzy Matter Element Method for Predicting Species Suitability Distribution Based on Environmental Data. *Sustainability* **2018**, *10*, 3444. [CrossRef]
50. Vaughan, I.P.; Ormerod, S.J. The Continuing Challenges of Testing Species Distribution Models. *J. Appl. Ecol.* **2005**, *42*, 720–730. [CrossRef]
51. Yu, F.Y.; Wang, T.J.; Groen, T.A.; Skidmore, A.K.; Yang, X.F.; Ma, K.P.; Wu, Z.F. Climate and land use changes will degrade the distribution of Rhododendrons in China. *Sci. Total Environ.* **2019**, *659*, 515–528. [CrossRef]
52. Touré, D.; Ge, J. The Response of Plant Species Diversity to the Interrelationships between Soil and Environmental Factors in the Limestone Forests of Southwest China. *J. Environ. Earth Sci.* **2014**, *4*, 105–122.
53. Ma, B.; Sun, J. Predicting the distribution of *Stipa purpurea* across the Tibetan Plateau via the MaxEnt model. *BMC Ecol.* **2018**, *18*, 10. [CrossRef]
54. Fielding, A.H.; Bell, J.F. A review of methods for the assessment of prediction errors in conservation presence/absence models. *Environ. Conserv.* **1997**, *24*, 38–49. [CrossRef]
55. Qin, A.; Liu, B.; Guo, Q.; Bussmann, R.W.; Ma, F.; Jian, Z.; Xu, G.; Pei, S. Maxent modeling for predicting impacts of climate change on the potential distribution of *Thuja sutchuenensis* Franch., an extremely endangered conifer from southwestern China. *Glob. Ecol. Conserv.* **2017**, *10*, 139–146. [CrossRef]
56. Liu, C.; Newell, G.; White, M. On the selection of thresholds for predicting species occurrence with presence-only data. *Ecol. Evol.* **2016**, *6*, 337–348. [CrossRef]
57. Wang, S.; Xu, X.; Shrestha, N.; Zimmermann, N.E.; Wang, Z. Response of spatial vegetation distribution in China to climate changes since the Last Glacial Maximum (LGM). *PLoS ONE* **2017**, *12*, e0175742. [CrossRef]
58. Xiang-Long, M.; Xing-Hua, Q.; Ze-Yuan, H.; Yong-Bin, G.; Ya-Nan, W.; Tong-le, H.; Li-Ming, W.; Ke-Qiang, C.; Shu-Tong, W. Latent Infection of Valsa mali in the Seeds, Seedlings and Twigs of Crabapple and Apple Trees is a Potential Inoculum Source of Valsa Canker. *Sci. Rep.* **2019**, *9*, 7738. [CrossRef]
59. Loo, Y.Y.; Billa, L.; Singh, A. Effect of climate change on seasonal monsoon in Asia and its impact on the variability of monsoon rainfall in Southeast Asia. *Geosci. Front.* **2015**, *6*, 817–823. [CrossRef]
60. Stejskal, V.; Vendl, T.; Li, Z.; Aulicky, R. Minimal Thermal Requirements for Development and Activity of Stored Product and Food Industry Pests (Acari, Coleoptera, Lepidoptera, Psocoptera, Diptera and Blattodea): A Review. *Insects* **2019**, *10*, 149. [CrossRef]
61. Shi, S.; Yu, J.; Wang, F.; Wang, P.; Zhang, Y.; Jin, K. Quantitative contributions of climate change and human activities to vegetation changes over multiple time scales on the Loess Plateau. *Sci. Total Environ.* **2021**, *755 Pt 2*, 142419. [CrossRef] [PubMed]
62. Gao, C.; Zhou, P.; Jia, P.; Liu, Z.; Wei, L.; Tian, H. Spatial driving forces of dominant land use/land cover transformations in the Dongjiang River watershed, Southern China. *Environ. Monit. Assess.* **2016**, *188*, 84. [CrossRef] [PubMed]
63. Wu, X.; Shen, Z.; Liu, R.; Ding, X. Land Use/Cover Dynamics in Response to Changes in Environmental and Socio-Political Forces in the Upper Reaches of the Yangtze River, China. *Sensors* **2008**, *8*, 8104–8122. [CrossRef] [PubMed]

64. Hu, W.; Wang, Y.; Dong, P.; Zhang, D.; Yu, W.; Ma, Z.; Chen, G.; Liu, Z.; Du, J.; Chen, B.; et al. Predicting potential mangrove distributions at the global northern distribution margin using an ecological niche model: Determining conservation and reforestation involvement. *For. Ecol. Manag.* **2020**, *478*, 118517. [CrossRef]
65. Bruelheide, H.; Jimenez-Alfaro, B.; Jandt, U.; Sabatini, F.M. Deriving site-specific species pools from large databases. *Ecography* **2020**, *43*, 1215–1228. [CrossRef]
66. Mas, J.F.; Kolb, M.; Paegelow, M.; Olmedo, M.T.C.; Houet, T. Inductive pattern-based land use/cover change models: A comparison of four software packages. *Environ. Modell. Softw.* **2014**, *51*, 94–111. [CrossRef]
67. Liu, X.P.; Liang, X.; Li, X.; Xu, X.C.; Ou, J.P.; Chen, Y.M.; Li, S.Y.; Wang, S.J.; Pei, F.S. A future land use simulation model (FLUS) for simulating multiple land use scenarios by coupling human and natural effects. *Landsc. Urban Plan.* **2017**, *168*, 94–116. [CrossRef]

MDPI
St. Alban-Anlage 66
4052 Basel
Switzerland
Tel. +41 61 683 77 34
Fax +41 61 302 89 18
www.mdpi.com

Agriculture Editorial Office
E-mail: agriculture@mdpi.com
www.mdpi.com/journal/agriculture

www.ingramcontent.com/pod-product-compliance
Lightning Source LLC
Chambersburg PA
CBHW061548220926
51850CB00005B/281
9783036568027